Third Edition
AIR POLLUTION
Volume III: Measuring, Monitoring, and Surveillance of Air Pollution

The third edition of this standard reference work has grown to keep pace with our expanding knowledge of air pollution, and will now consist of five volumes. Most chapters which appeared in the second edition have been revised. New chapters concern themselves chiefly with aspects of air quality management such as data handling, emission inventory, mathematical modeling, and control strategy analysis; global pollution and its monitoring; and more detailed attention to pollution from automobiles and incinerators.

Volume III describes the sampling, analysis, measurement, and monitoring of air pollution.

AIR POLLUTION

THIRD EDITION

VOLUME III

Measuring, Monitoring, and

Surveillance of Air Pollution

ENVIRONMENTAL SCIENCES

An Interdisciplinary Monograph Series

Editors: Douglas H. K. Lee, E. Wendell Hewson, and Daniel Okun

A complete list of titles in this series appears at the end of this volume.

AIR POLLUTION

THIRD EDITION

VOLUME III

Measuring, Monitoring, and Surveillance of Air Pollution

Edited by

Arthur C. Stern
Department of Environmental Sciences and Engineering
School of Public Health
University of North Carolina at Chapel Hill
Chapel Hill, North Carolina

ACADEMIC PRESS New York San Francisco London 1976

A Subsidiary of Harcourt Brace Jovanovich, Publishers

ACADEMIC PRESS, INC.
111 Fifth Avenue, New York, New York 10003

United Kingdom Edition published by
ACADEMIC PRESS, INC. (LONDON) LTD.
24/28 Oval Road, London NW1

Library of Congress Cataloging in Publication Data

Stern, Arthur Cecil.
Air pollution.

(Environmental sciences)
Includes bibliographical references and indexes.
CONTENTS: v. 3. Measuring, monitoring, and surveillance.—v. 4. Engineering control of air pollution.
I. Air–Pollution–Collected works. I. Title. II. Series. [DNLM: 1. Air pollution. WA754 A292]
TD883.S83 1976 363.6 76-8256
ISBN 0–12–666603–2

PRINTED IN THE UNITED STATES OF AMERICA

To Elizabeth and Mark, Amanda and Simon Michael

Contents

4. Sampling and Calibration of Gaseous Pollutants

Herman D. Axelrod and James P. Lodge, Jr.

5. Hydrocarbons and Carbon Oxides

Aubrey Paul Altschuller

6. Sulfur Compounds

Donald F. Adams

7. Nitrogen Compounds and Oxidants

Morris Katz

8. Odors

Austin H. Phelps

PART B AMBIENT AIR SURVEILLANCE

9. Ambient Air Quality Surveillance

Robert J. Bryan

10. Air Pollution Effects Surveillance

Charles W. Gruber and George Jutze

11. Radionuclide Surveillance

Harry F. Schulte

12. Air Quality Data Handling and Analysis

Charles E. Zimmer

13. Global Monitoring and Surveillance of Air Pollution

Richard D. Cadle

PART C SOURCE SURVEILLANCE

14. Stack Sampling

Harold J. Paulus and Raymond W. Thron

15. Source Monitoring

John S. Nader

16. Measurement, Testing, and Surveillance of Air Pollution from Mobile Sources

William F. Biller and John Harkins

List of Contributors

Numbers in parentheses indicate the pages on which the authors' contributions begin.

Donald F. Adams (213), Department of Civil and Environmental Engineering, Washington State University, Pullman, Washington

Aubrey Paul Altschuller (183), Environmental Science Research Laboratory, U.S. Environmental Protection Agency, Research Triangle Park, North Carolina

Herman D. Axelrod* (145), National Center for Atmospheric Research, Boulder, Colorado

William F. Biller (647), 68 Yorktown Road, East Brunswick, New Jersey

Robert J. Bryan (343), Pacific Environmental Services, Inc., Santa Monica, California

Richard D. Cadle (485), National Center for Atmospheric Research, Boulder, Colorado

Paul M. Giever (3), Stanford Research Institute, Arlington, Virginia

Charles W. Gruber (393), Department of Civil and Environmental Engineering, University of Cincinnati, Cincinnati, Ohio

James R. Hammerle (717), National Air Data Branch, Office of Air Quality Planning and Standards, U.S. Environmental Protection Agency, Research Triangle Park, North Carolina

John Harkins (647), 291 Charro Avenue, Thousand Oaks, California

* Present address: Coors Spectro-Chemical Laboratory, Coors Porcelain Co., Golden, Colorado.

George Jutze (393), PEDCo-Environmental Specialists, Inc., Cincinnati, Ohio

Morris Katz (259), Faculty of Science, York University, Downsview, Ontario, Canada

James P. Lodge, Jr. (145), National Center for Atmospheric Research, Boulder, Colorado

Walter C. McCrone (99), McCrone Associates, Inc., Chicago, Illinois

John S. Nader (589), Division of Chemistry and Physics, National Environmental Research Center, U.S. Environmental Protection Agency, Research Triangle Park, North Carolina

Harold J. Paulus (525), Environmental Health, School of Public Health, University of Minnesota, Minneapolis, Minnesota

Austin H. Phelps (307), Corporate Air Pollution Control, The Procter and Gamble Company, Cincinnati, Ohio

Harry F. Schulte (413), Industrial Hygiene Group, Los Alamos Scientific Laboratory, Los Alamos, New Mexico

Raymond W. Thron (525), Metropolitan Council of the Twin Cities, St. Paul, Minnesota

Philip W. West (51), College of Chemistry and Physics, Louisiana State University, Baton Rouge, Louisiana

Charles E. Zimmer (453), PEDCo-Environmental Specialists, Inc., Cincinnati, Ohio

Preface

This third edition is addressed to the same audience as the previous ones: engineers, chemists, physicists, physicians, meteorologists, lawyers, economists, sociologists, agronomists, and toxicologists. It is concerned, as were the first two editions, with the cause, effect, transport, measurement, and control of air pollution.

So much new material has become available since the completion of the three-volume second edition that it has been necessary to use five volumes for this one. Volumes I through V were prepared simultaneously, and the total work was divided into five volumes to make it easier for the reader to use. Individual volumes can be used independently of the other volumes as a text or reference on the aspects of the subject covered therein.

Volume I covers two major areas: the nature of air pollution and the mechanism of its dispersal by meteorological factors and from stacks. Volume II covers the effect of air pollution on plants, animals, humans, materials, and the atmosphere. Volume III covers the sampling, analysis, measurement, and monitoring of air pollution. Volume IV covers two major areas: the emissions to the atmosphere from the principal air pollution sources and the control techniques and equipment used to minimize these emissions. Volume V covers the applicable laws, regulations, and standards; the administrative and organizational strategies and procedures used to administer them; and the energy and economic ramifications of air pollution control. The concluding chapter of Volume II discusses air pollution literature sources and gives guidance in locating information not to be found in these volumes.

To improve subject area coverage, the number of chapters was increased from 54 of the second edition (and 42 of the first edition) to 71. The scope of some of the chapters, whose subject areas were carried over from the second edition, has been changed. Every contributor to the second edition was offered the opportunity to prepare for this edition either a revision of his chapter in the second edition or a new chapter if the scope of his work had changed. Since 8 authors declined this offer and one was deceased, this edition includes 53 of the contributors to the second edition and 46 new ones.

The new chapters in this edition are concerned chiefly with aspects of air quality management such as data handling, emission inventory, mathematical modeling, and control strategy analysis; global pollution and its monitoring; and more detailed attention to pollution from automobiles and incinerators. The second edition chapter on Air Pollution Standards has been split into separate chapters on Air Quality Standards, Emission Standards for Stationary Sources, and Emission Standards for Mobile Sources. Even with the inclusion in this edition of the air pollution problems of additional industrial processes, many are still not covered in detail. It is hoped that the general principles discussed in Volume IV will help the reader faced with problems in industries not specifically covered.

Because I planned and edited these volumes, the gap areas and instances of repetition are my responsibility and not the authors'. As in the two previous editions, the contributors were asked to write for a scientifically advanced reader, and all were given the opportunity of last minute updating of their material.

As editor of this multiauthor treatise, I thank each author for both his contribution and his patience, and each author's family, including my own, for their forbearance and help. Special thanks are due my secretary, Patsy Garris, and her predecessors, who carried ninety-nine times the burden of the other authors' secretaries combined, and Eleanor G. Rollins for preparing the Subject Index for this volume. I should also like to thank the University of North Carolina for permitting my participation.

Arthur C. Stern

Contents of Other Volumes

VOLUME I AIR POLLUTANTS, THEIR TRANSFORMATION AND TRANSPORT

VOLUME V AIR QUALITY MANAGEMENT

Part A

SAMPLING AND ANALYSIS

1

Particulate Matter Sampling and Sizing

Paul M. Giever

I. General

The collection of an accurate and adequate sample from a gaseous medium presents a variety of complicated situations. Meeting these situations so that the best possible sample is collected is the first essential step in an air pollution evaluating and control program. Analysis and subsequent calculations are important, but, if the sample itself is not representative, the end results are of little or no value.

A thorough knowledge of techniques, instrumentation, and the behavior of gases and airborne particulate matter is essential. Each investigator must assure himself that the sampling units and procedures he has selected will provide the best results possible.

Steps in Sampling

Four important steps in making a quantitative determination of atmospheric pollutants are: (a) collection of the sample using the appropriate instruments and techniques for a particular pollutant, (b) special treatment or refinement of the collected sample, (c) analysis of the sample, and (d) evaluation of the quantities of pollutants collected.

1. Size of Sample

The sample must be of sufficient size to accommodate the evaluation technique employed. It may be that trial samples of various sizes, 1, 5, 10 m^3 or even more, may be required in order to obtain sufficient material for analysis. In other instances 1–10 liters may be adequate. The quantity required is inversely related to the sensitivity of the analytical technique employed.

2. Rate of Sampling

The sampling rate is determined by the collecting equipment employed. Some systems are designed to sample milliliters per minute while others operate in the range of several cubic meters per minute. The instruments and the sampling rates should be selected to meet the requirements of the program being undertaken. The final criterion in selecting the sampling rate is the efficiency of collection at the rate required to provide an adequate sample.

3. Duration of Sampling

A sampling period will determine the average concentration for the time covered. In Figures 1, 2, and 3 the actual pollutant concentration is shown by the solid line. If a sampling period of 4 hours is selected, the average concentration will be that shown in Figure 1. If the duration of sampling is reduced to a series of four 1-hour periods, while maintaining a fixed sampling rate, it will produce the results shown by the dashed line in Figure 2; the effect of further reducing the duration of sample time to a series of 15-minute intervals is shown in Figure 3.

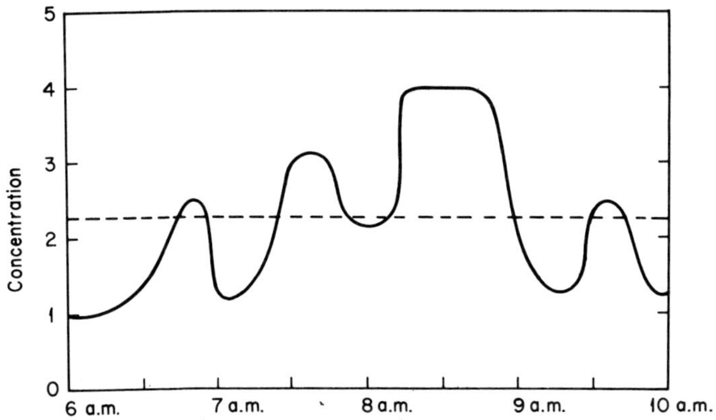

Figure 1. Apparent pollutant concentration resulting from a sampling period of 4 hours is shown by the dashed line.

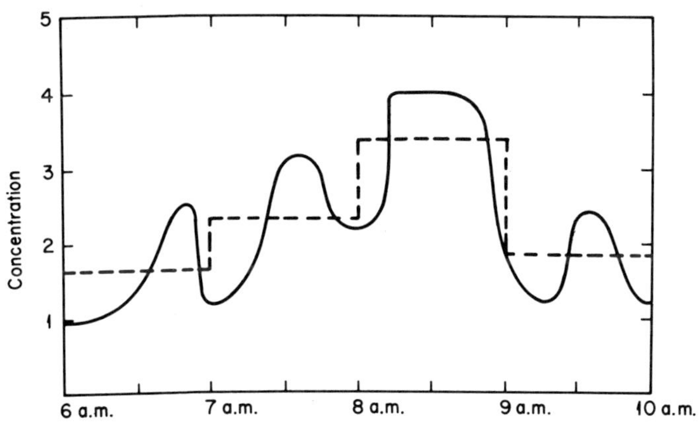

Figure 2. Apparent pollutant concentration resulting from a series of sampling periods of 1 hour is shown by the dashed line.

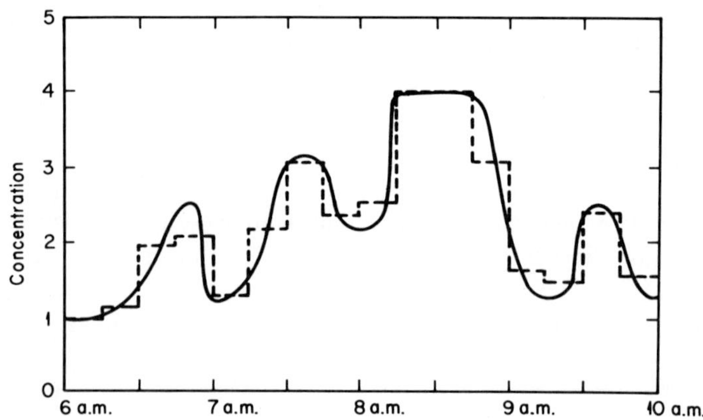

Figure 3. Apparent pollutant concentration resulting from a series of sampling periods of 15 minutes is shown by the dashed line.

Since peaks of fumigant concentrations may be the only significant values, the shortest sampling period generally reveals the most valuable information. The length of sampling period is not a purely arbitrary decision. Factors which affect the minimum possible duration include: (a) the expected significant concentration to be encountered; (b) permissible rate of sampling; and (c) the lower limit of the analytical procedure to be used. Thus the duration of sampling is related to the size of sample. Selection of an optimum sampling period requires consideration of the fact that very short durations result in a rapidly fluctuating record which may be difficult to read (*1*).

Consideration must be given to the intended use of the data sought, and the sampling time must be adjusted accordingly. If a 4- or a 24-hour average is adequate, the increased sampling and analysis time required to obtain 15-minute average data will be largely for academic purposes only.

4. Collection Limitations

a. General. Collection devices must be suited to the specific pollutant involved, and the collection efficiency as well as accuracy or precision should be known. An efficiency of below about 75% is usually unacceptable and a more efficient device should be employed (*2*).

In some instances it may be necessary to sample particulate matter in both liquid and solid form. It may also be necessary to sample a pollutant that is in both gaseous and particulate state, as, for example, fluorides. In these instances a filter is often used ahead of the gas absorbers.

Under such circumstances it is essential that the first collector units do not interfere with the operation of subsequent collector units. Again citing fluorides as an example, the filter material chosen is very important, as many types of filter media will react with gaseous components (*3*).

The collection efficiency of a sampling system is usually determined on the basis of either (a) the weight of particles collected, (b) a count of the particles removed, or (c) the changes in color of a filter. If the aerosol being measured is high in the number of large particles, the efficiency of its removal on a weight basis may be very high. Efficiency on a count basis usually gives much lower results if large particles are present. It may be possible to have an efficiency of 99.99% on a weight basis and only 50 or 60% on a particle count basis. The method of determining the efficiency will depend upon the ultimate use of the data. If fine particles are of significance, a count analysis may be indicated, but, if only the total weight of the particulate matter is of interest, a weight analysis should be made.

Particles may be altered by coalescence or disintegration during sampling and the particles collected may not be representative of those present in the atmosphere.

The effectiveness of a sampling system can be determined by trial on a reproducible synthetic atmosphere under controlled conditions. In no other way can accuracy of the elements of a sampling procedure be determined. Accuracy or precision determined in this manner can be used as an indication of the probable effectiveness or accuracy of that system when used in an uncontrolled atmosphere if the general range of conditions are somewhat comparable to the controlled conditions.

The use of two similar sampling devices in series is not a satisfactory method of determining the efficiency of a sampling unit since the device may be capable of collecting only particles of a certain size range, and any number of similar devices in series will still collect only particles within that size range.

Particles in the atmosphere and in the collected sample may undergo changes both as to composition and to size. It, therefore, may be impossible to identify in the atmosphere, or in the sample, the specific compounds originally discharged from a source. It will often be possible to select a specific sampling technique which will prevent, or at least minimize, alteration of the collected sample. For this reason the sampling method should be carefully selected to accommodate the material expected to be collected.

Transportation and storage, as well as time lapse between capture and analysis, may detract from the accuracy of a particulate matter measurement. It is usually a good policy to allow a minimum time lapse between collection and evaluation. A maximum of 24 hours between capture and measurement has been the recommended time for counting samples for hazardous health exposure.

Stored samples should generally be kept from heat and light and the possibility of interaction with storage container should be prevented. In general, glass containers are inert to particulate matter but some synthetic compounds such as polypropylene may serve equally well.

b. ACCURACY AND PRECISION. Precision refers to reproducibility of observations. Accuracy refers to the truth of the observation (*1*). The degree of accuracy required must be determined on the basis of the use of the data. Often a degree of accuracy and precision is specified that is unreasonably high, i.e., when the effects of the substance may not be evident in concentrations 10 or even 100 times the accuracy and precision required. An example of this is the reporting of COH (Coefficient of Haze) units to the first and even the second decimal place, and requiring a $\pm 1\%$ accuracy in sampling rate and light standardization. COH units are at best merely a measurement of the attenuation of transmitted light and are dependent on particles size, density, and color. COH units are calculated in terms of 1000 linear feet (304.8 m) from a measurement made in liters per minute, and the area represented is extrapolated from a 2.54-cm-diameter spot to square meters. Concentrations of atmospheric contaminants generally vary momentarily over a wide range resulting in an average value for the given period of time sampled.

c. SAMPLING METHODS FOR THE GENERAL ATMOSPHERE. Continuous sampling is essential to obtain the most representative information concerning variations in the concentration of pollutants in the atmosphere. In many instances, samples collected intermittently at a large number of locations will provide data of more value than will a continuous record at a single or a small number of locations.

Sequential sampling instruments employing automatic programmers serve to collect samples routinely at preset time intervals for extended periods of time. These units are usually mechanically or electronically operated and are a very valuable part of an ongoing sampling program.

Wind controlled operation of intermittent samplers, utilizing direction and/or velocity, has also been employed advantageously for many years in atmospheric sampling. For the most part these units are just samplers and the materials collected must be taken to the laboratory for analysis.

Continuously operating automatic instruments develop vast amounts of data and must be coupled with automatic data processing systems to make them practicable.

Table I Recommended Units for Reporting Air Pollution Measurements

Item measured	*Recommended units*	*Observed ranges*
Particle fallout		
Count	Number/cm^2/time interval	—
Weight	mg/cm^2/time interval	0.5 to 135 mg/ cm^2/month
Airborne particulates		
Count	Number/m^3	10×10^6 and up particles/m^3
Weight	$\mu g/m^3$	10–5000 $\mu g/m^3$
Gases and vapors	$\mu g/m^3$	Varies greatly
Instantaneous light transmission	% transmitted	0–100%
Visibility	km	—
Volume emission rates	m^3/minute	1.5 – 300 + m^3/minute
Sampling rates	m^3 (dm or cm)/minute	10 cm^3/minute to 3 m^3/minute
Temperature	Celsius scale	—
Time	0000 to 2400 hours/day	—
Pressure	mm of mercury	—
Velocity	m/second	0–100 m/sec
Gas volumes	Reported at 760 mm Hg and 21.1°C	—

d. UNITS OF EXPRESSION. A variety of units are used in the air pollution literature for expressing results of sampling, even in some instances mixing metric and English units. This practice should be discontinued and a strict adherence to the metric system, as shown in Table I, should be practiced (*4*).

II. Sample Collection

A. Introduction

The collection of accurate and precise atmospheric samples of particulate matter becomes increasingly difficult as their sizes, densities, and concentrations fluctuate.

B. Sampling Equipment

To obtain a meaningful sample of particulate matter it is usually necessary to use a positively operated source of air movement and an accurate means of measuring the amount of gas sampled. These two features combined with efficient sample collection constitute the major features of a sampling train or system. The collecting device is ideally the first unit in a sampling train, but it is often necessary to precede the collector with a probe and nozzle to obtain the sample from a desired location not suited to accommodate the collecting media.

Air movement is usually produced by a source of vacuum. Measurement devices are usually quantity or rate meters. Collection devices usually involve filtration, impingement, impaction, or precipitation.

1. Vacuum Sources

To operate a collection device it is necessary to draw a measured quantity of air through the system. Vacuum sources commonly employed include (a) vacuum pumps: positive displacement types (rotary, cycloid, piston, and diaphragm) and centrifugal types; (b) ejectors (aspirators): water, stream, compressed air, and compressed gas; (c) miscellaneous devices: liquid displacement, evacuated containers, automobile intake manifold, and expanding plastic bags, bellows, or syringes.

The two major types of vacuum pumps are positive displacement and centrifugal. The principal difference in their performance is that the relationship between suction pressure (head) and capacity is linear in the positive displacement and nonlinear in the centrifugal type (*5*). Both types are usually operated either by electric motors or manually.

Rotary and cycloid pumps are positive displacement pumps in which the vanes or impellers move rotationally, requiring no suction or discharge valving. Piston and diaphragm pumps are reciprocating positive displacement types utilizing suction and discharge valves and are prone to produce a pulsating type of gas flow. The diaphragm type has the advantage of nonreactive construction, which is desirable when the sample must pass through the pump before going to the collector.

In centrifugal pumps an impeller revolves in a casing. The gas being moved by the vanes of the impeller is propelled by centrifugal force into the casing and out the discharge.

Ejectors or aspirators operate on the venturi principle and require either water, steam, air, or some form of gas under pressure as the driving fluid to operate them. The movement of the high velocity driving fluid through the "throat" section of the venturi creates a pressure differential causing the sample air stream to enter the chamber. This type of suction source is particularly desirable where hazards from fire or explosion may exist.

Liquid displacement vessels, evacuated containers, automobile intake manifolds or expanding plastic bags, bellows, or syringes are among the variety of devices that can be used. These types of suction sources are most often used for grab samples or for unusual sampling conditions.

Sampling may be required for extended periods of time so that a heavy-duty motor-driven pump offers the most versatility and reliability. Battery- or line voltage-operated pumps are the most desirable but gasoline engine-driven pumps may also be used. If a gasoline engine-driven pump or generator is utilized care should be taken so as not to contaminate the atmosphere being measured with exhaust combustion products of the engine.

2. Air Metering Devices

Air metering devices are required to relate the quantity of pollutant collected to the amount of gas in the sample. Therefore, a precise and accurate measurement of the amount of gas drawn through the sampling device must be made. The flow measurement is usually made downstream from the collecting device, between it and the vacuum source. This arrangement introduces pressures other than atmospheric, so that pressure and temperature must be measured and used in calculating the correct volume. If flow is measured into the pump rather than out of it, no account need be made for gas leakage from the pump.

Two types of gas measurement devices are (a) ratemeters: orifices, nozzles, venturimeters, and Rotameters; and (b) volume (positive displacement) meters: wet test meters, dry test meters, and cycloid gas meters.

Ratemeters are measuring devices in which the gas passes through in a continuous stream. They are compact and inexpensive, measure instantaneous flow, but, even when calibrated, frequently are not considered to be as accurate as volume meters for obtaining total gas volume.

Orifices, venturis, and nozzles are typical "head" or pressure-loss type of measuring devices in which the pressure is measured by a manometer and the flow rate calculated from the manometer reading. These units must be calibrated. Their design, construction, installation, and flow rate computation have been standardized by the American Society of Mechanical Engineers (*6*).

Rotameters consist of a ball or float in a glass tube. The tube, mounted vertically, is slightly tapered in bore with the diameter decreasing downward. The flow to be measured is passed upward through the tube. The ball is propelled upward until it reaches a position in the tube where its weight is balanced by the force due to the gas flow. These units are used extensively for gas sampling measurements in instances where the degree of precision is satisfactory. They are not generally considered to be a precision type of instrument.

There tends to be an unsteadiness in the position of the float but the device can be used to check a prescribed sampling rate quickly (*7*). The flow rate through these meters for a given period of time is not usually considered to be an appropriate substitute for a volume meter.

Positive displacement meters are volume measuring devices designed to measure accurately the quantity of gas passing through the meter over the sampling period or for the time interval required. They are also referred to as volume meters and are designed so that the gas passing through the unit is captured in a specific area of fixed capacity, each filling of which is indicated on a meter or recorder.

A wet test meter consists of a cylindrical casing containing four horizontally mounted measuring compartments of equal volume partially submerged in water. The liquid level is such as to completely seal the compartments when in the lower half of the cylinder. Gas enters at the center and flows into a compartment causing it to rise, thereby producing motion. As the drum rotates each compartment fills alternately with gas and water so that it is completely filled and emptied in each revolution of the drum. This rotation is indicated by a dial on the face of the instrument. The volume measured is directly related to the water level and care must be taken to have the meter level filled with water to the indicated point. Wet test meters are accurate to $\pm 0.5\%$, are not generally considered suitable for field use but are used extensively as a secondary standard for calibrating other gas measuring devices.

The dry test meter consists of a pair of bellows interconnected by mechanical valves. Measurement is made by successive filling and emptying of the bellows with the number of such events indicated on a dial or meter. The dry gas meter is rugged and light weight. It can be used under relatively high pressures and is accurate to $\pm 1.0\%$. Meters routinely taken into the field should be calibrated at frequent intervals.

The cycloid or lobed impeller meter is examplified by the Roots meter in which twin contrarotating lobes operating within a casing measure the volume of gas with a reported accuracy of $\pm 1.0\%$. The revolutions of the lobes are indicated on a counter graduated in units of volume.

Volume meters are calibrated against a spirometer which is considered to be the primary calibrating unit.

C. Collecting and Evaluating

1. Settlement

Sampling particulate matter using gravimetric or sedimentation techniques involves the collection of particles that have sufficient mass to fall out of the atmosphere due to the force of gravity. The principle of collection is that particles settling through a medium will assume a constant velocity after they reach equilibrium between the molecular resistance of the medium and gravity. This type of sampling requires no vacuum source or metering system, and collection can be achieved in dust fall jars, plates, glass slides, dishes, or trays, with or without retention helpers in the form of gummed or grease surface coating. The method is suitable only for large particles of 20 to 50 μm minimum size. Being without suction or gas measurement, the method cannot provide data on volumetric concentration.

Gravitational collection has little or no application for particles 20 μm or smaller as they settle so slowly that very minor air movement prevents their collection.

This method of collection, while simple, is beset by the complicating effects of wind velocity, direction, turbulence, rain, and density of the settling material.

Variations in material, size and shape of collectors, and their placement in the field, have developed data that are not intercomparable. In order to obtain a basis for comparison of values from one location to another an American Society for Testing and Materials (ASTM) standard method was developed by Committee D-22 (D-1739—revised 1970) (*8*). The standard establishes the size of the collector and the relationship between its diameter and height. The collector must be a cylinder (Fig. 4) made of glass, plastic, or stainless steel (*9*). A holder should be provided to ensure safe and upright positioning of the collector.

The sample is deposited over a 30-day period and the material is dried and weighed. The results are expressed in terms of weight per unit area per month. The present trend is to report the weight collected in units of milligrams per square centimeter per month. Extrapolating results from the small collector area to a square mile, or square kilometer, as has been done routinely in the past, magnifies the inherent inaccuracies of the method.

Directional fallout containers have found some application as indicators of sources of pollution. Their opening to the atmosphere is activated by wind direction. They generally use eight containers, one for each cardinal wind direction.

In addition to the weight of particulate matter collected, dissolved solids may be evaluated by chemical analysis (*10*).

It must be kept in mind that this system is erratic, selective, nonpredictable, and usually nonrepeatable. It is an indicator that may show need for more precise sampling, or may be used to show trends over a period of years.

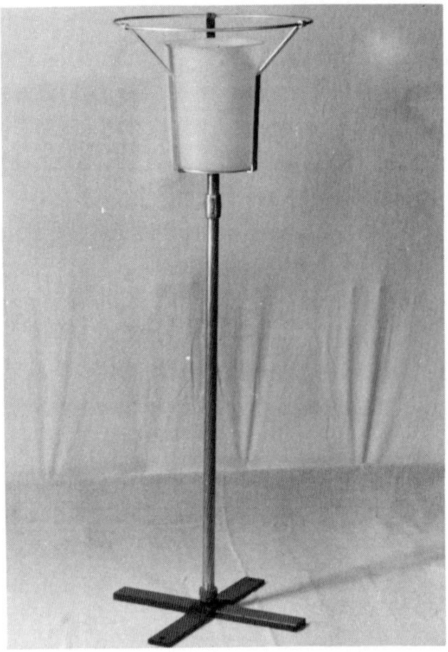

Figure 4. Fallout container. (Courtesy of Research Appliance Company.)

2. Filtration*

a. INTRODUCTION. Many procedures for the sampling of the atmosphere utilize filters as a collecting medium. Filters of appropriate composition and size may be used for qualitative and quantitative analysis, particle identification, and sizing.

The selection of the filter medium must be preceded by careful review of its characteristics and limitations and its relation to sampling requirements.

Most filter media now used in air sampling were developed for other applications, e.g., low ash content, high purity cellulose paper filters were developed for wet chemistry. They have the disadvantages of relatively low efficiency for collecting small particles, high flow resistance, and extreme sensitivity to moisture.

The filter serves to separate and retain suspended particles from the gas stream. It makes the separation by one or more of the following mechanisms: (a) sieving, (b) impaction, or (c) diffusion.

Filters generally employ not only all three of these mechanisms but also an electrostatic effect which is operative in some instances.

* See also Chapter 4, Vol. IV.

In sieving action any particle larger than the opening in the filter will be blocked and held mechanically on the surface of the filter or screen. Particles smaller than the pores penetrate the surface but may be retained within the filter as the pores may be offset thus actually providing a high degree of retention of particles smaller than the stated pore size (see also Section IV, A).

Collection by impaction occurs as the moving stream of particles flows around an obstruction or is forced to change direction abruptly. This type of collection is within the structure of the filter and is most applicable to particles of high density in high velocity flow. Collection by diffusion is a result of Brownian movement of the suspended particles. Low stream velocity aids in this type of collection. Impaction and diffusion are the predominant factors in collection of particles smaller than the pore size of the filter.

Applicability of typical filter media to atmospheric particulate matter sampling and analysis is shown in Table II (*11*).

b. Chemical Analysis. Filters used for chemical analysis of the particles collected should be carefully selected for their purity, low ash content, and the absence of interfering elements (Table II). Filter media having these characteristics generally have high pressure drop and are not desirable for applications where long sampling periods and large volumes of sampled gas are required. Where high efficiency collection of submicron particles is desired membrane (*12*) or Nucleopore (*13*) filters may be used.

c. Special Applications. Filter media in the form of tapes are used for sampling ambient and in-plant atmospheres with analysis being performed by measuring the attenuation of light by the tape and its collected particles. A typical filter tape sampler is shown in Figure 5. The reduction of transmitted light through the sampling spot is measured as COH (Coefficient of Haze) units and defined in terms of optical density. 1 COH unit = 0.01 O.D.—the quantity of particulate matter which produces an optical density of 0.01 on filter paper. The optical density of a sample is not proportional to the mass concentration of suspended particulate matter, unless the particle size, shape, color, and surface characteristics remain the same as that for which the optical density was calibrated. Another system measures light reflectance from samples collected on filter tape; the results are Reflectance Units of Dirt Shade (RUDS).

The filter medium most commonly used for these samplers is a cellulose paper of relatively low pressure resistance, sensitive to moisture changes

Table II Properties of Typical Filter Media Relative to Various Tests in Atmospheric Analysis[a,b]

Test purpose	*Filter type:* *Cellulose papers*	*Glass fiber high efficiency*	*Ester cellulose membranes*
Mass loading			
Gravimetric	X	2	3
Dust spot	4	4	7
Light scatter	X	5	5, 7
Microscope count and size distribution	X	X	8, 9
Particle identity	X	X	8, 9
Chemical analysis			
Wet methods	10	11	12
Instrumental	10	11	12
Biological assay	X	X	13
Radiological	14	14	3, 8, 15
Analysis for gaseous components	X	16	16

Remarks

1. Highly moisture sensitive; low efficiency
2. Good stability vs moisture; temp. resistant
3. Moisture sensitive; high flow resistance
4. Paper usually in tape form
5. Used to clean reference sample of air
6. Low collection efficiency
7. High flow resistance
8. Good particle distribution on top surface
9. Select proper pore size
10. Use chemical grades and low ash if necessary
11. Interfering materials may extract from filter
12. Possibility of interfering materials in filter
13. Standard filter for this use
14. Not applicable for a count
15. Suitable for all counts
16. Danger of reaction or sorption vs gaseous components of air sample

[a] Reprinted by permission of the American Society for Testing and Materials from Spec. Tech. Publ. 555, copyright 1974.
[b] Numbers in columns refer to itemized remarks; X's indicate applicability unsuitable in most cases.

Figure 5. Tape-type filter sampler. (Courtesy of Research Appliance Company.)

and at best somewhat inconsistent in optical density, i.e., ±5% as measured by light transmission. Available filters with less moisture sensitivity show greater flow resistance or unacceptable variations in optical density.

In some European countries, where pollutants originate largely from coal burning processes, relationships between measured spot reflectance and mass concentration have been published (*14*). Where the optical properties of particulate matter vary seasonally and diurnally such a relationship cannot be developed, and reflectance or transmittance must be used as independent measurements, rather than being converted to equivalent mass concentration values.

In order to add some degree of comparability to the spots from these sampling procedures, standardized methods of sampling, evaluation, and calibration, along with suggestions for locating stations and interpreting the data, have been developed by American Society for Testing and Materials (*15*) and the Air Pollution Control Association (*16*).

Mass monitoring using filters has many disadvantages. Slowness in determining the concentration by manual operation prevents collection of "real time" response and data. The use of beta gauging techniques for tape filter systems has lead to the development of fast responding, accurate, automatic mass monitors (*17*). Cellulose paper tape similar to that used in the previously described tape sampler is used. Mass determination is made by measuring the attenuation of beta radiation from a carbon-14 source (*18*). These monitors are generally computer operated and print-out data in standard units at standard conditions of temperature and pressure.

3. Impingement or Inertial Collection

Wet impingement is a commonly used method for collecting solid particulate matter (*19*). It has long been the standard method for collecting silica and other pneumoconiosis-producing dusts that are to be evaluated by a counting technique.

Collection occurs through abrupt changes in direction and velocity of the particles and the carrier gas stream which permits the retention of particles in a collector liquid. The standard impinger tube discharges the sample through an orifice having a bore of 2.3 mm onto a flat plate 5 mm distant. The discharge velocity is around 110 m/second and drops to a considerably lower velocity as it enters the impinger flask which has a diameter of 5.0 to 5.5 cm. The very high discharge velocity may cause shattering of large particles thus changing the particle size characteristics of the sample. Sampling rates are from 0.75 to 1.5 cubic feet per minute (cfm) [21.2 to 42.5 liters per minute (lpm)]. Collection efficiency is nearly 100% for particles down to 1 μm aerodynamic diameter at the optimum sampling rate of 1 cfm (28.3 lpm).

The midget impinger (*20*) and the microimpinger (*21*) are examples of smaller wet impingers utilizing the principle of the standard impinger. The midget impinger has been used extensively in dusty industrial atmospheres but is usually inadequate for cleaner ambient air. It may also be inadequate for sampling large particles due to the rather low intake velocity at the recommended sampling rate of 0.1 cfm (2.8 lpm).

4. Impaction (Dry Impingement)

When a gas stream containing solid particulate matter is deflected from its path of flow, the inertia of some of the particles will cause them to deviate from the line of flow to strike a deflecting surface upon which the particles adhere. The surface then is referred to as an "impaction surface" and the particles are said to be impacted (*22*). Particles of a specific size range will be retained on the surface while other particles will remain in the gas stream flowing away from the impaction surface. The size range of the particles retained on the impaction surface is a function of (a) velocity of the gas, (b) geometry of the flow, (c) viscosity of gas, (d) mean free path of the gas, and (e) particle density and shape.

Dry impactors in a series-stacked configuration have been used extensively for ambient air sampling. Impactors provide aerodynamic size separation so that mass evaluation of various particle size groups can be obtained. A large number of samplers of a variety of configurations and number of collecting stages have been developed.

The reliability of an impactor depends to a large extent on the accuracy of calibration of the collection characteristics of their impaction stages. Methods of calibration and interpretation of data are well developed (*23*).

5. Electrostatic Precipitation*

Electrostatic precipitation or electrical impaction collection techniques are used in a variety of different instruments (*24*). A high electrical potential difference is maintained between two electrodes causing a current to flow. Potential differences of from 12,000 to 45,000 V dc are used. Large numbers of ions are liberated and maintained in the proximity of the electrodes (*25*). The particles are collected on the inside of a cylinder housing the electrode, and may be washed or brushed out for analysis.

6. Thermal Precipitation

Collection of particles by thermal precipitation is based on the principle that small particles, under the influence of a temperature gradient, will move to the lower temperature (*26*). This phenomenon (thermophoresis) has been recognized for a considerable period of time, but is still not widely used as an ambient air sampling technique.

The thermal forces acting on the dust particles are relatively weak and a considerable temperature gradient is required for effective precipitation. Collection is most effective for particles below 10 μm and is approximately 100% for particles in size range of 0.01 to 10 μm.

Samplers generally utilize a slit, or a long narrow aperture to introduce the dust-laden gas to the space between an electrically heated wire or plate and a cold glass-collecting surface (*27*). Velocity of gas through

* See also Chapter 5, Vol. IV.

the sampler is low, and the volume of sample is correspondingly small. Air movement may be induced by aspiration using water at a rate of only around 5 to 7 ml/minute. Large particles may not be collected due to inefficient adhesion to the collection surface and subsequent reentrainment, since the precipitating forces are small.

Particle shattering is unlikely in this collection method making it very suitable for sizing and for study of small particles with the electron microscope. Samples can be collected directly on a grid or a thin microscope cover glass.

The method is rarely used for gravimetric analysis and has limited application in air pollution studies beyond the uses mentioned above.

7. Respirable Dust Sampling

If the objective of a sampling program is to obtain information on the nature and magnitude of the potential health hazard due to inhalation of airborne particles, the sampling technique must be capable of collecting those particles which will reach the critical respiratory organs to cause toxic action. The sampling method must be size selective for those particles which will be conveyed into the human respiratory tract.

The sampling of quartz and other toxic dusts by use of the standard impinger, discussed in Section II,C,3, had been the standard method for their evaluation for many years. In 1973 the American Conference of Government Industrial Hygienists (ACGIH) through its Committee on Threshold Limit Values (TLV), established the following TLV's for mineral dusts (*28*).

TLV for silica, SiO_2, crystalline quartz:

$$\frac{300}{\%\ \text{quartz} + 10}\ \text{million particles per cubic foot of air}$$

TLV for total dust, respirable and nonrespirable:

$$\frac{30}{\%\ \text{quartz} + 3}\ \text{mg/m}^3$$

The committee further specified that both concentration and percent quartz for the application of the limit are to be determined from the fraction passing a size selector with characteristics as shown in Table III.

The limits shown in Table III were also published as part of Safety and Health Standards for Federal Supply Contracts of the United States Department of Labor. By publication in the *Federal Register* (*28a*) these TLV's are now official Federal standards for industrial air.

Table III Criteria for Sampling Respirable Dust

Aerodynamic diameter (μm) (unit density sphere)	*% Passing selector*
2	90
2.5	75
3.5	50
5.0	25
10	0

In the sampling procedure established by the Aerosol Technology Committee of the American Industrial Hygiene Association (*29*), the particulate mater, which is first passed through a size selector (cyclone), must be analyzed for mass and not particle count. Filters which follow the cyclone are also evaluated by weight using a balance with a 0.01-mg sensitivity. The filter medium must be nonhygroscopic in nature, such as glass fiber, silver membrane, or polyvinyl chloride.

In developing these standards and sampling procedures these committees recognized two existing criteria which had been previously specified for "respirable" dust measurement: (a) the British Medical Research Council (BMRC) (*30*), also known as the "Johannesberg," criterion; and (b) that adopted at a 1961 Los Alamos Conference sponsored by the United States Atomic Energy Commission's Office of Health and Safety (*31*).

III. Collection and Evaluation for Particle Count and Size Determination

A. Particulate Matter Sampling and Sample Treatment for Analysis by Counting Methods

1. Introduction

Samples may be collected for particle count evaluation by any of the methods described in Section II. The techniques of sampling and counting that were originally developed for mineral dusts related to pneumoconiosis are the only ones standardized and bearing an official United States Public Health Service procedural description.

The standardized counting technique was published in United States Public Health Service Bulletin No. 217 in 1935. Discrepancies between the light-field counting method specified in Bulletin No. 217, and other techniques have been well documented in subsequent years, but because of the correlation between light-field counts and the incidence of human pneumoconiosis in the United States, the method has continued to be used for evaluation of these health hazards.

Where an evaluation of the concentration of pneumoconiosis-producing dust particles is to be made, the sample should be collected in a standard or midget impinger using dust-free alcohol and water; the count should be made by the standard light field counting technique using a Sedgwick–Rafter or Dunn counting cell (*32*).

Samples of asbestos or other fibers are collected on membrane filters, rather than by impinger, and are evaluated by techniques described in United States Public Health Service, National Institute for Occupational Safety and Health, Document No. TR-84 (*33*). The filter is transformed from an opaque solid membrane to a transparent, optically homogeneous gel, and the fibers are sized and counted by phase-contrast microscopy.

2. Samples Collected in a Liquid

a. Optical Microscopy. A standard biological microscope or its equivalent should be used for particle counting. An adjustable draw tube is preferable to a fixed tube type for ease of calibration. A microscope equipped with an Abbe condenser iris diaphragm, a 10× ocular and 3 objectives, 16, 4, and 1.8 mm, will provide the necessary range of magnification for evaluating particles in the 0.5–25-μm size range.

The counting area is defined by a "Whipple disk" (Fig. 6) or similar type of ocular grid, or by rulings on the counting cell. The portion of the grid to be evaluated must be calibrated with a stage micrometer. With the adjustable tube of the microscope the lines of the stage micrometer can be aligned with those of the Whipple disk thus establishing the calibration so that the exact area counted can be determined accurately.

Illumination for the light-field counting technique consists of a 15-W substage lamp placed under the condenser, or a 75-W microscope lamp positioned 10 in. in front of a plain mirror. A daylight filter is required for this type of counting (*34*).

b. Counting Cells. The committee on standard methods of the American Conference of Governmental Industrial Hygienists recommended that a standard counting cell have a liquid depth of 1.0 mm (*35*).

The Sedgwick–Rafter cell (Fig. 7) consists of a glass plate slightly larger than an ordinary microscope slide on which a rectangular area has been set off by cementing plate glass strips exactly 1.0 mm high to

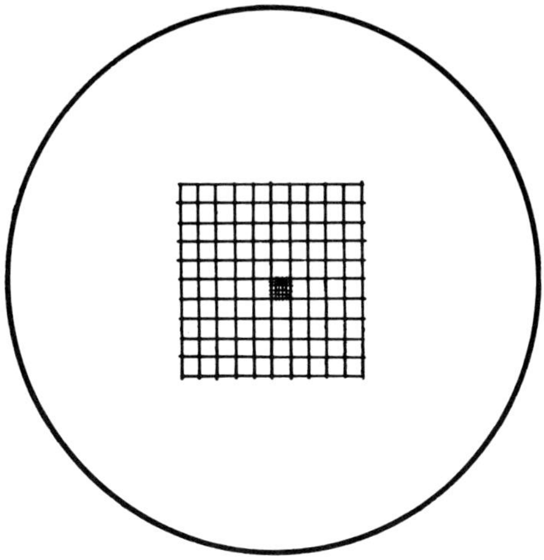

Figure 6. Whipple disk grid for dust counting.

Figure 7. Sedgwick–Rafter cell.

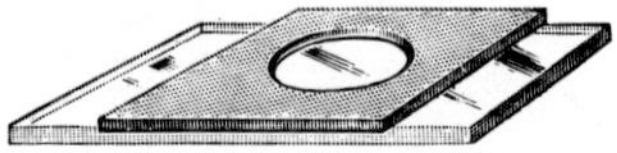

Figure 8. Dunn cell. (Courtesy of Arthur H. Thomas Co.)

the base. A cover glass completes the cell. The cell is durable and is easy to use, but the raised sides makes it difficult to clean.

The Dunn cell, first described by K. L. Dunn in 1934 (Fig. 8) is composed of three separate parts: (a) a rectangular glass plate 1 mm thick containing a circular hole for holding the sample liquid, (b) a thick base plate, and (c) a cover glass (*36*). Upon introduction of the liquid a seal is formed between the base plate and the sample container section allowing little or no sample to escape between them. This type cell can be easily cleaned.

A hemocytometer cell having a depth of 0.1 mm has also been used in dust counting. This type cell, which is designed primarily for blood counting, has a ruled grid on the cell base. The reduced depth decreases the time required for particles to settle, but other problems associated with it have been such that it is not used extensively.

3. Preparation of Sample

Samples collected in liquid are transferred from the impinger directly to the counting cell after any dilution that may be necessary. Samples in liquid suspension have generally shown greater repeatability if counted within 24 hours of collection. This, of course, depends on the material collected and the collecting medium.

The volume of the suspension should be increased by dilution with dust-free distilled water (or collecting medium) until the number of particles within any one counting field is less than 150. The sample counted should also contain four or more times the blank count, i.e., the count of the dust-free distilled water (or collecting medium). The blank count should be no more than 10 particles per field. Drinker and Hatch (*37*) suggest 50–75 particles per counting field while others prefer up to 100 particles. The number of particles that can be counted with greatest precision may be in the 75–100 range but is, of course, dependent to some extent on the range of particle size. With a wide range of particle sizes, the effects of superimposition may cause error since one large particle may obscure several smaller ones. Less error would be introduced when the particles are all somewhat nearer to the same size. Experience has shown that when a counting field has up to 150 particles and an aliquot of the sample is diluted by a factor of 3 (10–30 ml), the resulting particle count is likely to range from 60 to 70 particles, rather than the 50 one would expect. This indicates the potential error in counting samples containing an excessive number of particles.

The sample should be shaken vigorously to obtain uniform distribution of particles in the suspension. A portion of the sample should be taken up in a pipet and transferred immediately to a clean counting cell. If the sample has been collected in a midget impinger, the impinger tube may be used to transfer the sample instead of a pipet.

Two cells of each sample are to be prepared for counting. The second cell should be prepared with a lapse of time equal to that required for making a cell count so that when the first cell has been evaluated the second will have been settling for the appropriate time before counting. Dust in cells 1 mm in depth should be allowed to settle for 30 minutes before counting.

Where samples are collected by electrostatic precipitator, the particles are removed by washing and suspended in a fluid for counting. Wash the particles from the inside of the precipitator using a wash bottle and distilled water followed by swabbing the inside of the shell using a rubber policeman. The operation is repeated until the sample has been completely transferred to the liquid. Some types of electrical precipitators collect directly on a slide or foil on which the sample may be evaluated dry.

4. Evaluation by Optical Microscopy

After allowing the particles 30 minutes in which to settle, the cell is placed on the stage of the microscope and an area selected for counting. Counts are to be made in five separate sections of the cell, the sections are in the general area of the center and each of the four corners. If a circular cell is used, the corners are those of an "imaginary" square of the sample.

A uniform procedure should be developed and followed since accuracy of dust counting is greatly enhanced by physical skill, and the process can best be developed by repetitive actions. One specific quadrant of the Whipple disk (Fig. 6) should always be used for counting samples.

Begin the count at the upper left and proceed from left to right counting a set of five squares across the top, down to the next set of five squares and proceed from the right to the left edge of the grid, down to the third row and continue in this manner until all particles in the 25 squares of the calibrated quadrant have been counted. For particles that just touch the outer boundaries of the counting area, include all that touch the line along two sides and exclude those on the other two perimeters. All dust particles irrespective of size are counted. Each particle is tallied on a hand counter, and the number of particles is recorded. The second cell of the sample is treated in the same manner and a blank is counted for each set of samples.

The concentration of dust determined in this manner is expressed in millions of particles per cubic foot of gas sampled (mppcf) and reported to two significant figures (*38, 39*).

$$\text{Concentration} = \frac{(N_s - N_c) \times V_w}{V_a \times A \times D} \tag{1}$$

where

N_s = Mean count per field (suggested 10 fields)
N_c = Mean count per field for the blank
V_w = Total volume of liquid, ml
V_a = Gas volume sampled, ft^3 ($ft^3 \times 28.3$ = liters)
A = Area of counting field in square centimeters (determined by calibration)
D = Depth in cell in mm

Example

Duration of sample	10 minutes
Rate of sample	0.1 ft^3/m (2.8 lpm)
Volume in flask (V_w) total	100.0 ml
Cell depth	0.1 cm
Area counted	0.0025 cm^2 *

Cell 1	Cell 2
47	54
32	51
40	44
30	41
51	45
200	235

Total 200 + 235 = 435
N_s = 435/10 = 43.5 mean sample count
N_c = 5.0 mean blank count
$N_s - N_c$ = 38.5 mean net count per field
D = 0.1 cm

$$\text{Concentration} = \frac{38.5 \times 100}{0.1 \times 10 \times 0.0025 \times 0.1} = 15.4 \times 10^6 \text{ (ppcf)}$$

mppcf × 35.3 = million particles per cubic meter

* The counting area has been established and calibrated with the stage micrometer to be 0.05 cm × 0.05 cm.

B. Filter Samples

Although evaluation of filter samples by count is not directly comparable to the older wet collecting methods, it is a good method and should yield as good precision and accuracy as any other.

The collection of dust particles on filters for study and evaluation by electron microscopy is a recognized and established method for particles in the <0.3-μm sizes.

1. Membrane Filters

The membrane filter is widely used as a medium for collecting particulate matter for count analysis as well as for mass evaluation. Counting can be accomplished directly on the filter medium by rendering it translucent with an immersion oil of appropriate index of refraction.

With the membrane filter, dust particles are deposited in an essentially unaltered state without shattering or agglomerating. The individual particles in a representative area are counted using the light-field technique referred to previously. A blank count of unused membrane filters should also be made. United States Air Force Technical Order No. 00-25-203 recommends that for routine work an average background count may be obtained by counting four (4) filters per box of 100. If the background count approximates 10%, or more, of the sample count each individual filter should have its background count established. Alternatively, a larger sample volume can be used to minimize the significance of higher filter background counts.

The chemical homogeneity of the matrix of membrane filter material allows for almost complete transparency of the membrane if the pore volume (about 85% void space) is completely filled with a clear neutral liquid which has the same index of refraction as the membrane. In this manner, particles on the surface, and also those which have penetrated the membrane, become visible in transmitted, light-field illumination and can be counted or sized. Thus the preparation for immediate microscopic evaluation of the collected particles involves only placing the filter (or a portion of it) dust side down on a previously inspected clean microscope slide and adding a few drops of appropriate immersion liquid to the upper side.

If a permanent mount is desired for convenience of repeated handling, the filter may be wetted with a mixture of dimethyl formamide and ethyl alcohol which, upon drying, produces a permanent mount (*40*).

Acetone or some other organic solvent may be used to arrest the migration of particles toward the periphery if this occurs during the application of the immersion liquid. Condensing acetone vapors evenly over a filter on the microscope slide may serve to clear the membrane, without graininess or cloudiness, with the particles sealed in a single plane in a permanent mount (*41*). The use of acetone also provides a sufficient difference in refractive indexes between the cleared membrane and collected material having a refractive index close to 1.5, e.g., quartz.

The counting of particles collected on membrane filters is facilitated by ruled grids on the filter surface. The area of the grids or the deliniated counting area must be determined by calibration with the microscope at the magnification and tube length to be used.

Broad application of this technique in "Clean Room" monitoring has resulted in Technical Orders by the United States Air Force and the Society of Automotive Engineers (SAE) (*42*).

Evaluation of particles in the <1-μm range can be made with the use of oil immersion techniques. These techniques used at 1000× are applicable for evaluation of particles down to 0.3 to 0.5 μm in size. Numerical evaluation of samples is made by the following formula:

$$\text{Concentration} = \frac{N_T \times A_F}{n \times a \times V_a} \qquad (2)$$

where

N_T = Total number of particles counted in n areas
n = Number of unit areas counted
a = Unit area, mm^2
V_a = Air volume sampled, ft^3 ($ft^3 \times 0.0283 = m^3$)
A_F = Total effective filter area, mm^2

2. Evaluation by Electron Microscopy

The use of electron microscopic techniques for particle counting (*43*) has not found wide application probably because up to the present time there has been no urgent demand for numerical evaluation of particles in the submicron range.

Counting with the electron microscope does not impose any unusual problem. The number of particles enumerated must be related to a known area established by calibration and should also be representative of a known volume of the sample.

IV. Particle-Size Analysis*

A. *Introduction*

Particle size may be uniquely defined only for the special case of spherical particles, for which particle size is defined as the particle diameter.

* See also Chapter 3, this volume.

Natural aerosols are generally polydisperse with respect to particle size and encompass a range of approximately 6×10^{-4} to 10^3 μm. Except for liquid droplets, a few metallurgical fumes and certain natural air contaminants, such as spores and pollen, particulate matter in the atmosphere is nonspherical. The particle size of nonspherical particles may be defined in terms of equivalent diameters which depend on various geometrical or physical properties of the particles. It is evident that particle size for irregularly shaped particles will depend upon the physical principles employed for determining the average dimension or equivalent diameter. Generally, none of the methods can be expected to agree with each other with any appreciable precision. For this reason, it is essential that particle-size analysis be reported in terms of the method used. Thus, for instance, particles classified by geometrical similarity, as in sieving and microscopic sizing, and by hydrodynamic or aerodynamic similarity as in sedimentation or elutriation analysis, may show great differences in equivalent diameters. Manipulative skill is often essential for obtaining reliable results.

According to geometric similarity, particle size is related to the particle perimeter or its projected area. These measurements do not take into account particle density or shape. According to hydrodynamic or aerodynamic similarity particle size is related to particle motion through a liquid or a gas. These measurements do not take into account particle perimeter or projected area. Particles may vary so greatly in density that the application of Stokes' radius r_s is not without problems. It becomes more practical to define the equivalent aerodynamic size r_e or d_e, the radius or diameter of a sphere having the same falling velocity as the particle and a density equal to 1 gm/cm^3 (*44*).

The determination of size with reasonable precision is a somewhat difficult problem but it should be emphasized that the procurement of a representative sample for sizing is just as important and perhaps just as difficult.

B. Sieving

1. Dry Sieving

Dry sieving may be employed for evaluation of particles from approximately 0.25 in. (64 mm) to 74 μm and wet sieving is often used for particles down to 37 μm. The latter size corresponds to the aperture in the finest sieve commercially available and refers to the diameter of the sphere that will pass the aperture. Particle-size determination with sieves consists of placing the sample in the first of a nest of sieves and agitating the nest for a given period of time (*45*). Sieving may be done one sieve at a time; or with a nest of sieves, each having progressively smaller openings, so that by weighing the retained particles on each screen, particle-size distribution data may be obtained.

Sieving is a two-stage process. In the first stage, particles considerably smaller than the screen opening pass readily through the screen. The second stage involves the much slower passage of particles close to the size of the screen opening. The second stage is affected by the shape of the particle, moisture content, relative humidity, the method of shaking, and the amount of sample used for the test (*46*).

Usually from 50 to 150 gm of dry sample is placed in the top sieve of the series, and using a combination of shaking and tapping with the hand, or a mechanical shaker, the screens are agitated for from 10 to 15 minutes or until the rate of collection on a specific sieve indicates that particles of that specific size have been collected. The amount of material on each sieve is weighed.

The American Society for Testing and Materials Standard No. C-238-51 (*47*) stipulates:

> Procedure-5. (a) The sieving operation shall be conducted by means of a lateral and vertical motion of the sieve, accompanied by a jarring action so as to keep the sample moving continuously over the surface of the sieve. In no case shall fragments in the sample be turned or manipulated through the sieve by hand. If hand operation is employed, sieving shall be continued until not more than 1% by weight of the residue passes any sieve during 1 minute.
>
> If mechanical sieving is used, the device shall be such as to impart the type of agitation described in the preceding paragraph and the shaking shall be continued for a period of 15 minutes.

Sieves are used to determine size analysis of particles whose relative dimensions of length, breadth, and thickness may vary appreciably. Therefore, it is a process of statistical classification. A particle is classified according to its width. Whether a particle of near mesh dimension will or will not pass through a sieve is a matter of chance. Continued sieving will increase the chance, and degrade particle size by attrition. For these reasons there is no end point in a sieve analysis, and the point of termination must be fixed. Any of the following criteria may be used to establish an end point for a sieve analysis: (a) standard length of time for sieving, or (b) weight of material passing through a sieve per unit time is less than a specified percentage of the weight of the residue on the sieve.

Sieve size openings and wire diameters are standardized and certified in the United States by the National Bureau of Standards and the American Society for Testing and Materials (*48*).

Since 1910 many countries have adopted National Sieving Standards based on the same fixed ratio as the United States Series, $4\sqrt{2}:1$. This includes Canada, United Kingdom, Netherlands, New Zealand, Japan, India, Argentina, Chile, and Uruguay (*49*). These standards as designated correspond to the values for test sieve apertures recommended by the International Standards Organization, Geneva, Switzerland.

2. Wet Sieving

In the wet sieving technique, a liquid, usually water, is added to the sieves in which the finer fractions are to be separated, and sieve analysis conducted as previously described.

The wet method is useful for sieving materials below 74 μm, which is the smallest sieve size for dry sieving; or for samples originally collected in a liquid. The method is also valuable for samples in which the particles tend to agglomerate during dry sieving. Evaluation of the portions retained on each sieve can be made by gravimetric analysis after the liquid has been removed by evaporation.

The lowest size for sieve analysis is 37 μm. Classification of smaller sized particles is usually accomplished by microscopic techniques.

C. Size Analysis by Optical or Electron Microscopy

Evaluation of particle sizes by optical microscopy involves the measurement of a linear dimension using a micrometer or the comparison of the area of circles on an eyepiece graticule with the area of the particle. Samples collected by any of the previously described sampling procedures may be evaluated by optical or electron microscopy.

1. Preparation of Sample

If a dry or bulk sample is believed to contain a significant number of particles larger than 100 μm, it may be desirable to first perform a sieve analysis, in which case preparation of the sample would consist of drying as required and weighing the portion of the sample to be sieved. Particles less than a specified size, such as 100 μm, could then be further evaluated by use of microscopic techniques.

For microscopic observation a dry or bulk sample can be dispersed on a microscope slide. It is essential that the particles be dispersed on the slide in a random manner without bias or fractionation. Care must be taken so that individual particles are not shattered in the dispersion process; and agglomerates must be deflocculated. A method for producing a dust dispersion free of agglomerates is called the "rubbing out" method

(*50*). A drop of water or an alcohol, such as diacetone alcohol, may be put on a clean glass slide and a small amount of particulate matter added using the end of a toothpick or similar wooden device. This mixture is then spread across the slide in a thin film. The liquid evaporates, leaving a dispersion of the dust on the slide.

For a permanent slide, Canada balsam dissolved in xylene is used. Dust particles in alcohol are mixed with the Canada balsam and dispersed evenly on a glass slide. On evaporation of the solvent the dust remains embedded in the viscous residue. A clean cover slip is then placed over the dust preparation. The Canada balsam will normally harden in 1 or 2 hours, but examination of the slide can be started as soon as the slide is prepared. Quartz cannot be evaluated in a Canada balsam mounting because the index of refraction of quartz (1.54) is so close to that of the Canada balsam (1.535) that the particles cannot be seen.

Thermal precipitator samples which are collected dry on a glass cover slip may be mounted dust side down on a microscope slide and evaluated directly by microscopic observation. If 1000× magnification is desired, immersion oil is added at all glass–air interfaces. Sizing should be carried out across the dust strip to minimize any error that may result from size-selective deposition.

Samples collected on membrane filters may be placed on microscope slides, dust side down; immersion oil added to make the filter transparent; and size analysis conducted at the required magnification.

Sticky paper samples may be attached to a glass slide for evaluation using the same technique as described for counting samples.

Samples collected in a liquid or those suspended in a liquid may be placed on a microscope slide and analysis made after evaporating to dryness.

2. Techniques in Sizing

a. GENERAL. Since microscopic images allow only linear or area measurements, the size of irregularly shaped particles, when determined by microscopic measurement, is stated either in terms of the "projected" diameter, or in terms of one of the so-called "statistical" diameters. These can be ascertained by direct examination of the microscopic images of particles, by projection of the field of view on a screen, or by photomicrography.

The "projected" diameter is the diameter of a circle equal in area to the profile of the particle when viewed normal to its position of greatest stability (*46*). It is usually determined by comparing the particles with areas enclosed in calibrated circles of graduated size on an eyepiece reti-

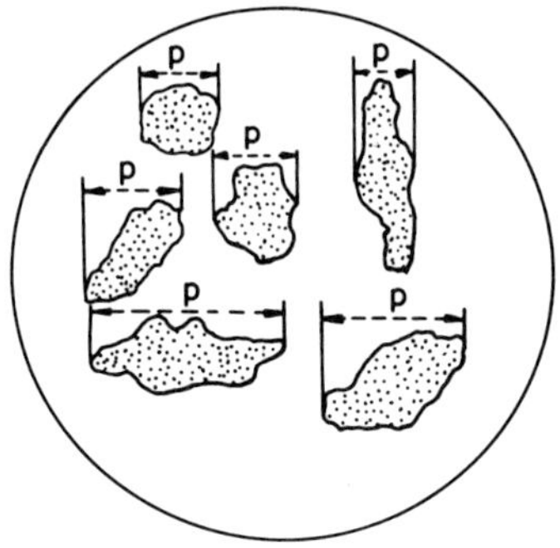

Figure 9. Statistical distribution of particles using Feret's diameter.

cle or by measuring directly with a filar micrometer. The areas may also be measured exactly with a planimeter, from projections on a screen, or on photomicrographs.

The "statistical" diameter (Fig. 9) concept follows from the consideration that some average linear measure of the projection of many particles in a fixed direction is statistically adequate for describing the size range or size distribution of particles (*51*). It requires the assumption that the particles of the sample studied are randomly oriented, and that the number of particles examined and measured is sufficiently large to be statistically adequate. There is little information to indicate how many particles should be sized, using statistical diameters, to achieve a given confidence level, although it has been found from numerous measurements that differences between repeated size frequency distribution determinations become smaller as the number of particles examined increases. For most purposes the sizing of 200 particles has been suggested as adequate (*52*).

Two of the most commonly used statistical diameters are Feret's and Martin's diameters (*53*) (Fig. 10). Feret's diameter is the mean distance between two tangents on opposite sides of the particle image, the tangents being drawn perpendicular to the direction of traverse. Martin's diameter is the length of a line that intercepts the particle image and divides it

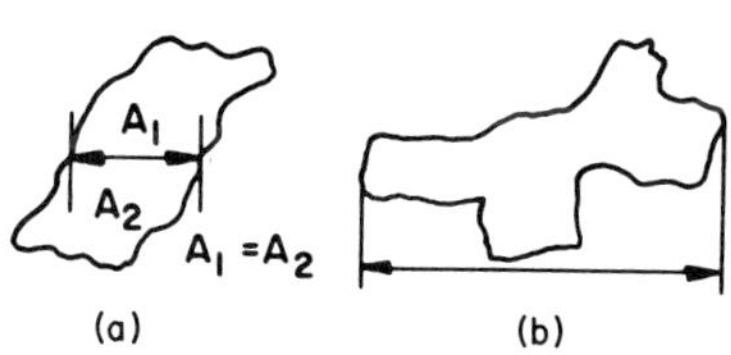

Figure 10. (a) Martin's diameter measuring mean length of line dividing particles into equal areas. (b) Feret's diameter measuring mean distance between tangents on opposite sides of particle image.

into two equal areas, with the bisecting line always being taken parallel to the direction of traverse.

Feret's diameter may be measured with a micrometer eyepiece having a movable cross hair, i.e., a filar micrometer. It may also be determined with an image-splitting eyepiece or, less frequently, has been estimated by means of an eyepiece reticle.

Experimental evidence indicates that, for a given particle distribution, Martin's diameter is slightly smaller than the mean projected diameter, while Feret's diameter is greater (*46*). The ratio of the two statistical diameters has been found to remain constant for a given material.

The projected diameter corresponds more closely to sedimentation measurements (*51*). Therefore, the manner in which the size of the particle was determined should always be given with the results. This is especially important when the size distribution by count is converted to a volume or mass distribution.

Sizing should be done in a randomly selected area of the microscope slide, but, in general, the evaluation should be made across the entire width of a selected area to compensate for any segregation of particles that may have occurred during the preparation of the sample; or during the actual sampling operation, if the sample to be sized is the collecting slide.

A monocular microscope used with either a filar micrometer or eyepiece reticle is preferred. The magnifications most commonly used are obtained with a 10× ocular and 16 mm (10×), 4.4 mm (44×), and 1.8 mm (95×) objectives. In some instances the filar micrometer is 12.5× instead of 10× thus achieving a higher magnification. These combinations will give 100, 450, and 950× with the 10× ocular and 125, 550, and 1187.5 with the 12.5× filar micrometer.

The 10× and 44× objectives are used dry with light transmitted through air, and the 95× objective is used in oil immersion with light transmitted through an oil of suitable refractive index. The general range of particle sizes evaluated using 1000× magnification and oil immerson is from 0.4 to 200 μm. The lower limit of sizing for 500× is 0.6 μm and for 100× is approximately 1.0 μm. The differentiation of particle sizes less than 1.0 μm is of a low order of reproducibility.

A microscope of lower magnification may offer greater ease and rapidity in evaluating particles of larger size. The instrument commonly used to size fallout or sticky paper samples has a magnification of from 7 to 60×.

b. Filar Micrometer. The filar micrometer (Fig. 11) consists of an eyepiece containing a movable hair controlled by a calibrated micrometer

Figure 11. Filar micrometer eyepiece for measuring particles. (Courtesy of American Optical Corporation.)

screw. In using the filar micrometer for particle-size determination the calibration is established by use of a stage micrometer. The distance set off by the markings on the micrometer screw, as the cross hair is moved from left to right, is directly related to the μm on the linear stage micrometer scale.

After calibration of the eyepiece, the sample to be evaluated is placed on the stage, the area for sizing is selected at random and all particles within the field are sized. If Feret's diameter is used, the cross hair is moved to the left edge of the particle and the micrometer reading is noted, then the cross hair is moved to the right edge of the particle and the micrometer reading again recorded. The total number of divisions traversed by the micrometer is determined by difference and converted to μm from the calibration data. The use of the filar micrometer requires two persons, one to adjust the cross hairs and the other to make the micrometer readings and take data.

c. Eyepiece Reticle. The eyepiece reticle consists of a glass plate which is placed in the focal plane of the ocular (Fig. 12). The reticle,

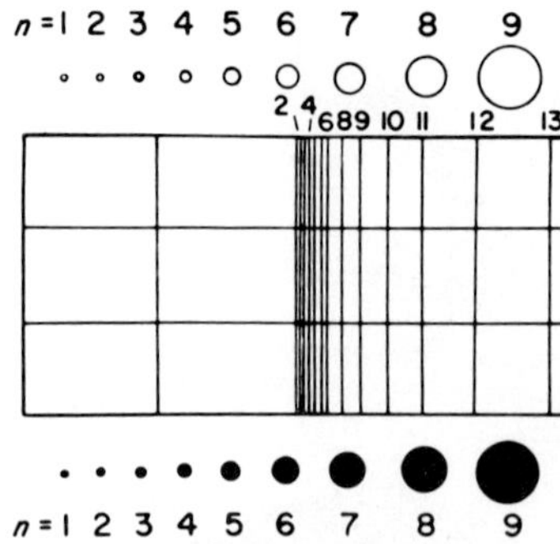

Figure 12. Eyepiece reticle for estimating particle size.

several types of which are available, has circles of various sizes which are compared to particles in the microscope field. One reticle has as indices a series of circles and a logarithmic scale for comparison with particles. The circles are numbered from 1 through 9 and the diameter (D) of each circle is defined in arbitrary units as $D = \sqrt{2^n}$, where n represents the number of the circle. The logarithmic scale shows the value of n for the measurement of particles larger than number 9 circle.

The large rectangle in this reticle is 100 D units in width and 200 D units in length. Calibration may be accomplished with a stage micrometer, by determining the dimensions of the rectangle in microns, and calculating the diameter of each circle.

An estimated size for each dust particle can then be assigned with reasonable speed and reproducibility by comparing the projected area or diameter of the particle to the area or diameter of the appropriate circle.

d. MICROPROJECTOR. A microscope with a projection prism may be used to project the images of the particles onto a screen. A suitable scale may be projected onto the screen along with the sample. This method requires more elaborate equipment but particle sizing is less tiring to the eyes.

D. Sizing with Electron Microscope

1. General

The electron microscope allows the direct observation of solid particulate matter down to around 0.001 μm in size. In contrast, the theoretical resolution of the light microscope is approximately 0.24 μm. Application of the electron microscope has been more for sizing and study of the morphology of particles than for counting. There is no single, widely accepted technique for sample collection for subsequent investigation by electron microscopy. Billings and Silverman (*54*) have outlined many of the applicable methods. The membrane filter, thermal precipitator, and electrostatic precipitator are the devices most commonly used for collecting particulate matter for evaluation with the electron microscope.

2. Techniques of Preparation and Measurement

If the sample has not been collected on an electron microscope specimen screen, it must be placed on one for introduction into the electron microscope (*55*). This is usually done by using a 0.7% Formvar (polyvinyl formal) solution in ethylene dichloride (*56*). A coated grid is placed, film side up, on a section of 10-mesh screen on the bottom of a petri dish. A section of filter is placed dust side down on the Formvar-coated grid. Ethyl acetate (or other suitable solvent such as acetone) is poured into the dish until the filter squares are wetted by capillary action. With the petri dish covered, the filter will dissolve in 10–20 minutes. The dust particles settle on the undissolved Formvar film. After drying, the specimens are ready for electron microscope examination.

A common method for determining size of particles is to project an electron photomicrograph on a screen ruled to show the size of a 1-μm particle. Because of the high magnification, and consequent small field, of the electron microscope, it is necessary to examine five or ten fields and size 200–400 particles for statistically acceptable results.

E. Sedimentation

The sedimentation method of particle-size analysis is based on the measurement of the terminal settling velocity at which particles fall under gravity through a gaseous or liquid sedimentation medium. Analysis by sedimentation techniques requires several grams of sample and, in general, yields a gross grouping of sizes rather than a specific size for individual particles.

If the terminal settling velocity U_t of the particle in a given sedimentation medium is known, an "equivalent" particle diameter d can be computed from Stokes' law for a sphere falling steadily under the influence of gravity:

$$d = \left[\frac{18 n U_t}{(P_P - P_f) g} \right]^{1/2} \tag{3}$$

where

P_P = Density of particle, grams per milliliter
P_f = Density of sedimentation medium, grams per milliliter
n = Absolute viscosity of sedimentation medium, poises
g = Acceleration of gravity, 981 cm/sec^2

The diameter d is commonly termed the Stokes' or sedimentation diameter, and may be defined in the case of irregular particles as the diameter of a sphere having the same settling velocity under gravity and the same density as the particle in question. Often it is convenient to use another term, called the reduced sedimentation diameter or the equivalent aero- (or hydro-) dynamic diameter. This is the diameter of a sphere of unit density having the same settling speed as the particle.

For Equation (3) to be strictly valid, even for spherical particles, the conditions of Stokes' law must be met. That is, the flow around the falling particle should be streamline and the particles should fall through the sedimentation medium remote from each other. Usually the settling speeds are low enough for the first condition to be satisfied. Cadle quotes experimental evidence that deviations from streamline flow are negligible for particle Reynolds numbers (Re) less than 1.2, where (*57*)

$$\text{Re} = P_f U_t d/n \tag{4}$$

Particles having diameters approximately equal to or less than the mean free path of the molecules of the sedimentation medium fall faster than Stokes' law predicts due to particle "slip" between the atoms or molecules of the fluid. To make allowance for this, the terminal settling velocity U_t must be multiplied by Cunningham's correction factor:

$$C = 1 + 2LA/d \tag{5}$$

where

L = Mean free path of the molecules in the sedimentation medium, centimeters
A = A numerical factor close to unity (0.9 is often used)
d = Diameter of particle, centimeters

The second requirement of Stokes' law, that of no mutual interference of particles in their downward motion, cannot be completely satisfied in practical sedimentation procedures. Usually a compromise is sought between the ideal concentration of particles (a single particle falling in an infinite medium) and one that an appropriate sensing device can detect accurately.

The sedimentation methods used for determination of particle-size distributions are of two types. The first of these, in which the changes of concentration at a particular level in a settling suspension are measured, is called incremental. Examples are pipet, hydrometer, and photoextinction methods. The second type, called cumulative, involves the measurement of the change in the overall concentration of the suspension, usually by a sedimentation balance or a manometric method. With the first, the fractional concentration at a chosen depth gives directly the weight fraction of those particles smaller than the size which would just fall from the surface to that depth in a given time from the commencement of settling. In the second type, a weight-deposited versus time curve, usually known as the sedimentation curve, must be obtained, which must be differentiated before the size distribution can be evaluated. The theory and procedures of the various sedimentation techniques have been described in detail by Orr and Dalla Valle (*51*).

Adequate dispersion of the particles in the suspension is essential in sedimentation analysis. The effects on the sample of the dispersion method used should be carefully considered. Thus, for example, high velocity gas jets, often used with gaseous sedimentation media, may impart electrostatic charges on the particles and affect their settling action. Similarly, mechanical agitators for liquid sedimentation systems are rarely adequate for uniform particle dispersion. Sedimentation systems are also quite sensitive to convective disturbances in the sedimentation medium, and precautions to ensure thermal stability of the system have to be devised. Even normal evaporation of liquid medium has been shown to cause convection currents (*58*).

F. Elutriation

Elutriation may be considered as the inverse of sedimentation. The same major problems of particle dispersion and dynamics are common to both methods.

Air elutriation devices consist of one or more interconnected chambers of decreasing diameter (*59*). Air is introduced into the dust sample holder and particles smaller than the separation size pass through a settling chamber to be collected by a filter or remain airborne in a subsequent chamber (Fig. 13). The size of cut can be adjusted by changing the velocity of the airstream through the instrument. A minimum quantity of several grams of sample is usually necessary to perform particle size classification. There is considerable agglomeration due to electrostatic effects which also causes particles to adhere to the walls and tubes of the system. Under these conditions accuracy of separation is poor and the total

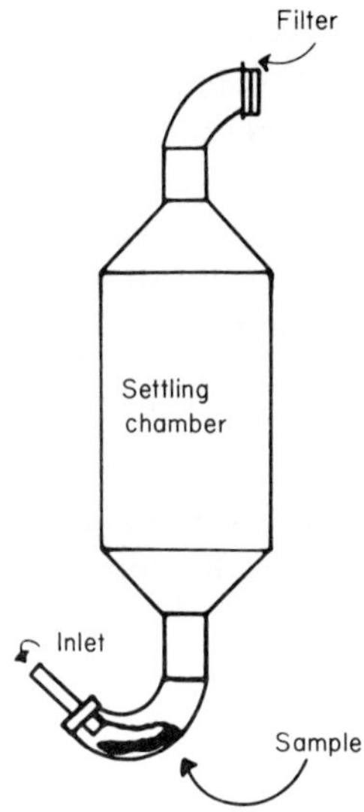

Figure 13. Simple form of an elutriation system for fractionating particles.

weight of fractions obtained may be more or less than original sample weight.

G. Centrifugal Classification

Centrifugal classification has the advantage of providing separation of smaller particles than can be achieved by gravitational techniques and at a more rapid rate. The technique has been found to be useful for particles in the submicron size ranges, down to as small as 0.025 μm. Formulas for calculating the increase in settling velocity and description of various centrifuge techniques have been presented by Orr and DallaValle *(51)*.

Particle size analyzers employing centrifuge techniques are commercially available *(60)*. In these, centrifugal force causes the particulate matter to deposit at specific locations in the analyzer peculiar to their settling velocity.

The Bahco microparticle classifier is a combination air centrifuge–elutriator. It has been specified as the standard instrument for particle-size evaluation by the American Society of Mechanical Engineers (ASME) Power Test Code *(61)*. The Bahco classifier separates dry, solid particulate matter (a sample of about 10 gm is necessary) into an elutriated or fine fraction and a "settled" or coarse fraction by subjecting the particles to a centrifugal force which is opposed by air drag on the moving particles. The weighed sample is introduced into a spiral air current, having suitable values of tangential and radial velocity, created by the vanes of the fan. Depending on the size, density, and shape of the particles, a certain "settled" fraction of the sample is accelerated by the centrifugal force toward the periphery, collected in the catch basin, and subsequently weighed. The fine fraction is carried out of the sifting chamber through the fan vanes, and its weight is taken as the difference in weight of the coarse fraction before and after separation. By varying the airflow through the sifting chamber by means of throttles or air orifice settings (eight are available), it is possible to change the terminal velocity limit of division, and thus divide the material into up to nine fractions by reclassifying the coarse fractions of the sample. If the weight and some measure of particle size of each sample fraction are known, a particle-size distribution curve can be plotted. Owing to variations in construction, no two Bahco classifiers perform identically, hence instrument calibration is necessary.

H. Impactors

Impactors, which were previously discussed as sampling instruments in Section II,C,4, are very useful in particle sizing. The cascade impaction principle was first described by May (*62*) in 1945, but a variety of design modifications have been made since then (Fig. 14). Some of the units

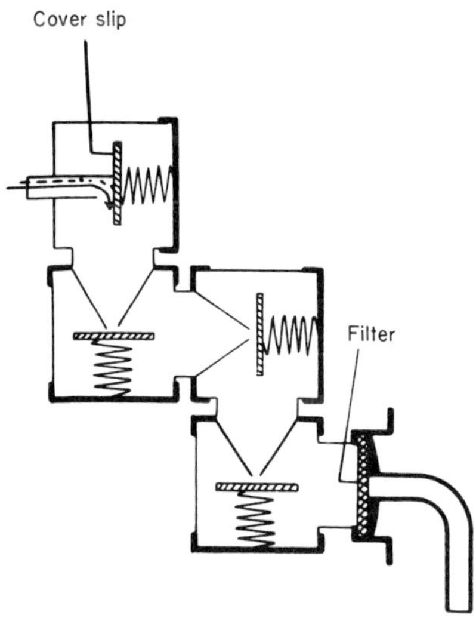

Figure 14. Schematic of a four-stage cascade impactor.

are comprised of seven or more collection stages followed by filters (*63*) (Fig. 15).

If the impactor functioned in an ideal manner, each stage would collect all particles above a certain "cutoff diameter," which could be predicted from theoretical consideration (*64*), and would collect no particles below that diameter. At each successive stage the cutoff diameter would be smaller than in the previous stage, and all particles collected on a given stage would have diameters between two fixed values (except the first stage, where the maximum size particle in the aerosol would determine the upper size limit). The particle collection characteristics of impactors are not ideal so that there is an overlap in sizes from stage to stage (*65*). Therefore, before cascade impactors can be used for estimating particle size or mass distributions of an aerosol, the various stages of the instrument must be calibrated. Manufacturers usually supply calibration curves for the various stages of their impactors for particles of unit density. After the particulate matter has been classified on the various stages

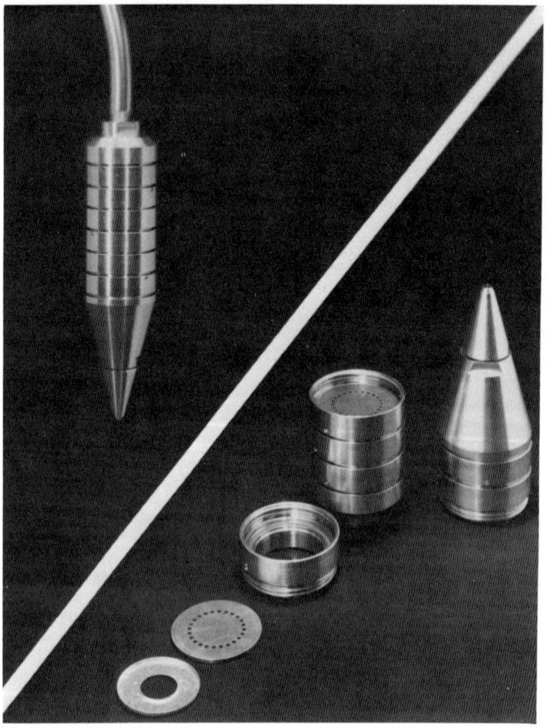

Figure 15. Inertial in-stack cascade impactor. (Courtesy of Meteorology Research Inc.)

of a calibrated cascade impactor, the amounts of material on each stage may be determined by chemical or physical analysis. The lowest particle size collected on the last stage of a typical commercial cascade impactor is about 0.24 μm (unit density), with smaller sizes retained by the filter stage (*66*).

The usefulness of an impactor in determining mass distributions of particulate matter depends to a large extent on the accuracy of its calibration. Numerous accounts of calibration procedures have been reported in the literature (*61*).

I. Automatic Instruments for Particle-Size Analysis

1. General

Numerous instruments have been developed for the analysis of particles by size classification and by enumeration. Many of these units should be designated as "classifiers" rather than sizers, as, in most instances, the particle separation is at best into groups or size ranges, and often the overlap may be rather extensive. To obtain identification of particles within size groups, optical or electron microscopy may be required.

The principal developments have been in sensing zone methods, where particle measurement is accomplished by causing a dilute suspension of particles to flow through a small, defined zone of electric, photic, or sonic energy so that either the passage of individual particles can be detected and particle size indicated or integrated data obtained with respect to all particles present. A suitable sensor produces either a series of electrical pulses, each proportional to the size of a particle, or an electrical signal proportional to the integral of such information. The rate of sensing is usually such that adequate numbers of particle pulses are produced in a few seconds or minutes.

2. Sensing-Zone Instruments

a. Electrolytic Sensing. A device utilizing electrolytic sensing was first described by Coulter in 1956 as a high-speed, high accuracy method (*67*). Since its initial development, a wide variety of applications and adaptive techniques have been worked out.

Electrolytic sensing detects particles suspended in a slightly conducting liquid as they are drawn from one chamber into another through a small orifice. For a given sample and electrolyte, the change in resistance is directly proportional to the volume of the particle if the particle diameter is not larger than 30% of the orifice diameter. The size of the electrical pulse produced is proportional to the change in the resistance as the particle passes through the orifice. A complete set of size distribution data may be derived from a single milliliter of suspension. A cumulative count of particles larger than a given size can be made over the entire particle-size range covered by the instrument.

Different sizes of orifices are used for each range of particle size to be measured. Calibration is with monodisperse particles of known volume (*68*).

This instrument (Fig. 16) has been used successfully as an advantageous substitute for laborious microscope counting. Both particle concentration and size distributions are obtained using a counting time of under 30 seconds. Individual particles are sensed one by one at rates up to 4000 per second. Human error is very low and statistical accuracies are high. The United States Bureau of Mines (*69*) has standardized this procedure as a replacement for their microprojection procedure. Industry (*70*) has used it for both ambient and source samples.

The procedure requires that the particles be collected in an electrolyte, or collected dry and transferred to the electrolyte. Both Greenberg–Smith and midget impingers are used for the former and surfaces such as membranes, glass slides, plates, and thimbles used for the latter. Surface-collected samples are transferred to the electrolyte by washing. In certain instances, ultrasonic energy is used to complete the surface washoff.

Figure 16. Coulter Counter Model TA. (Courtesy of Coulter Electronics, Inc.)

When solubility of sample particles is suspected, observations of the count rate stability will confirm or reject the suspicion. If solubility proves to be a problem, alcohol electrolytes are often substituted for aqueous electrolytes.

b. PHOTIC SENSING. The photic-sensing-zone method detects the passage of particles by forward light scattering, sideways scattering (Fig. 17), or shadowing. In all cases the zone and sensor are formed by an optical system. Either gaseous or liquid media may be used as suspension media, usually with some differences in mechanical structure of the instruments. Minimum sensing-zone sites practically achievable are on the order of 0.01 mm^3 while the minimum measurable particle sizes are in the order of 0.2 to 0.4 μm in gases and 3 to 5 μm in liquids. In liquids, agitation can be used to keep particles in uniform suspension provided that air bubble inclusion and sensing-zone turbulence are avoided. Agitation in gaseous media is relatively difficult, and measurement is practically limited to a maximum particle size of 10–70 μm depending on particle density.

Optical particle counters based on the scattering of light from single aerosol particles have the ability to determine the concentration and particle-size distribution of aerosols (*71*). Several types of instruments using this principle are now commercially available (*72*). The aerosol counter is designed so that individual particles are led into the sensing zone where they are illuminated by an intense beam of light. The light scattered at an angle is picked up by a photomultiplier, resulting in electrical pulses

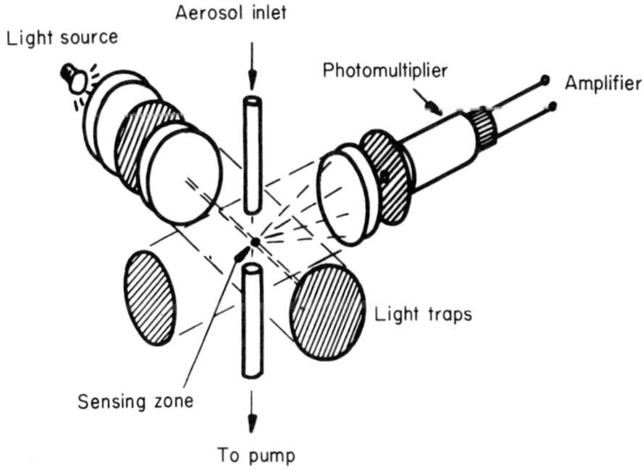

Figure 17. Photic sensing, utilizing forward or sideways light scattering or shadowing.

which give a measure of the particle-number concentration (*73*). If the instrument is to be used for particle-number versus size distribution determination, the individual pulses must be amplified, classified into amplitude groups by a pulse height analyzer, and counted.

The aerosol photometer receives light scattered from many particles illuminated in the sensing zone simultaneously. When the physical characteristics of the particles remain relatively constant, the intensity of scattered light is directly proportional to aerosol concentration, as long as the concentration is sufficiently low, and secondary scattering does not occur among the particles.

There are several advantages to using instruments of these types:

(a) The particles are counted and sized in the state in which they occur in the air, whereas most other methods evaluate particles after they are removed from the air.

(b) A change in particle concentration can be detected almost instantaneously. This makes it possible to pinpoint major problems in an industrial process or to serve as a quality control device for clean room atmospheres.

(c) Low aerosol concentrations, which may cause significant counts on theses instruments, could go undetected with other sampling and evaluating techniques.

The major drawback in the use of light scattering instruments for atmospheric particulate matter evaluation is that there is considerable difficulty in interpreting their measurements. Thus, light scattering methods do not give an objective measure of the airborne dust concentration independent of the composition and particle-size distribution of the aerosol. The intensity of scattered light, as a measure of particle size and concentration, varies in a complicated manner with the geometry of the illuminating and collecting lens systems and with the frequently unknown size, shape, refractive index, and bulk absorption coefficient of the aerosol particles. Some modifications in design principles to reduce or eliminate the influence of some of these variables have been suggested by Hodkinson (*74, 75*).

Because of the complexity of the scattering process and the optical and electronic components required, it is always necessary to calibrate these instruments against standardized particles of known size and optical properties. Studies have suggested that calibration of aerosol counters with only a few sizes of such standardized particles is not sufficient (*76*). Adequate calibration procedures should probably be designed to include tests on monodisperse, truly spherical, particles over the entire indicated size range as well as tests with monodisperse irregular and absorbing aerosols, and particles of different refractive indices.

c. Sonic Sensing. Detection in the sonic sensing zone is accomplished by reflection of an intermittent ultrasonic beam in a liquid medium (77). The liquid need be neither conductive nor clear.

In a liquid medium the theoretical particle size and sensing-zone size are nearly 10 μm and 0.3 mm^3, respectively, but practical operating levels are about 25 μm and 1 mm^3. Instruments which apply this technique to a gaseous medium have yet to be developed.

d. Coincidence Errors. Careful consideration must be given to the errors caused by coincident passage of particles through the sensing zone. Coincidence of randomly distributed particles increases approximately as the square of the concentration for particles at low concentration, and increases linearly with sensing-zone volume, following Poisson's law. Coincidence effects tend to shift the true distribution upward a little in size since pulses lost at a given size level appear in effect at a somewhat larger size level. The practical procedure to avoid significant levels of coincidence is to reduce particle concentration, sensing-zone size, or both.

e. Present Trend. There continues to be considerable interest by researchers in the development of instruments for particle size and count analysis. The development of a revolutionary device that can be obtained at a practical price, that will evaluate ambient atmospheres in the wide variety of sizes, densities and optical properties seems to be still somewhere in the future.

REFERENCES

1. E. R. Hendrickson, *in* "Air Pollution" (A. C. Stern, ed.), 2nd ed., Vol. II, p. 4. Academic Press, New York, New York, 1968.
2. L. Silverman, *Air Cond., Heat., Vent.* **52,** 88 (1955).
3. M. R. Pack and A. C. Hill, *J. Air Pollut. Contr. Ass.* **15,** 166 (1965).
4. F. P. Terraglio, J. P. Sheehy, and R. M. Manganelli, *J. Air Pollut. Contr. Ass.* **8,** 220 (1958).
5. P. M. Giever, ed., "Air Pollution Manual," 2nd ed., Part I, Chapter 9. Amer. Ind. Hyg. Ass., Dayton, Ohio, 1972.
6. "Flowmeter Computation Handbook." Amer. Soc. Mech. Eng., New York, New York, 1961.
7. E. Ower and R. C. Pankhurst, "The Measurment of Air Flow," p. 251. Pergamon, Oxford, England, 1966.
8. American Society for Testing and Materials Committee D-22. *1973 Book ASTM Stand.* Part 23 D-1739-70, p. 893. Amer. Soc. Testing Materials, Philadelphia, Pennsylvania, 1970 (revised).
9. J. S. Nader, *J. Air Pollut. Contr. Ass.* **8,** 35 (1958).

10. M. B. Jacobs, "The Chemical Analysis of Air Pollutants," p. 36. Wiley (Interscience), New York, New York, 1960.
11. W. J. Smith and A. L. Benson, Spec. Tech. Publ. No. 555, pp. 137–142. Amer. Soc. Test. Mater., Philadelphia, Pennsylvania.
12. M. Polydorova, K. Spurny, D. Paspa, and F. Benak, *Int. J. Air Water Pollut.* **9**, 23 (1965).
13. K. R. Spurny, J. P. Lodge, Jr., E. R. Frank, and D. C. Sheesley, *Environ. Sci. Technol.* **3**, 464 (1969).
14. "British Standard Smoke Calibration Curve," B.S. 1747, Part 2. British Standards Institution, London, England, 1964.
15. American Society for Testing and Materials Committee D-22, *1973 Book of ASTM Stand.,* Part 23 D-1899-68, p. 898. Amer. Soc. Testing Material, Philadelphia, Pennsylvania, 1973.
16. TR-2 Air Pollution Measurments Committee: Recommended Standard Method for Continuing Air Monitoring for Fine Particulate Matter by Filter Media Using the Low Volume Sampler (APM-2.2). *J. Air Pollut. Contr. Ass.* **13**, 397 (1963).
17. H. Dresia and F. Spohr, *Staub* **31**, 12 (1971).
18. R. B. Husar, *Atmos. Environ.* **8**, 183 (1974).
19. L. Greenburg and G. W. Smith, *U.S., Bur. Rep Invest. Mines,* **RI-2392** (1922).
20. J. B. Littlefield and H. H. Schrenk, *U.S., Bur. Rep. Invest. Mines,* **RI-3387** (1938).
21. A. L. Linch and M. Corn, *Amer. Ind. Hyg. Ass., J.* **26**, 601 (1965).
22. N. A. Fuchs, "The Mechanics of Aerosols." Macmillan, New York, New York, 1964.
23. T. T. Mercer, *Ann. Occup. Hyg.* **6**, 1 (1963).
24. E. C. Barnes, *Air Pollut., Proc. U.S. Tech. Conf. 1950,* p. 547 (1952).
25. H. J. White, "Industrial Electrostatic Precipitation." Addison-Wesley, Reading, Massachusetts, 1963.
26. L. Waldmann and K. H. Schmitt, *in* "Aerosol Science" (C. N. Davies, ed.), pp. 137, 162. Academic Press, New York, New York, 1966.
27. J. R. Ashford, *Brit. J. Appl. Phys.* **11**, 13–21 (1960).
28. Committee on Threshold Limit Values, "Threshold Limit Values for Chemical Substances and Physical Agents in the Workroom Environment with Intended Changes for 1973." Amer. Conf. Govern. Ind. Hyg., Cincinnati, Ohio, 1973.
28a. Federal Register, *Fed. Regist.* **36**, No. 105 (1971).
29. Aersol Technological Committee—Amer. Ind. Hyg. Assoc., "Guide for Respirable Mass Sampling Aerosol-Technology," *Amer. Ind. Hyg. Ass., J.* **31**, 133 (1970).
30. A. J. Orenstein, ed., "Proceedings of the Pneumoconiosis Conference held at the University of Witwaterstand, Johannesburg, 9–24 February, 1959," pp. 619–672. Little, Brown, Boston, Massachusetts, 1960.
31. T. Hatch and P. Gross, "Pulmonary Deposition and Retention of Inhaled Aerosols," p. 150. Academic Press, New York, New York, 1964.
32. F. A. Patty, "Sampling and Analysis of Contaminants, *Ind. Hyg. Toxicol*" (F. A. Patty, ed.), 2nd ed., p. 197. Wiley (Interscience) New York, New York, 1958.
33. N. A. Leidel, S. G. Bayer, and R. D. Zumwalde, "Membrane Filter Method for Evaluating Airborne Asbestos Fibers," NIOSH TR-84. Nat. Inst. Occ. Health Safety, Washington, D.C., 1974.
34. "The Industrial Environment, Its Evaluation and Control," Pub. Health Serv. Publ. No. 614, p. B-5-12. United States Dept. of Health, Education, and Welfare, Washington, D.C., 1965.

35. "The Industrial Environment, Its Evaluation and Control," Pub. Health Serv. Publ. No. 614, p. B-6-12. United States Dept. of Health, Education, and Welfare, Washington, D. C., 1965.
36. K. L. Dunn, *J. Ind. Hyg. Toxicol.* **16,** 29 (1934).
37. P. Drinker and T. Hatch, "Industrial Dust," 2nd ed., p. 152. McGraw-Hill, New York, New York, 1954.
38. C. E. Brown and H. H. Schrenk, *U.S., Bur. Mines, Inform. Circ.* **1c 7026,** 1–20 (1938).
39. C. E. Brown and H. H. Schrenk, *U.S., Bur. Mines, Rep. Invest.* **RI-3373,** 1–9 (1938).
40. "Dust Topics," Vol. 3, No. 1, pp. 18–19. Gelman Instrument Co., Ann Arbor, Michigan, 1966.
41. E. D. Myers, *Tenn. Ind. Hyg. News* **1,** 18 (1961).
42. "Procedure for the Determination of Particulate Contamination of Air in Dust Controlled Spaces by the Particle Count Method," Aerospace Recommended Practice. Soc. Automotive Eng., New York, New York, 1962.
43. C. E. Hall, "Introduction to Electron Microscopy," pp. 312–323. McGraw-Hill, New York, New York, 1953.
44. F. Stein, R. Quinlan, and M. Corn, *Amer. Ind. Hyg. Ass., J.* **27,** 39 (1966).
45. R. D. Cadle, "Particle Size Determination," p. 181. Wiley (Interscience), New York, New York, 1955.
46. G. Herdan, "Small Particle Statistics," 2nd ed. Academic Press, New York, New York, 1960.
47. "Standard Method of Test for Sieve Analysis of Magnesium Oxychloride Compositions, Aggregates, and Fillers," C238-51. *Book ASTM Stand.* Part 30, p. 158 (1964).
48. ASTM E 11-70 Standard Specifications for Wire Cloth Sieves for Testing Purposes, *Book ASTM Stand.* Part 30, pp. 97–101 (1970).
49. "Manual on Test Sieve Methods," STP 447A. Amer. Soc. Testing Materials, Philadelphia, Pennsylvania, 1972.
50. M. W. First, *Arch. Ind. Hyg. Occup. Med.* **7,** 58–60 (1953).
51. C. Orr and J. M. Dalla Valle, "Fine Particle Measurements, Size, Surface, and Pore Volume," pp. 83–91. Macmillan, New York, New York, 1960.
52. J. M. Dalla Valle, "Micromeritics, The Technology of Fine Particles," p. 69. Pitman, New York, New York, 1948.
53. "The Industrial Environment, Its Evaluation and Control," Pub. Health Serv. Publ. No. 614, p. B-7-6. United States Dept. of Health, Education, and Welfare, Washington, D.C., 1965.
54. C. E. Billings and L. Silverman, *J. Air Pollut. Contr. Ass.* **12,** 587–590 (1962).
55. D. A. Fraser, *Arch. Ind. Hyg. Occup. Med.* **8,** 412 (1953).
56. M. E. Kaine, "The Electron Microscope, The Present State of the Art," pp. 226–232. Spon, London, England, 1961.
57. R. D. Cadle, "Particle Size, Theory and Industrial Applications," p. 83. Van Nostrand-Reinhold, Princeton, New Jersey, 1965.
58. B. H. Kaye and C. R. G. Treasure, *Mater. Res. Stand.* **5,** 568–571 (1965).
59. W. H. Walton and J. Birt, *Appl. Phys., Suppl.* **3,** S29-S40 (1954).
60. A. Goetz and T. Kallai, *J. Air Pollut. Contr. Ass.* **12,** 479 (1962).
61. "Determining the Properties of Fine Particulate Matter," ASME Power Test Codes, PTC 28-1965. Amer. Soc. Mech. Eng., New York, New York, 1965.
62. K. R. May, *J. Sci. Instrum.* **22,** 187 (1945).

63. M. Lippman, *Amer. Ind. Hyg. Ass., J.* **20**, 406–416 (1959).
64. T. T. Mercer, *Amer. Ind. Hyg. Ass., J.* **26**, 236–241 (1965).
65. A. Renoux, G. Tymen, and G. Madelaine, *Atmos. Environ.* **8**, 41–55 (1974).
66. M. Pilat, P. S. Ensor, and J. C. Bosch, *Amer. Ind. Hyg. Ass., J.* **32**, 508–511 (1971).
67. W. H. Coulter, *Proc. Nat. Electron. Conf.* **12**, 1034 (1956).
68. R. N. Berglund and B. Y. H. Liu, *Environ. Sci. Technol.* **7**, 147–153 (1973).
69. F. G. Anderson, T. F. Tomb, and M. Jacobson, *U.S. Bur. Mines, Rep. Invest.* **RI-7105** (1968).
70. J. B. Dunson, Jr., Paper No. 5e, 63rd Annu. Meet., Chicago, Illinois. Amer. Inst. Chem. Eng., New York, New York, 1970.
71. B. Y. H. Liu, R. N. Berglund, and J. K. Agarwol, Particle Technology Laboratory Publication No. 209. Mechanical Engineering Department, University of Minnesota, Minneapolis, Minnesota, 1973.
72. J. S. Nader, G. C. Ortman, and M. T. Massey, *Amer. Ind. Hyg. Ass., J.* **20**, 42–48 (1961).
73. B. Y. H. Liu and D. Y. H. Pui, *J. Colloid Interface Sci.* **47**, 156–171 (1974).
74. J. R. Hodkinson, *Proc. Nat. Conf. Aerosols, 1st, 1962* pp. 181–194 (1965).
75. J. R. Hodkinson, *Appl. Opt.* **5**, 839 (1966).
76. K. T. Whitby and R. A. Vomela, *Environ. Sci. Technol.* **1**, 801–814 (1967).
77. R. H. Berg, *Mater. Res. Stand.* **5**, 119 (1965).

2

Analysis of Inorganic Particulates

Philip W. West

I. Introduction

The prime requisites for analytical methods that are to be used in determining air quality are reliability, accuracy, precision, and sensitivity. It is, of course, desirable to have methods that are rapid and simple, and it is usually important that equipment used be low in cost. The chemical analysis of inorganic pollutants may involve almost any of the techniques employed for quantitative or qualitative analysis. However, classic methods of qualitative analysis are now seldom employed, and quantitative determinations based on gravimetry and titrimetry are being replaced by newer analytical methods that offer advantages of sensitivity and selectivity. Colorimetry is still widely used in the determination of inorganic pollutants but even it is giving way to the more sophisticated spectrophotometric or fluorometric procedures. Basically, the analytical studies involve trace analyses but microchemical techniques are also widely used.

The simplicity, sensitivity, economy, and general reliability of many of the newer analytical techniques promise significant advances in the study of air pollution. In some instances, the more sophisticated methods may seem to require greater skills and impose greater cost. Actually, most of the new methods are simpler than the older techniques and even the cost of certain equipment, which may seem high at first appraisal, is quickly justified on the basis of the results obtained. Investments in equipment may soon be recovered through actual savings in the cost of manpower. It is extremely important that considered judgment be applied in selecting methods and equipment. All too often, procedures or instruments are proposed without proper warning regarding deficiencies such as the lack of specificity. Unfortunately some instruments and some procedures are used that measure only general properties or effects. If the results obtained are evaluated wisely, meaningful information can be obtained, but if the results are evaluated with blind faith the results may be worse than valueless.

Atomic absorption spectroscopy introduced by Walsh (*1*) in 1955 has become the most widely used technique for reference studies and critical measurements of metallic pollutant species. The technique is inherently specific, and it is often more sensitive than either flame photometry or emission spectroscopy. The technique is simple, and the equipment required is relatively inexpensive and readily obtainable.

The ring-oven technique introduced by Weisz (*2*) has been developed for use in air pollution investigations (*3–5*). Ring-oven methods are remarkably simple, reliable, and inexpensive. No special training is required to carry out the technique, and equipment can be shop made or obtained commercially for approximately $200. The methods now available are sensitive to microgram to nanogram amounts of material and have accuracies of 90% or better. The procedures are usually selective or even specific. The method is especially attractive where budgets are restricted or where it is desirable to have a method that can be applied in the field as well as in the laboratory.

Electroanalytical methods, such as anodic stripping and polarography, are useful techniques. Microscopy in its various forms is a basic analytic tool (see Chapter 3, this volume) and X-ray fluorescence spectroscopy is being touted as a method of the future for particulate monitoring.

II. Airborne Particulates

Airborne particulates or aerosols consist of either liquid or solid particles ranging in diameter from 0.01 μm or less, up to about 100 μm. The aerosols may be isolated by sedimentation, filtration, impingement, centrifugation, or by electrostatic or thermal precipitation. The evaluation of particulate burden is often based on the degree of soiling attributed to suspended dusts. Reduction in visibility, as measured photographically or by means of a transmissometer, gives a general measure of particulate loading. Such observations may be quite valuable in evaluating general air quality, but little specific information is provided. It is important that the identity of pollutant species be established and quantitative determinations of individual pollutants be made, especially if health effects are to be studied and toxic stresses are to be minimized.

A. Dust Stain Index (Smoke Stain Index)

There has been increased interest in measurements of the light transmitted or reflected by particulate deposits on paper or membrane filters. A measured volume of air is drawn through a small spot on a circle of filter paper or membrane filter, or samples are taken by means of automatic dust samplers which are now widely used for the collection of samples on filter paper tapes. Transmittance and reflectance measurements on the stains produced by the collected samples then serve for evaluation of the dust or soot burden of the atmosphere.

Hemeon has proposed that transmission measurements be made directly on the spots (*6*) and the results reported in terms of the "coefficient of haze" (COH). Gruber and Alpaugh have recommended that reflectance measurements be used (*7*) and the results expressed as "reflectance units of dirt shade" (RUDS). A COH unit has been defined as that quantity of particulate material which produces an optical density of 0.01 when measured by light transmission at 400 nm and when compared to the measurements on dust-free filter paper taken as 100%. Thus, a spot which passes 50% of the light that is transmitted by the clean reference area of the tape has an optical density of 0.301, or the sample has a value of 30.1 COH units. Smoke concentration may be expressed in terms of the number of COH units per 1000 linear feet of air drawn through the filter.

The soiling index or RUDS is the percentage reflectance obtained when

10,000 linear feet of air is drawn through the filter paper. The reflectance of the clean paper is used as the reference standard and is set at 100 on a Photovolt reflectance meter. With this system, a reflectance of 100 is "absolutely" clean, and a reflectance of 0 is termed "absolutely" black.

Considerable differences of opinion exist regarding the relative merits of transmittance versus reflectance measurements in evaluating dusts. It is of significance, however, that the report of the Working Party on Methods of Measuring Air Pollution and Survey Techniques of the Organization for Economic Cooperation and Development (OECD) has agreed that the estimation of smoke concentration from smoke stains can best be done by reflectometric measurements (*8*). The OECD Subgroup on smoke measurement has prepared a report which provides for differences in instruments and defines precise conditions for sampling. It was concluded that a general procedure for recording smoke stain and reflectometric readings is already a matter of agreement. The importance of procedural details was recognized in regard to the intercomparison of results but it was concluded that differences in instrumentation in different countries do not present difficulties and any standard reflectometer may be used. The Subgroup studied the relation between the weight of particulate caught and the darkness of the stain produced and pointed out that darkness of stain is dependent on the nature of the smoke or dust and cannot be used to measure the total weight. Where the stain is due mainly to the products of combustion, however, the darkness produced is less variable and, within limits, may be used as a measure of the material emitted.

The standardized procedures for evaluating smoke stains as evolved by the OEDC are recommended. The standard method for measuring particulate matter in the atmosphere and adopted as American Society for Testing and Materials designation No. D 1704-61 is also valuable. This procedure is essentially that which was first introduced by Katz, Sanderson, and Ferguson (*9*) and is based on the use of a tape sampler for filtration of the air followed by transmittance or reflectance measurements on the sample spot. A parallel sample is collected by filtration with a high volume sampler provided with a dried and tared high efficiency glass filter for determining the mass of particulate material and for providing a sample for subsequent chemical analysis.

B. Gravimetric Determination of Suspended Particulates

Measured volumes of air can be sampled by means of high volume samplers using previously dried and weighed glass filters (see Section III,C). The filter should be an 8 × 10-in. sheet of flash-fired fiberglass, and the usual sampling period is for 24 hours. At a flow rate of 25–50

ft^3/minute, the average sample collected will be from 50 to 250 mg. Balances (Mettler, Model H6T, with air pollution filter weighing chamber) are available with a special pan and case large enough to accommodate the large sheets. Alternately, high volume samplers with smaller filter heads can be used and the filters weighed on any conventional balance having the required sensitivity.

Samples collected on filter paper are ordinarily not weighed because of the difficulty of obtaining tares on paper. The absorption of water by the paper introduces serious variations for most work, although the errors due to such variations can be minimized when sufficiently large samples are collected. Large fluted filters have been used with high volume sampling *(10)*. The filters are dried at 110°C and are then stored in sealed containers ready for use. The quantity of dust collected with such filters is frequently so great that small variations in the weight of the filter due to the absorption of water is of little importance.

Molecular membrane filters manufactured from cellulose esters are very useful in dust studies. Available under the trade names of Millipore and Isopore filters, they exhibit remarkable uniformity of pore size and approach 100% efficiency for removing particles as small as 0.1 μm in size. Such filters are convenient for use in the gravimetric determination of dusts *(11)* because they have little tendency to absorb moisture, they are light, and they are uniform in weight. Although the filters are delicate and must be handled with care, suitable filter heads designed for their use are available. Not only can such filters be used for gravimetric determinations but they also lend themselves readily to the microscopic evaluation of collected dusts as well as for ring-oven studies.

Electrostatic precipitation provides an efficient means for collecting particles in the micron to submicron range *(12)*. Commercially available electrostatic precipitators will collect particles as small as 0.1 μm with efficiencies approaching 100%. Only relatively small amounts of sample can be collected with such equipment so that a semimicro- or microanalytical balance should be used for gravimetric measurements on the sample after it has been transferred from the collection tube to a suitable tared weighing vial. If subsequent analyses on the collected dusts are to be performed, it is usually necessary to resort to methods of high sensitivity such as emission spectroscopy, atomic absorption spectroscopy, spectrophotometry, or ring-oven techniques.

C. Preparation of Filters and Samples

Glass filter sheets used for gravimetric and associated studies of dusts should be flash-fired before taring. After collection, the sample should be equilibrated at 75°C and 50% relative humidity for at least 16 hours. After the equilibrated samples are weighed to determine the total dust, the sheets can be cut into appropriate sections for spectrographic, spectrophotometric, or other appropriate studies. It should be noted that glass filters are usually contaminated with substantial amounts of elements such as zinc, barium, sodium, and chlorine. If quantitative determinations are to be made, this may invalidate results on such elements and may even introduce errors for other elements because of variable matrix effects. It is usually advisable to consider the use of cellulose or membrane filters where quantitative studies are to be made.

D. Atomic Absorption Spectroscopy

Atomic absorption spectroscopy is a powerful tool for trace metal determinations in air particulate samples. In addition to being a very sensitive method, it provides selectivity, simplicity, and good reproducibility, in comparison with most other methods. Until recently, its application to chemical analysis had been confined to the study of solar and stellar atmospheres (*13*) and the determination of mercury vapors in the air (*14*). In 1955, Walsh first recognized the advantages of atomic absorption over emission methods, and devised a simple instrument for the analysis of solutions for a wide range of elements (*15*). Since then, considerable work has been done in the development and application of this technique to a variety of fields such as agriculture, biology, medicine, metallurgy, and geology. A number of review articles (*16–23*) and books (*24–26*) have appeared and should be consulted for more detailed information.

1. Sampling of Airborne Particulates for Atomic Absorption Spectroscopy

Proper selection of sampling procedure is necessary to obtain accurate results. Three major techniques are used for sampling: liquid impingers, filtration methods, and electrostatic precipitation. Because of its simplicity and efficiency, filtration is generally the method of choice for sampling airborne particulate matter. The filtering material must be chosen with great care. Glass fiber filters, which are most commonly used, happen to have particularly high levels of trace metal impurities (*27*). Cellular and polystyrene membrane filters have been used for sampling particulates. For a few metals such as selenium and mercury, liquid impingers are used for sampling. Mercury vapors from air are also very conveniently collected by amalgamation with silver (*28*).

2. Principle

Under most conditions ground-state atoms make up the bulk of an atomic population. These atoms absorb radiation at discrete wavelengths, which are characteristic for each element, and are thus raised to the excited state. The excited-state atoms, in turn, emit radiation on returning to the ground state. The amount of energy thus absorbed and emitted is a function of the number of atoms present and hence may serve as the measure of the concentration of the respective elements in the sample.

3. Instrumental

An atomic absorption spectrophotometer is essentially the same as an emission instrument except for the addition of a radiation source—usually a hollow cathode lamp which has an emitter cathode made of the element to be determined (Fig. 1). Electrodeless discharge lamps are also being used in place of hollow cathode lamps, particularly for volatile elements such as arsenic, mercury, cadmium, and selenium. An electrodeless discharge lamp consists of a small quartz tube containing vapors of the element of interest and an inert gas. The element is excited to produce radiation by means of an external electric field. Such lamps are now commercially available. The sample to be measured is generally atomized in a flame using conventional atomizers and burners. The radiation from the source transverses the flame into the entrance slit of a monochromator, which separates the desired resonance line from the other lines that may be present. Its intensity is measured by a conventional photomultiplier tube and amplifier. Usually the flame also emits radiation of the same wavelength as the source. This background radiation is compensated by modulating the radiation from the source, and feeding the signal from the detector to a tuned ac amplifier, which rejects the dc signal

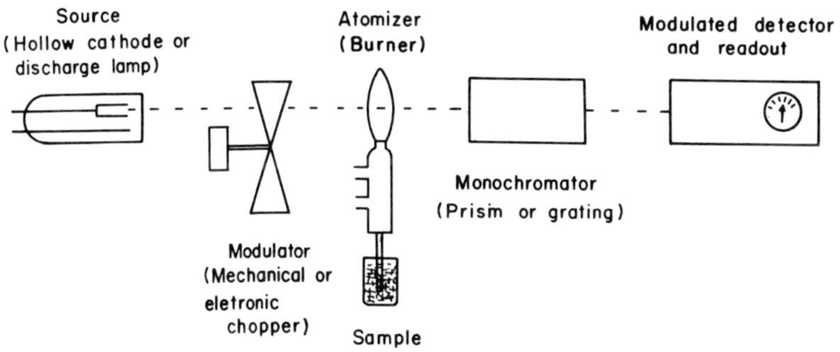

Figure 1. Schematic diagram of atomic absorption spectrometer.

contributed by emission from the flame. Modulation is accomplished by chopping the light beam mechanically, or the amplifier is tuned electrically to the chopped frequency of the source and therefore reads only the modulated radiation.

Most atomic absorption studies are made on solutions with a flame used as the means of sample atomization. Other methods for sample atomization include cathode sputtering (*29, 30*) by electrically heating a graphite tube, a graphite rod, or a tantalum ribbon in an atmosphere of argon or hydrogen.

4. Interferences

Atomic absorption spectroscopy is free of spectral interferences encountered in flame photometry and excitation interferences common in emission spectroscopy. Chemical interference, such as that of phosphate on calcium, affects absorption and emission methods equally, but may usually be overcome by the addition of a releasing agent such as EDTA (ethylenediaminetetraacetic acid) which serves to liberate the metal being determined from combination with an interfering species. A high temperature flame such as $N_2O–C_2H_2$ can also be used to overcome some of the interferences. It should be noted, however, that in the presence of a very high concentration of other salts or organic materials, the sensitivity may be reduced in some cases and enhanced in others. The decrease in absorption signal may be due to chemical interference such as formation of refractory compounds or due to change in the physical properties of the sample solution, such as its viscosity and surface tension, which reduces the rate of atomization. Enhancement of an absorption signal can take place by the reverse of the physical conditions mentioned above, as well as by scattering of light by solid particles. The latter can be corrected to some extent by the use of a deuterium background corrector now available with most of the commercial atomic absorption instruments. Most physical interferences can be overcome by an internal standardization technique.

5. Advantages and Disadvantages

Atomic absorption spectroscopy is distinguished from most other analytical methods by its sensitivity, speed, ease of operation, accuracy, and relative freedom from physical and chemical interferences. Its principal disadvantage is that it cannot be applied to nonmetals since their lines lie in the vacuum ultraviolet region (i.e., below 2000 Å).

Once a sample is brought into solution, no special technique or skill is needed and the analysis can be performed by semiskilled personnel with a little training. Usually no interferences are encountered while determining trace metals in dilute solutions. However, when analyzing solutions with higher salt concentrations, interference effects do occur. Such interferences can be overcome either by using one of the methods described under Section II,D,4 or by separating the metals of interest by the use of an extraction procedure.

Until recently, certain elements such as aluminum, barium, beryllium, silicon, titanium, tungsten, and vanadium could not be determined because they form refractory compounds. This can now be overcome by atomizing the sample in a nitrous oxide–acetylene flame.

6. Methods of Increasing Sensitivity

Atomic absorption methods, although sensitive enough for most purposes, must be made still more sensitive for the determination of certain trace metals in air such as Be, Mo, V, Ni, and Ti, etc. Such metals are present in air at a concentration level less than 10^{-3} $\mu g/m^3$ and with the conventional atomic absorption methods, a rather large volume of air (2000 m^3) must be sampled to obtain reasonably accurate results. Atomic absorption methods, however, can be made more sensitive by the following procedures:

a. Chemical Methods. The use of organic solvents such as alcohols, ketones, or esters, instead of water, has been found to enhance absorption and thus increase sensitivity. This is thought to be due to more efficient volatilization of droplets in the flame. Chelation followed by extraction into an organic solvent and atomization of the resulting organometallic species (*31*) serves not only to isolate and concentrate the metal from the sample system, but also enhance absorption due to atomization of an organic instead of an aqueous solution.

b. Optical Methods. The flame should have a long narrow light path so that as many atoms as possible are introduced into the optical path for the absorption measurement. The effective length of optical path may be increased by passing the light beam several times through the flame with the aid of a mirror system. Similar increase in path length may also be achieved by directing the flame through a long or T-shaped quartz tube (*32, 33*).

c. Flameless Atomization. Recently, flameless methods for atomization have been developed. Such methods increase the sensitivity from an order of magnitude to several orders of magnitude. Three types of flameless atomizing devices are commercially available: graphite tube furnace (Perkin-Elmer Corporation), carbon rod atomizer (Varian Techtron), and heated tantalum ribbon atomizer (Instrumentation Laboratories). Graphite tube and carbon rod atomizers provide good reducing conditions and very good sensitivity, but impurities have been reported in the carbon which can yield erroneous results. The tantalum ribbon atomizer has been reported to yield excellent sensitivity for several elements. However, it cannot be heated to temperatures above 2996°C (the melting point for Ta) which may not be sufficient for atomization of some elements.

Atomic absorption is at present suitable for determination of over 50 elements, 21 of which are listed in Table I. The precision, accuracy, and reliability is equal to or better than other commonly used methods of

Table I Atomic Absorption Spectroscopic Methods

		Sensitivities		
		ppm/% Absorption[a]		*μg/% Absorption*[b]
Metal	*Wavelength (Å)*	*Air–acetylene flame*	*Nitrous oxide–acetylene flame*	*Flameless atomization*
Ag	3281	0.06		2×10^{-5}
Al	3092		1.00	1×10^{-4}
Ba	5536		0.40	5×10^{-5}
Be	2349		0.025	7×10^{-6}
Bi	2231	0.40		1×10^{-4}
Ca	4227	0.08	0.05	3×10^{-6}
Cd	2288	0.03		7×10^{-6}
Co	2407	0.15		9×10^{-4}
Cr	3579	0.10		4×10^{-5}
Cu	3247	0.09		2×10^{-5}
Fe	2483	0.12		2×10^{-4}
Hg	2536			8×10^{-4}
K	7665	0.04		3×10^{-6}
Mg	2852	0.01		2×10^{-7}
Mn	2795	0.06		2×10^{-5}
Na	5890	0.015		3×10^{-6}
Ni	2320	0.15		4×10^{-4}
Pb	2170	0.50		3×10^{-5}
Sn	2246	2.50		4×10^{-6}
V	3184	1.70		1×10^{-3}
Zn	2138	0.02		4×10^{-6}

[a] Obtained with Perkin–Elmer Model No. 403.

[b] Sensitivities listed are by no means the maximum; sensitivities equal to or better than those listed can be obtained through proper control of parameters and impurities.

trace element determination, such as spectrophotometry, polarography, and emission spectrography. The monitoring of several potentially hazardous elements such as Be, Hg, Cd, and Mn, would require development of continuous analyzers. Atomic absorption has already been shown to have potential for this purpose (*34*).

E. Emission Spectrometry

Emission spectrometry entails the excitation of a sample to provide emission spectra of the metals present. This is usually accomplished by means of an ac arc or spark, or a dc arc. A Lundegarth flame atomizer has been used (*35*) to detect some 34 elements, among which are some of those of greatest interest in air pollution monitoring systems. In addition to the excitation unit, the spectrometer will include a system for dispersing and isolating discrete portions of the spectrum (a prism or grating) and a receptor, such as a photographic plate or a recording device for measuring the isolated radiant energy. The details of instrumentation may affect the cost and convenience but generally the method remains one of the best approaches for the routine identification and estimation of metals in particulate samples. It is sensitive (microgram to nanogram range for most metals), qualitatively free from interferences, rapid (once set up and applied routinely), and widely applicable. The equipment is relatively costly, and considerable skill and experience are required of the spectroscopist. Quantitative analyses may be subject to some error due to matrix effects.

Many important applications of emission spectroscopy are made in air pollution studies. The National Air Sampling Network applies emission and atomic absorption methods in studies of samples collected in various parts of the nation. Cholak, Schafer, and Hoffer (*36, 37*) have used spectrographic methods for broad studies of both airborne and settled dusts.

Cholak and Story (*38*) have applied emission methods for the determination of several metallic constituents of sootfall ash. Clayton (*39*) has investigated spectrographically the material trapped by accordion-pleated filters and obtained data on 24 elements. Keenan and Byers (*40*) employed spectrographic methods for the estimation of 21 elements in samples collected with high volume samplers. O'Neil (*41*) and Kennan and Holtz (*42*) have described a spectrographic technique for determining beryllium in air.

F. Flame Photometry

Flame photometry is now being more widely used than in the past because better methods of excitation have made more applications possible. Originally, flame photometry found use only in determining alkali or alkaline earth metals. With better methods of excitation it is now possible to use flame photometry in other determinations (*43, 44*). Flame photometry is sensitive and quite reliable, particularly when applied to air pollution problems where matrix effects are usually not serious. It is relatively inexpensive, rapid, and simple to apply.

The flame, although a low energy source as compared to an arc or spark, is very stable and is capable of exciting intense emission of radiant energy from elements with low excitation potentials. The method has been applied for the analysis of several water-soluble atmospheric pollutants such as sodium, potassium, calcium, magnesium, strontium, manganese, and iron.

Flame photometry may find wider application in the coming years for air pollution work, as improved instrumentation and hotter flames for excitation are becoming available. In addition, newly designed burners for introducing powdered samples may facilitate the direct study of dusts.

G. Ring-Oven Methods

The ring oven is a simple, inexpensive apparatus which permits the separation, concentration, and analysis of inorganic and organic substances at microgram and nanogram levels. Ring-oven methods offer great promise for the study of airborne particulates because they are so reliable, sensitive, and convenient (*4*). Results can be obtained in which the relative errors are no more than 5–10%, which at the microgram level compares very favorably with other applicable methods such as emission spectroscopy, polarography, spectrophotometry, and neutron activation. Such accuracies exceed present capabilities for the interpretation of the data. The necessary equipment can be obtained for approximately $100–$200, no special training is required for performing the analyses, and the methods can often be applied in the field.

The ring-oven technique was introduced in 1954 by Weisz (*2*). By means of this technique a complete scheme of separation and identification can be carried out on a single drop of unknown through application of classical chemical methods of precipitation and filtration (*45*). A scheme for the separation of 35 metal ions has been developed which is based on a combination of liquid–liquid extraction and the ring-oven technique (*46*). Recent developments of ring-oven methods have extended the applications to quantitative analysis, electrographic studies, and to the analysis of radioactive substances. These and many other applications have been summarized in the monograph by Weisz (*47*) dealing with the ring-oven technique.

1. Equipment

The only equipment needed for ring-oven studies is a ring oven, an electric hair drier or oven suitable for drying filter papers, and some micropipets and capillary pipets. An ultraviolet lamp is also desirable. The ring oven consists of a heated ring, ordinarily 22 or 33 mm i.d., which serves as a means of fixing solute materials in a sharply defined circle (Fig. 2). Although a suitable ring oven can be readily constructed in any good laboratory shop, commercial instruments are readily available. The Trace Oven manufactured by Arthur H. Thomas Co., Philadelphia, Pennsylvania is essentially the same as the ring oven originally designed by Weisz. The essential features of this unit are shown in Figure 3. The heating areas of the ring oven are usually made of aluminum, but various other materials can also be used, such as copper, steel, and glass. For extremely delicate analyses, gold- or platinum-plated equipment has been also used so as to avoid dangers from possible contamination.

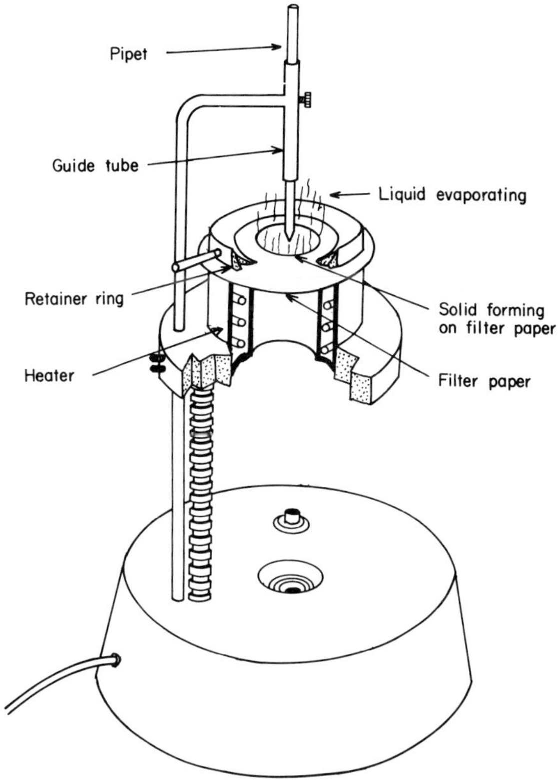

Figure 2. Essential features of ring oven.

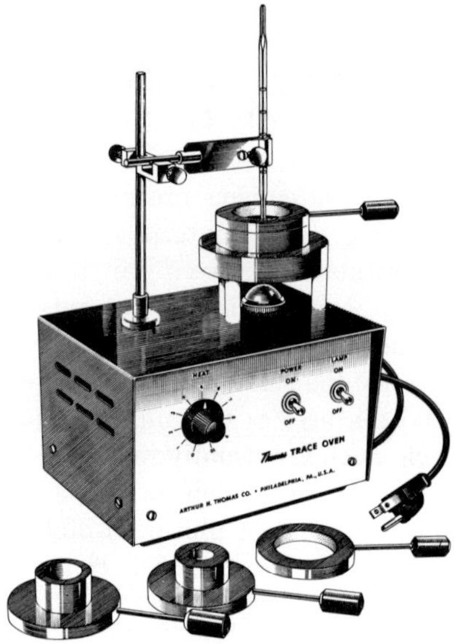

Figure 3. Ring oven (Trace Oven). (Courtesy of Arthur H. Thomas, Co. Philadelphia, Pennsylvania.)

2. Technique

Ring-oven methods are a special refinement of spot test procedures that are carried out on filter paper (Table II) (*48–64*). The detection and determination of the constituents of inorganic air pollutants is accomplished through the use of extremely sensitive organic reagents that are either specific or selective or which can be made highly specific or selective through the proper use of masking agents or through other "conditioning" processes. The critical innovation introduced by the ring-oven technique is the separation, transfer, and concentration of test materials which the method makes possible. The operations consist simply of placing a piece of filter paper on the heated surface of the ring oven and introducing soluble sample material at the exact center of the paper. The sample solution is then washed through the pores of the paper by the addition of 5–10 microdrops of an appropriate solvent. The sample solutes are thus transported or washed to the area of the heated ring surface. As the solution approaches the heated ring zone, the carrier solvent is evaporated, thus depositing the solute (sample) as a sharply defined ring. The filter paper with the ring of the deposited salts (sample) is finally removed from the surface of the ring oven and appropriate tests are then run on sectors of the sample ring. The filter paper is ordinarily cut into as many segments as there are tests or determinations to be run. If four determinations are to be run, the paper is cut into quandrants; if more determinations are to be run, each quandrant can be subdivided so as to provide as many ring segments as required. Copper, for example, can be as readily detected and/or determined on a small sector of the ring as it can on half or even the whole ring. One needs only to recognize the fact that the color of paint on a house can be matched as easily using one of the boards as it can by using a whole wall. The essential steps of a ring-oven analysis are depicted in Figure 4. Under step 3 the analyses are completed by adding appropriate reagents to the respective sectors to develop the definite test colors.

In an alternate procedure a microliter drop of the reagent is added to the center of the filter paper followed by a microliter drop of a solution containing the substance to be determined. The colored reaction product formed at the center can then be selectively washed to the ring zone by the use of a suitable solvent, thus intensifying (concentrating) the test color and isolating it from possible interfering stains.

For quantitative studies, the sample solution should, of course, be made up to a known volume and a suitable aliquot used for the determinations; and the final operation should be to match the individual sectors against standard rings obtained on known amounts of the species in question, carrying out all analytical operations under similar fixed conditions. Once the ring or ring sector has been processed to develop the test color, the quantity of material present is determined by matching the sample ring against a series of standard rings produced under similar conditions. The solutions used in preparing the standard rings should be made up to con-

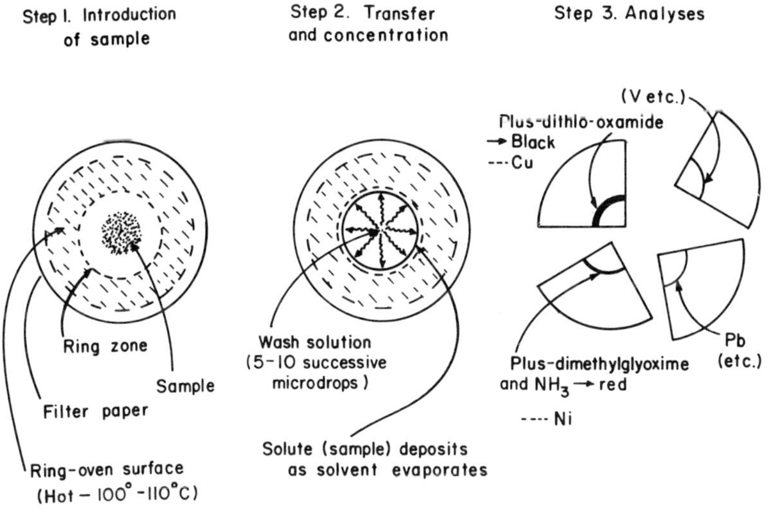

Figure 4. Ring-oven procedure steps.

Table II Ring-Oven Methods[a]

Determination	*Remarks*
Aluminum	
1. Spray ring with morin (3,5,7,2′,4′-pentahydroxyflavone, saturated in methanol).	Limit of identification: 0.02 μg Optimum range: 0.04–0.8 μg Fluorides prevent the reaction; Be^{+2}, Zr^{+4}, and Sb^{+3} interfere.
2. Dry.	Solutions and standard rings are stable.
3. Dip in HCl (2 *N*). 4. Observe under UV radiation; green-yellow fluorescence.	*Sample:* From sample tape, direct dissolution, and transfer using ten extractions with 30-μl portions HCl (2 *N*).[a] (Caution: When using filter paper without barrier rings, acid must not flood to ring oven surface as the metal would be attacked and Al contamination would result.) Aliquots of sample solutions may also be used.
Antimony	
1. Heat some phosphomolybdic acid (5% aqueous solution) to boiling.	Limit of identification: 0.1 μg Optimum range: 0.2–1.5 μg There are no significant interferences.
2. Dip test paper in hot reagent for 1 minute; remove excess reagent with distilled water; blue ring.	The reagent solution is stable 3–4 days and standard rings are stable for at least a month (*48*). *Sample:* From sample tape, direct dissolution and transfer. Dissolve by adding 30 μl H_2SO_4 (4 *N*). Let stand 2–3 minutes. Transfer by adding 30 μl KI (10%) and, after the spot has almost dried, extracting with 25–30, 25-μl portions of solvent (benzene–ethanol, 2:1). (Caution: solvent mixture should not diffuse closer than 1mm from the ring-oven surface, otherwise charring will occur.) The ring oven should be heated to approximately 80°C. Aliquots of sample solution may also be used (*49*).
Beryllium	
1. Spray ring with morin (3,5,7,2′,4′-pentahydroxyflavone, saturated in methanol) or rub with reagent crayon and heat with hot air.	Limit of identification: 0.01 μg Optimum range: 0.03–0.2 μg The method is specific. Reagent crayons or solutions are stable
2. Place the treated test paper in bath (NH_4OH–methanol, 1:1) for 5 minutes. Dry.	as are the test rings. Reagent crayons can be prepared by mixing 3.5 gm monoglyceryl stearate and 1.5 gm

Table II *(Continued)*

Determination	*Remarks*
3. Observe under UV radiation; yellow-green fluorescence.	paraffin wax in a test tube and heating in boiling water until a transparent melt is obtained. Fifty milligrams of morin are then dissolved and mixed in the melt which is then poured into waxed soda fountain straws and allowed to cool and solidify.
	Sample: From sample tape, direct dissolution and transfer. Dissolve by adding 30 μl NH_4 acetate (15%). Wash to ring zone with water. Followed by 30 μl EDTA (0.1 *M*). Aliquots of sample solutions may also be used *(50)*.
Cadmium	
1. Expose ring to NH_3.	Limit of identification: 0.08 μg
2. Add the reagent (ferrous dipyridyl iodide) to the ring using a capillary pipet. After 10 seconds, wash in a jet of distilled water for 30 seconds; pink ring.	Optimum range: 0.1–2.0 μg The method is specific for air pollution studies—exposure to NH_3 masks the interferences of Cu, Ni, Co, Zn; phosphate masks the interference of Pb. Some other colored products may be formed on addition of reagent but are washed off by water. The reagent ferrous dipyridyl iodide is prepared by dissolving 0.25 gm α,α-dipyridyl, 0.146 gm $FeSO_4{\cdot}7H_2O$, and 10 gm KI in 50 ml distilled water. The mixture is shaken vigorously for 30 minutes and filtered. The reagent is stable for 5 days. It becomes turbid on long standing. It should be filtered before use. *Sample:* From sample tapes, direct dissolution and transfer by a 30-μl portion of 15% NH_4 acetate washed to the ring zone with distilled water, followed by another 30-μl portion of NH_4 acetate (15%), 30 μl KCN (0.5%) and washed to ring zone with distilled water. A 30-μl portion of Na_3PO_4 (10%) is also added and washed to the ring *(51)*. Aliquots of sample solutions may also be used.
Cobalt	
1. Expose ring to NH_3.	Limit of identification: 0.05 μg
2. Spray or add Na_2HPO_4 (5%).	Optimum range: 0.1–3.0 μg

Table II *(Continued)*

Determination	*Remarks*
3. Spray or add 1-nitroso-2-naphthol (0.1% in acetone); red-brown ring.	The method is specific for air pollution studies—iron and other likely interferences are masked by phosphate and CN^-. Reagents and standard rings are stable for at least 1 month. *Sample:* From sample tapes, direct dissolution and transfer. A 30-μl portion of NH_4 acetate (15%) + 30 μl KCN (0.5%) washed to the ring zone with distilled water. Aliquots of sample solutions may also be used (*52*).
Copper	
1. Dry ring at room temperature. 2. Spray ring with reagent (saturated solution of dithiooxamide in 20% malonic acid in ethanol). 3. Dry in a jet of cold air; dark green-black ring.	Limit of identification: 0.08 μg Optimum range: 0.1–2.0 μg The method is specific for air pollution studies—potential interferences are masked by malonic acid and CN^-. Reagents and standard rings are stable for at least 1 month. *Sample:* From sample tape direct dissolution and transfer by adding 30-μl portion of NH_4 acetate (15%) + 30-μl portion of KCN (0.5%) and washing to the ring zone with distilled water. Aliquots of sample solutions may also be used (*53*).
Iron	
1. Expose ring to Br_2 fumes. 2. Spray with $K_4Fe(CN)_2$ (1%); blue ring.	Limit of identification: 0.01 μg Optimum range: 0.01–0.5 μg The method may be considered specific for air pollution studies. The reagent is stable and the standard rings may be kept indefinitely if protected from light and covered with a protective wax or acrylic spray. *Sample:* From sample tapes, direct dissolution and transfer. Dissolve by moistening sample spot with HCl (2 *N*) and washing to the ring zone with distilled water or HCl (0.1 *N*). Aliquots of sample solutions may also be used (*48*).
Lead	
1. Spray with dithizone (0.005% in CCl_4).	Limit of identification: 0.05 μg Optimum range: 0.1–1.0 μg

Table II *(Continued)*

Determination	*Remarks*
2. Immerse paper in wash solution (0.2% KCN and 2% NH_4OH) for 1 minute. 3. Wash paper in water and press between filter papers to remove excess water. Dry in warm air; brick red ring.	The method is specific for air pollution studies. The reagents and the standard rings are stable for 1 week if stored in the dark. Special care must be taken to insure the absence of any contamination by heavy metals in the reagents. This can be done by prior extraction of the reagents with a CCl_4 solution of dithizone, in order to reduce blanks. *Sample:* From sample tapes, direct dissolution and transfer. Dissolution by addition of 30 μl NH_4 acetate (15%) and 30 μl $Na_2S_2O_3$ (5%) and washing to the ring zone with distilled water. Another 30 μl NH_4 acetate (15%) + 30 μl KCN (0.5%) is added and washed to the ring zone with distilled water. Aliquots of sample solutions may also be used (*54*).
Manganese 1. Spray the ring with 0.1 *N* NaOH. 2. Dry in cool air. 3. Add benzidene solution (.01% in acetic acid) and dry; blue ring.	Limit of identification: 0.05 μg Optimum range: 0.1–2.0 μg The method is specific for air pollution studies. Reagents and solutions are stable for about 2 months. The test rings are stable for only $\frac{1}{2}$ hour. Therefore, a set of permanent standard rings is preferred in this method. *Sample:* From sample tapes, direct dissolution and transfer by addition of 30 μl of 2% H_2O_2 solution in 2 *N* HCl. Then add 30 μl NH_4 acetate (15%) + 30 μl KCN (0.5%) and wash to the ring zone with distilled water. Then, add enough 2 *N* HCl to diffuse to ring zone and dry. Aliquots of sample solutions may also be used (*55*).
Nickel 1. Expose ring to formaldehyde fumes for 2 minutes. 2. Spray with dimethylglyoxime (1% in ethanol).	Limit of identification: 0.08 μg Optimum range: 0.1–1.0 μg The method is specific for air pollution studies.

Table II *(Continued)*

Determination	*Remarks*
3. Expose to NH_3; brilliant red ring.	Reagents and the standard rings are stable. *Sample:* From sample tapes, direct dissolution and transfer by moistening the sample spot with 30 μl KCN (0.5%) and washing to the ring zone with distilled water. Aliquots of sample solutions may also be used (*48, 52*).
Nitrite	
1. Rub test area with reagent crayon.	Limit of identification: 0.01 μg
2. Let crayon melt, remove excess crayon by blotting with filter paper.	Optimum range: 0.025–0.25 μg Oxidizing agents interfere but can be destroyed by exposure to SO_2.
3. Wash in chloroform; pink-red ring.	Reagent crayons are stable for at least 2 months. Standard rings are stable if sealed between glass plates. Reagent crayons are prepared by heating 20 mg sulfanilamide, 20 mg α-naphthylamine, 5 gm monoglyceryl stearate, and 0.5 gm paraffin wax on a water bath. After the melt is thoroughly mixed, 3 drops concentrated HCl are added and the melt again stirred, poured into waxed soda fountain straws, and cooled. *Sample:* From sample tapes, transfer directly to ring zone by washing with ten 20-μl portions of NaOH (0.1 *N*). Allow to dry 1 minute on ring oven and then run test while paper is still on oven surface.
Phosphate	
1. Spray ring with orthodianisidine molybdate.	Limit of identification: 0.02 μg Optimum range: 0.05–2.0 μg
2. Dip treated ring 8–10 times in HCl (0.1 *M*) to remove excess reagent; brown rings.	The method is specific although Al^{3+}, Zr^{4+}, and Th^{4+} phosphates are so stable that they cannot readily be decomposed to release PO_4^{3-}. Reagent solutions and the standard rings are stable. Orthodianisidine molybdate: Orthodianisidine (0.125 gm) is dissolved in a solution consisting of 1 ml glacial acetic acid and 2.5 gm $Na_2MoO_4 \cdot 2H_2O$ in 20 ml of 3 *M* HCl.

Table II *(Continued)*

Determination	*Remarks*
	Sample: From sample tapes, direct dissolution and transfer. The sample spot is kept moistened for 5 minutes with HCl (6 *M*), after which the dissolved phosphate is washed to the ring zone with 20 μl portions of 0.1 *M* HCl. Aliquots of sample solutions may also be used (*56*).
Selenium 1. On purified filter paper add three 20-μl portions of $Na_2C_2O_4$ (1%). 2. Add 5-μl aliquots of sample solution (to provide 0.1 to 0.5 μg Se). 3. Add six 20-μl portions of $Na_2C_2O_4$. 4. Add 20 μl formic acid (2.5 *M*). 5. Add four 5-μl portions fresh reagent (1% 3,3′-diaminobenzidine hydrochloride). 6. Add 30 μl buffer, pH = 3.0 (prepared from potassium acid phthalate and tartaric acid).	Limit of identification: 0.08 μg Optimum range: 0.1–0.5 μg The method is free of interferences. The 3,3′-diaminobenzidine solution and the test rings are relatively unstable and should not be considered reliable for more than 12 hours. All filter papers tested were found to contain selenium impurities (as much as 2 μg per 22-mm diameter spot). Papers can be readily purified (*57*) by washing with 15 ml Na_2S (1%) which is added to the center of the filter paper by means of a capillary pipet. See reference for details. *Sample:* Samples are best collected by electrostatic precipitation or by bubbling air through distilled water. [Losses have been reported if collected on any kind of filter media (*58*).] Na_2S (1%) can also be used as an effective scrubbing solution.
Sulfate 1. Spray ring or dip it in reagent solution (mixture of 5% $BaCl_2 \cdot 2H_2O$ and 3% $KMnO_4$). 2. Dry 5 minutes at 110°C. 3. Wash ring in oxalic acid (3% $H_2C_2O_4 \cdot 2H_2O$ in 1 *M* HCl), until brown stains of $MnO_2 \cdot H_2O$ disappear; pink ring.	Limit of identification: 0.1 μg Optimum range: 0.1–1.0 μg Reagents and test rings are stable. Only SO_2 interferes (and it through oxidation to SO_4^{2-}). *Sample:* Samples are collected from sample tapes, direct dissolution and transfer by washing with ten 20-μl portions of HCl (0.1 *N*). Sulfur dioxide, adsorbed on particulate surfaces or deliberately trapped on filter surfaces (by moistening the paper

Table II *(Continued)*

Determination	*Remarks*
	with dilute NaOH prior to sampling) can be determined by adding three to four 20-μl portions of H_2O_2 prior to the transfer to the ring zone with HCl (*59*).
Sulfur 1. Add Cu(I) reagent to the ring zone. 2. After 3 minutes rinse in tap water; yellow-brown ring.	Limit of identification: 0.1 μg Optimum range: 0.2–2.0 μg The method is specific for air pollution studies (*60*). Copper (I) reagent: First a Cu(II) sulfate stock solution is prepared by dissolving 4 gm of $CuSO_4 \cdot 5H_2O$ in 100 ml of concentrated NH_4OH (29.1% ammonia by weight). To 5 ml of this stock Cu(II) solution add 5 ml hydroxylamine hydrochloride (20% w/v), dilute to 50 ml with distilled water and shake vigorously. The reagent is unstable and should be freshly prepared from the stock solutions each day. *Sample:* Samples are collected from the sample tapes, direct dissolution, and transfer by washing the spot with four 30-μl portions of acetone. The ring oven should be maintained at 50°C.
Sulfuric acid aerosol 1. Add 6 μl fluorescein solution (0.05% in water) and wash to the ring zone with methanol. 2. Add $KBrO_3$ (4%) around the ring with a capillary tube. 3. Dry paper on the ring oven for 3 minutes; pink ring.	Limit of identification: 0.1 μg H_2SO_4 Optimum range: 0.2–2.0 μg Acidity due to H_2SO_4 and hydrolyzed metal sulfates are measured (*61*). Volatile acids like HNO_3 and HCl are evaporated during sample pretreatment. The standard rings are stable when stored in dark and sealed containers. Florescein solution is stable for 10 days. *Sample:* Collect 0.1–2.0 m^3 of air (depending on acid levels) using a sequential tape sampler. The samples are placed in an oven at 100–110°C, for 3 hours to volatilize HCl and HNO_3. Then the tapes are centered on the ring oven and samples are directly

Table II *(Continued)*

Determination	*Remarks*
	dissolved and transferred using a 30-μl portion of KBr (4%) and washing to ring zone with distilled water.
Total protons 1. Add 30 μl KBr (4%) and enough distilled water to moisten the entire sample spot. 2. Add 30 μl fluorescein (0.05%) and 30 μl $KBrO_3$ (4%). 3. Wash the pink reaction product (eosin) to the ring zone using absolute alcohol; pink ring.	Limit of identification: 0.01 μg (H^+) Optimum range: 0.02–2.0 μg (H^+) Fluorescein solutions are stable for about 10 days. The standard rings are stable if stored in dark and in sealed containers. *Sample:* From sample tapes, direct dissolution, reaction with fluorescein and transfer (*62*). Aliquots of sample solutions may be used. The ring oven is maintained at 90°C.
Vanadium 1. Moisten sample spot (tape) with HCl (0.1 *N*). 2. Add three 20-μl portions of reagent. 3. Dry. Center on ring oven (40°–50°C) and wash test color to ring zone using $CHCl_3$. 4. Expose test ring to steam until color of excess reagent is washed out; gray-yellow ring.	Limit of identification: 0.01 μg Optimum range: 0.01–3.0 μg Reagent solutions and test rings are stable for at least 2–3 weeks. The reagent is prepared by dissolving 0.4 gm anthranilic acid in 20 ml ethanol and then adding 5 ml benzene. This is then added to a mixture of 12 ml salicylaldehyde in 4 ml ethanol. The reagent is ready after standing for 1 hour in a water bath held at 40–50°C. The method is specific. *Sample:* From sample tapes, moisten sample spot with HCl (0.1 *N*) and then run test reaction, extraction, and estimation (*63*).
Zinc 1. Add five or six 30-μl portions of $(NH_4)_2S_2O_3$ (5%) so that it covers the whole ring zone area. 2. Dry. Add 30 μl of dithizone (0.005% in CCl_4) to the center and wash.	Limit of identification: 0.05 μg Optimum range: 0.1–2.0 μg Dithizone is stable for about 1 week. The method is specific for air pollution studies. *Sample:* From sample tapes, direct dissolution and transfer by adding 30 μl of NH_4 acetate (15%) + 30 μl KCN (0.5%) and washing to the ring zone with distilled water. Aliquots of sample solution may also be used (*64*).

[a] Volumes specified throughout this table relate to ring ovens having a 33-mm aperture. If a unit with a 22-mm aperture is used, 15-μl volumes should be used.

tain 0.1 μg of pure substance per microliter of solution. A typical standard scale would be prepared by making 10 rings with 0, 1, 2, 3, 4, 5, 6, 7, 8, and 9 μl of standard solution. Once the colors are developed in the standard scale, they can ordinarily be preserved for future reference by spraying the rings with a clear acrylic spray, or simply by storing the rings in individual clear plastic envelopes or containers. Ottendorfer (*64*, *65*) has found that the visual matching of the rings is equal to, if not superior to, the results obtained with instrumental measurements of ring densities. With ordinary care most determinations can be carried out with sufficient accuracy by the above procedure. If more precise analyses are required, the statistical approach of Knödel and Weisz (*66*) should be used. This approach requires the preparation of three rings having three different aliquots of the unknown solution. These are then matched against standard rings and the weighted average used to determine the amount of unknown present. Accuracies of from 90 to 95% at the microgram range can be obtained with this approach.

3. Reagents

Most of the reagents used are organic reagents which provide great sensitivity and excellent selectivity. The reagents form intensely colored insoluble reaction products with the individual sample entities. The precipitate of reaction product is produced in the pores of the paper and normally becomes fixed and remains as a permanent record. Usually the reagents are kept in solution form and are applied to the test sectors simply by adding a drop or two of the reagent or by spraying the sector with the reagent solution. If the reagent is sprayed or the filter paper is dipped into the reagent solution, it may be necessary to wash out the excess reagent, especially when the reagent itself is colored. In such cases the reaction product should be insoluble in the wash solution. A recent innovation has been the development of reagent crayons. Many reagents can be put into solid solution by dissolving them in an appropriate melted matrix which can then be poured into a soda fountain straw and allowed to solidify. For use, the crayons are exposed by peeling off an appropriate length of the protective straw after which the crayon is rubbed across the surface to be tested. Gentle heat is usually required to melt the crayon and provide intimate mixing of the reagent with the sample.

4. Sampling

Because the ring-oven methods have such high sensitivity, sufficient sample can usually be obtained by means of automatic tape samplers.

In many cases various acid-soluble materials can be studied directly on a section of the tape used for sampling. A convenient approach for most routine studies is to obtain a sample spot having a diameter less than that of the desired ring (22 mm). The section of the tape containing the sample spot is centered on the preheated oven. The spot is then treated with NH_4 acetate and KCN, washed to the ring zone (*5*), and the ring developed by application of the appropriate reagent to the ring zone. For some metals treatment with acid is used instead of the acetate and cyanide. Sampling can also be accomplished by electrostatic precipitation, the use of membrane filters, or by means of high volume samplers. In such cases, the sample obtained can be dissolved in 6 *N* HCl or HNO_3 or some other suitable solvent, and the solution then analyzed using the ring-oven method.

5. General Remarks

Only quantitative or high quality qualitative filter paper should be used for ring-oven studies. All papers studied were found to contain traces of selenium (*58*). Except for studies of selenium itself, selenium impurities in the paper are probably not significant. Iron impurities, on the other hand, are more troublesome and papers with low iron content should be used as much as possible. For those cases where contaminated paper must be avoided, the paper can be pretreated to remove the unwanted impurities (*58*).

Particularly for quantitative work the washing technique should be carried out in such a way that the salts of the solute are deposited in sharply defined rings. The technique for producing sharp rings requires that the amount of solvent added for each washing and the interval between successive washings be such that the solvent diffuses through the pores of the paper all the way to the ring zone, but that it approaches the heated ring zone at such a rate that it does not produce overrunning or flooding.

As pointed out above, a certain amount of care is required in order to produce sharply defined rings. The principal source of difficulty is flooding of the ring zone which results from impatient adding of solutions. A recent innovation serves to eliminate some of the tedium of producing sharp rings. A physical barrier in the form of an impervious wax ring can be used to supplement the heat barrier provided by the ring oven (*67*). Impervious barrier rings may be prepared by melting a wax such as Bareco Polywax 2000 (Bareco Division, Petrolite Corporation, P.O. Drawer K, Tulsa, Oklahoma) in a shallow dish. As the wax melts to form a thin layer, a heated glass or metal tube is touched to the wax and then stamped on a piece of filter paper resting on a warm glass plate. The wax is thus transferred to the paper where it solidifies and forms the desired impervious barrier. The diameter of the stamped ring should be 1 or 2 mm less than the diameter of the ring-oven aperture.

As has been pointed out, ring-oven separations and development of test colors must be carried out on filter paper. Other materials, such as glass fiber filters, membrane filters, or filters made from polymeric fibers, cannot be used because they do not have the fibrous texture that is required for capillary absorption and desorption that is involved in the migration of solutions and the final fixing of colored reaction products. It is possible, however, to use various filter materials for collection of samples, provided these samples are transferred to filter paper for the final ring-oven processing.

If particulate samples are collected on membrane filters or other filter media besides filter paper tapes, an areal aliquot of the dust spot can be cut out with a sharp cork borer, and then placed dust side down on a piece of Whatman No. 41 filter paper. The spot is then centered exactly over the annular aperture of the ring oven and appropriate solvents and wash solutions are added in the customary manner to complete the transfer of the sample to the ring zone. If the ring oven has a 33-mm diameter hole, a 25–28-mm diameter sample spot should be used. If the ring oven has a 22-mm diameter hole, the dust spot taken should be 15–18 mm in diameter. After the analysis is complete, the calculations should take into account the fraction of the total sample represented by the spot actually analyzed.

6. Experiments for Practice and Demonstration

Two qualitative experiments are suggested which emphasize the sensitivity and simplicity of the ring-oven method. The simplest demonstration is to detect impurities of iron in various samples of filter paper. If various brands and grades of filter paper are tested with a common reagent such as potassium ferrocyanide, very few distinct tests for iron impurities will be obtained because of the small amount that is present. If the papers are placed on a ring oven, however, and the iron impurities washed to the ring zone, practically all papers will yield detectable amounts of iron. The iron distributed in the paper is so dispersed that it cannot be detected readily in place, but when it is dissolved and washed to the ring zone and deposited there, it has been concentrated approximately 100-fold into a sharp ring which is readily detected. The experiment gives convincing evidence of the elegance of the technique and its effectiveness for trace or microchemical studies. The experiment also emphasizes the importance of using pure papers for ring-oven studies, and it further demonstrates the importance of selecting papers having uniform textures—if the solvents and wash solutions do not migrate through the pores uniformly in all directions, the rings obtained will not be uniform. The experimental procedure is as follows:

(a) Place a circle of filter paper (5.5 cm diameter) on the hot surface (110°–120°C) of the ring oven and weigh it in position by means of the retaining ring. The temperature of the ring can be easily adjusted by means of a powerstat and as a general rule should be established at a temperature of about 15°C higher than the boiling point of the solvent used for washings.

(b) By means of a capillary pipet, add a small drop (15 μl) of dilute HCl to the center of the paper. This is accomplished by inserting the capillary pipet through the guide tube of the ring oven.

(c) The wash solution diffuses through the paper and it reaches the edge of the heated surface, the solvent evaporates, leaving behind the solute—in this case, the dissolved iron impurities. The time interval between successive additions of the solvent should be adjusted in such a way as to prevent the flooding of the ring zone which would result in the formation of diffused rings.

(d) Remove the filter paper from the ring oven and test for iron by adding a drop of dilute potassium ferrocyanide to the approximate area of the ring of the ring zone. A sharply defined blue ring should be formed.

A second demonstration that is readily performed is to mix a few milligrams of nickel sulfate or chloride with a few milligrams of iron(III) sulfate or chloride in 50 ml of distilled water. Place a drop of the mixture in the center of a circle of Whatman No. 41 or Schleicher and Schuell No. 595 filter paper positioned on the hot surface of a ring oven. Wash the sample to the ring oven with 15-μl portions of dilute HCl as described above. Remove the filter paper, cut it in half and test one-half with potassium ferrocyanide as described above. Add a drop of alcoholic solution of dimethylglyoxime to the other half and then expose it to ammonia fumes to test for nickel. The nickel dimethylglyoximate should appear as a sharply defined red line.

H. Other

Although optical methods of analysis and the ring-oven technique discussed above are sufficient for most analysis of particulate samples, some other methods find use and are considered below. The methods used, of course, depend on the analytical problem concerned as well as the expertise of the analyst and the availability of necessary equipment.

1. Gravimetric and Titrimetric Analysis

Because even the total amount of inorganic particulate materials that can be collected is quite small, the amount of any given constituent is generally in the nanogram to microgram range. At best, only a few milligrams of any given pollutant are available and, therefore, gravimetric procedures are seldom used for the quantitative analyses. Gravimetry on the microchemical scale is applicable but more convenient techniques are generally available. Likewise, titrimetry is finding less application, although coulometric titrations have sufficient sensitivity and accuracy to make them attractive for certain specific applications.

2. Colorimetry and Spectrophotometry

During the past two decades the number of specific, selective, and sensitive reagents for the determination of inorganic and organic compounds has increased rapidly. New reagents are available for colorimetric, spectrophotometric, fluorometric, and nephelometric determinations that are so sensitive and reliable that they are ideally suited for the analysis of inorganic particulates (Table III) (*68–76*). Sensitivities are sufficient even for the study of such small amounts of material as those collected on molecular membrane filters and paper filter tapes. Equally important, the selectivity of many of the reagents provides procedures that are simple and reliable. Where specific reagents cannot be found, it is often possible to use selective reagents and apply them in specific reactions through proper conditioning of the system by the use of masking agents or other conditioning techniques.

One of the most important determinations in the study of inorganic particulates is that of lead. Although lead has been determined by many different methods, the most widely used approach is the spectrophotometric determination employing diphenylthiocarbazone (dithizone). The reagent is very sensitive and when properly used is capable of giving excellent results. In addition to the determination of lead, dithizone can be used for the determination of a number of other heavy metals such as mercury, silver, bismuth, antimony, cadmium, and zinc. The great sensitivity of the reagent permits the determination of microgram amounts of these metals, and selectivity is established by the use of appropriate masking agents, fixing of oxidation states, and the control of pH.

An important determination that is best made spectrophotometrically is nitrate nitrogen. Nitrates are associated with aerosols through sorption of oxides of nitrogen with subsequent air oxidation to form nitric acid and nitrates. Chromotropic acid has been introduced for the spectrophotometric determination of nitrate. The method is significant because there are no interferences due to chloride. Nitrites can be prevented from reacting with the reagent or they may be permitted to react and thus be included with the nitrate values if both nitrate and nitrite are to be established as a combined value.

Chlorides in airborne particulates can be determined by direct spectrophotometry. The procedure is based on the use of iron(III) perchlorate, and the intense absorption by the chloro complexes of iron(III) at a wavelength of 340 nm provides a specific means for determining microgram amounts of this halide. Other halides do not interfere.

Although it is not a spectrophotometric measurement, sulfate is often determined by means of a spectrophotometer which is employed in the measurement of turbidities produced by barium sulfate precipitates. Sulfuric acid or its salts may be determined by adding a slight excess of barium chloride solution to an appropriate volume of acidic sample solution and after mixing, measuring the optical density at 500 nm.

3. Polarography and Anodic Stripping Voltammetry

The polarographic technique was introduced by Heyrovsky *(77)* in 1922. Since that time, the method has been applied extensively to solve a wide variety of analytical problems. A broad assortment of polarographic instruments have been designed and are commercially available. Many inorganic ions and organic compounds are easily oxidized or reduced and thus may be analyzed by polarography. The method is quite sensitive and provides both qualitative and quantitative information on as many as five or six species present in a sample and the results recorded on a single polarogram. Polarographic theory, techniques, and applications are discussed in detail in numerous texts, review articles, or monographs, so only a brief discussion of these will be included here. The reader is referred to the monograph *(78)* by Heyrovsky and Kuta for an authorative discussion of the subject.

The polarographic method is based on the fact that ions in solution diffuse at a constant rate when a potential is applied between two electrodes in that solution. The rate of diffusion is constant for a given system and thus serves to limit the current, i.e., the limiting current, that can be carried by any given species. The potential at which an oxidation or a reduction occurs is a thermodynamic property of the ion in solution and is different for different ions. This allows for the simultaneous determination of several ions in a single solution as illustrated by Figure 5.

Table III Some Colorimetric, Spectrophotometric, and Fluorometric Methods

Determination	*Remarks*
Arsenic	
1. Reduce to As^{3+}, convert to arsine, and distill using Gutzeit apparatus. Trapping is with 3 ml of silver diethyldithiocarbamate (0.5% in pyridine). 2. Measure absorbance at 560 nm.	Sensitivity: 0.1 μg All reagents must be As-free. Hydrogen sulfide interference can be avoided by use of a lead acetate plug. The method is tedious but accurate (*68*). *Sample:* Collect samples by impingement (30 liters/m) in NaOH (0.1 *N*), electrostatic precipitation (100 liters/m) or filtration on membrane filters (100 liters/m).
Beryllium	
1. Isolate by extracting beryllium butyrate into $CHCl_3$. 2. Transfer Be to a buffered aqueous system and add morin (2′,4′,3,5,7-pentahydroxyflavone). 3. Measure fluorescence of the morin complex in alkaline solution using excitation at 365 or 435 nm and fluorescence emission at 550 nm.	Sensitivity: 0.01 μg Large amounts of copper may interfere (*69, 70*). *Sample:* Collect samples with high volume sampler (500 liter/m). Wet ash with $HNO_3 + H_2SO_4$.
Cadmium	
1. To neutralized solution add sodium potassium tartrate, KCN, NaOH, and hydroxylamine hydrochloride. Extract with $CHCl_3$ solution of dithizone and read absorbance at 518 nm.	Sensitivity: 0.1 μg Only thallium interferes (*71*). *Sample:* Samples may be collected by impingement (30 liters/m) in dilute HNO_3, electrostatic precipitation, or by filtration. Cadmium and its salts are dissolved in HNO_3.
Chloride	
1. The aqueous solution of chlorides are treated with iron(III) perchlorate in $HClO_4$. The absorbance is measured at 350 nm.	Sensitivity: 10 μg No interferences: not even other halides interfere. Total chlorides may be determined by bubbling the sample through a 5 *N* $HClO_4$ solution of the reagent. Note that $HClO_4$ and its solution should be handled with the ordinary precautions (*72*). *Sample:* Particulate chlorides are best sampled on membrane filters (10 liters/m) or on glass filters using high volume samplers.
Lead	
1. Ash samples in HNO_3, then dissolve residue in 1% HNO_3 and treat solution with hydroxylamine hydro-	Sensitivity: 0.3 μg Bi and Tl interfere. Great care necessary to ensure purity of reagents and ad-

Table III *(Continued)*

Determination	*Remarks*
chloride. With careful adjustment of pH, add NH_4CN and extract with $CHCl_3$ solution of dithizone. Repeat extraction and measure absorbance at 510 nm.	herence to specified conditions (*73*). *Sample:* Collect particulate lead on membrane filter (100 liters/m) or by means of electrostatic precipitation. A standard impinger containing 75 ml of 1% HNO_3 may also be used at sampling rate of 30 liters/m.
Manganese	
1. Incinerate sample, then dissolve ash in a mixture of H_2SO_4–$H_2C_2O_4$. Evaporate to dryness, add H_2SO_4 (1:20), a drop of $AgNO_3$ (0.1 *N*) and 0.5 gm of $(NH_4)_2S_2O_8$. Heat, make up to volume, and measure absorbance at 525 nm.	Sensitivity: 1 μg Cr and large amounts of chloride interfere. The latter can be inhibited by the addition of $HgSO_4$ (*74*). *Sample:* Collect sample on membrane filter (100 liters/m) or by means of a high volume sampler or electrostatic precipitation.
Nitrate	
1. Dissolve nitrates in minimum volume of water (e.g., 2.5 ml), add 1 drop sulfite–urea solution, place flask in cold water and add a sulfuric acid solution of $Sb_2(SO_4)$. Add chromatropic acid solution, adjust volume, and measure absorbance at 410 nm.	Sensitivity: 0.5 μg No interferences (*75*). *Sample:* Collect sample by means of a high volume sampler or by electrostatic precipitation.
Sulfate	
1. Dissolve sulfate by refluxing for 90 minutes. Filter and adjust volume. 2. Acidify with HCl, add 4 ml glycerol–alcohol mixture and determine absorbance at 500 nm. Add 0.25 gm $BaCl_2$, mix, and let stand for 40 minutes. Read absorbance at 500 nm and subtract any correction for impurities.	Sensitivity: 50 μg turbidimetric (or 2 μg nephelometric). No interferences common for air pollution studies (*76*). *Sample:* Collect sample on glass fiber using a high volume sampler or use electrostatic precipitation.

The basis for quantitative analysis is indicated by the Ilkovic equation which states that for a given ion, the limiting current is proportional to concentration.

Figure 6 shows the design of a typical three-electrode polarographic cell. The working electrode is usually a dropping mercury electrode which has the advantage of giving a clean, reproducible mercury surface every few seconds, thus eliminating the likelihood of a buildup of contaminants. The reference electrode is usually a saturated calomel electrode and

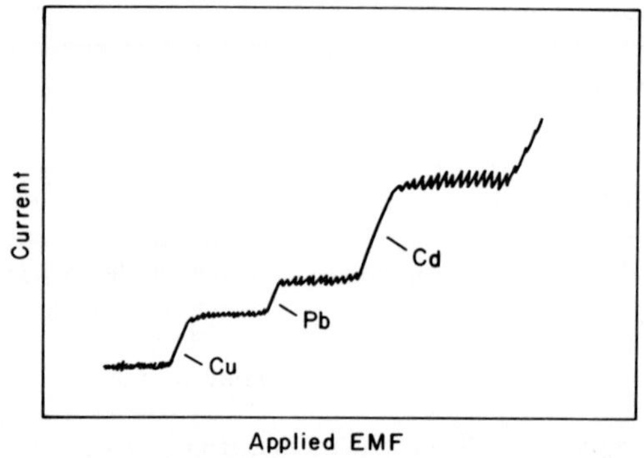

Figure 5. Typical polarogram showing multiple waves typical of samples having a number of metals.

serves to indicate the change in potential between the bulk solution and the working electrode. The auxillary electrode may be either a platinum electrode or a mercury pool electrode. In some applications the auxillary electrode is used as the reference electrode, thus providing a simple two-electrode system. Oxygen can interfere with many analyses, and should be removed by bubbling nitrogen or some other inert gas through the solution prior to the analysis.

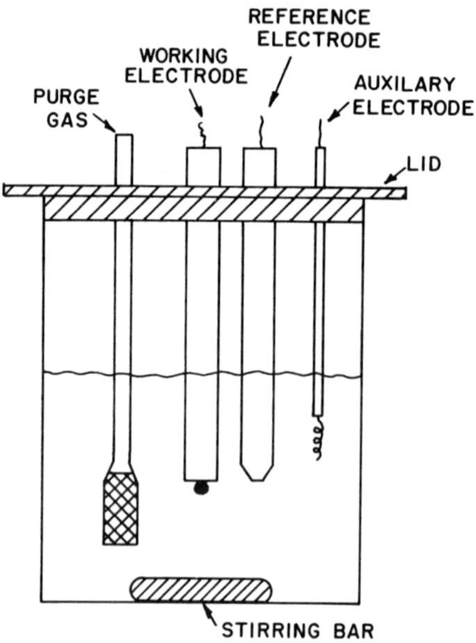

Figure 6. Three-electrode polarographic cell with a working electrode together with reference and auxillary electrodes.

A survey of the literature reveals that the behavior of some 80 elements in solution has been studied polarographically, including most of those of interest to workers in the field of air pollution. Many of these papers deal only with theoretical problems but certainly the adaptability of polarography to the problem of measuring air pollutants is great. The techniques involved have been the subject of several recent reviews (*79–81*). These and the references cited therein provide a more detailed description of these techniques than can be given here.

Most elements can be determined by conventional polarography to a level of about 10^{-6} *M* with as little as 10 ml of solution required. A few detection limits are as low as 10^{-8} *M*, which corresponds roughly to a sample requirement of 1 μg. A desirable feature of the conventional technique is that equipment costs can be kept very low, under $100 if a strip-chart recorder is available, by building one's own equipment using operational amplifier circuitry as shown in Figure 7.

As the potential sweep rate is increased, the time required to reach the oxidation potentials are decreased. In conventional polarography this creates some problems in the analysis. However, at the limit of the rapid sweep rate, the entire potential range can be swept during a portion of the life of a single mercury drop. This technique, in the form of cathode-ray polarography, has the effect of increasing the current flow and providing a smooth curve. By taking the derivative of the curve, the resolution can be greatly enhanced. Resolution of peaks of less than 50-mV separation is possible and detection limits are lowered to the 10^{-7} and 10^{-8} *M* range.

The development of differential pulse polarography has served to enhance the sensitivity of measurements of metal ions in solution. Figure 8 shows the relationship of applied potential and the points at which measurements are made. The pulsed signals are timed to coincide with the drop time of the dropping mercury electrode (DME). The current flowing is sampled immediately prior to the initiation and the completion

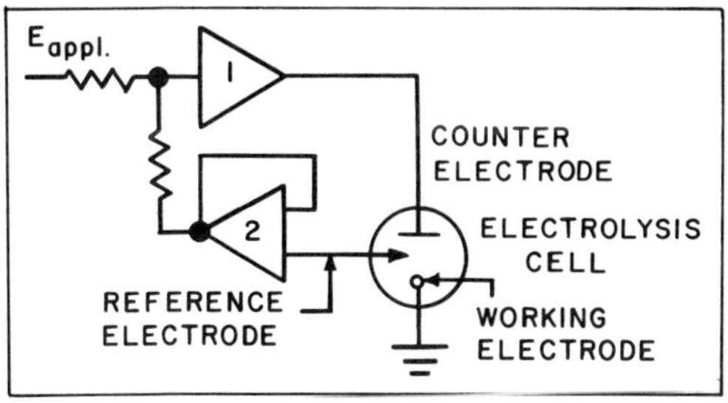

Figure 7. Three-electrode potentiostatic polarographic system.

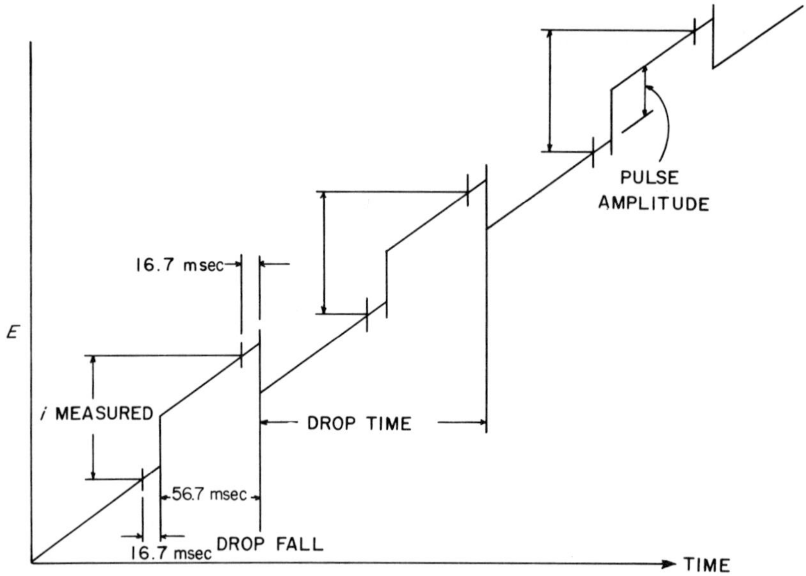

Figure 8. Potential excitation wave form employed in differential pulse polarography.

of the pulse. The change in current between the two times arises almost completely free from reduction of the electroactive species and occurs on a time interval such that it is larger than the diffusion-controlled current. The result of this approach is a peak for each species corresponding to the derivative of the standard form of the polarogram. This serves to enhance the resolution and sensitivity to levels approaching 10^{-9} *M*.

One of the most useful polarographically based techniques for trace analysis is anodic stripping voltammetry, much studied by Kemula (*82*). This technique can be applied to most metals and involves preconcentration of the metals by application of a large enough potential to the electrode system to plate out the metals on the electrode surface. The working electrode is usually a hanging mercury drop electrode (HMDE) or a graphite electrode with a thin mercury coating. Occasionally, a platinum electrode may be used. When essentially all of the metal ions of interest are removed from solution, the potential is scanned in the reverse direction and the current flow is recorded. Recently, other methods similar to those employed in differential pulse polarography have been used for the reverse reaction (*83*) giving better resolutions of the peaks present. Sensitivity, of anodic stripping voltammetry is very high (Fig. 9), permitting the analysis of solutions as dilute as 10^{-12} *M*.

As amplification technology and noise reduction systems have im-

Table IV Detection Limits For Some Common Elements

Element	*Polarography*[a] (μg)	*Anodic stripping*[a] (μg)
Aluminum	0.01	
Arsenic	0.75	
Bismuth	0.01	0.02
Bromine	1	
Cadmium	0.05	0.00005
Clorine	2	
Chromium	10	
Cobalt	0.01	0.002
Copper	0.02	0.00006
Erbium	0.01	
Gallium	0.01	
Germanium		1
Indium	0.01	
Iodine	1	
Iron	0.01	
Lead	0.01	0.00021
Molybdenum	10	
Nickel	0.01	
Niobium	4	
Selenium	5	
Silver	0.5	
Sulfur	10	
Tellurium	1	
Tin	0.05	
Tungsten	10	
Vanadium	0.01	
Zinc	0.01	0.00001
Zirconium	0.01	

[a] Detection limits for a 10-ml solution.

proved, there has been a considerable lowering of detection limits for polarographic analyses. At the current state of the art, polarography has much to offer the worker who must make routine daily determinations of several metallic pollutants in air. The high sensitivity and high selectivity make it very attractive and the speed of analysis is certainly no less than more widely accepted methods. Table IV gives some of the metals recently determined at trace levels and their detection limits. This list is by no means complete, but it includes most inorganic pollutants of interest. Figure 10 indicates differences in half-wave potentials of some common metal ions in different supporting electrolytes.

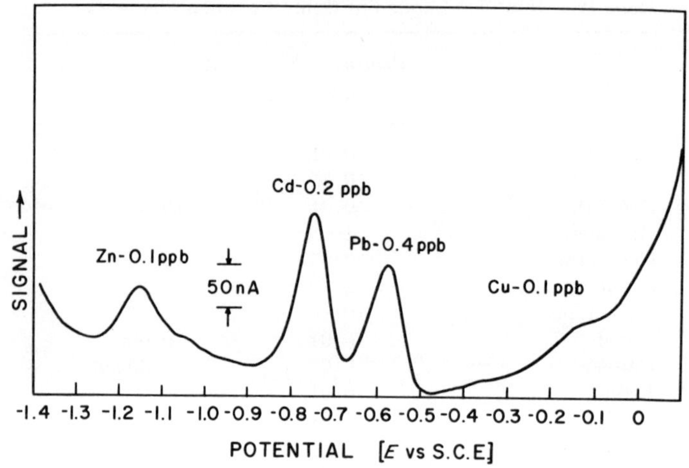

Figure 9. Pulsed stripping voltammogram typical of samples having a number of metals present.

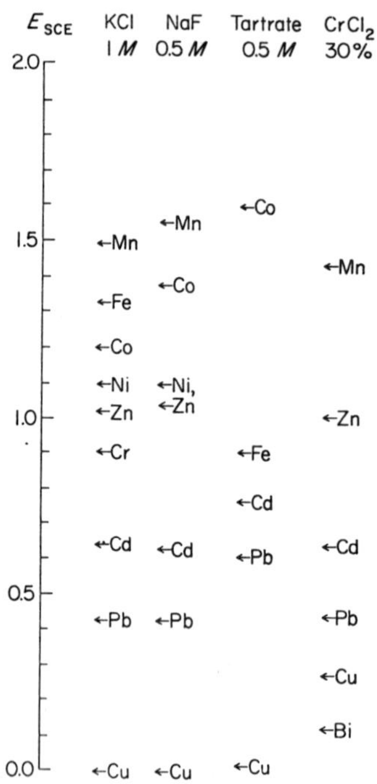

Figure 10. Some half-wave potentials in different supporting electrolytes.

4. Potentiometry and Coulometry

Potentiometric and coulometric methods have limited applications in the analysis of airborne particulates. Potentiometry is limited mainly to the use of ion selective electrodes. The ammonium and fluoride selective electrodes are good examples of what can be accomplished by this approach. As other electrodes are introduced or improved, it can be expected that more use will be made of potentiometry. In the case of coulometry, a number of applications can be anticipated. The applications, however, are likely to be more in the form of continuous monitoring equipment, and the development will be mainly by equipment manufacturers. The users of coulometric methods will be primarily the purchasers of monitoring equipment and any necessary background information will be part of the associated instrument manuals.

5. Neutron Activation

Neutron activation is a powerful, versatile tool for the study of airborne particulate samples. The great cost of the facility required for activation of samples limits the use of the technique. Likewise, the cost of counting equipment and the highly specialized training required for the applications of neutron activation dictates that the method be used only in special cases. Detailed discussion of the methods involved cannot be justified in this presentation.

6. X-Ray Fluorescence

X-ray fluorescence spectroscopy is being promoted as a very promising method for the study of airborne particulate samples. Work done at the Naval Research Laboratory under contract with the Environmental Pro tection Agency has demonstrated the potential usefulness of this technique for particulate pollutant samples.

X-ray emission analysis is a nondestructive method which requires little or no sample preparation. It is sufficiently sensitive that trace element determinations can be made on a milligram or so of sample. The technique is especially attractive because it has the inherant capability of identifying and determining a multitude of elements in a single sample with a scan rate as high as one element per minute.

X-ray fluorescence spectrometry is not without defect. In application, the ideal sample should be as thin as possible so as to minimize matrix effects. The matrix effects associated with X-ray absorption and scatter reduced the analytical sensitivity and thus introduce quantitative errors. Some special limitations apply to the collection of the sample because glass fiber and vinyl membrane filters cannot be used because of the high concentrations of contaminating elements likely to be present. The principal deterrent to the development of X-ray fluorescence spectroscopy is the cost and complexity of the equipment required. Like neutron activation, it is not a tool that can be casually incorporated into a program. The high cost can be justified only on the basis of an assured flow of thousands of samples. Under such circumstances, specially qualified personnel and the costly equipment can be justified.

7. Sulfuric Acid Aerosol, Sulfur Trioxide, and Total Sulfate

A new approach to the problem of determining sulfuric acid aerosol and total sulfates has been developed (*83a*) which solves the problem of interferences that have been inherent in previous methods. In the past, samples collected for the determination of sulfuric acid were a concentrated mixture of metal oxides, carbonates, silicates, sulfates, and general atmospheric debris, together with droplets of sulfuric acid. The highly reactive acid aerosol, in intimate contact with antagonistic oxides and carbonates, was neutralized by the antagonists and thus final measurements represented only the net sulfuric acid after the interfering reactions had been completed. The new approach is based on the stabilization of sulfuric acid at the moment of collection, thus eliminating or minimizing its reaction with possible interferng species.

Perimidylammonium bromide reacts with sulfate to form a stable and insoluble perimidylammonium sulfate. When this organic sulfate is pyrolyzed, it decomposes to form stoichiometric amounts of sulfur dioxide which can then be determined by the West–Gaeke method as a measure of the original sulfate. The obvious problem of collecting and stabilizing the acid sulfate and differentiating it from sulfate salts was solved by collecting aerosol samples on glass fiber filters impregnated with perimidylammonium bromide. Droplets of sulfuric acid react topochemically with the reagent at the moment of impact and are thus inhibited from reactions with collected co-pollutants. Salts such as calcium and sodium sulfate do not interfere, although they contain sulfate, because unlike liquid sulfuric acid, they are solids and thus cannot react topochemically with the reagent. It should be noted, however, that total sulfate can be determined by a simple modification of procedure. By moistening the collected sample with additional reagent, all sulfates are sufficiently dissociated that they can react and are thus converted to perimidylammonium sulfate. Subsequent pyrolysis and determination of the evolved sulfur dioxide provide quantitative values for total sulfates.

The impregnated filters are substantially more efficient than other filters used for collecting sulfuric acid. They also are effective for collecting sulfur trioxide as well as salts of sulfuric acid. The detection limit for sulfuric acid or its soluble salts is 0.3 μg as SO_4^{2-} and the working range is 1 to 50 μg.

III. Dustfall and Sootfall

The amount of dust that settles from the atmosphere or that is washed from the air by rainfall is a fundamental measurement in the study of air pollution. Most of the particles involved will range in size from approximately 1 to 200 μm. Although the measurement of dustfall or sootfall is relatively crude, the information can be quite meaningful. Also, the collected fallout materials may be analyzed and considerable information obtained regarding the nature of the offending particulates.

A. Gravimetric Determination of Total Solids

The determination of dustfall and sootfall is made gravimetrically on samples that have been collected for a given period of time. A container for collecting dustfall may be any type in which dust will settle by gravity and become entrapped. However, to prevent the reentrainment of the collected dust by wind action, the containers may contain a liquid or may have a greasy coating. Depending on local conditions, or the requirements of the studies being made, the collecting vessel may contain an algicide (e.g., 2 mg of quaternary ammonium chloride), or it may contain a volatile antifreeze such as isopropyl alcohol. In areas of heavy rainfall, precautions must be taken to prevent the containers from being filled to overflowing with collected rainwater. A convenient means for collecting dustfall or sootfall samples is shown in Figure 11. A perforated polished aluminum open box is used to support a polyethylene cup which is em-

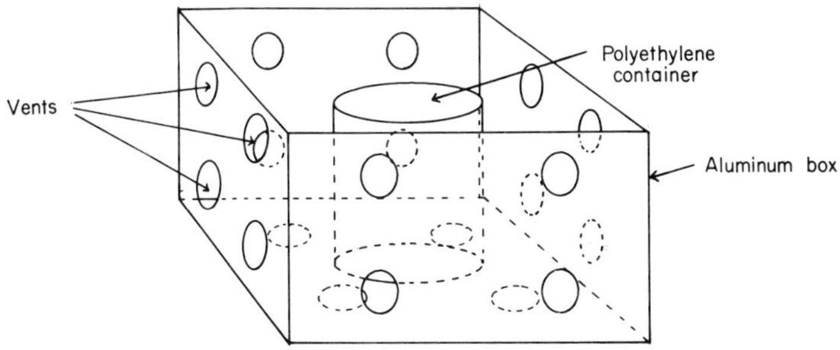

Figure 11. Dustfall collector assembly showing reflecting box to enhance evaporation and plastic collection vessel.

ployed for the actual measurement of the collected solids. The aluminum box has a dual function: it reflects solar heat to evaporate rainwater that may collect in the sampling cup; and it minimizes the loss of samples due to wind action. The perforations provide for the drainage of rainwater, and they minimize wind eddies that would otherwise cause sample losses. The polyethylene vessels are sufficiently light that they can be weighed directly on an analytical balance. Furthermore, they are not damaged by freezing conditions nor are they subject to corrosion and consequent changes in weight. The tared plastic container is placed in the aluminum box and left exposed for an appropriate period of time (usually a month) after which the container is covered with a matching lid and returned to the laboratory for subsequent study. Any moisture in the container can be evaporated on a water bath or by exposure to infrared radiation. Nonsignificant debris such as large bugs, leaves, or twigs should be removed and the amount of residue then determined by heating the container and its contents at 110°C for 1 hour, reweighing, and getting the dustfall and sootfall by difference. The results should be expressed either as milligrams per square centimeter per month, metric tons per square kilometer per month, or tons per square mile per month:

$$\text{mg/cm}^2 = \frac{1000\ (\text{gm of dust})}{0.7855\ (\text{diameter of container, cm})^2} \tag{1}$$

$$\text{tons/mile}^2 = \frac{5660\ (\text{gm of dust})}{(\text{diameter of container, in.})^2} \tag{2}$$

B. Soluble and Insoluble Solids

For some studies it is advantageous to differentiate between water-soluble and water-insoluble materials. After the total fallout has been determined, the amount of water-soluble material can be measured by washing down the inner sides of the sample container with a jet of hot distilled water. The washings (and insoluble residue) are quantitatively transferred and filtered through ashless filter paper (e.g., Whatman No. 42). The filtrate should be caught in a previously tared evaporating dish and subsequently evaporated and dried for 1 hour at 110°C. The residue is determined by reweighing the evaporating dish and obtaining the amount of residue by difference.

C. Chemical Analysis

It is often desirable to analyze the collected dust. The residue from the determination of soluble solids can be put back into aqueous solution, and its acidity or basicity determined by means of a pH meter. The total amount of acidic or basic material can, of course, be determined titrimetrically. Sodium, potassium, and calcium can be determined by flame photometry or atomic absorption spectroscopy. Other metals of special interest can be determined by appropriate spectrophotometric methods, polarography, or emission or atomic absorption spectroscopy. Fluorides are conveniently determined colorimetrically (*84*) or titrimetrically, but sulfates are best determined by the turbidimetric procedure after precipitation as barium sulfate (*85*). Nitrates should be determined by the chromotropic acid method (*75*) or the brucine procedure (*86*).

Table V Intersociety Committee Methods for Inorganic Particulates *(88)*

Analysis for	*Method number*
Suspended particulate matter in the atmosphere (high-volume method)	11101-01-70T
Atmospheric soiling index by transmittance (paper tape sampler method)	11201-01-72T
Antimony content of the atmosphere	12102-01-69T
Arsenic content of atmospheric particulate matter	12103-01-68T
Beryllium content of atmospheric particulate matter	12105-01-70T
Cadmium content of atmospheric particulate matter	12110-01-73T
Copper content of atmospheric particulate matter by atomic absorption spectrophotometry	12114-01-73T
Iron content of atmospheric particulate matter (bathophenanthroline method)	12126-01-70T
Iron content of atmospheric particulate matter (volumetric method)	12126-02-70T
Inorganic lead content of the atmosphere	12128-01-71T
Manganese content of atmospheric particulate matter	12132-01-70T
Molybdenum content atmospheric particulate matter	12134-01-73T
Molybdenum content of atmospheric particulate matter by atomic absorption spectrophotometry	12134-02-73T
Nickel content of atmospheric particulate matter by atomic absorption spectrophotometry	12136-01-73T
Selenium content of atmospheric particulate matter	12154-01-69T
Fluoride content of the atmosphere and plant tissues (manual method)	12202-01-68T 12202-01-72T
Fluoride content of the atmosphere and plant tissues (semi-automated method)	12202-02-68T
Gaseous and particulate fluorides in the atmosphere (separation and collection with sodium bicarbonate coated-glass tube and particulate filter)	12202-03-72T
Gaseous and particulate fluorides in the atmosphere (separation and collection with a double paper tape sampler)	12202-04-72T
Chloride content of the atmosphere (manual method)	12203-01-68T
Nitrate in atmospheric particulate matter (2,4-xylenol method)	12306-01-70T
Nitrate in atmospheric particulate matter (brucine method)	12306-02-72T
Dustfall from the atmosphere	21101-01-70T

Table VI Summary of Methods

Method	*Applications*	*Sensitivity*	*Reliability*	*Remarks*
Titrimetry				
General	General (ions)	mg^{+}	Nonspecific; technical skill required unless automated	Slow to rapid, depending on the system
Coulometric	General (ions)	mg–μg	Nonspecific; usually automated systems are quite precise	Quite rapid; equipment cost, \$300–3000
Gravimetry	Limited (ions and total mass)	mg^{+}	Selective methods but dependent on technical skill	Slow; equipment cost, \$300–1500
Colorimetry and filter photometry	General (ions)	mg–ng	Selective or even specific; quite reliable	Usually rapid and simple; equipment cost, \$30–5000; often automated
Turbidimetry and nephelometry	Limited (ions)	mg–μg	Selective	Usually rapid and simple; equipment cost, \$300–5000
Spectrophotometry	General (ions)	mg–ng	Selective or even specific; good reliability	Usually rapid and simple; equipment cost, \$2000–15,000; may be automated
Fluorimetry	Limited (ions and molecules)	μg–ng	Quite selective but may be subject to interferences	Usually rapid; equipment cost, \$5000–16,000
Microscopy				
General	General (ions and molecules)		May be very reliable, depending on skill and experience	Quite rapid; equipment cost, \$150–500
Polarized light	General (molecular species)		Very reliable but is dependent on training and skill	Quite slow; equipment cost, \$600–2500
Electron	General (molecular species)		Requires training and skill	Quite slow; equipment cost, \$30,000–110,000
Microprobe	General (elements)	<pg	Very reliable but dependent on skill	Slow; equipment cost, \$30,000–35,000

Polarography	Metals	mg–μg	Quite accurate but requires some skill	Rather slow but 4 or 5 metals may be determined on 1 sample; not automated; equipment cost, $500–3000
Flame photometry	Many metals	mg–μg	Essentially specific with few interferences; quite accurate	Simple and rapid; may be partially automated; equipment cost, $500–3000
Emission spectrometry	Metals	mg–ng	Specific; some interferences due to matrix effects; skill required	Training necessary; rapid for routine studies of many samples; may be automated; equipment cost, $5000–50,000
Atomic absorption spectrometry	Metals	μg–ng	Specific and very reliable	Simple and rapid; not as rapid as emission methods if more than 3 or 4 metals are run per sample; may be partially automated; equipment cost, $2500–8000
Neutron activation	General (elements)	μg–pg	Specific and very reliable	Simplicity and speed dependent on facilities and equipment; equipment cost, $60,000 and up
Ring oven	General (ions)	μg–ng	Highly selective and quite reliable	Simple and quite rapid; equipment cost, $150–200
X-ray fluorescence	Elements (atomic 11 and above)	μg–ng	Specific; subject to some matrix effects	Excellent for multielement determinations on a routine basis; expensive

It is often of interest to determine the free carbon present in fallout samples. Free carbon values may be correlated with the soiling properties of atmospheric dusts and may also be significant in physiological studies because of the well-known adsorptive capacity of carbon and the possibility that gaseous pollutants may be concentrated on particle surfaces. This determination can be accomplished (*87*) by digesting the dustfall sample with 25 ml of 70% nitric acid. After the mixture has been boiled for 20 minutes, it is diluted with 125 ml of 6 *N* nitric acid and allowed to stand overnight. The nitric acid digestion destroys the organic matter, and the residue is then collected by filtration though a tared porcelain filter crucible. The crucibles and contents are dried for 2 hours at 140°C, cooled and weighed. They are ignited for 2 hours at 700°C, cooled, and again weighed. The loss in weight is reported as free carbon.

IV. Intersociety Committee Methods of Air Sampling and Analysis

Many of the methods discussed in this chapter in more general terms are described in great detail—in "cookbook" fashion, as some people call it—in the published methods of the Intersociety Committee (ISC) for a Manual of Methods of Air Sampling and Analysis. The methods for inorganic particulates published at the time of preparation of this chapter are listed in Table V. Methods published subsequently will be found in more recent issues of the journal *Health Laboratory Science* or in more recent editions of the compilation of ISC methods (*88*).

V. Conclusion

This discussion has been restricted to certain basic determinations together with selected special methods. Considerable emphasis has been placed on atomic absorption spectroscopy and ring-oven methods because they are particularly applicable to the study of airborne particulates. They are especially attractive because of their specificity, sensitivity, convenience, and general reliability. Table VI gives a general summary of the various analytical methods now used in studying inorganic particulates.

ACKNOWLEDGMENTS

The preparation of this chapter represents the suggestions and assistance of a number of colleagues. The contributions of S. L. Sachdev, Arun D. Shendrikar,

F. K. West, George Lyles, James W. Robinson, Leonard Nelms, Ray Maddalone, Ronnie L. Thomas, G. L. Lundquist, and V. Dharmarajan are gratefully acknowledged.

REFERENCES

1. A. Walsh, *Spectrochim. Acta* **7,** 108 (1955).
2. H. Weisz, *Mikrochim. Acta* p. 140 (1954).
3. P. W. West, H. Weisz, G. C. Gaeke, Jr., and G. Lyles, *Anal. Chem.* **32,** 943 (1960).
4. P. W. West, *J. Air Pollut. Contr. Ass.* **16,** 601 (1966).
5. P. W. West, "Determination of Air Quality" (G. Mamamtov and W. D. Shults, eds.), pp. 131–142. Plenum, New York, New York, 1972.
6. W. L. Hemeon, G. F. Haines, Jr., and H. M. Ide, *J. Air Pollut. Contr. Ass.* **3,** 22 (1953).
7. C. W. Gruber and E. L. Alpaugh, *J. Air Pollut. Contr. Ass.* **4,** 143 (1954).
8. Report of the Working Party, "Methods of Measuring Air Pollution," No. 17913. Organ. Econ. Coop. Develop., Paris, France, 1965.
9. M. Katz, H. P. Sanderson, and M. B. Ferguson, *Anal. Chem.* **30,** 1172 (1958).
10. L. Silverman and F. J. Viles, *J. Ind. Hyg. Toxicol.* **30,** 124 (1948).
11. W. M. First and L. Silverman, *Arch. Ind. Hyg. Occup. Med.* **7,** 1 (1953).
12. C. F. Gottschlich, "Removal of Particulate Matter from Gaseous Wastes: Electrostatic Precipitators." Amer. Petrol. Inst., New York, New York, 1961.
13. S. Rosseland, "Theoretical Astrophysics." Oxford Univ. Press (Clarendon), London, England and New York, New York, 1936.
14. T. T. Woodson, *Rev. Sci. Instrum.* **10,** 308 (1939).
15. A. Walsh, Louisiana State Symp. Mod. Methods Anal. Chem. 1962.
16. J. E. Allan, *Spectrochim. Acta* **18,** 605 (1962).
17. J. B. Willis, *Proc. Roy. Aust. Chem. Inst.* p. 245 (1962).
18. D. J. David, *Analyst* **85,** 779 (1960).
19. B. M. Gatehouse and J. B. Willis, *Spectrochim. Acta* **17,** 710 (1961).
20. J. W. Robinson, *Anal. Chem.* **32,** 17A (1960).
21. W. Leithe, *Angew. Chem.* **73,** 488 (1961).
22. A. C. Menzies, *Anal. Chem.* **32,** 898 (1960).
23. J. W. Robinson, *Ind. Chem.* **38,** 226 and 362 (1962).
24. A. Walsh, "Advances in Spectroscopy," Vol. II. Wiley (Interscience), New York, New York, 1961.
25. W. T. Elwell, and J. A. F. Gidley, "Atomic Absorption Spectrophotometry." Pergamon, Oxford, England, 1961.
26. J. W. Robinson, "Atomic Absorption Spectroscopy." Dekker, New York, New York, 1966.
27. J. Y. Hwang, *Anal. Chem.* **44,** 20A (1972).
28. S. L. Sachdev, private communication.
29. B. J. Russel and A. Walsh, *Spectrochim. Acta* **15,** 883 (1959).
30. B. M. Gatehouse and A. Walsh, *Spectrochim. Acta* **16,** 602 (1960).
31. S. L. Sachdev and P. W. West, *Environ. Sci. Technol.* **4,** 751 (1970).
32. J. W. Robinson, *Anal. Chim. Acta* **27,** 465 (1962).
33. S. L. Sachdev, J. W. Robinson, and P. W. West, *Anal. Chim. Acta* **38,** 499 (1967).
34. H. P. Loftin, C. M. Christian, and J. W. Robinson, *Spectrosc. Lett.* **3,** 111 (1970).

35. J. Cholak and D. M. Hubbard, *Ind. Eng. Chem., Anal. Ed.* **16,** 728 (1944).
36. J. Cholak, L. J. Schafer, and R. F. Hoffer, *Arch. Ind. Hyg. Occup. Med.* **2,** 443 (1950).
37. J. Cholak, L. J. Schafer, and R. F. Hoffer, *Arch. Ind. Hyg. Occup. Med.* **6,** 314 (1952).
38. J. Cholak and R. V. Story, *J. Opt. Soc. Amer.* **31,** 730 (1941).
39. G. D. Clayton, "Determination of Atmospheric Contaminants." Amer. Gas Ass., New York, New York, 1953.
40. R. G. Keenan and D. H. Byers, *Arch. Ind. Hyg. Occup. Med.* **6,** 226 (1952).
41. R. L. O'Neil, *Anal. Chem.* **34,** 781 (1962).
42. R. G. Keenan and J. L. Holtz, *Amer. Ind. Hyg. Asso., J.* **25,** 254 (1964).
43. J. A. Dean, "Flame Photometry." McGraw-Hill, New York, New York, 1960.
44. R. Herrmann and C. T. J. Alkamade, "Flame Photometry" (transl. by P. T. Gilbert, Jr.). Wiley (Interscience), New York, New York, 1952.
45. H. Weisz, *Chem. Age (London)* **70,** 1039 (1954).
46. P. W. West and A. K. Mulkherji, *Anal. Chem.* **31,** 947 (1959).
47. H. Weisz, "Microanalysis by the Ring Oven Technique." Pergamon, Oxford, England, 1961.
48. P. W. West, H. Weisz, G. C. Gaeke, Jr., and G. Lyles, *Anal. Chem.* **32,** 943 (1960).
49. P. W. West and A. J. Llacer, *Anal. Chem.* **34,** 555 (1962).
50. P. W. West and E. Jungreis, *Anal. Chim. Acta* **45,** 188 (1969).
51. V. Dharmarajan and P. W. West, *Anal. Chim. Acta* **57,** 469 (1971).
52. V. Dharmarajan, Ph.D. Dissertation, Louisiana State University, Baton Rouge, Louisiana, 1972.
53. P. W. West and L. R. M. Pitombo, *Anal. Chim. Acta* **37,** 374 (1967).
54. A. D. Shendrikar and P. W. West, *Anal. Chim. Acta* **61,** 43 (1972).
55. M. McDaniel and P. W. West, unpublished studies.
56. S. L. Sachdev and P. W. West, *Atmos. Environ.* **2,** 331 (1968).
57. A. D. Shendrikar and P. W. West, unpublished studies.
58. P. W. West and C. Cimmerman, *Anal. Chem.* **36,** 2013 (1964).
59. C. Huygen, *Mikrochim. Ichnoanal. Acta* p. 6 (1963).
60. R. Rowland and P. W. West, *Anal. Chim. Acta* **61,** 152 (1972).
61. A. D. Shendrikar and P. W. West, unpublished studies.
62. A. D. Shendrikar and P. W. West, unpublished studies.
63. E. Jungreis and P. W. West, *Anal. Chim. Acta* **44,** 440 (1969).
64. L. J. Ottendorfer, DECHEMA (*Deut. Ges. Chem. Apparatewesen*) *Monogr.* **44,** 227 (1962).
65. L. J. Ottendorfer, "Analytical Chemistry," Proc. Feigl Anniv. Symp., p. 100. Elsevier, Amsterdam, 1962.
66. W. Knodel and H. Weisz, *Mikrochim. Acta.* p. 417 (1957).
67. F. K. West, V. Dharmarajan, and P. W. West, *Anal. Chim. Acta* (in press).
68. V. Vasák and V. Sedivec, *Chem. Listy* **46,** 341 (1952).
69. S. Banarjee, A. K. Sundaram, and H. D. Sharma, *Anal. Chim. Acta* **10,** 256 (1954).
70. C. W. Sill and C. P. Willis, *Anal. Chem.* **31,** 598 (1959); J. Walkley, *Amer. Ind. Hyg. Ass., J.* **20,** 241 (1959).
71. B. E. Saltzman, *Anal. Chem.* **25,** 593 (1953).
72. P. W. West and H. Coll, Monograph No. 3. Amer. Geophys. Union, 1959.
73. L. J. Snyder and S. R. Henderson, *Anal. Chem.* **33,** 1175 (1961); H. B. Elkins and B. P. W. Ruotolo, *Amer. Ind. Hyg. Ass., Quart.* **14,** 111 (1953).

74. "Standard Methods for the Examination of Water and Waste Water," 12th ed. Amer. Pub. Health Ass., New York, New York, 1965.
75. P. W. West and T. P. Ramachandran, *Anal. Chim. Acta* **35,** 317 (1966).
76. Interbranch Chemical Advisory Committee, "Selected Methods for the Measurement of Air Pollutants." U.S. Pub. Health Serv., Washington, D.C., 1965.
77. J. Heyrovsky, *Chem. Listy* **16,** 256 (1922).
78. J. Heyrovsky and J. Kuta, "Principles of Polarography." Academic Press, New York, New York, 1966.
79. J. Maienthal and J. K. Taylor, *Advan. Chem. Ser.* **73,** 172 (1968).
80. J. B. Flato, *Anal. Chem.* **44,** 75A (1972).
81. W. D. Ellis, *J. Chem. Educ.* **50,** A131 (1973).
82. W. Kemula, *Pure Appl. Chem.* **15,** 283 (1967).
83. H. Siegerman and G. O'Dom, *Amer. Lab.* **4,** 59 (1972).
83a. R. L. Thomas, V. Dharmarajan, G. L. Lundquist and P. W. West, *Anal. Chem.* (in press).
84. P. W. West, G. Lyles, and J. L. Miller, *Environ. Sci. Technol.* **4,** 487–91 (1969).
85. Interbranch Chemical Advisory Committee, *U.S., Pub. Health Serv., Publ.* **999-AP-11** (1965).
86. D. Jenkins and L. L. Medsker, *Anal. Chem.* **36,** 610 (1964).
87. R. McCarthy and C. E. Moore, *Anal. Chem.* **24,** 411 (1952).
88. Intersociety Committee, "Methods of Air Sampling and Analysis." Amer. Pub. Health Ass., Washington, D.C., 1972.

3

Microscopy and Pollution Analysis*

Walter C. McCrone

Every microscopist is well aware of pollution. Every sample he looks at contains a few, or often many, particles that have settled out of the air. Recent measurements in one of our own laboratories (not one of the clean rooms) showed nearly 1000 particles larger than 5 μm settling every hour on each square centimeter of bench surface.

* Much of the material in this chapter has been drawn from: W. C. McCrone and John G. Delly, "The Particle Atlas," 2nd ed. Ann Arbor Science Publishers, Ann Arbor, Michigan, 1973.

Microscopists studying pollution know that microscopy is the way to study particulate pollutants. Single particles of almost any size can be identified by a trained microscopist. His procedures are the ultimate in ultramicroanalysis, i.e., the identification of very small samples. His lower limit is about 1 picogram (10^{-12} gm) with the light microscope, about 1 femtogram (10^{-15} gm) with the scanning electron microscope and the electron microprobe, 1 attogram (10^{-18} gm) with the transmission electron microscope, and even less with the ion probe.

The microscopist can recognize many particles at sight—cement, coffee chaff, brass machining chips, zinc oxide foundry fume, paper fibers, pollens, dog hair, calcite, pumice, potato starch, etc. Recognition is just as rapid and certain, for most particles, as recognition of a next door neighbor. The identifying characteristics include size, shape, color, transparency, homogeneity, polarization colors, extinction, and refractive indices.

There will always be some subnanogram particles that cannot be quickly identified from memory or by comparison with pictures, descriptions, or known samples. If important enough to justify the often expensive further analytical study, such particles can nearly always be identified. The procedures require careful isolation of the particle, followed by measurement of suitable additional physical properties: refractive indices, diffraction pattern, or elemental composition. Procedures for micromanipulation as well as use of the necessary analytical tools, e.g., polarizing microscope, scanning and transmission electron microscopes, X-ray diffraction camera and electron and ion microprobes, are described in "The Particle Atlas" (*1*).

I. Sampling Methods

Sampling purely for particle identification is usually quicker and easier than when a sample for particle size, dust loading, or percentage composition is required. This is because it is not essential that the sample be representative in terms of size and composition. It is only necessary that these criteria be reasonably well met and that all compositions be present. Settled dust samples are often sampled by pressing a sticky transparent tape of the desired surface. A Kleenex, or better a cloth wipe, may also be used. Suspended dusts are sampled by pump and membrane filter or by impactor. The Andersen type is preferred by microscopists over the cascade impactor.

The microscopist always identifies particles with as little sample treatment as possible. If he has to remove any particles for closer study, he often does this one particle or a few particles at a time. If he is studying a membrane filter sample by reflected light under the stereobinocular microscope, he may remove specific particles by touching each with an Aroclor-coated, fine-point tungsten needle. The needle is coated by touching the tip to a 1% benzene solution of Aroclor, allowing the solvent to evaporate. This coated tip with a 1–5 μm radius will pick up one specific particle when directed by a steady hand under the stereobinocular microscope. The particles are then transferred to a small Aroclor drop on a clean microscope slide. After adding a cover slip, they can be examined in the usual way.

A. Suspended Dust

1. Membrane Filter Samples

The simplest procedure for the collection of suspended dust is certainly a membrane (or Nucleopore) filter and air pump. This usually involves a standard membrane filter size, but a simple micro sampler using a 6-mm filter can be used (*2*). If the desired particles cannot be picked off the filter a portion of it can be mounted in a liquid medium between a slide and a cover slip. The proper mounting liquid will eliminate visual interference from the filter itself. The filter may also be dissolved in a drop of solvent on the slide, or the filter may be dissolved in a few milliliters of solvent in a centrifuge tube and the particles recovered after centrifuging and decanting several times.

2. Impactor Samples

The cascade impactor is used by the microscopist by using a very thin layer of adhesive film and a short collection time to minimize sample size. Every particle should impact the adhesive film rather than another particle. The Andersen sampler will collect more sample with less pile-up of particles. The particles are usually studied *in situ* on the impacting surface.

B. Settled Dust

A quick sampling procedure for dust settled on surfaces is the applicaton of transparent sticky tape. When applied to a surface and removed, loose particles adhere to the adhesive layer of the tape. This tape can then be applied to a clean microscope slide and labeled. The particles, sandwiched between the slide and the tape backing, are protected, and they can be studied microscopically through the tape backing. If necessary, the particles are removed by lifting the tape, applying a small drop of benzene and using a fine needle to make a small ball of adhesive containing the desired particle. A larger portion of particles can also be removed in the same way, placed on a clean slide, and covered with a cover slip for more careful microscopic study.

Collections of settled dust in the field are often made by allowing dust to settle onto a thin Aroclor (1260) film on a microscope slide. Particles adhere to the film and cannot be blown away. Transparent sticky tape does the same thing and has the added advantage of being flexible. A piece can be wrapped around a tree or pole and a mark indicating north can be made. When the tape is examined, the direction from which the particle originated can be determined from the reference mark. Collection efficiency is dependent on particle size and wind speed; nevertheless, sticky tape offers a simple inexpensive way to obtain windblown particles. If it is necessary to remove the particles, it is best to use a tape having an adhesive that is soluble in some organic solvent. The adhesive is dissolved and washed along with the particles into a centrifuge tube, and the particles are spun to the bottom. The solvent (with dissolved adhesive) is decanted, and the particles are washed and centrifuged several times to remove all traces of adhesive.

II. Particle Analysis

A. Morphological Analysis

If you can recognize an apple or a rose at sight, you should be able to recognize single particles of coal, cement, or gypsum when the microscope magnifies them to an equivalent size. This is the basis for morphological analysis microscopy. No other microanalytical method identifies such small samples of such diverse substances so quickly.

Single particles of almost any substance—animal, vegetable, mineral, industrial by-product, raw material, corrosion product, fly ash—can be identified in a few seconds to a few minutes by a microscopist trained in particle identification (Fig. 1). The particle must be >1 μm in diameter and 1 picogram, or 10^{-12} gm in weight for identification by light microscopy, or 2–3 nm in diameter and 1 attogram, or 10^{-18} gm, in weight for identification by electron microscopy.

What steps must be taken to identify these tiny samples? Since nearly all are invisible to the naked eye, they must be made visible by using an appropriate microscope. To be able to identify any of the thousands of particles that may be present requires a good polarizing microscope and a trained microscopist. The equipment required can cost as little as $1000, although any microscopist skilled in this highly specialized field certainly deserves more sophisticated equipment. A total cost of perhaps $8000–10,000 is more realistic.

In addition, few laboratories are clean enough to permit work of this kind. The laboratory itself is dirty; the air in the room is dust laden; fingers carry particles; slides and cover slips are never clean. In short, there are millions of particles in any laboratory just as large, just as visible, and just as interesting microscopically, as the particle to be identified. It is essential that the sample be opened and examined under conditions such that contamination is avoided. This requires a clean bench if not a clean laboratory (Fig. 2). Finally, each microscopist must have a stereo binocular microscope and a polarizing microscope, the latter fitted with top lighting, and a few accessories. These include a photomicrographic camera, Cargille mounting liquids, and a library.

Now, what properties can be measured on small particles that will aid identification? The first is shape: Is it a very flat plate of mica, highly fibrous asbestos, pitted wood pulp fiber, conchoidally fractured quartz or ground glass, or a perfect sphere of fly ash from a pulverized coal boiler? Next, size may be important for some substances, e.g., pollens, starches, and fibers. Color is very important, and both transmitted and reflected light must be used. Homogeneity and transparency should also be noted. Crossed polars should be used to differentiate between isotropic substances such as glass, pumice, fluorite or sphalerite, and birefringent substances such as quartz, talc, nylon, fur fibers, and starch. The degree of birefringence often permits recognition of limestone, foraminifera, Orlon, quartz, and so forth. The refractive indices are usually not determined precisely, although an estimate of the value relative to the mounting liquid is often useful. For example, calcite and siderite which look very much alike can be quickly and easily differentiated when mounted in the usual mounting medium, Aroclor, which has a refractive index of about 1.66. Particle identification with these techniques is called morphological analysis.

Difficulties begin, however, when microscopic examination fails to identify the particle because the microscopist has not seen the particle before. He then must have recourse to electron microscopy, X-ray diffraction, the electron microprobe, or the ion probe. To do this the particle must be removed from the mounting medium used for light microscopic examination, cleaned, and remounted for the particular tool to be next applied. Few ultramicroanalysts are capable of carrying out these steps, and specialized "particle-pickers" are usually required for single subnanogram particle manipulation. Special tools are used but no mechanical micromanipulators.

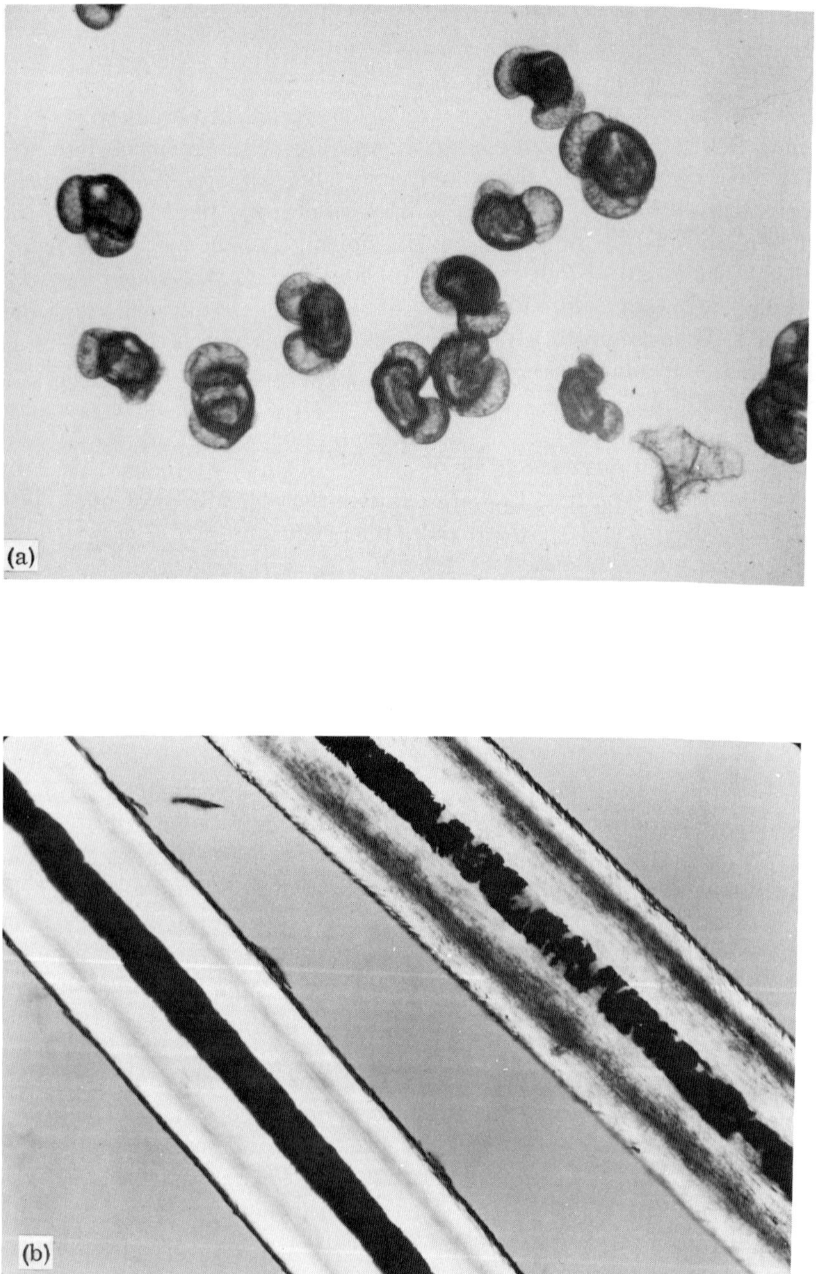

Figure 1. Light micrographs of a few representative atmospheric pollutants: (a) pine pollen, (b) human hair.

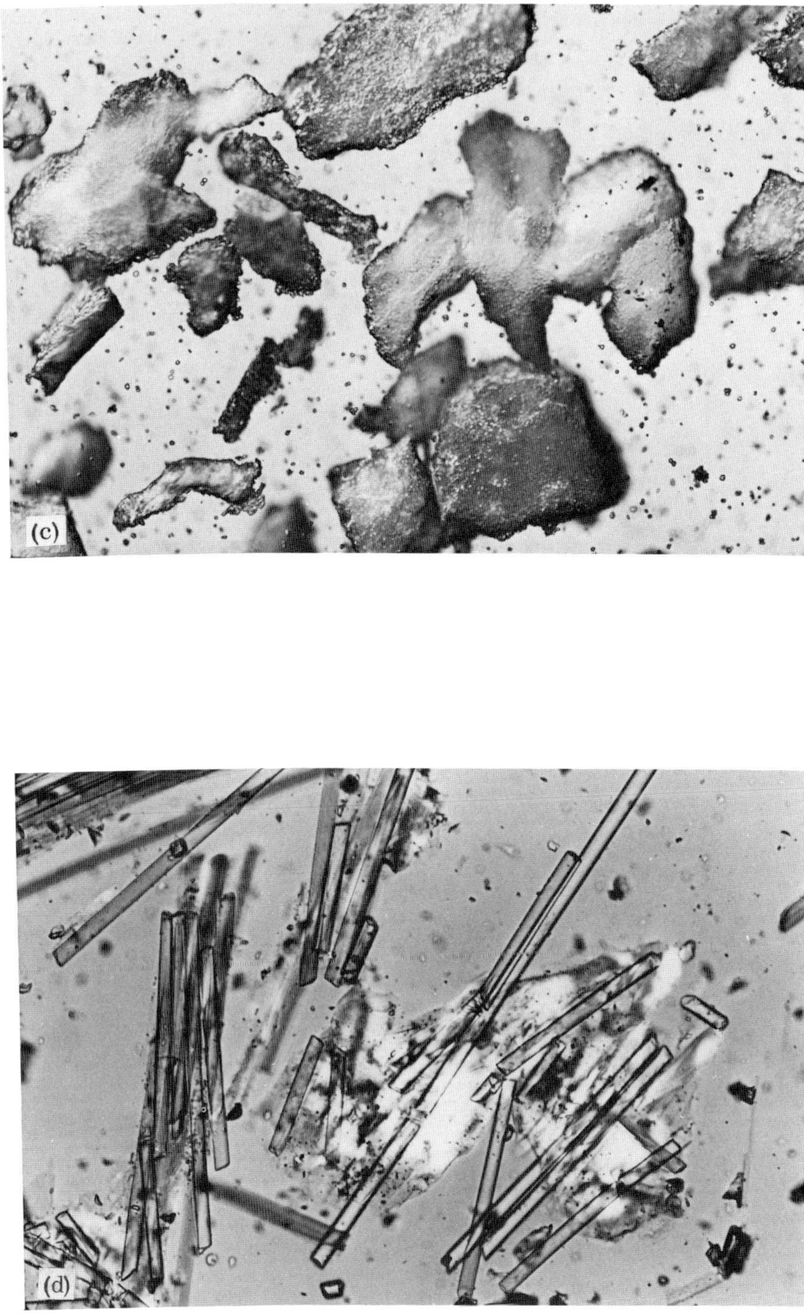

Figure. 1. (c) Corn meal, (d) glass fiber reinforced plastic.

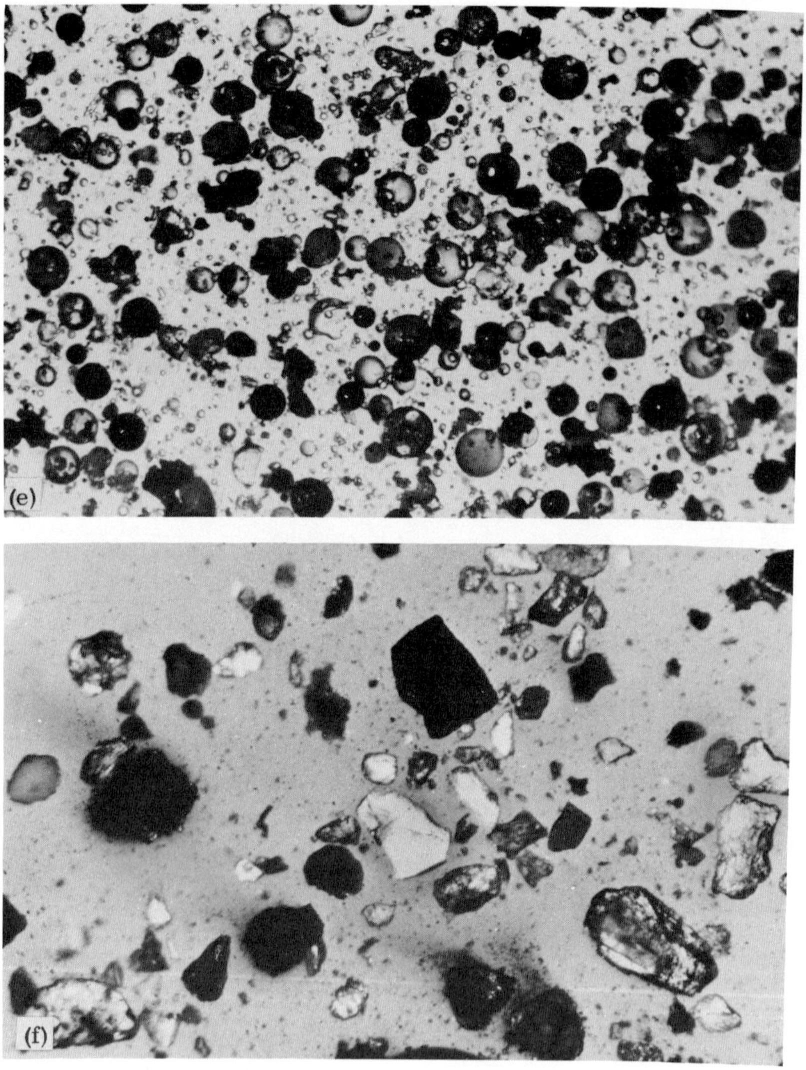

Figure 1. (e) Power plant pulverized fly ash, (f) road dust.

The following tools are useful for successful handling of very small particles.

1. Tungsten Needles

Good tungsten needles are essential for particle handling. Three different tip diameters are used for different purposes:

Needle 1 has a short blunt tip (50 μm diameter) and is very sturdy. It is used mainly for scribing marks on metal surfaces.

Needle 2 is not so sturdy as needle 1. It has a finer tip (10 μm diameter), and is used mainly for cutting particles from collodion.

Needle 3 has a very fine tip (1 μm diameter). It is used most frequently for single particle manipulation. Because the tip is rather fragile, it bends or breaks frequently. Many of these needles should be kept on hand since more than one are often necessary to complete manipulation of a single particle.

There are two ways to make tungsten needles—electrolytically and with $NaNO_2$. The $NaNO_2$ method is quicker, but the electrolytic procedure affords more control over the tip diameter.

a. Sodium Nitrite Methods. Tungsten reacts exothermically with sodium nitrite ($NaNO_2$) to form sodium tungstate. This serves to chemically etch the tungsten, and, when applied to a wire, a fine-tip needle is quickly produced.

Hold a stick of sodium nitrite ($NaNO_2$) and a tungsten needle over a Bunsen, or equivalent, burner. When the needle is red hot and the tip of the stick begins to melt, stroke the very end ($\sim$5 mm) of the needle gently through the $NaNO_2$ melt. One to three strokes are usually sufficient for a No. 1 needle. Only a few additional passes through the $NaNO_2$ are necessary for a No. 2 or No. 3 needle (Fig. 3). Needles can usually

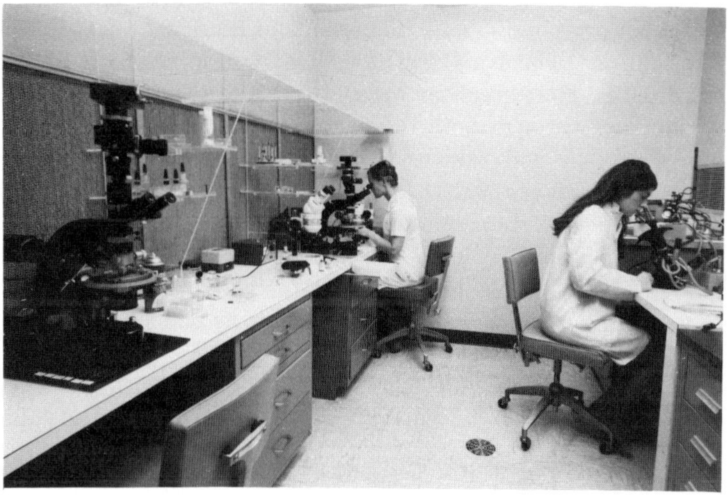

Figure 2. A clean room laboratory for particle microscopy.

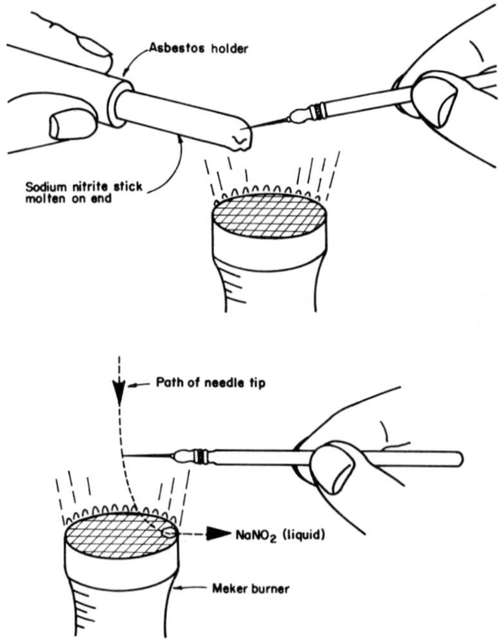

Figure 3. Preparation of tungsten needles using $NaNO_2$.

be resharpened with a single stroke. After microscopic examination, wash the tip gently under the water tap and dry.

An alternate procedure begins by placing a small lump of $NaNO_2$ powder, about $\frac{1}{4}$ in. in diameter, to one side of the grid of a Meker burner. Light the burner. The $NaNO_2$ will melt and form a small droplet supported on the grid surface. Starting at the top of the flame, quickly bring the needle down through both the flame and the drop of molten $NaNO_2$ (Fig. 3). Only the end (about a 5 mm length) of the tungsten needle should pass through the $NaNO_2$ drop and for less than 1 second. Examine the needle under a stereomicroscope after the first pass through the $NaNO_2$ drop. If the needle is not sharp enough or its tip is not long enough, repeat the process. Hold the tungsten wire in the flame for as short a time as possible, and keep the flame as low as possible; otherwise, the tungsten will become brittle.

To clean the $NaNO_2$ from the grid, turn the Meker burner upside down and hold the head under the water tap. The $NaNO_2$ will quickly dissolve.

b. Electrolysis Method. Two to six needles in their aluminum holders are suspended in a 15% KOH solution to a depth of 1–5 mm (Fig.

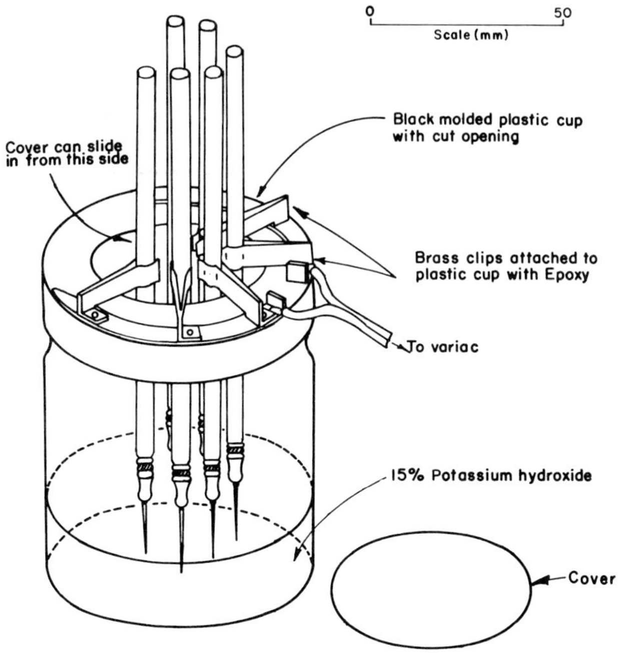

Figure 4. Arrangement for electrolysis procedure for making metal needles.

4). The needle holders are held in position by small brass clips which are in turn connected in series (the same number of clips on each lead). With the variable voltage transformer set at 10 V, 2–5 minutes of etching should produce a No. 1 or No. 2 tip of suitable diameter. The electrolysis method usually produces No. 1 and No. 2 needles with a somewhat rough surface. Fine No. 3 needles with a smooth surface are easier to make using $NaNO_2$ (Fig. 5). If more than one or two needles are to be made, the operation is best carried out in a chemical hood or other well-ventilated area.

2. Polyethylene Needles

Occasionally, a very soft needle is needed to pick up a contaminant particle without scratching the surface or picking up any of the substrate. Suitable soft needles are prepared from polyethylene as follows. A 50-mm length of $\frac{1}{4}$-in. diameter polyethylene tubing is heated over an alcohol lamp whose heat is focused into a very narrow band by placing a metal plate with a narrow slit about 3 in. above the flame. When the tubing becomes transparent and soft, it is slowly pulled out to one-third of its

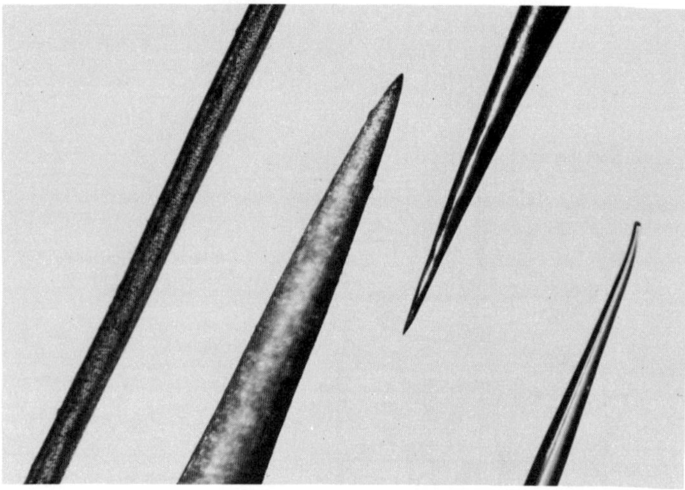

Figure 5. Tungsten and polyethylene needles.

original diameter. A tiny portion of the thinned-out tubing is then reheated and pulled out until you lose sight of the fiber. The fiber is cut, and the tip warmed for a fraction of a second well above the flame to round off the cut tip. See Figure 5 for a photomicrograph of a finished polyethylene needle. A good needle should be about 2 cm long and 0.5 mm in diameter. To have strength, it should taper to 10 to 20 μm at the tip within the last 2–3 mm. Steps 1–4 in the procedure for making polyethylene pipettes (Fig. 7) describe an equivalent procedure to that outlined here for formation of a polyethylene needle.

3. Microdroplets of Liquid

It is very important to be able to deliver tiny droplets of liquid onto any surface. There are two general types of apparatus for this: a glass fiber brush sealed into a capillary, and a polyethylene micropipette. Both devices are capable of delivering droplets as small as 100 μm in diameter.

a. Glass Fiber Brush. We have borrowed this clever idea from Fred Schneider of IBM (private communication). A bundle of 20 to 30 glass fibers, 5 to 8 mm long and about 20 μm in diameter, is sealed with a microflame into one end of a glass capillary. A melting point capillary

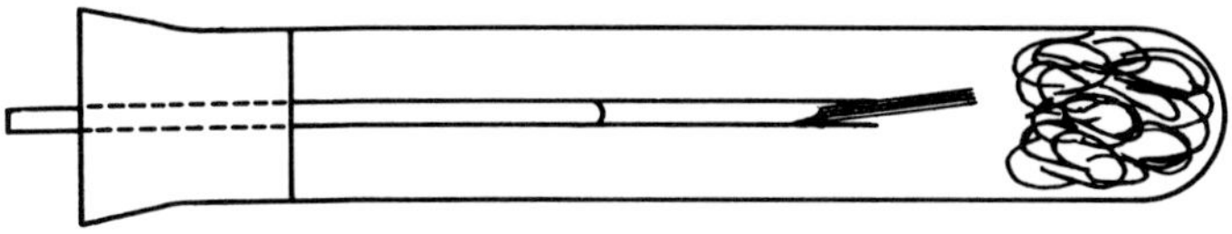

Figure 6. Protective container for glass fiber brush.

1 to 2 mm in diameter and open at both ends is convenient (Fig. 6). To use for small drop deposition, the capillary is partly filled (2- to 3-cm-long column of the desired liquid) so that the sealed-in end of the glass fiber bundle is fully immersed. Droplets can then be "painted" onto any surface by touching the glass fiber "brush" to the desired spot. Drop size depends on the number and size of glass fibers, the liquid head in the capillary, the splaying of the fiber brush on the surface, and any movement of the brush across the surface.

Normally, the glass fiber brush is used with pure liquids because evaporation concentrates a solution on the brush and may clog the fibers. This effect can be alleviated if the solution-filled capillary is stored with its brush tip in a bottle or test tube containing a little of the appropriate solvent soaked into a cotton ball at the bottom (Fig. 6). Evaporation from the open end of the capillary is very slow, negligible over a period of several hours, and can be minimized further by drawing down the open end of the capillary to a very fine opening. In this way, a dilute solution of, say, collodion can be kept ready for instant use for days and even weeks. It is essential that the brush end of the capillary be stored in the saturated solvent vapor corresponding to the solution used. Strictly speaking, one avoids small concentration changes due to differences in vapor pressure if the brush tip is stored over the same solution rather than over the pure solvent.

b. Polyethylene Pipettes. These pipettes will deliver drops of organic solvent as small as 100 μm diameter on any surface. They are unbreakable and so flexible and soft that they do not scratch even the softest surfaces. Polyethylene is inert and does not dissolve or swell in most organic solvents. They are made as follows (Fig. 7):

Step 1. Rotate a piece of polyethylene tubing over a small flame (use a microburner or an alcohol lamp). The tubing should be held at least 2–3 in. above the flame.

Step 2. As the tubing slowly becomes transparent, pull it apart very gently, continuing the rotation. As the tubing becomes thinner, slowly raise it away from the flame. Work in a draft-free area so that the flame

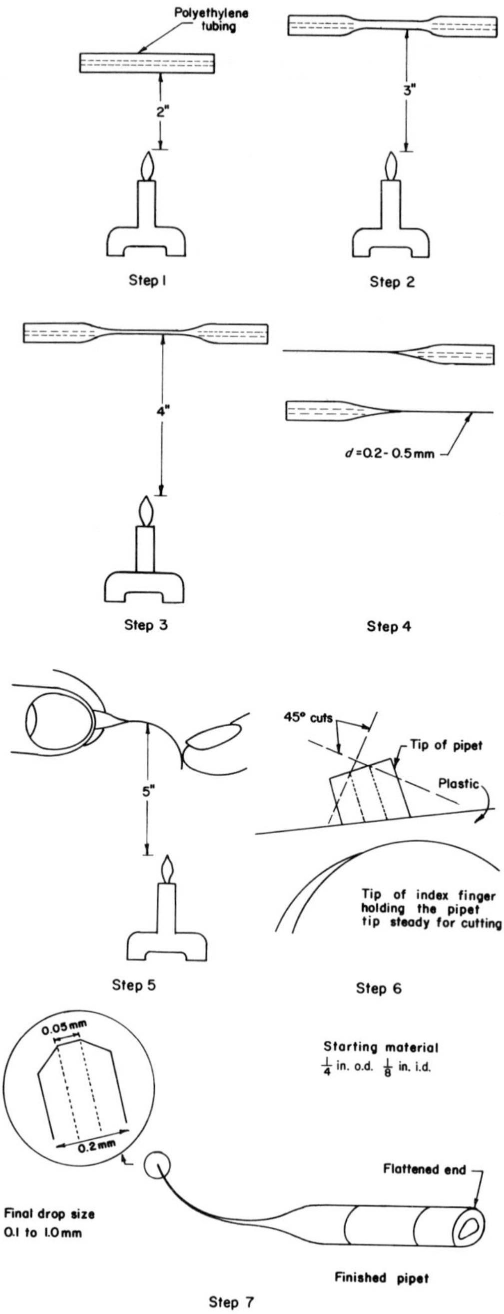

Figure 7. Steps in making polyethylene pipets.

is very steady. After the tubing is about half its original thickness, let it cool (it will again become translucent).

Step 3. Repeat steps 1 and 2, heating only a small central portion of the thinned-out area. Continue the operation 3–5 in. above the flame until the central portion looks like a thread.

Step 4. Let the tubing cool and cut it into halves at the center of the threadlike part.

Step 5. Bend one tip slightly in your fingers. Hold it about 6 in. above the flame, and, when your fingers just begin to feel uncomfortable, remove your hands and let the pipette cool.

Step 6. Focus on the pipette with the stereomicroscope. Hold the tip steady with a thin piece of plastic or cardboard and make two cuts with a razor blade. This tip shape makes uniform drop delivery reliable, an essential factor in mounting very small particles.

Step 7. Heat the opposite end of the micropipette and flatten it on one side. The pipette will then rest on the bench-top with the tip up. This avoids contamination. To fill, touch the tip with a drop of solvent. At least five micropipettes should be prepared, each delivering a different size drop ranging from 100 to 1000 μm diameter.

c. Sealed-Off Water Dropper. Small droplets of water are conveniently applied to a surface by sealing off the tip of a glass dropper bottle and drawing the tip out into a small loop. The size of the loop (1–3 mm) determines the size of the drop (2–5 mm) deposited by contact of the flat side of the loop on the desired surface.

4. Glass Flakes

Glass flakes, small cover slips, and membrane filter squares should all be kept on hand for refractive index determination of small single particles. A supply of glass flakes can be made from four full-size cover slips. Wash them and hold them tightly together, between thumb and forefinger, over a very fine sieve, e.g., U.S. standard sieve No. 400 with 37-μm openings. Tap the edges with a hard piece of metal, e.g., metal holder for carbide stylus. Many flakes will break off and fall into the sieve. Wash the fragments on the screen with water and acetone to remove flakes smaller than 37 μm and some small metal fragments. There should be several hundred flakes about 20 μm thick and 35 to 500 μm in diameter. Collect and store them in a small petri dish. They are convenient markers for small particles placed on any substrate and which are more difficult to find. They are also useful for supporting the small cover slips above such particles during refractive index determination.

5. Small Cover Slips

Crush three clean No. 1 cover slips in a mortar until most of the pieces are smaller than 2 to 3 mm in diameter. Place the ground-glass fragments on an 18 mesh, U.S. standard sieve with 1 mm openings and wash them with water and acetone to remove most of the pieces less than 1 mm in diameter. Store the remaining pieces in a small petri dish.

6. Membrane Filter Squares

Place a membrane filter, 150 μm thick, on a glass slide. With a sharp blade, cut the filter into $\frac{1}{4}$-in. squares. Make all cuts at a 45° angle to the vertical in order to obtain a good beveled edge on each side. Store in a small petri dish.

B. X-Ray Diffraction

There are relatively few ultramicro analytical techniques capable of identification of the solid phase of a chemical compound. X-ray diffraction is such a one (polarized light microscopy is the only other general one). Microprobes, mass spectrometers, emission spectrographs, and chemical methods, in general, give only the constituent atoms. How they are combined and their arrangement in the solid state can only be determined by microscopy and by diffraction.

The identification of the polymorphic form is particularly important. Sulfur, for example, can exist in several different crystal forms, each with distinctive crystal properties. Titanium dioxide is available commercially in three crystal forms (rutile, anatase, and brookite). Quartz, tridymite, and cristobalite are different forms of silica. Calcite and argonite are both calcium carbonate. Nearly every substance shows one or more metastable crystalline forms, each with distinctive properties and uses. The importance of diffraction as a method for identifying the specific crystal form of a substance is therefore apparent. This usefulness is enhanced by the ability to recover the sample unchanged and by the sensitivity of the procedure (about 10^{-11} gm).

X-ray powder diffraction measures the interplanar spacings of atoms in a crystal. The Bragg equation, $n\lambda = 2d \sin \theta$ (where n is an integer, λ the X-ray wavelength, d the interplanar spacing, and θ the diffraction angle), is used to convert observed θ angles to corresponding d values. If the X-ray powder diffraction data for a given compound have been determined and tabulated, then subsequent samples of that compound can be identified by comparison of the measured data with the tabulation.

The ASTM data file (*3*) holds powder diffraction data for well over 20,000 compounds and is being constantly expanded.

Many materials are obvious candidates for X-ray powder analysis; minerals, abrasives, corrosion products, pigments, fertilizers, etc. Most biological materials, polymers, and combustion products are noncrystalline and thus do not give X-ray diffraction patterns interpretable by means of the Bragg equation.

1. Sample Size

Analysis by X-ray powder diffraction is best carried out on fine powders. The X-rays diffracted by the sample are usually recorded either by a film camera or a counter diffractometer. The camera collects the X-rays for a longer period of time; thus it, rather than the counters, must be used for ultramicro samples because of the low X-ray flux in the diffracted beams. Samples as small as 100 ng, e.g., single particles 40 to 50 μm in diameter are usually adequate for powder diffraction. With smaller samples, good powder diffraction patterns are not as easily obtained. The increase in exposure time, required for small samples, produces darkening of the film due to scatter of X-rays by air and the sample mount. This increase in background reduces contrast, making it more difficult to see and measure the diffraction lines.

Various steps may be taken to reduce background when the sample size is limited: (1) critically align the collimation system; (2) reduce air scatter by removal of air from the camera; (3) reduce air scatter by reduction of the air path (reduction of camera diameter); and (4) reduce support scatter by reducing the size of the sample support.

For success with single subnanogram particles all of these steps must be taken.

2. Equipment

X-ray diffraction can be successful for subnanogram particles with a small diameter, vacuum camera (Fig. 8) with the single particle mounted on the tip of a very fine glass fiber with a minimum of glass and adhesive (Fig. 9). The glass fiber must have a very narrow diameter, typically 1–3 μm. The particle must, of course, be polycrystalline or ground to a very fine crystalline powder.

3. Grinding Single Small 5–50-μm Particles

A powder X-ray diffraction pattern is far easier to interpret than a single crystal pattern. However, a small particle is likely to be a single

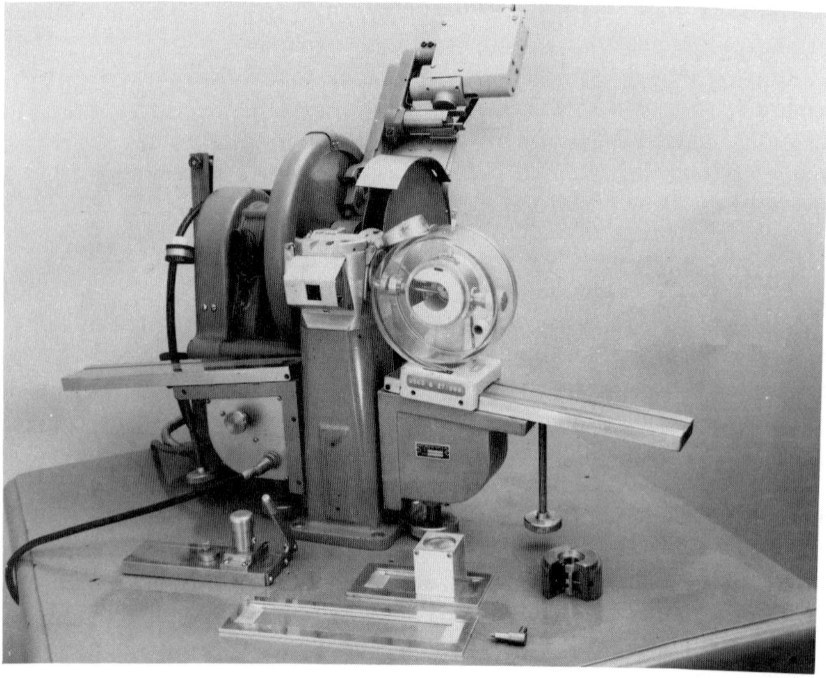

Figure 8. The 57.3-mm-diameter Philips camera with the 28.65-mm insert in place in the vacuum box.

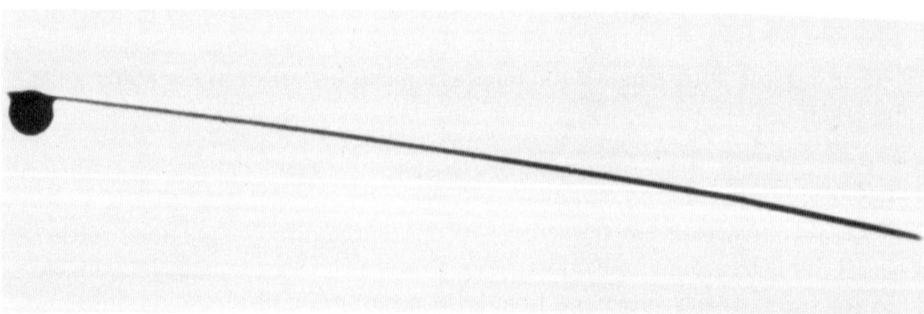

Figure 9. A 5-μm particle mounted on the tip of a 1-μm glass fiber for X-ray diffraction.

crystal and, if so, it must be crushed to a fine powder before mounting for X-ray diffraction. The procedure for crushing a single particle, 5–50 μm in diameter and mounting the powder for X-ray diffraction is as follows (Fig. 10):

Step 1. Place the particle to be crushed on a clean glass slide. Use a stereomicroscope at 25× to 50× to monitor placement of a small drop of dilute collodion next to the particle and, with a No. 2 tungsten needle, slowly spread the solution around and over the particle. When the film dries, it should not be thicker than the particle.

Step 2. Place a small, 2 to 3 mm fragment of No. 1 cover slip over the particle and, with a microspatula or other convenient tool, press down firmly. Check with the polarizing microscope to see if enough fragments are present. If the particle cannot be crushed, the collodion film around the particle is probably too thick. Thin the film with a drop of solvent before repeating the crushing procedure. If the particle presses out into a flake it is probably biological or polymeric and should not be X rayed.

Step 3. Transfer the slide to the stereomicroscope and remove the

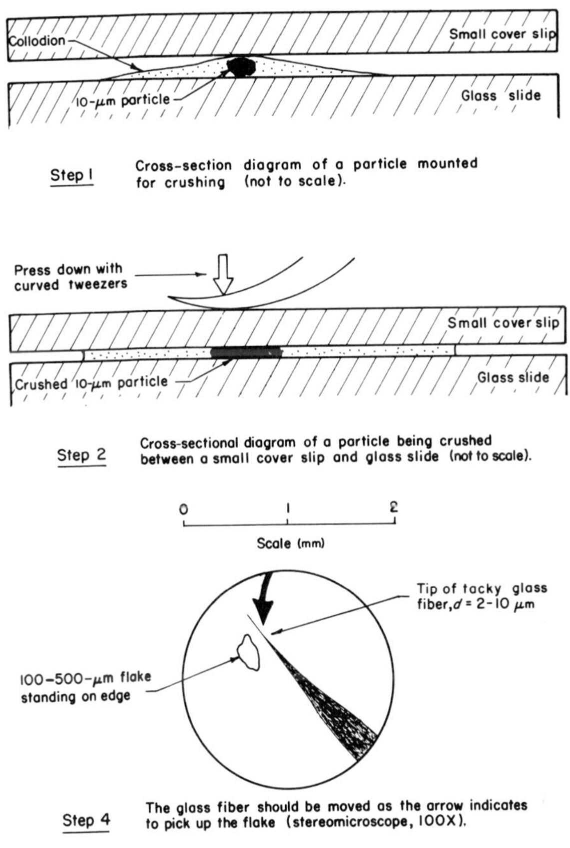

Figure 10. Crushing of a small particle between slide and cover slip to obtain a powder for X-ray diffraction.

cover slip. Place a small drop of amyl acetate on the slide and, with a tungsten needle, bring a small amount of the solvent over the collodion film and crushed particle. The softened collodion will tack down all the loose fragments. With the remaining solvent, try to wash as much collodion as possible from the crushed particle without dispersing the fragments too much. With a No. 2 tungsten needle, cut a neat circle just large enough to include all the fragments. Place a small drop of H_2O (~200 μm) next to the circled area. With a tungsten needle, bring it into the field of view and float off all the excess collodion around the circular flake.

Step 4. Now float the flake off and tip it on edge to minimize adhesive forces between it and the glass slide. Flakes larger than 500 μm across should be rolled or folded before they are picked up. Place the X-ray pin, bearing a drawn-out glass fiber, in a holder. Using the stereomicroscope to follow the operation, dip the tip of the glass fiber in a solution of amyl acetate and rubber cement. Before the cement dries, touch it lightly to the collodion flake and pick it up. If the flake does not adhere to the fiber, try adding more rubber cement to the fiber tip or change the position of the flake on the glass slide with a needle.

Step 5. Lift the fiber and the flake and hold them for about 30 seconds in the vapor space of a freshly opened wide-mouth bottle containing amyl acetate.

Step 6. Examine the fiber tip with the polarizing microscope (the 16× objective has enough working distance to permit this). The flake should have formed a ball adhering to the fiber.

For polycrystalline particles which do not need to be crushed proceed directly with Step 4.

Under previously accepted standard conditions, samples 100 μm in diameter (1 μg) were considered to be the minimum size for routine analysis. By careful control, choice of apparatus, and technique, polycrystalline samples as small as 2 μm can be identified, and 5-μm samples can be considered routine.

C. Scanning Electron Microscope

Although the starting point for the identification of small particles is the polarizing microscope there are limitations to polarized light microscopy. First, subpicogram particles, e.g., some foundry fumes, clays, and pigments, cannot usually be identified. Second, rapid identification of any given substance depends on the analyst's ability to remember the identifying characteristics for that substance or to refer to existing atlases or tables that include those characteristics. If he cannot identify the sample he must then use other ultramicroanalytical tools, starting usually with the scanning electron microscope (SEM).

Use of the SEM is a logical second step for three good reasons. First, it shows the shape and surface of even large particles (1 to 100 μm) better than the polarizing microscope. This is due to superior depth of field (about 300 μm compared with about 1 μm for the polarizing microscope) and better contrast. The SEM image is best compared with that obtained using reflected light with a stereobinocular light microscope; however, resolution with the latter is unfortunately only about 2000 nm. One can also compare the SEM image with reflected light images in the polarizing microscope, but here, with maximum resolution, the depth of field is very low and continuous focusing is necessary to visualize the entire particle.

A second advantage of the SEM over the light microscope is superior magnification and resolution. The realistic or useful magnification for the SEM is about 30,000× compared with 1500× for the light microscope. This is very helpful for visualizing and identifying small particles.

The third advantage of the SEM is its ability to determine semiquantitatively the chemical composition of even tiny single particles. This is done by adding an energy dispersive X-ray analyzer (EDXRA) capable of detecting any element in the periodic table above nitrogen (Fig. 11)

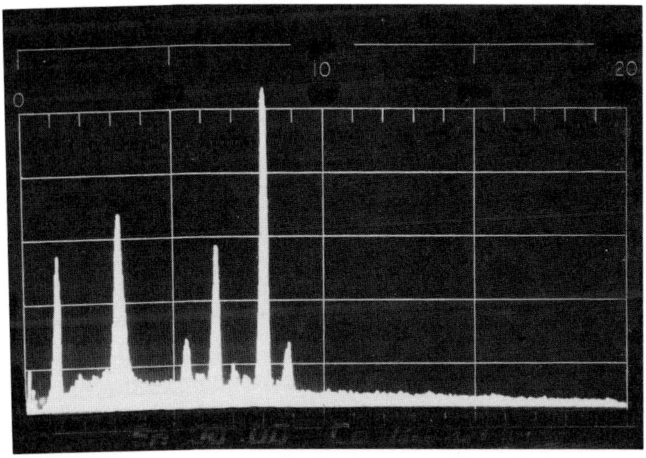

Figure 11. EDXRA pattern showing copper (0.93, 8.04, 8.90), silver (2.98, 3.15), chromium (5.41, 5.95), iron (6.40, 7.06), and nickel (7.47).

An elemental analysis of particle is obviously more generally valuable than, say, a refractive index or a diffraction pattern since these latter procedures identify substances only if the data have been previously measured and tabulated. Furthermore, it requires only seconds and minimal training to obtain an EDXRA analysis, whereas measurement of indices or diffraction patterns requires much more training and much more time.

We have cited certain advantages of the SEM for particle identification, but unfortunately, one should not expect to replace the light microscope and other physical analytical tools with the SEM. To put the SEM in its proper perspective, we need to consider its limitations. In addition to the high vacuum requirement, it also takes longer to mount the sample for SEM than for light microscopy. In addition, most common particulate substances can be quickly identified by light microscopy. Without EDXRA, the SEM is inferior to the light microscope. Most substances easily identified by SEM (without EDXRA) are also easily identified by light microscopy. Generally, these include biological substances such as natural and synthetic fibers and pollens, spores, and protozoa. Most other particles may show interesting shape and surface characteristics under the SEM but these may not be sufficiently distinctive to effect identification (Fig. 12). This is equally true for light microscopy, but additional characterization data are quickly obtained by crossing the polars, rotating the stage, focusing up and down, and using transmitted or reflected light. These procedures may or may not identify the particle, but no such flexibility is possible with the SEM. Only EDXRA makes generally possible the identification of particles not identifiable by polarized light microscopy. Some microscopists argue that SEM without EDXRA can identify some pigments, foundry fumes, and clays not identifiable by light microscopy. This, however, is only partly true. Most, if not all, pigments can be quickly identified by polarized light microscopy, and few clays or fumes can be identified by SEM alone. Generally, however, their shape and surface are simply not distinctive enough, and other data must be obtained.

For SEM it is better to mount small particles directly on a highly polished beryllium surface. The surface of the standard aluminum sample holder has too many grooves, and the aluminum substrate interferes with EDXRA elemental analysis. Therefore, a highly polished beryllium plate is mounted on top of the aluminum holder with Duco cement and scribed for quick location of particles. Since only deep scribe marks are visible on the beryllium surface, Roman numerals, which are formed with straight lines, are used instead of rounded Arabic numerals to label the columns and rows on the plate (Fig. 13). The following steps are required for mounting small single particles for SEM:

Step 1. Transfer the isolated particle in a thin (~2 μm) collodion flake to an SEM mount (keeping the flake flat) using a No. 3 tungsten needle.

Step 2. Wash essentially all of the collodion from the particle as follows. With a polyethylene micropipette, place a drop of amyl acetate next to the flat flake. With a No. 3 tungsten needle, gently move the solvent toward, over, and away from the particle. If this is done correctly, most of the collodion will be carried by the solvent away from the particle, and the particle will remain in place. If some of the collodion still remains, producing interference colors around the particle in reflected light, repeat the washing until only the shiny beryllium surface is visible around the particle. There will still be a few nanometers of collodion holding the particle.

Step 3. Draw a pattern around the particle as shown in Figure 14. However, make the marks with a No. 1 tungsten needle as *deep* as possible. This makes it easier for the SEM operator to locate the particle.

To get good data on the morphology of small particles, mount them first for SEM; they can be submitted later for microprobe, X ray, etc. Keep in mind, however, that the 10^{-20} nm of gold or other conductive material coated over the particle to eliminate charging effects in the SEM will be detected in subsequent microprobe analysis. Small single particles develop negative charges in the electron beam which tend to distort the beam and interfere with the observation of surface detail. Once the particles are properly mounted for SEM, however, no further mounting is necessary to get elemental composition with an EDXRA. Also the SEM sample pin can be transferred to a microprobe for analysis.

D. Transmission Electron Microscope

The visible light microscopist is limited by the resolution of his instrument to the study of particles larger than about 1 μm (1000 nm). This limitation is imposed by the relationship between resolution and wavelength of visible light; high resolution is possible only with shorter wavelengths. Since the detection and focusing of significantly shorter wavelengths of electromagnetic waves is impractical, a different kind of radiation source and different ways of focusing and presenting the image must be found. The transmission electron microscope (TEM) fulfills these requirements.

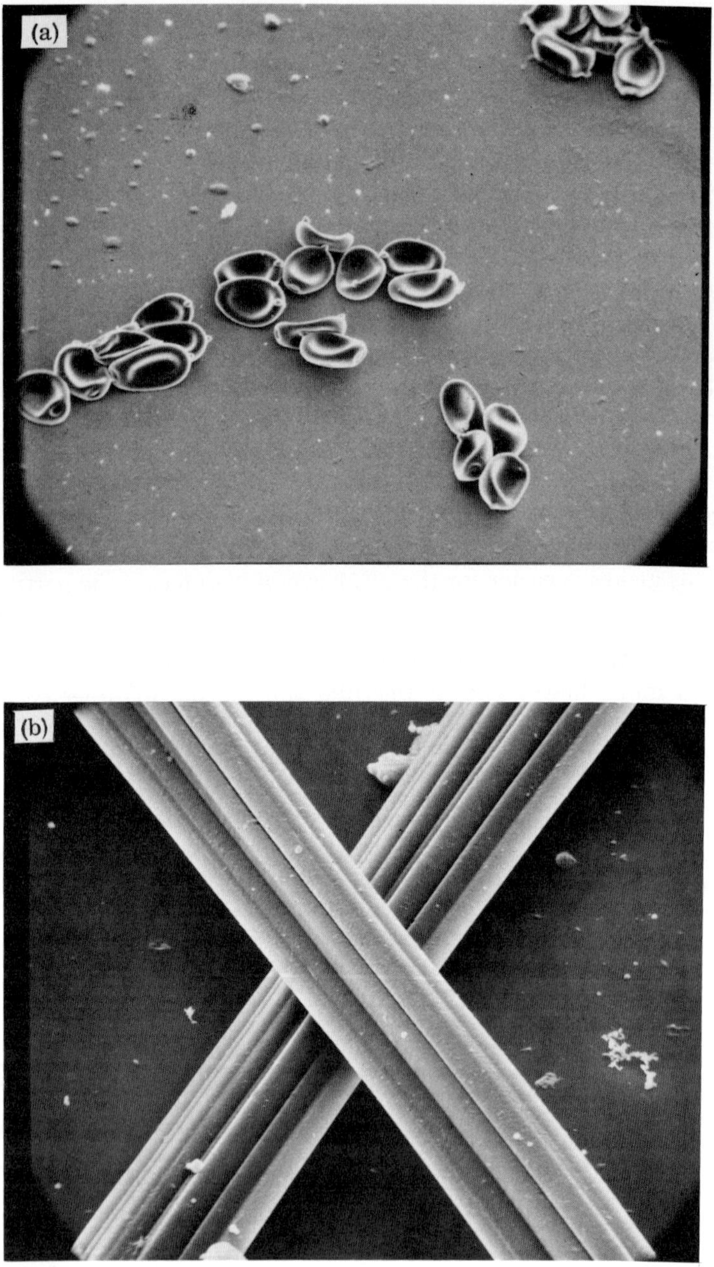

Figure 12. SEM micrographs of a few representative atmospheric pollutants: (a) mushroom spores, 1000×; (b) vicose rayon fibers, 1000×. (Micrographs courtesy of John A. Brown, McCrone Associates.)

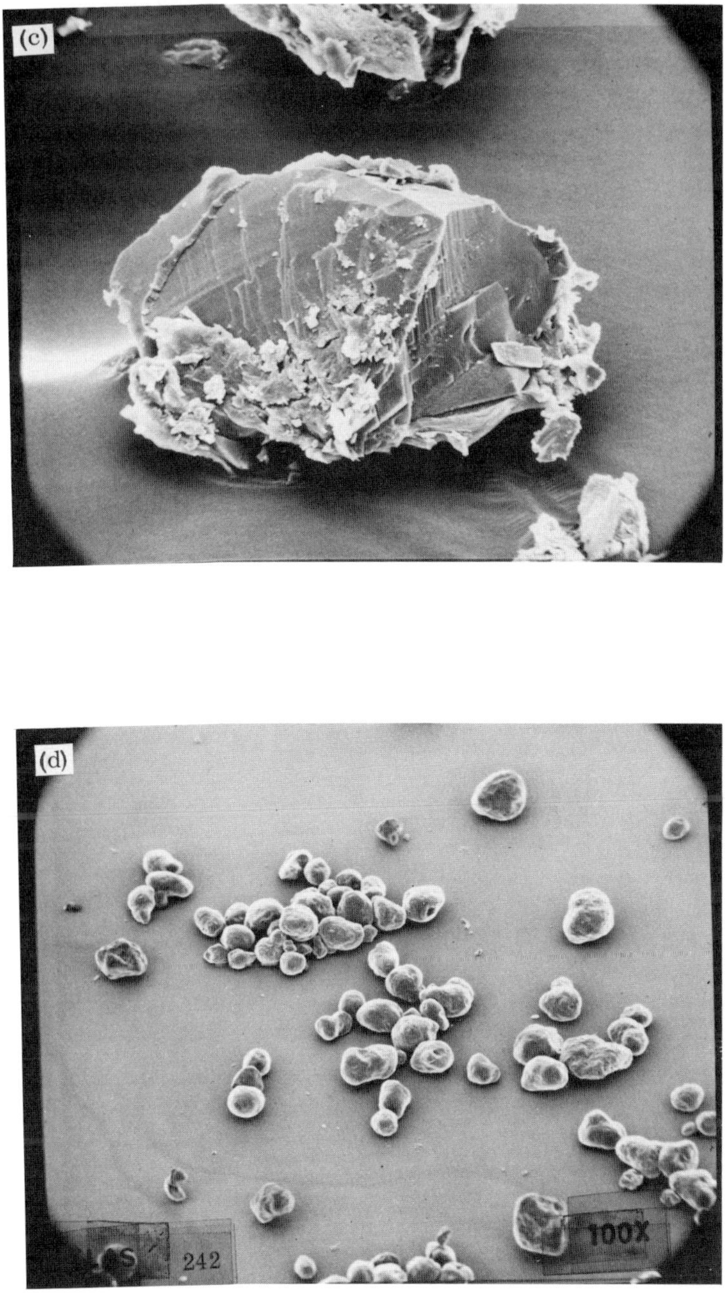

Figure 12. (c) Corundum, 100×; (d) oil refinery catalyst, 100×. (Micrographs courtesy of John A. Brown, McCrone Associates.)

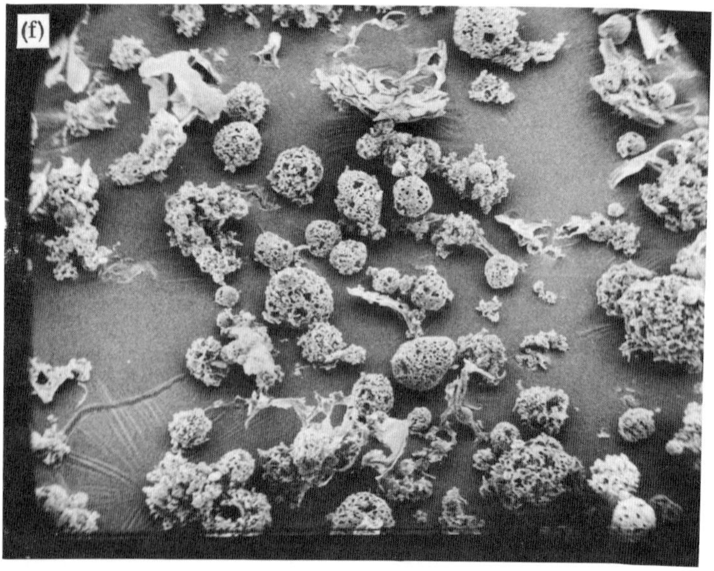

Figure 12. (e) Mica paint extender, 100×; (f) oil soot, 100×. (Micrographs courtesy of John A. Brown, McCrone Associates.)

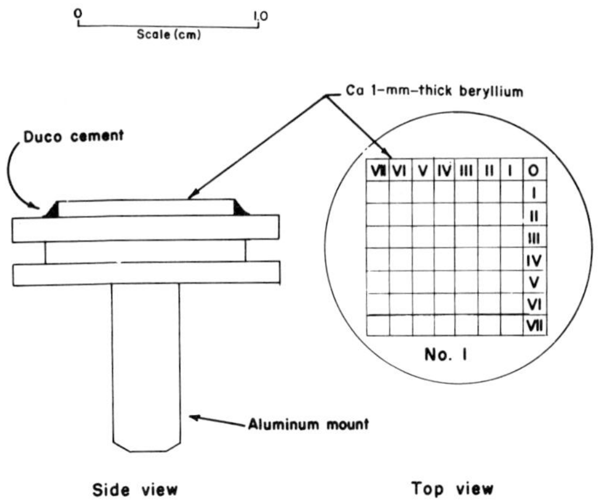

Figure 13. Aluminum stub with beryllium plate for single particle mounting.

Figure 14. Pattern drawn in the beryllium metal by the light microscopist to describe particle location and ensure analysis of the correct particle. The house is 0.3 mm high and seven particles are mounted in door, 2 windows, on chimney, and in each of 3 billows of smoke.

The wavelength of a high energy electron beam is extremely short, 3.7 pm (0.0037 nm) for a 100-kV beam compared with 400–700 nm (0.4–0.7 μm) for a visible light beam. Most important, the electron beam can be focused by magnetic lenses. In this way, it has been possible to achieve point-to-point resolution of the order of 0.2 nm (2 Å). Coupled with this increased resolution a greater depth of field and greatly increased depth of focus are obtainable. Thus the depth of field, the depth of the sample simultaneously in focus, is between 0.5 and 2.0 μm, and the depth of focus, the existence of a focused image, is approximately 1×10^3 cm. The large depth of field means that all particles in a preparation will be in focus throughout a range of about 1000 nm (1 μm). The tremendous depth of focus permits placement of a focusing screen at virtually any point behind the final lens, thus permitting the image recording camera to be positioned by design convenience rather than by optical restrictions. Furthermore, the wavelengths of electrons are of the order of the interatomic spacing of the atoms present in the sample, hence diffraction conditions exist just as with X rays. We can, therefore, diffract electrons and identify crystals in the same way we use X-ray diffraction.

Obviously, then, we have better resolution (about 0.5 nm) and better depth of field (about 1 μm even at very high magnifications, say 100,000×) than the light microscope. Unfortunately, it is difficult to prepare the particle(s) for TEM, and, again, identification by shape alone is not likely. Fortunately, the TEM can be used to obtain selected area electron diffraction (SAED), and this parameter is nearly as useful as EDXRA. Incidentally, the installation of EDXRA on the TEM will certainly increase in the future in order to permit elemental analysis.

The advantages of the TEM are useful, and often necessary, for the identification of subpicogram particles. The TEM is far more useful when fitted with SAED, and the combination of SAED and EDXRA on the TEM is best of all.

There are limitations to the TEM as with all techniques. Most serious, perhaps, is high vacuum; the electron beam and the sample must be at 10^{-4} to 10^{-5} Torr. As a result, all samples must be stable in the high vacuum and to the thermal action of electron beam bombardment. Another limitation is the size of the specimen imposed by its position inside the objective lens. In order to maintain the area of interest in an undistorted magnetic field, the specimen area must not exceed 3 mm in diameter. Thus, when examining particles from a bulk sample, extreme care must be taken to ensure that the observed portion is representative; several samples from the same material should usually be examined. Finally, due to low penetrating power of the electron beam, only particles thinner than about 50 nm can be observed in any detail with a 100-kV electron beam and give selected area electron diffraction (SAED). Thicker particles may be seen in the beam but give only silhouettes. TEM with higher accelerating voltages are now becoming available, thereby increasing penetration to as much as 3 μm for some materials at high voltages, >1

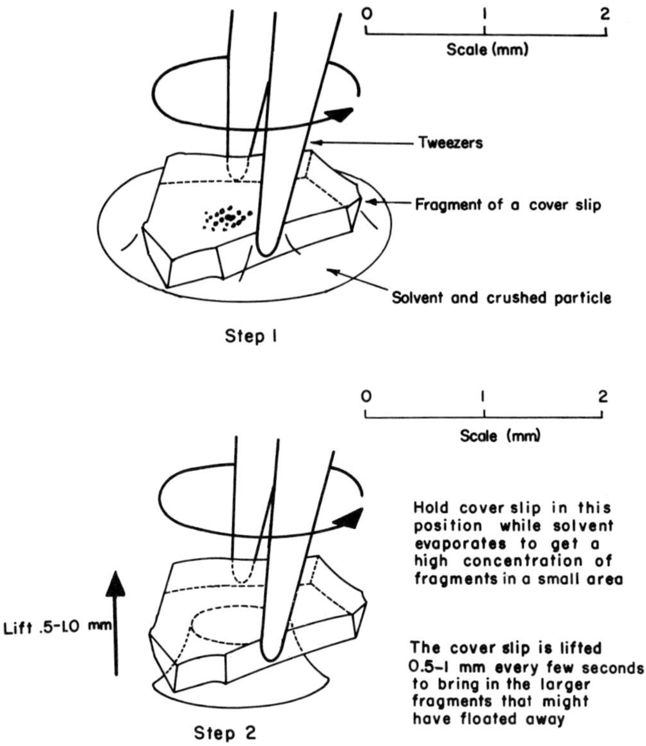

Figure 15. Use of a cover slip to crush small single particles.

MV. We do not recommend these very expensive high voltage instruments to identify particles. Other better ways to do this are discussed throughout this chapter.

1. Mounting >10-μm Particles

Particles greater than 10 μm give more information in the TEM if they are first crushed into submicrometer fragments. Good SAED patterns can be obtained on heavy concentrations of <0.1-μm particles. The <10-μm particle can be crushed as shown in Figure 10. However, even if one repeats Steps 1–3 in Section II,B,3 a few times, there may still not be enough <0.1-μm fragments.

A large particle, especially one greater than 10 μm, can be dispersed into submicrometer fragments better by using the following procedure with a stereomicroscope at 50× magnification (Fig. 15):

Step 1. The particle is placed on a clean glass slide. All mounting medium is washed off. A small cover slip 1 to 2 mm on edge is put on top. A 2-mm drop of solvent is placed next to the cover slip. A number of solvents will work well, e.g., amyl acetate, benzene, xylene, and H_2O. Sometimes good results can be obtained without any solvent. The solvent should have a tendency to redisperse the particles during the grinding procedure, and not aggregate them. The solvent is then flowed under the cover slip with a tungsten needle. The cover slip is gripped firmly with a pair of tweezers and pressed down onto the particle with a circular grinding motion. It is best to dip the tweezer tips in dilute rubber cement to coat them before gripping the cover slip. This improves the grip and helps prevent chipping glass from the cover slip into the sample.

Step 2. Every few seconds, the cover slip is lifted a fraction of an inch to decrease the size of the drop and bring the larger uncrushed fragments that have floated away back under the cover slip. After the solvent evaporates, the particle size range is determined, using the polarizing microscope at high magnification. If many fragments are greater than 0.5 μm, the procedure is repeated with another 2-mm drop of solvent.

The concentration of fragments in a given area can be regulated by the amount of solvent remaining when the cover slip is removed. If a very high concentration of fragments is desired, the cover slip is held 1 mm or less over the crushed fragments until *almost* all the solvent has evaporated. When the cover slip is lifted, the volume of the remaining liquid and fragments is so small that it usually will not spread into a circle greater than 1–2 mm in diameter (see Fig. 15, Step 2). The smaller fragments will migrate toward the circumference, and the larger ones will remain in the center; however, ~10% of the fragments will adhere to the cover slip, the tweezers and other areas of the glass slide.

Step 3. The TEM grid can now be positioned over the crushed particles (Fig. 16).

Step 4. Same as Steps 5–8 described below and shown in Figure 17. The particles mounted in a 50- to 100-nm Formvar film on a TEM grid are then carbon coated and the grid is introduced into the TEM. If the particles are to be examined by an electron microprobe analyzer, the grid is first mounted in a brass holder.

2. Mounting <10-μm Particles

The methods described above for mounting samples for transmission electron microscopy cannot be used for a single particle 1 to 10 μm in diameter. Since there is only one particle to work with, it must be placed in a known position on the TEM grid so that it can be located afterward. The following modified procedures permit electron microscopy on a single

Figure 16. A portion of the TEM grid positioned over the particles to be examined.

small particle; with practice, these procedures can be carried out in 5 to 10 minutes.

Step 1. Place the isolated particle in a collodion film on a clean glass slide.

Step 2. Center the particle in the field of view of a stereomicroscope. *Use stage clips to hold the slide in place.*

Step 3. Place a drop of amyl acetate, 5 to 10 mm in diameter, next to the particle. With a No. 1 tungsten needle, wash almost all collodion away by spreading the solvent very gently over the particle and away from it. Do not lose sight of the particle (especially if it is less than 3 μm in diameter) until the solvent evaporates and the particle is in a known, fixed position in the field of view.

Step 4. (Steps 4 through 8 are illustrated in Figs. 16 and 17.) Gently place a center-marked TEM grid over the particle, and position it so that the particle lies in the center of a "known" square. This is done by pushing on the sides of the grid with a No. 3 tungsten needle; the grid might "stick" to a No. 1 or No. 2 needle.

Step 5. Place a drop of 0.5% Formvar in ethylene dichloride about 1 cm from the grid. With a tungsten needle, gently pass some of the solution over the grid. Usually one slow pass of the solvent over the grid will deposit a 20-nm Formvar layer. Draw excess solvent away with the

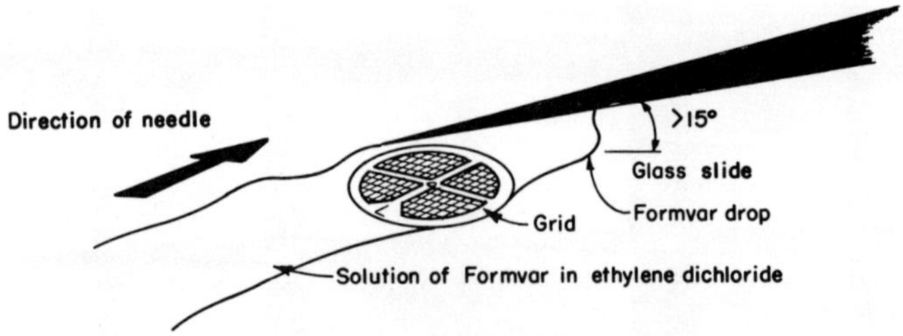

Step 5 With a tungsten needle spread a thin film of Formvar in ethylene dichloride over and under the grid.

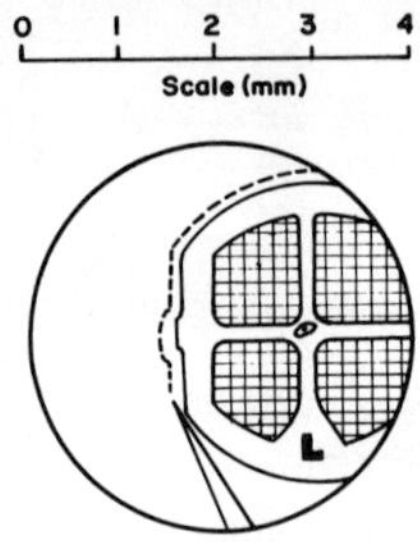

Step 6 Cut the Formvar film all around the grid (stereomicroscope, 50×).

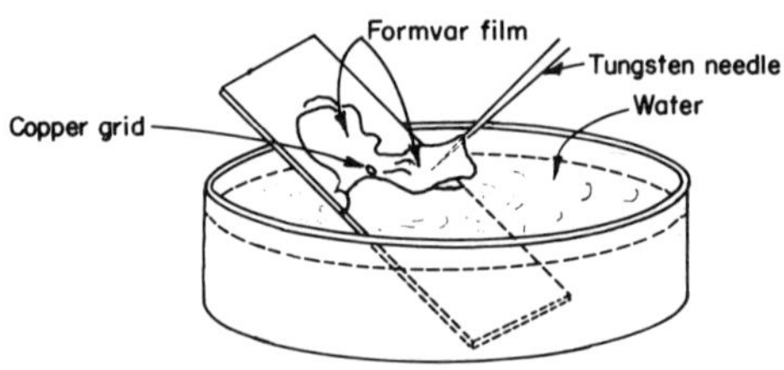

Step 7 Remove excess Formvar film from water surface and glass slide.

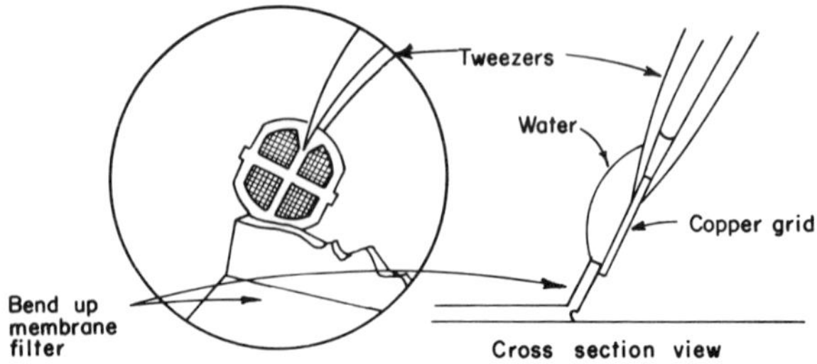

Step 8 Remove excess water from grid (stereomicroscope, 25×)

Figure 17. Steps in mounting particles on a TEM grid.

needle. Film thickness will be greatly influenced, however, by room and slide temperature, humidity, and the length of time the drop is allowed to remain on the slide before it is used. To test the film thickness, scribe a few lines in adjoining squares with a No. 3 needle. At 100 × magnification, the lines should be *just visible*. If the film is too thick, pass a drop of ethylene dichloride over the grid and test the thickness again.

Step 6. Make a cut in the Formvar film around the grid.

Step 7. Breathe on the slide to deposit a film of moisture; slowly dip it at a 45° angle into a dish of distilled H_2O until the grid is almost submerged. Remove excess Formvar film with the needle. Gently touch the side of the grid, which should pop to the water surface.

Step 8. Pick up the grid with tweezers and remove excess water with a membrane filter square.

Step 9. Place the grid, film side up, on a clean glass slide and use a polarizing microscope to examine the film around the particle. Use either a 16× or a 45× objective and reflected light. Determine the retardation color of the film (Michel–Lévy birefringence chart). Figure 18 shows the relationship between retardation, interference color, and thickness of the Formvar film. A dark gray film with a retardation of less than 150 nm is too thin and will most likely break under the electron beam. First-order light gray to white interference colors are ideal for electron microscopy. Yellow-orange films are usable. Films with retardation greater than 420 nm are too thick. If the film is too thin, place the grid, film side down, on a glass slide and pass the 0.5% Formvar in

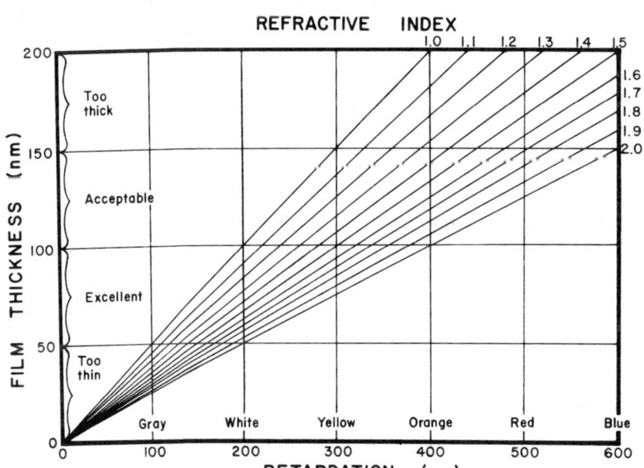

Figure 18. Relationship between retardation, interference color, and thickness of thin films.

ethylene dichloride solution once again over the grid. If the film is too thick, pass a drop of ethylene dichloride 2 mm in diameter over the grid, then proceed with Step 6.

Step 10. The grid is carbon coated before submission to the TEM lab. If elemental analysis is also required, the grid is also mounted in a grid holder for the electron microprobe.

3. Mounting Very Fine Particles

A very fine precipitate on a smooth surface can sometimes be mounted directly on an electron microscope grid for TEM or microprobe analysis just as described above for crushed particles. If the surface on which the grid is resting is very small (e.g., a silicon integrated circuit chip), use a polyethylene micropipette to deliver the Formvar solution. However, if the surface is over $\frac{1}{4}$ in. in diameter (e.g., ball bearing race), a polyethylene needle (Section II,A,2) can be dipped in Formvar solution and used to deposit a somewhat larger drop.

The collodion film is cut with a No. 2 tungsten needle all around the grid, and, if the object being replicated can be dipped in H_2O, the grid should pop to the surface. However, if one cannot submerge the object in water, a small drop of H_2O can be delivered to one side of the grid with a pipet or, for a smaller water drop, with a sealed-off H_2O dropper (Section II,A,3,c). The grid is then removed with the help of the water drop manipulated with a No. 1 or No. 2 tungsten needle.

Although the Formvar replica in a few grid openings may be broken, many other intact grid openings will give sufficient particles for study. The film thickness is now checked (Fig. 18), and, if suitable, the grid may be carbon coated and submitted for either TEM or microprobe analysis.

E. Electron Microprobe Analyzer

The electron microprobe analyzer (EMA) is an ultramicroanalytical tool combining light microscopy, scanning electron microscopy, X-ray fluorescence, and cathodoluminescence. Samples as small as 1.0 μm (10^{-13} gm) can be analyzed for most elements present to 1% or greater. Detection limits are as low as 10^{-15} gm for most elements, hence 0.1 μm particles of pure elements can be identified. This is achieved by focusing an electron beam to a spot smaller than 1 μm^2 on the sample. The characteristic X rays thus emitted are analyzed by wavelength, or energy, dispersing systems for both quantitative and qualitative analysis.

The ability to analyze very small samples quantitatively and nondestructively makes the EMA eminently suitable for small particle identification. About the only disadvantages of the EMA, aside from high cost, are the inability to detect Li (and usually Be), even when it exists in high concentration in the sample, and trace concentrations in a sample. One hundred parts per million is about the lower limit of detectability for most elements.

1. Sample Preparation

In the early days of EMA, flat highly polished samples were required if quantitative analysis were to be possible, because surface topography of the samples strongly affected X-ray intensity. Therefore, accurate quantitative analysis of unpolished samples was almost impossible. Even though small particles extend above the substrate and less absorption correction is needed, most quantitative analyses depend on the comparison of sample count rates with the count rate on a flat standard. Absorption of emitted X rays is greater for the polished standard than for the particle; hence, quantitative analysis of particles is less accurate when low take-off angles cause greater differences in absorption between the bulk standard and the particle. Modern EMA instruments have a take-off angle ranging from 50° to 75°, which means that surface topography affects X-ray intensity to a lesser degree. This is fortunate for small particle analysis because sectioning and polishing small particles is impractical and, in many cases, virtually impossible.

Small particles are usually mounted on a substrate for EMA analysis. The two substrates most commonly used are beryllium and carbon. A beryllium substrate is suitable for particles down to 1 μm in diameter, while a carbon film on an electron microscope grid is usually used as a substrate for particles less than 1 μm. Beryllium metal is hard, easy to polish, has a very low level of white radiation, and gives emission lines undetected by the usual microprobe. However, even high purity beryllium contains some submicrometer or even micrometer size inclusions. Fortunately, these inclusions are not very frequent and the probability of an inclusion being close to the mounted particle is low. The surface for mounting small particles must be fairly easy to polish, and, even more important, it should have low white X radiation when bombarded by an electron beam. Such requirements rule out the heavy elements. Small particles, fixed to a beryllium plate with a very thin collodion film, are usually analyzed without any further preparation.

The beryllium plate used for EMA differs from the one used for SEM. The thin, highly polished centimeter square plate is mounted on a centimeter long section of a 25-mm cylinder brass or aluminum. Each plate has an inscribed rectangular grid of 1-mm squares. The squares are numbered in columns and rows, and the location of particles is recorded using this numbering system. A particle in a specific square on the beryllium plate may be located by several methods; the most frequent method is by light microscopy in the probe. If the optical characteristics of a particle do not make location easy, secondary or back-scattered electron signals can be used. The secondary electron picture is closely related to the topography and is equivalent to the optical image with very oblique illumination. The back-scattered electron signal depends on the mean atomic number of each sample point and is useful when the substrate and the particle have a large difference in atomic number. If any element in a small particle is known, the easiest way to locate such a particle is to set one spectrometer for this element and scan the suspected location electronically. When the beam hits the particle, the signal on the spectrometer will rise well above the background. This can even be done manually as the operator listens to the audible clicks of the counter.

An EMA sample should be electrically conductive to carry away charge buildup from the impinging electrons; however, small particles, even insulators, become quickly conductive in the probe due to carbon buildup at the points of beam impact. Only nonconducting particles larger than 30 to 50 μm require a conductive coating, and even then only for more precise work.

a. Mounting Single 1–10-μm Particles

Step 1. Make an accurate drawing of the particle on the microprobe analysis sheet, listing all dimensions. This information is very important to make certain the correct particle is mounted and later analyzed by the microprobe operator.

Step 2. Mount the particle in a collodion flake. Before the flake is floated off the glass slide, transfer the slide to the polarizing microscope. With a 45× objective, check to see if the right particle is inside the flake; use the drawing from step 1 to make this check.

Step 3. Place the glass slide bearing the collodion flake on a taped stack of glass slides, or other support, equal in height to the beryllium mounting blocks. To observe the flake, use the stereomicroscope at 50× total magnification with the 2× auxiliary objective lens, transmitted light and *no* polarizing filters.

Float the flake with a drop of water onto a No. 3 tungsten needle. Carefully lift the needle and slide the microprobe mounting block into the field of view. Find an empty square and gently place the flake in the center. If the flake does not come off the needle readily, roll the needle slowly between your fingers, keeping the needle tip on and nearly parallel to the block. Refocus, and increase the magnification to 100× if necessary. The flake should lie flat on the block. If you cannot see the particle clearly in the collodion flake or if there is more than one particle, examine the flake on the polished beryllium surface with a 45× objective and bright-field top light. Locate the correct particle by its size and shape, and mentally note the particle position in relation to the shape of the collodion flake.

Step 4. Return the holder to the stereomicroscope. The surface of the mounting block should be very brightly illuminated. If it is not, replace the ground-glass stage plate with a clear glass plate. All the pits and scratches in the surface should show up as dark spots and lines on an almost white background. With the collodion flake in the center of the field of view, bring both a polyethylene micropipette and the tip of a No. 3 tungsten needle into the field of view, one on either side of the collodion flake.

Deposit a 500- to 1000-μm drop of amyl acetate near the flake and spread it quickly over the whole flake with the No. 3 needle. Keeping your eye on the particle, spread the drop over as large an area as possible. In 5 to 10 seconds, the solvent will evaporate, hopefully leaving a first-order yellow interference color in the collodion film around the particle. A film with brightly colored rings around the particle is much too thick. To thin such a film, place another drop of solvent next to the particle and repeat the procedure. If it has been properly mounted, the particle should be clearly visible with middle first-order colors in the surrounding collodion film. No other particle should be within 15 μm.

Step 5. With a No. 1 tungsten needle, make an identifying outline of marks around the particle (Fig. 14). Working directly from the stereomicroscope, precisely draw the area, including the identification marks, on the microprobe analysis sheet. Also indicate the particle position by listing the column and row of the sample mounting block and the block number.

b. Mass Scanning. Frequently a sample containing a large number of small particles must be characterized by chemical composition. In such cases, the only practical approach is so-called mass scanning, the analysis of a large number of small particles. Since the elements in the sample are unknown, the approach is to analyze every particle for all elements. Obviously, such an approach is practical only with a fully automatic system since one has to analyze at least a few thousand particles to characterize a sample of a large number of small particles. Even a few thousand particles represent a statistically reliable sample only when the variety of compositions is small.

For typical mass scanning, many particles are dispersed on a beryllium substrate. The electron microprobe for mass scanning is necessarily automated, at least the specimen stage, scanning beam, and spectrometer drives. The specimen stage is initially manually driven to the starting position, then the automated system takes over control of the probe. The electron beam moves until it hits a particle, which is recognized by the change in specimen current. All possible chemical compositions except those containing H, He, Li, and Be have a higher average atomic number than the beryllium substrate, so there is, with these very minor exceptions, always a lower current on the particle than on the substrate. When the beam hits a particle, the computer stops the beam scan, centers the beam on the particle by noting the position of minimum specimen current, and then starts scanning the spectrometers through the whole range of elements. Element intensities well above background are printed out, and, when all spectrometer ranges have been covered, the beam moves to the next particle and repeats the procedure.

When all the particles within that field have been analyzed, the specimen stage is moved one step onto a new field and the procedure starts again. Obviously, the speed of such mass scanning is very slow, and the number of particles analyzed is between 4 and 50 per hour, the higher rate by the energy dispersive system. If some elements can be skipped (assuming that they are not present in the sample from the data obtained by earlier bulk analysis), the mass scanning procedure by spectrometer may be significantly shortened.

Mass scanning can be made quantitative since each particle is sized as well as analyzed. It is not, however, practical to do this when an important component is present in percentages less than 1–5%. It is too difficult to analyze a sufficient number of particles to yield a representative sample. This has obvious implications for needle-in-the-haystack analysis which is discussed in the next section. It is also obvious that the computer can be programmed not only to print out the elemental composition for each particle but also to classify them into different mineralogical or other arbitrary groups, such as silicates, oxides, and metals.

c. Needle-in-the-Haystack Analysis. A corollary of mass scanning is the situation where one knows the particle composition of interest, e.g., quartz in mine atmospheres, lead or asbestos particles in urban atmospheres (*4*). Particles of interest may be present as parts per million, and often many thousands of particles must be scanned to find a significant number of "needles." It is not usually practical by this method to measure the percentage of the trace component but the detection of specific trace components can be done and the composition of particles containing a specific element is also feasible.

This procedure is similar to mass scanning except that the spectrometer is set for an element, or combination of elements, present in the particles of interest. If, for example, one wants to find and analyze all lead particles, the spectrometer scans all particles until it senses lead. The stage then stops, and that particle is quantitatively analyzed before the stage continues looking for other lead particles.

The approach is adequate as long as the number of "needles" is not too small (<0.05 to 0.1%) relative to the "haystack." In samples containing a few ppm or less of the desired phase, the location step is time consuming.

2. Automatic Analysis

Sample preparation for automatic analysis has some unique aspects, and therefore requires special attention. Since the operator does not have direct control of the sample through the light microscope or through any of the imaging systems, it is essential that the sample be evenly dispersed on the substrate. There must be no agglomeration, since this would seriously hamper automatic analysis; an agglomerate of particles will be regarded as a single particle by the computerized system. In such cases, a very unusual particle composition should be expected. Since automatic focusing of X rays from the particles has yet to be introduced for EMA it is important to have all particles on the same level. The most reliable method for positioning samples on the correct level is to have a perfectly flat substrate and particles whose sizes do not exceed the normal focusing tolerance of the X-ray system (usually 5–10 μm). Particles larger in vertical dimension than this result in large errors unless the operator intervenes.

3. Quantitative Analysis

Quantitative analysis of solid samples is a well-established but nevertheless difficult technique. It requires, besides superior instrumentation, an experienced operator with a knowledge of the numerous variables and sources of error encountered in such work. With such knowledge, it is possible, under favorable conditions, to produce routine results with errors as low as 1% relative. However, in small particle analysis, such a low error is difficult to achieve because the total X-ray intensity is reduced by lack of the particle mass necessary for full excitation of X rays. A large fraction of the high speed electrons and primary X rays passes

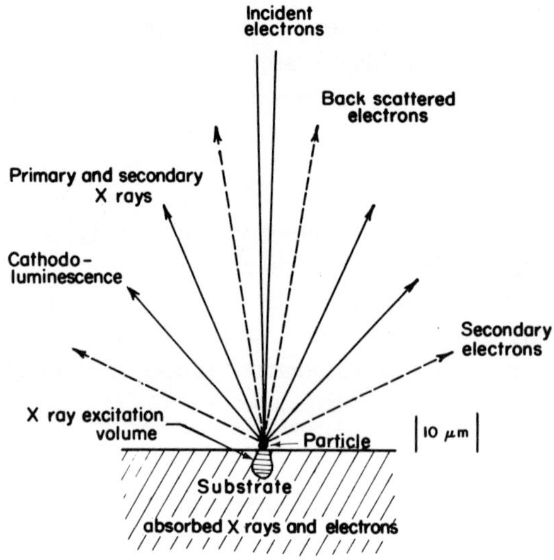

Figure 19. Interaction of the electron beam with a small particle.

through small particles and is absorbed in the substrate (Fig. 19). The effect of particle size on the intensity of X rays emitted is shown in Fig. 20. Besides the mass effect, there are a number of other effects (not fully investigated, such as particle shape, defocusing of the X rays because of the particle size, etc.) which adversely affect X-ray intensity.

F. Ion Microprobe Analyzer

The ion microprobe analyzer (IMA; Fig. 21), is a device for performing sputter source mass spectrometry on small particles or small areas

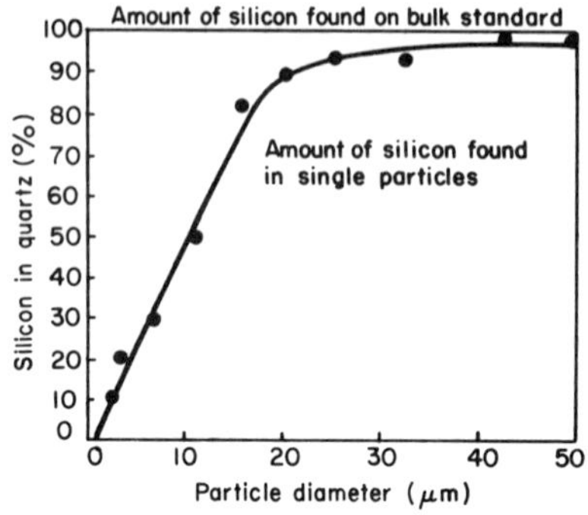

Figure 20. The effect of particle size on the observed percentage of silicon in quartz particles. A conducting particle would give a better X-ray yield, e.g., a 5-μm stainless-steel particle would lie at 50%.

Figure 21. The Applied Research Laboratory ion microprobe.

of "bulk" samples. As such, it is one of the most sensitive tools available for small particle analysis. It is sensitive to every element in the periodic table and can, under ideal conditions, detect as little as 10^{-20} gm of some elements and 10^{-19} gm of most elements. It is fully capable of trace analysis (ppm) on ultramicro samples (<1 μm or 10^{-12} gm) and will obtain ppb of some elements in appropriate samples. The spatial resolution is of the order of 1 μm in x–y and a few 10's of angstroms in the z direction. Samples considerably smaller than the 1 μm x–y resolution limit can be readily analyzed. The time required for semiquantitative analysis may be as short as 4 seconds but is usually about 40 seconds.

1. Instrumentation

The IMA has an ion rather than electron source, a column of two electrostatic lenses, a light microscope to locate the sample, and a mass spectrometer to perform the analyses. A duoplasmatron is used as ion source. The ions thus formed are extracted through a small differential pumping aperture into the high vacuum of the main system. The source will work with various gases, either inert or active. For most analytical work, oxygen seems to be the best choice. For some work, inert gases such as argon or neon can prove of value, while, in other special cases, fluorine may be desirable.

After extraction, the ions are accelerated to potentials between 5 and

20 kV and sent through a primary magnet to isolate single ion species all having the same velocity. With the usual oxygen source, one gets positively charged mono- or diatomic ions. O_2^+ is the ion normally selected for routine work. After passing through the primary magnet, the ions are focused on the surface of the sample by two electrostatic lenses (condenser and objective). With a two-lens combination, a 1-μm spot size on the sample can be achieved with a current between 10^{-9} and 10^{-10} A. For most types of work with small particles, this current density is more than adequate. Typical count rates on particles vary from 10^6 counts/second for good ionizers, such as aluminum, to about 5×10^3 counts/second for poor ones, such as gold. A light microscope concentric with the ion optical column allows viewing of the sample while under ion bombardment.

When a beam of ions strikes the sample, atoms from the sample are sputtered off—a certain fraction as ions. These ions are collected and sent through a conventional mass spectrometer. The ions of the desired species are selected by setting the magnetic field to the proper value. They then pass through a slit and are collected on a conventional Daly detector. The detector consists of a target which emits secondary electrons when bombarded by ions and a scintillator–photomultiplier arrangement to preamplify this signal. This type of preamplification allows gains of 10^6 with essentially no noise.

2. Applications of IMA

The applications of this instrument are as many and varied as those of the electron microprobe analyzer. Any airborne, waterborne, or contaminant particles can be analyzed with this tool. Micrometeorites, lead particles from auto exhaust, and contaminants on integrated circuits are three good examples. Though the IMA operates in much the same manner as the electron probe, it has several advantages over that instrument. In general, the count rates and peak-to-background ratio are about $1000\times$ better than the electron probe. This means that we can either get the same type of data as EMA $1000\times$ faster (or on $1000\times$ less sample) or we can look at trace levels $1000\times$ smaller. These attributes can be very useful and, in some cases, indispensable for solving practical problems.

3. Particle Analysis

The analysis of small particles by the IMA is fairly straightforward. The samples must first be placed on a suitable substrate. Since the solid solution sensitivity of the IMA is of the order of 1 ppm down to 1 ppb, the substrate must be reasonably pure. In addition, to avoid confusing the background spectrum, the substrate should have a single isotope if possible. Also, the substrate should be one that ionizes quite poorly with the bombarding gas chosen. Although beryllium or graphite are usually used, ultrapure gold or tantalum have worked well with oxygen as the bombarding gas. Where it is desirable to analyze single particles, they are best mounted on polished gold or tantalum having a scribed grid of approximately 1-mm squares. Beryllium or graphite may also be used for convenience. The squares are lettered in columns and numbered in rows and are thus quite easy to locate with the light microscope of the IMA.

4. Particle Handling Techniques

A procedure somewhat similar to that used for EMA is followed for mounting particles for the IMA. Particles for the ion probe are usually mounted on polished beryllium or polished pyrolytic graphite, and these plates are mounted, in turn, in an aluminum mount. The particles are "tacked" to the surface with a film of collodion no more than about 20 nm thick.

a. MOUNTING <5-μM PARTICLES. Particles <5 μm in diameter cannot be mounted in a thin film of Formvar on a TEM grid as described in the EMA procedure because the IMA would burn holes in the film. Instead, particles must be mounted for IMA using the procedure described in Section II,E,1,a. That is, transfer the particle in a flake of collodion to the metal surface, then wash away most of the collodion with amyl acetate. Finally, mark its position clearly within the 1-mm square.

b. MOUNTING >5-μM PARTICLES. Particles >5 μm should be first crushed because large particles will collect enough charge in the IMA to interfere with the results. These large particles could be carbon coated to eliminate charging; however, there are two good reasons why this is avoided.

First, it takes only 2 to 5 minutes to crush a large particle and mount the fragments for analysis; much more time is needed to carbon coat it.

Second, once the metal substrate has been carbon coated, it has to be cleaned before it can be used again for mounting. Since each sample-mounting block can accommodate 100 to 200 particles, by not cleaning them off, even when all squares are filled, they provide an excellent permanent sample storage arrangement.

Therefore, if a single particle greater than about 10 μm in diameter is to be mounted for IMA, it is first mounted in collodion, then crushed as described in detail in Section II,B. The best IMA data come from 2- to 5-μm particles, so do not crush the particle into very fine fragments. The flake with the crushed particle is then placed in the center of a square on the polished metal surface. A 0.5-mm drop of amyl acetate is added to one side, and most of the collodion is washed off. The fragments should be kept within a 100-μm area. This area is then clearly marked with a No. 1 tungsten needle.

Fortunately, the IMA is extremely fast. Since the sample is being continuously burned, it is absolutely imperative that data be collected rapidly. It is possible to scan the entire mass range from 1 to 300 in a few tens of a second and still get sensitivities of the order of 0.1 of 1% In general, a 1-μm particle can have a survival time of as long as $\frac{1}{2}$ hour under conditions suitable for trace analyses down to about 10 ppm. In general, it is possible to get good quantitative data on, at least, the major components of particles as small as 0.1 to 0.2 μm before they are completely lost.

G. *Automated Analysis*

Application of computers to analytical instrumentation has distinct advantages when justified. In many cases of electron microprobe analysis, the time required for the preparation of programs for running a particular sample would be greater than the operator time involved in the analysis. Therefore, the applications of computer control to the electron microprobe analyzer are limited to those cases in which a particular routine operation is frequently performed. In our particle analysis laboratory, computers are regularly applied to electron microprobe analyses. Some particles can even be recognized automatically by size and shape, e.g., chromosomes. By the time this chapter appears in print, we will be using a computer to control our IMA, and it seems likely that this combination will eventually solve the toughest of particle analysis problems—identification of subnanogram organic particles.

The advantages of the computer over the human operator are (1) tireless performance, even with continuous 24-hour days; (2) no mistakes; and (3) more rapid and accurate data reduction.

The wide availability of low cost computers makes certain their rapid application to analytical instrumentation. The choice of computer is between a small dedicated one and a larger instrument that can service the entire analytical range of a given laboratory.

The suitability of the electron microprobe analyzer for automation has long been recognized. Most probe manufacturers, in fact, supply some degree of automation in their instruments, be it spectrometer presetting or step-scanning facilities for the specimen. All probes, however, require some operator control although not necessarily continuously.

Automation may be readily applied to the analysis of small particles. First, a search program is required to locate the particle, and specific instructions must be given to the microprobe to set the spectrometer for the elements of interest. The computer may also make decisions as to which particle should be analyzed; for example, particles containing particular elements can be either accepted or rejected. In more sophisticated programs, the computer can be made to count particles containing a specific combination of elements, measure particle size and, finally, perform the data reduction necessary to give quantitative elemental composition.

III. Conclusion

The light microscopist occupies a special place in small particle technology. This is especially true, as happens increasingly, when the sample is a single tiny particle or if the sample consists of many small particles of nearly as many different substances. A trained light microscopist can identify in seconds any one of several thousand single substances. They can be single particles weighing as little as a picogram and as small as a micrometer in diameter. If he cannot identify a given particle at sight he can, if he chooses, make further measurements or tests using only the optical microscope and its accessories and usually identify it. On the other hand, if he feels another analytical method might speed the analysis, he can remount even single subnanogram particles for electron microscopy, X-ray diffraction, and electron or ion microprobe analysis. Occasionally, several of these sophisticated tools must be used, and again the light microscopist can recover the tiny particle from one mount, then clean and remount it for a second or third analysis. Having isolated the particle from its original sample location and followed it through successive analytical stages, he is in the best position to integrate the findings and write the final report.

ACKNOWLEDGMENTS

The methods described are those in use in the author's laboratory. The successful application of ultramicroanalytical techniques to small particle identification is due to the excellent staff of this laboratory. It is difficult to single out one person as more deserving of credit than another but most of us would agree that the techniques of particle handling and mounting developed by Anna Teetsov are a prerequisite for success in this field. Without these micromanipulation techniques the microprobes and other ultramicroanalytical tools would be far less effective.

REFERENCES

1. W. C. McCrone and J. G. Delly, "Particle Atlas," 2nd ed. Ann Arbor Sci. Publ., Ann Arbor, Michigan, 1973.
2. W. C. McCrone and J. G. Delly, "Particle Atlas," 2nd ed., p. 239. Ann Arbor Sci. Publ., Ann Arbor, Michigan, 1973.
3. Joint Committee on Powder Diffraction Standards, Swarthmore, Pennsylvania.
4. M. Bayard and C. Terhaar, *Nature (London)* **232**, 553 (1971).

4

Sampling and Calibration of Gaseous Pollutants

Herman D. Axelrod and James P. Lodge, Jr.

I. Sampling

A. Introduction

Sampling is the segregation of a portion of matter—in the present case, air with its impurities—for analysis. Some optical methods require only that the sampled air be kept changing, but more often sampled air must be moved at a defined rate for a defined time through the sampling device, or a defined volume of air must be isolated. Because of the low concentrations of most air pollutants, it is also frequently necessary to concentrate them in the act of sampling in order to obtain enough of them for subsequent examination.

When considering when and how to sample, the intended application of the experimental result must be defined. The sampling results might be used to provide data for mean daily concentrations in a region, highest attained concentrations in a city, or long-term trends in a particular locality. Consequently, the experiment must be designed to provide the correct information.

1. Type of Sampling: Integrated or Real Time

Sampling for atmospheric constituents can be separated into several time scales. If the experimenter is interested in measuring mean daily levels, then techniques that can take an integrated sample over long time periods are desirable. The equipment and technique is usually quite simple since resolution of the concentration variations is not needed.

The extreme opposite is when the experimenter wishes to observe the many concentration fluctuations that might occur during a short time period. The equipment required is usually quite sophisticated since real-time measurements are required.

Obviously it would be a waste of money to purchase real-time instrumentation when a mean daily value is sufficient to answer the question at hand. Indeed, the volume of output data presented by the real-time instrument might overwhelm the scientific staff.

The question of automatic versus manual sampling must also be based upon the type of data needed for the study. Intermittent data requirements can be met effectively by manual operation while continuous or semicontinuous information is best provided by automatic equipment interfaced with digital recording systems, the last mentioned used for eventual computer data reduction. Lodge (*1*) discusses when not to use automatic air pollution monitoring instruments.

Although many of the parameters of a given technique may be known, the gas concentration pertinent to the study might be unknown, making it necessary to determine the optimum sample size and sampling time. If possible, the amount of air sampled should provide a healthy signal-to-noise ratio without overloading the sampling or analysis system.

2. Storage

The period of time from the collection of the sample to the time of analysis can be an important factor in the quality of the data. The longer the sample is in storage the more opportunities for contamination, loss due to reaction or sorption, or the sample container being damaged or misplaced.

The ideal sample would be infinitely stable, but since we are dealing with reality the technique should be checked for storage stability before data gathering, and appropriate corrections should be made if there has been sample decay. The storage container, the material of the container, the concentration range of the stored constituent, the other chemical species present, and environmental conditions can all affect storage stability. In general, storage is best in low-temperature lightless areas. Since a universal storage container cannot be recommended, each technique must be examined for its own merits.

3. Efficiency of Collection and Analysis

In order to calculate accurately the desired gas concentration, the collection efficiency and the analysis efficiency must be known.

The collection efficiency can be defined as the ratio of the amount of the desired material collected during sampling to the total amount actually present in the parcel of sampled air. This efficiency depends upon the technique used, the gas under investigation and the associated chemistry, the concentration of the gas, the sampling rate, and environmental conditions and interferences. For completeness each parameter should be studied under controlled conditions that best describe the sampled environment.

If the sample is taken by drawing a known volume of air into an evacuated container, then the efficiency of its isolation should be considered 100%, if wall effects are ignored. However, concentrating techniques such as bubblers, impingers, and adsorptive surfaces usually show efficiencies less than 100%. The chemistry involved can greatly affect the percentage of isolation.

One of the traditional techniques for measuring isolation efficiencies is to place two or more sampling units in series. Air containing the desired species is passed through the series units, and the amounts of gas present in each unit are determined. Assuming that efficiency is independent of gas concentration (*2*), the efficiency is calculated from the amount isolated by the first unit divided by the total amount isolated by all the units. This procedure is at times subject to large errors as demonstrated by Urone *et al.* (*3*). They showed that the amount of SO_2 captured by the first sampling unit was a function of atmospheric concentration even though the absolute amount present in the air sample in all cases was the same.

From the work of Urone *et al.* (*3*) and of others, it is apparent that a threshold concentration for isolation is present and that if the first unit removes a fraction of the material, the gas concentration passing through the first unit might be at the isolation threshold and so pass through succeeding units. This would show that no gas had been isolated by the subsequent units, and a 100% collection efficiency would be falsely assigned to the method.

An alternate procedure for the determination of collection efficiency would be to measure the amount isolated from a wide range of standard gas concentrations. Inefficient isolation should begin to appear as lower and lower known concentrations are sampled or as the isolation threshold is approached.

The analysis efficiency, on the other hand, can be related to loss of the sample during the analysis procedure. In other cases, the chemistry may be different for the standard and the samples. The risk of this increases when standard and sample differ greatly. While it is not our intention to discuss individual methods, it might be worthwhile to point out one example of the last mentioned problem. In the Saltzman (*4*) method for nitrogen dioxide, a sodium nitrite standard is used to calibrate the reagent solution. The debate on this subject continues; however, it has been shown in several circumstances that in this procedure a given amount of nitrogen dioxide gives a response equivalent to 75% of that of the sodium nitrite standard containing the same amount of nitrogen.

Inlet tubing, sampling manifolds, etc., can decrease apparent concentrations due to sorption. Desorption from tubing and manifolds can also lead to erroneously higher values when sampling for low concentrations. A number of authors (*2, 5–9*) have clearly demonstrated that a poor choice of tubing material for sample procurement can cause severe sample losses. A specific tubing material must be matched to the constituent to be measured, and factors such as constituent concentration, temperature, relative humidity, tube length and diameter, sampling flow rate, and tubing preconditioning must be considered. Proper material selection will minimize losses and can even make it possible to sample for carbon monoxide (*10*), ozone (*11, 12*), and sulfur dioxide (*10*) at the discharge side of a pump.

At times it is advantageous to remove particulate matter prior to gas sampling. This is usually done by placing a filter ahead of the sampling unit. This technique requires caution because there is always the danger that the gas under investigation will be retained by either the filter, the filter holder, or the particles accumulated on the filter, and not enter the sampling unit. As with the sampling tube, losses occur with prefilter systems and are subject to the same kinds of environmental, chemical, and sampling variables. Uptake of sulfur dioxide by prefilters, for example, is documented by Byers and Davis (*13*) and Axelrod *et al.* (*14*).

4. Interferences

Inaccurate findings can be the result of interferences from a variety of sources. The typical interference is a constituent isolated during the sampling which subsequently reacts with the material under investigation or with the analytical reagents employed. While the gas under investigation and the interferent might coexist in the atmosphere, sampling (usually a concentration procedure) tends to increase both constituent concentrations and to provide a medium (liquid absorption or solid adsorption) for their reaction. This is especially true for cryotrapping, in which larger amounts of air are sampled and only certain materials in highly concentrated form are retained. If the sample eventually reverts to room temperature many interactions may occur, resulting in sample losses. It is obvious that for best results the potential interferent should not be retained by the sampling system; or the gas under investigation should be converted into a stable form.

5. Reproducibility and Accuracy

The reproducibility of a method is the measure of its repeatability, while accuracy indicates how close a method comes to the true value. Techniques can be both very reproducible and very accurate; some can be either one or the other; and still other techniques may be neither. Whether it is necessary for a method to possess both of these qualities is dependent upon how the data will be used.

If the experimenter wishes to know whether the present atmospheric gas concentration is in compliance with specific values set by a government agency, then clearly the accuracy of the method must be known. For studying trends it is more important to have a reproducible method than an accurate one.

6. Units of Measurement

Atmospheric gas concentrations are reported in the literature in a variety of ways. Authors have presented data as mixing ratios by volume or by weight, or as concentrations in units of per unit volume. It is important that the reported units be defined. In addition, mixing ratios of 10^{-6}, 10^{-9}, and 10^{-12} mean, respectively, parts per million, billion, and trillion in the United States, and parts per million, millard, and billion in some European countries. There has been a recent effort to encourage reporting in units of micrograms per cubic meter, ostensibly because it is "metric."

This makes little sense as both volumetric and gravimetric mixing ratios (i.e., parts per million by volume or weight, etc.) are dimensionless, and so are neither metric nor nonmetric. In fact, mixing ratios have the advantage of not requiring gas-law corrections to or from standard conditions. On the other hand, it seems inelegant to present concentrations of particulate species as mixing ratios since, in the molecular sense at least, they are not mixed. So for the present, one is likely to find sulfate concentrations in micrograms per cubic meter and sulfur dioxide in parts per billion.

B. Sampling Systems

1. Grab Sampling

Grab sampling is a technique commonly used in air pollution investigations. The procedure is to rapidly acquire a volume of air that can later be analyzed for specific pollutants. The containers are usually made from flexible plastics, steel, glass, or hypodermic syringes.

The use of hypodermic syringes is an easy way to obtain a grab sample for future analysis. Syringes constructed of all glass, polyethylene, and glass and Teflon are available in various sizes. Careful selection of the syringe material and repeated preconditioning treatments may be necessary to avoid irreversible adsorption of the pollutant (*15*). After sampling, syringes can be sealed and returned to the laboratory for analysis of the contents. Syringes can be used dry, or they can contain a specific absorption or analysis solution so that the procedures of grab sampling and analysis can be accomplished in one operation with the syringe (*4, 15, 16*).

Plastic bags have been widely used for grab sampling and sample storage before analysis. The bags can be filled in one of several ways. Air samples can be introduced into the bag from the discharge side of a vacuum pump. (Obviously, the pump must not remove the pollutant being sought nor contaminate the sample.) Another procedure is to place the bag inside an air-tight container with only the inlet exposed. The air surrounding the bag is removed causing the bag to inflate (Fig. 1). A similar technique would be to use water instead of air as the surrounding jacket and allow the water to escape.

Schuette (*17*) reviewed previous work in bag sampling and concluded that the following factors were important for maintaining the integrity of the sample: surface area of bag, concentration of the gas, relative humidity, temperature, physical and chemical properties of the bag material, and the presence of other reactive compounds. Despite these poten-

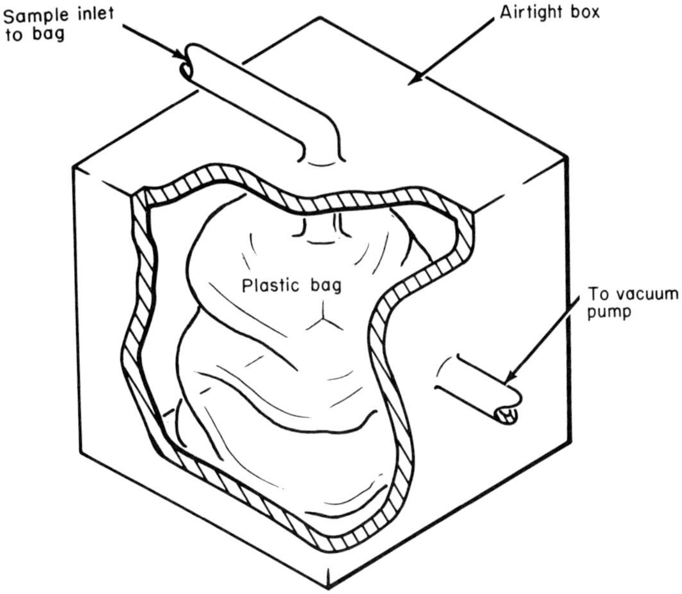

Figure 1. A plastic bag being used for grab sampling.

tial problems plastic bags have been successfully used by many investigators (*18–27*).

Rigid containers made from stainless steel or glass can also be used for grab sampling. With rigid containers, samples are usually gathered with an evacuated container by opening a valve, and allowing air to enter and be captured. An alternative to evacuation is to have two valves, one at each end, and to flush the container with the sampled air prior to sealing the container so that the walls will have an opportunity to equilibrate with the incoming air. The problems with storage in plastic bags described earlier can also occur with inflexible containers.

The use of an oxide coating on stainless steel in the sampling for hydrogen and methane (*28*) showed that preconditioning or pretreatment of the container can help maintain integrity of the sample. Glass containers fitted with greased or Teflon stopcocks are also good containers. Evacuated rigid walled containers, too, can hold absorption solutions and are used in a way similar to that discussed in syringe sampling.

2. Adsorption Sampling

Adsorbents are used to isolate a variety of pollutant gases. Unfortunately this technique has not been developed to the point at which specific adsorbents can be used. As a result, a mixture of gases is usually retained, which can cause analytical difficulties.

Some of the adsorbents commonly used are activated aluminas, silica gel, molecular sieve, charcoal, and Celite. The sample is drawn through a container with the adsorbent maintained at ambient or subambient temperatures (see Section I,B,4). After sampling, the container is returned to the laboratory for analysis.

The adsorbed material can be removed either by heating the adsorbent to volatilize the trapped material—which can then be flushed by an inert gas into another system—or by washing the adsorbent with a suitable solvent. Analysis can be made in a number of different ways.

There are several problems associated with this technique. If water vapor is present in the sample stream, the adsorbent can be deactivated. Since the adsorbent is nonspecific, gases concentrated on the surface can potentially react with each other. Because of the nature of certain gases and adsorbents, it is not always possible to remove the adsorbed gas without decomposition once it is attached to the adsorbent. In addition, certain adsorbents are known to cause isomerization, and so alter the sample (*21, 29, 30*).

Despite the limitations discussed above, adsorptive sampling has been used as a first step in the measurement of a wide variety of gases. Some representative examples are: organics (*21, 29–39*); radioiodine (*40*); volcanic gases (*41*); nitrous oxide (*42–44*); ammonia and amines (*45*); carbon monoxide (*46, 7*); carbon dioxide (*48*); ozone (*49*); and lindane pesticide (*50*). Adsorptive sampling for ethylene has been done using brominated charcoal to react with the ethylene to form, and be retained as, ethylene dibromide (*51*).

3. Absorption Sampling

Absorption can be described as the process of transferring one or more gaseous components into a liquid or solid medium in which they dissolve. Many techniques based on absorption are presently used in air pollution sampling.

a. Bubblers and Impingers.* These items are kinds of equipment designed to trap specific atmospheric gases in a solution that can later be analyzed. The gas to be sampled is drawn through a tube, the downstream end of which is below the surface of a liquid. The gas forms bubbles

* See also Chapter 1, this volume.

and is dispersed into the liquid. If the rate of gas dissolution, or gas reaction with the solution, is faster than the time needed for the bubbles to leave the liquid, the gas will be retained.

The dispersion tube can be an open-ended tube or one with a frit (a material with many 50- to 100-μm holes). The open tube is best for strong oxidants, such as ozone, while the frit is suitable for dispersing most other gases. Frits can be prepared in various shapes and sizes; glass, plastic, and ceramic have been used as materials. An advantage of the ceramic is that the holes are uniform in size (*52*).

Another alternative is to use glass tubing with small holes that can be prepared by sealing copper wires into a glass tube and then dissolving the wire with nitric acid (*53*). Since dispersion devices tend to vary in number and size of holes, frit pressure drop can be excessive. For this reason it is advisable to measure the pressure drop of each frit before use (*54*). It has also been shown that there is no advantage to using very fine dispersing devices since the bubbles formed at the frit tend to coalesce rapidly (*54, 55*). Coarse frits, then, are recommended.

A number of designs for bubblers and impingers are available (Fig. 2). The differences between them are usually based on convenience rather

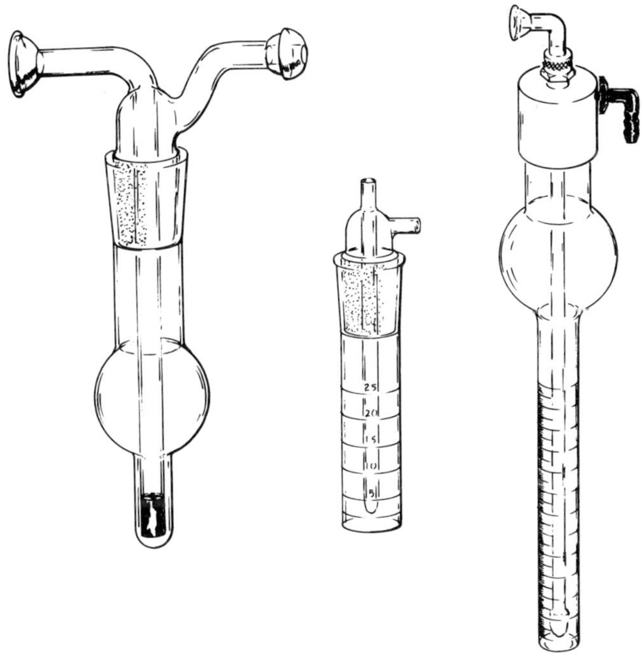

Figure 2. Typical bubbler and impinger sampling equipment.

than efficiency. Most systems are constructed entirely from glass (*56–64*); others have "O" ring seals to minimize breakage (*65*); still others have interchangeable parts of glass and Teflon (*66*), or a threaded Teflon top (*52*).

Saltzman (*54*), Calvert and Workman (*55*), and Gage (*67*) studied the theory of bubbler-impinger operations and the many factors influencing efficiency of isolation. The efficiency of the various types has been determined and cross correlated by Roberts and McKee (*68*). Procedures for checking efficiencies have been suggested (*54*), some of which use radioactive gases (*3, 69*). Impingers usually have only one dispersing device, but if efficiency is a problem, several dispersing devices can be placed in series in a single unit to increase the capability for isolation (*70–72*). This was successfully demonstrated by Bergshoeff (*71*).

A serious limitation in the use of bubbler-impinger samplers for gases is the evaporation rate of the solvent. In dry climates, the evaporation rate can be high, limiting the amount of air to be sampled for a given liquid volume. However, a system to maintain a constant solvent level by sensing the impinger weight has been developed by Kats and Thompson (*73*). For volatile organic solvents, bubblers may be maintained in an appropriate subambient temperature bath (*74*).

b. SOLID ABSORPTION. The "lead peroxide candle" method, a traditional technique for sulfur dioxide monitoring, is an example of solid absorption. The candle is a ceramic cylinder coated with lead dioxide which is exposed to the atmosphere for lengthy time periods (usually 1 month). The sulfur dioxide comes in contact with the candle forming lead sulfate. At the appropriate time the candle is returned to the laboratory and is analyzed for sulfate (*75–77*). Unfortunately, this technique is not a good indication of sulfur dioxide levels since the amount of absorption is dependent upon factors such as wind velocity during the exposure period. The use of various types of coated plates has been proposed to simplify sampling (*78, 79*).

c. ION-EXCHANGE RESINS. Cole and Shulman (*80*) demonstrated the use of ion-exchange resins for the clean up of industrial gas. This technique could be applied to atmospheric sampling of pollutants since it offers simplicity and portability. Not much work has been done in this area, but Paez and Guagnini (*81*) used ion-exchange resins to collect hydrogen sulfide for subsequent analysis.

d. FILTER SAMPLING. Filter material impregnated with reagents for specific gases can be used for atmospheric sampling. One of the more common techniques of this kind is the paper tape sampler. A roll of paper tape is impregnated and dried. The tape is then placed in a mechanical unit which sequences a tape section into position for sampling. After a known volume of air has been drawn through it, the tape is removed and the sample analyzed. This system has been successfully used for the measurement of hydrogen sulfide (*82–84*).

Individually impregnated filters have also been used for air pollution sampling. These filters offer several advantages; they can sample large quantities of air without the problems of liquid evaporation; they can sample at high flow rates; and they are light weight, nonfragile, inexpensive, and easily transported to the analysis laboratory (*85*). Paper, glass fiber, and membrane filters have been used for such purposes. Sampling procedures have been given for sulfur dioxide (*86, 86a*), hydrogen sulfide (*87–91*), ammonia (*92–94*), nitrogen dioxide (*94a*), and mercaptans (*95*), all using impregnated filters.

e. Gas Chromatographic Column Packings. An easy, convenient method of isolating a particular gas for sampling uses gas chromatographic columns. In a study made by Eggersten and Nelsen (*96*), air is drawn through a column at ambient or subambient temperatures and the gas under investigation is absorbed by the liquid phase coating. The column is then returned to the laboratory for analysis. This procedure has had a variety of applications (*97–102*), including pesticides (*100*), and appears to be much more efficient than isolating with open tubes at cryogenic temperatures.

A variation of the liquid-coated gas chromatographic column method is the use of liquid-coated platinum wire. Air is passed over the wire (in its appropriate housing) trapping the desired gases. The wire is then returned to the analysis laboratory, heated by electrical current, and the volatile material is introduced into a gas chromatograph (*103*).

Absorption of atmospheric constituents by liquid coatings on solid supports is also useful for isolating gases over long periods of time. Levaggi *et al.* (*104*) made 24–96-hour integrated measurements of nitrogen dioxide by trapping this gas on a triethanolamine coating on molecular sieve.

f. Diffusion-Absorption. This technique is the reverse of the permeation tube technique used in gas calibrations. In a procedure developed by Reiszner and West (*105*) for sulfur dioxide, absorbing solution is placed in an inverted glass test tube with a membrane covering the open end. The sulfur dioxide permeates the membrane at a known rate and is trapped by the absorbing solution. The solution is then returned to the laboratory for analysis.

4. Cryogenic Sampling

Cryogenic sampling or cold trapping is treated here separately, even though it includes adsorptive processes. The cryosampling technique is designed to preconcentrate chemical species so that an accurate analysis can be made. The procedure is to place a sampling trap at a subambient temperature and to draw air through the trap causing the vapor under investigation to condense and be retained. The trap is then brought to the analysis laboratory, the sample is removed, and an analysis is performed. This basic procedure has been used by many scientists to condense and isolate various organic materials for future analysis (*23, 97, 106–113*). Williams (*97*) used a cryosampler to help identify and quantitate over 30 organic compounds.

Although cryosampling is very useful, it is not an easy technique. Coolants frequently used for the trap are ice water, solid carbon dioxide—isopropyl alcohol, liquid oxygen, liquid nitrogen, and liquid helium. A more complete selection of solutions and their respective equilibrium temperatures is shown in Table I (*114*). Some of the coolants listed have

Table I Summary of Cold Bath Solutions[a]

Coolant	*Temperature (°C)*
Ice and water[b]	0
Ice and NaCl	−21
Carbon tetrachloride slush[b,c]	−22.9
Chlorobenzene slush[b,c]	−45.2
Chloroform slush[b,c]	−63.5
Dry ice and acetone[b]	−78.5
Dry ice and cellosolve[b]	−78.5
Dry ice and isopropanol[b]	−78.5
Ethyl acetate slush[b,c]	−83.6
Toluene slush[c]	−95
Carbon disulfide slush[b,c]	−111.6
Methyl cyclohexane slush[b,c]	−126.3
N-Pentane slush[c]	−130
Liquid air	−147
Isopentane slush[c]	−160.5
Liquid oxygen	−183
Liquid nitrogen	−196

[a] Hendrickson (*114*).
[b] Adequate for secondary temperature standard.
[c] The slushes may be prepared by placing solvent in a Dewar vessel and adding small increments of liquid nitrogen with rapid stirring until the consistency of a thick milkshake is obtained.

serious disadvantages, particularly for field use. Liquid nitrogen tends to condense oxygen in the trap. This liquid oxygen can be dangerous; also it will react with many of the other condensibles. Liquid oxygen as a coolant is dangerous, too, and requires careful handling. All sub-0°C coolants tend to condense large quantities of water that can eventually react with other condensibles, plug the trapping system, or interfere with the analysis. This problem can be minimized by using a desiccant filter at the entrance to the trap (*107*). A properly selected desiccant will retain most of the water and quantitatively pass the compound of interest. However, this is not always possible; for example, most desiccants retain ammonia.

Since different materials have different boiling points the sample air can be fractionated by using several traps placed in series and at successively colder temperatures. This technique helps to decrease water problems and interactions between compounds, and performs a helpful separation prior to analysis.

The cryotrapping technique has a threshold at which collection is made. For example, if the saturation vapor pressure of the test substance at the trap temperature is 1 mTorr, and its partial pressure in the sampled air is 100 mTorr, then 99% of the material will be trapped. However, if its partial pressure is 2 mTorr, then only 50% will be collected. For partial pressures below 1 mTorr, no material will be collected. Vapor pressure considerations for sampling are discussed by Weaver *et al.* (*115*) and Cave (*116*).

These limitations can be overcome with a "noninert" trap, in which an adsorbent is placed in the sampling tube. With the proper adsorbent, the material under investigation will condense and tightly adhere, and vapor pressure is no longer important. Typical adsorbents are molecular sieve, activated charcoal, silica gel, and chromatographic column materials. Altshuller (*22*) discusses the application of cryosampling with adsorbents.

Proper handling of the isolated material after sampling is very important. There is always a danger that, since this technique tends to concentrate materials, the condensed species will react with each other, especially if the sample is allowed to reach room temperature before analysis. In addition the trap walls are often good catalytic surfaces for reactions. For this reason, the sample should be kept at its cryo-temperature up to analysis time. Adsorbents in the trap tend to minimize interactions between compounds although this depends on both the compounds and adsorbents.

Since air is frequently trapped along with the constituents under investigation, some experimenters flush the trap with helium to remove the air before analysis. Some of the condensibles are also removed in the process (*108*).

C. Sampling Equipment*

1. Pumps, Sequential Samplers and Operating Power

Proper selection of the prime air mover can ensure effective sampling. Gas sampling—and particle sampling with membrane filters—usually involve pumps. Some fibrous filters are open enough for fans or blowers to be used instead.

The variety of pumps available makes a discussion of each one impractical here. However, choice of a pump should be based on: power requirements (operating voltage and current), pumping capacity, weight, ruggedness, chemical inertness, environmental operating conditions, total operating time between repairs, and, perhaps, operating noise.

The efficiency of the pump is an important consideration for sampling with batteries in remote areas. Smaller size pumps tend to be more efficient than larger ones. Taylor (*117*) has examined the problem of pump operation and has designed an efficient pump for air sampling.

A sequential sampler is often used for sampling operations that require intermittent sample taking over long time periods. This device consists of a timer, a switching valve, and a set of sampling containers so that successive samplers are operated at successive time intervals during the sampling period. Since a wide variety of systems is available it should be fairly easy to obtain one that meets specific requirements. Not all of them are equally dependable. Therefore, for high durability, it may be necessary to test several before making a selection.

Typical power to operate the above units is ac electricity from a local power line. At times, however, one might wish to operate on batteries. A 12-V automobile battery will usually perform satisfactorily and is readily available. Two other batteries which also have been used are the Ni–Cd battery and the lead–acid gel electrolyte battery. All of the above are rechargeable and are rugged.

Our experience has shown that the best one of the group is the lead–acid gel electrolyte. This battery has the rechargeability characteristics of the other batteries but it is priced at about one-tenth of the cost of a Ni–Cd battery. In addition, the gel electrolyte allows the battery to be used in any position without fear of the electrolyte spilling.

Samples can also be taken by devices requiring no electrical power.

* See also Chapter 1, this volume.

For example, one can operate by hand a bellows to draw or force air into the sampling system. Another approach is to use a siphon or aspirator bottle to draw air into the sampling system. Siphons and aspirator bottles are further discussed in Section I,C,3.

2. Volume and Flow Rate Measurements

Depending upon the method chosen for sampling, the experimenter must know the volume of the sample container or the flow rate of the sampling system.

When obtaining a sample with a rigid system and subsequently using the entire volume of the sample for analysis, the volume of the container must be accurately known. One way of determining the volume of the container is to weigh the container empty and then fill it completely with a liquid of known density. The container is weighed again, and the volume calculated from the weight of liquid and its density.

Another technique is to evacuate the container to a very low pressure. The container is then connected to a known volume filled with air and attached to a pressure gauge. The air-filled volume is opened and allowed to expand into the evacuated container. The resulting lower pressure in the gauge is recorded. With the initial and final pressures known, and the volume of the air-filled container and gauge known, the volume of the evacuated container can be determined. The evacuated container can be attached to a volumetric device (wet or dry test meter) in a similar way. The container is then opened, and the volume of air drawn in measured by the meter.

There are several procedures for calibrating flow control devices. The techniques are not all covered here but American Society for Testing and Materials Method D 1071-55 (*118*) gives a good description of most of them. Wet or dry gas meters are frequently used for volume or flow rate measurements. The dry test meter is a device containing two bellows of known volume and a sliding valve. Each bellows is alternately filled and emptied; the motion drives an indicator of total volume. The wet gas meter consists of a drum with a paddle wheel, the wheel half submerged in water. The air coming in fills the open section of the paddle wheel and makes it rotate. In this way one section is being filled while the other is being emptied. Complete descriptions are given by Nelson (*119*). When using the wet gas meter the experimenter must be concerned with dissolution of the measured gas into the water until saturation is reached. On the other hand, prolonged use of the wet gas meter with a dry gas tends to evaporate the water and cause errors.

The gas meters along with other equipment can be calibrated using an aspirator bottle (*118*). The gas exiting from the device to be calibrated is fed into a carboy containing a liquid of known density. The liquid is forced out through a siphon tube and the amount weighed. The volume removed must equal the volume passing through the test device.

A soap bubble flowmeter can be used for low flow rate calibration. This unit is a buret of appropriate volume fitted with a rubber bulb at the bottom and a side arm above it. The bulb is filled with soap solution. Air is either forced into the buret through the side arm or drawn through from the buret top. The bulb is then squeezed, causing a soap bubble to enter the buret and traverse the barrel. The time it takes the bubble to travel a known volume is used to calculate the flow rate. Several flowmeter designs have been presented by Nelson (*119*) and James and Martin (*120*); a more accurate system with a plastic float and mercury seal has been developed (*121*).

One of the most commonly used flow rate measuring devices is the rotameter. This item is a tube of V-shaped bore with a float or floats inside. A gas passing through the tube will cause the float to rise until its downward force is equal to the force of the gas. The height of the float is proportional to the flow rate. Some units contain several floats of different densities in order to increase the useful range. If possible the Rotameter metering scale should be calibrated under the conditions of use since the float height attained is also a function of gas density (*122*). Many Rotameters are equipped with needle valves for flow rate adjustment. Since the Rotameter is pressure sensitive it is recommended that the needle valve be placed downstream from the Rotameter (*123*). Rotameters can also be used in continuous readout systems (*124*).

The use of orifice flowmeters is an easy, inexpensive method for measuring flow rates (*125*). The meter is made by placing a restricting capillary in the flow line and measuring the pressure drop across the capillary with a manometer. The relationship between the pressure drop and the flow rate can be determined using one of the previously described procedures. Care must be taken not to partially clog or damage the orifice during use since this will alter the calibration. Orifice flowmeters are quite simple and can be used with corrosive gases (*126*).

Thermal mass flowmeters can also be used for flow rate measurements. A number of styles have been designed, but they all function by measuring the amount of cooling which results from a gas passing over a heated surface. Some designs place a heated thermistor in the gas stream while others pass the gas through a stainless-steel tube externally heated to a specific temperature. The cooling effect of the gas versus a reference is indicative of the gas flow rate and heat capacity. Consequently, this system must be calibrated for the specific gas to be measured. A review of the various thermal mass flowmeters has been made by Benson *et al.* (*127*).

A flowmeter based upon the measurement of ionized gas has been developed by Castle and Sewell (*128*). They use an α source to ionize the flowing gas, and measure the resulting ions flowing past a given point. This system can be used as a volume or mass flowmeter or an anemometer.

3. Constant Flow Rate Sampling

Depending upon the situation, atmospheric samples can be taken either instantaneously or in an integrated way. An instantaneous sample is air trapped at a particular moment, so it may not be a representative sample. Conversely, a sample taken over a fixed time interval is more representative, but averages the high and low values. Integrated samples are generally inherent in preconcentration techniques where large air volumes are sampled to obtain sufficient amounts of the test gas for analysis.

To obtain a proper integrated sample, the volume of the sample must be accurately known. This is done most easily if the sampling rate is constant. The simplest approach is to place a valve and flowmeter between the sample collector and a vacuum pump. As the flow rate changes during the sampling period, the experimenter adjusts the valve until the flow rate is back to the desired volume. This procedure requires the constant attention of the experimenter, and can result in errors made both by the experimenter and by the instruments. The valve alone will not maintain a constant flow rate because of the changes in the back pressure of the sample collector during the time period.

The sampling pump can be made to maintain a uniform sampling rate. One procedure is to place a restricting orifice on the discharge side of the pump (*23*), and another is to sense the air flow rate and alter the pump speed (*129*).

One of the oldest techniques uses the aspirator or siphon bottle. In this procedure a sample collector is connected to a bottle filled with a known quantity of water. The water is allowed to flow out at a known rate, drawing air into the bottle and through the sample collector at a rate equal to the flow rate of the water. The total volume sampled will equal the water displaced. The aspirator system is bulky but it does have the advantage of not requiring electrical power for operation. This technique has been used extensively (*130–133*). Hermann (*134*) shows that an 8-liter sample can be obtained over a 24–48-hour period. For constant sampling rate the aspirator system must maintain a constant head of pressure (*134*).

Constant flow rate sampling can also be maintained with critical orifices. These consist of a small constriction or orifice placed in the sampling stream between the sampler and the vacuum pump. If the downstream pressure is lowered, a point is reached at which the air passing through the orifice is at sonic velocity (critical orifice) and the flow rate will be constant and independent of further drops in the downstream pressure. The critical point for air is reached when the downstream pressure is less than 53% of the upstream pressure (*135*). Although with critical orifices the flow rate is independent of downstream pressure changes, it is dependent on the upstream pressure (*135*). For this reason if the critical orifice is controlling the flow rate for an impinger or filter sampler, changes in upstream pressure due to solution evaporation or filter clogging can affect the sampling flow rate. Since contamination of the orifice will change the flow, too, an orifice protector (usually a filter) of low pressure drop should be placed directly upstream of the orifice (*136*).

Several kinds of critical orifices have been designed for field sampling. One of the easiest and least expensive is the hypodermic needle (*136, 137*). These needles are placed on the sampling line; by selecting the proper gauge and needle length, flow rates from 0.3 to 12 liters per minute can be maintained (*136*). Other approaches are to use watch jewels (*138*) or glass capillaries (*139*).

Another technique is to use a float in conjunction with an orifice (*140, 141*). In this case changes in the pressure drop across an orifice cause the float to alter position and so maintain a constant flow.

A more sophisticated device for keeping flow rates constant is a differential flow controller (*142*) in the same position occupied by the critical orifice. The advantage of the differential flow controller is that it only requires a 180-Torr pressure drop for good operation, as opposed to approximately $\frac{1}{2}$ atm for critical orifices. In addition, a wide range of flow rates can be selected by adjusting the needle valve on the flow controller. Still, flow controllers are more expensive than critical orifices and the needle valve can be damaged if an orifice protector is not used.

Mass flowmeters connected in servo fashion to an adjustable valve are commercially available (*121, 143*). The sensing element of the flowmeter regulates the valve opening so that the flow rate through the system should be constant. Although the equipment is expensive, it makes excellent control possible.

4. Sampling from Manifolds

A common procedure is to sample and measure several gases from a common manifold. Inherent in this technique are the problems of sample loss on the intake manifold walls. The gases to be sampled can be irreversibly adsorbed or decomposed by wall materials and preferential losses can occur to gas mixtures thereby altering the gas ratios.

Sampling from manifolds also tends to smooth out abrupt changes in gas concentrations. This is because the manifold has a dead volume which buffers abrupt concentrations of incoming gases. This effect can be overcome by using a small volume manifold or by having high flow rates of air passing through the manifold.

D. When to Sample

The sampling of air contaminants in such a way that the collected sample has a known relationship to the airborne concentrations of the tested substances is certainly not simple. On the other hand, the scheduling of a program of sampling so that it will answer the questions that motivate the program is at least as complex. The mathematics of optimal program design have not been worked out. However, it is possible to get a feeling for the general requirements of sampling design by considering extreme cases.

If we ignore the measurement error, the concentration of a completely inert and constant ingredient of the atmosphere like neon can be measured definitively with a single determination at a single point, with no point on the earth's surface preferable to any other. At the other extreme, to completely map the concentration of an infinitely variable component would require an infinite number of measuring stations. Actually, most substances relevant to air pollution are quite variable, and their distribution is accurately described by classical logarithmic normal statistics (*144*).

The question of how many samples must be taken at how many individual measuring stations in order to determine an aerial and time mean value of a pollutant concentration within a stated tolerance interval of the true mean can be answered, theoretically, from classic statistics. This argument is well developed in a series of reports from an air pollution study of Nashville, Tennessee (*145–150*). However, classical statistics is predicated on the assumption that the input data are randomly ordered, as though one took all the possible values and shook them in a hat. In fact, the data are highly autocorrelated. Adjacent stations are likely to yield similar values, and, because of meteorological persistence, high or low values at any given station are likely to be followed by similarly high or low values in the next measurement. The Nashville experimenters themselves point out that, in general, two suitably located stations, each sampling on a random day every 2 weeks, give essentially as good a determination of mean levels as their far more numerous network. However, the dichotomy between experiment and theory was never faced in the published reports.

Hence, the problem of specifying station number, location, and sampling frequency for optimum operation remains unresolved. This is for the comparatively simple case of determining (for example) the annual mean air pollution level of an entire city. While this sort of information is very useful in the study of trends in their pollutant concentration, it is not particularly applicable to the estimation of the air pollution dosage of a single individual (whether plant, animal, or human) or to the problem of studying air pollutant distributions within a given city.

Individual doses can, of course, be determined by attaching a continuously integrating sampler to the individual (or, if the individual is a plant, locating it immediately adjacent to it). Obviously, in studies of large population groups this can be unwieldy, and the individual may object to taking a frequently noisy sampling device to bed with him. There are also technological problems about integrating samplers small enough to carry that will operate for the length of time this sort of experimental design necessitates. It would certainly be preferable to have an accurate map of concentration isopleths throughout the city, and merely ask the individual to report his daily itinerary. However, the whole statistical problem of establishing isopleths within a designated accuracy has not yet been attacked. Some data exist on the relationship of weather map accuracy to weather station spacings, and should be applicable. In the meantime, it seems desirable to decide in advance on the order of isopleth spacing that will be sought, and to space the sampling network at no greater intervals. Unfortunately, in nearly all air pollution experiments to date, the spacing and number of stations has been decided by budget limitations rather than by any real attempt at experimental design from first principles.

II. Calibration

A. Introduction

Calibration relates the data an instrument or technique produces to the real concentrations of atmospheric pollutants. *Calibration* generally means comparison with a reference measure that is constant but may or may not itself be a measure of real concentrations. *Standardization* means that the reference measure has a one-to-one correspondence with true concentrations. Accurately produced test atmospheres are used for both purposes.

Producing an atmosphere of precisely known composition is one of the most difficult tasks facing the air pollution researcher, whether this atmosphere is for calibrating an analytical instrument or for exposing test vegetation, animals, or humans. The difficulty in comingling air with pollutants in precise proportions when the respective volumes are in a ratio of several million to one is intensified by the possibility of subsequent reaction of the pollutants with compounds already in the air, with each other, and with the materials that compose the confining walls of the system. It seems likely that many of the discrepancies in experimental results in air pollution research are the result of underestimating the difficulty of this problem.

These considerations create some degree of uncertainty about the actual concentrations produced by any system. However, once the interreactions described above are understood, they can be modified by an optimal design developed for each particular experiment.

It is not our purpose to report on all the different techniques used for calibration in the air pollution field. Nelson (*119*) has covered almost all the different calibration techniques and information. Hersch (*151*) has discussed some of the lesser known techniques.

1. Basic Considerations

Whenever possible the concentration of substances in the test atmosphere should be established by calculation from established physical constants and the parameters of the system. If calibration from first principles cannot be made, the controlled atmosphere must be analyzed instead by chemical or physical means. To avoid circular reasoning, these means must be independent of the system for which the controlled atmosphere is being prepared. Any errors in the analytical method will be perpetuated in all its subsequent applications.

Systems for producing test atmospheres may be categorized by the method used to determine the concentration. Those in which the concentrations of trace substances are computed from flow rates and physical constants are called *primary systems*. Those in which the final concentrations are measured as they appear in the system are called *secondary*.

Ideally a controlled atmosphere used to test a field analytical method should duplicate the pertinent qualities of the air to be measured in the field. This is impossible in practice because it presupposes a knowledge of field conditions that would make further field measurements superfluous. There are very few cases at present in which we can specify the pertinent features of the atmosphere. We must first work with simplified systems to isolate the effects of pure pollutants or simple mixtures, then work our way by successive approximations toward controlled models of the complex mixture that is our urban atmosphere. It is hoped that the complex mixture will include all known and potential interferences.

The degree to which the same considerations apply to atmospheres in animal and other studies will depend on the design of the experiment. Amdur's (*152*) studies on the effect of aerosols on the toxicity to animals of certain gases suggest that our view of the pertinent features of atmospheres for toxicity studies has been too simplified. Thus, the coexistence of particles with the trace gases may lead to underestimation of effective dosage, since inhaled aerosols may deliver high local concentrations of sorbed gaseous toxicants to small areas of the lungs, despite a low mean concentration.

Another consideration is that the test gas produced should be the same as the one to be sampled, or to provide a similar response. According to Altshuller (*22*), the use of a few organic compounds to indicate the true response towards all organics by a chromatographic flame ionization detector might be incorrect.

2. Components of the System

The first requirement of a controlled atmosphere is a source of the major gas or gases that are to comprise this atmosphere. Air is used most frequently, argon less so, and other gases are sometimes used for special purposes. The purity of these gases is extremely important when realistic air pollution concentrations are to be produced. A gas certified to contain 99.99% nitrogen, for example, may still contain 100 ppm of impurity, compared with the 1 ppm or less to be introduced in the test atmosphere. "Pure air" presents a problem of definition, since the purest air available on earth in the natural state contains hundreds of substances in varying trace amounts. The result is an operational definition of "pure air" meaning air freed of anything known to interfere with the study being made.

Air containing a minimum of impurities can be isolated with a system of several filters of increasing efficiency, a charcoal bed for the removal of organic matter and polar gases, a furnace for combustion of the last traces of organics, an air conditioner for controlling temperature and humidity, and a final filter to remove any particulate matter introduced by the conditioner or the furnace. Depending on the specific requirements of the study some of the stages of purification can be omitted, pure compressed oxygen and nitrogen can be blended in a 1:4 ratio, or specific chemical methods of purification can be substituted.

For example, Sleik and Turk (*153*) propose the specific removal of unsaturated hydrocarbons by means of a bromine-impregnated charcoal.

A small self-regenerating mechanical scrubber for hydrocarbons can also be used *(154)*. A moisture scrubber version of the same device is described by Skarstrom *(155)*. An air purifier designed for gas chromatography *(155a)* can also be used. Small amounts of acidic gases may be removed, together with carbon dioxide, by soda lime or an equivalent solid alkali. Hydrogen fluoride can be removed on charcoal impregnated with sodium silicate *(156)*.

The other components of the controlled atmosphere system are a source of the test substance or substances; a container, duct, or other boundary for the system; and devices to limit, define, or measure the flux or amounts of air, and test substances and final mixture. Regardless of the type of system employed the physical hardware should not retain any of the diluted gas. Smith and Hochstettler *(157)* have used radioactive chemicals to check the reliability of dilution systems and thus correct for adsorptive effects.

B. System Types

The motion of gases within all controlled atmosphere systems determines whether they are *static* or *dynamic* systems. Static and dynamic systems each have capabilities and limitations that must be evaluated when a system is chosen for a particular purpose. No one system meets all the requirements of all possible applications.

1. Static Systems

In a true static system, appropriate quantities of the test substances are added to a fixed and isolated volume of air. All tests are conducted within this fixed volume. Depending on the size and geometry of the volume and the nature of the test substances, some sort of stirring may be provided.

Fully static systems are frequently considered primary, and sometimes are. However, dilute gaseous mixtures in static systems are analogous to highly dilute liquid solutions, and it can never be assumed *a priori* that they are free of significant errors caused by gas phase reactions and by interaction with the walls of the system. Consequently, if a system is to be used as a primary standard, there has to be assurance that such losses will not occur. This limitation is less important for secondary static systems but even there the experimental design must be examined to be sure that the constancy of composition of the static system is not a requirement. If it is, then the same tests are necessary as for primary static systems.

a. FLEXIBLE WALLS. Static systems can be divided into those with flexible and those with rigid walled containers; each has its own assets and liabilities. The flexible systems are containers or bags constructed from a polymeric material that changes shape during filling or emptying operations. The containers are made from a wide variety of materials such as Teflon, Tedlar, Mylar, aluminized-Mylar, polyethylene, rubber, and cellophane. The bags are made from sheets of material and sealed by heat or special tapes.

After the bags have been constructed and checked to see that there are no leaks, the bags are filled with a measured volume of the diluting gas and the test gas or vapor is introduced—usually by injecting a small volume of the test gas with a syringe. (An alternate way is to place a sealed ampoule inside the bag during its construction and break the ampoule at the appropriate time. The experimenter must be careful not to pierce the wall of the bag during this operation.) The bag is then massaged to produce thorough mixing. If the bag is to be taken into the field where there is a possibility of damage, it is recommended that it is transported inside a cardboard box (*158*).

The biggest problem with this system is the loss of standard gas during storage. The trapped gas can sorb onto walls, diffuse through the walls, react with the walls or other contaminants present, or decompose. In some cases exposure of the bags to sunlight can accelerate standard gas losses (*158*). For this reason it is imperative that the standard gas be matched with the wall material for optimum stability.

Baker and Doerr (*18*) studied many combinations of stored gases and container materials and concluded that there was no universal system. Moisture permeating the bag wall appeared to play an important role in the decomposition of some gases. They recommended a double bag system: an inner one that was impermeable to the stored gas, and an outer aluminized–Mylar that was impermeable to moisture.

Clemons and Altshuller (*19*) showed that aliphatic hydrocarbons could be stored for long periods in Mylar bags, while aromatic hydrocarbons were unstable under the same conditions. Conner and Nader (*159*) and Sawicki *et al.* (*20*) have reported similar findings. Stability of some halocarbons in bags and steel cylinders was also reported by Saltzman *et al.* (*160*).

Preconditioning of the bags is important for maintaining the integrity of the standard. The usual procedure is to fill the container several times with the desired or higher concentrations of the test gas (*21, 159*). After this has been done, the bags are ready for use. With sulfur dioxide, however, the preconditioning appears to deteriorate when the bag is not being used (*21, 159*).

Despite these precautions, preparations of standard gases by identical means in similarly treated containers can vary as much as 25% from each other (*158*).

b. Rigid Walls. The other kind of static system uses rigid containers usually constructed of glass, steel, or an inflexible plastic. Rigidity has the advantage of ruggedness and the disadvantage of difficulty in removing the sample from the container. Rigid containers have the same problems of surface sorption and reactions as flexible ones.

Perhaps the simplest example of a rigid static system is the syringe. A sample of test gas can be drawn into a syringe, partially filling it, and then a diluting gas can also be drawn in, further filling the syringe to the desired volume. The contents can be injected into the system being tested. In this procedure the dead volume of the needle must be considered along with the volume of the test gas drawn into the syringe.

The dead volume of the needle can be used to advantage by drawing a quantity of test gas into the syringe. This gas is quickly expelled leaving the test gas only in the needle. Next, diluting gas is drawn in, diluting the needle volume of test gas. Again, the syringe gas can be expelled, leaving the needle filled, and further dilutions can be made until the desired concentration is reached.

Another approach (*161–164*) is to place sealed ampoules containing known quantities of the test substance in the container. The container is then filled with the diluting gas. The ampoule is broken and the contents escape, mixing with the surrounding atmosphere.

Two other means of filling rigid containers are by weight (*165*) and by partial pressure (*28*).

Another very practical approach to the problem of maintaining known concentrations in a rigid system has been discussed at length by Setterlind (*164*). In this method a number of flasks of carboys of identical size, each filled with the test substance at known concentration, are connected in series by means of suitable glass or plastic tubing. The sample is withdrawn from the flask at the end of the line. A compensating volume of fresh air enters the first flask and mixes with the test mixture there; it is this mixture which moves to the second flask. In this way, the effect of the pure air added decreases with the number of flasks.

Setterlind showed that if mixing were perfect, for a system of five flasks almost $2\frac{1}{2}$ times the volume of any individual flask could be withdrawn from the fifth flask before its residual concentration of added material decreased to 90% of its initial value. By contrast, only 10% of the contents of a single flask can be withdrawn before this decrease takes place. Since these relationships are independent of flask size, it is evident that almost five times as much usable sample can be withdrawn from the five-flask system as from an equal volume of air in a single container. For the general case of a series of n flasks connected in this fashion, Setterlind showed that the residual concentration in the nth flask is given by the equation

$$C_n = C_0 \exp(-V/V_0)\left[1 + \frac{1}{1!}\left(\frac{V}{V_0}\right) + \frac{1}{2!}\left(\frac{V}{V_0}\right)^2 + \cdots + \frac{1}{(n-1)!}\left(\frac{V}{V_0}\right)^{n-1}\right]$$

where

C_0 = initial concentration
C_n = residual concentration in the nth flask
V_0 = the volume of the flask
V = volume withdrawn

The expression in brackets is a well-known series expansion for $\exp(V/V_0)$. Thus, for an infinite number of flasks, it yields $C_n = C_0$, as in fact it should. Setterlind also discussed at length methods of initial production of known concentrations in the flasks.

Urone *et al.* (*3*) developed an elegant adaptation of the classical laboratory vacuum system in which isotopically tagged sulfur dioxide could be purified by trap-to-trap distillation, measured in volume by a modified McLeod gauge, and then added to determined volumes of air. The system was used both in a study of analytical methods and for the determination of the nature of storage losses in Mylar bags.

Another approach of considerable usefulness is the volume expansion technique (*28, 166*). In this method, a small known volume is filled with pure test gas. This container is then attached to a larger known volume container filled with diluting gas, and the contents of the two are allowed to mix with each other. The diluted gas can now be used, or the small volume can again be diluted by attaching it to another larger volume containing only diluting gas and repeating the process. This kind of system is rather slow, but then all rigid static systems are slow because of the equilibration time needed.

2. Dynamic Systems

If the linear flow rate through the system is substantially constant throughout, the system is called dynamic. Whenever possible, dynamic systems are used for studies of the stability of analytical instruments and in other situations where it is desirable not to have an effective limit to the volume of gas mixture that can be supplied. The necessary parts for such a system are a constant supply of air and of the trace gas, at rates that are in the correct ratio, that are accurately known, and that can be kept constant over an extended period of time. It is not difficult to provide an accurately measured flow of the diluting air. Accurate measurement of the flow rate, one-millionth as great, of the trace gas, is far more difficult. Unless the added test gas is spontaneously unstable like ozone, a system meeting these criteria is primary.

a. DIFFUSION. One of the systems most commonly used for the production of test gases, especially when the gases have rather high boiling points, is open tube diffusion. In this procedure, first proposed by McKelvey and Hoelscher (*167*), a liquified quantity of the test gas or vapor is placed in a thermostated open tube and a diluting gas is passed over the tube mouth, drawing the vapor of the liquified gas into the diluting gas stream.

Altshuller and Cohen (*168*) studied this technique thoroughly and showed that the final concentration was dependent on variables such as the liquid vapor pressure, liquid temperature, tube diameter, and the diluting gas flow rate. They also designed an apparatus that yielded results in much better accord with the theory (Fig. 3). Other apparatus designs are described by Dravnieks and Krotoszynski (*169*) and Lovelock (*170*).

Saltzman and Gilbert (*171*) keep the test gas to be added under a very slight, constant positive pressure as it diffuses through a somewhat porous asbestos plug in a three-way stopcock with a T-bore. Nitrogen dioxide enters the horizontal arm of the T, and the bulk of it leaves through the vertical arm, which leads to a tube submerged in a few centimeters of water. The depth of submersion controls the back pressure of the system, and is maintained manually or automatically (*172*) to compensate for the output pressure. The remaining horizontal arm of the stopcock plug is gently stuffed with purified "Gooch" asbestos. The flow rate through the plug is determined, as a function of back pressure, by means of a soap-bubble flowmeter. Flows as low as 0.01 ml/minute can be produced. This flow apparently remains constant for an extended period of time. The very small flow that results can be blended with a large flow of purified air to produce a primary test mixture. This system has been used by Boström (*69*) with radioactive sulfur dioxide in a demonstration of the collection efficiency of various bubblers and trapping solutions for that gas; and by Saltzman (*54, 173*) for ammonia and ethylene oxide dilutions.

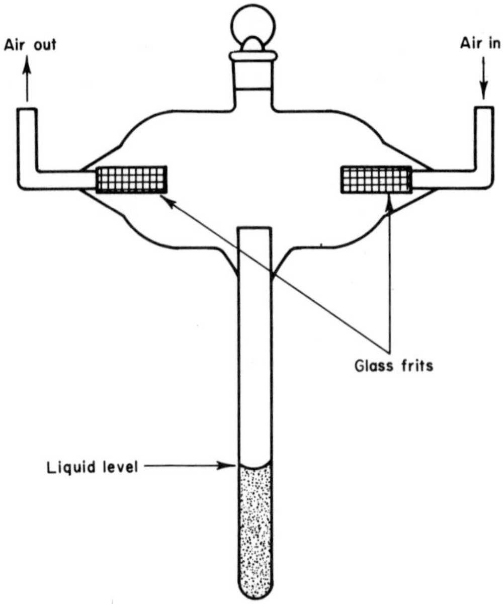

Figure 3. Diffusion cell (*168*). [A. P. Altshuller and I. R. Cohen, *Anal. Chem.* 32, 802 (1960). Copyright (1960) by the American Chemical Society. Reprinted by permission of the copyright owner.]

Mass spectrometer leaks and other narrow capillaries allow small quantities of gas to enter a diluting gas stream. Hales *et al.* (*174*) show that packed capillaries can provide a small gas flow for dilution. The flows can be calculated from theoretical equations. The system has the advantage of being usable for a wide variety of gases. Axelrod *et al.* (*175*) have used porous ceramic plugs attached directly to a gas cylinder. The pressure behind the plug is varied, altering the output flow of the contaminant gas into the diluting stream. By changing the cylinder pressure behind the plug and the diluting gas flow rate, a number of standard concentrations can be provided.

Plastic materials are permeable to gases under pressure. This is the basis for one of the most important methods of preparation of standard gas mixtures, the permeation tube (Fig. 4) (*175a*), first developed by O'Keeffe and Ortman (*176, 177*). Their procedure involved keeping a liquid substance inside a sealed plastic tube, usually Teflon. The vapor above the liquid diffuses through the tube wall at a rate dependent upon tube temperature, liquid substance, tube material, wall thickness, and exposed surface area. The emitted vapor is diluted into a gas stream which is passed over the tube. Gas concentration can be calculated from the diffusion rate and the diluting gas flow rate. The theoretical principles

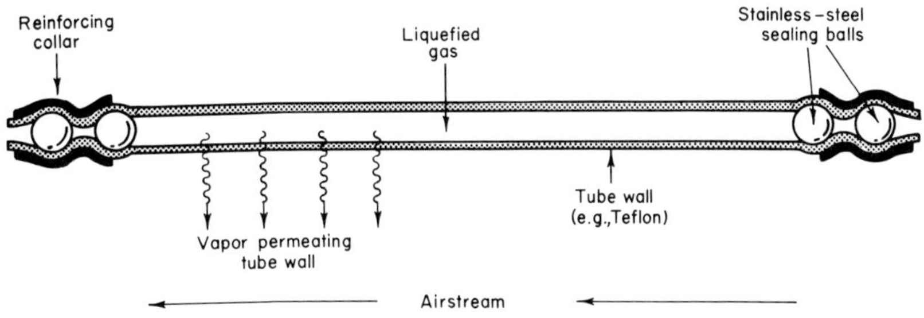

Figure 4. Permeation tube (*175a*). [Reprinted by permission of Willard-Grant Press, copyright, 1972.]

involved are given by Lucero (*178*); the procedures for the tube preparation are reported by Scaringelli *et al.* (*179*).

The tubes can be calibrated by measuring their weight loss as a function of time and tube temperature (*176*). To obtain a measurable weight loss using laboratory balances, repeated weighings over a several week period are necessary. The calibration time can be markedly shortened by the use of a microbalance (*180*). Calibration can also be accomplished by measuring the volume of gas evolved (*181*).

Since the diffusion rate is dependent on temperature, a tube must be calibrated and used under controlled temperature conditions. A $\pm 1°C$ variation in temperature can result in a $\pm 10\%$ diffusion rate change (*182*). Therefore, $\pm 0.1°C$ temperature control is recommended.

The permeation tube has been used to prepare sulfur dioxide standards as well as a wider range of other standard gas mixtures (*179, 183–185*), but the technique is not universal for all gases since some have critical pressures high enough to rupture the tube walls. The experimenter must be cautious about assuming that the tube material is inert towards the liquid substance. It has been reported that nitrogen dioxide causes blistering of the Teflon tube wall (*176*) and that moisture present in the diluting gases tends to accelerate this phenomena (*186*). [Lists of various tube materials and compatible gases are included in Nelson's (*119*) book.]

Moisture tends to affect the diffusion rate of some gases, so that calibration should be made under conditions of future use (*186*). When air is used as the diluting gas for hydrogen sulfide or mercaptan permeation tubes, a yellow-white film is formed inside the tube, probably due to oxygen permeating into the tube and oxidizing the sulfide to elemental sulfur (*184*). For this reason, the use of an inert diluting gas is recommended.

When the diffusion rate through a permeating tube is higher than desirable, O'Keeffe and Ortman (*187*) recommend the use of a glass bottle filled with the liquid substance and a Teflon plug forced into the mouth.

This device has to be calibrated by analysis of the emitted gas. Other container designs have been suggested by Saltzman *et al.* (*186*).

Permeation devices are lightweight and, if they can be maintained at a constant temperature, gas standards can be prepared in the field (*188*). Certified standard sulfur dioxide tubes are available (*189*).

Recently workers have used a Teflon wafer to control the diffusion rate of a test gas into a diluting stream (*189a*). This technique has the advantage of being useful for both condensible and noncondensible gases. In addition, the calibration time for this system as compared to that for a Teflon permeation tube was greatly reduced through the use of a very sensitive differential pressure gauge.

b. DISPLACEMENT. One of the most common direct displacement techniques is a mechanically driven glass syringe used to inject vapors into a diluting gas stream. Saltzman (*190*) uses a motor-driven syringe for micrometering of gases. Smith and Grant (*191*) uses a motorized microburet of the Gilmont type with suitable modifications to make it gas tight and to permit easy flushing and filling with gas. This system produces approximately 1 liter per minute of gas mixture having a concentration of the order of 100 ppm. It is difficult to drive a large syringe sufficiently slowly by this means to achieve much lower concentrations. Smaller syringes have the disadvantage that their ultimate capacity is limited.

Goetz and Kallai (*192*) have described a device in which a length of fine thermometer capillary is sealed to the tip of a syringe. The distal end of the capillary extends into the air flow of a dynamic system. The syringe is supported vertically with the plunger up. To the end of the plunger is attached a small "squirrel cage" impeller driven by a precisely horizontal air jet. At sufficient rotational speed, the plunger of the syringe is almost frictionless in the barrel, and so exerts a force equal to its own weight on the gas contained in the syringe. This gas escapes through the capillary in a manner analagous to Saltzman's asbestos plug technique. In fact, there seems to be no reason why an asbestos plug cannot be used with a spinning syringe to provide flow rates smaller than those obtainable with thermometer capillary, and with less danger of plugging. The flow rate can be conveniently monitored by timing the vertical movement of the plunger over the graduations on the syringe barrel. Additional types of syringe systems have been reported by Kuczynski (*193, 194*), Hill and Newell (*195*), Thomas and Amtower (*196*), and Nelson and Griggs (*197*).

A major disadvantage of syringe techniques is the dependence of total gas production on syringe volume. When the syringe is empty, the system must be stopped until the syringe can be refilled. This objection was overcome by Fultyn (*198*) when he used two syringes, one filling, while the other is emptying. Another problem is that the diluting air passes over the syringe needle and thereby drags out the test gas. Kuczynski (*193*) overcame this by placing glass wool around the needle.

Another technique involves electrolytically or chemically generating a gas which exerts pressure to force a column of liquid or gas through a capillary into a diluting gas stream. Still another is to quantitatively and slowly inject water onto a mercury column, causing the column to push against a gas or liquid, thus forcing it through a capillary into a gas stream. There are many variations on these ideas.

A rather obvious technique is the use of dynamic serial dilution on the microscale. In such a system, a small flow of the test gas is metered into a larger flow of air. After the gas and air are mixed, an aliquot of this mixture is metered into a still larger flow of air. In theory, the process may be repeated a number of times to produce very low concentrations, the excess mixture at each stage of dilution being discarded. A major problem with serial dilution systems is the supply of the large volume of "clean" diluting gas that is required since much of the diluting gas is wasted. Several systems have been constructed in which the division between the portion of the gas mixture passed on for further dilution and that which is discarded is controlled by needle valves or orifices. General experience with systems of this kind seems to indicate that they are prey to a variety of resonances, instabilities, and ambiguous corrections that make them extremely hard to control.

This basic technique has been refined by using mixing ratios dependent upon the metering capabilities of thermal mass flowmeters (*143*). This system is able to produce 10^5 dilutions with each stage with a reported reproducibility of 2%.

Another device for blending gases involves the use of a gas doser (*199*). In this technique, a known volume bore of a stopcock is flushed with the test gas. It rotates into a second position in which the bore is flushed with diluting gas, the bore dose entering a diluting gas stream. If the stopcock volume bore and the diluting gas flow rate are known along with the number of times per unit time the bore is filled with test gas and flushed into the diluting stream one can calculate the diluted test gas concentration. Since the system provides test gas pulses, hold-up flasks are needed to smooth out the test gas concentration. Two units can be used in series, each stage providing a 10^5 dilution. For many gases, a wide range of concentrations can be produced; however, a larger amount of plumbing is required and equilibration times for steady output concentrations can be up to several hours.

c. Exponential Dilution. The exponential dilution flask is an alternate means of preparing gas standards. As originally conceived by Lovelock (*170*), this system is simple to operate and very low concentrations can be attained. The flask (Fig. 5) (*200*) is cylindrically shaped with a paddle wheel inside rotated by an external motor. A known flow rate of diluting gas is passed through the flask and at a precise time an injection is made of known volume of test gas. The diluting gas continues to flow, and if there is thorough mixing the test gas concentration should decay exponentially with time. The test gas can be introduced into the flask by syringe injection (*170*) or by allowing the diluting gas to pass over a test gas permeation tube (*201*). The latter technique is reported to be more reproducible but requires a period of time for the flask to reach a steady-state concentration prior to the start of the dilution procedure.

The technique has the disadvantage of an output concentration continually decreasing so that the time must be carefully noted to sample the output at a desired concentration. Of the different designs reported, one was made of miscellaneous materials (*200*) while another was constructed totally from Teflon (*202*). The choice of material is dependent on adsorption effects. It has been reported that the initial adsorption and the resulting desorption can cause nonexponential decay, especially at very low

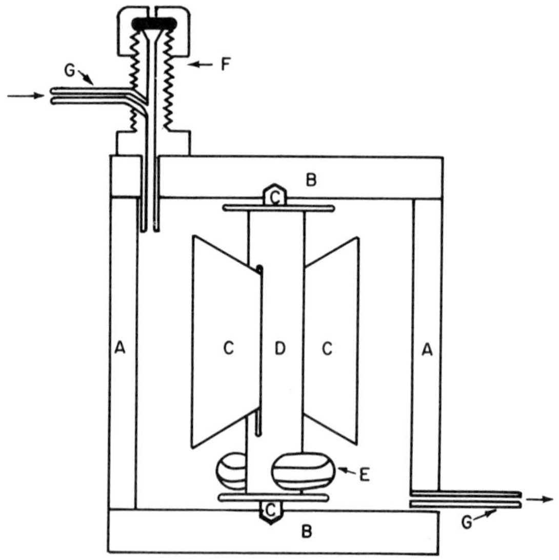

Figure 5. Exponential dilution flask (*200*). Key to symbols: A, phenolic cylinder; B, Lucite end pieces; C, Teflon; D, plastic roll film spools; E, magnetic stirring bar; F, injection port; G, copper tubing.

concentrations. Fowlis and Scott (*203*) minimized adsorption effects by using an inert solvent in the flask.

REFERENCES

1. J. P. Lodge, Jr., *Instr. Tech.*, 28–29, Dec. (1971).
2. R. O. McCaldin and E. R. Hendrickson, *Amer. Ind. Hyg. Ass., J.* **20,** 509 (1959).
3. P. Urone, J. B. Evans, and C. M. Noyes, *Anal. Chem.* **37,** 1104 (1965).
4. B. E. Saltzman, *Anal. Chem.* **26,** 1949 (1954).
5. H. C. Wohlers, H. Newstein, and D. Daunis, *J. Air Pollut. Contr. Ass.* **17,** 753 (1967).
6. H. C. Wohlers, N. M. Trieff, H. Newstein, and W. Stevens, *Atmos. Environ.* **1,** 121 (1967).
7. W. H. Perry and E. C. Tabor, *Arch. Environ. Health* **4,** 254 (1962).
8. A. F. Welch and J. P. Terry, *Amer. Ind. Hyg. Ass., J.* **21,** 316 (1960).
9. A. P. Altshuller and A. F. Wartburg, *Int. J. Air Pollut.* **4,** 70 (1961).
10. H. C. Wohlers, H. Newstein, and W. E. Jackson, *Atmos. Environ.* **4,** 435 (1970).
11. W. D. Komhyr, *Rev. Sci. Instrum.* **38,** 981 (1967).
12. W. D. Komhyr, *Ann. Geophys.* **25,** 203 (1969).
13. R. L. Byers and J. W. Davis, *J. Air Pollut. Contr. Ass.* **23,** 236 (1970).
14. H. D. Axelrod, R. J. Teck, and J. P. Lodge, Jr., *J. Air Pollut. Contr. Ass.* **21,** 218 (1971).
15. M. C. Meador and R. M. Bethea, *Environ. Sci. Technol.* **4,** 853 (1970).
16. P. R. Averell, W. F. Hart, N. T. Woodberry, and W. R. Bradley, *Anal. Chem.* **19,** 1040 (1947).
17. F. J. Schuette, *Atmos. Environ.* **1,** 515 (1967).
18. R. A. Baker and D. C. Doerr, *Int. J. Air Water Pollut.* **2,** 142 (1959).
19. C. A. Clemons and A. P. Altshuller, *J. Air Pollut. Contr. Ass.* **14,** 407 (1964).
20. E. Sawicki, R. C. Corey, A. E. Dooley, J. B. Gisclard, J. L. Monkman, R. E. Neligan, and L. A. Ripperton, *Health Lab. Sci.* **7,** Suppl. 1, 23 (1970).
21. A. P. Altshuller, A. F. Wartburg, I. R. Cohen, and S. F. Sleva, *Int. J. Air Water Pollut.* **6,** 75 (1962).
22. A. P. Altshuller, *In* "Advances in Chromatography" (J. C. Giddings and R. A. Keller, eds.), pp. 229–262. Dekker, New York, New York, 1968.
23. R. J. Gordon, H. Mayrsohn, and R. M. Ingels, *Environ. Sci. Technol.* **2,** 1117 (1968).
24. T. A. Bellar, J. E. Sigsby, C. A. Clemons, and A. P. Altshuller, *Anal. Chem.* **34,** 763 (1962).
25. A. P. Altshuller, T. A. Bellar, C. A. Clemons, and E. VanderZander, *Int. J. Air Water Pollut.* **8,** 29 (1964).
26. R. S. Ajemian and N. E. Whitman, *J. Air Pollut. Contr. Ass.* **20,** 310 (1970).
27. R. Jeltes and E. Burghardt, *Atmos. Environ.* **6,** 793 (1972).
28. L. E. Heidt and D. H. Ehhalt, *J. Chromatogr.* **69,** 103 (1972).
29. A. Turk, J. I. Morrow, and B. E. Kaplan, *Anal. Chem.* **34,** 561 (1962).
30. P. W. West, B. Sen, and N. A. Gibson, *Anal. Chem.* **30,** 1390 (1958).
31. F. R. Cropper and S. Kaminsky, *Anal. Chem.* **35,** 735 (1963).
32. H. Buchwald, *Occup. Health Rev.* **17,** 14 (1965).

33. B. Fallentin, *Dan. Kemi* **46,** 133 (1965); *Chem. Abstr.* **64,** 2654h (1966).
34. A. Dravnieks, A. O'Donnell, and T. Burgwald, *Environ. Sci. Technol.* **5,** 1220 (1971).
35. C. L. Fraust and E. R. Herman, *Amer. Ind. Hyg. Ass., J.* **27,** 68 (1966).
36. N. E. Whitman and A. E. Johnston, *Amer. Ind. Hyg. Ass., J.* **25,** 464 (1964).
37. S. B. Smith and R. J. Grant, Report. Res. Develop. Dept. Pittsburgh Coke and Chem. Co., Pittsburgh, Pennsylvania, 1958.
38. N. Brennen, E. Cieplinski, L. S. Ettre, and V. J. Coates, *J. Chromatogr.* 3, 230 (1960).
39. D. L. Brooman and E. Edgerly, Jr., *J. Air Pollut. Contr. Ass.* **16,** 25 (1966).
40. G. Pearson, J. W. R. Dutton, and F. Elliot, *U.K. At. Energy Auth., Prod. Group,* **PG Rep 408,** (1962); *Chem. Abstr.* **58,** 7583a (1963).
41. J. J. Naughton, E. F. Heald, and I. L. Barnes, Jr., *J. Geophys. Res.* **68,** 539 (1963).
42. R. Bock and K. Schütz, *J. Anal. Chem.* **237,** 32 (1968).
43. M. D. LaHue, H. D. Axelrod, and J. P. Lodge, Jr., *Anal. Chem.* **43,** 1113 (1971).
44. J. Hahn, *Anal. Chem.* **44,** 1889 (1972).
45. T. Ohira, M. Takeda, T. Ishiguro, and O. Koyama, *J. Jap. Soc. Air Pollut.* **2,** 48 (1967).
46. L. Dubois and J. L. Monkman, *Mikrochim. Acta* p. 313 (1970).
47. L. Dubois and J. L. Monkman, *Anal. Chem.* **44,** 74 (1972).
48. G. J. Ferguson, *Rev. Sci. Instrum.* **34,** 403 (1963).
49. E. Briner, *Helv. Chim. Acta* **21,** 1218 (1938).
50. I. Hornstein and W. N. Sullivan, *Anal. Chem.* **25,** 496 (1953).
51. A. Turk, J. Morrow, P. F. Levy, and P. Weissman, *Int. J. Air Water Pollut.* **5,** 14 (1961).
52. H. D. Axelrod, A. F. Wartburg, R. J. Teck, and J. P. Ldoge, Jr., *Anal. Chem.* **43,** 1916 (1971).
53. J. R. Branham and E. D. Sperling, *J. Res. Nat. Bur. Stand.* **22,** 701 (1939).
54. B. E. Saltzman, *Anal. Chem.* **33,** 1100 (1961).
55. S. Calvert and W. Workman, *Amer. Ind. Hyg. Ass., J.* **22,** 318 (1961).
56. National Institute of Health, U.S. Public Health Service, Eng. Univ. Div., *Ind. Eng. Chem., Anal. Ed.* **16,** 346 (1944).
57. R. Kieselbach, *Ind. Eng. Chem., Anal. Ed.* **16,** 538 (1944).
58. L. Silverman, Ind. Hyg. Found., *Chem. Toxicol. Serv. Bull.* No. 1, pp. 1–72 (1947).
59. W. N. Carson, Jr., *Anal. Chem.* **21,** 538 (1949).
60. A. Ehmert, *Meteorol. Rundsch.* **4,** 64 (1951).
61. A. A. W. Russell, *Analyst* **79,** 524 (1954).
62. L. Silverman, *Air Cond., Heat., Vent.* **52,** 88 (1955).
63. H. A. Kamphausen, *Chem. Ind.* (*London*) p. 816 (1962).
64. F. A. Bell, Jr., *J. Air Pollut. Contr. Ass.* **13,** 127 (1963).
65. A. L. Linch and M. Corn, *Amer. Ind. Hyg. Ass. J.* **26,** 601 (1965).
66. A. F. Wartburg, J. B. Pate, and J. P. Lodge, Jr., *Environ. Sci. Technol.* 3, 767 (1969).
67. J. C. Gage, *Analyst* **85,** 196 (1960).
68. L. R. Roberts and H. C. McKee, *J. Air Pollut. Contr. Ass.* **9,** 51 (1959).
69. C. E. Boström, *Int. J. Air Water Pollut.* **9,** 333 (1965).
70. G. Bergshoeff, *Chem. Weekbl.* **57,** 637 (1961).
71. G. Bergshoeff, *Int. J. Air Water Pollut.* **10,** 629 (1966).

72. I. Skare, *Int. J. Air Water Pollut.* **9**, 129 (1965).
73. G. Kats and C. R. Thompson, *J. Air Pollut. Contr. Ass.* **19**, 265 (1969).
74. T. Okita, *Bull. Inst. Pub. Health, Tokyo* **16**, 59 (1967).
75. H. R. Hickey and E. R. Hendrickson, *J. Air Pollut. Contr. Ass.* **15**, 409 (1965).
76. E. N. Lawrence, *Int. J. Air Water Pollut.* **8**, 381 (1964).
77. S. Omichi, *Bunseki Kagaku* **13**, 339 (1964); *Chem. Abstr.* **61**, 1373 (1964).
78. M. Boulerice and W. Brabant, *J. Air Pollut. Contr. Ass.* **19**, 432 (1969).
79. N. A. Huey, *J. Air Pollut. Contr. Ass.* **18**, 610 (1968).
80. R. Cole and H. L. Shulman, *Ind. Eng. Chem.* **52**, 589 (1960).
81. D. M. Paez and O. A. Guagnini, *Mikrochim. Acta* p. 220 (1971).
82. J. P. Paré, *J. Air Pollut. Contr. Ass.* **16**, 325 (1966).
83. H. P. Sanderson, R. Thomas, and M. Katz, *J. Air Pollut. Contr. Ass.* **16**, 328 (1966).
84. S. Hochheiser and L. A. Elfers, *Environ. Sci. Technol.* **4**, 672 (1970).
85. J. P. Lodge, Jr., J. B. Pate, and H. A. Huitt, *Amer. Ind. Hyg. Ass., J.* **24**, 380 (1963).
86. C. Huygen, *Anal. Chim. Acta* **28**, 349 (1963).
86a. A. M. Chamberland, P. Bourbon, and R. Malbose, *Int. J. Environ. Anal. Chem.* **2**, 303 (1973).
87. D. A. Falgout and C. I. Harding, *J. Air Pollut. Contr. Ass.* **18**, 15 (1968).
88. A. F. Smith, D. G. Jenkins, and D. E. Cunningworth, *J. Appl. Chem.* **11**, 317 (1961).
89. D. F. S. Natusch, H. B. Klonis, H. D. Axelrod, R. J. Teck, and J. P. Lodge, Jr., *Anal. Chem.* **44**, 2067 (1972).
90. T. Okita, J. P. Lodge, Jr., and H. D. Axelrod, *Environ. Sci. Technol.* **5**, 532 (1971).
91. C. Huygen, *Anal. Chim. Acta* **30**, 556 (1964).
92. T. Okita and S. Kanamori, *Atmos. Environ.* **5**, 621 (1971).
93. S. Kanamori, *J. Jap. Soc. Air Pollut.* **5**, 84 (1970).
94. S. Kanamori and T. Okita, *J. Jap. Soc. Air Pollut.* **6**, 277 (1971).
94a. J. Alary, P. Bourbon, P. Chovin, C. Delaunay, J. Esclasson, and J.-C. Lepert, *C. R. Acad. Sci.* **278**, 651 (1974).
95. T. Okita, R. Sugai, and I. Kifume, *J. Jap. Soc. Air Pollut.* **4**, 118 (1969).
96. F. T. Eggersten and F. M. Nelsen, *Anal. Chem.* **30**, 1040 (1958).
97. I. H. Williams, *Anal. Chem.* **37**, 1723 (1965).
98. A. P. Altshuller, *J. Gas Chromatogr.* **1**, No. 7, 6 (1963).
99. S. Araki and T. Kato, *Bunseki Kagaku* **12**, 1027 (1963); *Chem. Abstr.* **58**, 10702 (1963).
100. W. A. Aue and P. M. Teli, *J. Chromatogr.* **62**, 15 (1971).
101. H. Shoji, T. Yamamoto, and K. Nishida, *Trans. Jap. Soc. Civ. Eng.* **91**, 25 (1963); *Air Pollut. Contr. Ass. Abstr.* No. 5316 (1963).
102. J. Novák, V. Vasák, and J. Janák, *Anal. Chem.* **37**, 660 (1965).
103. A. Linenberg, E. Yanis, D. Shamir, and D. Hagel, *Pittsburgh Conf. Anal. Chem. Appl. Spectrosc., Cleveland, Ohio, 1973.*
104. D. A. Levaggi, W. Siu, and M. Feldstein, *J. Air Pollut. Contr. Ass.* **23**, 30 (1973).
105. K. D. Reiszner and P. W. West, *Environ. Sci. Technol.* **7**, 526 (1973).
106. M. Shepherd, S. M. Rock, R. Howard, and J. Stormes, *Anal. Chem.* **23**, 1431 (1951).
107. P. S. Farrington, R. L. Pecsok, R. L. Meeker, and T. J. Olson, *Anal. Chem.* **31**, 1512 (1959).

108. M. Feldstein and S. Balestrieri, *J. Air Pollut. Contr. Ass.* **15,** 177 (1965).
109. R. A. Rasmussen, *Amer. Lab.* **4,** 19 (1972).
110. T. A. Bellar, M. F. Brown, and J. E. Sigsby, Jr., *Anal. Chem.* **35,** 924 (1963).
111. W. A. Lonnemann, T. A. Bellar, and A. P. Altshuller, *Environ. Sci. Technol.* **2,** 1017 (1968).
112. E. R. Stephens and F. R. Burleson, *J. Air Pollut. Contr. Ass.* **17,** 147 (1967).
113. E. R. Stephens and F. R. Burleson, *J. Air Pollut. Contr. Ass.* **19,** 929 (1969).
114. E. R. Hendrickson, *in* "Air Pollution" (A. C. Stern, ed.), 2nd ed., Vol. II, pp. 3–53. Academic Press, New York, New York, 1968.
115. E. R. Weaver, E. E. Hughes, S. M. Gunther, S. Schuhmann, N. T. Redfearn, and R. Gorden, Jr., *J. Res. Nat. Bur. Stand.* **59,** 383 (1957).
116. G. C. B. Cave, *Tappi* **46,** 1 (1963).
117. R. D. Taylor, Report UCRL-51239, TID-4500, UC-41. Lawrence Radiation Lab., University of California, Livermore, California, 1972.
118. American Society for Testing and Materials Method D 1071-55, Part 19. ASTM, Philadelphia, Pennsylvania, 1971.
119. G. O. Nelson, *in* "Controlled Test Atmospheres," Ann Arbor Sci. Publ., Ann Arbor, Michigan, 1971.
120. A. T. James and A. J. P. Martin, *Biochem. J.* **50,** 679 (1952).
121. "Product Bulletins." Brooks Instruments, Hatfield, Pennsylvania, 1973.
122. J. M. Mansfield and J. D. Winefordner, *Anal. Chim. Acta* **40,** 357 (1968).
123. C. Veillon and J. Y. Park, *Anal. Chem.* **42,** 684 (1970).
124. S. Lowell and S. Karp, *Anal. Chem.* **43,** 492 (1971).
125. H. M. Lemon and H. Wise, *Science* **99,** 43 (1944).
126. A. B. Waugh and P. W. Wilson, *Anal. Chem.* **44,** 2118 (1972).
127. J. M. Benson, W. C. Baker, and E. Easten, *Instrum. Contr. Syst.* **43,** 85 (1970).
128. G. S. P. Castle and M. R. Sewell, *Proc. IEEE 6th Annual Meeting,* p. 177 (1971).
129. D. M. Allen, U.S. Patent 3,501,899 (1970).
130. L. F. Kachor, *Gig. Sanit.* 1950, 48 (1950).
131. M. A. Pinigin, *Predel'no Dopustimy Kontsent. Atmos. Zagryaz.* p. 143 (1960).
132. L. Silverman and W. B. Wardlow, *Ind. Eng. Chem., Anal. Ed.* **12,** 682 (1940).
133. W. Stanilewicz, *Przem. Chem.* **44,** 641 (1965); *Chem. Abstr.* **64,** 12208f (1966).
134. G. Hermann, *Chem. Tech. (Leipzig)* **17,** 97 (1965).
135. M. B. Jacobs, "The Analytical Toxicology of Industrial Inorganic Poisons." Wiley (Interscience), New York, New York, 1967.
136. J. P. Lodge, Jr., J. B. Pate, B. E. Ammons, and G. A. Swanson, *J. Air Pollut. Contr. Ass.* **16,** 197 (1966).
137. M. Corn and W. Bell, *Amer. Ind. Hyg. Ass., J.* **24,** 502 (1963).
138. D. L. Brenchley. *J. Air Pollut. Contr. Ass.* **22,** 967 (1972).
139. C. Huygen, *J. Air Pollut. Contr. Ass.* **20,** 675 (1970).
140. A. Setterlind, *Amer. J. Pub. Health* **34,** 863 (1944).
141. H. Scholz and O. J. Stebel, *Gas-Wasserfach* **91,** 127 (1950); *Chem. Abstr.* **44,** 7593d (1950).
142. A. F. Wartburg, H. D. Axelrod, R. J. Teck, M. D. LaHue, and J. P. Lodge, Jr., *Anal. Chem.* **45,** 423 (1973).
143. "Product Bulletins." Tylan Corp., Torrance, California, 1973.
144. R. I. Larsen, C. E. Zimmer, D. A. Lynn, and K. G. Blemel, *J. Air Pollut. Contr. Ass.* **17,** 85 (1967).
145. R. I. Larsen, W. W. Stalker, and C. R. Claydon, *J. Air Pollut. Contr. Ass.* **11,** 529 (1961).

146. D. M. Keagy, W. W. Stalker, C. E. Zimmer, and R. C. Dickerson, *J. Air Pollut. Contr. Ass.* **11**, 270 (1961).
147. W. W. Stalker and R. C. Dickerson, *J. Air Pollut. Contr. Ass.* **12**, 111 (1962).
148. W. W. Stalker and R. C. Dickerson, *J. Air Pollut. Contr. Ass.* **12**, 170 (1962).
149. W. W. Stalker, R. C. Dickerson, and G. D. Kramer, *J. Air Pollut. Contr. Ass.* **12**, 361 (1962).
150. W. W. Stalker, P. A. Kenline, and H. J. Paulus, *J. Air Pollut. Contr. Ass.* **14**, 469 (1964).
151. P. A. Hersch, *J. Air Pollut. Contr. Ass.* **19**, 164 (1969).
152. M. O. Amdur, *Int. J. Air Pollut.* **1**, 172 (1959).
153. H. Sleik and A. Turk, *in* "Air Conservation Engineering," 2nd ed., p. 9. Connor Eng. Corp., Danbury, Connecticut, 1953.
154. W. H. King, Jr., *Anal. Chem.* **43**, 984 (1971).
155. C. W. Skarstrom, U.S. Patent 2,944,627 (1960).
155a. F. W. Williams and H. G. Eaton, *Anal. Chem.* **46**, 933 (1974).
156. A. Turk, *Pub. Health Rep.* **75**, 1187 (1960).
157. H. O. Smith and A. D. Hochstettler, *Environ. Sci. Technol.* **3**, 169 (1969).
158. G. A. Jutze and R. J. Lewis, *J. Air Pollut. Contr. Ass.* **15**, 323 (1965).
159. W. D. Conner and J. S. Nader, *Amer. Ind. Hyg. Ass., J.* **25**, 291 (1964).
160. B. E. Saltzman, A. I. Coleman, and C. A. Clemons, *Anal. Chem.* **38**, 753 (1966).
161. S. W. Comer and A. V. Jensen, *Anal. Chem.* **36**, 799 (1964).
162. M. Fugas, R. Pauković, and Z. Topolnik, *Arch. Hig. rada* **10**, 155 (1959); *Chem. Abstr.* **54**, 14847e (1960).
163. R. C. Paule, *Anal. Chem.* **44**, 1537 (1972).
164. A. Setterlind, *Amer. Ind. Hyg. Ass., Quart.* **14**, 113 (1953).
165. J. E. Miller, A. J. Carroll, and D. E. Emerson, *U.S., Bur. Mines, Rep. Invest.* RI-6674 (1965).
166. W. D. Komhyr, "Technique for Preparing CO_2, CO and Other Accurately Calibrated Reference Gases." Report. Air Resources Lab., National Oceanic and Atmospheric Administration, Boulder, Colorado, 1972.
167. J. W. McKelvey and H. E. Hoelscher, *Anal. Chem.* **29**, 123 (1959).
168. A. P. Altshuller and I. R. Cohen, *Anal. Chem.* **32**, 802 (1960).
169. A. Dravnieks and B. K. Krotoszynski, *J. Gas Chromatogr.* **6**, 144 (1968).
170. J. E. Lovelock, *Anal. Chem.* **33**, 162 (1961).
171. B. E. Saltzman and N. Gilbert, *Amer. Ind. Hyg. Ass., J.* **20**, 379 (1959).
172. C. B. Avera, Jr., *Rev. Sci. Instrum.* **32**, 985 (1961).
173. B. E. Saltzman and A. F. Wartburg, *Anal. Chem.* **37**, 1261 (1965).
174. J. M. Hales, D. A. Snyder, and J. L. York, *Amer. Ind. Hyg. Ass., J.* **32**, 299 (1971).
175. H. D. Axelrod, R. J. Teck, J. P. Lodge, Jr., and R. H. Allen, *Anal. Chem.* **43**, 496 (1971).
175a. S. E. Manahan, *in* "Environmental Chemistry," p. 379. Willard-Grant Press, Boston, Massachusetts, 1972.
176. A. E. O'Keeffe and G. C. Ortman, *Anal. Chem.* **38**, 760 (1966).
177. A. E. O'Keeffe and G. C. Ortman, *Anal. Chem.* **41**, 1598 (1969).
178. D. P. Lucero, *Anal. Chem.* **43**, 1744 (1971).
179. F. P. Scaringelli, A. E. O'Keeffe, E. Rosenberg, and J. P. Bell, *Anal. Chem.* **42**, 871 (1970).
180. L. J. Purdue and R. J. Thompson, *Anal. Chem.* **44**, 1034 (1972).
181. B. E. Saltzman, C. R. Feldman, and A. E. O'Keeffe, *Environ. Sci. Technol* **3**, 1275 (1969).

182. B. E. Saltzman, *J. Air Pollut. Contr. Ass.* **18**, 326 (1968).
183. A. L. Linch, R. F. Stalzer, and D. F. Lefferts, *Amer. Ind. Hyg. Ass., J.* **29**, 79 (1968).
184. R. K. Stevens, A. E. O'Keeffe, and G. C. Ortman, *Environ. Sci. Technol.* **3**, 652 (1969).
185. F. P. Scaringelli, S. A. Frey, and B. E. Saltzman, *Amer. Ind. Hyg. Ass., J.* **28**, 260 (1967).
186. B. E. Saltzman, W. R. Burg, and G. Ramaswamy, *Environ. Sci. Technol.* **5**, 1121 (1971).
187. A. E. O'Keeffe and G. C. Ortman, *Anal. Chem.* **39**, 1047 (1967).
188. J. H. Blacker and R. S. Brief, *Amer. Ind. Hyg. Ass., J.* **32**, 668 (1971).
189. National Bureau of Standards, Washington, D.C.
189a. R. N. Dietz, E. A. Cote, and J. D. Smith, *Anal. Chem.* **46**, 315 (1974).
190. B. E. Saltzman, *Amer. Ind. Hyg. Ass., Quart.* **16**, 122 (1955).
191. S. B. Smith and R. J. Grant, *Robert A. Taft Sanit. Eng. Cent., Tech. Rep* **A 59-3** U.S. Public Health Service, Cincinnati, Ohio. (1959).
192. A. Goetz and T. Kallai, *J. Air Pollut. Contr. Ass.* **12**, 437 (1962).
193. E. R. Kuczynski, *J. Air Pollut. Contr. Ass.* **13**, 435 (1963).
194. E. R. Kuczynski, *Environ. Sci. Technol.* **1**, 68 (1967).
195. D. W. Hill and H. A. Newell, *J. Sci. Instrum.* **42**, 783 (1965).
196. M. D. Thomas and R. E. Amtower, *J. Air Pollut. Contr. Ass.* **16**, 618 (1966).
197. G. O. Nelson and K. S. Griggs, *Rev. Sci. Instrum.* **39**, 927 (1968).
198. R. V. Fultyn, *Amer. Ind. Hyg. Ass., J.* **22**, 49 (1961).
199. H. D. Axelrod, J. B. Pate, W. R. Barchet, and J. P. Lodge, Jr., *Atmos. Environ.* **4**, 209 (1970).
200. H. P. Williams and J. D. Winefordner, *J. Gas Chromatogr.* **4**, 217 (1966).
201. F. Bruner, C. Canulli, and M. Possanzini, *Anal. Chem.* **45**, 1790 (1973).
202. F. Bruner, A. Liberti, M. Possanzini, and I. Allegrini, *Anal. Chem.* **44**, 2070 (1972).
203. I. A. Fowlis and R. P. W. Scott, *J. Chromatogr.* **11**, 1 (1963).

5

Hydrocarbons and Carbon Oxides

Aubrey Paul Altshuller

I. Introduction

The analysis of organic gaseous pollutants has been the subject of several reviews (*1, 2*). A review of analysis of hydrocarbons, oxygenated organics, organic nitrogen, sulfur, and halogen compounds by gas chromatography is available, covering the analytical literature to 1966 (*3*). A broader, but less comprehensive, review of analytical applications to air pollution to 1971 also is available (*4*).

Many of the analytical techniques of earlier years have not been utilized in recent years. In particular, colorimetric, electrometric, infrared, and mass spectrometric methods now receive little application to atmospheric hydrocarbons. Flame ionization analyzers, along with subtractive and gas chromatographic analysis using flame ionization detectors have been used almost exclusively for hydrocarbons. Gas chromatography has found application to analysis of other volatile organics, but colorimetric methods continue to be important for aldehyde analysis. Carbon monoxide and carbon dioxide are analyzed by several optical spectroscopic techniques and gas chromatographic and mass spectrometric procedures.

II. Carbon Oxides

A. Carbon Monoxide

Two types of needs have emphasized the importance of improving measurement capability for carbon monoxide. A primary air quality standard based on health considerations has been established in the United States setting maximum 1-hour concentrations at 35 ppm, 40 mg/m^3, and 8-hour concentrations at 9 ppm, 10 mg/m^3, not to be exceeded more than once a year (*5, 6*). Because of the short time intervals involved instrumental techniques are required. The first need, therefore, is to relate urban carbon monoxide concentrations to vehicular exhaust emissions. The second need is to better define global or background levels of carbon monoxide. Atmospheric measurements are considered important to determine whether or not the input from pollution sources can increase global or background concentration levels of carbon monoxide. New measurement techniques have been necessary to determine the distribution of carbon monoxide and its sources and sinks on a global scale.

The nondispersive infrared (NDIR) analyzer is the instrument ordinarily selected for monitoring carbon monoxide concentrations in urban situations. It is the reference method specified for the United States air quality standard (*5*). Gas chromatographic carbon monoxide analyzers provide an alternate method for monitoring (*7*). The gas chromatographic technique offers greater sensitivity, a wide linear range, and a high degree of specificity.

For measurement of carbon monoxide in nonurban or remote sampling locations the nondispersive infrared analyzer has inadequate sensitivity.

Concentrations of carbon monoxide in such areas usually range from 0.1 to 1 ppm. The gas chromatographic technique or the ultraviolet mercury replacement method has been applied successfully to measurement of low concentrations of carbon monoxide in the ambient atmosphere (*7*, *8*).

1. Nondispersive Infrared Spectrometry

Almost all air quality data for carbon monoxide were obtained by NDIR analyzers (*6*). Because the NDIR method has been selected as the United States reference method for measurement of carbon monoxide, it will continue to be of importance, and the performance characteristics of NDIR analyzers will be crucial in assessing improvements in air quality.

The principle involved is selective absorption by carbon monoxide of infrared radiation. The infrared radiation from the emitting source is split into parallel beams and passed through a sample cell, containing the atmospheric sample, and through a reference cell. Subsequently, the net radiation in the two beams are passed into matched selective carbon monoxide detectors. When carbon monoxide is present in the sample cell, energy is absorbed, and the temperature and pressure in the corresponding detector cell is reduced relative to that in the detector cell on the reference side of the analyzer. A diaphram between the two cells is displaced, and the amount of displacement is detected, electronically amplified, and an output signal produced. This particular method is only one of the infrared techniques which can be utilized, but it is the approach in common use.

The NDIR method has been subjected to a collaborative study of its characteristics (*9*). Test gas mixtures containing 8, 30, and 53 mg/m^3 of carbon monoxide in air were prepared and distributed to collaborative laboratories. Because of the water vapor response of NDIR analyzers, testing also was done for water vapor before introducing the mixture into the inlet system of the analyzers. Replication, repeatability, reproducibility, sensitivity, accuracy, and water vapor interference were investigated. The reproducibility between laboratories on the same mixtures was variable over the measured concentration range as follows: 2.5 mg/m^3 at 10 mg/m^3 level, 2.3 mg/m^3 at 20 mg/m^3, 2.6 mg/m^3 at 30 mg/m^3, 3.0 mg/m^3 at 40 mg/m^3, 3.7 mg/m^3 at 50 mg/m^3, 4.3 mg/m^3 at 60 mg m^3. The minimal detectable sensitivity was 0.3 mg/m^3 (0.27 ppm). Water vapor interference is adequately compensated for by using a drying agent or refrigeration. However, use of narrow-band optical filters may not be adequate to compensate for water vapor interference. Accuracy is completely dependent on calibration standards. Even with considerable effort in preparation and analysis the results were estimated to be 2.5% high. This type of carefully conducted collaborative test probably represents the best results obtainable at the time of the test. Routine operation of such analyzers is not necessarily capable of yielding such performance. For example, inadequate compensation for water vapor interference can and does occur. Under such circumstances monitoring results would be substantially less acceptable than those of these tests. A general discussion of the characteristics of NDIR analyzers, their sampling requirements, and some examples of monitoring applications is available (*10*).

Carbon monoxide measurement by NDIR analysis has been reported at sites representing a variety of traffic conditions in Detroit, Michigan; New York, New York; and Los Angeles, California (*11*). The diurnal variations in carbon monoxide concentration at each site, including weekday–weekend and seasonal effects is presented and correlated to traffic densities. Some 500 samples were collected in Toronto, Ontario by use of 100-ml syringes and returned to the laboratory for analysis in a NDIR unit with a short path cell (*12*). Samples were collected indoors and in passenger cars and subway cars, as well as on sidewalks. Empirical equations were developed relating pedestrian and driver exposure to local traffic density. The method for calibration of the NDIR analyzer used appears marginal for the lower range of concentrations reported.

2. Gas Chromatography

Gas chromatographic analysis has been used since the early 1960's to measure carbon monoxide (in the 0.1 to 10 ppm range) produced by simulated solar irradiation of ethylene–nitrogen oxide, propionaldehyde, and propylene–nitrogen oxide in air (*13–15*). In 1967 and 1968 carbon monoxide measurements were made on ambient air samples collected in Tedlar film bags in Los Angeles, California (*16*). A sequential sampler for collecting 12 air samples per 24-hour period has been combined with a gas chromatographic procedure capable of analyzing and computing the concentration of the 12 samples in $2\frac{1}{2}$ hours by this procedure (*17, 18*). Carbon monoxide is separated from methane on a molecular sieve column. After separation the carbon monoxide is quantitatively converted to methane followed by analysis as methane with a flame ionization detector. This procedure offers great potential for atmospheric analysis of carbon monoxide from 0.05 ppm to the highest urban concentration levels. The response to carbon monoxide is linear over the entire concentration range and the method is specific to carbon monoxide. In addition to its use in urban measurement the method can be used for measuring carbon monoxide at suburban, rural, and global background levels (*19, 20*).

In one automated system for measuring carbon monoxide in polluted air (*21*), the gas chromatographic column consists of a 12 ft length of $\frac{1}{4}$-in. o.d. stainless steel packed with 5A molecular sieve, 60/80 mesh, maintained at 115°C. The helium gas flow is 200 ml/minute, and the hydrogen and air flow to the flame ionization detector maintained at 150°C are 60 and 400 ml/minute, respectively. The catalytic reactor consists of a 6 in. length of $\frac{1}{4}$-in. o.d. stainless-steel tube packed with 10% nickel on 42/60 mesh C-22 firebrick maintained at 360°C with a hydrogen flow of 30 ml/minute. For repetitive analysis 24 hours a day a stripper column is essential to prevent water vapor, carbon dioxide, and hydrocarbons other than methane from reaching the molecular sieve column. The stripper column consists of 12 in. length of $\frac{1}{4}$-in. o.d. stainless-steel tube packed with 5 in. of 10% Carbowax 400 on 60/80 mesh chromosorb W-HD, 5 in. of 60/80 mesh silica gel, and 2 in. of malcosorb maintained at 25° ± 5°C. The sample air passes through the sample loop at 100 ml/minute, and a sample is injected into the analyzer every 10 minutes. The sample flows through the stripper column, then into the molecular sieve column, or into the catalytic reactor and through the flame ionization detector. Samples collected in evacuated 1-liter steel cylinders can be analyzed. The system has proved capable of measuring carbon monoxide concentrations ranging from 0.3 to 10 ppm.

Measurements by both a commercial gas chromatographic analyzer (*22*) and an NDIR analyzer were carried out concurrently in St. Louis, Missouri, between October 14 and December 21, 1971 (*23*). On the average the diurnal concentration levels of carbon monoxide measured by the NDIR analyzer were 3 to 6 ppm higher than those measured by the gas chromatographic analyzer. These differences were attributed to drift or interferences in the NDIR method.

3. Other Methods

Another method for measuring higher atmospheric carbon monoxide concentrations is the potassium sulfite indicator tube technique (*24*).

Several other analytical methods have been developed in recent years for measurement of carbon monoxide but these techniques, if utilized, have been involved in atmospheric research investigations. The instruments have required more or less continuous operator attention, and they have not been evaluated for routine monitoring.

Infrared radiation has been utilized to vibrationally excite the $C^{16}O$ and $C^{18}O$ molecules at approximately 8 Torr each, diluted up to 1 atm pressure with argon in a sealed fluorescent cell (*25*). When the gases are of extremely high purity, the vibrationally excited molecules are deexcited by reemission of fluorescent radiation. The emitted fluorescent radiation from both the $C^{16}O$ and $C^{18}O$ molecules is chopped alternately by negative chopper cells containing impure $C^{16}O$ and $C^{18}O$ which absorb the $C^{16}O$ or $C^{18}O$ fluorescent radiation. The detector receives alternate pulses of $C^{16}O$ and $C^{18}O$ fluorescent radiation which has passed through the sample cell. Only the $C^{16}O$ radiation is significantly attenuated by the atmospheric carbon monoxide since the $C^{18}O$ isotopic abundance is only 0.2%. The output is proportional to $S(C^{18}O)-S(C^{16}O)$ or to the $C^{16}O$ concentration in the sample cell. This type of analyzer has improved sensitivity when compared to the normal NDIR analyzer and has high rejection ratios for carbon dioxide, water vapor, and hydrocarbon interferences. It has been utilized in measurement from aircraft.

Gas chromatographic frontal analysis for continuous determination of carbon monoxide (*26*) requires one adsorbent column to remove nitrogen and oxygen to prevent these gases from passing into the flame ionization detector. The eluted gases are passed onto a nickel catalyst to convert carbon monoxide to methane prior to the flame ionization detection. A complex set of solenoid valves are used to control the flow through the four columns used. The averaging time is 6 minutes. Because of base line drift the first derivative is taken but sensitivity is reduced to 2 ppm. Continuous measurement of carbon monoxide appears to be possible but at the penalty of some increase in system complexity and reduced sensitivity. A sensitivity of 2 ppm is inadequate outside of urban or high traffic density areas.

Another carbon monoxide analyzer is based on the oxidation of carbon monoxide with hot mercuric oxide to form mercury, followed by photometric measurement of the mercury vapor (*8*). Partial oxidation of hydrogen is compensated for by use prior to the HgO cell of an Ag_2O oxidizer for carbon monoxide which will not oxidize hydrogen. Since water vapor causes an increase in mercury vapor concentration, it also must be removed before the HgO. Olefins and carbonyl compounds will interfere by reaction with the HgO. Instruments of this type have found use in measuring background concentrations of carbon monoxide and in measurements from aircraft (*27, 28*).

Another technique involves use of an isotope mass spectrometer to measure the $^{18}O/^{16}O$ ratio, $^{13}C/^{12}C$ ratio, and total carbon monoxide concentration (*29*). This technique requires a large sample (e.g., 87 liters) which must be processed to remove water vapor, CO_2, N_2O, NO, and NO_2, followed by oxidation of CO to CO_2, purification, and manometric measurement of CO_2. While this method is extremely valuable in relating the isotopic forms of carbon monoxide for formation and decay processes, it is not a convenient measurement method for total carbon monoxide analysis.

Gas chromatography and the mercury analyzer have been used to determine carbon monoxide in remote areas including polar sites, over the ocean, and the upper troposphere (*8, 19, 20, 27, 28*). Although concentration of carbon monoxide as low as 0.04 ppm has been reported in remote sites, the concentration levels near the surface in "clear air" ordinarily range between 0.1 and 0.3 ppm. Oxidation of atmospheric methane, probably by hydroxyl radicals, is a major source of a global background of 0.10 to 0.15 ppm (*29*). The ocean and other waters have been identified as a source for carbon monoxide (*19*). Some carbon monoxide is emitted from vegetation and from degradation of chlorophyll in plants and trees (*29*). The contributions at rural sites from anthropogenic CO is detectable particularly in the winter in northern latitudes (*29*). Sinks for carbon monoxide include soil microorganisms (*30*) and consumption by reaction with hydroxyl radicals (*29, 31*). Carbon monoxide decreases rapidly above the tropopause (*28, 32*). The overall net effective residence time for carbon monoxide appears to be about 0.1 year based on the measured levels and the very large production rates from natural sources (*29, 31*). These measurements have lead to the conclusion that natural sources and sinks on a global scale greatly exceed man-made contributions. It now seems unlikely that anthropogenic sources can perturb the global background levels of carbon monoxide.

B. Carbon Dioxide

There have been only a few publications primarily concerned with the measurement of carbon dioxide in urban areas. A nondispersive infrared analyzer has been utilized to obtain measurements of carbon dioxide at urban sites in New Orleans, Louisiana, and St. Louis, Missouri and at both urban and nonurban sites in Cincinnati, Ohio (*33*). The contribution of natural sources of carbon dioxide to total carbon dioxide concentration levels is significant at urban sites. It was concluded that carbon dioxide could be used as an indicator of total pollution during all hours only in the winter and, during the other seasons, only in the afternoon hours (*33*).

The direct mass spectrometric technique for carbon dioxide requires high inlet pressure sample gas and removal of oxygen with phosphorus (*34*). Water vapor also has to be removed before the inlet because water vapor causes desorption of organic species from prior samples with interference at mass numbers 40 and 44. Water vapor also causes deactivation of the phosphorus by reacting with it to form phosphoric acid. Accuracy of better than 1% is claimed for the method. Analyses of carbon dioxide samples taken within minutes of each other at an isolated rural location showed substantial concentration variations with the measurements over 6 days ranging from 311 to 339 ppm (*34*).

III. Total Analysis for Organic Substances

Organic analysis is necessary because of the participation of hydrocarbons in the atmospheric photochemical reactions producing ozone and other oxidants, eye irritants, phytotoxicants, and inorganic and organic submicrometer aerosols. One of the six ambient air standards promulgated by the United States Environmental Protection Agency (EPA) is concerned with hydrocarbons (*35*). Vehicular emission standards for hydrocarbons also have been established in several countries. The United States ambient air quality standard of 100 $\mu g/m^3$ (0.24 ppm) applies to the time period from 6 to 9 a.m. The measurement is for all atmospheric hydrocarbons (and other organics) except methane. This standard is associated with the control approach for oxidant/ozone, since it is the oxidant/ozone not the organic reactants which have been associated with health effects (*35*). The instrumental technique selected as the United States reference method involves the use of the hydrogen flame ionization detector (*35*). The flame ionization analyzer also is used to determine compliance with the United States vehicular emission standard for hydrocarbons.

Infrared techniques and mass spectrometry have been applied in earlier work (*1*). However, these analytical approaches have been displaced by the flame ionization method. Infrared spectrometers have had some continued use in laboratory photochemical research on model hydrocarbon–nitrogen oxide systems. Some measurements have been made on atmospheric samples using Fourier transform spectrometers (*36*).

IV. Compositional Analysis of Volatile Organic Substances

A. Total Hydrocarbons

Most measurements of atmospheric hydrocarbons involve the use of flame ionization analyzers. The flame ionization analyzer responds to all hydrocarbons and most other organic compounds. There are small differences in response among various hydrocarbons (*37*). Carbon atoms with linkages to oxygen, nitrogen, sulfur, halogen, or other polar groups cause reduced or negligible signals with flame ionization analyzers. The hydrogen flame ionization analyzer shows negligible response to such molecules as formaldehyde or carbon monoxide.

Because of the lack of photochemical reactivity of methane, techniques were developed to measure methane and compute nonmethane hydrocarbons by difference—total hydrocarbons minus methane. The original technique involved use of an activated carbon adsorption column to hold back hydrocarbons other than methane (*38, 39*). Breakthrough of light hydrocarbons other than methane can occur unless the carbon column is replaced at appropriate intervals. A modification makes it possible to use one hydrogen flame ionization analyzer to alternately measure total hydrocarbon and methane (*40*).

In a selective combustion technique for use with the hydrogen flame ionization detector (*41*), the combustion temperature is varied by adjusting the power to a platinum wire in a quartz tube to either burn all hydrocarbons to establish a zero reading, burn all hydrocarbons except methane, or pass all hydrocarbons. The combustor also is used with a water-sensing sorption detector or carbon dioxide detector to measure the water or carbon dioxide formed from the hydrocarbon oxidation. As used, the techniques are slow so that total hydrocarbons and methane are determined during different portions of hourly operating intervals. The procedure utilizing the water-sensing detector is reported to be capable of differentiation of hydrocarbon classes (*41*).

A commercial analyzer for total hydrocarbons, methane, and carbon monoxide was evaluated as part of an instrument evaluation study in St. Louis, Missouri during October, November, and December, 1971 (*23*). The average nonmethane hydrocarbon between 6 and 9 a.m. was 0.77 ppm and exceeded the United States EPA standard of 0.24 ppm 42 out of 65 days of valid measurements. High purity hydrogen and air, which had been through a catalytic scrubber to remove trace hydrocarbon contaminants, were found essential for optimum performance (*23*).

B. Methane

As already mentioned, methane is photochemically unreactive on the time scale of interest in urban reactions. Methane is oxidized slowly to carbon monoxide on a global scale. Methane is known to be a permanent geophysical constituent of the earth's atmosphere with a ground-level concentration of 1.3 ± 0.1 ppm (*19, 20, 39, 42–44*). In urban areas methane can occasionally reach concentration levels three to five times greater than geophysical concentration levels (*39, 42, 45*). This variability of urban concentration levels makes it necessary to monitor methane at frequent intervals.

The early procedures for measuring methane were based on differential or subtractive techniques. In the United States Environmental Protection Agency reference method for hydrocarbons, discrete volumes of air are analyzed periodically (4 to 12 times an hour) by the hydrogen flame ionization analyzer method (*35*). A portion of each volume of air is passed through a stripper column into a gas chromatographic column which separates methane from carbon monoxide. After separation, the methane is analyzed with the hydrogen flame ionization detector. The combination of column separation and detector makes the procedure specific for methane. The stripper or precolumn protects the analytical column against deterioration by removing water, carbon dioxide, and higher molecular weight hydrocarbons.

A gas chromatographic column which has been found to be satisfactory for the methane determination is a 12 ft length of $\frac{1}{4}$-in. o.d. stainless steel packed with 5A molecular sieve, 60/80 mesh (*21*). A stripper column which has been found to be satisfactory is a 12 in. length of $\frac{1}{4}$-in. o.d. stainless-steel tube packed with 5 in. of 10% Carbowax 400 on 60/80 mesh chromosorb W-HD, 5 in. of 60/80 mesh silica gel, and 2 in. of malcosorb (Mallinckrodt). The gas chromatographic column is operated at 115°C and the stripper column at 25 ± 5°C (*21*). An analyzer using these columns was used to measure methane and carbon monoxide continuously for several months in Cincinnati, Ohio. For continuous use of the analyzer the stripper column is needed. With use of the stripper column, the properties of the analytical column remain constant for at least three months. Backflushing the stripper column assists in maintaining it in usable form for 6 to 8 weeks (*21*). Samples collected in evacuated rigid containers or in flexible containers such as plastic bags can be analyzed by this method.

C. Other Hydrocarbons

Atmospheric hydrocarbon composition is of interest for several reasons. Investigators have been concerned with relating atmospheric hydrocarbon composition to source emissions of hydrocarbons. For example, atmospheric composition can be compared to automobile exhaust emission composition to determine the quantitative contribution of this class of sources to overall atmospheric loadings of hydrocarbons. Hour by hour analyses provide diurnal patterns which can be related to source emissions. Ratios of individual hydrocarbons can be used to obtain additional relationships. Isolation of atmospheric samples in containers transparent to solar radiation, followed by irradiation, provides results on relative rates of reaction under realistic reaction conditions. Concurrent analyses for nitrogen oxides, ozone, and other products provide additional insight into atmospheric photochemical reactions.

Gas chromatography has made it possible to analyze an atmospheric sample for many individual aliphatic and aromatic hydrocarbons (*1*). Emphasis has been on analyzing a sufficient number of samples to obtain an improved knowledge of atmospheric composition.

Almost all of the samples whose analyses have been published (Table I) (*44, 46–52*) were collected in the Los Angeles, California Basin area near ground level. Concentration by freeze-out techniques of hydrocarbons from 70 to 500 ml of air has been used by some of the analysts to obtain detection limits between 0.1 and 1 ppb. Flame ionization detectors were used by all of the investigators reported. Samples were collected in Tedlar plastic bags (*16, 45, 48, 49*) or glass containers (*43, 46, 50*). The contents were passed into freeze-out traps, the hydrocarbon components concentrated, and injected into the chromatographs operating under isothermal rather than temperature programmed conditions. As many as

Table I. Gas Chromatographic Column Separations of Hydrocarbon in Atmospheric Samples

	Hydrocarbons analyzed			
Columns used	*Paraffins and acetylene*	*Olefins*	*Aromatics*	*Reference*
Porapak N	C_1—C_2	C_2	—	*44, 46, 47*
Alumina	C_2—C_3	C_2—C_3	—	*44, 46, 47*
Dimethyl sulfolane on C-22 firebrick	C_2H_2, C_3—C_6	C_3—C_5	—	*44, 46, 47*
Open-tubular copper (OTC) *m*-bis-*m*-phenoxyphenoxyl benzene plus Apiezon grease	—	—	C_6 C_{10}	*48*
Didecylphthalate on 42/60 mesh firebrick (mercuric perchlorate precolumn to remove unsaturates)	C_2—C_5	C_2—C_4	—	*49*
13× Molecular sieve column	CH_4, CO	—	—	*50, 51*
Dibutylmaleate + bis-2-(methoxyethyl) adipate on 60/80 mesh gas chrom Z	C_2—C_5	C_2—C_5	—	*50, 51*
Silica gel	C_2—C_5	C_2—C_4	—	*52*
OTC-Coated dibutylmaleate	C_4—C_7	C_4—C_6	—	*52*
OTC-Coated *m*-bis-*m*-phenoxyphenoxyl benzene	—	—	C_7—C_{10}	*50–52*

three different isothermal chromatograms have been needed to permit adequate separation of components.

V. Applications to Atmospheric Analysis

Fifty-three individual hydrocarbons have been reported by one or more investigators (Table II). They include 18 paraffinic hydrocarbons, two acetylenic hydrocarbons, 16 olefinic hydrocarbons, and 17 aromatic hydrocarbons. The retention times for two and sometimes three hydrocarbons are the same, so some peaks cannot be identified uniquely with a single hydrocarbon. When irradiations are used, the substantial differences in rates of olefins compared to paraffinic hydrocarbons provide some guidance. However, as many as 10 of these 53 hydrocarbons may not be considered as positively identified. The peaks identified do account for at least 95% of the total hydrocarbons measured by a flame ionization analyzer. The ten hydrocarbons—ethane, propane, isobutane, *n*-butane, isopentane, *n*-pentane, acetylene, ethylene, toluene, and *m*-xylene—account for 75% of the nonmethane hydrocarbon on a ppm by volume basis, and 55% of the nonmethane hydrocarbon on a carbon ppm basis. These ten hydrocarbons plus methane account for 95% of the total hydrocarbon on a ppm by volume basis, and 70% of the total hydrocarbon on a carbon

Table II. Hydrocarbons Analyzed in Atmospheric Samples

Paraffins	*Acetylenes*	*Olefins*	*Aromatics*
Methane	Acetylene	Ethylene	Benzene
Ethane	Methyl-acetylene	Propylene	Toluene
Propane		1-Butene	Ethylbenzene
n-Butane		Isobutene	*p*-Xylene
Isobutane		*trans*-2-Butene	*o*-Xylene
n-Pentane		*cis*-2-Butene	*m*-Xylene
Isopentane		1,3-Butadiene	Styrene
Cyclopentane		1-Pentene	*n*-Propylbenzene
n-Hexane		*cis*-2-Pentene	Isopropylbenzene
2-Methylpentane		*trans*-2-Pentene	1,2,4-Trimethylbenzene
3-Methylpentane		2-Methyl-1-butene	1,3,5-Trimethylbenzene
2,2-Dimethylbutane		3-Methyl-1-butene	*p*-Ethyltoluene
2,3-Dimethylbutane		2-Methyl-2-butene	*o*-Ethyltoluene
Cyclohexane		2-Methyl-1,3-butadiene	*m*-Ethyltoluene
n-Heptane		Cyclopentene	Isobutylbenzene
3-Methylhexane		1-Hexene	*sec*-Butylbenzene
2,4-Dimethylpentane			*t*-Butylbenzene
Methylcyclohexane			

ppm basis (*49*). These results are particularly applicable to the atmospheric hydrocarbon composition during high morning vehicular traffic conditions. Under lower vehicular traffic conditions, and both earlier and later in the day, paraffinic hydrocarbons become even more dominant and the olefins even less significant on a concentration basis (*45*).

Solar irradiation of atmospheric samples result in photochemical reactions with the following order of increasing rates of disappearance of hydrocarbons: 2-olefins $>$ 1-olefins (except ethylene) $>$ trialkylbenzenes $>$ ethylene $\geqq$ dialkylbenzenes $\geqq$ monoalkylbenzenes $\geqq$ butanes and higher molecular weight paraffins $>$ ethane, acetylene, propane (*16, 49*). The rates of disappearance of paraffins vary widely. For solar irradiation periods of 6 or 8 hours, the percent consumptions for ethane, acetylene, and propane cannot be distinguished from zero (*49*). Comparisons of these results for solar irradiation of individual atmospheric hydrocarbon samples with corresponding laboratory results with simulated solar irradiation of various hydrocarbon samples show reasonable agreement in relative rates of reaction (*49*).

One can compare the atmospheric hydrocarbon composition observed with the composition expected if automobile exhaust completely determined atmospheric composition. The largest differences observed are in the methane, ethane, propane, isobutane and, to a lesser extent, *n*-butane concentrations relative to the other hydrocarbon compounds (*45*). These paraffinic hydrocarbons are present in larger amounts and show a marked difference (except for *n*-butane) in diurnal behavior. These differences have been associated with natural gas losses, diffusion of hydrocarbon gases through soil into air from petroleum deposits, and, to some extent, petroleum evaporation losses (*45, 46, 48, 50*). In contrast, such hydrocarbons as ethylene, acetylene, propylene, 1-butene plus isobutene, *n*-pentane, isopentane, toluene, and xylenes show very similar diurnal behavior in their concentration patterns (*45*). This result verifies a close similarity to motor vehicle emission patterns. The differences in the diurnal patterns for the butanes compared to the pentanes, suggest one or more additional significant sources of these paraffins. Evaporative losses are one possibility. It should be noted that dynamometer driver cycles are only an approximation to the compositional mix in actual traffic on city streets (*48*). Therefore, ratios obtained from analyses of such exhaust samples are more likely only to provide guidance. Compositions and even ratios for hydrocarbons should be extrapolated with considerable caution to atmospheric conditions. However the paraffinic hydrocarbon anomaly is so large and characteristic as to be totally inconsistent with motor vehicle exhaust compositional patterns. Comparison of the hydrocarbon compositions at two different sites within the Los Angeles, California Basin clearly shows this effect. Azusa, California with between one-third and one-half the traffic density of downtown Los Angeles, California based on the ratios of hydrocarbons associated with vehicular exhaust, has from two-thirds as much to about equal concentrations of ethane, propane, and isobutane depending on the time of day.

On a percent consumption basis, as a class (excluding methane), the paraffins disappear one-fifth as fast, and the aromatics one-half as fast, as the olefins (*49*). Because of the relative atmospheric abundances, the aromatics contribute as much or more carbon atoms than either the olefins or paraffins (*49, 51*). Over 4- to 6-hour midday irradiation periods 20–25% of the total initial carbon atoms (excluding methane) have been found to be converted to products (*16, 51*) with about 33% of the reactive carbon atoms in the products identified as formaldehyde and peroxyacetyl nitrate (*16*). If source strengths and mixing volumes are constant throughout the day, the older irradiated air packets will be greatly depleted in the faster reacting components. In particular, olefins should be depleted significantly. Comparisons of the decreases of slowly reacting butanes with rapidly reacting 2-butenes indicate that the decreases in the 2-butenes by a factor of 4 between morning and midafternoon in the central Los Angeles, California Basin area result about equally from increased mixing volumes and photochemically induced reactions (*45*). Such changes in composition with irradiation time would be maximized for long trajectories over water or land areas of low-emission source strengths of hydrocarbons. Conversely, samples with relative olefin-to-paraffin or olefin-to-acetylene ratios close to auto exhaust composition must have been taken very close to vehicular traffic or represent little photochemical activity. The later situation would be characteristic of lower sunlight intensity conditions in the late fall, winter, or early spring.

VI. Analysis of Aliphatic Oxygenated Compounds

Analysis for volatile organic substances containing oxygen in the atmosphere or in emissions usually is done by colorimetric or "wet" methods, rather than by instrumental methods. In principle, instrumental methods can be used to analyze a number of classes of oxygenated substances. In practice, applying instrumental techniques to the analysis of oxygenates has been limited to applications of gas chromatography to emission source samples.

Many of the oxygenated organic substances present in emissions and atmospheric samples can be lost, owing to any or all of the following processes: (a) removal with water vapor in traps, (b) condensation along with water vapor as emission samples cook, (c) adsorption on surfaces of sampling or analytical equipment, and (d) removal along with water vapor and carbon dioxide in the use of many types of chemical adsorbents. Although these factors can adversely affect all methods of analysis, sampling prior to instrumental analyses often is remote from the laboratory analytical instrumentation. Losses in sampling, storage, and transfer of samples to instruments can be greater than in sampling directly into liquid reagents for colorimetric analyses.

Use of mass spectrometry has been very limited in analyses for oxygenated organic substances in air pollution studies (*52, 53*). Aldehydes and ketones contribute to the same mass peaks as do paraffinic hydrocarbons. Although variations in intensities of parent mass peaks to fragmentation product mass peaks suggest the presence of carbonyl compounds (*52*), no attempt has been made to quantitatively analyze for aldehydes or ketones. Alcohols have been tentatively identified and estimated in some combustion emissions and in the atmosphere.

The paraffinic chains on oxygenated substances do contribute to the responses of both nondispersive and dispersive infrared methods for total hydrocarbons or total organics. Total aldehydes have been estimated in irradiated automobile exhaust samples from the absorbance at 3.6 μm. Methanol and acetone have been determined and several other oxygenates identified in incinerator effluents by the dispersive infrared method.

The analysis of carbonyl compounds by the bisulfite procedure (sulfoxylate method) has been widely applied in air pollution and industrial hygiene. An addition complex is formed between the carbonyl compound and the bisulfite, the excess bisulfite is destroyed, the addition compound decomposes, and the bisulfite ion released is titrated with standard iodine solution. Although the collection efficiency for lower molecular weight aldehydes is satisfactory, higher molecular weight aldehydes and ketones can be partially lost. Such losses occur at room temperature and at flow rates higher than 3 liters/minute (*54*). Use of several impingers, an ice bath, and lower flow rates improves collection efficiencies significantly (*54*). Since the sensitivity of the method is not high, large samples must be collected for analysis of atmospheric samples. This method has been used for analysis of atmospheric samples (*55, 56*), automobile exhaust (*57*), diesel exhaust (*58, 59*), and incinerator effluents (*60*).

In early work Ellis (*61*) used various chemical methods to analyze carbonyl compounds in automobile exhaust emissions. Carbonyls react with hydroxylamine hydrochloride to form oximes and hydrochloric acid. The amount of hydrochloric acid formed is determined titrimetrically. Carbonyls also were determined by forming the 2,4-dinitrophenylhydrazones and measuring the optical absorbance of the products at 420 nm.

This latter procedure was further evaluated and applied by Papa (*62*) for analysis of carbonyls in automobile exhaust gases. Papa found that use of a solvent medium 70 and 30% methanol–methylene chloride in pyridine stabilized the colored derivatives, particularly the formaldehyde derivative which decomposed readily. At 440 nm the molar absorptivities of the aldehyde and ketone derivatives were within a ±10% range of each other.

Early efforts were made to analyze oxygenated compounds in automobile exhaust emissions by gas–liquid partition chromatography (*63–65*). The procedure utilized physical condensation followed by vaporization into gas–liquid partition columns. Because of the lack of specificity involved in collection and analysis, this procedure was effective only on emissions from engines used with individual hydrocarbons as fuels. Much of the automobile exhaust research has emphasized collection of carbonyls into reagent solutions so as to form products and separate the carbonyls from the more abundant and numerous hydrocarbons in automobile exhaust emissions.

Levaggi and Feldstein (*66, 67*) went back to collection of the carbonyls in sodium bisulfite solution contained in two impingers in parallel. Most of the carbonyls were analyzed by gas chromatography although formaldehyde and acrolein had to be analyzed by specific colorimetric procedures. This technique was applied to effluents from Webb press, litho oven, brake debonding core oven, formaldehyde plant, phthalic acid plant, coffee roasting, and paint spray booths (*66, 67*).

In practice this procedure is closer to a measurement of various lower molecular weight aldehydes because bisulfite solution does not efficiently collect ketones or other oxygenates. Therefore, the procedure has been proposed as a tentative method of analysis for formaldehyde, acrolein, and other low molecular weight aliphatic aldehydes in the atmosphere by the Intersociety Committee (*68*).

Fracchia and co-workers (*69*) used gas–liquid partition chromatography to analyze the 2,4-dinitrophenylhydrazones of carbonyl compounds in automobile exhaust. The carbonyls in auto exhaust were collected in two bubblers containing acidified 2,4-dinitrophenylhydrazine and precipitated as hydrazones. The hydrazones were filtered, extracted, dissolved in carbon disulfide, separated using a 6-ft × 0.085-in. i.d. stainless-steel column packed with 10% SF 96 on 60 to 80 mesh acid-washed chromosorb W and analyzed with a flame ionization detector. Calibration, collection efficiencies, and recovery efficiencies for the hydrazones were investigated. Formaldehyde, acetaldehyde, acrolein, propionaldehyde, crotonaldehyde, butyraldehyde, acetone, 2-butanone, and benzaldehyde were identified as present in the automobile exhaust using a regular grade of gasoline as the fuel. Formaldehyde accounts for about two-thirds of the carbonyls present in these emissions.

A gas chromatographic analysis technique for lower molecular weight oxygenated compounds in emissions was developed by Bellar and Sigsby (*70*). These authors indicated a number of the problems associated with forming and handling derivatives of oxygenated compounds. Their system involves use of a cutter column to retain oxygenates, but pass hydrocarbons; a trap to condense the oxygenates from the long backflush peak, and an analytical column followed by a flame ionization detector. The cutter column consisted of a 12-ft-long, 0.25-in. o.d., #304 stainless-steel tube packed with 20% 1,2,3,-tris-(2-cyanoethoxy)propane on 45–60 mesh (not acid washed) chromosorb W operated at 22°C with a 100 cm^3/minute helium flow rate. The trap consisted of a 16-in.-long, 0.125-in. o.d., #304 stainless-steel tube packed with 12 in. of 10% sucrose octaacetate on 60–80 mesh gas chrom Z and 2 in. on each end of siliconized glass wool. The trap in the trapping mode was operated at —55°C and in desorbing into the analytical column was programmed from —55°C to 100°C at 12°C/second. The analytical column consisted of a 12-ft-long, 0.105-in. i.d., #304 stainless-steel tube packed with 50–80 mesh parapak Q and operated at 156°C with a 50 cm^3/minute helium flow rate. Analysis was done with a flame ionization detector. This technique was applied to samples of automobile exhaust, a wood burning incinerator, and ambient air. Eight lower molecular weight oxygenates were determined in automobile exhaust and incinerator emissions including methanol, acetaldehyde, acrolein, propanol, acetone, 2-methylpropanal, butanal, and methylethyl ketone. It was possible to measure methanol, acetaldehyde, acrolein, and acetone in an ambient air sample near traffic. Formaldehyde was not detected by this technique.

Vogh (*71*) investigated the isolation and analysis of oxygenates as oximes and applied his procedure to both gasoline engine and diesel exhaust emissions. Oximes could be formed from aliphatic and aromatic aldehydes and ketones. The samples were collected in methanol–hydroxylamine solution made basic by adding sodium hydroxide. Hydrocarbons were extracted with *n*-pentane. After neutralization the oximes were extracted with ethyl ether. Phenols also were carried along with the oximes through the extraction procedure. In the auto exhaust, sample formaldehyde and three other aliphatic aldehydes, several aromatic aldehydes, acetone and 2-butanone, benzylalcohol, phenol, and the cresols were identified. These separations were made with a 42-in.-long, 4-mm i.d. borosilicate tube packed with 0.1% Carbowax 20 M on glass beads. Analysis was done with a flame ionization detector.

Another technique for analysis of oxygenates in automobile exhaust was developed by Seizinger and Dimitriades (*72*). The exhaust was collected in a heated (55°C) Teflon bag. Samples from the Teflon bag flowed through a chromosorb bed consisting of a 10-in. × 0.375-in. o.d. stainless-steel tube packed with 60–80 mesh chromosorb 102 (porous polymer). The contents of the chromosorb bed were purged with helium at 55°C. The retained organics were backflushed from the bed into the first of a two-stage chromatographic system. The first stage column consisted of a 30-in. × 0.25-in. o.d. stainless-steel tube packed with 10% 1,2,3-tris-(2-cyanoethoxy)propane and programmed from —20 to 110°C at 4°C/minute. Helium flow was 350 cm^3/minute. The resultant chromatographic fractions were frozen out in a packed loop trap and revaporized into the second-stage column. This column consisted of a 150-ft × 0.02-in. i.d. Carbowax 20 M support coated open tubular column programmed to 210°C at 10°C/minute with a helium flow of 8 cm^3/minute. The effluent was split to permit parallel measurement with a flame ionization detector and a mass spectrometer. This technique was inadequate for analysis of the light oxygenates which were separated by a method similar to that of Bellar and Sigsby (*70*). Analysis of an engine operated on simple hydrocarbon fuels permitted identification of over 30 products. The products included a number of aliphatic and aromatic aldehydes, ketones and alcohols, phenols, furans, an ether, methyl formate, and nitromethane. Formaldehyde was not measured.

Subsequently, Seizinger and Dimitriades (*73, 74*) using simple hydrocarbon fuels in engines with and without an oxidation catalyst measured a very wide range of oxygenated organic compounds by gas–liquid partition chromatography on auto exhaust samples. In addition to the usually identified saturated aliphatic aldehydes and ketones, they reported unsaturated ketones, benzaldehyde, tolualdehyde, ethylbenzaldehyde, salicyaldehyde, and acetophenone. Other oxygenates measured include methyl formate, nitromethane, and furans (*73, 74*).

Paper and thin-layer chromatography techniques have been developed to analyze carbonyls in automobile exhaust emissions (*75, 76*). Ascending paper chromatography of the 2,4-dinitrophenylhydrazones of the aliphatic aldehydes and ketones was adapted to apply to auto exhaust emissions collected in bubblers containing 15% sodium bisulfite in ice water (*75*). The carbonyls identified included formaldehyde, acetaldehyde, propionaldehyde, acrolein, crotonaldehyde, acetone, and methyl ethyl ketone. Various losses in collection, conversion, and decomposition limited the quantitative applicability of this chromatographic procedure. Subsequently, Barber and Sawicki (*76*) investigated the paper and thin-layer chromatographic properties of the 4-nitrophenylhydrazones of the aromatic carbonyls. Auto exhaust emissions were collected into a series of three impingers containing 1.0 *N* sodium hydroxide at ice water temperature. Benzaldehyde and acetophenone were identified in auto exhaust by their R_f values, by ultraviolet spectra, and by fluorescence and phosphorescence characteristics. *Ortho*- and *m*-hydroxyl benzophenones also were identified. The concentrations of benzaldehyde, acetophenone, and the hydroxybenzophenones in the exhaust were 0.5, 0.6, and 1 mg/m^3, respectively. Propionaldehyde and acetone also were identified but were not quantitatively determined.

A. Aldehydes

1. Total Aldehydes

Colorimetric methods of analysis have been used extensively for the analysis of aldehydes. However, a number of the available colorimetric methods respond much more to formaldehyde than to other aldehydes (*77*). Such methods include those involving Schiff's reagent, chromotropic acid reagent, phenylhydrazine reagent (Schryver's method), and 2-hydrazinobenzothiazole reagent. These colorimetric techniques cannot be used for the quantitative determination of total aldehydes in atmospheric or emission source samples.

A method based on the 3-methyl-2-benzothiazole hydrazone developed by Sawicki and co-workers (*78*) responds to aliphatic aldehydes. The method has been evaluated and applied to atmospheric and emission source analysis (*78–80*). The azine resulting from the reaction of the aldehyde with this reagent is oxidized by a ferric chloride–sulfamic acid solution to form a blue cationic dye in acid media. The blue dye formed with formaldehyde has molar absorptivities at 635 or 670 nm which are about 25% greater than the dyes formed with acetaldehyde and other straight-chain aliphatic aldehydes and about 300% higher than the dyes formed with branched-chain and unsaturated aldehydes (*78*). Therefore, the 3-methyl-2-benzothiazole hydrazone method tends to underestimate the concentration of total aldehydes if expressed in terms of formaldehyde. While formaldehyde has been shown to constitute one-half to two-thirds of the aldehydes in many atmospheric or emission source samples (*79–82*), the deviation still can be significant. A correction factor has been suggested for analyses of atmospheric samples from photochemical reactions (*80*). The method has been reported to be applicable to both short duration (*79, 80*) and 24-hour sampling procedures (*83*). Collection efficiencies have been determined for acetaldehyde and propionaldehyde (*84*). The 3-methyl-2-benzothiazole method has been applied to atmospheric analysis (*79, 81*), smog chamber analysis (*80, 82*), automobile exhaust emissions (*73, 74*), and diesel exhaust emissions (*85*). The procedure has been proposed as a tentative method for formaldehyde with application to other aldehydes by the Intersociety Committee (*86*). However, the procedure in the Intersociety Committee manual is somewhat ambiguous in not making it clear that other aldehydes besides formaldehyde react appreciably with the reagent, and these other aldehydes are present in appreciable quantities in the atmosphere and in many kinds of source emissions.

All of the gas chromatographic procedures for oxygenates can be used to estimate total aldehydes by summing up the gas chromatographic peaks associated with individual aldehydes. The procedures involving formation of derivatives measure the formaldehyde as well as the other aldehydes (*69, 71*). Those procedures which measure the aldehydes directly have not been able to measure formaldehyde; so a separate analysis for formaldehyde must be made (*66, 67, 70, 72*).

Two titrimetric methods for total aldehydes were applied to automobile exhaust analysis (*61*). One of these methods involves oxidation of the aldehydes to the corresponding acids by hydrogen peroxide in alkaline solution followed by acid–base titration. The original aqueous solution must be neutralized before oxidation of the aldehydes. In the other method, silver oxide is reduced to metallic silver by the aldehydes. The metallic silver is removed by titration, and the unreacted silver ion is titrated with ammonium thiocyanate solution. These titrimetric methods are of too low sensitivities to be useful in atmospheric analysis for aldehydes.

The C—H stretch vibrations of CHO groups in the 3.5- to 3.7-μm region should be useful for infrared analysis of aldehydes. However, infrared techniques have found little application to analysis of aldehydes in atmospheric or source emission samples. Infrared spectroscopy has been used to analyze aldehydes produced by ultraviolet irradiation of hydrocarbon–nitrogen oxide mixtures in smog chamber investigations (*87–90*).

2. Formaldehyde

Formaldehyde is the most abundant aldehyde found in the atmosphere and in emissions. Most of the analyses of formaldehyde in atmospheric, smog chamber, and emission source samples have been carried out by colorimetric methods. The chromotropic acid method has received extensive use in colorimetric analysis of formaldehyde in air pollution applications.

Formaldehyde will react with chromotropic acid (4,5-dihydroxy-2,7-naphthalenedisulfonic acid) in sulfuric acid solution to form a purple chromogen with an absorption maximum at 580 nm (*91–93*). The method is sensitive, and involves a simple and rapid reaction. Although other aldehydes do not react significantly with chromotropic acid, alcohols, olefins, aromatic hydrocarbons, and phenols constitute negative interferences (*93*). Nitrogen dioxide also can constitute an interference under some circumstances (*85, 94*). The formaldehyde can be collected into water, aqueous bisulfite solution, or sulfuric acid (*93, 95*). Cares (*96*) has suggested that collection in aqueous bisulfite solution followed by heating to remove nitrates and volatile organics will minimize the interferences particularly by nitrogen oxides. Collection in sulfuric acid solution provides a tenfold increase in sensitivity compared to collection in aqueous solutions. The choice of collection media also can be important in minimizing interferences depending on the distribution of interferences in atmospheric and emission source samples (*95*). Collection in acid solution is satisfactory for atmospheric irradiated dilute automobile exhaust, diesel exhaust, incinerator effluent, but not for automobile exhaust (*95*). Collection in aqueous solution is satisfactory for atmospheric samples, irradiated diluted automobile exhaust and raw automobile exhaust, but not for diesel exhaust or incinerator effluent.

The Intersociety Committee recommended use of the chromotropic acid method for atmospheric analysis of formaldehyde (*97*). Only the collection into water is mentioned.

Collection in sulfuric acid solution actually is the method of choice for atmospheric analysis because of its higher sensitivity (*79, 93, 95*). The chromotropic acid method with either aqueous or acid collection has been used for atmospheric samples (*79, 81*), diluted or irradiated diluted automobile exhaust (*80, 82, 95, 98*), raw automobile exhaust (*74, 77, 93*), diesel exhaust (*58, 85, 94, 99*), incinerator effluents (*100, 101*), and other source effluents (*66, 67*). Several variations of the chromotropic acid reagent have been developed by using the related 6-amino-1-naphthol-3-sulfonic acid (J-acid) and phenyl J-acid (*102*).

In earlier work considerable difficulty with slow color formation and incomplete collection was experienced with acidified Schiff's reagent (*103*). Two modified procedures were proposed for quantitative analysis of formaldehyde in air with this reagent (*103, 104*). Use of pararosaniline, rather than basic fuchsin, increased sensitivity, and much more rapid color formation was obtained (*104*). This latter modification was applicable to an instrumental analysis of formaldehyde (*104*).

The Schryver method—reaction of formaldehyde with phenylhydrazine, followed by oxidation with potassium ferricyanide to form a red

derivatives permit analysis of formaldehyde along with other carbonyls. Although the product can be unstable, modifications have been developed that lead to a more stable product (*107, 109*). This method has been applied to the determination of formaldehyde in automobile (*61, 110*) and diesel exhausts (*110*).

The substitution of 2-hydrazinobenzothiazole for phenylhydrazine results in a blue chromogen of high color intensity and good stability (*111*). The 2-hydrazinobenzothiazole and chromotropic acid methods have been compared in analyses of diluted and diluted, irradiated automobile exhaust (*77*).

As already discussed, gas chromatographic procedures for oxygenates involving formation of 2,4-dinitrophenylhydrazone (*69*) or oxime (*71*) derivatives permit analysis of formaldehyde along with other carbonyls.

Automobile exhaust has been analyzed by a titrimetric potassium cyanide procedure of relatively low sensitivity (*61*). The technique depends upon the addition of formaldehyde to potassium cyanide to form a cyanohydrin. Excess cyanide ion is measured by titration with silver nitrate.

Polarographic techniques for formaldehyde in aqueous potassium hydroxide (*105*) and as the semicarbazone (*112*) have been developed. However, polarographic techniques have not been utilized to analyze formaldehyde in atmospheric or emission samples. Other methods including mass spectroscopic and infrared spectroscopy also have received little use in analysis of formaldehyde in air.

3. Acrolein

Analyses of acrolein in atmospheric and emission samples has been of particular interest because of the high eye irritation potential of acrolein (*56, 77, 82, 87*). Acrolein is present at much lower concentration levels than formaldehyde in the atmosphere and in most emission sources. Therefore, a very sensitive method of analysis is essential. Most of the acrolein measurements reported since the early 1960's have been obtained by the 4-hexylresorcinol reagent procedure developed by Cohen and Altshuller (*113*). This method is the only colorimetric method available with sufficient sensitivity for air pollution applications (*113, 114*). Acrolein reacts with 4-hexylresorcinol in an ethyl alcohol–trichloroacetic acid solvent medium in the presence of mercuric chloride resulting in a blue chromogen of high color intensity and adequate stability (*100*). Interferences are minimal (*94, 113*). The applicability of the method to automobile exhaust emissions and simulated photochemical atmospheres in smog chambers was demonstrated in the original work (*113*). The 4-hexylresorcinol method received rapid application within several years to samples from the Los Angeles, California, atmosphere (*56, 79*), smog chamber studies (*82*), and automobile (*75, 77, 82, 113*) and diesel exhaust emissions (*58, 85, 94, 99*). More recently additional applications have been made to atmospheric analysis (*81*) and various stationary combustion sources (*66, 67*).

The Intersociety Committee recommended the use of the 4-hexylresorcinol reagent procedure for atmospheric analysis of acrolein (*115*). The collection into the 4-hexylresorcinol solution of 50 liters of atmospheric samples at 2 liters/minute will result in an absorptivity of 0.05 units at 0.01 ppm of acrolein (*113, 115*). Concentrations down to 0.001 to 0.002 ppm of acrolein have been estimated (*79*).

Acrolein has been separated and analyzed in auto exhaust by means of gas–liquid partition chromatography directly and as a derivative (*63–65, 69, 70, 72*). Only Bellar and Sigsby (*70*) reported acrolein in an atmospheric sample by gas–liquid partition chromatography. The precision, accuracy, and sensitivity of these procedures for acrolein are not established. Polarographic (*112, 116*) and paper chromatographic (*75*) procedures also have been used to identify acrolein. A polarographic technique was used to estimate acrolein concentrations in the atmospheres of paint and varnish plants (*116*).

B. Ketones

In early work, Ellis (*61*) measured aliphatic ketones in automobile exhaust by chemical methods. He reported ketones at much lower concentrations than aldehydes in automobile exhaust (*61*). Subsequently, several aliphatic ketones were identified by gas–liquid partition chromatographic methods (*63–65, 72*) in automobile exhaust formed by burning individual hydrocarbons as fuels. Barber and co-workers identified acetophenone in automobile exhaust (*117*). Beller and Sigsby (*70*) using a more involved but direct gas–liquid partition chromatographic procedure measured acetone and 2-butanone in both incinerator emissions and automobile exhaust using gasoline as the fuel. Fracchia and co-workers (*69*) by gas–liquid partition chromatography were able to identify the 2,4-dinitrohydrazones of acetone and 2-butanone from automobile exhaust using a regular grade of gasoline as the fuel. Acetone and 2-butanone in automobile exhaust were measured as oxime derivatives by Vogh (*71*) using gas–liquid partition chromatography. Seizinger and Dimitriades (*73, 74*) reported measurement of a number of saturated and unsaturated ketones and acetophenone in exhaust from burning individual hydrocarbons.

Very few measurements of ketones in atmospheric samples have been reported. Acetone has been measured by a polarographic method (*112*). Bellar and Sigsby (*70*) analyzed acetone in an ambient air sample near traffic by a gas–liquid partition chromatographic procedure.

C. Organic Acids

Total organic acids have been identified in incinerator effluents (*60*) and in coal-fired power plant emissions (*118*). Formic acid has been detected in diesel exhaust emissions by infrared spectroscopy (*94*).

Formic acid was reported by Mader and co-workers in samples collected from the Los Angeles, California, atmosphere based on chemical analyses (*119*). Hanst has analyzed formic acid formation with time of day by Fourier transform spectrometry on samples collected in Pasadena, California (*120*).

D. Alcohols

Methanol has been determined in incinerator effluents by infrared analysis (*60*) and gas–liquid partition chromatography (*70*). Methanol also has been identified in diesel exhaust by gas chromatography (*94*). A number of aliphatic alcohols in automobile exhaust from burning of individual hydrocarbons as fuels have been identified and estimated by gas–liquid partition chromatographic procedures (*63–65, 70, 72*). Alcohols have been determined in automobile exhaust by mass spectrometry (*53*). Total alcohols were estimated in automobile exhaust by converting the alcohols to the nitrite esters, which were determined by a titrimetric procedure (*61*).

In early work using mass spectrometric analysis alcohols were identified in freeze-out samples collected in the vicinity of an industrial area in Louisville, Kentucky (*52*). Methanol was analyzed by gas–liquid partition chromatography in ambient air collected near traffic in Cincinnati, Ohio (*70*).

E. Phenols

A number of investigators have concerned themselves with measuring phenols in automobile exhaust. In early work Smith and co-workers (*121*) and Hoffmann and Wynder (*122*) showed that phenols were significant components of automobile exhaust. Barber *et al.* (*117*) separated and analyzed phenols by use of both paper chromatography and gas–liquid partition chromatography. They reported phenol; *o*-cresol; *p*-cresol;

m-cresol; 3,5-xylenol; 2,5-xylenol; 3,4-xylenol; and 2,3-xylenol in automobile exhaust (*117*). A group of spectrophotometric methods including a piperonal chloride procedure, a *p*-nitroaniline procedure, 4-amino antipyrine procedure A, 4-aminoantipyrine procedure B, nitrous acid procedure, and ultraviolet procedure were intercompared for analysis of phenols in automobile exhaust (*123*). Samples were collected in a series of impingers containing aqueous alkaline solutions cooled in ice water (*123*). The methods of choice for analysis of the phenols where simplicity and routine usability are desired are the *p*-nitroaniline and the 4-aminoantipyrine B procedures. The 4-aminoantipyrene procedure used by Smith and co-workers has been proposed as tentative Intersociety method for determination of phenolic compound in air (*124*).

REFERENCES

1. A. P. Altshuller, *in* "Air Pollution" (A. C. Stern, ed.), 2nd ed., Vol. II, pp. 115–145. Academic Press, New York, New York, 1968.
2. M. Katz, *in* "Air Pollution" (A. C. Stern, ed.), 2nd ed., Vol. II, pp. 53–114. Academic Press, New York, New York, 1968.
3. A. P. Altshuller, *Advan. Chromatogr.* **5,** 229–262 (1968).
4. A. P. Altshuller, *Nat. Bur. Stand. (U.S.), Spec. Publ.* **351,** 245–286 (1972).
5. Environmental Protection Agency, National Primary and Secondary Air Quality Standards, *Fed. Regist.* **36,** No. 84, 8186–8201 (1971).
6. U.S. Dept. of Health, Education, and Welfare, Air Quality Criteria for Carbon Monoxide, *Nat. Air Pollut. Contr. Admin. (U.S.), Publ.* **AP-62,** 6-1 to 6-16 (1970).
7. R. K. Stevens and A. E. O'Keeffe, *Anal. Chem.* **42,** 143A (1970).
8. R. C. Robbins, K. M. Borg, and E. Robinson, *J. Air Pollut. Contr. Ass.* **18,** 106 (1968).
9. H. C. McKee and R. E. Childers, "Collaborative Study of Reference Method for the Continuous Measurement of Carbon Monoxide in the Atmosphere (Non-Dispersive Infrared Spectrometry)," Contract CPA 70-40. Division of Chemistry and Physics, National Environmental Research Center, United States Environmental Protection Agency, Research Triangle Park, North Carolina, 1972.
10. P. C. Wolf, *Environ. Sci. Technol.* **5,** 212 (1971).
11. J. M. Colucci and C. R. Begeman, *Environ. Sci. Technol.* **3,** 41 (1969).
12. G. Goodwin, G. Wright, and R. J. Shepard, *Arch. Environ. Health* **25,** 305 (1972).
13. A. P. Altshuller and I. R. Cohen, *Int. J. Air Water Pollut.* **8,** 611 (1964).
14. A. P. Altshuller, I. R. Cohen, and T. C. Purcell, *Can. J. Chem.* **44,** 2973 (1966).
15. A. P. Altshuller, S. L. Kopczynski, W. A. Lonneman, T. L. Becker, and R. Slater, *Environ. Sci. Technol.* **1,** 899 (1967).
16. A. P. Altshuller, S. L. Kopczynski, W. A. Lonneman, and F. D. Sutterfield, *Environ. Sci. Technol.* **4,** 503 (1970).
17. R. S. Ajemian and N. E. Whitman, *J. Air Pollut. Contr. Ass.* **20,** 310 (1970).
18. K. Porter and D. H. Volman, *Anal. Chem.* **34,** 748 (1962).

19. J. W. Swinnerton, V. J. Linnerton, and C. H. Cheek, *Environ. Sci. Technol.* **3,** 836 (1969).
20. C. E. Decker, J. R. Smith, and G. C. Ortman, "Evaluation of Techniques for Measurement of Low Concentrations of Trace Gases in Atmosphere." Contract No. CPA-22-69-109. National Air Pollution Control Administration, U.S. Dept. of Health, Education, and Welfare, Washington, D.C., 1970.
21. R. K. Stevens, A. E. O'Keeffe, and G. A. Ortman, *Proc. 11th Conf. Ind. Hyg. Stud.,* Pap. No. 8 (1970).
22. R. Villalobos and R. L. Chapman, *in* "Air Quality Instrumentation" (J. Scales, ed.), Vol. 1, pp. 114–128. Instrum. Soc. Amer., Pittsburgh, Pennsylvania, 1972.
23. J. R. Ramsey, *Arch. Environ. Health* **13,** 44 (1966).
24. E. A. McClatchie, "Development of an Infrared Fluorescent Gas Analyzer." Contract CPA 70-152. Division of Chemistry and Physics, National Environmental Research Center, United States Environmental Protection Agency, Research Triangle Park, North Carolina, 1972.
25. L. Dubois and J. L. Monkman, *Anal. Chem.* **44,** 74 (1972).
26. L. A. Cavanagh, C. F. Schadt, and E. Robinson, *Environ. Sci. Technol.* **3,** 251 (1969).
27. W. Seiler and C. Junge, *J. Geophys. Res.* **75,** 2217 (1970).
28. C. M. Stevens, L. Krout, D. Walling, A. Venters, L. Ross, and A. Engelkemeir, *Joint Meet. Amer. Geophys, Union Amer. Meteorol Soc. 1972* p. 4-1 (1972).
29. J. Pressman and P. Warneck, *J. Atmos. Sci.* **27,** 155 (1970).
30. B. Weinstock, and H. Nike, *Science* **176,** 290 (1972).
31. J. F. Clark and R. B. Faoro, *J. Air Pollut. Contr. Ass.* **16,** 212 (1966).
32. R. E. Inman, R. B. Ingersoll, and E. V. Levy, *Science* **172,** 1229 (1971).
33. E. E. Hughes and W. D. Darko, *Anal. Chem.* **40,** 866 (1968).
34. Environmental Protection Agency, National Primary and Secondary Air Quality Standards, *Fed. Regist.* **36,** No. 84, 8186–8201 (1971).
35. P. Hanst, A. Lefohn, and B. Gay, *Appl. Spectrosc.* **27,** 188 (1973).
36. W. A. Dietz, *J. Gas Chromatogr.* **5,** 68 (1967).
37. G. C. Ortman, *Anal. Chem.* **33,** 644 (1966).
38. A. P. Altshuller, G. C. Ortman, B. E. Saltzman, and R. E. Neligan, *J. Air Pollut. Contr. Ass.* **16,** 87 (1966).
39. R. G. McCann, *J. Air Pollut. Contr. Ass.* **21,** 502 (1971).
40. W. H. King, Jr., *Environ. Sci. Technol.* **4,** 1136 (1970).
41. R. K. Stevens, T. A. Clark, C. E. Decker, and L. F. Ballard, 65th Annu. Meet. Air Pollut. Contr. Ass. (1972).
42. A. P. Altshuller, T. A. Bellar, C. A. Clemons, and E. VanZander, *Int. J. Air Water Pollut.* **8,** 29 (1964).
43. E. R. Stephens and F. R. Burleson, *J. Air Pollut. Contr. Ass.* **19,** 929 (1969).
44. R. A. Lamontagne, J. W. Swinnerton, V. J. Linnenbon, and W. Smith, *Joint Meet. Amer. Geophys. Union Amer. Meteorol. Soc., 1972* p. 9-1 (1972).
45. A. P. Altshuller, W. A. Lonneman, F. D. Sutterfield, and S. L. Kopczynski, *Environ. Sci. Technol.* **5,** 1009 (1971).
46. E. R. Stephens, E. F. Darley, and F. R. Burleson, *Proc. Amer. Petrol. Inst., Div. Refining* **47,** 466 (1967).
47. W. A. Lonneman, T. A. Beller, and A. P. Altshuller, *Environ. Sci. Technol.* **2,** 1017 (1968).
48. R. J. Gordon, H. Mayrsohn, and R. M. Ingels, *Environ. Sci. Technol.* **2,** 1117 (1968).

49. S. L. Kopczynski, W. A. Lonneman, F. D. Sutterfield, and P. E. Darley, *Environ. Sci. Technol.* **6**, 342 (1972).
50. E. R. Stephens and F. R. Burleson, *J. Air Pollut. Contr. Ass.* **17**, 147 (1967).
51. A. P. Altshuller, S. L. Kopczynski, W. A. Lonneman, F. D. Sutterfield, and D. L. Wilson, *Environ. Sci. Technol.* **4**, 44 (1970).
52. E. R. Weaver, E. E. Hughes, S. M. Gunther, S. Schuhmann, N. T. Redfearn, and R. Gordon, Jr., *J. Res. Nat. Bur. Stand.* **59**, 383 (1957).
53. J. K. Walker and C. L. O'Hara, *Anal. Chem.* **26**, 352 (1954).
54. K. W. Wilson, *Anal. Chem.* **30**, 1127 (1960).
55. N. A. Renzetti, L. H. Rogers, and R. Tice, *in* "An Aerometric Survey of the Los Angeles Basin, August-November, 1954" (N. A. Renzetti, ed.), Rep. No. 9. Air Pollut. Found., Los Angeles, California, 1956.
56. N. A. Renzetti and R. J. Bryan, *J. Air Pollut. Contr. Ass.* **11**, 421 (1961).
57. G. P. Larson, J. C. Chipman, and E. K. Kauper, *J. Air Pollut. Contr. Ass.* **5**, 84 (1955).
58. M. C. Battigelli, *J. Occup. Med.* **10**, 5 (1968).
59. C. E. Schumann and C. W. Gruber, *J. Air Pollut. Contr. Ass.* **14**, 53 (1964).
60. J. E. Yocum, G. M. Hein, and H. W. Nelson, *J. Air Pollut. Contr. Ass.* **6**, 84 (1956).
61. C. F. Ellis, *U.S. Bur. Mines, Rep. Invest.* **RI-5822** (1961).
62. L. J. Papa, *Environ. Sci. Technol.* **3**, 397 (1969).
63. K. J. Hughes and R. W. Hurn, *J. Air Pollut. Contr. Ass.* **10**, 367 (1960).
64. K. J. Hughes, R. W. Hurn, and F. G. Edwards, *Gas Chromatogr., Proc. Int. Symp., 2nd, 1959* p. 171 (1961).
65. C. F. Ellis, R. K. Kendall, and B. H. Eccleston, *Anal. Chem.* **37**, 511 (1965).
66. D. A. Levaggi and M. Feldstein, *J. Air Pollut. Contr. Ass.* **19**, 43 (1969).
67. D. A. Levaggi and M. Feldstein, *J. Air Pollut. Contr. Ass.* **20**, 312 (1970).
68. Intersociety Committee, "Methods of Air Sampling and Analysis" (Tentative Method of Analysis for Low Molecular Weight Aliphatic Aldehydes in the Atmosphere, No. 43501-01-71T), Part II. Amer. Pub. Health Ass., Washington, D.C., 1972.
69. M. F. Fracchia, F. J. Schuette, and P. K. Mueller, *Environ. Sci. Technol.* **1**, 915 (1967).
70. T. A. Bellar and J. E. Sigsby, Jr., *Environ. Sci. Technol.* **4**, 150 (1970).
71. J. W. Vogh, *Anal. Chem.* **43**, 1618 (1971).
72. D. E. Seizinger and B. Dimitriades, *J. Air Pollut. Contr. Ass.* **22**, 47 (1972).
73. D. E. Seizinger and B. Dimitriades, *U.S., Bur. Mines, Rep. Invest.* **RI-7675** (1972).
74. B. E. Seizinger and B. Dimitriades, *U.S., Bur. Mines, Rep. Invest.* **RI-7837** (1973).
75. E. D. Barber and J. P. Lodge, *Anal. Chem.* **35**, 348 (1963).
76. E. D. Barber and E. Sawicki, *Anal. Chem.* **40**, 984 (1968).
77. A. P. Altshuller, I. R. Cohen, M. E. Meyers, and A. F. Wartburg, Jr., *Anal. Chim. Acta* **25**, 101 (1961).
78. E. Sawicki, T. R. Hauser, T. W. Stanley, and W. Elbert, *Anal. Chem.* **33**, 93 (1961).
79. A. P. Altshuller and S. P. McPherson, *J. Air Pollut. Contr. Ass.* **13**, 109 (1963).
80. A. P. Altshuller and L. J. Leng, *Anal. Chem.* **35**, 1541 (1963).
81. Scott Research Laboratories, "Atmospheric Reaction Studies in the Los Angeles Basin," Final Report on Phase I, Vol. II, Contract CPA 22-69-19. National

Air Pollution Control Administration, U.S. Public Health Service, Washington, D.C., 1969.
82. P. W. Leach, L. J. Leng, T. A. Bellar, J. E. Sigsby, Jr., and A. P. Altshuller, *J. Air Pollut. Contr. Ass.* **14,** 176 (1964).
83. T. R. Hauser and R. L. Cummins, *Anal. Chem.* **36,** 679 (1964).
84. I. R. Cohen and A. P. Altshuller, *Anal. Chem.* **38,** 1418 (1966).
85. L. B. Reckner, W. E. Scott, and W. F. Biller, *Proc. Amer. Petrol. Inst., Sect. 3* **45,** 133 (1965).
86. Intersociety Committee, "Methods of Air Sampling and Analysis" [Tentative Method of Analysis for Formaldehyde Content of the Atmosphere (MBTH-Colorimetric Method-Applications to Other Aldehydes), No. 43502-02-70T], Part II. Amer. Pub. Health Ass., Washington, D.C., 1972.
87. E. A. Schuck and G. J. Doyle, "Photooxidation of Hydrocarbons in Mixtures Containing Oxides of Nitrogen and Sulfur Dioxide," Rep. No. 29. Air Pollut. Found., San Marino, California, 1959.
88. C. S. Tuesday, *in* "Chemical Reactions in the Lower and Upper Atmosphere" (R. D. Cadle, ed.), p. 1–49. Wiley (Interscience), New York, New York, 1961
89. E. R. Stephens and W. E. Scott, *Proc. Amer. Petrol. Inst., Sect. 3* **42,** 665 (1962).
90. S. L. Kopczynski, *Int. J. Air Water Pollut.* **8,** 107 (1964).
91. C. E. Bricker and H. R. Johnson, *Ind. Eng. Chem., Anal. Ed.* **17,** 400 (1945).
92. P. W. West and B. Sen, *J. Anal. Chem.* **153,** 177 (1956).
93. A. P. Altshuller, D. L. Miller, and S. F. Sleva, *Anal. Chem.* **33,** 62 (1961).
94. R. H. Linnell and W. E. Scott, *Arch. Environ. Health* **5,** 616 (1962).
95. A. P. Altshuller, L. J. Leng, and A. F. Wartburg, *Int. J. Air Water Pollut.* **6,** 381 (1962).
96. J. W. Cares, *Amer. Ind. Hyg. Ass., J.* **29,** (1968).
97. Intersociety Committee, "Methods of Air Sampling and Analysis" [Tentative Method of Analysis for Formaldehyde Content of the Atmosphere (Colorimetric Method), No. 43502-01-69T], Part II. Amer. Pub. Health Ass., Washington, D.C., 1972.
98. A. P. Altshuller, S. L. Kopczynski, W. Lonneman, and D. Wilson, *J. Air Pollut. Contr. Ass.* **17,** 734 (1967).
99. R. H. Linnell and W. E. Scott, *J. Air Pollut. Contr. Ass.* **12,** 510 (1962).
100. R. L. Stenburg, R. P. Hangebrauck, D. J. von Lehmden, and A. H. Rose, Jr., *J. Air Pollut. Contr. Ass.* **11,** 376 (1961).
101. R. L. Stenburg, R. P. Hangebrauck, D. J. von Lehmden, and A. H. Rose, Jr., *J. Air Pollut. Contr. Ass.* **12,** 83 (1962).
102. E. Sawicki, T. R. Hauser, and S. P. McPherson, *Anal. Chem.* **34,** 1460 (1962).
103. A. C. Rayner and C. M. Jephcott, *Anal. Chem.* **33,** 627 (1961).
104. G. R. Lyles, F. B. Dowling, and V. J. Blanchard, *J. Air Pollut. Contr. Ass.* **15,** 106 (1965).
105. E. C. Barnes and H. W. Speicher, *J. Ind. Hyg. Toxicol.* **24,** 9 (1942).
106. R. W. Kersey, J. R. Maddocks, and T. E. Johnson, *Analyst* **65,** 203 (1940).
107. M. Tannenbaum and C. E. Bricker, *Anal. Chem.* **23,** 354 (1951).
108. N. Zurlo and A. M. Griffini, *Med. Lav.* **45,** 692 (1954).
109. M. B. Jacobs, "The Chemical Analysis of Air Pollutants," p. 267. Wiley (Interscience), New York, New York, 1960.
110. M. A. Elliot, G. J. Nebel, and F. G. Rounds, *J. Air Pollut. Contr. Ass.* **5,** 103 (1955).

111. E. Sawicki and T. R. Hauser, *Anal. Chem.* **32,** 1434 (1960).
112. D. M. Coulson, *Anal. Chim. Acta* **19,** 284 (1958).
113. I. R. Cohen and A. P. Altshuller, *Anal. Chem.* **33,** 726 (1961).
114. A. P. Altshuller, *Int. J. Air Water Pollut.* **6,** 169 (1962).
115. Intersociety Committee, "Methods of Air Sampling and Analysis," [Tentative Method of Analysis for Acrolein Content of the Atmosphere (Colorimetric), No. 43505-01-70T], Part II. Amer. Pub. Health Ass., Washington, D.C., 1972.
116. W. A. van Sandt, R. J. Graul, and W. J. Roberts, *Amer. Ind. Hyg. Ass., J.* **16,** 221 (1955).
117. E. D. Barber, E. Sawicki, and S. P. McPherson, *Anal. Chem.* **36,** 2442 (1964).
118. S. T. Cuffe, R. W. Gestle, A. A. Orning, and C. H. Schwartz, *J. Air Pollut. Contr. Ass.* **14,** 353 (1964).
119. P. P. Mader, G. R. Cann, and L. Palmer, *Plant Physiol.* **30,** 318 (1955).
120. P. L. Hanst, W. E. Wilson, R. K. Patterson, B. W. Gay, Jr., L. Chaney, and C. S. Burton, *167th Meet. Amer. Chem. Soc.* Environmental Chemistry Paper, No. 55 (1974).
121. R. G. Smith, J. E. McEwen, and R. E. Barrow, *Amer. Ind. Hyg. Ass., J.* **20,** 149 (1959).
122. D. Hoffmann and E. L. Wynder, *Nat. Cancer Inst., Monogr.* **9,** 91 (1962).
123. T. W. Stanley, E. Sawicki, H. Johnson, and J. D. Pfaff, *Mikrochim. Acta* p. 48 (1965).
124. Intersociety Committee, "Methods of Air Sampling and Analysis" [Tentative Method of Analysis for Determination of Phenolic Compounds in the Atmosphere (4-Aminoantipyrine Method), No. 17320-01-70T], Part II. Amer. Pub. Health Ass., Washington, D.C., 1972.

6

Sulfur Compounds

Donald F. Adams

I. Introduction

Most of the methods applicable to the measurement of the sulfur-containing gases in ambient air must be considered in a state of evolution. However there has been significant progress from the early application of wet chemical analytical techniques to the more recent development of sophisticated gas chromatographic separation and photoluminescent detection procedures which permit sulfur-specific identification within the spectrum of sulfur containing compounds present in the atmosphere. Information on the accuracy, precision, and reliability of these methods is very limited. Only one method, the modified West–Gaeke sulfur dioxide procedure (*1–4*), has been collaboratively tested, the lowest concentration range studied being well above the levels most frequently present in rural and global background air.

Atmospheric sulfur dioxide concentrations of interest generally range from fractional parts per billion ($\mu g/m^3$) to one or two parts per million (mg/m^3). The primary ambient air quality standards for sulfur dioxide of the United States Environmental Protection Agency (USEPA) are (a) 60 $\mu g/m^3$ (20 ppb), annual arithmetic mean and (b) 260 $\mu g/m^3$ (100 ppb), maximum 24-hour concentration, not to be exceeded more than once per year (*4*). Interest has developed in measuring (a) the background sulfur dioxide concentrations in remote areas where new installations of large industrial sources, such as fossil fuel-fired electric power generating facilities, have been proposed, (b) the long-range effect of all sulfur dioxide emissions on the chemical composition of the global atmosphere, and (c) nature's contributions to the global atmospheric sulfur gas background, including sulfur dioxide, hydrogen sulfide, dimethyl sulfide (DMS), and dimethyl disulfide (DMDS) produced by biogeochemical processes.

Lovelock *et al.* (*5*) postulate that DMS is a direct product of organic decomposition in contrast to the previously accepted theory that the primary product of vegetative decomposition is hydrogen sulfide. Robinson and Robbins (*6*) estimate that the global background sulfur dioxide concentration is approximately 0.52 $\mu g/m^3$ (0.2 ppb). This estimate is based upon a limited number of background measurements using various modifications of the West–Gaeke method. Their review (*6*) reports the limits of detection of the West–Gaeke method to be 0.78 $\mu g/m^3$ (0.3 ppb) sulfur dioxide. Other studies of this analytical method have stated the lower limits of the method to be 26 $\mu g/m^3$ (10 ppb) in an air sample of 30 liters.

II. Sulfur Dioxide

A. Colorimetric Methods—West–Gaeke

In the modified West–Gaeke colorimetric method sulfur dioxide is collected in 0.1 *M* sodium tetrachloromercurate(II) (TCM) to form a dichlorosulfitomercurate complex (DCSM) which resists oxidation by either the oxygen in the air or that dissolved in the absorbing solution. The addition of ethylenediamine tetraacetatic acid disodium salt (EDTA) to the TCM absorbing solution complexes heavy metals that could oxidize sulfur dioxide before DCSM formation is complete (*7, 8*). Once DCSM is formed it is stable to such oxidants as ozone and oxides of nitrogen. Any ozone which remains in the absorbing solution after absorption is allowed to decay before adding the colorimetric reagents. Any nitrite ion present in the absorbing solution from the absorption of oxides of nitrogen from the sampled air is destroyed by the addition of sulfamic acid (*9*).

The colorimetric determination of sulfur dioxide is based upon the measurement of the red-violet color produced by the reaction of DCSM with hydrochloric acid, pararosaniline, and formaldehyde. The intensely colored pararosaniline methyl sulfonic acid acts as a two-color pH indicator. The pH of the final solution may be adjusted to 1.6 ± 0.1 by addition of 3 *M* phosphoric acid and the absorbance is read at 548 nm with a molar extinction of 47.7×10^3. A greater quantity of phosphoric acid may be added to yield a final pH of 1.2 ± 0.1. The wavelength of maximum absorbance under these conditions is 575 nm with a molar extinction coefficient of 37.0×10^3.

Several authors have considered the influence of dye purity on the analytical procedure (*10, 11*). Procedures for purification of commercial quality dye by recrystallization have been published (*2, 4, 11*). Purity of the pararosaniline dye affects the sensitivity of the colorimetric procedure, although not necessarily the accuracy or reproducibility when a standard curve of absorbance versus concentration is prepared for each dye lot, and all other analytical conditions are held constant. Dye purified especially for the colorimetric analysis of sulfur dioxide is commercially available (*12*).

The Sulfur Subcommittee of the Intersociety Committee (ISC) has published a tentative method for sulfur dioxide which specifies two procedures for the colorimetric determination of sulfur dioxide collected in a tetrachloromercurate absorption solution (*2*). "Method A" gives the highest sensitivity and requires careful control of the pH used in the color development. "Method B," which uses a larger quantity of phosphoric acid, does not require as close pH control during color development, and has the advantage of a lower blank. Method A is described as a "research" method because of its greater sensitivity, whereas Method B is considered by some to be more suitable for routine use by control agency laboratories because of the less demanding conditions required for color development. ISC Method A corresponds most closely with the USEPA manual reference method published in the United States Federal Register (*4*).

Fourteen laboratories participated in a collaborative test program of the EPA's method (*4*). Test atmospheres were generated from United States National Bureau of Standards permeation tubes and purified air. The laboratory work was conducted in the home laboratory of each collaborator. Prior to the final test program, however, a familiarization session was held to review the entire procedure with the participants. The collaborative test contractor concluded that "the overall average deviations from the expected values for each concentration tested were not significant, and therefore, no systematic error, bias, or inaccuracy was detectable (*13*).

ISC "Method A" also corresponds closely to American Society for Testing and Materials (ASTM) Method D2914-T (*3*). Method D2914-T was collaboratively tested as a part of ASTM's Project Threshold (*14*). In this test, all collaborators conducted their sampling and analysis independently, but at the same time and place. All samplers were attached to a single, large sampling manifold through which ambient air was drawn, thus minimizing the possibility that the various collaborators might sample different test atmospheres. This test protocol was repeated in three cities, New York City (Manhattan), New York; Los Angeles, California; and Bloomington, Indiana. In addition, for some tests the ambient air was also spiked with known concentrations of sulfur dioxide to provide a measure of the recovery of the added sulfur dioxide from the three types of "dirty" air represented by the three cities. Simultaneous analyses for other atmospheric pollutants present in the sampled ambient air of the three test cities were not made. Thus these data cannot be used to evaluate the possible existence or extent of interference(s) from the unknown, but simultaneously present, air pollutants which vary from one sampling site to another. It cannot be tacitly assumed that the possible effects of various interferents are linear with their concentration, or that, once the interferent "demand" for sulfur dioxide has been satisfied, any further effect will be linear. Although interferents (especially nitrogen dioxide, ozone, and metallic particulates) may be added back to a purified airstream to evaluate their effect upon the analytical method, it is extremely difficult to produce the original air mixture minus sulfur dioxide. Both types of collaborative tests are obviously needed to confirm the overall reliability of the TCM–pararosaniline sulfur dioxide procedures (or any other ambient air analysis procedure for that matter).

Different statistical procedures were used to evaluate the data from the USEPA and ASTM collaborative tests. At the time of writing this chapter, no collaborative tests had been undertaken to resolve the relative superiority of the USEPA reference method, and by close equivalence ISC "Method A," and ASTM Method D-2914-T, over the ISC "Method B" which has been considered by some to be more suitable for routine use. The results from EPA's and ASTM's collaborative tests of essentially the same method indicate differing lower limit or detection (18 $\mu g/m^3$ vs 4 $\mu g/m^3$ sulfur dioxide). The lower limit of detection reported by ASTM may be related to the longer sampling time used in their test. A negative bias of the measurement of sulfur dioxide recovered from spiked-ambient samples was —22, —6, and —4% at Los Angeles, Bloomington, and Manhattan, respectively. Although the bias was not concentration dependent, it may represent a possible oxidant interference.

A major potential source of error associated with the analysis of sulfur dioxide by the West–Gaeke colorimetric procedure is the widely differing collection efficiency reported for Greenburg–Smith and midget impingers when atmospheric concentrations decrease to the 26 $\mu g/m^3$ (10 ppb) level; and the as yet unestablished but implied drastic decrease in collection efficiency in the global background range of 0.05 to 13 $\mu g/m^3$ (0.02–5 ppb). Urone *et al.* (*15*) investigated the collection efficiency in TCM solution of microgram quantities of sulfur dioxide tagged with ^{35}S. The tagged SO_2 was followed from its original pure state to its mixture at high dilution in pure air, to the scrubbing solution, and finally to the quantitation step. This tracer technique showed that the sampling methods lost efficiency below 260 $\mu g/m^3$ (100 ppb). A second bubbler in series did not capture the $^{35}SO_2$ lost by the first bubbler. This study also showed that series bubblers cannot be used to determine absorber collection efficiency. Boström found a 99% collection efficiency for sulfur dioxide in TCM solution within the range of 139–1390 $\mu g/m^3$ (100–1000 ppb) (*16*).

White and Hill (*17*) conducted a field evaluation of the pararosaniline method for the determination of parts per billion concentrations of atmospheric sulfur dioxide in a remote desert canyon in Utah because of the probable low background sulfur dioxide there. Known amounts of sulfur dioxide were generated from permeation tubes and added to filtered and unfiltered desert air. The stability of DCSM solutions was also checked by aerating solutions prepared from known concentrations of sodium sulfite and TCM reagent. These authors concluded that the method lost collection efficiency for known sulfur dioxide concentrations below 52 $\mu g/m^3$ (20 ppb). The authors suggested that it might be possible to apply correction factors to measured values down to 6.8 $\mu g/m^3$ (2.6 ppb) (Fig. 1). It is stated that below this concentration random errors make the use of correction factors unfeasible. Aeration with unfiltered desert air containing approximately 58.9 $\mu g/m^3$ (30 ppb) ozone, NO_2 at 0–9.4 $\mu g/m^3$ (0–5 ppb), and particulate in the range of 10–20 $\mu g/m^3$ produced more loss of DCSM than did aeration with purified air. The authors concluded that "measurements of SO_2 below 13 $\mu g/m^3$ (5 ppb), where *in situ* proof of sampling efficiency was not shown, are questionable."

Stevens *et al.* (*18*) compared a modified, 24-hour West–Gaeke proce-

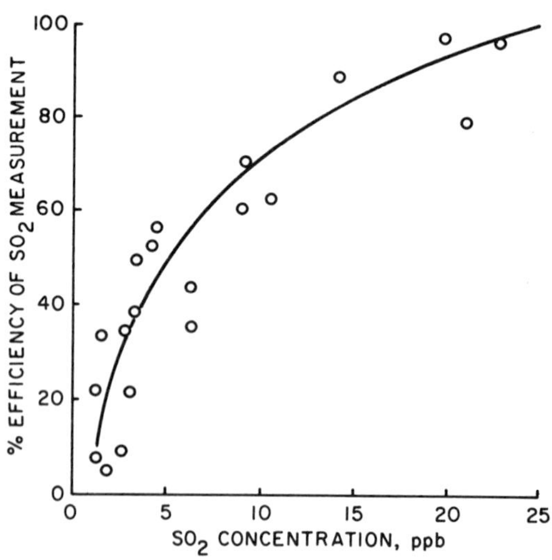

Figure 1. Overall efficiency of sulfur dioxide measurement by the pararosaniline method under field conditions vs SO_2 concentration.

dure against a total sulfur flame photometric detector (FPD), a gas chromatograph–FPD (GC–FPD), and a coulometric analyzer. In this field comparison only the data from the total sulfur FPD and coulometric analyzers correlated with the West–Gaeke method above $r = +0.874$. The lower correlation obtained with the GC–FPD analyzer was attributed to sporadic instrumental difficulties associated with the instrument electronics and gas flow controllers.

The coulometric and flame photometric methods are accepted as "equivalent" to the manual reference colorimetric procedure (*4*) by USEPA (*19*).

B. Colorimetric Methods—Other

Attari (*20*) absorbed sulfur dioxide into a solution of ferric ammonium chloride, perchloric acid, and phenanthroline dye. A color complex with an absorbance at 510 nm was formed. The color developed within 10 minutes, but tended to fade with time. Hydrogen sulfide interfered seriously.

Barium chloranilate will react with sulfate ion releasing the red-colored acid chloranilic ion. Kawai (*21*) used this reaction for the direct analysis of sulfur dioxide in an absorption solution containing hydrogen peroxide and barium chloranilate. The method appears satisfactory for flue gas analysis but lacks the sensitivity required for ambient air analyses.

C. Conductivity Methods

The most common sulfur dioxide monitoring method historically is exemplified by the Thomas Autometer (*22*), in which the conductivity of a dilute sulfuric acid–hydrogen peroxide reagent changes due to the absorption of pollutants. The change in conductivity is assumed to result primarily from sulfur dioxide absorbed from the sampled air and oxidized to sulfuric acid. The data are interpreted in terms of equivalent sulfur dioxide concentrations. The limit of sensitivity of these instruments generally averages 131 $\mu g/m^3$ (50 ppb); however, they are nonspecific, since conductivity change may result from any material which is collected by the absorbing solution and alters its conductivity. In many field situations, sulfur dioxide is the major pollutant present. If other pollutants are also present, their solubility, or their collection efficiency in the scrubber may be significantly lower than for sulfur dioxide.

Numerous field comparisons have been conducted which indicate that conductivity agrees fairly well with other sulfur dioxide procedures (*23–25*), although the observed difference may be either positive or negative depending upon the characteristics of the indigenous air pollution. Hydrochloric acid gas, ammonia, and chlorine significantly increase conductivity. Shikiya and MacPhee (*25*) found unexplained two- to fourfold differences between different conductivity analyzers and between conductivity and colorimetric analyzers. Terabe *et al.* (*26*) found that in Kawasaki, Japan, the conductivity method gave higher sulfur dioxide values than did the colorimetric method. Although the conductivity procedure may be acceptable for isolated areas having point sources of sulfur dioxide, its high potential for interferences, both negative and positive, limits its application.

D. Iodometric Methods

Iodometric methods were among the first adapted for air pollution analysis from the industrial hygiene literature. In the iodine method, the sulfur dioxide is collected in an impinger containing a standard solution of sodium hydroxide. After the sample has been collected, the absorbing solution is acidified and the liberated sulfurous acid is titrated with standard iodine solution (*27*). This method is not specific for sulfur dioxide since the method responds to the total reducing gases present in the sample, including hydrogen sulfide and methyl mercaptan. It is not particularly suitable for field application, unless laboratory facilities are available for immediate titration of the samples. The maximum storage time for low concentration samples is thought to be 24 hours.

Another iodometric method of historic interest uses a standard iodine–potassium iodide absorbing solution (*28*). Both sulfur dioxide and hydrogen sulfide react with iodine; the remaining iodine is subsequently titrated with standard sodium thiosulfate solution. Although potassium iodide is added to produce a KI_3 complex to minimize the stripping of iodine from the absorbing solution, measureable loss is nevertheless encountered. In practice the iodine–potassium iodide scrubber is followed by a second impinger containing a known volume of standard sodium thiosulfate to collect the stripped iodine vapor. The contents of both impingers are combined and the remaining iodine titrated with standard thiosulfate. Iodometric methods generally suffer from lack of sensitivity required for most ambient air and background air studies. Katz (*29*) used a dilute starch–iodine solution in a continuous, colorimetric instrument for the determination of reducing gases including sulfur dioxide and hydrogen sulfide.

E. Adsorption Sampling Methods

Stratmann (*30*) describes a method wherein sulfur dioxide is sorbed on silica gel. After sampling a known volume of air, the sulfur dioxide is then desorbed and reduced to hydrogen sulfide at 700–900°C over a platinum catalyst. The resultant hydrogen sulfide is absorbed in a 2% solution of ammonium molybdate and the equivalent sulfur dioxide determined colorimetrically. The method is fairly specific for sulfur dioxide, but the final colorimetric determination of the resulting hydrogen sulfide by molybdenum complex does not utilize the most sensitive method available for the determination of sulfur dioxide.

F. Filtration Methods

Pate *et al.* (*31, 32*) collect sulfur dioxide by filtering air through potassium bicarbonate impregnated types HA and AA Millipore filters. The filters are then analyzed for collected sulfate. Impregnated filters have been used by others to collect sulfur dioxide (*33*). The collection efficiency of impregnated filters is dependent upon humidity, temperature, and the atmospheric concentration of sulfur dioxide. Huygen (*34*) showed that collection efficiency decreases rapidly below 25% relative humidity. No controlled experiments have been reported which define the performance of these filters below the parts per billion concentration range.

The duration of the sampling period determines whether the technique would be classified as "short-term batch" or "cumulative." The technique can provide air concentrations of sulfur dioxide if the volume of air sampled is measured. Such quantification is not possible with static samplers such as the lead peroxide candle and Huey sulfation plate described below.

The sulfur dioxide absorbed by sodium carbonate–glycerine impregnated membrane filters has been measured directly by nondispersive X-ray fluorescence (*35*). The 2.3-keV sulfur X ray was counted with a beryllium window proportional counter using an iron-55 source deposited on a platinum disc. The minimum sensitivity is stated to be 30 μg sulfur dioxide with a $\pm 5\%$ standard deviation. The major potential interference is chlorine.

G. Static Collectors*

The lead peroxide candle static collector was developed in England by Wilsdon and McConnell (*36*) in 1932 as an inexpensive method for measuring the relative "sulfation" of the atmosphere at numerous selected sampling sites within a prescribed study area. The sulfur dioxide collection efficiency is dependent upon temperature, relative humidity, wind

* See also Chapter 10, this volume.

speed, atmospheric concentration of sulfur dioxide, and the length of exposure period. Huey (*37*) developed a more convenient procedure utilizing an inverted 48-mm petri dish containing the lead peroxide paste. This technique eliminates the tedious job of coating the 100-cm^2 gauze surface used in the candle technique. The simplified petri dish holder bracket may be attached to a convenient utility pole, thus eliminating the square or round louvered shelters used with the candle. The performance of the plates has been compared with the candles at several different locations. The plates were approximately 20% more reactive than the candles (*37*).

Numerous investigators (*38–40*) have reported varying relationships between sulfation and sulfur dioxide concentration in parts per million as representing the sulfur dioxide concentration over the period of the peroxide candle exposure. The method will not, however, provide short-term concentration data because of its relative insensitivity.

Harding and Kelley (*41*) used a 44-station network of 3 × 4-in. mild steel plates to demonstrate monthly corrosion patterns. The pattern of corrosion gave a reasonably good description of the sulfur dioxide dispersion from several nearby sources. The correlation coefficient between sulfation and corrosion was +0.60, indicating that less than half of the variance could be accounted for by the relation to sulfation. Sea salt gave high corrosion rates near the Jacksonville beach, but dropped off to background within 1 mile from the beach.

Ikeda (*42*) determined the relative sulfur dioxide index in the ambient air by exposing chemically impregnated filters for specified time periods such as 30 days. An active carbon filter paper method was used for sulfur dioxide. Hydrogen sulfide was collected on filter paper impregnated with gum tragacanth. Following exposure, the filters were extracted with distilled water and titrated with barium chloranilate.

III. Hydrogen Sulfide

Much of the literature prior to 1968 (and even some more recent publications) on the determination of hydrogen sulfide, methyl mercaptan, the alkyl sulfides, and disulfides present conflicting, questionable, and sometimes erroneous conclusions, some in error by as much as one to three orders of magnitude.

The primary sources of error associated with these data are (a) errors in the published calibration curves for the lead acetate tape method, (b) the unrecognized significance of the photodecomposition of collected cadmium sulfide in the ppb concentration range which resulted in losses up to at least 80% prior to colorimetric determination of the collected sulfide by the methylene blue procedure, and (c) the as yet unestablished collection efficiency for these gases in Greenburg–Smith or midget impingers in the low and submicrogram per cubic meter range.

A. Cadmium Hydroxide–Methylene Blue

In the Jacobs *et al.* (*43*) method, hydrogen sulfide is absorbed in an aqueous, alkaline suspension of cadmium hydroxide prepared by dissolving 4.3 gm of cadmium sulfate ($2CdSO_4 \cdot 8H_2O$) and 0.3 gm of sodium hydroxide in 1 liter of distilled water. The methylene blue is then formed by reaction between the absorbed sulfide and *p*-amino-*N*,*N*-dimethylaniline in the presence of ferric chloride. The molar absorptivity is reported to be about 34,000 in the range of 5–50 μg S^{2-}/100 ml, thus providing an exceptionally sensitive system. Although color development is influenced by temperature, time, and acidity, it is quite reproducible using ordinary laboratory precautions.

Jacobs *et al.* emphasize the problem of the "oxidation" of sulfides by the relatively large volumes of air which are aspirated through an impinger and suggest that "oxidation" is minimized by using an alkaline cadmium hydroxide absorbing mixture. They prefer this absorbent to those employing zinc acetate or ammoniacal cadmium chloride (*44*). Hasselhuhn (*45*) in 1966 reported that the zinc slurry is superior to the widely accepted cadmium hydroxide slurry. This report appears to contradict the earlier (1956–1957) data of Jacobs *et al.* (*43*) and Marbach and Doty (*46*).

In 1956, Marbach and Doty reported minimum sulfide losses when the cadmium was either completely precipitated with an equimolar addition of sodium hydroxide (pH 9.6) or when an excess of sodium hydroxide was present (pH 13.0). Marbach and Doty used the equivalent of 4.3 gm of $3CdSO_4 \cdot 8H_2O$ and 1.3 gm NaOH to prepare their fully precipitated $Cd(OH)_2$ sulfide collection media. Subsequently, in 1957 Jacobs *et al.* recommended only 0.3 gm sodium hydroxide per liter (pH 7.5). The preponderance of opinion favors the cadmium hydroxide slurry absorbing reagent.

The possibility that the observed loss could be related to low collection efficiency had been eliminated by the work of Boström (*47*) wherein the collection efficiency for isotopic H_2S^* in cadmium hydroxide suspension was found to be in the range of 93–98% within the range of 1.4–40 $\mu g/m^3$ (1–29 ppb). The CdS* collected in the absorption solution was oxidized to $S^*O_4^{2-}$ with hydrogen peroxide in slightly acid solution. Unfortunately, the analytical method did not provide any information concerning the fate of the excited S^{2-} during aspiration and storage. Analysis was based upon the amount of excited $BaSO_4$ precipitated from the excited SO_4^- produced from the complete oxidation of CdS* rather than a direct determination of the sulfide ion. The collection efficiency may decrease at lower atmospheric concentrations.

Since Jacobs *et al.* did not discuss the reasoning behind their reduction of the 1:1 molar ratio of Cd/2NaOH—recommended earlier by Marbach and Doty—to the Jacobs ratio of 1:0.25, a series of Cd/NaOH ratios between 1:1 and 1:0.25 were reexamined (*48*) to determine the influence of alkalinity. Samples of higher pH gave consistently higher sulfide recoveries, however, the improvement was not considered to be significant. Variable losses of 30–70% were observed over the pH range of 7.5 to 12.0.

To test the Jacobs *et al.* (*43*) suggestion that sulfide losses are caused by oxidation, replicated impingers containing 15 ml of $Cd(OH)_2$ suspension and added standard sulfide were purged with nitrogen, air, and oxygen. Sulfide losses were not significantly greater for the air and oxygen treatments than for the nitrogen treatment, indicating that the observed sulfide losses are not related to oxidation. However, it was observed that the sulfide loss appears to be related to the length of exposure to laboratory light indicating a possible photodecomposition of the sulfide. A parallel study of a number of common antioxidants and protective colloids showed that adding 1% STRactan 10 (*48, 49*) to the cadmium hydroxide slurry, significantly improves sulfide recovery both in the light and in the dark.

The modified cadmium hydroxide slurry with 1% added STRactan (*49*) provides reasonable protection from photodecomposition. Recoveries are in the order of 80% ± 3% in the low parts per billion concentration range. This method has been field tested (*50*), selected as the method of choice by the Bay Area (California) Air Pollution Control District, and published as a tentative method by the Intersociety Committee (*51*). Thomas *et al.* studied the effect of potential interfering compounds (*52*). The percent hydrogen sulfide recovery with each interferent is shown in Table I.

Falgout and Harding (*53*) believed that their inability to find hydrogen sulfide in the air of Jacksonville, Florida, by the methylene blue method, even though the odor of hydrogen sulfide was detected, was due to the presence of an interfering substance such as ozone. Based upon present knowledge it seems more reasonable that the collected cadmium sulfide photodecomposed prior to the determination of the sulfide by the methylene blue colorimetric procedure. Prescher and Lahmann (*54*) collected hydrogen sulfide in cadmium hydroxide suspension kept in the dark from the moment of air intake until the analysis was accomplished. These

Table I Cadmium Hydroxide–STRactan–Methylene Blue Method Interference Studies

Interferent gas	Concentration interferent added (ppb)	Concentration range of H_2S studied (% H_2S recovered)		
		21–33 ppb	*3.5 ppb*	*0.5–1.5 ppb*
SO_2	924	95	102	
	211			128
CH_3SH	43	93	107	
	37			103
$(CH_3)_2S$	186	108	112	
	228			84
$(CH_3S)_2$	32	99		
	145			102
	139			96
CH_3OH	1031–4469	101	107	
	450–550			88
$CH_3 \cdot CHO$	436–614	95	118	
	161			102
$(CH_3)_2CO$	365		107	
	443			94
α-Pinene	277		117	
NO_2	2560			41
	300			89
	42			96
O_3	668		64	
	57			85
	26			104
Cl_2	1129		51	
	42			69

authors also compared the Buck and Stratmann (*55*) molybdenum blue method with the Mecklenburg and Rosenkranzer (*56*) methylene blue method for determination of hydrogen sulfide and concluded that the reaction of hydrogen sulfide with methylene blue produces a more intense color and is therefore to be preferred.

B. Adsorption Sampling

Buck and Gies (*57*) collected atmospheric hydrogen sulfide in dry sorption tubes packed with glass beads treated with $KHSO_4$ and Ag_2SO_4. The collected sulfide is dissolved with a solution of $ZnCl_2$ in concentrated hydrochloric acid and the liberated hydrogen sulfide determined by the molybdenum blue method. The concentration range over which the authors tested this method, 278–417 $\mu g/m^3$ (200–300 ppb), is well above values normally found in the atmosphere.

Although the authors carefully considered a wide variety of possible interferences from other air pollutants, the validity of their hydrogen sulfide test atmosphere was not adequately established. Air oxidation, light sensitivity, and storage delays which have been shown to affect adversely the stability of cadmium sulfide and lead sulfide, were not evaluated. These are possible sources of similar errors when collecting hydrogen sulfide with the silver sulfate sorption tube, particularly since silver salts are known to be light sensitive.

A relative detection limit of 9 $\mu g/m^3$ (6.5 ppb) hydrogen sulfide for a 30-minute sample and 19.5 $\mu g/m^3$ (14 ppb) hydrogen sulfide over a 15-minute sample cycle is claimed for the Ag_2SO_4 sorption method. Extrapolated to a 2-hour midget impinger cycle, this would be equivalent to a sensitivity of 3.1 $\mu g/m^3$ (2.2 ppb) hydrogen sulfide. Buck and Gies conclude that the above sensitivities are "below the odor threshold (approximately 0.05 mg/m^3)." This conclusion is questionable, since the odor threshold for hydrogen sulfide is actually of the order of 0.56–5.6 $\mu g/m^3$ (0.4–4 ppb).

Natusch *et al.* (*57a*) described the collection of trace levels of hydrogen sulfide as low as 0.007 $\mu g/m^3$ (5 ppt) on a silver nitrate impregnated filter. The collected Ag_2S is dissolved with NaCN solution and analyzed fluorometrically using dilute fluorescein mercuric acetate. Air is sampled for several hours at 6.6 liter/minute at these background concentrations. For higher hydrogen sulfide concentrations, shorter sampling times or slower flow rates could be selected, or an aliquot of the filter washings could be taken for analysis.

C. Chemically Treated Filter Tape Samplers

1. Lead Acetate Tape

In lead acetate hydrogen sulfide tape samplers air is drawn through a lead acetate impregnated tape for a predetermined time, leaving a darkened area when hydrogen sulfide is present in the air. The exposed tape is then returned to the laboratory and the optical density determined. The hydrogen sulfide concentration is reported to be proportional to the optical density of the spot (*58*).

Two-hour average hydrogen sulfide concentrations have been determined in many localities using the American Iron and Steel Institute (AISI) sequential tape sampler (*59*) or other similar instruments (*60–62*). In field applications where the tape is returned to the laboratory for optical density determination there is a time lapse of as long as several days between the first and the last sample on the roll of tape. There is conflicting opinion concerning the stability of the lead sulfide color with storage either on the roll in the tape sampler, or subsequently in the laboratory, when analysis is delayed or when the tapes are exposed to air or light (*63, 64*).

Improved tape instrumentation provides either continuous photometric readout of the developing spot throughout the sampling period, or a single automatic readout upon completion of the sampling period (*59–62*). With continuous readout, short-term fluctuations in hydrogen sulfide concentration can be calculated from the slope of the recorder tracing. Immediate readout should minimize fading of the lead sulfide spots, since no more than 2 hours of storage is involved. This approach, however, has an inherent source of error because of the reported instability of the lead sulfide color under continuous instrument photometer illumination (*63–65*).

According to one instrument manufacturer's calibration curve, the minimum 2-hour average hydrogen sulfide concentration detectable by the lead acetate tape method is approximately 1.4 $\mu g/m^3$ (1.0 ppb) (*59*). However, two laboratories have been unable to produce lead acetate tape darkening until a minimum threshold concentration of 70–83 $\mu g/m^3$ (50–60 ppb) hydrogen sulfide has been exceeded, regardless of the exposure time (*66, 67*). In these tests, one laboratory utilized a well-designed gas dilution panel; the other a hydrogen sulfide permeation tube. Therefore, it is presumed that the dilution techniques used for the original instrument calibration were less precise than those that later became available.

Sanderson *et al.* (*63*) reported that lead sulfide spots fade seriously when they are exposed to light and laboratory air and are bleached by ozone and sulfur dioxide. They conclude that the fading of the lead sulfide spots during sampling, and in the interval between sampling and analysis, reduces the status of the method to a "qualitative indicator of the presence of hydrogen sulfide."

A report (*68*) that lead acetate tape gave negligible response when the odor of hydrogen sulfide was detected is undoubtedly related to subsequent observations that the spots fade on storage and under light, that the color formation is modified when sulfur dioxide is present, and that lead acetate tape will not discolor until at least 42–83 $\mu g/m^3$ (30–60 ppb) hydrogen sulfide is present. Another cause for misinterpretation of these data is that while the earlier literature reports the odor threshold for hydrogen sulfide of the order of 42–139 $\mu g/m^3$ (30–100 ppb), more recent studies utilizing more refined techniques for production and chemical determination of low concentrations, indicate its odor threshold to be in the order of 0.56–5.6 $\mu g/m^3$ (0.4–4 ppb) (*69–71*).

These data provide reasonable explanations for much of the paradoxical data provided by the lead acetate tape technique. The difficulty in providing rational explanations for the ability of trained personnel to detect the odor of hydrogen sulfide while conducting field sampling which did not reveal its presence was further confounded by reports, such as High and Horstman (*72*), of reasonably good agreement between data obtained by lead acetate paper tape samplers and bubblers using the unstabilized cadmium hydroxide–methylene blue method. The agreement obtained was apparently coincidental, since both methods were subsequently shown to report concentrations lower than those actually present. When two diverse sampling and analytical methods provide comparable concentration data, which agreed with the then available data for odor threshold it seemed logical at the time to conclude that the available methods and the threshold value were all reliable.

Siu *et al.* (*67*) suggest the following changes in the continuous readout lead acetate tape system to minimize some of these deficiencies: decrease the lamp voltage to 60 V to reduce photodecomposition of the lead sulfide spots; increase the relative humidity of the air being sampled to 80% by first bubbling it through a 2% sulfuric acid solution; eliminate sulfuric dioxide interference with a barium acetate prefilter. Although a net increase in sensitivity to 28 $\mu g/m^3$ (20 ppb) is achieved, this improved detection threshold is still nearly tenfold above the human olfactory threshold for hydrogen sulfide.

Presher and Lahmann (*54*) conclude that neither lead acetate nor silver nitrate impregnated tapes are sufficiently sensitive or accurate for hydrogen sulfide monitoring.

2. Mercuric Chloride Tape

Paré (*65*) has confirmed many of the Sanderson *et al.* (*63*) observations and suggested the use of a mercuric chloride impregnated tape. The resulting spots are stable when exposed to 19.6 mg/m^3 (10 ppm) of ozone for three days. The spots are also stable in the presence of nitrogen oxides. Although Paré reports the spots to be stable in the presence of sulfur dioxide, Dubois and Monkman (*73*) found that the color threshold of the mercuric chloride tape is significantly increased when appreciable sulfur dioxide is present in the sampled air.

Hochheiser and Elfers (*74*) found that in the range 0.7–139 $\mu g/m^3$ (0.5 to 100 ppb) hydrogen sulfide, the colored spots are relatively stable with only a 10% color decay over a 7-day period following color development.

They reported that redevelopment of the tape with ammonium hydroxide solution prior to optical evaluation restores the samples to their original color. Addition of glycerine to the impregnating solution helps maintain humidity of the tape and improves collection efficiency. When methyl mercaptan is present at higher concentrations it produces yellow spots and interferes with the measurement of hydrogen sulfide. Chlorine bleaches the spots. No interference was noted with nitrogen dioxide, sulfur dioxide, organic sulfur gases, or ammonia. This lack of interference from sulfur dioxide may not contradict the Dubois and Monkman observation that the response threshold for the mercuric chloride tape is enhanced by sulfur dioxide. The mercuric chloride tape method is complicated by the need to develop the color of the spots in the laboratory just prior to their photometric measurement. Although mercuric chloride corrodes the metal parts of presently available tape samplers, substitution of more resistant metals, such as stainless steel, might eliminate this problem, but would most likely raise the cost of the instruments.

D. Silver Membrane Filter

Falgout and Harding (*53*) describe a dynamic method for the determination of hydrogen sulfide and methyl mercaptan by measuring the decrease in reflectance of silver membrane filters through which air had been drawn at a known rate for a selected time period. The method depends upon the reaction between metallic silver and hydrogen sulfide. The authors conclude that the silver membrane response to hydrogen sulfide and methyl mercaptan is not affected by nitrogen dioxide or ozone, and that the filters can readily be protected from ultraviolet light by wrapping the filter holder with black tape and from atmospheric particulates by placing a cellulose filter ahead of the silver membrane filter. As applied by the authors, air is sampled at the rate of 0.9 liters/minute and for 24 hours. Average concentrations of sulfide and mercaptan are reported as $\Delta R/m^3$, where ΔR is the change in reflectance. Further evaluation of the silver membrane technique should be undertaken under controlled conditions in which ΔR is related to known concentrations of hydrogen sulfide generated by the permeation technique.

E. Fluorimetry

Axelrod *et al.* (*75*) describe a fluorescence analysis method for hydrogen sulfide using a fluorescein–mercuric acetate reagent. The method is reported to have a detection limit of 1.4 $\mu g/m^3$ (1 ppb) hydrogen sulfide.

A slight quenching effect is observed with 10^5-fold excesses of Br^-, SO_3^{2-}, and NO_2^-. Hardwick *et al.* (*76*) also determine hydrogen sulfide by the degree of fluorescence quenching of a 0.01 *N* sodium hydroxide solution containing 2.5 to 60 mg/liter of fluorescein–mercuric acetate, depending upon the desired range of application from 0.3 to 60 μg hydrogen sulfide.

Andrew and Nichols (*77*) describe a continuous analyzer for hydrogen sulfide. The fluorescence of tetraacetoxymercurifluorescein is quenched linearly between 7–1390 μg/m³ (5–1000 ppb) hydrogen sulfide. A short-term sensitivity of approximately 7 μg/m³ (5 ppb) is claimed. This reagent appears to have useful application for continuous automatic hydrogen sulfide analysis. However, extensive studies will be required to establish possible interferences from other air pollutants and to confirm the reported sensitivity.

F. Selective Ion Electrodes

More than 20 ion-selective electrodes are commercially available. Light (*78*) used three electrodes—fluoride, sulfide, and cyanide—as detectors for process monitoring instruments for use in the pulp and petroleum industries. There is insufficient evidence to establish the limits of sensitivity and hence the potential usefulness of the sulfide electrode in an atmospheric monitor, although considerable progress has been reported in the application of the fluoride electrode for atmospheric monitoring in the ppb concentration range in the vicinity of alumina reduction plants (*79*).

G. Sensory Evaluation

No presently available instrument or chemical method has the required sensitivity to replace fully the human nose. In situations where a single source is responsible for odor complaints, a continuing community odor survey may be a valuable technique to evaluate the effectiveness of sulfur gas abatement methods. To be of value, the survey must be conducted on a daily basis over a period of years to establish trends. The results of an occasional survey will have little significance because of day-to-day ground-level odor intensity variation because of meteorological variables even though there be a constant level of odor emission.

The value of an organized, continuing community odor survey is perhaps best exemplified by the many years of observations by a volunteer odor panel living in the vicinity of the Peña Pobre kraft mill near Mexico City (*80*).

Semiquantitative community odor evaluation can be achieved with the

Scentometer, a simple, inexpensive portable device developed by Huey and co-workers (*81*) which is discussed in greater detail in Chapter 8, this volume.

H. Effects Methods

Various effects methods have been used to provide a relative measure of sulfides in the air. Chanin *et al.* (*82*) suggest the use of ceramic tiles soaked in an aqueous solution of lead acetate, acetic acid, and glycerol to provide a semiquantitative measure of hydrogen sulfide level by visual estimate of the darkening of tiles. Gilardi and Manganelli (*83*) have provided a comprehensive evaluation of the influence of exposure variables upon the lead sulfide color developed and have concluded that such tiles have a useful upper exposure limit of 8 hours and that tiles have to be exposed in a light-proof chamber designed to minimize air movement.

Qualitative evaluations of atmospheric hydrogen sulfide have also been made by measuring the darkening of test panels coated with lead-base paint (*84*). Following 30-day field exposure, the amount of paint darkening is reported as the percent decrease in light reflectance.

IV. Mercaptans

The odor threshold for methyl mercaptan is in the range of 0.6–1.0 $\mu g/m^3$ (0.3–0.5 ppb) (*69–71*). Extremely sensitive analytical techniques, which are not presently available, are required if one is to measure ambient levels of mercaptans down to the human odor-perception threshold.

Mercaptans in air are sampled by absorbtion in an aqueous solution of mercuric acetate–acetic acid in midget impingers (*85*). Its spectrophotometric determination depends upon reaction between *N*,*N*-dimethyl-*p*-phenylenediamine and mercaptan in dilute nitric acid in the presence of ferric chloride.

The mercuric acetate must be free of mercurous ion. If mercurous ion is present turbidity will result when the chloride ion-containing reagents are added in the last step of the analytical procedure.

This method is intended for the determination of mercaptans in the range below 196 $\mu g/m^3$ (100 ppb). The minimum detectable amount of methyl mercaptan is 0.04 μg per ml in a final liquid volume of 25 ml. When sampling air at the maximum recommended rate of 1 liter/minute for 2 hours, the minimum detectable mercaptan concentration is 3.9 $\mu g/m^3$ (2.0 ppb).

Moore *et al.* (*85*) report that hydrogen sulfide does not interfere with this reaction when present in the range of 0–150 μg. Figure 2 compares the absorbance of the complexes produced by hydrogen sulfide, mercaptan, and alkyl disulfides with the Jacobs and Moore reagents and confirms that hydrogen sulfide should not interfere with this phase of the determination of mercaptan. The triangles in Figure 2 indicate the wavelengths at which the absorbance measurements are made, i.e., 670 nm for hydrogen sulfide and 500 nm for methyl mercaptan. One-hundred μg hydrogen sulfide gives a mercaptan color equivalent to 1.5–2.0 μg mercaptan. However, if hydrogen sulfide is present in the sampled air, it may cause a precipitate in the sample absorbing solution which must be removed before proceeding with the analysis.

Approximately equimolar response is obtained from the hydrolysis products of dimethyl disulfide—the extinction coefficient being 4.4×10^3 for the amine–mercaptan reaction product and 5.16×10^3 for the amine–dimethyl disulfide reaction product. In practice the collection efficiency for dimethyl disulfide in aqueous mercuric acetate is low, and the actual interference negligible. Sulfur dioxide up to 250 μg does not influence color development even when sampling a test atmosphere containing 786

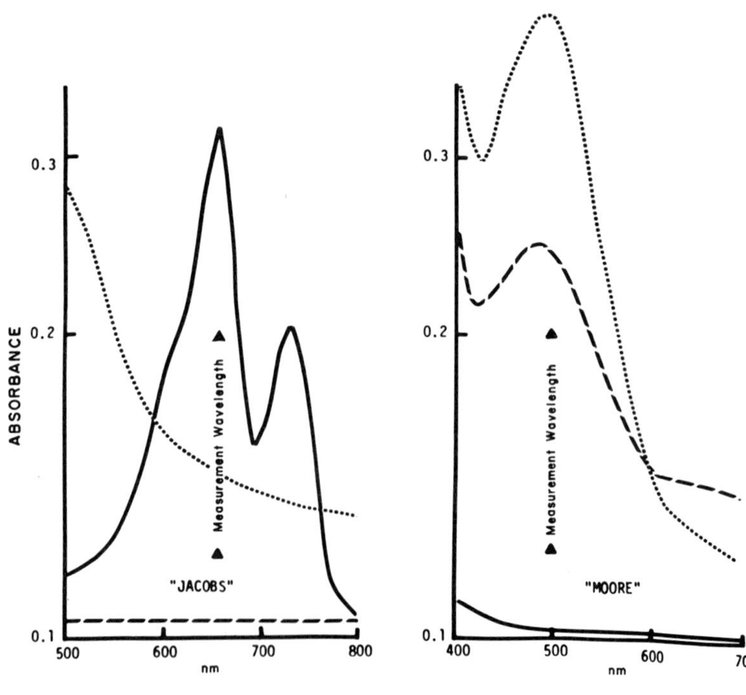

Figure 2. Absorbance—"Jacobs" and "Moore" reagents; μg of reagent added: solid line, 25 μg H_2S; dotted line, 100 μg CH_3SH; dashed line, 103 μg dimethyl disulfide.

mg/m^3 (300 ppm) sulfur dioxide. Nitrogen dioxide does not interfere up to 700 μg nitrogen dioxide when sampling a test atmosphere at 6 ppm. Higher concentrations of nitrogen dioxide cause a positive interference when mercaptans are present, but no interference in the absence of mercaptans.

In several instances field interference has been reported from unknown compounds when using this method in the vicinity of kraft mills. In one instance a light-colored precipitate was formed which was not believed due to emissions from the kraft mill, since in most instances when the precipitate occurred the wind was not blowing from the mill.

The coefficient of variation for four mercaptans from methyl to hexyl mercaptan ranges from 0.0–2.5%. The coefficient of variation increases with increasing molecular weight of the mercaptan.

Okita (*86*) reports use of Gelman A glass fiber filters impregnated with 3% mercuric cyanide to collect mercaptans, followed by the analytical procedure described below.

V. Alkyl Sulfides and Disulfides

Satisfactory wet chemical analytical methods for alkyl sulfides and disulfides sensitive to concentrations usually found in the ambient air, were not available at the time of writing this chapter. Okita (*86*) uses Gelman A glass fiber filters impregnated with 5% mercuric chloride to absorb both dimethyl sulfide and mercaptans. A sampling rate of 1 liter/minute through 5-cm filters is used. The collected gases are liberated by adding 20% hydrogen chloride to the filters and purging with nitrogen. The released gases are collected in a refrigerated organic solvent, and the collected compounds are determined by gas chromatography using a stainless-steel column packed with tricresyl phosphate on Celite and a flame ionization detector. Scaled-up filters, 8 × 10 in. using 100-liters/minute flow rates permit detection of fractional parts per billion for 10–30-minute sampling periods.

VI. Multifunctional Instrumental Methods

As the number of trace gases of interest increases an urgent need for multifunctional methodology and detectors capable of simultaneously monitoring a variety of species is generated. Gas chromatography has been utilized with various detectors and/or columns to achieve the necessary selectivity and separations. Coulometry, with preselective filters or scrubbers, provides another such approach.

These methods usually utilize detection techniques which are not specific for individual sulfur compounds. Coulometric (amperometric) or flame photometric detectors (FPD) can be utilized to measure the total sulfur gas content of the air. When used in conjunction with scrubbers, chemically impregnated filters, or gas chromatographic columns to give initial separation of individual sulfur gases in a complex mixture, they can provide a quantitative measurement for each sulfur gas present.

A. Coulometry

Prior to the development of FPD, several modifications of coulometric (amperometric) detectors were used extensively for the determination of total bromine-titratable compounds (mostly sulfur containing) in the atmosphere. Coulometry and flame photometry have been defined in the *Federal Register* as equivalent instrumental methods for the determination of sulfur dioxide (*19*), that is, continuous techniques which will provide results similar to those obtained by the manual reference modified West–Gaeke procedure (*4*).

Coulometry is based upon electrically generating a selected titrant (ion or element) in a titration cell. The current required to generate sufficient titrant is a linear measure of the concentration of reactable compounds in the gas sample being passed through the coulometric cell. In practice the titration cell has two sets of electrodes—a generating pair and a reference pair. A constant low level of titrant is continuously generated to produce a "zero level" on the instrument. When reactable compounds enter the cell the preset level of titrant is decreased. An associated feedback amplifier responds to the change in sensed titrant level and generates sufficient additional titrant to maintain the initially selected "zero level" concentration. The additional generating current is related stoichiometrically to the reactant by the following equation:

$$\text{Microequivalents} = \frac{Q \text{ Coulombs} \times 10^6}{(96{,}000 \text{ Coulombs/equivalent})} \tag{1}$$

Instrumentation for coulometric titration with electrolytically generated bromine has been used for more than two decades for the determination of bromine-oxidizable sulfur gases (*87*). The method is not sulfur dioxide specific since any compound which will react with bromine can be interpreted as equivalent sulfur dioxide. The Titrilog (*88, 89*), one of the first commercially available instruments embodying this principle, had a sensitivity of approximately 260 μg/m^3 (100 ppb) for sulfur dioxide. The sensitivity of the Titrilog was improved approximately tenfold to 26 μg/m^3 (10 ppb) by addition of a "R-C" network (*90, 91*).

The Barton "Titron" (*92*) represents a similar instrument using bromine coulometry and solid-state electronics. This instrument has a limiting sensitivity of approximately 35 $\mu g/m^3$ (25 ppb) for hydrogen sulfide. Theon (*93*) has reported a 7 $\mu g/m^3$ (5 ppb) sensitivity for "total sulfur" in the ambient air with this instrument by substituting a potentiometric recorder with a 10-in.-wide chart for the $2\frac{1}{2}$-in. galvanometric recorder provided by the manufacturer. The Titrilog II (*89*) has a reported sensitivity of 52 $\mu g/m^3$ (20 ppb) sulfur dioxide.

The Philips coulometric analyzer (*94*) indicates a standard full scale sensitivity of 2990 $\mu g/m^3$ (1.15 ppm) sulfur dioxide with a possibility for achieving 299 $\mu g/m^3$ (0.115 ppm) full scale sensitivity. The reported response time is 1.5 minutes to 63% of final value and 3 minutes to 95% of final value. The instrument is provided with a built-in sulfur dioxide calibration source which may represent the first sulfur dioxide instrument that can automatically provide an instrument calibration. Beckman's (*95*) coulometric instrument is reported to have a slower response time (*96*).

Microtitration cells utilizing silver–silver chloride electrodes have been used for amperometric analysis of hydrogen sulfide and mercaptans (*97*). The cell will not respond to sulfur dioxide or alkyl sulfides and disulfides. The sensitivity of this detector is reported to be approximately 260 $\mu g/m^3$ (100 ppb).

An iodine microtitration cell (MCT) (*98, 99*) and a Dohrmann coulometer (*100*) have been used for the analysis of sulfur-containing pesticides after their separation by gas chromatography and either oxidation to sulfur dioxide or reduction to hydrogen sulfide in an appropriate furnace (*101–103*). This detector has also been used for the direct analysis of sulfur dioxide in air; the sensitivity of the iodine microtitration cell for direct atmospheric analysis is 260 $\mu g/m^3$. The cell response is relatively slow and the electrolyte too short lived for practical application.

The microtitration cell, as modified and converted to bromine coulometry by Adams *et al.* (*104*), has a limiting sensitivity of approximately 7.8–13 $\mu g/m^3$ (3–5 ppb) for hydrogen sulfide, thereby approaching the sensory detection threshold. Selective impregnated membrane prefilters (Table II) developed for use in sequence ahead of the bromine MCT cell, provide separate analyses for sulfur dioxide, hydrogen sulfide, mercaptans, and alkyl sulfides, and disulfides (*105*). The filters must be placed in the order shown in Table II for optimum results. When mercuric nitrate or silver nitrate impregnated filters are used to retain hydrogen sulfide or methyl mercaptan, the nitric acid reaction product interferes with MCT response. This interference can be eliminated by using a bicar-

Table II Selective Filtration Series, Percent Retention

Impregnant	Test gas (% retained by filter)				
	SO_2	*H_2S*	*CH_3SH*	*Dimethyl sulfide*	*Dimethyl disulfide*
None	0	0	0	0	0
$NaHCO_3$	100	10	4	5	3
Lead acetate	100	100	4	5	5
Mercuric acetate	100	100	100	5	3
Hg acetate + tartaric acid and $NaHCO_3$	100	100	100	85	10
$AgNO_3 + H_3BO_3$ + tartaric acid and $NaHCO_3$	100	100	100	100	100

bonate impregnated backup filter. These filters (25 mm) have a useful life of at least 24 hours at concentrations well above those normally observed in the field when the air flow rate is maintained below 200 ml/minute.

The filters may be preceded by an all plastic and stainless-steel electrostatic precipitator to prevent buildup of particulates on their faces. The electrostatic filter must be operated at a sufficiently low voltage to prevent ozone production which would yield a negative response from the MCT.

In an automatic sequential samping system, air is sequentially sampled through a series of these filters. A six-cam timer automatically controls the 5-minute sequence of the five prefilters and a 5-minute instrument zero over a 30-minute cycle. The sample air then passed through the bromine MCT, thereby providing a stepwise recording of the concentrations of the various sulfur-containing gas fractions present in the air (Fig. 3). Electrolyte in the microtitration cell is automatically replaced to maintain a drift-free condition. A sensitivity of 26 $\mu g/m^3$ (10 ppb) sulfur dioxide is achieved and the response time for 90% of final value is attained in 15 seconds.

B. Flame Photometry

In the flame photometric detector (FPD) (*106, 107*) the sulfur-containing species are burned in a hydrogen-rich flame, and the sulfur is converted to excited S_2 which is at a higher energy state. Upon return of the excited S_2 molecule to the ground state, light is emitted at character-

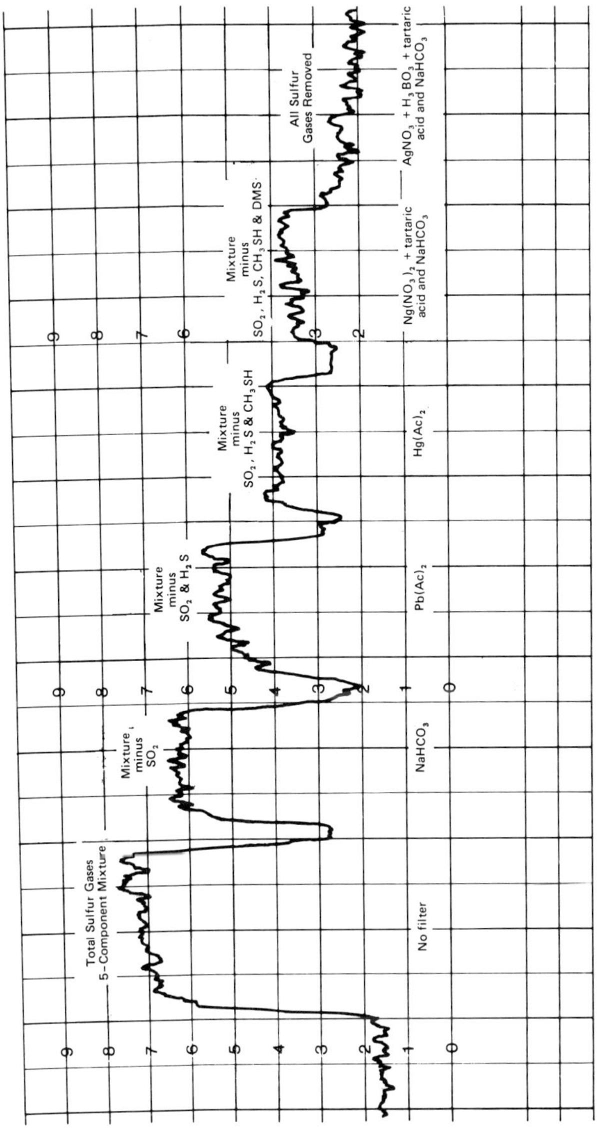

Figure 3. Stepwise separation and analysis of 5-component, sulfur-containing gas mixture.

istic wavelengths between 300 and 420 nm. An optical filter (394 ± 5 nm) isolates one of the sulfur waveband's emission from other sulfur species wavelengths and extraneous light. Sulfur emission bands at 374 and 383.8 nm are of almost equal intensity and could be used for the quantification of sulfur.

Phosphorus is a potential source of interference because it produces a photofluorescence in the 394-nm region. However, phosphorus compounds are not generally found in the ambient air, except in the vicinity of agrichemical spray applications of phosphorous-containing pesticides, and, even if present, are not usually within the concentration range of sulfur compounds. The FPD discrimination ratio is approximately 10,000:1 for hydrocarbons and at least 1000:1 for other gases (*108*). Air produces a spectral continuum yielding a measurable signal at 394 nm, which contributes to the background noise. These potential interferents need be considered only when the FPD is used as a continuous total sulfur gas analyzer. When FPD is used with gas chromatographic separation, these interferences are not considered to be significant. The response of the detector is nonlinear. A linear relationship may be obtained by plotting the detector response against concentration on a log scale or by using a log-linear amplifier. With either of these techniques, the linear dynamic range is approximately 130–500 $\mu g/m^3$ (50 to 200 ppb). The precision or repeatability of the method depends upon close control of sample and hydrogen flow to the detector. Changes in flow of 1% changes detector response approximately 2%. Adams (*109*) suggested that "it should be possible to use this detector for the direct analysis of the atmosphere for total sulfur gases." Harding (*110*) reports varying detector sensitivity to hydrogen sulfide and methyl mercaptan within the range of 85–170 $\mu g/m^3$ (50–100 ppb).

Sugiyama *et al.* (*110a*) examined the effect of bond structure of sulfur atoms on the emissive response of sulfur-containing compounds with a FPD. The FPD response varied widely with the type of sulfur compound. For example, butyl sulfide, the most FPD-sensitive compound studied, was approximately 100 times more responsive than pentyl disulfide.

Shiffman and Frank (*110b*) discussed the different emissivities observed from each sulfur-containing compound, and indicated the need for a standard calibration curve for each anticipated sulfur-containing compound when the FPD is used in the gas chromatographic mode. If the FPD is used as a total sulfur gas analyzer, it must be calibrated for the major compound in the mixture and a varying degree of error anticipated depending upon the preponderance of the major compound in the mixture.

Greer and Bydalek (*111*) characterized the response of the FPD for hydrogen sulfide and sulfur dioxide, the two most likely sulfur-containing gases to be found in urban air. They showed that the relative loss of detector sensitivity at higher sulfur masses is due to self-absorption rather than photomultiplier tube fatigue. Maximum detector sensitivity is expected between 0 and 60 ng of sulfur. Above this mass the FPD response becomes nonlinear.

Stevens and co-workers (*112, 113*) report the use of an improved FPD for the direct analysis of total, volatile sulfur compounds in the atmosphere. The detector is reported to have a specificity ratio for sulfur to nonsulfur compounds of 40,000:1 and a sensitivity to sulfur dioxide down to 13 $\mu g/m^3$ (5 ppb).

C. Gas Chromatography

Since its first applications to the analysis of kraft source gases (*114–117*), the use of gas chromatography (GC) for the separation and determination of individual sulfur compounds has progressed through several generations of detectors and column packings, with ever improving sensitivity and specificity. In GC the air sample was first dried by passing through a drying agent and then concentrated in a cooled section of stainless-steel tubing packed with the same column packing used for separation. The samples are then desorbed and analyzed by a detector with flame photometric (FPD), flame ionization (FID), or electron capture (ECD) detectors. The ECD has been found to be unsatisfactory because of its loss of sensitivity when water vapor is eluted, and its apparent temporary poisoning by methyl mercaptan (*118*). Although the FID has been used extensively for gas chromatographic analysis of atmospheric organic gases and vapors, it has been displaced for sulfur gas analysis by the more specific FPD discussed above. Using GC analysis, Williams (*119*) has identified more than 30 organic compounds including dimethyl disulfide in ambient air.

Eads and Cooper (*120*) report that "odor producing sulfur compounds . . . including sulfur dioxide, have been detected at very low levels . . . by modifying a gas chromatograph coupled to a microcoulometer." The primary difference between this system and those previously reported (*121, 122*) is the use of a sample concentration step wherein the air is drawn through an impinger containing methanol. Proper microliter quantities of the methanol are then injected into the GC system. This procedure overly complicates the technique without achieving greater sensitivity. The claimed sensitivity is questionable since the GC column used is 10% Triton X-305 on Fluoroport T which is known to retain appreciable quantities of sulfur-containing compounds.

Brody and Chaney (*123*) have reported the use of the FPD for the GC analysis of sulfur- and phosphorous-containing compounds for air and water pollutants, pesticide residues, and natural gas.

Although the FPD detector now provides adequate sensitivity and compound specificity, automated GC analysis of ambient air concentrations of sulfur gases has had to wait for development of nonreactive partition liquids and column packings. The literature contains conflicting information concerning the ability of various GC columns, e.g., Triton X-305 (*114, 115*), Deactigel (*124, 125*), and polyphenyl alcohol (*126*) to separate and/or release specific sulfur-containing gases. These anomalies may be due in part to individual differences in column preparation.

Stevens and co-workers (*121*) report two techniques for the direct gas chromatographic analysis of sulfur gas mixtures in the low ppb range. The first involves "differential response chromatography" (*127*), in which known concentrations of the gases to be determined are first added to the chromatographic carrier gas by passing the carrier gas over appropriate permeation tubes. A frontal elution results and the eluate composition assumes a steady-state value. Unknown samples are then introduced into the same column; when they emerge they will generate signals corresponding, in elution time to the minor components, but, in magnitude and algebraic sign, to the difference in concentration from the stream. Using this technique, Stevens *et al.* separated from other sulfur gases, and determined chromatographically, sulfur dioxide concentrations as low as 52 $\mu g/m^3$ (20 ppb).

The second, and more useful, GC method involves minimizing scavenging and reaction by the column packing. Polyphenyl ether containing 0.05% phosphoric acid, coated at 5% on 30–40 mesh Teflon, permits passage and separation of highly reactive sulfur-containing gases down to concentrations of 26 $\mu g/m^3$ (10 ppb).

The direct chromatographic technique has been automated (*18*) with air samples periodically obtained with a rotating multiport valve. The limit of detection is 6 $\mu g/m^3$ (2 ppb).

When Stevens and O'Keeffe (*128*) compared total atmospheric sulfur-containing gases, as measured by the FPD procedure against sulfur dioxide as measured by the modified West–Gaeke procedure, good correlation was found between these two methods; it was concluded that sulfur dioxide was the major sulfur-bearing gas in Cincinnati, Ohio air.

The automated, semicontinuous FPD-GC method described by the Sulfur Subcommittee of the Intersociety Committee (*129*) includes an automatic, multiport switching valve combining sampling and calibrating functions. The sampled air and calibration gas(es) are alternatively drawn through the valve sample loop. On a preset cycle the valve sample loop is switched to the carrier gas, and the sample in the loop is purged into the gas chromatographic column for resolution. On alternate cycles the calibration gas(es) is (are) purged through the sample loop into the GC column to provide standard reference peak(s) and to maintain column conditioning. The basic components of the system are shown in Figure 4. The automated 10-port rotary sampling valve for air and calibration gas sampling volume measurement and injection is shown in Figure 5.

The FPD, when coupled with a suitable GC column, offers sulfur-selective potential through separation of the sulfur-containing compounds which may be simultaneously present in community air; whereas preselective filtration techniques are limited to separation by compound class. To extend FPD–GC system detection limits to the 1 $\mu g/m^3$ (sub-ppb) range required to establish the rural and global background levels, it will be necessary (*130*) to (a) optimize the flow characteristics of the hydrogen-rich flame, (b) optimize the photomultiplier bias voltage with suppression of the high frequency photomultiplier tube signal/noise ratio input to the electrometer, (c) apply constant temperature cryogenic cooling to the photomultiplier tube, (d) upgrade GC electrometers, and (e) develop an automatic high-ratio sample enrichment freeze-out loop capable of handling large volumes of air for subsequent slug injection onto the column head.

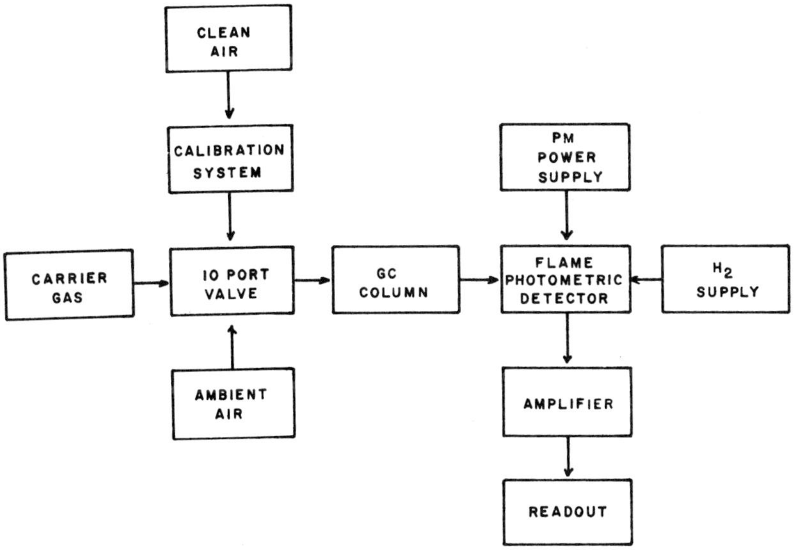

Figure 4. Automated Flame Photometric Detector–Gas Chromotography system.

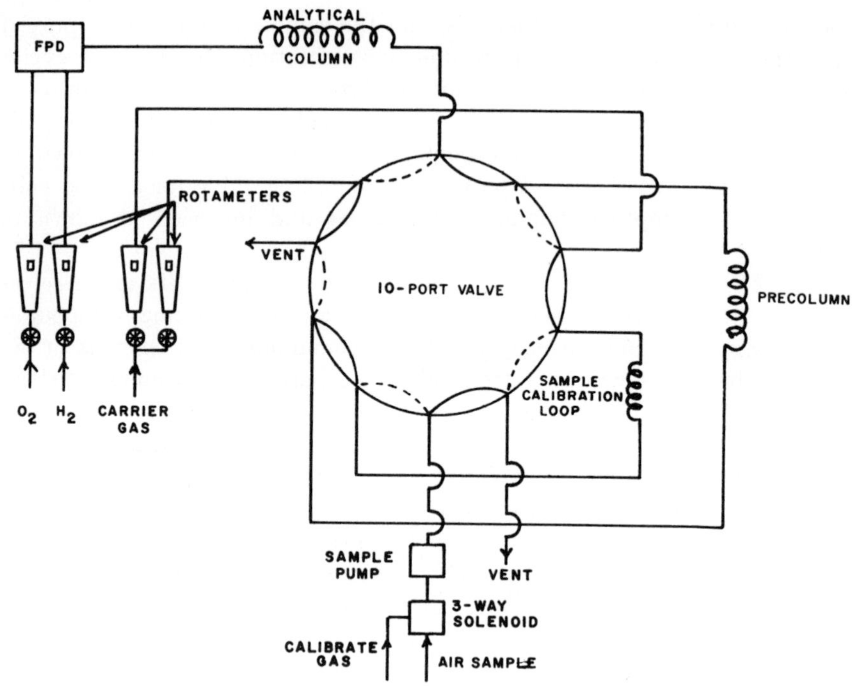

Figure 5. Automated 10-port rotary gas sampling valve.

D. Dual Flame Photometry/Flame Ionization

Stahl (*131*) has used a FPD flame as a FID flame to measure hydrocarbon levels in the atmosphere. To accomplish this an ion collection probe is inserted into the FPD burner chamber. Although the FPD flame is normally operated in a hydrogen-rich condition and the FID flame normally burns in an air-rich condition, this author found that the hydrocarbon ionization characteristics of the FPD were suitable for hydrocarbon analysis. Furthermore, the FPD gave a measure of the number of hydrocarbon molecules contained in the sample, whereas the typical FID analyzer response is in terms of methane equivalents. Apparently in the cooler hydrogen-hyperventilated flame the larger hydrocarbon molecules do not fragment to the precursor CH radical of the primary CHO^+ ion (*132–134*). The single hydrogen-rich flame FPD/FID system provides a continuous readout of total sulfur and total hydrocarbon concentrations when the FPD flame is fed ambient air. When associated with a GC column, or an appropriate catalytic hydrocarbon oxidizer and a sulfur dioxide scrubber, the FPD/FID may also provide a continuous measurement of hydrogen sulfide and methane. This dual system appears to have the potential for measuring methane and total hydrocarbons on one channel and hydrogen sulfide and total sulfur gases on the other, if equipped with suitable switching valves.

E. Other Chemiluminescent Methods

Snyder and Wooten (*135*) have investigated the use of O atom chemiluminescence for the detection of sulfur dioxide. The reaction of sulfur dioxide and O atoms produces sulfur dioxide chemiluminescence in the ultraviolet

$$O + O + SO_2 \rightarrow O_2 + SO_2 + h\nu \qquad (2)$$

The sulfur dioxide sensitivity using this reaction is 2.6 $\mu g/m^3$ (1 ppb). It was necessary to use an interference filter at 280 nm since potentially interfering molecules produce chemiluminescence with O atoms in the ultraviolet. Development of this reaction into a practical instrument for measuring ambient sulfur dioxide has been hindered by lack of a reliable source of O atoms.

Kummer *et al.* (*136*) observed chemiluminescence from the gas-phase reactions of ozone with a variety of sulfur compounds including hydrogen sulfide and organic sulfides. The emission observed from both hydrogen sulfide and organic sulfides is due to electronically excited sulfur dioxide. This suggests that the use of ozone and of a second reacting gas could lead to a nonspecific detector for gaseous sulfur compounds. Since the reaction rate of ozone with different sulfur compounds varies considerably, the sensitivity of such a detector would depend upon the compounds measured.

Okabe *et al.* (*137*) have measured the intensity of the ultraviolet fluorescence of sulfur dioxide produced by the absorption of the Zn 2138 Å line. A linear response from 26 to 1310 mg/m^3 (0.1 to 500 ppm) with a reported detection limit of 26 $\mu g/m^3$ (10 ppb) is claimed. The method is unaffected by water vapor and other chemicals. An early commercial prototype instrument has a sensitivity of 1.3–2.6 mg/m^3 (5 to 10 ppm) on the 130 mg/m^3 (500 ppm) low range (*138, 139*). These authors recognize the need to extend the sensitivity to a lower ppb range. Use of a more intense light source such as a laser would be suitable, but a laser at the required wavelength has not yet been developed.

Schwarz *et al.* (*139a*) modified the above instrument to extend the detection limit to the low ppb range by means of photon counting of the sulfur dioxide fluorescence excited and an improved cell design. At 22.4 $\mu g/m^3$ (8.6 ppb), the standard deviation is 29% for a counting time of

1 minute. The detector response is linear from 22.4 $\mu g/m^3$ to 4688 mg/m^3 (8.6 ppb to 1.8 ppm).

F. Electrochemical Transducers

Electrochemical transducers (ECT) represent a recent development of sealed detectors. The gaseous pollutant to be measured diffuses into a cell and reacts quantitatively. This technique avoids the requirement for closely calibrated and regulated liquid reagent and air flows associated with the automation of wet chemistry techniques for sulfur gases such as conductimetry, colorimetry, and coulometry. Reacting gases diffuse through a thin semipermeable membrane into the ECT where an electrochemical reaction occurs depending upon the transducer electrolyte, sensor and counter electrode composition, and an externally imposed "retarding" potential. The resulting current is proportional to the concentration of the reacting gas in the sample. Selectivity of the ECT is dependent upon the type of semipermeable membrane, the electrolyte, the electrode materials, and the retarding current (*140*). The useful life of the ECT in large part is determined by the rate of counter diffusion of water vapor from the electrolyte through the membrane. The more sensitive detectors have thinner membranes and hence a shorter useful life. Sample air flow is not critical since the diffusion driving force is proportional to the partial pressure of the reactive gas in the sample air. The sulfur dioxide ECT appears to have adequate sensitivity and stability for source monitoring and for ambient monitoring in SO_2 polluted urban areas. Selectivity (rejection ratio for other sulfur-containing compounds including hydrogen sulfide, methyl mercaptan, etc.) appears to be adequate for most ambient sampling situations. However, field and laboratory evaluation of several ambient sulfur dioxide ECT units below the 131 $\mu g/m^3$ (50 ppb) range have shown unacceptable temperature and baseline drifts which complicate data evaluation and raise questions regarding data validity.

G. Solid-State Detectors

Reynolds (*141*) reports the use of a N—P solid-state semiconductor Taguchi Gas Sensor (TGS detector) for detection of gases and organic vapors in an air carrier. The gases replace the surface concentration of O_2 and disturb the Fermi level of electrons thereby causing a current to flow. Reynolds stated that the TGS detector sensitivity is equal to or better than the FPD for sulfur-containing gases. The TGS is nonspecific, but could be used with a gas chromatograph to measure ambient air mixtures following separation on a suitable GC column. Because of its nonspecificity the resultant chromatograms would be more complex than would be obtained with an FPD under similar circumstances.

Bechtold (*142*) used a solid-state electrochemical cell Pt/Ag/AgI/Ag_2S/Pt to measure concentration changes in sulfur- (and halogen-) containing compounds in a carrier gas stream. The sulfur (and halogen) compounds must be first oxidized over platinum at 800°C and then passed over the electrode surface within an operating temperature range of 200–400°C. The electrode is reported to be linear over a range of four orders of magnitude with a detection limit for carbon disulfide of approximately 1 ng. Although this detector was first used as a GC detector, it has a potential use in a continuous monitor for sulfur gases.

LaRoy *et al.* (*143*) have investigated the electrical properties of thin-film solid-state electrochemical cells using a lanthanum fluoride electrolyte. The cells have been found to be quantitative sensors for reducible gases, including oxygen, carbon dioxide, sulfur dioxide, nitrogen dioxide, and nitric oxide. The cells are formed on glazed alumina substrates by ultra high vacuum evaporation. An open gridlike structure is used to permit access of gas to the cathode-electrolyte interface. The relative selectivity of the cells to various reducible gases is not yet clear.

Bélanger (*143a*) reported the determination of sulfur dioxide by anodic oxidation on a gold–lead dioxide electrode. Air is bubbled through a 1 *N* $HClO_4$ electrolyte at 240 ml/minute and the sulfur dioxide oxidized in contact with the electrode. Air containing 167 $\mu g/m^3$ (0.12 ppm) produced an anodic current of 2.4 μA at an electrode polarization voltage of +950 mV. A signal equal to 10% of the background noise was obtained for a sulfur dioxide concentration of 70 $\mu g/m^3$ (0.05 ppm). Considerable improvement in sensitivity will be required to match the sulfur dioxide sensitivity of the flame photometric detector.

H. Other Methods

Recent developments in microwave, correlation, and laser spectroscopy indicate their high potential for measuring certain air pollutants without prior separation or actual sampling of the air. The unique spectral lines occurring at high microwave, ultraviolet, and infrared frequencies provide specificity for component identification and measurement. The primary drawbacks at the present state of the art are cost, size of equipment, sensitivity of the equipment to vibration or movement, and inadequate sensitivity for the concentration range required for many ambient and background air studies. Further developments in instrument optics and solid-state components and their application may well lead to less expensive compact analyzers with greater sensitivity for determining sulfur-containing gases in nonurban and background atmospheres. Correlation spectroscopy has been used successfully in source monitoring and plume tracing. The contribution of the earth's albedo to SO_2 profile measurements obtained with vertical correlation spectroscopy from aircraft has not yet been adequately defined.

I. Plasma Source Emission Spectrometry

McCormack *et al.* (*144*) have reported a sensitive emission spectrometric detector with excellent sensitivity for certain chemical bonds including C—S and C—N. This detector was successfully used by Bache and Lisk (*145*) for the detection of organophosphorous insecticide residues following GC separation. Characteristic spectral lines for these and other organic bonds produced in a microwave-excited argon plasma are measured in the ultraviolet light region using a grating spectrometer.

Moye (*146*) increased the sensitivity of this detector by systematic study of the discharge tube, cavity geometry, microwave power, discharge pressure, carrier and discharge gas, and carrier gas flow rate. Use of an argon–helium carrier gas mixture also increased the detector sensitivity limits.

An automatic instrument probably could be constructed using a relatively low-cost, multichannel quantometer-type spectrometer by substituting a microwave-excited argon plasma emission source for the usual spark or arc source. The airborne sulfur compounds would first have to be collected and concentrated in a liquid absorbent and then periodically injected into the argon plasma. A concentration step is required to separate the sulfur-, phosphorus-, and nitrogen-containing compounds from air, since the argon plasma will not form in the presence of air. The sensitivity of this technique for C—S bonds should approximate 18 $\mu g/m^3$ (10 ppb), if present for 5 minutes. Specific instrumentation utilizing this detector is not commercially available and would probably be more expensive than the usual first-generation air monitors.

Natusch and Thorpe (*146a*) reviewed the spectrum of selective gas chromatographic detectors available for analyzing complex environmental or biological samples. These authors state that the disadvantage of the microwave plasma detector is due to widely distributed spectral interferences from CN, C_2, and CH which necessitate the use of a monochrometer in place of the simple FPD narrow bandpass filter. These interferences can be minimized as noted above (*146*) but its use requires vacuum equipment.

J. The D² Spectrometer

The D² spectrometer measures the transmission versus wavelength function of a scanning single beam spectrometer which is electromechanically processed to produce a signal proportional to the second derivative of this function (*147*). Several advantages of the D² signal are (a) at low concentrations it is directly proportional to the total mass of the gas in the analysis path, (b) it is independent of intensity fluctuations of the excitation light source, (c) measurements of trace concentrations of the desired compounds can be made even though the spectral band of interest lies within a broad absorption band of another gas of considerably greater strength, and (d) conversely a narrow band of an interference gas has a negligible effect upon the measurement of a broader absorption band of a gas of interest.

The D² spectrometer is sensitive to the ultraviolet absorption spectra (210–315 nm) for several important pollutant gases including sulfur dioxide, nitric oxide, nitrogen dioxide, and ozone. The reported limits of detection for these four compounds are 7.9, 3.7, 37.7, and 23.5 $\mu g/m^3$ (3, 3, 20, and 12 ppb), respectively. The absorption cell is 1 m in length and the minimum detection limits stated are for 20 passes of the internally reflected light. The commercial instrument is set for a 72-second scan for each of four pollutants, and the complete scan time is 7 minutes in the automatic mode.

K. Long-Path Infrared Analysis Technique

Intramolecular bonding may be studied by infrared analysis technique. The usual multiple-reflection gas cell of 10–40-m path length would have a sensitivity in the ppm level. The Franklin Institute 430-m long-path infrared gas cell used in the study of the complex photochemical reactions of the Los Angeles-type smog has a sensitivity of approximately 67 $\mu g/m^3$ (100 ppb) (*148*). Infrared instrumentation is considered to be the least practicable of the physical methods for direct automatic air analysis.

L. Mass Spectroscopy

The mass spectrograph has not found extensive use for the measurement of gaseous pollutants, although it represents a sensitive multifunctional detector system. The quadrupole design reduced the instrumental size and power requirements. Despite favorable sensitivities, the conventional 70-V electron impact ionization mode is unsuitable because ions formed from major components of air, i.e., the oxygen, O^+ fragment, obscures the parent peak of methane at 16 amu. Reduction of the ionizing energy of the electron below the threshold of major constituent ionization results in inadequate sensitivity. Photoionization will overcome many of the problems inherent to electron impact ionization. The photoionizing energy may be controlled with an ultraviolet monochrometer or appropriate light source, to minimize ion fragmentation. When applied to trace gas analysis in air, major component ion fragments can be eliminated and still ionize many air pollution-related trace gases.

The major drawback of photoionization is its lowered sensitivity when using an ultraviolet monochrometer. Driscoll and Warneck (*149*) studied four different ionizing energies below the nitrogen ionization threshold. Hydrogen sulfide gave appreciable fragmentation and the potential of its 32-amu parent peak with the $^{16}O^{18}O$ isotope of oxygen was avoided by using the 104.8-nm ionizing wavelength. Detection limits for hydrogen sulfide and sulfur dioxide were reported to be in the 7.8–10 mg/m^3 (3–4 ppm) range, which is not adequate for ambient air analysis. Driscoll and Warneck have considered techniques for achieving the required sensitivity gain of 10^4 by direct ionization from an argon or hydrogen light source.

M. Condensation Nuclei Formation

Condensation nuclei (CN) and larger aerosols are first removed from an air sample by filtration. The aerosol-free air then enters a sample reaction chamber where its sulfur dioxide is converted to sulfuric acid aerosol by ultraviolet light in the presence of water vapor (*150*). The resultant sample containing the aerosol is then expanded in a CN monitor to produce a water vapor cloud. The optical transmittance of the cloud (CN/cm^3) is approximately linear in a log–log plot versus the initial sulfur dioxide concentration. It has been observed that the pH of the initial filter can exert a profound influence upon the measured sulfur dioxide concentration.

N. Remote Sensing Methods

Urban air quality is generally measured by point sensors. Long path averaging sensors, while unable to provide concentration information at points along the line of sight, have the advantage of providing a more representative average between two points, than could be provided by arithmetically averaging the readout from two point sensors, one located at the source and the second at the reflector of a long path averaging sensor. Long path sensors have also demonstrated their utility in (a) source surveillence through total flux data obtained by circling the source with a mobile instrument and (b) ambient air quality measurements vertically through the mixing layer.

Long-path sensors employ classic optical as well as recently developed laser methods. Classic methods include selective optical filters, spectrographs, interferometers, optical correlation, and second derivative devices. Specificity is obtained with either narrow or broad bandwidths and oscillating replica spectra. Sensitivity is gained by long-path reflection or multireflection.

Laser methods use high spectral purity to achieve specificity. Raman backscatter, resonance fluorescence, thermal radiation, and resonance absorption principles have been evaluted by several researchers for use in air monitoring. Although resonance absorption requires retroflectors or scatterers, it can be operated at safe energy levels to provide high sensitivity over a longe range and has the further advantage of being the only method capable of detecting pollutants in their natural airborne state—"long path"—as contrasted with "point" measurements of traditional "ingested" samples. Collimation problems are not considered with resonance absorption methods because a wide angle specular reflector can be used.

Snowman *et al.* (*151*) are developing an infrared laser atmosphere monitoring system (ILAMS) for multipollutant measurement of integrated concentrations over an optical path. The ILAMS uses a cooperative reflector for long-path monitoring at low, safe power levels, i.e., 1 W or less in the 3–12 μm range. Laser power densities at the transmitter are kept within safe densities (below 0.01 W/cm^2) by beam expansion through the system optics. Pollutants are identified by absorption spectroscopy in the middle infrared by their strong absorption bands.

The system employs a unique spectral scanning laser configuration for rapid sequential transmission of different wavelengths to obtain maximum spectral information from returned energy. This is achieved by introducing elements into the laser which create a separate optical cavity for each of the selected wavelengths. Means are also provided for switching rapidly from one wavelength to another. The beam from the plasma tube is spectrally dispersed by reflection onto a grating. Positioned mirrors intercept selected wavelengths and direct them through a chopper wheel to the end mirrors. Properly spaced holes in the rotating chopper wheel permit only one laser wavelength at a time to pass through to the end mirrors. The narrow energy wavebands of lasers as coherent high resolution energy sources typically provide two orders of magnitude greater resolution specificity than is available in instruments using broad spectrum radiation. Signal processing techniques separate pollutant absorption patterns from interference patterns.

Snowman and his co-workers predict that a similar instrument using two or three lasers would offer sufficient lasing lines to identify ozone, sulfur dioxide, ammonia, nitric oxide, nitrogen dioxide, ethylene, carbon dioxide, carbon monoxide, and representative hydrocarbons.

Tannenbaum *et al.* (*152*) have field tested a Remote Raman System (RRS) using an ultraviolet lasar to stimulate Raman scatter from airborne compounds. The scattered frequencies are then monitored for identification and analysis. Their system employs a doubled ruby laser at 3472 Å and a 36-in. diameter collimating telescope. The reported sensitivities for the prototype RRS are in the range of 5.2 mg/m^3 (2 ppm) for sulfur dioxide, 6 mg/m^3 (4 ppm) for ethane and propane, 4.2 mg/m^3 (6 ppm) for ammonia, and 294 mg/m^3 (240 ppm) for nitric oxide. Their technique, however, requires improvement in size, cost, and sensitivity.

VII. Sample Lines and Particulate Removal

Except for the remote sampling methods just discussed, the rest of these ambient air analytical methods require transport of an air sample from a sample inlet, external to the room or equipment shelter, through a sample delivery tube to a sampler or analytical instrument. A filter is generally used at the probe inlet or in-line to prevent introduction of particulate matter into the sampling or analytical device. Both the sample tube and the filter present surfaces on which loss of reactive gases such as sulfur dioxide and hydrogen sulfide can potentially occur.

A. Sample Lines

Wohlers *et al.* (*153*) studied the loss of sulfur dioxide in 30.5-m lengths of 1.3-cm diameter tubing of glass, Teflon, tygon, polypropylene, polyvinyl chloride (PVC), aluminum, and stainless steel. Under the conditions of their study, adsorption or desorption of sulfur dioxide was observed only for tygon, PVC, and aluminum tubing. These conclusions were confirmed on a 305-m length of 1.9-cm diameter stainless-steel tubing. Yamada and Charlson recommend use of a 4- to 6-in. diameter stainless-steel sample manifold equipped with an auxillary blower to move large volumes of sample air within close proximity of the sampling device, followed by close coupling to the sampler or analyzer with $\frac{1}{4}$-in. stainless-steel tubing (*154*). The sampled air is removed from the manifold by separate sample or instrument pumps. Slowik and Sansone have examined mild carbon steel tubing and found that a 10-ft length of this material is satisfactory for use with sulfur dioxide monitors (*155*). In usual practice, from 3 to 7 m of sample tubing is required at fixed or mobile field sampling stations. Longer runs of sampling tubing are required only for sampling from an elevated location, such as a tower.

B. Particulate Filters

Byers and Davis (*156*) studied adsorption of sulfur dioxide on filter media and filter holders to determine the degree of sulfur dioxide loss obtained within the ambient range of 130–260 $\mu g/m^3$ (50–100 ppb). Within this range, membrane filters adsorb very little sulfur dioxide, but glass fiber filters absorb appreciable sulfur dioxide. Teflon and polyvinyl chloride filter holders are relatively inert to sulfur dioxide, but stainless steel and aluminum are more reactive. The best combination of filter holder and filter reported consists of a Tenite plastic holder fitted with a Nuclepore membrane filter. Adams *et al.* (*105*) found that one brand of membrane filters retained significant quantities of sulfur dioxide, whereas another brand did not. The excessive sulfur dioxide retention of the one type of filter was attributed to soluble sodium salts apparently introduced into the membrane in the manufacturing process. These filters did not retain sulfur dioxide after leaching in distilled water and drying, and compared favorably with the other brand. Each batch of filters purchased should be checked in-line between a permeation source and the detector to establish its suitability for its intended use.

Burke *et al.* (*156a*) studied the interaction of sulfur dioxide with airborne particulate matter by using frontal analysis chromatography. Particles were removed from the filter media and packed as chromatographic columns. The effect of organic particulate matter was also investigated by subjecting the column packings to different thermal pretreatments. A stream of known concentration of sulfur dioxide was passed through the column and the uptake of sulfur dioxide monitored for 35 minutes after breakthrough of the sulfur dioxide front. A stream of pure nitrogen was then passed through the column and the sulfur dioxide desorption monitored. The column was then heated at 115°C and the release of sulfur dioxide noted. The authors concluded that the interaction of sulfur dioxide and particulate matter is strongly associated with the degree of organic matter present. It is also significant that 30–40% of the adsorbed sulfur dioxide was volatilized from the particulate matter present in Tucson, Arizona air which is indicative of an unreacted sulfur dioxide fraction. Much work remains to be done in characterizing particulate matter–sulfur dioxide interaction in other urban environments.

The adsorption and desorption of sulfur dioxide and other reactive sulfur-containing gases by sampling lines, particulate filters, and filter holders can cause 3–5-minute maximum sulfur dioxide peaks to be lost from the record or at least partially attenuated.

The Philips sulfur dioxide coulometric analyzer is available with a heated filter-holder to protect the coulometric cell from particulate deposits. By heating the cellulose filter, sulfur dioxide adsorption is thought to be minimized. The manufacturer's instructions recommend filter replacement at least every 3 months. Although this analyzer is equipped with a permeation bulb to provide a known concentration of sulfur dioxide on demand, the instrument plumbing is designed to check only the instrument detection cell response. There is no convenient way to determine filter and sample line losses when instrument calibration is checked. This calibration deficiency is common to all presently available sampling and automatic analysis systems and few field operators conduct a two-mode calibration of (a) the sampler or instrument and (b) the filter, sample line, or manifold, plus the sampler or instrument to determine the degree of sample loss incurred by the sample air filtration and delivery system.

To overcome this common deficiency the Sulfur Subcommittee of the Intersociety Committee (*157*) has proposed a four-mode calibration procedure to be used for all instrumental methods for sulfur-containing gases. The system includes a sulfur gas permeation dilution system and a multiport valve which will deliver in sequence a known concentration of calibration gas within the anticipated ambient air range of the sulfur-containing gas(es) to be measured (a) directly to the instrument using clean air, (b) to the instrument through the entire sample handling system including the particulate filter if one is used to remove airborne particulates, (c) directly to the ambient air being sampled by the instrument, and finally (d) by addition to the ambient air at the sample line probe to the instrument through the entire sample handling system. This four-mode calibration technique will provide essential sample validation information including (a) instrument calibration, (b) degree of sample loss in the sample handling system, and (c) degree of interference from other pollutants present in the ambient air at the time of calibration. If, as proposed, this four-mode calibration is automated to provide at least three or four daily calibrations, significant additional information will be provided on instrument baseline drift relative to diurnal temperature changes.

REFERENCES

1. P. W. West and G. C. Gaeke, *Anal. Chem.* **28,** 1816 (1956).
2. D. F. Adams, M. Corn, C. I. Harding, J. P. Pate, A. L. Plumley, F. P. Scaringelli, and P. Urone, "Methods of Air Sampling and Analysis" [Tentative

Method of Analysis for Sulfur Dioxide Content of the Atmosphere (Colorimetric), No. 42401-01-69T], p. 447. Amer. Pub. Health Ass., Washington, D.C., 1972.

3. "ASTM Standards—Water; Atmospheric Analysis," ASTM Designation D2914-70T, Part 23. Amer. Soc. Test. Mater., Philadelphia, Pennsylvania, 1972.
4. United States Environmental Protection Agency, *Fed. Regist.* **36,** No. 158, 15492 (1971).
5. J. E. Lovelock, R. J. Maggs, and R. A. Rasmussen, *Nature (London)* **237,** 452 (1972).
6. E. Robinson and R. C. Robbins, "Sources, Abundance, and Fate of Gaseous Atmospheric Pollutants," Final Rep., SRI Project 6755 Supplement. Stanford Research Institute, Palo Alto, California, 1968.
7. N. Zurlo and A. M. Griffini, *Med. Lav.* **53,** 330 (1962).
8. F. P. Scaringelli, L. Elfers, D. Norris, and S. Hochheiser, *Anal. Chem.* **42,** 1818 (1970).
9. J. B. Pate, B. E. Ammons, G. A. Swanson, and J. P. Lodge, Jr., *Anal. Chem.* **37,** 942 (1965).
10. J. B. Pate, J. P. Lodge, Jr., and A. F. Wartburg, *Anal. Chem.* **34,** 1660 (1962).
11. F. P. Scaringelli, B. E. Saltzman, and S. A. Frey, *Anal. Chem.* **39,** 1709 (1967).
12. Hartman-Leddon, Philadelphia, Pennsylvania.
13. H. C. McKee, R. E. Childers, and O. Saenz, Jr., "Collaborative Study of Reference Method for Determination of Sulfur Dioxide in the Atmosphere (Pararosaniline Method)," Contract CPA 70-40. Southwest Research Institute, Houston, Texas, 1971.
14. Anonymous, *Amer. Ass. Test. Mater. Stand. News* p. 38 (1973).
15. P. Urone, J. B. Evans, and C. H. Noyes, *Anal. Chem.* **37,** 1104 (1965).
16. C. E. Boström, *Int. J. Air Water Pollut.* **9,** 333 (1965).
17. K. L. White and A. C. Hill, *J. Air Pollut. Contr. Ass.* (in press).
18. R. K. Stevens, T. A. Clark, C. E. Decker, and L. P. Ballard, 65th Annu. Meet., Pap. No. 72-13 Air Pollut. Contr. Ass., Pittsburgh, Pennsylvania, 1972.
19. United States Environmental Protection Agency, *Fed. Regist.* **36,** No. 67, 6684 (1971).
20. A. Attari, T. P. Igielski, and B. Jaselskis, *Anal. Chem.* **42,** 1282 (1970).
21. Kawai, T., *Netsu Kanri* **22,** 20 (1970).
22. M. D. Thomas, J. O. Ivie, and T. C. Fitt, *Ind. Eng. Chem., Anal. Ed.* **18,** 383 (1946).
23. J. E. Yocum, R. L. Richardson, I. M. Saslaw, and S. Chapman, Proc. 49th Annu. Meet. Air Pollut. Contr. Ass., Pittsburgh, Pennsylvania, 1956.
24. E. R. Kuczynski, *Environ. Sci. Technol.* **1,** 68 (1967).
25. J. M. Shikiya and R. D. McPhee, 61st Annu. Meet., Pap. No. 68-72. Air Pollut. Contr. Ass., Pittsburgh, Pennsylvania, 1968.
26. M. Terabe, S. Oomichi, F. E. Benson, V. A. Newill, and J. E. Thompson, *J. Air Pollut. Contr. Ass.* **17,** 673 (1967).
27. M. B. Jacobs, "The Chemical Analysis of Air Pollutants," Chapter 8. Wiley (Interscience), New York, New York, 1960.
28. M. B. Jacobs, "The Analytical Chemistry of Industrial Poisons, Hazards, and Solvents," Chapter 9. Wiley (Interscience), New York, New York, 1949.
29. M. Katz, *Anal. Chem.* **22,** 1040 (1950).
30. H. Stratmann, *Mikrochim. Acta* **6,** 688 (1954).
31. J. B. Pate, J. P. Lodge, Jr., and M. P. Neary, *Anal. Chim. Acta* **28,** 341 (1963).

32. J. P. Lodge, Jr., J. B. Pate, and H. A. Huitt, *Amer. Ind. Hyg. Ass., J.* **24**, 380 (1963).
33. J. Forrest and L. Newman, *Atmos. Environ.* **7**, 561 (1973).
34. C. Huygen, *Anal. Chim. Acta* **28**, 349 (1963).
35. J. Harding and B. Schlein, "Nuclear Techniques in Environmental Pollution," SM-142a/9. IAEA, Vienna, Austria, 1971.
36. B. H. Wilsdon and F. J. McConnell, *J. Soc. Chem. Ind.* **53**, 385 (1934).
37. N. A. Huey, *J. Air Pollut. Contr. Ass.* **18**, 610 (1968).
38. F. W. Thomas and C. M. Davidson, *J. Air Pollut. Contr. Ass.* **11**, 1 (1961).
39. W. W. Stalker, R. C. Dickerson, and G. D. Kramer, *Amer. Ind. Hyg. Ass., J.* **24**, 68 (1963).
40. H. R. Hickey and E. R. Hendrickson, *J. Air Pollut. Contr. Ass.* **15**, 409 (1965).
41. C. I. Harding and T. R. Kelley, *J. Air Pollut. Contr. Ass.* **17**, 545 (1967).
42. T. Ikeda, *J. Hyg. Chem.* **13**, 41 (1967).
43. M. B. Jacobs, M. M. Braverman, and S. Hochheiser, *Anal. Chem.* **29**, 1349 (1957).
44. J. Budd and H. A. Berwick, *Anal. Chem.* **23**, 1536 (1952).
45. B. Hasselhuhn, *in* "Atmospheric Emissions from Sulfate Pulping" (E. R. Hendrickson, ed.), p. 373. U.S. Public Health Service, National Council for Stream Improvement and University of Florida, Sanibel Island, Florida, 1966.
46. E. P. Marbach and D. M. Doty, *J. Agr. Food Chem.* **4**, 881 (1956).
47. C. E. Boström, *Air Water Pollut. J.* **10**, 435 (1966).
48. W. L. Bamesberger and D. F. Adams, *Environ. Sci. Technol.* **3**, 258 (1969).
49. STRactan, Trademark Brand Arabinoglactan. St. Regis Paper Co., New York, New York.
50. W. L. Bamesberger and D. F. Adams, *Tappi* **52**, 1302 (1969).
51. D. F. Adams, D. Falgout, J. O. Frohliger, J. B. Pate, A. L. Plumley, F. P. Scaringelli, and P. Urone, "Methods of Air Sampling and Analysis" (Tentative Method of Analysis for Hydrogen Sulfide Content of the Atmosphere, No. 72402-01-70T), p. 426. Amer. Pub. Health Ass., Washington, D.C.. 1972.
52. B. Thomas, W. L. Bamesberger, and D. F. Adams, Preprints, Division of Water, Air, and Waste Chemistry, *157th Annu. Meet.*, Pap. No. 63, p. 239. Amer. Chem. Soc., Washington, D.C., 1969.
53. D. A. Falgout and C. I. Harding, *J. Air Pollut. Contr. Ass.* **18**, 15 (1968).
54. K. E. Prescher and E. Lahmann, *Gesundh.-Ing.* **87**, 351 (1966).
55. M. Buck and H. Stratmann, *Staub* **24**, 241 (1964).
56. W. Mecklenburg and F. Rosenkranzer, *Z. Anorg. Chem.* **86**, 143 (1914).
57. M. Buck and H. Gies, *Staub* (*Engl. Ed.*) **26**, 27 (1966).
57a. D. F. S. Natusch, H. B. Klonis, H. D. Axelrod, R. J. Teck, and J. P. Lodge, Jr., *Anal. Chem.* **44**, 2067 (1972).
58. J. D. Sensenbaugh and W. C. L. Hemeon, *Air Repair* **4**, 5 (1954).
59. Research Appliance Co., Gibsonia, Pennsylvania.
60. D. V. Moses and L. T. Jilk, U.S. Patent 2,232,622 (E. I. DuPont DeNemours & Co.) (1941).
61. Rubicon Company, Philadelphia, Pennsylvania.
62. Gelman Instrument Co., Ann Arbor, Michigan.
63. H. P. Sanderson, R. Thomas, and M. Katz, *J. Air Pollut. Contr. Ass.* **16**, 329 (1966).
64. U.S. Department of Health, Education and Welfare, "A Study of Air Pollution in the Interstate Region of Lewiston, Idaho and Clarkston, Washington," Pub.

Health Serv. Publ. 999-AP-8. U.S. Dept. of Health, Education, and Welfare, Cincinnati, Ohio, 1964.
65. J. P. Paré, *J. Air Pollut. Contr. Ass.* **16,** 325 (1966).
66. W. L. Bamesberger and D. F. Adams, *Tappi* **52,** 1302 (1969).
67. W. Siu, D. A. Levaggi, L. Potter, R. Martin, and M. Feldstein, *J. Air Pollut. Contr. Ass.* **21,** 636 (1971).
68. J. P. Sheey, J. J. Henderson, C. I. Harding, and A. L. Davis, U.S. Pub. Health Serv., Publ 999-AP-3. U.S. Dept. of Health, Education, and Welfare, Cincinnati, Ohio, 1963.
69. G. Leonardos, "Research on Chemical Odors, Part I." Arthur D. Little, Inc., Cambridge, Massachusetts, 1968.
70. D. F. Adams, F. A. Young, and R. A. Luhr, *Tappi* **51,** 62A (1968).
71. T. Lindvall, *Nord. Hyg. Tidskr., Suppl.* **2,** 1–181 (1970).
72. M. D. High and S. W. Horstman, *Amer. Ind. Hyg. Ass., J.* **26,** 366 (1965).
73. L. Dubois and J. L. Monkman, "The Analysis of Airborne Pollutants," Conference Background Paper D25-3. Prepared for the National Conference on Pollution and Our Environment, Canadian Council of Resource Ministers, Montreal, Quebec, 1966.
74. J. Hochheiser and L. A. Elfers, *Environ. Sci. Technol.* **4,** 672 (1970).
75. H. D. Axelrod, J. H. Cary, J. E. Bonelli, and J. P. Lodge, Jr., *Anal. Chem.* **41,** 1956 (1969).
76. B. A. Hardwick, D. K. B. Thistlethwayte, and R. T. Fowler, *Atmos. Environ.* **4,** 379 (1970).
77. T. R. Andrew and P. N. R. Nichols, *Analyst* **90,** 367 (1965).
78. T. S. Light, *Nat. Bur. Stand. (U.S.), Spec. Publ.* **314** (1969).
79. T. Okita, M. Mori, S. Ito, and H. Ogino, 64th Annu. Meet., Pap. No. 71-3. Air Pollut. Contr. Ass., Pittsburgh, Pennsylvania, 1971.
80. W. Lenz and A. A. Tirado, *in* "Atmospheric Emissions from Sulfate Pulping" (E. R. Hendrickson, ed.), p. 31. U.S. Public Health Service, National Council for Stream Improvement and University of Florida, Sanibel Island, Florida, 1966.
81. N. A. Huey, L. C. Boering, G. A. Jutze, and C. W. Gruber, *J. Air Pollut. Contr. Ass.* **10,** 441 (1960).
82. G. A. Chanin, J. R. Elwood, and E. H. Chow, *Sewage Ind. Wastes* **26,** 1217 (1954).
83. E. R. Gilardi and R. M. Manganelli, "A Laboratory Study of A Lead-Acetate-Tile Method for the Quantitative Measurement of Low Concentrations of Hydrogen Sulfide," At. Pollut. Tech. Bull. No. 15. National Council for Stream Improvement, New York, New York, 1962.
84. H. C. Wohlers and M. Feldstein, *J. Air Pollut. Contr. Ass.* **16,** 19 (1966).
85. H. Moore, H. H. Helwig, and R. J. Graul, *J. Ind. Hyg.* **21,** 466 (1960).
86. T. Okita, *Atmos. Environ.* **4,** 93 (1970).
87. H. W. Washburn and R. R. Austin, *in* "Air Pollution" (L. C. McCabe, ed.), Chapter 72. McGraw-Hill, New York, New York, 1952.
88. Consolidated Electrodynamics Corp. (Bell & Howell Electronic Instrumentation Group), N. Sierra Madre Villa, California. (No longer available).
89. Process Analyzers, Inc., Houston, Texas.
90. J. S. Nader and J. L. Dolphin, *J. Air Pollut. Contr. Ass.* **8,** 336 (1959).
91. H. C. McKee and W. L. Rolliwitz, *J. Air Pollut. Contr. Ass.* **8,** 338 (1959).
92. ITT-Barton Process Instruments and Controls, Monterey Park, California.
93. G. Thoen, G. G. DeHaas, and R. R. Austin, *Tappi* **51,** 246 (1968).

94. Philips Electronic Instruments, Mt. Vernon, New York.
95. Beckman Instrument Co., Fullerton, California.
96. L. B. Potter, *Pap., 10th Anal. Methods Conf., 1969* (1969).
97. E. M. Fredricks and G. A. Harlow, *Anal. Chem.* **36**, 263 (1964).
98. D. M. Coulsen and L. A. Cavanagh, *Pap., Pittsburgh Conf. Anal. Chem. Appl. Spectrosc.* (1961).
99. P. J. Klaas, *Anal. Chem.* **33**, 1851 (1961).
100. Dohrmann Division of Envirotech Corp., Mountain View, California.
101. H. W. Bialkowsky and G. G. DeHaas, *Pulp Pap. Mag. Can.* **53**, 99 (1952).
102. E. Lieber and R. Rosen, *Ind. Eng. Chem., Anal. Ed.* **4**, 90 (1932).
103. J. Martin and J. A. Grant. *Anal. Chem.* **37**, 644 (1965).
104. D. F. Adams, G. A. Jensen, J. P. Steadman, R. K. Koppe, and T. J. Robertson, *Anal. Chem.* **38**, 1094 (1966).
105. D. F. Adams, W. L. Bamesberger, and T. J. Robertson, *J. Air Pollut. Contr. Ass.* **18**, 145 (1968).
106. B. Draeger and Heinrich Draegerwerk, West German Patent 1,133,918 (1962).
107. W. L. Crider, *Anal. Chem.* **37**, 1770 (1965).
108. R. E. Pescar and C. H. Hartmann, *J. Chromatogr. Sci.* **11**, 492 (1973).
109. D. F. Adams, "Chemistry of Odors and Flavors," Gordon Res. Conf. (discussion). Crystal Mountain, Washington, 1966.
110. C. I. Harding, personal communication (1966).
110a. T. Sugiyama, Y. Suzuki, and T. Takeuchi, *J. Chromatogr. Sci.* **11**, 641 (1973).
110b. S. Shiffman and C. W. Frank, *Anal. Chem.* **46**, 1804 (1974).
111. D. G. Greer and T. J. Bydalek, *Environ. Sci. Technol.* **7**, 153 (1973).
112. R. K. Stevens, *Pittsburgh Conf. Anal. Chem. Appl. Spectrose., 1967*.
113. R. K. Stevens, A. E. O'Keeffe, and G. C. Ortman, *Environ. Sci. Technol.* **3**, 652 (1969).
114. D. F. Adams and R. K. Koppe, *Tappi* **41**, 366 (1958).
115. D. F. Adams and R. K. Koppe, *Tappi* **42**, 601 (1959).
116. D. F. Adams and R. K. Koppe, *Tappi* **43**, 602 (1960).
117. G. C. B. Cave, *Tappi* **46**, 1 (1963).
118. D. F. Adams, R. K. Koppe, and W. N. Tuttle, *J. Air Pollut. Contr. Ass.* **15**, 31 (1965).
119. I. H. Williams, Annu. Meet. Pac. Northwest Int. Sect. Air Pollut. Contr. Ass., Pittsburgh, Pennsylvania, 1965.
120. E. A. Eads and W. H. Cooper, Lamar State College of Technology, Beaumont, Texas, personal communication (1969).
121. R. K. Stevens, A. E. O'Keeffe, J. D. Mulik, and K. J. Krost, *Pap., 157th Meet. Amer. Chem. Soc. Div. Anal. Chem. 1969* Paper No. 56 (1969).
122. R. K. Koppe and D. F. Adams, *Environ. Sci. Technol.* **1**, 479 (1967).
123. S. S. Brody and J. E. Chaney, *J. Gas Chromatogr.* **4**, 42 (1966).
124. C. H. Hartmann, *Pap., Joint Conf. Sensing Environ. Pollutants, 1971*.
125. W. L. Thornsberry, *Anal. Chem.* **43**, 452 (1971).
126. R. K. Stevens, J. D. Mulik, A. E. O'Keeffe, and K. J. Krost, *Anal. Chem.* **43**, 827 (1971).
127. A. A. Zhukhovitskii and N. M. Turkel'taub, *Dokl. Akad. Nauk SSSR* **143**, 646 (1961).
128. R. K. Stevens and A. E. O'Keeffe, *Anal. Chem.*, **42**, 143A (1970).
129. D. F. Adams, J. O. Frohliger, D. Falgout, A. M. Hartley, J. B. Pate, A. L. Plumley, E. P. Scaringelli, and P. Urone, *Health Lab. Sci.* **10**, 241 (1973).

130. R. A. Rasmussen, *Amer. Lab.* **4,** No. 12 (1972).
131. Q. R. Stahl, private communication, Meloy Instruments, Springfield, Virginia.
132. J. Peeters, C. Vinckier, and G. Van Tiggelen, *Oxid. Combust. Rev.* **4,** 133 (1969).
133. T. Fueno, N. R. Murkherjee, T. Ree, and H. Eyring, "Eighth (International) Symposium on Combustion," p. 222. Williams & Wilkins, Baltimore, Maryland, 1962.
134. W. J. Williams, "Eleventh (International) Symposium on Combustion," p. 311. Combustion Institute, Pittsburgh, Pennsylvania, 1967.
135. A. D. Snyder and G. W. Wooten, National Technical Information Service, Publ. No. 188 103. U.S. Dept. of Commerce, Springfield, Virginia, 1969.
136. W. A. Kummer, J. N. Pitts, Jr., and R. P. Steer, *Environ. Sci. Technol.* **5,** 1045 (1971).
137. H. Okabe, P. L. Splitstone, and J. J. Ball, *J. Air Pollut. Contr. Ass.* **23,** 514 (1973).
138. Celesco Industries, Costa Mesa, California.
139. Annonymous, *Chem. Eng. News* **51,** 47 (1973).
139a. F. P. Schwarz, H. Okabe, and J. K. Whittaker, *Anal. Chem.* **46,** 1024 (1974).
140. R. Chand, U.S. Patent 3,763,025.
141. R. A. Reynolds, Electro/Mass Laboratory, St. Louis, Missouri, personal communication (1972).
142. E. Bechtold, *Ber. Bunsenges. Phys. Chem.* **70,** 713 (1966).
143. B. C. LaRoy, A. C. Lilly, Jr., and C. O. Tiller, *Pap., Electrochem. Soc. Meet.,* (1972).
143a. G. Bélanger, *Anal. Chem.* **46,** 1576 (1974).
144. A. J. McCormack, S. C. Tong, and W. D. Cooke, *Anal. Chem.* **37,** 1470 (1965).
145. C. A. Bache and D. J. Lisk, *Anal. Chem.* **37,** 1477 (1965).
146. H. A. Moye, *Anal. Chem.* **39,** 1441 (1967).
146a. D. F. S. Natusch and T. M. Thorpe, *Anal. Chem.* **45,** 1184A (1973).
147. D. Williams and B. C. Kolitz, *Appl. Opt.* **7,** 607 (1968).
148. E. R. Stephens, P. L. Hanst, R. C. Doerr, and W. E. Scott, *Ind. Eng. Chem.* **48,** 149B (1956).
149. J. N. Driscoll and P. Warneck, *J. Air Pollut. Contr. Ass.* **23,** 858 (1973).
150. Environment/One Corp., Schenectady, New York.
151. L. R. Snowman, M. L. Noble, and R. J. Gillmaster, 65th Annu. Meet., Pap. No. 72-10. Air Pollut. Contr. Ass., Pittsburgh, Pennsylvania, 1972.
152. H. Tannenbaum, D. Tannenbaum, H. DeLong, F. R. Schildkraut, S. Kleiner, and T. Hirschfeld, 65th Annu. Meet., Pap. No. 72-11. Air Pollut. Contr. Ass., Pittsburgh, Pennsylvania, 1972.
153. H. C. Wohlers, N. Newstein, and D. Daunis, *J. Air Pollut. Contr. Ass.* **17,** 753 (1967).
154. V. M. Yamada and R. J. Charlson, *Environ. Sci. Technol.* **3,** 483 (1969).
155. A. A. Slowik and E. B. Sansone, 66th Annu. Meet., Pap. No. 73-153. Air Pollut. Contr. Ass., Pittsburgh, Pennsylvania, 1973.
156. R. L. Byers and J. W. Davis, *J. Air Pollut. Contr. Ass.* **20,** 236 (1970).
156a. M. F. Burke, R. K. Baker, and J. L. Moyers, *J. Chromatogr. Sci.* **11,** 575 (1973).
157. D. F. Adams, D. Falgout, J. O. Frohliger, J. B. Pate, A. L. Plumley F. P. Scaringelli, and P. Urone, *Health Lab. Sci.* **11,** 140, (1974).

7

Nitrogen Compounds and Oxidants

Morris Katz

I. Nitrogen Compounds

A. Introduction

The principal compounds of nitrogen and oxygen that form gases of interest in air pollution sampling and analysis arc nitric oxide, NO, and nitrogen dioxide, NO_2. Nitric oxide is formed to the extent of less than 0.1% to about 0.5% at flame temperatures, along with much smaller amounts of nitrogen dioxide, during the combustion of all types of fossil fuels. When discharged to the atmosphere, nitric oxide oxidizes at a measurable rate to nitrogen dioxide. The latter gas is the most reactive of the gaseous oxides of nitrogen and is a primary absorber of sunlight in photochemical atmospheric reactions that produce photochemical smog. Both NO and NO_2 are important contaminants for measurement in relation to emission and ambient air quality standards.

Nitrous oxide, N_2O, is also present in the atmosphere, as a natural constituent, to the extent of about 0.25 to 0.50 ppm by volume. The major source of this gas is the anaerobic fermentation of nitrates present in the soil. Ordinarily this gas is of no significance in air pollution analysis as it is relatively inert and biologically nontoxic even at high concentrations.

Ammonia occurs in the atmosphere in low concentrations in the vicinity of natural decomposition processes of organic matter containing nitrogen. Such sources include sewers, lavatories, animal feed lots, and manure pits. Industrial emission sources include coking plants, nitric acid production, fertilizers, chemicals, and plastics. This gas is washed out of the atmosphere readily by precipitation.

Information on the physical and chemical properties of these nitrogen compounds and the range of atmospheric concentrations is presented in Chapter 2, Vol. I.

B. *Nitrogen Oxides*

1. Manual Methods for NO and NO_2

a. Intersociety Committee Method of Analysis for Nitrogen Dioxide (*1*). This method is based on the Griess–Saltzman colorimetric, azo dye-forming reagent (*2*) and is probably the most sensitive of several manual methods available for atmospheric analysis of NO_2. The absorbing reagent consists of a solution of 5 gm of anhydrous sulfanilic acid, 140 ml of glacial acetic acid, and 20 ml of a 0.1% stock solution of *N*-(1-naphthyl)ethylenediamine dihydrochloride per liter of distilled water.

Three methods are available for gas sampling. For sampling air containing less than 5 ppm of NO_2, the sample at a measured flow rate of 0.4 liter/minute is passed through a fritted glass bubbler containing 10.0 ml of absorbing reagent for 10 to 30 minutes (or longer) until a sufficient final color has been developed. The maximum pore diameter of the frit in the bubbler is 60 μm.

Higher concentration ranges up to 100 ppm may be sampled by the evacuated bottle method, using bottles of appropriate size containing 10.0 ml or other convenient volume of absorbing reagent. For 1-cm spectrophotometer cells, a 5:1 ratio of air sample volume to reagent volume will cover a range up to 100 ppm. A 25:1 ratio is sufficient to measure down to 2 ppm. The bottles are evacuated to the vapor pressure of the absorbing solution, the pressure and temperature being recorded. About 15 minutes, with occasional shaking, is allowed for complete absorption and color development.

A glass syringe of 50- or 100-ml capacity, containing 10.0 ml of absorbing reagent may be employed to draw in an air sample. Absorption of nitrogen dioxide is completed by capping and shaking the syringe vigorously for 1 minute, after which the remaining air is expelled. Additional air may be drawn in and the process repeated several times, if necessary, to develop sufficient final color for measurement.

The reaction of NO_2 with the absorbing reagent produces a red-violet color. Color development is complete within 15 minutes at room temperature. Measurement of the color is accomplished by transfer to stoppered cuvettes and determination of the absorbance in a spectrophotometer at 550 nm, using the unexposed reagent as a reference.

The analytical method may be calibrated by the use of standard solutions of sodium nitrite. A graduated series of standard solutions are diluted with absorbing reagent to a total volume of 25 ml in each case. After mixing and standing for 15 minutes to complete the color development, the absorbance is read with a spectrophotometer at 550 nm. Standardization by nitrite is based upon the empirical relation that 0.72 mole of sodium nitrite produces the same color as 1 mole of NO_2 (*1, 3*). For calibration purposes, a plot of the absorbances of the standard colors against microliters of nitrogen dioxide per milliliters of absorbing reagent is constructed.

i. *Calculation.* The volume of air samples is corrected to 25°C and 760 Torr. Quantities of nitrogen dioxide may be expressed as microliters. The molar volume of gas at the above standard conditions is 24.47 liters, and the molecular weight of $NaNO_2$ is 69.00.

$$1\ \mu l\ NO_2 = \frac{10^{-6}}{24.47} \times 0.72 \times 69.0 = 2.03 \times 10^{-6}\ gm\ NaNO_2 \qquad (1)$$

The results for samples are computed as follows:

$$NO_2\ (ppm) = (\text{corrected absorbance}) \times K/V \qquad (2)$$

where

K = standardization factor or μl NO_2 at absorbance of exactly 1.0; for 1-cm cells the value of K is about 0.73

V = volume of air sample at standard conditions, in liters per milliliter of absorbing reagent for a standard solution containing nitrite equivalent to 0.4 μl NO_2 per ml

If V is a simple multiple of K, calculations are simplified. Thus, if exactly 7.3 liters of air are sampled through the bubbler containing 10 ml of absorbing reagent, $K/V = 1$, and the absorbance yields ppm directly.

Greater accuracy may be achieved by standardization with accurately known gas samples in a precision flow dilution system (*3–5*) or by use of the permeation tube technique (*6*). If this gaseous method is employed, the stoichiometric factor of 0.72 is eliminated from the calculation of concentration.

ii. *Application.* This method is intended for the determination of atmospheric nitrogen dioxide in the range of 0.005 to about 5 ppm by volume or 0.01 to 10 μg/liter, when sampling is conducted in fritted bubblers and when high sensitivity is required. A precision of 1% of the mean is possible. The limiting factors are the measurements of the volume of the air sample and the absorbance of the color.

Interferences from other nitrogen oxides and other gases, with the exception of sulfur dioxide, which might be found in the polluted atmosphere, are negligible. A tenfold ratio of sulfur dioxide to nitrogen dioxide produces no effect and a 30-fold ratio slowly bleaches the color of the azo dye-forming reagent to a slight extent only. Addition of 1% acetone to the reagent before use will retard the fading. However, if the evacuated bottle or syringe method is used to sample high concentrations above 5 ppm, interference from NO is possible, due to oxidation to NO_2. A fivefold ratio of ozone to nitrogen dioxide causes a small interference as the reagent may assume a slightly orange tint instead of the stable pink color. Peroxyacyl nitrate (PAN) can yield a response of approximately 15–35% of an equivalent molar concentration of NO_2 (*7*). However, concentrations of PAN in the ordinary ambient air are too low to cause any significant error.

b. United States Environmental Protection Agency (USEPA) Method for NO_2. The USEPA Reference Method (*8*) is a 24-hour sampling method for nitrogen dioxide in the atmosphere based upon the technique devised by Jacobs and Hochheiser (*9*). The gas sample is passed through an absorber containing a gas dispersion tube with a fritted end of porosity about 70–100 μm maximum pore diameter. A 0.1 N solution of sodium hydroxide (50 ml) is employed as absorbing reagent. The flow rate of 0.20 liter/minute is measured before and after sampling by means of a calibrated flowmeter to determine the air sample volume. Analysis is performed by replacing water lost by evaporation during sampling and pipeting 10 ml of the absorbing reagent into a test tube. The following reagents are added: 1.0 ml of dilute hydrogen peroxide, 10.0 ml of sulfanilamide–phosphoric acid solution, and 1.4 ml of *N*-(1-naphthyl)ethylenediamine dihydrochloride solution. After an interval of 10 minutes for color development, the absorbance is measured at 540 nm against a blank. The measurements are compared against a standard calibration curve.

The range of the analysis is 0.04 to 1.5 μg NO_2/ml. With 50 ml of absorbing reagent and a sampling rate of 200 ml/minute for 24 hours, the range is 20 to 740 $\mu g/m^3$ (0.01–0.4 ppm) nitrogen dioxide. Interference of sulfur dioxide is eliminated by conversion to sulfuric acid by hydrogen peroxide, prior to analysis. The relative standard deviations are found to be 14.4 and 21.5% at nitrogen dioxide concentrations of 140 and 200 $\mu g/m^3$, respectively, based on an automated analysis of samples collected from a standard test atmosphere. Over the above concentration range, an average efficiency of only 35% was obtained from test atmospheres.

It is obvious that the low efficiency of this method is due principally to the use of sodium hydroxide solution as the absorbing reagent in contrast to the high efficiency of the Griess–Saltzman reagent. Studies have been made of the efficiency of the sodium hydroxide absorption method. Using four simple nozzle bubblers in series, an absorption efficiency of 60–70% was obtained (*10*) on the basis of the following reaction:

$$2\,NO_2 + 2\,OH^- \rightarrow NO_2^- + NO_3^- + H_2O \tag{3}$$

With a single nozzle bubbler the efficiency was only about 30% (*11*). The use of two polyethylene coarse fritted tip bubblers of 30–50 μm maximum pore diameter in series yielded an absorption efficiency of about 62% at flow rates of 0.35 to 0.50 liter/minute, as compared to 36% efficiency for two nozzle tip bubblers in series (*12*). The assumption in the above equation that 1 mole of NO_2 yields the same colorimetric value as 0.5 mole of nitrite ion has not been clearly established.

c. METHOD OF LEVAGGI *et al.* (*13*) FOR NITROGEN DIOXIDE (24-HOUR AVERAGE). This is a much more accurate method than the one just described. Nitrogen dioxide in the air sample is absorbed in a solution of aqueous triethanolamine. This solution contains 15 gm of triethanolamine and 3 ml of *n*-butanol per liter. Subsequent analysis is performed by use of the azo dye-forming Griess–Saltzman reagent and measurement of the resultant color in a spectrophotometer at 540 nm.

This method may be used with confidence in the range of 10 to 1000 $\mu g/m^3$, or 0.005 to 0.50 ppm NO_2, based on a 24-hour sampling period. For a 1-cm spectrophotometer cell path, an absorbance of 0.1 is equivalent to 0.14 μg/ml of NO_2 in the absorbing solution. There is no interference from sulfur dioxide in concentrations up to 2000 $\mu g/m^3$ (0.7 ppm) if hydrogen peroxide is added to the absorbing solution after sampling and before color development. Ozone causes no interference at concentrations up to 1000 $\mu g/m^3$ and nitric oxide does not interfere at concentrations up to 800 $\mu g/m^3$ (0.6 ppm) as a 24-hour average (*13*). Absorption efficiency is in the range of 95 to 99%, using 50 ml of absorbing solution in the plastic USEPA bubbler (*8*). Accuracy of the method can be considered as comparable to that obtained in short-term sampling with the Griess–Saltzman reagent, or less than 2% when the nitrite factor of 0.764 is employed (*14*).

A modification (*14a*) of the above Levaggi *et al.* method involves bubbling air through 50 ml of an absorbing reagent containing 20 gm triethanolamine, 0.5 gm *o*-methoxyphenol, and 0.25 gm sodium *m*-bisulfite per liter of distilled water. A sampling rate of 200 ml/minute is maintained for 24 hours in the field. Subsequent colorimetric analysis is performed in the laboratory by reacting the exposed reagent with hydrogen peroxide to prevent SO_2 interference and then with sulfanilamide and 8-anilino-1-naphthalene sulfonic acid.

The range of the method is 20 to 700 $\mu g/m^3$ nitrogen dioxide or 0.025 to 4 μg NO_2^-/ml of absorbing reagent. Beer's law is obeyed throughout this range. A concentration of 0.025 μg NO_2^-/ml produces an absorbance of 0.013 in 1-cm cells. No interferences were found in tests of the following pollutants at a NO_2 concentration of 100 $\mu g/m^3$: ammonia, 205 $\mu g/m^3$; formaldehyde, 750 $\mu g/m^3$; nitric oxide, 734 $\mu g/m^3$; phenol, 150 $\mu g/m^3$; ozone, 400 $\mu g/m^3$; and sulfur dioxide, 439 $\mu g/m^3$.

Measurements with standard atmospheres of nitrogen dioxide, prepared by the use of permeation tubes, showed a collection efficiency of 93% and a relative standard deviation of 2% throughout the range of the method. The absorbing reagent shows no deterioration for a period up to 3 weeks prior to sampling and collected samples are stable for 3 weeks after sampling.

d. Sodium Arsenite Procedure for Nitrogen Dioxide. Nitrogen dioxide is collected by passing the air sample at about 200 ml/minute through 50 ml of an absorbing solution containing 4.0 gm sodium hydroxide and 1.0 gm sodium arsenite per liter of distilled water (*14b, 14c*). The method is suitable for the collection of 24-hour samples in the field and subsequent analysis in the laboratory. Nitrite ion produced in the absorbing solution is reacted with dilute hydrogen peroxide, sulfanilamide, phosphoric acid and *N*-(1-naphthyl)ethylenediamine dihydrochloride to form an azo dye and is then determined colorimetrically (*14b, 14c*).

The method has a range of 0.04 to 2.0 μg NO_2^-/ml, equivalent to 20 to 750 $\mu g/m^3$ of nitrogen dioxide at a sampling rate of 200 ml/minute for 24 hours through 50 ml of absorbing reagent. Beer's law is obeyed throughout this range. A concentration of 0.04 μg NO_2^-/ml will produce an absorbance of approximately 0.02 with 1-cm cells.

Nitric oxide in the air sample constitutes a positive interferent as it can increase the NO_2 response by 5 to 15% of the NO_2 actually present in the sample. The addition of hydrogen peroxide to the absorbing solution prior to colorimetric analysis with the azo dye-forming reagents serves to eliminate sulfur dioxide interference by conversion to sulfate ion. However, Beard and Margeson (*14d*), after ruggedness test evaluation of the arsenite procedure, found that carbon dioxide in excess of ambient air concentration constituted a negative interference and method response may also be affected by variations in sample flow rate much above 200 ml/minute. They reported a mean collection efficiency of 82.2% with a relative standard deviation of 4.5%. Collected samples remain stable over a storage period of 6 weeks.

e. PHENOLDISULFONIC ACID METHOD FOR OXIDES OF NITROGEN. This procedure has been described fully by the Intersociety Committee (*15*) and a comparable method has been adopted by the American Society for Testing and Materials (ASTM) (*16*) for the analysis of combustion products containing 2 to 500 ppm NO_2. It is a long-established method and was first employed for the determination of nitrate in water by Chamot *et al.* (*17*). It was later adapted to the measurement of oxides of nitrogen in air by Cook (*18*).

Nitric oxide (NO), nitrous anhydride (N_2O_3), nitrogen dioxide (NO_2), nitrogen tetroxide (N_2O_4), vapor or mist of nitric acid (HNO_3), and nitrous acid (HNO_2), but not nitrous oxide (N_2O), may be collected and oxidized to the nitrate ion in an evacuated flask containing sulfuric acid and hydrogen peroxide. The yellow compound resulting from reaction of the nitrate ion with phenoldisulfonic acid is measured colorimetrically at 400 nm.

The chemical reactions that take place are as follows:

$$2\,NO_2 + H_2O_2 = 2\,HNO_3 \tag{4}$$

$$2\,NO + 3\,H_2O_2 = 2\,HNO_3 + 2\,H_2O \tag{5}$$

$$HNO_3 + HOC_6H_3(SO_3H)_2 = HOC_6H_2(NO_2)(SO_3H)_2 + H_2O \tag{6}$$

Application of this method to ambient air is useful only in industrial locations where the concentrations of oxides of nitrogen are in the range of about 25 ppm by volume or greater. At lower levels of concentration, the absorption of the gas sample by the reagent in the evacuated flask is very slow and incomplete, even after several days. Only 90% recovery has been reported (*19*) in the range of 12 to 94 ppm, with an average deviation of 6.8%. Lower recoveries were obtained by Gill (*20*) in the concentration range below 16 ppm.

Inorganic nitrates, nitrites, and organic nitrogen compounds cause interferences by yielding high results. Chlorides and other halides, if present in relatively high concentration, must by removed by addition of silver nitrate, as they cause negative interference.

f. XYLENOL METHOD FOR NITROGEN OXIDES. In this method the air sample is collected in an evacuated flask containing 62.5% sulfuric acid as the absorbing reagent. After standing for 2 hours, a drop (0.05 ml) of 2% potassium permanganate is added to the flask and thoroughly mixed with the sulfuric acid in order to oxidize any nitrous acid present to nitric acid. 0.02 ml of 1% 2,4-xylenol dissolved in propylene glycol is then added to form the nitrated xylenol compound. After a 10-minute period of nitration, the mixture is diluted with 100 ml of distilled water and the nitroxylenol is distilled into a water-cooled 25-ml graduate containing 1 ml of 2% sodium hydroxide. The resultant yellow colored distillate is made up to 10 ml and then analyzed in the spectrophotometer at 435 nm (*21*). The method is applicable in the range of about 2 to 500 ppm of nitrate.

This method possesses a number of advantages over the more cumbersome phenoldisulfonic acid method. All of the operations can be performed in one flask and the distillation step removes most of the interfering substances to yield a clear solution for colorimetric estimation. The initial reaction mixture is diluted considerably prior to distillation and, thus, charring of organic matter is avoided, in contrast to the evaporation step that is required when employing phenoldisulfonic acid.

Interferences are limited to hydrogen peroxide, large amounts of halogens, and metallic and sulfide dusts. These substances must be removed, if present in the gas sample.

Instead of the distillation step, the nitroxylenol may be extracted from the acid solution with toluene (*22*). The toluene extract is shaken with 10 ml of 0.4 N sodium hydroxide and the resultant colored layer is removed and determined photometrically as above at 435 nm. A variation of this procedure is described in the Intersociety Committee method for the determination of nitrate in atmospheric particulate matter. The nitrates in the sample are removed first by extraction with water (*23*).

g. Nitric Oxide—Oxidation to NO_2. The Intersociety Committee has published a manual method for the analysis of nitric oxide in air in the range of 0.005 to about 5 ppm by volume or 6 to 6000 $\mu g/m^3$, with a sensitivity of 0.01 μg NO per 10 ml of absorbing solution (*24*). The sampling train consists, in order, of a Rotameter, absorber for nitrogen dioxide, humidity regulator, oxidizer, fritted absorber, and pump. The nitrogen dioxide absorber removes chemically the NO_2 that is normally present in the atmosphere. This absorber consists of a 20-mm i.d. $\times$ 50-mm long polyethylene tube with connecting caps at both ends. It is filled with pellets of 10–20 mesh, porous, inert material, such as firebrick, alumina, or zeolites that have been soaked in 20% aqueous triethanolamine, drained and dried for 30–60 minutes at 95°C in an oven (*25*). The dried pellets should be free flowing.

The humidity regulator contains a mixture of 50% anhydrous 50% hydrated sodium acetate coarse-grained crystal pellets packed into a 20-mm i.d. $\times$ 50-mm long polyethylene tube to provide steady 40–70% relative humidity for the oxidizer (*26*). To prepare the oxidizing material for almost total conversion of NO to NO_2 (efficiency 98–100%), glass, firebrick, or alumina of 15–40 mesh size is soaked in a solution 17 gm chromium trioxide in 100 ml of distilled water. This mixture is drained, dried in an oven at 105–115°C and exposed to 70% relative humidity (by exposing a thin layer of pellets in a petri dish to a saturated solution of sodium acetate contained in a desiccator) (*27–30*). The oxidizer pellets are packed to a depth of 10 ml between two glass wool plugs in a 15-mm i.d. glass tube with connecting ends. Finally, for absorption of the NO_2, a fritted absorber containing 10 ml of Griess–Saltzman reagent is employed. The air sample is drawn through the train at a rate of about 0.4 liter/minute or less in order to develop sufficient final color in the above reagent (about 10 minutes).

Sulfur dioxide does not interfere in this method as it is removed by the chromium trioxide in the oxidizer (*30*). Depletion of the oxidizer is indicated by a color change from orange to brown. Strongly oxidizing gases such as PAN and halogens, may interfere by affecting the stability of the color of the Griess—Saltzman reagent. To avoid this interference, the absorbance of the solution should be determined within 1 hour (*2*).

2. Continuous Monitoring Methods for NO and NO_2

Various types of continuous analyzers have been developed for the measurement of NO and NO_2 in the ambient atmosphere. Earlier developments in such analyzers, using colorimetric estimation of the azo dye-forming reaction, have been described by Katz *et al.* (*31*), Thomas *et al.* (*32*), and Cholak and co-workers (*33*). More recent efforts have overcome the difficulties connected with the efficient oxidation of nitric oxide to nitrogen dioxide. Glass, firebrick, or alumina pellets of 15–40 mesh size coated with chromium trioxide and equilibrated at 70% relative humidity have been found to have high conversion efficiency for the oxidation of NO to NO_2 (*27*).

Continuous colorimetric analyzers available for the measurement of nitric oxide and nitrogen dioxide in ambient air usually operate in the range of 0.01 to 2 ppm ($\sim$12–2450 μg NO/m^3 or 19–3760 μg NO_2/m^3) (*34*). A metered sample flow is contacted with a metered liquid reagent flow to produce a color corresponding to the NO_2 content. The nitric oxide is converted to NO_2 by a solid oxidant in the sampling train upstream from the absorber. This oxidant consists of pellets impregnated with chromium trioxide as described in the manual methods mentioned earlier. The liquid absorbing solution consists of the Griess–Saltzman reagent. A recording photometer measures the change in absorbance of the liquid reagent or azo dye at 550 nm. The recorder scale can be calibrated in terms of NO_2 concentration.

Another approach involves the reaction of NO with ozone to produce activated NO_2 and the measurement of the chemiluminescent emission from the electronic transition.

$$NO_2^* \rightarrow NO_2 + h\nu \tag{7}$$

The emission spectrum covers the range of 0.6 to 3 μm, with a maximum intensity near 1.2 μm. This method of analysis for NO provides a linear relation over an extremely wide concentration range from 0.004 ppm to more than 1000 ppm. Fontijn *et al.* (*35*) have evaluated the use of photomultipliers that respond to chemiluminescent emissions in the range of 0.6 to 0.9 μm. For optimum sensitivity, the operating temperature of the photomultiplier should be lowered to $-20°$C. The method may be employed conversely as a procedure for the measurement of ozone. It is highly specific since no interferences have been observed for a wide range of other gases present in polluted air.

Nitric oxide reacts with atomic oxygen to produce excited states that show emission lines commencing at 350 to 450 nm and increasing to a maximum near 650 nm (*36*). This reaction can also be employed to determine NO and NO_2, since atomic oxygen will react with NO_2 to form NO on a one-to-one basis. A rapid response instrument for measuring nitrogen oxides may be constructed on this basis. A prototype continuous chemiluminescent NO monitor for ambient air analysis by reaction with ozone has been studied by Ronco and Fontijn (*37*). An instrument for the analysis of oxides of nitrogen by reaction with atomic oxygen, as described above, has been investigated by Iitis and Mueller (*38*).

Several instruments have been developed to utilize molecular correlation spectrometry for the analysis of NO_2 as well as SO_2. In this technique the molecular absorption of a band in the ultraviolet or visible spectrum containing vibrational or rotational spectra is compared with a replica of this spectral band stored within the spectrometer. Such spectrometers with an optical path length of 2.5 m have been tested in point source sampling trials in Los Angeles, California and Toronto, Ontario (*39*, *40*).

Another type of spectrometer that is designed to process the second derivative of the transmission wavelength function has been developed to analyze for nitric oxide, nitrogen dioxide, ozone, and sulfur dioxide in the wavelength region of 190 to 400 nm (*41*). For a light cell of 1-m path with 20 passes, the sensitivities that have been reported are 0.003 ppm for NO and 0.02 ppm for NO_2.

Both correlation spectroscopy and the technique of second-derivative optics possess advantages of greater specificity and sensitivity in relation to details of vibrational–rotational band structure or transmission–wavelength function. However, for purposes of ambient air monitoring, such instruments are expensive due to the complicated, advanced optical and electronic systems that are required, in contrast to the much less costly colorimetric or chemiluminescent techniques.

A new approach involves the use of infrared lasers and infrared spectroscopy for the monitoring of gaseous air pollutants (*42*). Hinkley and Kelley (*43*) have described the operation of tunable diode lasers for the detection of gaseous air pollutants, using a sensitive, mercury–cadmium–telluride detector, cooled in liquid nitrogen. For the analysis of nitric oxide in the ppm range, Kreuzer and Patel have employed a laser providing infrared radiation in an optoacoustic spectrometer instrument (*44*). Various types of point-sampling and remote types of instrumentation developments have been reviewed by Altshuller (*45*).

A dispersive infrared analyzer for the measurement of nitric oxide and NO_x, utilizing a single-beam, dual-wavelength ratioing system has been described by Lord *et al.* (*46*). The beam passes through the sample cell and is focused on the entrance slit of a monochromator. Incident radiation is dispersed by a plane grating. Two adjacent wavelengths, one where the sample absorbs and one where it does not, are measured alternately and then ratioed. The concentration of NO is proportional to this ratio, I/I_0, and follows Beer's law. This type of instrument has been evaluated by the United States Environmental Protection Agency for the measurement of exhaust emissions from piston and diesel engines and showed a correlation of 0.97 or better with chemiluminescence analysis. For the measurement of NO_2 and NO_x, the instrument is coupled with an NO_2 thermal converter.

The use of NO_2 converters for application with the chemiluminescent monitor has been described by Sigsby *et al.* (*47*). Most converters employ either low temperature surface reactions (225°C) to convert the NO_2 to NO or the thermal decomposition of NO_2 to NO at the higher temperature of about 600°C.

3. Calibration of Continuous Colorimetric Analyzers for NO and NO_2

A method for calibration of such continuous colorimetric analyzers has been published by the Intersociety Committee (*48*). The calibration procedure is divided into two phases: static and dynamic. Static calibration is a test of the optical measuring system. Reagent solutions are used to simulate actual flow and reagent conditions encountered during normal operation. A static calibration curve is a plot of analyzer responses versus reagent solutions equivalent to specific NO_2 concentrations. Adjustment of the instrument variables, e.g., air and liquid flow rates, is performed after static calibration to make output response conform to predetermined concentrations in parts per million or micrograms per cubic meter of pollutant.

Reagent solutions equivalent to specific NO_2 concentrations are prepared by reacting known quantities of sodium nitrite solutions with the azo dye-forming reagent and introducing these calibrating solutions into the measuring cell of the photometer at a rate that is commensurate to that of the analyzer.

Dynamic calibration is a performance test of the complete instrument when sampling a gas of known concentration. The efficiency of the absorbing system and the correctness of air and liquid flow rates are specifically tested. A flow dilution panel is used to generate calibrated concentrations of gases which are fed into the instrument. The mixture is analyzed for NO_2 by the manual method. The nitric oxide content is determined similarly after first removing any NO_2 with the Griess–Saltzman highly selective absorber and then oxidizing the NO to NO_2 with high efficiency solid oxidant. The calibrating gases should be essentially free of interfering compounds. Relative humidities in excess of 90% for extended periods inactivate the NO oxidizer. Operating the oxidizer 5°C above ambient temperatures (e.g., by heating with a 7-W lamp) eliminates this interference.

a. Static Calibration Procedure. This procedure involves the use of a solution of sodium nitrite containing 0.0203 gm of $NaNO_2$ per liter of (nitrite-free) distilled water. This solution is equivalent to 10 μl of NO_2/ml or 20 μg NO_2/ml. Standardization is based upon the empirical observation that 0.72 mole of $NaNO_2$ produces the same color as 1 mole of NO_2 gas. Solutions equivalent to 0.04, 0.08, 0.12, and 0.20 μl NO_2/ml, respectively, are employed for the absorbance measurements and the recorder readings are plotted against concentration of NO_2 to obtain a calibration graph. In this procedure the reagent flow rate is maintained constant at the same rate as the analyzer utilizes in its actual operation (between 1 and 3 ml/minute).

b. Dynamic Calibration Procedure. Permeation tubes are available for NO_2 and may be used in conjuction with a thermostat and a dilution panel for mixing the NO_2 with a measured volume flow of clean air to prepare a series of test gas concentrations (*49*). The test gas concentrations are passed through the continuous analyzer to obtain a series of photometer recorder readings. Similar test gas concentrations are also analyzed by the manual azo dye method to derive a spectrophotometer calibration curve.

Nitric oxide is not available in permeation tubes because of its low boiling point, —152°C. However, 10–20 ppm NO in nitrogen may be obtained in commercially available cylinders. Any traces of NO_2 that may be present in this gas must be removed by prior passage through a bubbler containing azo dye reagent. The purified nitric oxide is mixed with the required amount of clean air by means of a dilution panel and then passed through the humidifier and chromium trioxide oxidizer to convert the NO to NO_2. The rest of the calibration follows the procedure used for NO_2.

Calculation of results from the dynamic calibration may be made as follows: By definition

$$\text{ppm (v/v)} = \frac{\text{microliters of pollutant}}{\text{liters of sample}} \tag{8}$$

In a dynamic system, such as an oxide of nitrogen analyzer, a material balance gives

$$\text{ppm NO}_2\text{ (v/v)} = \frac{(\text{microliters NO}_2\text{/ml of solution})(\text{ml of solution/minute})}{\text{liters of air/minute}} \tag{9}$$

If a direct readout on the chart is desired, that is, the chart reading in absorbance units is equivalent to ppm NO_2, a suitable reagent flow rate is chosen and the air flow is calculated using the static calibration curve. Often a scale factor f_s is used so that the output reading is a simple fraction of input concentration.

$$\text{Chart reading } A = f_s \text{ ppm } NO_2 = f_s \text{ (microliters } NO_2/\text{ml}) \times \frac{Q_r}{Q_a} \quad (10)$$

or

$$Q_a = f_s(\text{microliters } NO_2/\text{ml}) \times \frac{Q_r}{A} \quad (11)$$

but

$$(\text{microliters } NO_2/\text{ml})/A = 1/b \quad (12)$$

therefore

$$Q_a = Q_r \times f_s/b \quad (13)$$

where

A = absorbance units
Q_r = reagent flow rate (ml/min)
Q_a = air flow rate (liter/min)
f_s = a fractional number
b = slope of static calibration curve

C. Ammonia

In general, air sampling methods for the determination of ammonia involve absorption of the gas in dilute sulfuric acid. Subsequent analysis may be performed by reaction with colorimetric reagents such as Nessler solution, alkaline hypochlorite, and phenol or transformation to nitrite and conversion to an azo dye.

1. Nessler Reagent Method

Buck and Stratmann (*50*) have described a method in which the air sample is passed through a special impinger of 275 ml, containing an outer ground-glass joint, from which the absorbed ammonia can be distilled directly into a condenser and receiver. This impinger contains 50 ml of 0.01 *N* H_2SO_4 and 10 ml of 0.4% potassium permanganate. The air sample is aspirated at a measured rate of about 30 liters/minute for 30 minutes.

In the distillation step 30 ml of 30% NaOH are added and the liquid in the impinger is boiled by heating with a liquid bath at 138–140°C. The condensate containing the ammonia is collected in a receiver containing 30 ml of 0.01 N H_2SO_4, heated to about 120°C. After all of the ammonia has been collected, the contents of the receiver are transferred to a 50-ml volumetric flask, 4 ml of Nessler reagent are added and the volume made up to the mark with distilled water.

After standing for 20 minutes the absorbance of the colored solution is determined in a spectrophotometer at 450 nm, using a 2-cm cell. The method can be calibrated with solutions of ammonia in the range of 10 to 100 μg, subjected to steam distillation and treated as above. Substantially complete absorption of ammonia can be achieved for air sample NH_3 concentrations in the range of 17 to 400 $\mu g/m^3$.

2. Indophenol Method*

The air sample at a measured flow rate of 1 to 2 liter/minute is aspirated in a fritted bubbler or midget impinger containing 10 ml of 0.1 N sulfuric acid for about 1 hour. The contents of the absorber are transferred to a 25.0-ml glass-stoppered graduate cylinder, 2 ml of a buffer, (prepared by dissolving 5.0 gm of $Na_3PO_4 \cdot 12H_2O$ and 7.4 ml of 6.75 M NaOH in 100 ml of distilled water), 5 ml of a phenol solution (prepared by mixing 20 ml of 45% phenol solution with 1 ml of 2% sodium nitroprusside and diluting to 100 ml with distilled water), and 2.5 ml of hypochlorite solution (from a mixture of 30 ml of 0.1 N sodium hypochlorite and 30 ml of 6.75 M sodium hydroxide, diluted to 100 ml) are added, mixed, and diluted to 25 ml. All solutions and sample are maintained at 25°C and the graduated cylinder is stored in the dark for 30 minutes for color development of the blue dye. The absorbance of the solution is measured at 630 nm against a reagent blank, using 1-cm cells.

Calibration of the method can be achieved by use of ammonium chloride solutions containing 5, 10, and 15 μg of ammonia per 25 ml of solution. At the above sampling rate and sampling time of 1 hour, a concentration range of 20 to 700 $\mu g/m^3$ may be determined (0.025 to 1.0 ppm). Ammonium compounds in suspended particulate matter will be determined by this method if they are not removed by prefiltration of the sample. Nitrite and sufite may cause interference if present in a 100-fold excess. Formaldehyde may also interfere to the extent of 10–15%. Replicate samples collected with glass prefilters show a relative coefficient of variation of 30% in the 0.7- to 21-$\mu g/m^3$ (1–30 ppb) range. This coefficient decreases at higher concentrations and is as low as 5% in the 1 ppm range (700 $\mu g/m^3$) (*52*).

* See Ref. (*15*).

3. Nitrite Method*

The air sample is passed through a prefilter to remove particulate matter and then aspirated in a fritted bubbler or impinger containing 0.1 *N* H_2SO_4 at a rate of 1 to 2 liters/minute for 10 minutes to 2 or more hours to collect an adequate sample. The resultant ammonium sulfate solution is reacted with hypochlorite in the presence of bromide as a catalyst, in an alkaline medium, to form nitrite. After destruction of the excess hypochlorite with sodium arsenite, the nitrite is transformed into an azo dye by addition of a buffered color-forming reagent consisting of sulfanilamide, citric acid, and naphthylethylenediamine hydrochloride in solution in 5 *N* H_2SO_4. The color is measured by absorbance at 545 nm, using an appropriate cell length.

The method may be standardized with pure ammonium sulfate solutions. A constant 65% of the absorbed ammonia is converted to nitrite (*53*). This method is less subject to interferences in urban atmospheres such as formaldehyde than the indophenol method. The sensitivity in molar absorptivity of ammonia is 38,000. At the above sampling rate for 1 hour, the range of ammonia concentration is 14 to 220 $\mu g/m^3$. Formaldehyde up to 100 ppm does not interfere. Nitrite and hydrolyzable amino compounds may cause interference. The sampling efficiency of glass impingers is better than 95% at levels above 400 $\mu g\ NH_3/m^3$. Absorption efficiency decreases at low concentration levels in the range of 10 to 50 $\mu g/m^3$.

4. UV Spectrophotometry

The determination of ammonia in air by UV absorption at 204.3 nm in 10-cm quartz cells has been described by Gunther *et al.* (*56*). At this wavelength the molar extinction coefficient is 2790. Ammonia exhibits several strong absorption bands in the range of 190 to 230 nm. The lower concentration limit is about 7 ppm. A UV procedure for the continuous automatic determination of ammonia in air at concentration levels above 10 ppm has been reported by Kolbezen, Eckert, and Wilson, applicable to the environment of citrus fruits in storerooms (*57*).

5. One-Step Method for Ammonia and Aliphatic Amines

A direct, one-step method for the analysis of ammonia and aliphatic amines in air has been described by Kramer and Sech (*57a*). Ammonia

* See Refs. (*53–55*).

in the air sample is collected in dioxane containing a new reagent, *o*-(benzenesulfonamido)-*p*-benzoquinone. The reaction between ammonia and this reagent is so rapid that instantaneous determinations may be made of the ammonia concentrations being sampled. Measurement of the concentration is determined in the spectrophotometer by absorption at 480 nm. The molar extinction coefficient is 3.7×10^3.

D. Nitrous Oxide, N_2O

Although nitrous oxide is of no importance as an air pollutant, it occurs naturally in the atmosphere in concentrations up to about 0.25 ppm. It is formed by the decomposition of inorganic and organic nitrogen compounds and in the waste gases of industrial processes involving the oxidation of ammonia to produce nitric acid and of oxidation processes utilizing nitric acid. The gas is readily soluble in water but does not undergo specific chemical reactions at ordinary temperatures. Mass spectrometry and infrared analysis may be employed for the detection of N_2O. Its infrared spectrum shows a strong absorption band at 1275 cm^{-1}. However, a sensitive method for low concentrations involves the use of gas–solid or gas–liquid chromatography as described by Leithe (*58*).

Analysis for N_2O by Gas Chromatography

A silica gel column, activated by heating for 5 hours at 160°C, is employed to concentrate the N_2O in the gas sample. To remove carbon dioxide, a soda lime cartridge is inserted between the sample inlet and the silica gel column. During sampling of air at 500 ml/minute, the gel concentration column (U-tube, 30 cm $\times$ 6 mm) is immersed in a Dewar flask filled with acetone and dry ice (−70°C). The N_2O is desorbed from the concentration column by first flushing with helium and then immersing the concentration column in water at room temperature and connecting to a separation column.

The separation column (10.5 $\times$ 6 mm) contains a fixed phase of 19% propylene carbonate and 16.5% glutaronitrile on 0.2–0.3 mm Sterchamol. Helium at a flow rate of 60 ml/minute is employed as the carrier gas at 20°C. The air peak appears after 6 minutes and the N_2O peak after 13 minutes. When concentrated from a 10-liter air sample, the detection limit for N_2O is 0.05 ppm.

LaHue, Axelrod, and Lodge (*582*) describe a method for the measurement of atmospheric nitrous oxide (N_2O) which is present normally to the extent of about 250–300 ppb (0.25–0.30 ppm). For air sampling purposes, a molecular sieve 5A trap is employed to adsorb the N_2O. Upstream of this trap, water vapor and CO_2 are removed by passage of the air over anhydrous calcium sulfate and ascarite, prior to reaching the molecular sieve. To desorb the N_2O from the molecular sieve, a stream of helium gas saturated with water vapor is employed. The water-displaced N_2O is then trapped on a silica gel column submerged in a dry ice–isopropyl alcohol bath. This silica gel column is then heated to transfer the N_2O into a gas chromatograph for analysis.

E. Primary and Secondary Aliphatic Amines

Ninhydrin Method*

The air sample is passed at a measured flow rate of 1 to 2 liters/minute for 30 minutes through a midget impinger equipped with a fritted gas dispersing tube, containing 10 ml of absorbing solution, consisting of 1% v/v concentrated hydrochloric acid in isopropanol. To compensate for evaporative losses, the volume of the sampling solution is made up to 10 ml with isopropanol. An aliquot of 3 ml or less is transferred to a glass-stoppered tube. A blank is also run for each series of samples.

To each tube, 5 ml of pyridine and 2 ml of ninhydrin (1,2,3-triketo-hydrindene) are added and made up to volume with absorbing solution. The tubes are stoppered and placed in a water bath at 85°C for 7 minutes. After removal of the tubes from this bath, they are immersed in cold water for 10 minutes. The solution in each tube is then transferred to 1-cm measuring cells and the absorbance determined against that of the blank in a spectrophotometer at 575 nm.

The method is calibrated and standardized against solutions of *n*-butylamine containing 2 to 100 μg/ml. The reaction with ninhydrin of primary and secondary amines produces a purple dye (Ruhemann's purple) (*59*). Tertiary amines do not react. The results are expressed as equivalent molar amounts of *n*-butylamine (*61*). A 30-liter gas sample will provide a limit of detection 300 μg/m^3 (0.1 ppm) of equivalent butylamine when using 10 ml of absorbing solution. Lower concentrations require larger gas samples. The absorption efficiency is quantitative when 30 ml of absorbing solution is used for concentrations in the range of 16 to 300 mg/m^3 in air (*59*). Ammonia and alkaline substances, other than primary and secondary amines, do not interfere but some amino acids may interfere on a molar basis.

* See Refs. (*59–61*).

F. Hydrazine, NH_2–NH_2

This substance is used as a rocket propellant, and also for the removal of dissolved oxygen in supply water for high pressure boilers and as a reactant in the plastics industry.

Colorimetric Analysis

Low concentrations of hydrazine in air may be determined by scrubbing the air sample with 10 ml of a dilute solution of sulfurous or sulfuric acid in a fritted impinger or bubbler at a rate of 5 liters/minute for about 20 minutes or longer, as required. The solution is transferred to a 50-ml volumetric flask, 10 ml of dimethylaminobenzaldehyde reagent (5 gm of aldehyde added to 5 ml of absorption solution and diluted to 100 ml with ethanol) are added and made up to the mark with distilled water. After standing for 20–30 minutes for color development, the absorbance is determined at 455 nm in 1-cm cells against a similarly treated blank (*62*). An effective absorption solution is 40 ml sulfur dioxide diluted to 500 ml with distilled water.

Calibration of the method is accomplished by means of hydrazine sulfate solutions in the range of 0 to 25 μg hydrazine in 50 ml in accordance with the above procedure. The molar extinction coefficient is 64,000 and the detection limit for a 100-liter air sample is about 3 ppb.

A continuous colorimetric analyzer for the determination of asymmetric dimethylhydrazine in the ppm range has been described by Buck and Eldridge (*63*). It is based on the coulometric generation of bromine as a reagent, with a sensitivity of $\pm$0.1 ppm.

II. Oxidants and Ozone

A. Introduction

The role of ozone and oxidants as by-products of atmospheric photochemical reactions involving nitrogen dioxide, hydrocarbons, and organic free radicals was first noted during photochemical smog episodes in the Los Angeles, California area. Ordinarily, ozone in the lower atmosphere is present only in very low concentrations in the range of less than 1 to about 2 or 3 parts per hundred million (pphm). During a photochemical smog incident the total oxidant concentration may attain values as high as 50 pphm, of which about 90% consists of ozone. Measurements of the diurnal variation of primary and secondary contaminants in Los

Angeles show that in the early morning hours the concentration of NO increases first, followed by a rise in the level of NO_2. In the late morning hours as the nitric oxide concentration disappears, ozone levels commence to increase and reach a maximum about noon. The NO_2 declines from its peak as the level of ozone is increased. Values of ozone decline to near zero at night.

The presence of higher than normal levels of ozone or oxidant in urban and semirural atmospheres is considered to be an indicator of photochemical atmospheric reactions. These have been reported in an increasing number of cities and other areas in the United States, Canada, and other countries. The need for accurate analytical methods for the determination of ozone and oxidants has assumed, therefore, considerable importance. Oxidants include ozonides and peroxides of hydrocarbons and substances such as peroxyacyl nitrates.

B. Analysis of Photochemical Oxidants—Ozone

1. United States Federal (USEPA) Reference Method*

The basis for this method is the gas-phase chemiluminescent reaction between ambient air containing ozone and ethylene to produce an excited species which emits light in the visible range (*65–68*). The resulting reaction is specific for ozone and takes place in a chamber of borosilicate glass construction which has a light transparent, flat window coupled to a photomultiplier tube. The signal produced by the phototube is proportional to the ozone concentration. This signal may be further amplified and read directly on a recorder or converted into a digital display. The method is applicable to the continuous measurement of ozone in ambient air.

The range of the method is from 9.8 $\mu g/m^3$ of ozone to greater than 1960 $\mu g/m^3$ (0.005 to more than 1 ppm O_3). The method is claimed to be accurate within $\pm 7\%$, and the average deviation from the mean of repeated measurements does not exceed 5%.

Test atmospheres are generated by means of an ozone source consisting of a quartz tube into which ozone-free air is introduced and irradiated with a stable low pressure mercury lamp. The level of irradiation is controlled by an adjustable aluminum sleeve which fits around the lamp. At a fixed level of illumination, ozone is produced at a constant rate. Levels of ozone in the test atmospheres are determined by the neutral

* See Ref. (*64*).

buffered potassium iodide method, 1% solution. A calibration curve is prepared by measuring the absorbance of a series of diluted standard iodine solutions against calculated equivalents of ozone.

2. Intersociety Committee Chemiluminescent Method*

This method involves also the gas-phase reaction between ozone and ethylene, leading to emission of visible light and is substantially similar to the USEPA Reference Method described above. Chemiluminescent instruments for this method are available in several measurement ranges in the same instrument, for example, from 0 to 40, 0 to 400, and 0 to 4000 μg ozone/m^3 (0–0.02, 0–0.2, and 0–2 ppm, respectively, by volume). The lower limit of detection has been reported as 6 to 8 $\mu g\ O_3/m^3$ in air (0.003 to 0.004 ppm). When operating this type of instrument at the lowest measurement range, a slight dependence on humidity has been observed but other substances normally present in ambient air do not interfere (*67, 68*).

Carroll *et al.* (*68*) define precision in terms of stability and repeatability of ozone response for such instruments. Instability due to baseline drift should be equivalent to no more than $\pm 20\ \mu g\ O_3/m^3$. Noise level should not exceed $\pm 16\ \mu g\ O_3/m^3$, and repeatability of response to a standardized ozone–air mixture should not exceed a coefficient of variation of 1% as the composite error associated with both the ozone measurement and the standardization operations.

The air inlet should contain a Teflon filter capable of removing all particles greater than 5 μm in diameter. Ethylene flow to the reaction chamber is controlled by a meter capable of measuring flows between 0 and 80 ml/minute. The photomultiplier tube should be of a high-gain low dark current type (about 10^{-9}) having its maximum gain at 430 nm. A high voltage power supply of up to 2000 V is required. The direct current amplifier should be capable of full scale amplification of currents from 10^{-10} to 10^{-7} A. Commonly, an electrometer with an output of 0–10 mV or greater is employed. Sampling lines should be constructed of Teflon or borosilicate glass only. Ozone is destroyed by contact with rubber and polyvinyl chloride tubing and some metals show a constant small adsorption which affects the measurement of ozone (*70*).

a. CALCULATION. The concentration of ozone may be calculated as follows:

$$V_R = F \times T \times (P/760) \times 298/(t + 273) \qquad (14)$$

* See Ref. (*69*).

where

V_R = volume of air at standard conditions, in liters
F = flowmeter air rate, liters/minute
T = time, in minutes
P = barometric pressure in Torr
t = sampling temperature, in °C

The total micrograms of ozone are read from the spectrophotometric calibration curve. The concentration of ozone in the gas phase is given by

$$\text{Ozone, } \mu g/m^3 = \frac{\text{total } \mu g \text{ } O_3 \text{ per 10 ml}}{V_R} \times 10^3 \tag{15}$$

Conversion between $\mu g/m^3$ values for ozone and ppm can be made as follows:

$$\text{ppm ozone} = \frac{\mu g \text{ } O_3/m^3 \times 24.45}{48.00} \times 10^{-3} = \mu g \text{ } O_3/m^3 \times 5.1 \times 10^{-4} \tag{16}$$

where

24.45 = molar volume of a gas at 25°C and 760 Torr
48.00 = molecular weight of ozone in grams

3. Buffered Potassium Iodide—Manual Method

Byers and Saltzman (*71*) studied the reaction of a 1% phosphate buffered, potassium iodide solution (0.1 *M* KH_2PO_4 and 0.1 *M* Na_2HPO_4) for the quantitative absorption of ozone in air (*71*), with liberation of an equivalent amount of iodine. This reaction forms the basis of an Intersociety Committee Method (*72*) for the manual analysis of oxidizing substances in the atmosphere. The 1% KI solution is buffered at pH 6.8 ± 0.2, and the liberated iodine is determined by spectrophotometric measurement of the absorbance of triiodide ion at 352 nm, according to the following stoichiometric reaction:

$$O_3 + 3KI + H_2O = KI_3 + 2KOH + O_2 \tag{17}$$

The method is applicable to the determination of oxidant concentrations in the range of 0.01–10 ppm as ozone. When 10 ml of absorbing solution is used and 1 to 10 μl of ozone are collected, the corresponding absorbances of the solution range between 0.1 and 1.0 in a 1-cm cell. The interference of sulfur dioxide can be eliminated without loss of oxidant by incorporating a chromic acid impregnated paper absorber in the sampling train upstream from the midget impinger (*73*). This chromic acid absorber will also oxidize nitric oxide to NO_2. When sulfur dioxide is less than 10% of the nitric oxide concentration, the use of this absorber is not recommended. In this case a concurrent determination of SO_2 should be made and this concentration added to the oxidant value as a correction.

Nitrogen dioxide yields a response in 1% KI, equivalent to 10% of that of an equimolar concentration of ozone. Nitrogen dioxide interference can be corrected by concurrent analysis for NO_2 and subtracting one-tenth of this concentration from the total oxidant value. Peroxyacetyl nitrate, if present, yields a response equivalent to 50% of that of an equimolar ozone concentration (*74*). Concentrations of this compound in the atmosphere may range up to 0.1 ppm. Other substances, besides ozone, that will liberate iodine with this method include peroxy compounds, hydroperoxides, hydrogen peroxide, organic nitrites, and halogens (*75, 76*). Hydrogen sulfide causes a negative interference.

The method is calibrated by the use of dilute, standardized iodine solutions. Exactly 4.09 ml of 0.001 *M* iodine solution is diluted with absorbing solution (1% KI in 0.1 *M* phosphate buffer) just before use to 100-ml volume. This concentration of calibrating iodine solution is equivalent to 1 μl of O_3 per ml. Graduated amounts of calibrating solution (up to 10 ml) are added to a series of 10-ml volumetric flasks and diluted to volume with absorbing solution. The absorbances of the colored solutions are plotted against concentrations of ozone in μl 10 ml absorbing reagent to yield a straight line plot that follows Beer's law.

4. Amperometric Instruments—Continuous Monitoring*

The principle of this electrical monitoring method is the reaction of ozone and other oxidants with iodide ions present in a sensing reagent, to produce iodine which then forms triiodide ion. The triiodide ion is then converted electrically back to iodide ion. Amperometric conversion of the triiodide to iodide ion is accomplished through cathodic reduction utilizing either an electrolytic cell (*78, 79*) or a galvanic cell (*80, 81*).

In the electrolytic procedure an iodide-containing electrolyte is employed and a potential of 0.025–0.30 V is applied between the electrodes. The initial current polarizes the cathode with a layer of hydrogen which prevents further current flow (*82*). A fraction of the triiodide ion in solution reacts with some of the hydrogen. Consumption of hydrogen causes flow of polarization current to replace the reacted hydrogen and reestablish polarization. The polarization current is proportional to the triiodide ion concentration according to the configuration of the electrodes.

In the galvanic cell cathodic reduction of iodine occurs spontaneously

* See Ref. (*77*).

along with parallel oxidation of the carbon anode. The galvanic oxidant detector operates as an electrochemical fuel cell in which the fuels are oxidant (in the sample) and carbon (in the anode).

a. RANGE AND SENSITIVITY. Amperometric oxidant detectors operate most reliably at atmospheric oxidant concentration ranges equivalent to 0.01–0.2 ppm of ozone (0.02–0.4 μg/liter). Usable response may be obtained from oxidant levels equivalent to as much as 1 ppm ozone. In the range 0.2–1 ppm results in comparison to colorimetric procedures show some scatter and differences.

b. INTERFERENCES. Amperometric oxidant detectors respond to substances that are capable of either oxidizing iodide ion to iodine (positive response) or reducing iodine to iodide ion (negative response) in the sensing solution. The most common oxidizing interference is nitrogen dioxide, whereas sulfur dioxide, and hydrogen sulfide are among the most common negative interferences.

Presence of nitrogen dioxide in the sample causes positive interference equivalent to 0.03–0.20 moles of ozone/mole of nitrogen dioxide (*81, 83, 84*). The exact level of the detector's specific response to nitrogen dioxide depends on design of detector, composition of sensing solution, operating conditions, and unknown factors. Because of these unknown factors, the specific response to nitrogen dioxide for each detector must be determined experimentally rather than taken from literature data. The oxidant measurements may be corrected by subtracting from the total detector response the amount of response caused by nitrogen dioxide.

Presence of sulfur dioxide in the sample causes negative interference equivalent to approximately 1.0 mole of ozone/mole of sulfur dioxide. The following methods can be used for eliminating the sulfur dioxide interference.

i. *First method.* Add to the total detector response an increment equal in value to the reduction caused by the sulfur dioxide. To determine the latter, the sulfur dioxide level in the sample and the specific response of the detector to sulfur dioxide must be determined previously. This correction method is recommended only when sulfur dioxide levels do not exceed those of oxidants.

ii. *Second method.* Remove sulfur dioxide from the sample stream before it enters the oxidant sensor cell. Such removal, without affecting the oxidant, can be accomplished using a solid absorber consisting of chromic acid impregnated quartz chips. Performance for this system must be established experimentally rather than from literature data. This correction method is recommended when sulfur dioxide levels exceed those of oxidant.

Presence of hydrogen sulfide and/or other reducing compounds in the sample causes negative interference of varying magnitude. To correct oxidant measurement results for these interferences, add to the total detector response an increment equal in value to the reduction caused by the reducing compounds present. The latter are determined from both the total concentration of the reducing compounds in the sample and the specific response of the detector to these compounds. If a sulfur dioxide absorber is used, correction should be applied for the presence of reducing compounds in the sample stream that emerges from the sulfur dioxide absorber.

c. PRECISION AND ACCURACY. Because of the nonspecific nature of atmospheric oxidant, precision of the analytical method is defined in terms of stability and reproducibility of the detector's response to ozone. In continuous monitoring, the hourly variation in response to ozone present at a constant level between 0 and 0.50 ppm should be equivalent to ±0.01 ppm (*82*) or less (*81*). Day-to-day variability is considerably larger and its magnitude varies with individual instruments (*82, 85*). Accuracy of the method has not been established.

d. CALIBRATION OF AMPEROMETRIC DETECTORS. Because of the nonspecific nature of atmospheric oxidant, the response of an amperometric detector to atmospheric oxidant is expressed in terms of an ozone concentration level (ppm) that causes equal response. Therefore, the amperometric detectors must be calibrated to determine their specific response to ozone. Theoretically, the specific response to ozone is fixed and is equivalent to 2 electrons/ozone molecule. In practice, this response has values that are below the theoretical level and vary from instrument to instrument.

Therefore, each detector must be calibrated experimentally, using standard gaseous mixtures of ozone. Such mixtures are prepared by passing pure, filtered air or oxygen through an ozonizer and receiving the ozonized mixture into a properly conditioned plastic bag. Ozone concentration of varying levels can be obtained either by diluting the ozonizer output with pure, filtered air to varying degrees or by varying the airflow rate through the ozonizer. In this manner a series of mixtures can be obtained that covers the ozone concentration range of 0–2 ppm. Following preparation, the concentrations of the ozone mixtures in the bags are determined by the neutral phosphate buffered potassium iodide method (*72*). For calibration, the detector is first zeroed using a stream of air that has been freed from oxidizing and/or reducing contaminants by passing through a filter and an activated charcoal column. The detector is then connected to the bag that contains the standard ozone–air mixture, using a piece of conditioned Teflon (or other inactive material) tubing of minimum length. The resultant deflection data are treated mathematically as follows to define the detector's response as a function of ozone concentration.

e. Calculation of Detector Response. The instruments are calibrated by the manufacturers in accordance with the theoretical coulometric yield of 2 electrons/molecule of reacted ozone. The theoretical calibration relationships are given by

$$V = rI = r\frac{P \times F \times C \times 2 \times f}{R(t + 273)} \tag{18}$$

$$V = 5.16 \times 10^{-5} \times \frac{r \times P \times F \times C}{t + 273} \tag{19}$$

where

- I is detector response, milliamperes
- V is detector response, millivolts
- F is sample flow rate, ml/minute
- r is resistance of resistor used to convert signal current to voltage drop, ohms
- P is barometric pressure, Torr
- t is ambient temperature of the constant volume pump or the Rotameter, °C
- C is concentration of ozone in sample, ppm
- f is the amount of electricity produced from coulometric conversion of one electrochemical equivalent of ozone (96,494 C)
- R is the universal gas constant in appropriate units

In the electrolytic cell, response is affected by liquid flow rate although the latter is not included in the theoretical calibration equation. Also, response seems to be affected by temperature to a greater extent than expected from the perfect gas law.

In practice, the response is a linear function of ozone concentration at ozone levels only below 1 ppm. This linear function is different from that described by Equations (18) and (19) and is determined from the calibration data in the following manner: Plot response values against ozone concentrations in the standardized ozone mixtures. Apply the method of least squares to obtain the slope and intercept of the straight line that fits the data best. If the intercept value is significantly different from zero, apply the defined linear function only within the range of ozone concentrations used in calibration and obtain additional calibration points using ozone concentrations closer to zero.

To convert response data into ozone concentration, Equation (20) is used. If response is proportional to ozone concentration, the value of k_1 is zero.

$$C = (t + 273)/FP \qquad [k_1 + k_2 V] \tag{20}$$

The values of k_1 and k_2 are calculated from the intercept and slope, respectively, of the calibration line.

If the recorder has appropriate zero and span controls, it can be adjusted to give corrected readings under calibration conditions. If sample temperature and pressure deviate significantly, correct the sample flow rate to calibration conditions according to the perfect gas law.

5. Colorimetric Instruments—Continuous Monitoring

Most colorimetric instruments operate on the principle of the liberation of iodine and formation of the triiodide ion by ozone and other oxidants from a 10% neutral, buffered potassium iodide solution. The change in color of the KI reagent solution due to the presence of triiodide ion is measured photometrically at 352 nm (*75*) and recorded continuously on a strip chart recorder. Air samples are scrubbed in a wetted wall absorber by the reagent solution. A constant air-to-liquid ratio is maintained and the instrument can be calibrated to read directly in ozone concentrations as a reference oxidant. Calibration may be based on the Intersociety Committee Manual method for the analysis of oxidizing substances (*72*). A complete description of the operation and calibration of colorimetric instruments is presented in an Intersociety Committee Method (*86*).

a. Range and Sensitivity. Colorimetric oxidant detectors operate most reliably at atmospheric oxidant levels ranging from 0.01 to 1.0 ppm of ozone (20–2000 $\mu g/m^3$) or other equivalent oxidant (*87*).

Usable response may be obtained from oxidant levels equivalent to as much as 100 ppm ozone. At such high oxidant levels the stability and speed of detector response are markedly inferior to those attained in the 0.01–1.0 ppm range.

b. Interferences Sulfur dioxide produces a negative interference equivalent to 100% of an equimolar concentration of oxidant (*83*). To eliminate interferences either of the following two procedures may be used.

Up to 100-fold ratio of sulfur dioxide (and probably H_2S) to oxidant may be eliminated without loss of oxidant by incorporating a conditioned chromic acid paper absorber in the sampling train upstream from the wetted wall absorber *(73)*.

Alternatively, the effect of sulfur dioxide (or H_2S) on the oxidant reading can be corrected by concurrently analyzing for sulfur dioxide (or H_2S) and adding this concentration to the total oxidant value. This procedure is not applicable when sulfur dioxide (or H_2S) levels exceed those of the oxidant.

Nitrogen dioxide is known to give a response in 10% KI equivalent to about 20% *(88)* of that of an equimolar concentration of ozone. The contribution of the nitrogen dioxide to the oxidant reading can be eliminated by measuring the instrument response to a known stream of NO_2 and by concurrently analyzing for nitrogen dioxide and subtracting the contribution of this concentration from the total oxidant value. The exact level of the specific response to nitrogen dioxide depends on design of the absorbing column, composition of absorbing solution, operating conditions, and unknown factors.

Nitric oxide does not interfere unless an SO_2 absorber as described above is used. The SO_2 absorber will oxidize nitric oxide to nitrogen dioxide. Therefore, when sulfur dioxide is less than 20% of the nitric oxide concentration, the use of chromic acid paper is not recommended.

Hydrogen sulfide, reducing dusts, or droplets can act as negative interferences.

c. Precision and Accuracy. In continuous monitoring, the hourly variation of the instrument response to a constant level of ozone set between 0 and 0.5 ppm should be equivalent to ±0.01 ppm or less. Day-to-day variability of instrument response to a constant level of ozone should be less than ±0.03 ppm. This performance is achieved by available instruments.

The accuracy of this method has not been established because of the difficulty in preparing known concentrations of ozone within the range of this method.

d. Apparatus for Continuous Colorimetric Instruments. Air sampling lines and all parts of the instrument that come in contact with the airstream should be constructed of borosilicate glass, Teflon, or stainless steel. Fluorocarbon or fluorosilicon grease may be used sparingly, if necessary. Ozone for dynamic calibration purposes may be stored for short periods up to 1 hour in Teflon or Tedlar bags that have a surface-to-volume ratio of less than 20 m^2/m^3.

A suitable air pump consists of a constant volume, self-lubricating carbon-vane pressure-suction pump with vacuum gauge and relief valve that can operate continuously at sample flow rates between 2 and 5 liters/minute. It should be equipped with a bleeder inlet to the suction line and a needle valve for flow adjustment and for protection of the pump from overheating and corrosion.

The reagent lines and all parts of the instrument that come in contact with the unexposed potassium iodide solution should be made of borosilicate glass because of the corrosive nature of free iodine. For the reagent solution, a self-priming constant delivery pump should be employed that will operate continuously at a precisely known flow rate within 1% variation in any 24-hour period. A needle valve and calibrated glass Rotameter are used to measure and control the air flow, upstream from the contact column. The latter is constructed from a 500–600-mm straight glass tube of 6–10 mm i.d. A 4-mm pitch helix of 1-mm glass rod is fitted snugly inside this tube to increase the area of absorbing surface and to ensure that this surface is completely wetted with reagent. The holdup of liquid in this absorber should be minimized for rapid response.

The photoelectric sensor consists of a dual beam type, photometer system with an air path or a flow-through reference optical cell and a flow-through measuring cell. The optical cells should be transparent in the UV at 352 ± 20 nm, with a maximum deviation during the calibration procedure of 1% of full span. Baseline drift over any 24-hour period should be less than 2% of full span.

An iodine absorbing column is employed to remove the free iodine formed in the reagent solution. This consists of a cylinder filled either with purified 8–20 mesh activated charcoal or with nylon wool (*79*). The column should be capable of removing the free iodine for up to a period of 2 months. To obtain a zero reading on the instrument, a cartridge packed with granular activated charcoal and soda lime may be employed to remove traces of oxidants, NO_2, and SO_2 from the air sample.

The ozonator consists of a 4-W ozone lamp (such as General Electric G4511 or similar), a voltage stabilizer, a ballast transformer, a rheostat, and a 0–1000-mA ac-meter in series. It is powered by a 115-V ac source. The ozone chamber is fabricated from a 45/50 standard taper joint with the lamp socket and the inlet and outlet gas stream tubes sealed in. Varying levels of ozone can be obtained by adjusting lamp current and flow rate of gaseous streams.

e. Absorbing Reagent. Potassium iodide in the absorbing reagent must be of highest purity, free from iodine-reducing substances such as hypophosphite. All reagents must be made up with double distilled water; deionized water is not recommended for this purpose. The absorbing reagent consists of a 0.5 *M* neutral buffer solution of potassium dihydrogen phosphate and disodium hydrogen phosphate and 10% potassium iodide absorbing solution (2 liters of phosphate buffer in 10 liters of 10% KI).

For static calibration purposes, a series of dilute, standardized iodine solutions are prepared, containing the equivalent of 0.1, 0.2, 0.3, and 0.4 μl O_3/ml in 10% buffered KI absorbing solution. The corresponding photometer readings are recorded and plotted on coordinate paper. The linearity of response of the instrument is established by determining the concentration in microliters O_3/milliliter at which the curve intersects an absorbance of 1.0 or 10% transmittance (90% of full scale on the chart). The recorder is adjusted to this reading.

f. DYNAMIC CALIBRATION. The ozone for calibration is generated in an ultraviolet ozonator, as described above. The concentrations of the ozone mixtures so generated are determined by the Intersociety Committee method (*72*) for analysis of oxidizing substances. Under recommended flow conditions, the detector's initial response (lag time) is reached within 60 seconds. Longer response time may be caused by oxidant destruction on unconditioned surfaces of sample lines. Conditioning can be accomplished by operating the instrument on air samples containing about 1 ppm ozone until response time is minimized. Following each startup, the instrument can be stabilized by drawing ambient air for at least 24 hours before it is calibrated or used for recording purposes.

The analyzer operating parameters are adjusted approximately, so that the chart reading will give the concentrations directly in ppm O_3. The liquid flow rate (f_1) in milliliters/minute is set according to the manufacturer's specifications. Then, the airflow rate (f_a) in liters/minute is determined by multiplying the liquid flow rate f_1 by the inverse average slope value

$$1/b = c_s \qquad (21)$$

where c_s is the concentration in microliters O_3/milliliter of absorbing solution which will give an analyzer response of absorbance equal to 1.0:

$$f_a = c_s \times f_1 = f_1/b \qquad (22)$$

The instrument is zeroed by switching a charcoal filter into the sampling line. The analyzer is connected to one leg of the ozone dilution panel. The other leg is connected to the measuring column containing buffered KI absorbing reagent for determining the ozone concentration. When the trace on the recorder is stabilized, the measuring column with absorbing solution is switched into place, and the ozone concentration is then determined colorimetrically.

If the response of the dynamically determined points deviate by more than 10% from the static calibration curve, either the efficiency of the contact column is low or other operating parameters are off. Check the following parameters until the cause is discovered: liquid flow rate, airflow rate, change in static response, recorder malfunction.

The maximum deviation of any individual sampling point from the best-fit curve drawn through the dynamic calibration points is generally not more than 0.025 ppm. Those points that deviate more than the stated amount are questionable and should be repeated.

Determine the lag and time to 95% of ultimate response of the analyzer under simulated operating conditions as follows. Disconnect the ozone sample line. Allow the analyzer to sample filtered air until the analyzer trace is on baseline. Then reconnect to the ozone generator. Record the lag time by determining the interval between the time the ozone generator was reconnected and the observed start of the analyzer response on the chart. Then record the time to 95% of ultimate response by measuring the interval from the input concentration change to 95% of the ultimate steady-state concentration.

6. Miscellaneous Colorimetric Methods—Organic Reagents

A number of colorimetric methods for the estimation of ozone have been reported that involve reactions with organic bases and dyes. In some cases, the color produced by interfering substances, such as NO_2, differs from that produced by ozone.

a. 4,4′-DIMETHOXYSTILBENE—BRAVO AND LODGE (*89*). This procedure is based on the reaction of ozone with $CH_3O \cdot C_6H_4 \cdot CH{=}CH \cdot C_6H_4 \cdot OCH_3$ to form anisaldehyde by cleavage of the dimethoxystilbene. In the presence of fluoranthene and trifluoroacetic acid, the reaction mixture forms an intense blue dye. The air sample is passed at a rate of 0.1 to 0.15 liter/minute for 15 to 60 minutes through a fritted glass impinger containing 3 ml of a solution of 5 mg 4,4′-dimethoxystilbene in 100 ml of sym-tetrachloroethane. A 1-ml aliquot of this solution is mixed with 1 ml fluoranthene solution (5% in chloroform) and 0.8 ml trifluoroacetic acid in a 10-ml volumetric flask. After standing for 5 minutes, the mixture is made up to volume with trifluoroacetic acid. Photometric determination of the absorbance is carried out at 610 nm, using a reagent blank as the reference solution. The molar extinction coefficient is 35,000 and the detection limit of the method is 10 ppb for an 8-liter air sample.

The method is calibrated by the use of weighed amounts of anisaldehyde and determination of the absorbance of the color produced by reaction with fluoranthene and trifluoroacetic acid in the proportions given above. Calibration curves are plotted for corresponding ozone concentrations and absorbances on the basis that 1 mole of ozone is equivalent to 1 mole of aldehyde in the reaction with stilbene. Nitrogen dioxide, sulfur dioxide, peroxyacetyl nitrate, and methyl hydroperoxide do not interfere in concentrations below 1 ppm.

b. DIPHENYLAMINESULFONATE. According to Bovee and Robinson (*88*), this redox indicator develops a turquoise color in the presence of ozone. Air is sampled at 2.8 liters/minute through a midget impinger containing 10 ml of absorbing reagent until the appearance of a distinct color. The reagent consists of a solution of 1% sodium diphenylaminesulfonate in 0.02% perchloric acid. Losses due to evaporation during sampling are made up to volume with the reagent. The resulting absorbance is determined photometrically at 593 nm, using a 75-mm celi.

The method is calibrated by passage of ozonized air produced by irradiation with a mercury vapor UV lamp, the ozone concentrations being measured by reaction with standard neutral, buffered potassium iodide solution. The authors claim a detection limit of about ± 0.2 μg O_3 but 400 μg NO_2 causes interference equivalent to 20 μg O_3. Peroxides and chlorine gas produce a violet color with the reagent and NO_2 yields a yellowish-green color.

c. *N*-TETRAMETHYLDIAMINODIPHENYL METHANE. This base was employed by Arnold and Mentzel (*90*) as a qualitative reagent by impregnation of paper for the detection of ozone. It reacts in dilute acetic acid solution to produce a violet color with ozone, no color with hydrogen peroxide, a blue color with chlorine, and a yellow color with NO_2. Galster (*91*) and Fuhrmann (*92*) found that this reagent in citric acid solution is suitable for the continuous colorimetric determination of ozone at 565 nm. Fuhrmann employed this method in his Imcometer, a special instrument for measuring air pollutants continuously. The molar extinction coefficient of this reagent is about 40,000.

d. 1,2-DI-(4-PYRIDYL)ETHYLENE. The reaction of ozone with 1,2-di-(4-pyridyl)ethylene in glacial acetic acid solution to form pyridine-4-aldehyde has been employed by Hauser and Bradley (*93*) as an analytical method for atmospheric sampling. The aldehyde reaction product is determined by the method of Sawicki *et al.* (*94*) using 3-methyl-2-benzothiazolone hydrazone which is a very sensitive reagent for the aldehyde group.

Air is sampled at a rate of 0.5 liter/minute through a fritted impinger containing 15 ml of the absorption solution. This solution is prepared by dissolving 0.5 gm of 1,2-di-(4-pyridyl)ethylene in 100 ml glacial acetic acid. After sampling, any loss of solution in the impinger is made up to 15 ml volume by addition of absorption solution reagent and the contents transferred to a test tube. One milliliter of the color-forming reagent is added and the mixture heated for 20 minutes in a boiling water bath. After cooling, the absorbance of the intense yellow reaction product is determined photometrically at 442 nm. The color-forming reagent is prepared by dissolving 0.2 gm of 3-methyl-2-benzothiazolone hydrazone hydrochloride in 100 ml of distilled water.

Calibration of this method is performed by adding measured amounts of pyridine-4-aldehyde in glacial acetic acid solution and reacting with the color-forming hydrazone as described above. Based on the assumption that 2 moles of pyridine-4-aldehyde are formed per mole of ozone and dipyridyl ethylene and a theoretical yield of 61%, 2.75 μg of pyridine-4-aldehyde is equivalent to 1 μg ozone. The limit of ozone concentration that can be detected is about 20 ppb in a 15-liter air sample.

e. PHENOLPHTHALIN METHOD. Haagen-Smit and Brunelle (*95*) developed this method in investigations of the oxidant content of the atmosphere in the Los Angeles, California area. This test is based on the development of a characteristic pink color when the reduced form of phenolphthalein is oxidized by hydrogen peroxide or other oxidants in the presence of copper sulfate. The method is calibrated against standards of dilute hydrogen peroxide and the results are higher by a factor of 2 or more compared with equivalent amounts of ozone. The pink dye can be reduced readily by sulfur dioxide which constitutes a serious interference in the method, especially in urban areas where the levels of this gas are comparatively high. Also, the reagent is affected by nitrogen dioxide and many other oxidizing agents. The halogens, acid gases, and hydrogen sulfide can interfere as reducing agents.

i. *Phenolphthalin reagent.* Prepared by refluxing 1 gm of phenolphthalein with 10 gm of sodium hydroxide, 5 gm of zinc dust, and 20 ml of water for 2 hours, until colorless. The mixture is filtered through a sintered glass filter and diluted with 50 ml of distilled water. This concentrated reagent is stored in a dark-colored stoppered bottle containing granulated zinc. A 0.01 *M* coper sulfate solution is also required.

ii. *Absorbing reagent.* Prepare by diluting 10 ml of the concentrated phenolphthalin reagent with 30 ml of water. Add 1.0 ml of this dilute reagent and 1.0 ml of 0.01 *M* copper sulfate solution to a 100-ml volumetric flask and make up to volume with water. This mixture is shaken and used for the air sample as the test reagent.

The air sample at a rate of about 1 liter/minute is passed through a midget impinger or bubbler containing 10 ml of the above test reagent. After the development of the characteristic pink color, the absorbance is determined in a colorimeter, using a green filter. The concentration of the oxidant is determined by reference to a standard calibration curve. This standard curve is prepared from dilute solutions of hydrogen peroxide containing 0.3–3.0 μg of H_2O_2 per 10 ml of solution, to which 0.1 ml each of the phenolphthalin and copper sulfate reagents have been added. The color developed after standing 15 minutes is read in the colorimeter.

f. Miscellaneous Reagents. A method for the photometric determination of ozone, similar to that of Hauser and Bradley described above, except for the substitution of chloroform for glacial acetic acid as the solvent and the use of a different hydrazone color-forming reagent has been proposed by Amos (*96*). This color-developing reagent is 2-diphenylacetyl-1,3-indandione-1-hydrazone as a 2.5×10^{-4} M solution in chloroform. The air sampling fritted glass bubbler contains 5 ml of absorbing solution consisting of 0.25% w/v solution of 1,2-di-(4-pyridyl)ethylene in chloroform. During sampling the bubbler is immersed in an ice/salt bath at −18°C in a Dewar flask. Another bath at −18°C containing a glass coil upstream of the bubbler is used to freeze out water vapor.

When sampling is completed, the volume of absorbing solution is adjusted to 5 ml. One ml of the solution is pipetted into a 5-ml volumetric flask, 1 ml of color-developing reagent and 5 μl of concentrated HCl are added. The flask is stoppered, shaken, and then heated in a water bath at 65°C for 30 minutes. After cooling, the solution volume is made up to 5 ml with chloroform and the intensity of fluorescence measured at 536 nm, using an excitation wavelength of 468 nm.

The method was calibrated by passing microgram quantities of pyridine-4-aldehyde through the analytical procedure. For standardization purposes, ozone was generated to prepare various concentrations of ozone in air which were analyzed by the above procedure and simultaneously determined by means of an ozone analyzer (OREC Model No. MSA-3). It was found that 1.0 μg of ozone liberates 2.9 μg of pyridine-4-aldehyde from 1,2-di-(4-pyridyl)ethylene in chloroform solution. The method is claimed to be extremely sensitive and can be used to determine ozone concentrations as low as 0.02 ppm.

Hendricks and Larsen (*97*) have studied and compared seven methods for the determination of ozone and oxidants (including peroxides) in air such as buffered potassium iodide, phenolphthalein, diphenylamine sulfonate, fluorescein, dimethoxystilbene, rubber cracking, and the oxidation of NO to NO_2.

C. *Peroxyacetyl Nitrate (PAN)*

PAN was first discovered to be present in the Los Angeles, California atmosphere during photochemical smog episodes that caused the characteristic silvering, glazing, or bronzing of the undersurface of leaves of sensitive plant species. It is considered to be the primary phytotoxicant which causes this "smog" type injury (*98–100*). When isolated in the pure state PAN is an extremely unstable compound. It is one of the eye irritants produced in photochemical smog reactions.

1. Gas Chromatographic Analysis for PAN

Gas chromatography, which is used in this method for measuring peroxyacetyl nitrate (*98, 101, 102*) in ambient air, exploits the exceptional ability of PAN to capture free electrons to provide a method for measuring ppb concentrations of PAN in a 2- or 3-ml sample of ambient air. Higher homologues in the series such as peroxypropionyl nitrate can be detected but procedures for calibration have not been worked out since the amounts present are much smaller than the amounts of PAN found. Peroxybenzoyl nitrate (*103*) might be measured by an adaption of the present technique.

Three major components are required: (a) a chromatographic column to separate PAN from the many other compounds present; (b) a sample introduction system, in this case a gas sample valve activated by a timer and solenoid; (c) an electron capture detector with associated amplifier and recorder. The instrument injects a sample every 15 minutes and about 1.5 minutes are required to develop the chromatogram. The minimum detectable quantity is about 1 ppb although this could by lowered if there were need.

a. RANGE AND SENSITIVITY. Since the detector used in this method is based on the capture of electrons, large amounts of substances which have a high affinity for electrons can saturate the detector by capturing all the electrons. As saturation is approached, the response becomes very nonlinear. To maintain linear response it is necessary to restrict the sample so that the steady current through the detector is reduced by no more than about 25%. This limits the upper end of the concentration range.

The maximum depends on sample size, column length, temperature, and carrier gas flow rate, and is usually less than 1 ppm by volume.

On the other hand maximum sensitivity permits detection of concentrations of less than 1 ppb (10^{-9} by volume or 4.95 $\mu g/m^3$ at 25°C and 760 Torr).

b. INTERFERENCES. There are no known interferences with this PAN method. An interfering substance would have to meet three conditions: (a) it must have high electron affinity; (b) it must have an elution time very close to that of PAN; and (c) it must be present in the sample at a concentration detectable by this procedure. These stringent conditions eliminate virtually everything else.

c. PRECISION AND ACCURACY. Repetitive sampling of standard samples containing about 60 ppb of PAN gave a relative standard deviation of less than 2% for the amount present. Accuracy of course is dependent on the validity of the calibration which in turn depends on the accuracy of the infrared method. This has been compared with a colorimetric method based on hydrolysis to nitrite ion with good agreement (*104*). It is probable that the overall accuracy of the method is within 5%.

d. APPARATUS. The procedures for measuring PAN described by Darley *et al.* (*102*) have been modified and used in the automated system described by the Intersociety Committee (*105*). A gas chromatograph with an electron capture detector or any similar instrument is equipped with an automatic system which injects a sample of ambient air into the following column every 15 minutes. A 9-in.-long column of 1/8-in. i.d. Teflon tubing packed with 5% Carbowax E400 on 100–120 mesh hexamethyldisilazane treated chromosorb W is used. It is operated at 25°C oven temperature with a nitrogen carrier gas flow of 40 ml/minute. Under these conditions the retention time for PAN is 60 seconds.

A stainless-steel hexaport valve with an external 2-ml stainless-steel sample loop is used. With this system there is no detectable decomposition of PAN.

The automatic sampling mechanism attached to the outside of the chromatograph case consists of (a) a timing mechanism; (b) sample valve; (c) solenoid activator; and (d) time delay relay. The timing mechanism consists of a micro switch, a cam to operate the microswitch, and a standard Synchron 4 RPH motor. The cam is designed to energize the solenoid for 90 seconds during each 15-minute period. The shaft of the hexaport valve is attached rigidly to the solenoid shaft and positioned so that the air sample flows through the 2-ml sample loop during the time the solenoid is energized. A coil spring of proper tension installed between the solenoid and sample valve returns the valve to injection position when the solenoid is deenergized.

Calibration requires reference to a known PAN mixture which is most easily measured using a 10-cm path-length gas cell and a spectrometer capable of scanning the 1161 cm^{-1} band. Since this system cannot measure concentrations less than a few hundred ppm, provision must be made for quantitative dilution by a factor of about 1000.

e. REAGENTS. The only reagents required for this determination are the column packing, nitrogen carrier gas, and PAN for calibration. Dry "pre-purified" nitrogen can be obtained in 220-ft^3 cylinders from various suppliers.

PAN is synthesized, purified as described by Stephens *et al.* (*106*) and stored in 34-liter stainless-steel cylinders for use in the dynamic calibration system for the chromatograph (*107*). The cylinders, pressurized with nitrogen to 100 psi, contain 500 to 1000 ppm PAN and are stored at 60°F. An infrared spectrophotometer with a 10-cm cell is employed to determine the concentration of PAN in the cylinders, using the absorptivities of 13.9×10^{-4} ppm^{-1} m^{-1} at 8.6 μm (1161 cm^{-1}) (*108*).

A simplified alternate method which requires much less apparatus may also be used. Small quantities of PAN are prepared directly in a 10-cm infrared cell with a Pyrex glass body by photolyzing traces of ethyl nitrite, CH_3CH_2ONO, in pure oxygen. The PAN formed is then measured in the cell and is not separated from unreacted nitrite or from by-products. However it must be diluted by a factor of several thousand to reach the electron capture concentration range. Blacklight fluorescent lamps provide a good energy source for the necessary 300–400-nm radiation but it might well be possible to use the sun for this photolysis. The cell is first flushed with pure oxygen and then small amounts (50 μl) of vapor from the space above the liquid in the ethyl nitrite bottle are added at 15-minute intervals while irradiating. The ethyl nitrite concentration must be kept low to maximize yield.

f. PROCEDURE. Dry nitrogen carrier gas is flushed continuously through the 2-ml sample loop of the hexaport valve for 13.5 minutes of each period when the solenoid is deenergized. When the solenoid is energized the sample loop is open to the atmosphere to be sampled and a pump draws a continuous flow of air through the loop for 90 seconds. When the solenoid is again deenergized, the carrier gas flushes the 2-ml sample into the column.

The time delay relay energizes the chart drive motor on the strip chart recorder at the end of the 90-second sample period when the solenoid is deenergized. The chart drive motor is allowed to run for 90 seconds which is ample time to record the chromatogram. The column is too short to separate peroxypropionyl nitrate (PPN) and peroxybutyryl nitrate (PBN), although a shoulder on the PAN peak, assumed to be PPN, appears occasionally.

g. CALIBRATION AND CALCULATION. The chromatograph is calibrated with a flow dilution panel from a supply containing PAN at a concentration high enough to be measured in a 10-cm infrared cell (about 1000 ppm).

In the first dilution of the dynamic system *(107)*, 1 part PAN from the storage cylinder is diluted with 100 parts activated charcoal-filtered air. A similar 1:100 dilution with filtered air in the second step of the dilution system reduces the concentration of PAN in a constant flow of air to the ppb range. Calibration curves are plotted from calculated concentrations of PAN and peak height.

The tank concentration is calculated from the infrared spectrum *(108)* by applying the Beer–Lambert equation to the absorbance at 8.6 μm (8.6×10^{-6} m):

$$C\ (\text{ppm}) = A/(0.1\ \text{m}) \times (13.9 \times 10^{-4}\ \text{ppm}^{-1}\ \text{m}^{-1}) \qquad (23)$$

where C is concentration in ppm and A is absorbance $= \log_{10}(I_0/I)$. For accurate measurement this concentration must be of the order of a 1000 ppm. This is about 10,000 times higher than the concentration required for calibration so dilution is carried out in two steps of about 100-fold using a dilution panel. The final concentration is then calculated by the formula:

$$C_e = C[f_1/(F_1 + f_1)] \times [f_2/(F_2 + f_2)] \qquad (24)$$

in which f_1 and f_2 are rates in the small streams and F_1 and F_2 are flow rates in the large streams.

Calculated concentrations of PAN and peak height show a linear relationship in the range from 1 to about 50 ppb; therefore, concentrations within this range can be calculated by multiplying the peak height by a constant. The constant varies from one instrument to another and varies inversely with changes in standing current of a particular instrument.

Contamination of the electron capture detector causes a slow decline in standing current and a concomitant reduction in sensitivity. It is desirable to calibrate the instrument about once a week to compensate for the gradual reduction in sensitivity. When the detector is thoroughly cleaned the original sensitivity is regained.

III. Discussion

A. Chemiluminescent Reactions

1. Ozone

Stevens and Hodgeson (*109*) have reviewed the applications of chemiluminescent reactions to measure oxides of nitrogen, ozone, and sulfur compounds continuously. Probably, the first chemiluminescence detector for atmospheric ozone concentrations was developed by Regener (*101, 111*) who utilized the relatively intense emission from the reaction between the organic dye, Rhodamine B, adsorbed on an activated silica gel surface, and ozone. Improvements in stability and lifetime of the Regener detector were made by Hodgeson *et al.* (*112*). An internal ozone source was found to be necessary to provide the frequent calibrations of the dye–silica gel surface to compensate for changes in sensitivity. Chemiluminescent emission with this detector permits detection of less than 1 ppb (10^{-3} ppm) of ozone in the atmosphere and yields a linear response for measurements up to 0.4 ppm. Moisture has a desensitizing effect on the chemiluminescent surface but this can be eliminated by means of a silicone resin coating technique.

A number of commercial monitors for ozone have been developed on the basis of the gas-phase chemiluminescent reaction between ozone and ethylene, reported earlier by Nederbragt *et al.* (*65*). This reaction yields an emission in the 300–600-nm region, with a maximum at about 435 nm. The emission intensity is directly proportional to the ozone concentration. A detector of this type responds linearly to ozone concentrations between 0.003 and 30 ppm, without any observable interferences.

Extensive field studies in Los Angeles, California (*113*) and St. Louis, Missouri (*114*) by Stevens *et al.* established good correlations between the Regener and the ozone–ethylene methods. These studies led to the adoption of the latter method as the USEPA Reference Method cited in the *Federal Register* of April 30, 1971.

2. Oxides of Nitrogen, NO and NO_2

The chemiluminescent reaction between nitric oxide and ozone is the basis for a number of detection instruments. Clough and Thrush (*115*) have described the kinetics and mechanism of this reaction in accordance with the following steps:

$$NO + O_3 \rightarrow NO_2^* + O_2 \tag{25}$$
$$NO + O_3 \rightarrow NO_2 + O_2 \tag{26}$$
$$NO_2^* \rightarrow NO_2 + h\nu \tag{27}$$
$$NO_2^* + M \rightarrow NO_2 + M \tag{28}$$

where NO_2^* represents the excited electronic state that emits radiation between 600 and 3000 nm, with a maximum at 1200 nm. Only a fraction of the $NO + O_3$ collisions leads to excited nitrogen dioxide molecules and quenching of the NO_2^* allows only a fraction of the possible photons to be emitted. The emission of red and infrared light in this reaction is a modification of the afterglow effect in the atmosphere that is observed during electrical discharges through the air.

Since the direct emission reaction is useful only for the detection and analysis of NO, the determination of a mixture of NO and NO_2 (NO_x) is possible only if the NO_2 present is converted first to NO by catalytic or thermal means. Results of initial development of a chemiluminescent NO detector were reported by Fontijn *et al.* in 1970 (*35*), and this work was extended by Stedman *et al.* (*116*). A monitor based on this approach led to the development of commercial instruments for application largely to measurements of motor vehicle exhaust gases. Conversion of NO_2 to NO by passage through a stainless-steel tube heated to 600°C or higher occurs quantitatively, according to Sigsby *et al.* (*47*). However, ammonia is also oxidized to NO under these conditions and constitutes a potential interference, unless it is removed by prior scrubbing with dilute acid. Catalytic reduction of NO_2 at 240°C to NO, over various metal surfaces, may be accomplished without oxidation of the ammonia.

Atomic oxygen reacts with NO to form excited NO_2^* molecules, as follows:

$$NO + O \rightarrow NO_2^* \tag{29}$$
$$NO_2^* \rightarrow NO_2 + h\nu \tag{30}$$
$$NO_2^* + M \rightarrow NO_2 + M \tag{31}$$
$$NO_2 + O \rightarrow NO + O_2 \tag{32}$$

However, there is a rapid reaction between NO_2 and oxygen atoms to form NO. Emission of light from the excited nitrogen dioxide molecules [Equation (30)] is in the wavelength range of 400 to 1400 nm, representing the air afterglow reaction (*117*). The possibility of using the above oxygen atom reactions for the measurement of NO_x in air pollution has been insufficiently explored, as yet.

Air exerts a quenching effect on the chemiluminescent reaction between nitric oxide (NO) and ozone. Consequently, some models of detectors employ a vacuum pump to supply the air sample and the ozone mixture to the reaction chamber at a pressure of only 1 to 5 Torr. The air sample is passed into the reaction chamber at flow rates of 50 to 100 cm^3/minute at STP and similar rates are employed for the mixture of about 0.5% ozone in oxygen. A light cutoff filter is placed between the reaction chamber and the phototube to absorb any emission below 600 nm, to remove any interference from emissions at lower wavelengths due to the reaction of ozone with olefins or other molecules present as pollutants.

Increased sensitivity may be obtained by cooling the photomultiplier tubes of NO detectors down to —25°C or lower. The concentration limit of NO detection by this means is 0.001 ppm and the linear dynamic range is as high as 10^7. Other pollutants that may be present in the ambient air cause no interference (*116*).

The feasibility of measuring ambient NO concentrations in a luminescent reactor cell at atmospheric pressure is under investigation (*118*). In spite of the quenching effect of air at 1 atm pressure, the decreased light emission rate is still significant and can be partially compensated for by an increase in sample flow rates to 1–2 liters/minute, using a reduced volume of reaction chamber. Some commercial instruments use small diaphragm air pumps and operate at about 200 Torr or greater pressure, with sample flows of 100 to 200 cm^3/minute. Under these operating conditions, the minimum detectable concentration is between 0.002 and 0.005 ppm.

A paper by McClenny, Hodgeson, and Bell (*119*) describes a method for the detection of nitrogen dioxide in ambient air concentration levels by photolysis of the NO_2 at reduced pressures. The resultant oxygen atoms are detected by the chemiluminescence of their gas-phase reaction with NO. The intensity of chemiluminescence is a linear function of NO_2 concentration over the range of 1.0 ppb (0.001 ppm) to 10^3 ppb (1.0 ppm), with an optimum photomultiplier response of 0.1 mA/ppb. Owing to the fast reaction between O atoms and NO_2, concentrations greater than 10^3 ppb of NO_2 show a nonlinear response. For typical ambient air level concentrations of hydrocarbons, negligible interferences are found from oxygen atom–hydrocarbon reactions that could occur in this system.

This chemiluminescent type of photofragment detection involves: (a) the selective photolysis of a trace gas in an irradiation chamber; (b) transfer of the photofragments through a light trap and into a reaction chamber; (c) recombination of the photofragments by a gas-phase reaction with a pure mixer gas, resulting in the subsequent emission of a characteristic chemiluminescent spectrum; and (d) selective detection of all or a portion of this spectrum by a filter-photomultiplier combination.

Specifically, NO_2 molecules in a sample airstream are photolyzed at reduced pressures to produce O atoms:

$$NO_2 + h\nu \rightarrow NO + O \tag{33}$$

The continuously flowing sample airstream carries the O atoms from the irradiation chamber through a light trap and into a closely coupled reaction chamber. The O atoms are detected by the chemiluminescence which occurs when they are mixed with nitric oxide. Pure NO is introduced into the reaction chamber at a rate typically less than a factor of 10^{-2} of the sample stream flow rate. A portion of the O atoms collide with NO molecules, resulting in a chemiluminescent emission. The chemiluminescent spectrum for this reaction extends from 400 nm into the infrared with a maximum intensity at approximately 650 nm.

The use of flame chemiluminescence for the detection of oxides of nitrogen is described by Krost, Hodgeson, and Stevens (*120*). Excitation of characteristic nitrogen compound bands was achieved by means of the usual oxygen–hydrogen flame. The emission observed in the reaction between atomic hydrogen and NO was viewed by means of a scanning monochromator, above the flame, at 690 nm. This chemiluminescence results probably from the biomolecular reaction between H atoms and NO, according to the following mechanism:

$$H + NO \rightarrow HNO^* \tag{34}$$
$$HNO^* \rightarrow HNO + h\nu \text{ (660 to 770 nm)} \tag{35}$$

The intensity of this reaction is proportional to the product of H atom and NO concentrations

$$I = I_0[H][NO] \tag{36}$$

The detection limits of the apparatus described in this study are 0.15 ppm for nitrogen oxides and 0.004 ppm for sulfur compounds (the sulfur emission being viewed at 394 nm).

B. Miscellaneous Methods for Nitrogen Compounds

The use of a nitrate-specific ion electrode has been proposed by Di Martini (*121*) for the determination of nitric oxide and nitrogen dioxide in the parts per million range in flowing gas mixtures. In this method, the NO_x compounds are oxidized in the gas phase by ozone to nitrogen pentoxide (N_2O_5). Hydrolysis of the nitrogen pentoxide leads to the formation of nitric acid and this is followed by the quantitative determination of the nitrate ion with a selective ion electrode. The author states that electrode response is relatively stable over wide pH and temperature ranges, with few chemical interferences. However, variable efficiencies were obtained in the range of 0.06 to 5.7 ppm NO_x.

Another modification of the above method has been reported by Kneebone and Freiser (*122*) for the determination of NO_x in ambient air. The air sample is passed through gas-washing bottles containing 2% hydrogen peroxide solution to absorb and oxidize the NO_x to nitric acid. Subsequent treatment of the absorbing solution with MnO_2 destroys the remaining peroxide, and the nitrate concentration is then measured potentiometrically, utilizing a coated-wire nitrate ion selective electrode. The method can be used in the presence of at least a 40-fold excess of SO_2 or SO_3.

Gas-phase electron paramagnetic resonance (EPR) has been applied by Uehara and Arimitsu (*123*) to the detection of NO and NO_2 in air. The only gaseous paramagnetic species in polluted air are O_2, NO, and NO_2. Although all of these gases, in principle, are observed by EPR, it is possible to detect NO and NO_2 separately, since the NO_x gases have an electric dipole moment. The oxygen molecule is nonpolar and, therefore, is not observed at all by Stark-modulated EPR. Furthermore, NO is a linear molecule, whereas NO_2 is nonlinear. In the proposed method, NO_2 in the air sample is detected by Zeeman modulation of EPR at 1 atm of sample pressure, without interference from O_2 and NO. Nitric oxide is determined by Stark modulation under a reduced sample pressure of less than 0.1 Torr, without any interference. The sensitivity of the method is claimed to be better than 30 ppb for NO and NO_2 with a sample volume of 1.5 liters, using a novel low temperature trapping technique.

An increasing trend has developed toward displacement of wet chemical methods of analysis for gaseous pollutants by instrumental techniques of coulometry, potentiometry, gas and other types of chromatography and chemiluminescence. It is anticipated that this progress will be accelerated in order to develop more reliable, rapid, and specific methods that depend upon unique physical or chemical properties of molecules.

REFERENCES

1. Intersociety Committee, "Methods of Air Sampling and Analysis," p. 329. Amer. Pub. Health Ass., Washington, D.C., 1972.
2. B. E. Saltzman, *Anal. Chem.* **26,** 1949 (1954).
3. B. E. Saltzman and A. F. Wartburg, Jr., *Anal. Chem.* **37,** 1261 (1965).
4. M. D. Thomas and R. E. Amtower, *J. Air Pollut. Cont. Ass.* **16,** 618 (1966).
5. B. E. Saltzman, *Anal. Chem.* **33,** 1103 (1961).
6. A. E. O'Keefe and G. C. Ortman, *Anal. Chem.* **38,** 760 (1966).
7. P. K. Mueller, F. P. Terraglio, and Y. Tokiwa, "Chemical Interferences in Continuous Air Analysis," Proc. Conf. Methods Air Pollut. Stud. 7th, Department of Public Health, State of California, Berkeley, California, 1965.
8. United States Environmental Protection Agency, *Fed. Regist.* **36,** 8200 (1971).
9. M. B. Jacobs and S. Hochheiser, *Anal. Chem.* **30,** 426 (1958).

10. E. C. Tabor and C. C. Golden, *J. Air Pollut. Contr. Ass.* **15,** 7 (1965).
11. W. H. Perry and E. C. Tabor, *Arch. Environ. Health* **4,** 254 (1962).
12. F. L. Meadows and W. W. Stalker, *Amer. Ind. Hyg. Ass., J.* **27,** 559 (1966).
13. D. A. Levaggi, W. Siu, and M. Feldstein, *J. Air Pollut. Contr. Ass.* **23,** 30 (1973).
14. F. P. Scaringelli, E. Rosenburg, and K. A. Rehme, *Environ. Sci. Technol.* **4,** 924 (1970).

14a. J. D. Mulik, R. G. Fuerst, J. R. Meeker, M. Guyer, and E. Sawicki, "A new twenty-four hour manual method of collection and colorimetric analysis of atmospheric NO_2," 165th Nat. Meet., American Chemical Society, Washington, D.C., 1973.

14b. A. A. Christie, R. G. Lidzey, and D. W. F. Radford, *Analyst* **95,** 519 (1970).

14c. E. L. Merryman, C. W. Spicer, and A. Levy, *Environ. Sci. Technol.* **7,** No. 11, 1056 (1973).

14d. M. E. Beard and J. H. Margeson, "An Evaluation of the Arsenite Procedure for the Determination of Nitrogen Dioxide in Ambient Air." Methods Standardization Branch, United States Environmental Protection Agency, Research Triangle Park, North Carolina, 1974.

15. Intersociety Committee, "Methods of Air Sampling and Analysis," p. 337. Amer. Pub. Health Ass., Washington, D.C., 1972.
16. American Society for Testing and Materials Committee D-22, ASTM Designation D 1608–60 (reapproved 1967), "Annual Standards," Part 23, p. 461. ASTM, Philadelphia, Pennsylvania, 1968.
17. E. M. Chamot, D. S. Pratt, and H. W. Redfield, *J. Amer. Chem. Soc.* **31,** 922 (1909); **32,** 630 (1910); **33,** 336 (1911).
18. W. A. Cook, *Amer. J. Pub. Health* **26,** 80 (1936).
19. F. A. Patty and G. M. Petty, *J. Ind. Hyg. Toxicol.* **25,** 361 (1943).
20. W. E. Gill, *Amer. Ind. Hyg. Ass.* **21,** 87 (1960).
21. H. Yagoda and F. H. Goldman, *J. Ind. Hyg. Toxicol.* **25,** 440 (1943).
22. H. Barnes, *Analyst* **75,** 388 (1950).
23. Intersociety Committee, "Methods of Air Sampling and Analysis," p. 322. Amer. Pub. Health Ass., Washington, D.C., 1972.
24. Intersociety Committee, "Methods of Air Sampling and Analysis," p. 325. Amer. Pub. Health Ass., Washington, D.C.
25. D. A. Levaggi, W. Siu, E. L. Kothny, and M. Feldstein, *Environ. Sci. Technol* **6,** 250 (1972).
26. I. C. Huygen, *Anal. Chem.* **42,** 407 (1970).
27. D. A. Levaggi, E. L. Kothny, T. Belsky, E. R. de Vera, and P. K. Mueller, *17th Ann. Meet.,* Institute of Environmental Science, Berkeley, California, 1971.
28. D. A. Levaggi, E. L. Kothny, T. Belsky, E. R. de Vera, and P. K. Mueller, *Environ. Sci. Technol.* **8,** 348 (1974).
29. T. Belsky, *Amer. Ind. Hyg. Lab., Rep.* No. 85 (1970).
30. B. E. Saltzman and A. F. Wartburg, *Anal. Chem.* **37,** 779 (1965).
31. G. E. Moore, A. F. W. Cole, and M. Katz, *J. Air Pollut. Contr. Ass.* **7,** 25 (1957).
32. M. D. Thomas, J. A. MacLeod, R. C. Robbins, R. C. Goettleman, R. W. Eldvidge, and L. H. Rogers, *Anal. Chem.* **28,** 1810 (1956).
33. J. Cholak, L. J. Shafer, D. Yeager, and W. J. Younker, *AMA Arch. Ind. Health* **15,** 198 (1957).
34. American Society for Testing and Materials Committee D-22, Method D2012-69, "Annual Standards," Part 23, p. 620. ASTM, Philadelphia, Pennsylvania, 1970.
35. A. Fontijn, A. J. Sabadell, and R. J. Ronco, *Anal. Chem.* **42,** 575 (1970).

36. A. D. Snyder and G. W. Wooten, Contract CPA 22-69-8. U.S. Dept of Health, Education, and Welfare, National Air Pollution Control Administration, Durham, North Carolina, 1969.
37. R. J. Ronco and A. Fontijn, Contract CPA 70–79. United States Environmental Protection Agency, Air Pollution Control Office, Durham, North Carolina, 1971.
38. R. Iitis and J. C. Mueller, Contract CPA 70-70. United States Environmental Protection Agency, Air Pollution Control Office, Durham, North Carolina, 1971.
39. D. T. Williams and B. L. Kolitz, *Appl. Opt.* **7**, 607 (1968).
40. Barringer Research, Contract PH 22-68-44. U.S. Dept. of Health, Education, and Welfare, National Air Pollution Control Administration, Durham, North Carolina, 1969.
41. Spectrometrics of Florida, Technical Bulletin on Model III d^2 Analyzer. Pinellas Park, Florida, 1971.
42. P. L. Hanst, *Appl. Spectrosc.* **24**, 161 (1970).
43. E. D. Hinkley and P. L. Kelley, *Science* **171**, 635 (1971).
44. L. B. Kreuzer and C. K. N. Patel, *Science* **173**, 45 (1971).
45. A. P. Altshuller, *Nat. Bur. Stand. (U.S.), Spec. Publ.* **351**, 245 (1972).
46. H. C. Lord, D. W. Egan, F. L. Johnson, and L. D. McIntosh, *J. Air Pollut. Contr. Ass.* **24**, 136 (1974).
47. J. E. Sigsby, F. M. Black, T. A. Bellar, and D. L. Klosterman, *Environ. Sci. Technol.* **7**, 51 (1973).
48. Intersociety Committee, *Health Lab. Sci.* **9**, 327 (1972).
49. Intersociety Committee, *Health Lab. Sci.* **7**, 4 (1970); "Methods of Air Sampling and Analysis," p. 447. Amer. Pub. Health Ass., Washington, D.C., 1972.
50. M. Buck and H. Stratmann, *Z. Anal. Chem.* **213**, 241 (1965).
51. P. J. Rommers and J. Visser, *Analyst* **94**, 653 (1969).
52. Intersociety Committee, *Health Lab. Sci.* **10**, 115 (1973).
53. V. W. Truesdale, *Analyst* **96**, 584 (1971).
54. Intersociety Committee, *Health Lab. Sci.* **11**, No. 2 (1974).
55. J. E. Harwood and A. L. Kuhn, *Water Res.* **4**, 805 (1970).
56. F. A. Gunther, J. H. Barkley, M. J. Kolbezen, R. C. Blinn, and E. A. Staggs, *Anal. Chem.* **28**, 1985 (1956).
57. M. J. Kolbezen, J. W. Eckert, and C. W. Wilson, *Anal. Chem.* **36**, 593 (1964).
57a. D. N. Kramer and J. M. Sech, *Anal. Chem.* **44**, 395 (1972).
58. W. Leithe, "The Analysis of Air Pollutants," p. 176. Humphrey Sci. Publ., Ann Arbor, Michigan, 1970.
58a. M. D. LaHue, H. D. Axelrod, and J. P. Lodge, Jr., *Anal. Chem.* **43**, 1113 (1971).
59. R. F. Scherberger, F. A. Miller, and D. W. Fassett, *Amer. Ind. Hyg. Ass., J.* **21**, 471 (1960).
60. D. J. McCaldin, *Chem. Rev.* **60**, 39 (1960).
61. Intersociety Committee, *Health Lab. Sci.* **11**, No. 2 (1974).
62. W. Pilz and E. Stelzl, *Z. Anal. Chem.* **219**, 416 (1966).
63. R. P. Buck and R. W. Eldridge, *Anal. Chem.* **37**, 1242 (1965).
64. United States Environmental Protection Agency, *Fed. Regist.* **36**, No. 84, 8195 (1971).
65. G. W. Nederbragt, A. Van Der Horst, and J. Van Duijn, *Nature (London)* **206**, 87 (1965).
66. G. J. Warren and G. Babcock, *Rev. Sci. Instrum.* **41**, 280 (1970).
67. J. A. Hodgeson, B. E. Martin, and R. E. Baumgardner, "Comparison of Chemiluminescence Methods for Measurement of Atmospheric Ozone." Eastern Anal.

Symposium New York, New York, 1970, Reprint No. 77. National Environmental Research Center, United States Environmental Protection Agency, Research Triangle Park, North Carolina, 1970.
68. H. Carroll, B. Dimitriades, and M. Ray, *U.S., Bur. Mines, Rep. Invest.* **RI-7650** (1972).
69. Intersociety Committee, *Health Lab. Sci.* **11,** No. 2 (1974).
70. A. P. Altshuller, and A. F. Wartburg, *Int. J. Air Water Pollut.* **4,** 70 (1961).
71. D. H. Byers and B. E. Saltzman, *Amer. Ind. Hyg. Ass., J.* **19,** 251 (1958).
72. Intersociety Committee, "Methods of Air Sampling and Analysis," p. 351. Amer. Pub. Health Ass., Washington, D.C., 1972.
73. B. E. Saltzman and A. F. Wartburg, *Anal. Chem.* **37,** 779 (1965).
74. P. K. Mueller, F. P. Terraglio, and Y. Tokiwa, "Chemical Interferences in Continuous Air Analysis," Proc. 7th Conf. Methods Air Pollut. Stud. Department of Public Health, State of California, Berkeley, California, 1965.
75. B. E. Saltzman and N. Gilbert, *Anal. Chem.* **31,** 1914 (1959).
76. A. P. Altshuller and A. F. Wartburg, *Int. J. Air Water Pollut.* **4,** 70 (1961).
77. Intersociety Committee, "Methods of Air Sampling and Analysis," p. 341. Amer. Pub. Health Ass., Washington, D.C. 1972.
78. G. M. Mast and H. E. Saunders, *Instrum. Soc. Amer. Trans.* **1,** 325 (1962).
79. F. Schulze, *Anal. Chem.* **38,** 748 (1966).
80. Beckman Instruments, Inc., Fullerton, California.
81. P. Hersch and R. Deuringer, *Anal. Chem.* **35,** 897 (1963).
82. L. Potter and S. Duckworth, *J. Air Pollut. Contr. Ass.* **15,** 207 (1965).
83. I. Cherniak, and R. J. Bryan, *J. Air Pollut. Cont. Ass.* **15,** 351 (1965).
84. A. F. Wartburg, A. W. Brewer, and J. P. Lodge, Jr., *Int. J. Air Water Pollut.* **8,** 21 (1964).
85. E. R. Stephens, "Infrared Calibration of Coulometric Analyzers," Proc. 7th Conf. Methods Air Pollut. Stud. Department of Public Health, State of California, Berkeley, California, 1965.
86. Intersociety Committee, "Methods of Air Sampling and Analysis," p. 356. Amer. Pub. Health Ass., Washington, D.C., 1972.
87. F. E. Littman and R. Benoliel, *Anal. Chem.* **25,** 1480 (1953).
88. H. H. Bovee and R. J. Robinson, *Anal. Chem.* **33,** 1115, (1961).
89. H. A. Bravo and J. P. Lodge, *Anal. Chem.* **36,** 671 (1964).
90. C. Arnold and C. Mentzel, *Ber. Deut. Chem. Ges.* **35,** 1324 (1902).
91. H. Galster, *Z. Anal. Chem.* **186,** (1962).
92. H. Fuhrmann, *Staub* **25,** 266 (1965).
93. T. R. Hauser and D. W. Bradley, *Anal. Chem.* **38,** 1529 (1966).
94. E. Sawicki, T. R. Hauser, T. W. Stanley, and W. Elbert, *Anal. Chem.* **33,** 93 (1961).
95. A. J. Haagen-Smit and M. F. Brunnelle, *Int. J. Air Pollut.* **1,** 51, (1958).
96. D. Amos, *Anal. Chem.* **42,** 842 (1970).
97. R. H. Hendricks and L. B. Larsen, *Amer. Ind. Hyg. Ass., J.* **27,** 81 (1966).
98. E. R. Stephens, E. F. Darley, O. C. Taylor, and W. E. Scott, *Int. J. Air Water Pollut.* **4,** 79 (1961).
99. J. T. Middleton, *Annu. Rev. Plant Physiol.* **12,** 431 (1961).
100. R. A. Glater, R. A. Solberg, and F. M. Scott, *Amer. J. Bot.* **49,** 954 (1962).
101. E. R. Stephens, P. L. Hanst, R. C. Doerr, and W. E. Scott, *Ind. Eng. Chem.* **48,** 1498 (1956).
102. E. F. Darley, K. A. Kettner, and E. R. Stephens, *Anal. Chem.* **35,** 589 (1963).

103. J. M. Heuss and W. A. Glasson, *Environ. Sci. Technol.* **2**, 1109 (1968).
104. E. R. Stephens and M. A. Price, "Colorimetric Analysis of PAN," 8th Conf. Methods Air Pollut. Ind. Hyg. Stud. Department of Public Health, State of California, Berkeley, California, 1967.
105. Intersociety Committee, "Methods of Air Sampling and Analysis," p. 215. Amer. Pub. Health Ass., Washington, D. C., 1972.
106. E. R. Stephens, F. R. Burleson, and E. A. Cardiff, *J. Air Pollut. Contr. Ass.* **15**, 87 (1965).
107. R. L. Plata, "Calibration and Comparison of Coulometric and Flame Ionization for Monitoring PAN in Experimental Atmospheres," 9th Conf. Methods Air Pollut. and Ind. Hyg. Stud. Department of Public Health, State of California, Berkeley, California, 1968.
108. E. R. Stephens, *Anal. Chem.* **36**, 928 (1964).
109. R. K. Stevens and J. A. Hodgeson, *Anal. Chem.* **45**, 443A (1973).
110. V. H. Regener, *J. Geophys. Res.* **65**, 3975 (1960).
111. V. H. Regener, *J. Geophys. Res.* **69**, 3795 (1964).
112. J. A. Hodgeson, K. J. Krost, A. E. O'Keeffe, and R. K. Stevens, *Anal. Chem.* **42**, 1795 (1970).
113. R. K. Stevens, J. A. Hodgeson, L. F. Ballard, and C. E. Decker, *in* "Determination of Air Quality" (G. Mamamtov and W. D. Shults, eds.), p. 83. Plenum, New York, New York, 1970
114. R. K. Stevens, T. A. Clark, L. F. Ballard, and C. E. Decker, Annu. Meet., Pap. No. 72-13. Air Pollution Control Association, Pittsburgh, Pennsylvania, 1972.
115. P. N. Clough andn B. A. Thrush, *Trans. Faraday Soc.* **63**, 915 (1967).
116. D. H. Stedman, E. E. Daby, F. Stuhl, and H. Nikki, *J. Air Pollut. Contr. Ass.* **22**, 260 (1972).
117. F. Kaufman, *Progr. React. Kinet.* **1**, 1 (1961).
118. J. A. Hodgeson, K. A. Rehme, B. E. Martin, and R. K. Stevens, Annu. Meet. Pap. No. 72-12. Air Pollution Control Association, Pittsburgh, Pennsylvania, 1972.
119. W. A. McClenny, J. A. Hodgeson, and J. P. Bell, *Anal. Chem.* **45**, 1514 (1973).
120. K. J. Krost, J. A. Hodgeson, and R. K. Stevens, *Anal. Chem.* **45**, 1800 (1973).
121. R. Di Martini, *Anal. Chem.* **42**, 1102 (1970).
122. R. B. Kneebone and H. Freiser, *Anal. Chem.* **45**, 449 (1973).
123. H. Uehara and S. Arimitsu, *Anal. Chem.* **45**, 1897 (1973).

8

Odors

Austin H. Phelps

I. Introduction

There are several degrees of odor measurement sophistication. The most delicate, the blending of perfumes or whiskeys, requires a particularly acute sense of smell, very delicate appreciation of even the slightest deviation from a desired standard, and imagination in estimating the effects of combinations of different compounds. In the second category of odor measurement, food science or food production, odor is one of several organoleptic tests used for evaluating products and determining probable causes of problems. In dairy science, for example, smelling and tasting of milk may be used to determine the quality of the raw material and the effectiveness of its cooling. These judgments can be made only by carefully selected people who have received proper training.

Air pollution odor measurement is less demanding in both the accuracy required and the training necessary. The basic problem in air pollution odor measurement is to determine if something is present in the air which does not belong there. Absolute identification of all the ingredients or all the odor bodies is rarely necessary. In this context, many of the refinements and problems encountered in the more accurate odor measurements necessary for food or perfumery disappear.

Odor is a physiological response of individuals. Only when this response can be replicated, can we say that odor measurement has been placed on an objective basis. Instruments which measure the physical or chemical characteristics of an odor-laden gas, as an indirect means of measuring odor, are of questionable value in describing the odor potential of an unknown mixture. To date, the only reliable and accurate istrument for measuring odor is the human nose.

There are no simple obvious physiological responses which correlate directly with odor (*1*). Dravnieks and Laffort indicate that odor response to pure compounds (both the threshold level and the change in intensity with concentration) can be predicted quite well by using data from gas chromatography (*2*).

At the third Karolinska Institute Symposium on Environmental Health (*3*) the following definitions were drawn up:

Odor — A product of the activation of the sense of smell, an olfactory experience.

Odorant — Any chemical compound which can stimulate the olfactory (smell) sense.

Intensity — The suprathreshold perceived intensity of an odor experience. It can be assessed through scaling or by comparison to reference standards.

Quality — The property which permits identification of a given odor's characteristics through perception of the unique balance of qualities comprising the experience.

Acceptability — The continuum with judgments of positive response (acceptance or pleasure or liking) at the one extreme and a

> negative response (nonacceptance, displeasure, or rejection) at the other extreme. The position on the continuum at which a given stimulus is placed may differ with the individual, his condition, his environment, attitudes, experience, and the like.

It was reported at the symposium that the most important dimension of an odor is probably acceptability, e.g., what percentage of a population is annoyed by a particular odor. Unfortunately, there are no physical methods for determining acceptability and one must resort to sociological inquiry methods.

Some industries incorrectly assume that they will have no odor problem in their neighborhood because their particular effluent is "pleasant." Bakers, coffee producers, perfumers, and distillers sometimes consider their discharges unobjectionable. If we compare those odors with the odors emanating from a rendering plant, this opinion has some merit. However, the presence of any odor which persists, and does not normally belong there, will be a source of annoyance to the neighborhood. The result will be complaints. A check of the records of any air pollution control official, who has areas where residences and industry approach each other, will attest to this fact. Complaints have been received about coffee roasters, bakeries, and many other normally pleasant odors. Kerka and Kaiser many years ago compiled a list of the most frequent odorant complaints reported by municipal air pollution control bureaus (*4*). Their work shows that several odors which might be considered pleasant are causes of complaint. While it may be easier for an employee to accustom himself to coffee roasting than to putrefaction in a rendering plant, a neighborhood considers day-in, day-out persistent odors—even of baking bread—to be a nuisance. Therefore, the material to follow will not dwell overly long on the description of odors, since it is the author's belief that any odor may become objectionable.

II. Classification of Odor Measurement Problems

For air pollution, odor measurement problems occur in four different pairings; effluent source or ambient air, and relatively pure compounds or mixtures of known or unknown components in varying concentrations.

Odor measurement at the source is the easiest and most accurate for obtaining useful data for odor abatement, whether for pure compounds or mixtures. The concentration of the odor in the stack is usually high, and its presence is assured by its continual outflow. It is undisturbed by wind or atmospheric conditions, and a sample may readily be drawn. Ambient odor, on the other hand, is often weak, even when at the complaint level, and may be several thousand dilutions below that in the stack. Furthermore, it is very fleeting in nature, and, at any one point on the ground, can vary by several orders of magnitude as the wind shifts, or as puffs come down to the ground. Variation in wind direction, velocity, turbulence, humidity, and temperature complicate both the dilution of the odor and the sense of smell. A steady-state condition, ideal for measurement of any parameter, rarely exists for odor in ambient air. There are cases, of course, where massive odor emissions float to the ground and result in a persistent cloud of odor nearby. However, more generally, the ambient odor has to be chased up and down the street to find sufficient concentration to identify the source. Detecting the odor and determining the wind direction is about all that can be done with any accuracy.

If the odor is a relatively pure single compound, which can be isolated, its odor threshold and behavior downwind can be determined by simple analytical and mathematical techniques, even though it is wise to remember that minor impurities can cause great changes in the odor of "pure" compounds. Pure single compound exhausts are usually more uniform than exhausts containing mixtures. Odor concentration can be approximated by measuring the compound itself by physical or chemical methods. Odor threshold values can be determined once and for all and, in many cases, may be found in the literature. These can be used in conjunction with the emission rate of a compound to yield a value representing the odor emission in units which can be used in diffusion equations. Pure compounds, unfortunately, are not very common in nuisance complaints. Production of a pure compound usually involves many processing steps whose only purpose is to ensure the purity of the compound. Thus, complicated mixtures will usually be handled and emitted during the manufacture of a pure compound.

Although the discharge from a coffee roaster might be considered a single compound, since only coffee is handled, the organic flavors and other compounds emitted cover a wide spectrum. In the case of a rendering plant, even less can be said about what specifically is emitted, although several compounds, such as mercaptans and amines, are certainly easy to predict. For mixtures of compounds, the best approach for measuring the nuisance effect is by odor dilution techniques. Collection of a sample at the source is important to ensure that the mixture which is being measured is the one which will be received by the neighborhood. Many conpounds have effects on each other, either synergistic or antagonistic. Even if it were possible to measure analytically each of the compounds present and its concentration, the effect of each on the odor of the mixture would still be unknown. The total odor effect from a mixture of compounds cannot be determined by adding the effects of the individual components. In some cases, it is possible to define an objectionable component sufficiently well to analyze for it and make reasonably good prediction of the odor effect on a community. For example, total reduced sulfur (TRS) is often taken as the measure of kraft mill odor. Even though such emissions contain mercaptans, methyl sulfide, dimethyl sulfide, and other reduced sulfur compounds, their predominant component is hydrogen sulfide. By assuming that all the reduced sulfur compounds have approximately the same odor threshold as hydrogen sulfide, it is possible to make reasonably good estimates of total odor emission from uncontrolled kraft mills with massive TRS emissions. As kraft mills accomplish better control, it may become necessary to analyze for the specific components, rather than the total reduced sulfur to estimate odor emission.

In similar fashion, it might be possible to sample for total hydrocarbon emissions from a coffee roasting operation and make a reasonably good prediction of total odor emission, or to find some compound or class of compounds in rendering plant exhaust which could be analyzed and used to define its total odor effect.

The method for odor measurement at the source recommended here is the syringe technique involving successive dilutions. It is particularly useful for mixtures of unknown components, and is also the method of choice for pure compounds. A wider variety of methods are available for ambient odors.

III. Methods of Describing Odors

Odors are described in terms of both intensity and character. One system for describing intensity, dating back at least as far as 1920 (*5*) uses the following scale of odor intensities: 0 = no odor; 1 = very faint; 2 = faint; 3 = easily noticeable; 4 = strong; 5 = very strong. This number system is subjective and represents the opinion of each member of a panel.

Stevens (*6*) has shown that most physiological responses follow the equation:

$$\text{Sensation} = k(\text{stimulus})^n$$

which is called the Stevens power law. Stevens has shown that it applies to loudness, brightness, taste, visual length, duration, finger span, and numerous other physiological responses as well as to odor. Dravnieks (7) has shown that the law holds quite well for the sensation of odor, and that exponent n varies from 0.2 to 0.8 depending upon the specific odorant. The exponent also varies among panelists on an odor panel, but the averages of several panelists are reasonbly stable (Fig. 1).

Defining only a threshold (i.e., zero intensity) value for a compound is insufficient for understanding its odor as an air pollutant, because the slope of the concentration–intensity line is equally important in determining its effect on a community. The slope defines the rate of decay of odor sensation with successive decreases in concentration, or with successive dilutions. The range in intensity from 0 to 5 covers 8 log cycles of concentration for ethyl mercaptan, 6 for butyl thioether, and 4 for crotonaldehyde. Both industrial pollution control engineers and control officials want to design for "no odor," rather than "faint" or "noticeable" intensity. The safest number to use for atmospheric diffusion or downwind calculations is the threshold concentration. An odor intensity of 5 measured in one location does not give us much information as to how far the parcel of air must travel or how much it must be diluted before its intensity is reduced to 2, 1, or 0. The descriptions of the numbers on the odor scale are easily understood by the rawest recruit and may be used for community odor patrols by untrained observers. Within each individual's own frame of reference, the results seem to be reproducible

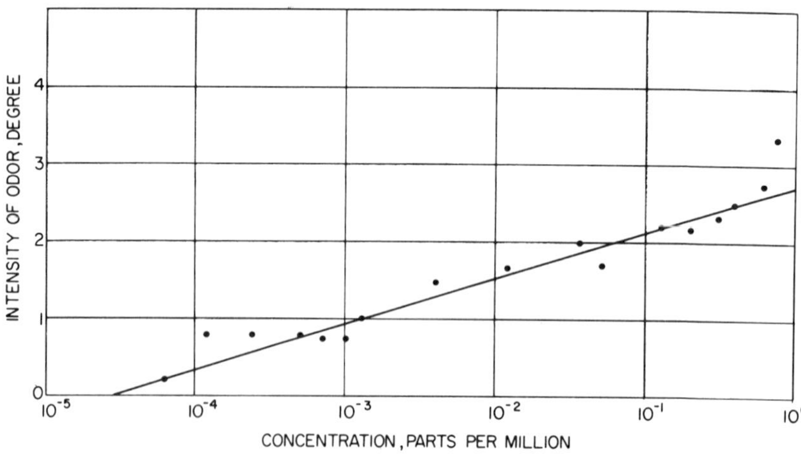

Figure 1. One of several tests of odor intensity vs concentration for ethyl mercaptan. Intensity (sensation) closely approximates logarithm of concentration (stimulus).

enough to be useful for a community problem—at least to determine whether or not an odor is present, and whether or not it is strong.

One of the disadvantages of any method that requires measurement of strong odors, as well as faint ones, is that each attempt to measure a strong concentration may cause olfactory fatigue in a subject, and subsequent measurements may not have the same reliability as those prior to exposure to higher concentrations. Purging the equipment and the nasal passages, with odor-free air and possibly resting the panelists adds to the time required for evaluation.

The character of an odor is described in terms similar to those used in describing taste. Most flavors can be described as varying concentrations of sweet, sour, salt, or bitter. Several classification schemes have been offered for odor using such descriptions as fruity, burnt, flowery, and so on (*8*). With careful training, sensitive individuals can be taught to classify an odor using these descriptive terms, and a scale of intensity for each, with enough accuracy to identify a compound by its odor description. This method obviously has merit in the field of food science or the blending of compounds where scent is important.

Dravnieks (*9*), building on a speculation by Zwaardemaker (*10*), suggests a possibility for odor titration and a scheme for odor description. His surmise is that, if the odors of two odorants are similar, a large amount of the first must be added to change the odor of the second; but if the odors are different, a small amount of the first will change the odor of the second. Nitrobenzene benzaldehyde and benzonitrile, have an almondlike odor, whereas *n*-butanol does not. A relatively small amount of nitrobenzene added to *n*-butanol changes the character of the *n*-butanol odor, whereas it takes a much larger amount of nitrobenzene added to the benzonitrile to change the benzonitrile odor. With an appropriate selection of compounds having different odor characteristics, it may be possible to perform an odor titration to numerically define an odorant by quantifying the amount of each standard required to change the character of the "unknown."

The best classification scheme for community ambient odors is to use descriptive terms which relate and can help identify sources of odors. In most air pollution control situations the probable odor sources in a community are known to both industry and regulatory agencies. For example, an odor can smell like coffee, a rendering plant, or sulfur dioxide. The description of kraft mill odors as being similar to rotten cabbage is one which most observers seem to agree upon. Anyone in a pulp mill community will identify the source from that description. One difficulty with odor characterization is that it sometimes is so subjective as to preclude communication among panel members. The author recalls experimenting with an odor measuring device and having one subject respond that the odor reminded him of "little green balls." This reference may have meant something to the subject, but had no meaning to anyone else.

IV. Units of Measurement—The Odor Unit

The relative strength of an odor is determined by dilution. The greater the number of dilutions with odor-free air necessary to bring an odorous gas sample to the threshold, or barely detectable, level, the stronger the odor.

Using the odor threshold as a basis, we must place some quantitative value on odor in order to apply the various atmospheric diffusion equations for predicting odor intensities at various downwind distances. This has been done—the odor unit has been established and defined *(11–13)*. It is the amount of odor necessary to contaminate 1 ft^3 of clean (odor-free) air to the threshold, or barely perceptible, level.

For example, to say that a sample of a gaseous emission contains 10 odor units/ft^3 means that when one part of the sample is diluted with nine parts of odor-free air, the resultant mixture will have a barely detectable odor. The 10 odor units/ft^3 is analogous to a dust loading figure. It is a concentration, but it states nothing in terms of pounds per minute. Therefore, as in the case of particulate matter discharges, it is necessary to know the air flow rate in the duct or stack in cubic feet per minute (cfm). The product of flow rate times its odor intensity gives the odor units emitted. To be specific, if an odor source has a concentration of 10 odor units/ft^3, and a flow rate in the vent or stack of 10 cfm, the odor output would be

$$10 \text{ odor units/ft}^3 \times 10 \text{ cfm} = 100 \text{ odor units/minute}$$

It is this value, 100 odor units/minute, which can be used in atmospheric diffusion equations in much the same manner as pounds of SO_2 per minute is used in these equations. While this approach is not strictly dimensionally correct *(14)*, it does allow for numerical computation of odor emission by diffusion equations.

Since the nose is so sensitive, the number of odor units in plant discharges will often be very large. A large number does not necessarily mean that a community odor problem will result. Discharges of 50,000 to 100,000 odor units per minute, which are not at all unlikely in a given industrial discharge, may not be perceptible outside the plant boundary line. With a few thousand feet distance between source and receptor even a million odor units per minute may not give a detectable response in a neighborhood.

V. Techniques of Odor Measurement by Dilution

The reason for our preference for the nose as a measuring device for odor is the fact that it is attached to a head stuffed with some brains. It is the mental ability to single out an odor which does not belong, from the ones which might normally be expected, which makes the human nose and mind so useful.

There are four major procedures for the use of the nose for this purpose: (a) the odor-containing air and odor-free diluent air are mixed in a container, the nose is brought to an opening in the container, and the measurement made by inhalation from the container; (b) the odor-containing and odor-free air are mixed in a container and part or all of the contents of the container are injected from the container into the nose; (c) the odor-containing and odor-free air are mixed in a duct which supplies a flow of the mixture to a nose, or facepiece, or to a hood or chamber enclosing the observer's head or body; and (d) the act of inhalation is used to provide the energy to induce odor-free air to mix with odor-containing air in a measurable dilution ratio. Some precautions and elements of technique apply generally to all methods. These have to do with the selection and use of panel members.

Qualification of panel members for odor measurement is the subject of much concern. As a quick and rough method of identifying good panel members, their eating practices might well be observed. It has been our experience that simple "meat and potatoes" men make very poor odor observers. The connection between taste and odor is well known and in general those who are not fussy about their food and do not, when given the opportunity, select complicated dishes on a menu will often be found to be off by one or two orders of magnitude in threshold measurement. Actually, in either industrial or control organizations, a panel will necessarily be drawn from the personnel of the organization. After the first few panel tests those who can smell reliably at low concentration, and those who cannot, may be readily identified.

Those panelists who consistently give results markedly different from the rest of the group may simply be dropped from the test panels and replaced by new panelists. In a reasonably stable industrial organization or air pollution control agency this approach is adequate. However, if a panel has to be set up for a specific problem, where there is no previous history as to the panelists' osmic ability, a triangle approach (*15, 16*) may be used to screen potential panelists. In this approach, two odorous compounds having significantly different quality are dissolved in water at low concentrations and presented to members of the panel as three samples—two of which contain identical concentrations of one odorant, the third containing the other odorant. The test for a potential panelist is to find the odd sample among the three. Sniffing from the flask is used rather than more-refined techniques for presenting the odor and persons who consistently cannot find the odd sample are excluded from the panel.

Since air pollution odor problems do not require the fine discrimination of a whiskey judge or perfume blender, a two- or preferably three-person panel, whose odor judgment has been demonstrated to be reasonably good, should suffice. Larger panels are obviously to be preferred, particularly if court action and proof are liable to be required. Even so, reasonably accurate estimates of a community's response to an odor can be made if necessary by a single individual who has experience with odor pollution, just as an experienced chef can determine reasonably well all by himself whether the soup is too salty for the customers.

An excellent study of the variability of panel members was made by Prince and Ince (*17*). In their study, different observers showed appreciably different sensitivities to a given odor. The day-to-day variation for a given observer is small while he is in good health, and such variations as are observed may be due to changes in the environment. Careful selection and training of observers will materially reduce variance. The standard error for a particularly good observer could easily be $\pm 40\%$ in concentration. If the average results are taken for a group of four good observers, the standard error will be halved, and results reproducible to about $\pm 20\%$ in concentration are obtainable. There are significant variations between observers for two different odors. For example, the difference between odor thresholds found by two observers was slight for benzophenone, but to detect *p*-methoxyacetophenone one observer required five times the concentration required by the other. This implies that detection of a specific reference odor at a certain concentration cannot be the sole criterion for selection of panel members. The use of a standard odor can, however, be used to tell whether panel members are in good osmic health.

The odor samples should be prepared by a nonparticipating member of the panel, and the concentrations presented to panel members should be unknown to them, and in random order. To avoid odor fatigue, it is preferable to start with an extremely low concentration and try to approach the threshold from the bottom of the scale rather than from the top. Too frequent high concentration may deaden odor discrimination and give a false threshold value. A high concentration of a compound which can irritate the nasal surfaces can destroy a panel member's effectiveness for the duration of the test. Once there is some indication of the threshold value, several samples above and below this value should be tried for the panel members to try to zero-in on the threshold limit value. There will, of course, be no concentration which simultaneously fulfills both of the following requirements: Any concentration below X will never be detected by a panel member; and any concentration above X will always be detected by a panel member. These actually are two different points, and there is a band of undefinable concentration between. While attempting to determine odor thresholds in this zone, several blank samples with no odor materials should be presented to the panel at random to detect any imaginary responses on the part of panel members—to keep them loose and honest. In this zone, close to threshold values, samples should not be handed to panel members in ascending or descending order or they will catch on to the pattern and try to second-guess the samples. The observers should smell, not think!

The above assumes that for each panelist a true momentary sensory threshold actually exists. This assumption is adequate for use by regulatory agencies or those in industry trying to determine which sources of odor must be controlled and how thoroughly they must be controlled. Other assumptions may be preferred for more precise laboratory determinations, such as the theory of signal detectability. Methods based on either of these two assumptions are discussed by Lindvall (*18*).

If threshold values are developed by small panels in either industry or a control office, the question of which concentration to use as a threshold seems valid. Should we use the concentration above which all members of the panel always report an odor or should we use the concentration below which no member reports an odor as the threshold value? The values obtained in either case will not be exact, but this is a common engineering problem. Perhaps the best value to be used for engineering purposes is that concentration for which only one member, of a panel of three or more, gives a positive value. This is not the zero threshold, but considering the variation in people, atmospheric diffusion, wind direction, and other meteorological factors, this seems like a reasonable value to use. An emission whose downwind concentration after atmospheric dilution would be detectable by only one member of the panel should not be a major neighborhood nuisance.

The Los Angeles County, California Air Pollution Control District (*19*) defines the threshold value as that point where 50% of the panel can distinguish that an odor is present. Hemeon (*14*) using the Stevens' power law, determines threshold value by having panelists define odor presented to them in increments of strength. Thus a panelist may indicate a number 2 intensity at one concentration and a number 3 intensity at another. He draws a line on semilog paper through these intensity responses, and defines the threshold value as the point where this line intercepts the ordinate.

We prefer designating the result of each test only in "yes" or "no" terms. Since we are looking for the threshold, we are relatively unconcerned about whether a particular injection is strong, faint, or overpowering. No attempt to define those concentrations is advised. Nor do we want qualitative observations such as "ye gods!" We are not trying to determine whether the odor is pleasant or unpleasant—we merely want to know if it is there. In practice, it is difficult to prevent members of the panel from given subjective responses over and above a simple yes or no. An occasional "ugh!" or "wow!" or "maybe" must be expected. These may be used as a clue as to whether the next dilution should be several times more or less dilute.

There undoubtedly is a difference between injecting an odor into the nose, and immersion of the nose into a container of the odorous gas. Stone (*20*) reports an effort to measure the difference between two analogous methods. That analogous to injection involved uncorking a sample solution and inhaling above the level of the liquid. This was compared with piping a known concentration into a hood where the subject was completely enclosed in the odorous gas. Subjects appeared more sensitive to odor concentration by the former method, but their responses were more variable. Variation occurred not only markedly between observers, but also for an observer at different times. Measurement by nasal immersion in odor has several complications. All equipment, including the hood, piping, and diluting devices, must themselves be odorless, must not retain any odor on surface walls, or rob the sample of any material, and the dilutions to be tested must be capable of being changed easily and rapidly with no retention.

Difficult or not, we would expect any immersion technique to give better results than injection because exact control of the dilution rates and the rate of odor delivery is under control of the operator and can be uniformly maintained. With the injection technique the panel member himself controls the rate of release, the rate of injection, and the amount of inhalation. Dilution in the nose by the eductor effect of the injected jet of air occurs. Also, the air in the nose which must be displaced by the injected sample effects some dilution which is impossible to measure.

One rationale for categorizing odor measuring techniques is the manner of presentation of the odor to the panelist: sniffing techniques where samples are presented to either one or both nostrils under control of the panelist; techniques where the odor is presented in a hood or face mask at a set rate; and odors presented in a room where total immersion in the odorant occurs. Discussions of these techniques are preceded by a brief presentation of some odor measuring techniques which are primarily of historical interest.

A. Techniques of Historical Interest

The Katz and Allison Odorometer developed in the 1920's is of interest because it was used to measure odor threshold values, some of which are still valid and useful (*21*). In fact, many odor threshold tables still cited in the literature are based on this original reference, although not always with proper citation. The odorometer device presented the observer with a mixture containing measured amount of odorous compound in air so that odor thresholds could be expressed in parts per million (Fig. 2). Room air is passed through a filter of charcoal and cotton wool, to remove any odorous material; then through calcium chloride, to remove moisture; through a flow measuring device using tubing as an orifice; and then to the vaporizer. The vaporizer is generally a U-tube, and the pure air either passes over the surface of the compound of interest or is bubbled through the compound. The U-tube is maintained at a constant temperature and rate of evaporation determined by loss in weight. This

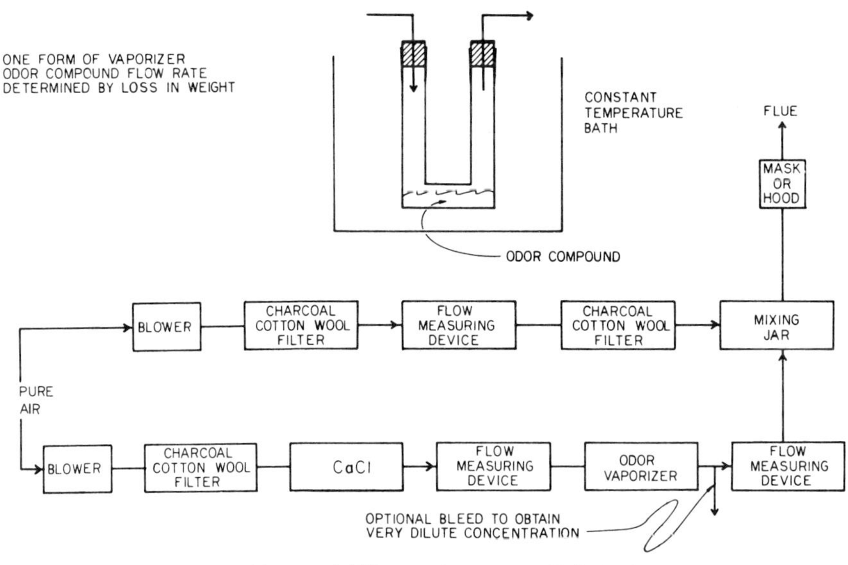

Figure 2. Schematic of Katz and Allison odorometer. Relatively pure air is rendered odor free by charcoal and filtered. Air for vaporizing is dehumidified and passed over compound in vaporizer tube. Measured flows of odorous and pure air are mixed and evaluated by panelists.

mixture of air and evaporated odor compound passes through another flowmeter. The mixture flows to a mixing jar into which air, which has been metered and purified by charcoal filters, flows. The diluted, odorous air with known concentration of odor compound passes to a mask or hood from which the observer inhales and evaluates the odor. The observer's head is positioned in a ventilated area to maintain relatively odor-free conditions in the test room. The Fair–Wells osmoscope (*22*) of somewhat later vintage consists of two telescoping tubes. Along the middle section of the inner tube there are specially arranged holes which are closed or opened by adjusting the outer tube. The inner tube conveys the odor to be tested and terminates on one end in a special nosepiece. The opposite end of the tube is inserted into the odorous environment. The outer tube is positioned to progressively open or close holes, allowing, with each inhalation, odor-free room air to dilute the odor traveling through the inner tube. This device was improved by Prince and Ince (*17*). Since odor perception varies in logarithmic increments, one of their useful improvements was to arrange the holes in a geometric series so that the odor sensation was approximately doubled or halved at each step.

In the Procter & Gamble Osmo (*23*), developed in 1950, the central element is a cylinder with many equally sized holes. Surrounding the cylinder is an outer cylinder enclosing a mixing chamber. A piston divides the inner cylinder into two sections so that the ratio of holes exposed in each can be varied by an external crank. By varying the ratio of holes exposed to odorous air and odor-free air, dilutions can be measured and the sample presented to the panelists.

B. Sniffing Techniques

Among the techniques in which the sample is presented directly to a nostril for sniffing, the one recommended by the author is ASTM Method D 1391-57 (*24*). The equipment necessary for this method are a few graduated 100- and 2-ml syringes and hypodermic needles. The syringes are standard medical equipment which can be obtained at any medical supply house; for example, Luer hypodermic syringes, 100 cc Ideal-Lok, with syringe adapters for covering the opening. Only a small sample of odorous gas is required.

The sample is taken from a small hole in the duct by withdrawing the syringe plunger to the 100-cc mark. $1\frac{1}{2}$- to 3-in.-long needles are required so that the odorous material can be withdrawn from well within the duct without risk of dilution with outside air. After the sample is taken from the duct, the needle is removed and a cap placed on the syringe to avoid losing any sample. The sample syringe is taken to an odor-free room where clean air is available for making dilutions. The ground joints of a glass syringe are extremely tight. Only negligible amounts of the sample can leak to the atmosphere.

Measured quantities of the odorous gas are injected into the dilution syringe, either directly from the sample syringe, or indirectly through the use of a transfer syringe (Fig. 3). The remainder of the dilution syringe volume is filled with odor-free air and the odorous air mixture is then injected into the nostril by depressing the syringe plunger (Fig. 4). The ratio

$$\frac{\text{ml odor-free air} + \text{ml odorous air}}{\text{ml odorous air}}$$

gives the number of odor units per cubic foot, or per cubic meter, in the source gas. For very dilute concentrations of very odorous compounds, a small amount of odorous gas is first placed in the sample syringe. The remaining volume is then filled with odor-free air. Most of this mixture is then expelled, i.e., 90 ml and more free air drawn in; and this dilution process continued until the desired dilution is reached. It is possible to make reasonably accurate dilutions to the order of 1 ppm.

Some complication with the syringes can occur in obtaining samples from hot or saturated gases. A technique used by the Los Angeles County, California Air Pollution Control District (*19*) to partially solve this problem is to draw the hot or saturated gas into a known volume

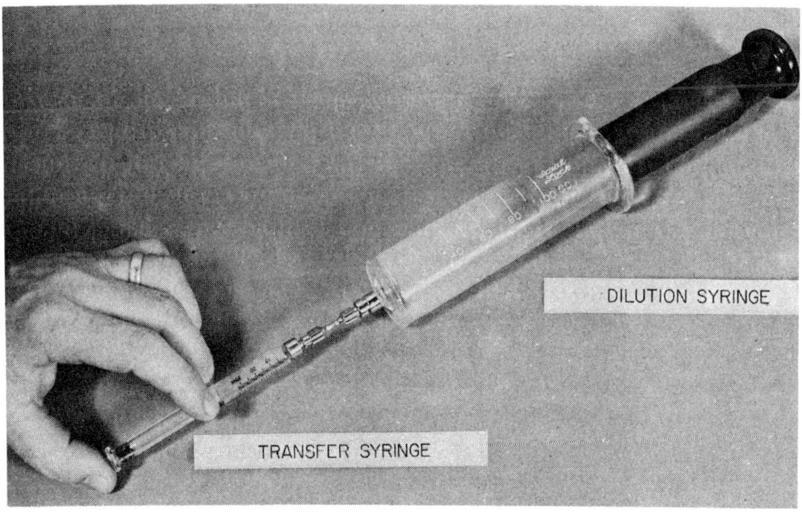

Figure 3. Transfer of odor sample to dilution syringe. Sample in transfer syringe was measured out of a sample syringe, which is identical with dilution syringe.

Figure 4. Evaluation of diluted odor sample. Panelist is injecting contents into nostril to determine if any odor is present at this dilution.

of odor-free air during sampling. That is, a 100-ml syringe is initially filled with 90 ml of odor-free air, and only 10 ml of sample is drawn in. This avoids overheating the syringes and prevents condensation from subsequent cooling. Of course, this initial dilution must be considered in calculating the odor intensity of the sample.

The ASTM method has been questioned because of its high surface-to-volume ratio, presentation of the sample to only one nostril, lack of control over the rate of presentation of the nostril, possible dilution effects, and difficulties in reproducing threshold values for pure chemicals requiring a large number of dilutions. It is being revised to better specify the method of pretesting panelists, presenting different strength dilutions to panelists and computing the panel average. With experience on the part of the panel in the use of this technique, successful results in quantifying odor emissions and the degree of odor control have been reported (*15*, *19*).

Dravnieks (*7*) proposes an eight-step, *n*-butanol scale prepared in a simple vapor diluting constant flow system. Panelists are given an unknown odor concentration of the odorant to be investigated from one inhalation port and are allowed to compare it with the concentration of *n*-butanol at any of the eight butanol air dilution ports. While comparisons are best when the unknown odorant is similar in smell to *n*-butanol, judgments are reported to be satisfactory even when the quality of the unknown is different.

In the Dravnieks and Prokop (*25*) olfactometer one diluted odor sample and two nonodorous air blanks are presented dynamically at each dilution level. Panelists are asked to judge which of the three sniffing ports is odorous and to signal their choice (Fig. 5). The tubing and bag

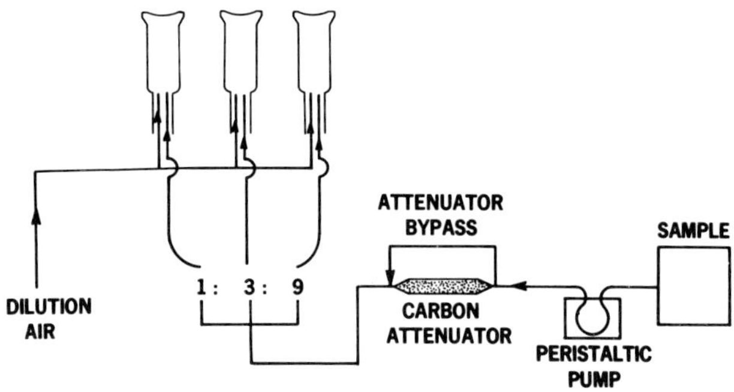

Figure 5. Schematic of Dravnieks–Prokop device developed for the National Rendering Association. Odorous sample is delivered in ratio of 1:3:9 to each of 3 sniffing ports along with a volume of dilution air which is the same for each of the sniffing ports, but may be varied to give higher or lower dilution at the sniffing ports.

are flushed with several volumes of the sample to be evaluated, the preflush is expelled by collapsing the bag, and the bag is filled with the sample to be evaluated. The sample from the bag is divided by stainless-steel capillary tubing to produce three flows having a ratio of 1:3:9. Dilution air is supplied by a separate pump. Provision is made for double blind testing so that neither the panelist nor the sample provider knows which port is receiving the odor until after judgment has been made. Ten effluent samples from rendering plants were evaluated two or more times by this method. Five of these were evaluated twice by the syringe method. No statistically significant difference between the two measuring techniques was found, but the triangle technique was found to be faster and more convenient for the panelists.

C. Hood and Face Mask Techniques

The Institute of Hygiene of the National Institute of Public Health, Sweden, collects samples of gases from industrial exhausts (*18, 26*) in special plastic bags from which the sample is pumped by means of a diaphragm pump into a dosing system consisting of a number of calibrated steel capillary tubes (Fig. 6). The sample is then mixed with charcoal filtered air and conducted to an exposure hood where the face is inserted. This method permits sampling at an odorous site with subsequent odor measurement in an odor-free room. A most impressive test of this method was that samples were drawn from the same point in the process from two entirely different plants, and odor threshold values were found to be approximately the same, dilutions to threshold very near 10^5 for both plants.

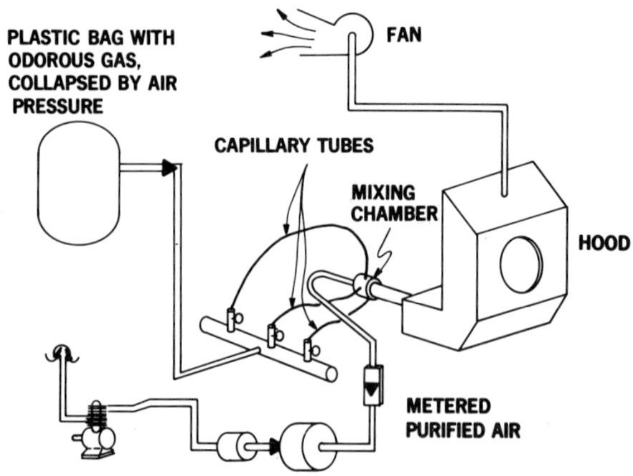

Figure 6. Schematic of Swedish apparatus. Plastic bag sample is obtained at plant, diluted, and evaluated at laboratory.

In the use of an exposure hood it is important to provide for an even distribution of the air stream and good mixing prior to reaching the face port. Where the hood is used outdoors, it is acceptable to allow the test sample to be fed to the hood at a rate greater than normal respiratory volume. In a closed space this is less acceptable, since the excess odorant is being pumped into the room. However, a flow to the hood greater than respiratory volume ensures that additional dilution from around the interface between hood and breathing zone will not occur. Lindval (*18*), in a modification of this method, uses a fan to maintain the hood at a slight negative pressure with respect to the exposure chamber (approximately 1 mm H_2O) to prevent the odorant from leaking into the exposure chamber. Another refinement is to compress the plastic bag to deliver the sample rather than risk its contamination by passing it through a pump. Dilution is controlled by the use of calibrated steel capillary tubes each of which connects a receiver to the hollow stem of a dosage valve. Odorant flows continually through each capillary and stem and out through small peripheral holes in the opposite end of the stem. The dosage valve stems are moved forward to deliver this flow to the main airstream by compressed air-operated pistons or are retracted by the pistons to deliver the odorant flow to an exhaust channel. Thus each of the six dilutions can be quickly delivered to the main airstream.

Hemeon (*14*) also uses a hood which covers the nose (Fig. 7). Samples are collected in a throw-away polyethylene sample bag, used only once, inside which an expandable wire frame is placed before sealing off the neck with a sampling tube. After assembly, the bag interior is evacuated by collapsing the wire frame. The sample is collected by slowly expanding

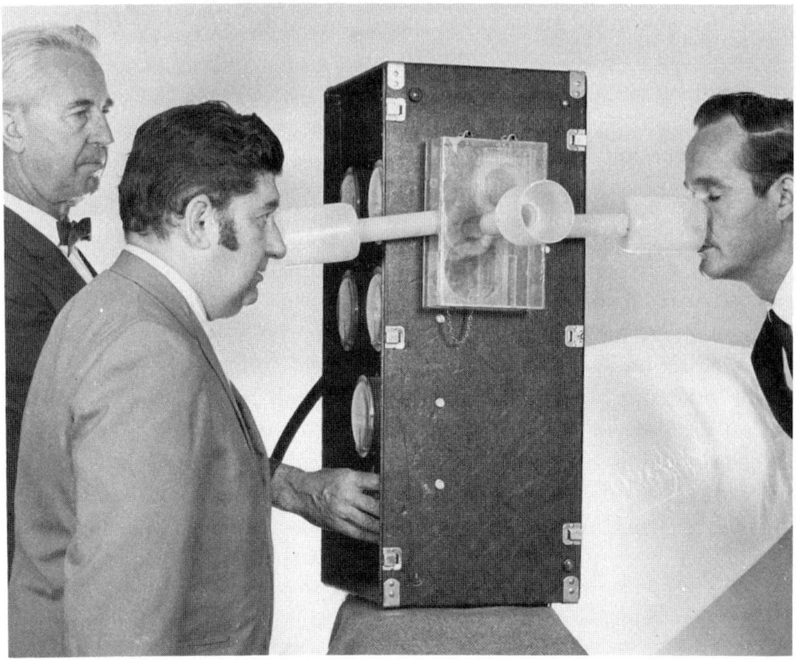

Figure 7. A sample (3–5 ft^3) of the odor bearing air stream is collected in a plastic bag visible at right rear, now connected to the odorometer. Dilution streams are delivered through the distribution box to face pieces. (Courtesy of Hemeon Associates.)

the wire cage to an open position. The odor sample is then transported to an odor meter at a convenient test location and the bag connected to the odor meter inlet in preparation for the test. Control of flow of the odor-bearing gas stream from the bag is provided by a primary valve and the control of the dilution air stream by other valves which extend through the instrument panel of the odor meter. Two stages of dilution are available. The first allowing for dilution from 30 to 400, and the second from 400 to 150,000. For the second stage, an aliquot is drawn from the primary airstream discharge and delivered to a secondary dilution air system. The sample for presentation to the panel is delivered at a rate of about 5 cfm, well in excess of breathing requirements, to avoid hood dilution effects.

Wilby (*27*) describes another form of odor comparator which delivers the samples to be tested out of a hooded opening of about 100 in.^{2} at a low velocity. The observer's face is inserted in the hood exhaust stream. Wilby's comparator was designed to be portable enough to be used in the field by one man. His device was designed to test the odor threshold in natural gas transmission lines to determine whether it has adequate odorization for the users to be able to detect leaks.

D. Total Immersion Techniques

Under the sponsorship of the Manufacturing Chemists' Association (MCA), Arthur D. Little, Inc. made a study of the odor threshold of 53 chemicals, by use of an odor test room of 500 ft^3 volume (*28*). The chemical to be studied was introduced quantitatively in amounts calculated to achieve an appropriate concentration level by a variety of methods depending on its physical properties. Electric fans were used to circulate the odor for several minutes, after which time the fans were shut down and the air in the test room allowed to come to rest. Panelists who had been waiting in an odor-free antechamber, entered the test room, then recorded their odor observations by describing the odor and evaluating its intensity. Upon completion of the analysis, the test room and antechamber were purged for 20 minutes with activated carbon-treated air. To extend this study, a smaller (5 ft $\times$ 4 ft $\times$ 3 ft) aluminum test chamber was constructed for the presentation of odorants on a dynamic mode (Fig. 8). The panelists sit while an airstream passes continuously from front to rear. Intake air is drawn through a bank of activated carbon-filled panels, and odorized by means of a syringe pump remotely

Figure 8. Dynamic test chamber used by A. D. Little, Inc. Test stream is delivered via grill in chamber (visible through window in door) to seated panelist.

controlled to give a variety of concentrations and durations of exposure. The odorant compounds are dissolved in a low-odor solvent which serves both as a carrier and a tracer for the odorant dose. A total hydrocarbon analyzer with flame ionization detector is used to measure the concentration of the carrier solvent in the airstream prior to its entry into the chamber. The carrier solvent is present in a concentration well below its odor threshold, but above its limit of detection by the analyzer.

E. Gas Chromatographic Techniques

Dravnieks *et al.* (*29*) and Dravnieks (*30*) have reported on methods for collecting odor samples on highly porous hydrophobic organic polymers and studying the collected samples by temperature programmed gas chromatography. With collection of the odorous compounds in a fluidized bed, the vapors simultaneously reach all the particles of the polymeric adsorbent, and the maximum collectable amount of each component is rapidly accumulated. As an alternate, a fixed air volume can be passed through a packed collector. The latter technique presents analytical difficulties if some compounds reach equilibration sooner than others and begin to pass through. The identification of individual odorous components is still possible, but the quantification of their concentration is much more difficult.

In sampling, air is drawn through a container holding about 5 gm of the adsorbent. Experience indicates that most organic species will be completely collected from a 10-liter air sample. Most samples are analyzed by a 200-ft open tubular Carbowax 20M column with programmed temperature rise. The exit carrier gas from the column is split equally, with one portion going to a flame ionization detector and second portion going to a sniffing port (Fig. 9). With each peak on the detector, an observer can present his nose to the sniffing port and determine whether, for that particular compound, an odor is present.

With a multicomponent mixture, it is thus possible to find those fractions which exhibit a significant odor response at the sniffing port and by chemical logic to identify each odorous compound in the mixture. Even though a single investigation can result in an impressive resolution of odorants, it is possible that many compounds will still remain unresolved or be masked by particularly large peaks. Another column using a different packing can be used to resolve peaks of particular interest. Since the Carbowax column has the best resolution for compounds which are more or less polar, a nonpolar packing is used in the other column. When a peak of interest begins, the flow can temporarily be changed and directed to the second column for further resolution.

Since there is concentration of the sample on the polymeric adsorbent,

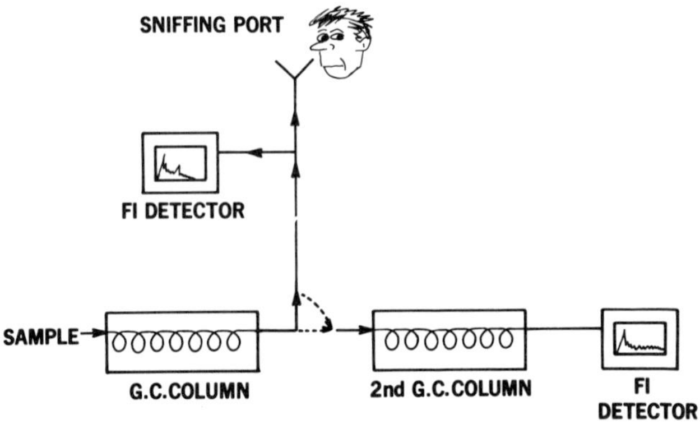

Figure 9. Schematic of gas chromatograph arrangement for investigating odor. Odorous peak can be further resolved when desirable by switching to second column.

individual compounds are presented at the sniffing port approximately one order of magnitude more concentrated than they were in the gas volume from which they were originally sampled. Thus, by this technique, it may be possible to identify odorous compounds which, at the particular time they are present, are below threshold, or are unnoticed because other odorous compounds temporarily mask them. By the use of this technique it may be possible to take "fingerprints" of the odor-producing compounds in the ambient air of a community, and to later sample suspected sources to determine whether that source's fingerprint was found at the scene of the complaint.

Dravnieks and Laffort (*2*) have also investigated ways of predicting the odorous properties of pure compounds from gas chromatographic and physicochemical data. When square and product terms were included among the predictive factors, models were developed which give good estimates of threshold and intensity slope values and qualitative data. While the polynomial algebra on which the analysis rests may not be the most suitable form for combining factors to describe olfactory effects, the analysis indicates that correlations exist. Future work may lead to a model by which the odor properties of pure compounds can be predicted without the use of a human panel.

VI. Ambient Odors

A simple odor patrol is probably the best means of detecting existing or potential nuisance from odorous discharges (*31*). An odor patrol consists of having at least one observer make a regular patrol around the plant or through the district every day (or at some other time interval) noting any odors observed and the wind direction. With a little experience almost any individual with a normal sense of smell can become expert enough in detecting and distinguishing separate odors to locate the offending source. A record of such odor patrols in also useful as an indication of where odor control funds might best be spent or where pollution control regulatory efforts might best be directed. Obviously, industry should expend its funds in odor control on those sources which are most frequently noticed on the street. This may not necessarily be identical with the magnitude of discharges within the plant. Various stack heights, terrain or building features, chemical interactions, and potential diffusion in an industrial complex confound the prediction of which emissions will cause nuisance problems. The odor patrol method has the disadvantage of not providing numbers which can be used for calculating back to the probable original odor intensity.

To some extent numerical data can be developed by providing the odor patrol with instrumentation. In general, the instruments and measuring techniques, such as the syringe technique, described for measuring the odor strength of sources are too complex and cumbersome to use on odor patrol. The nearest approach to a portable instrument for quantifying an observer's response is the scentometer (*32*) (Fig. 10). The device has

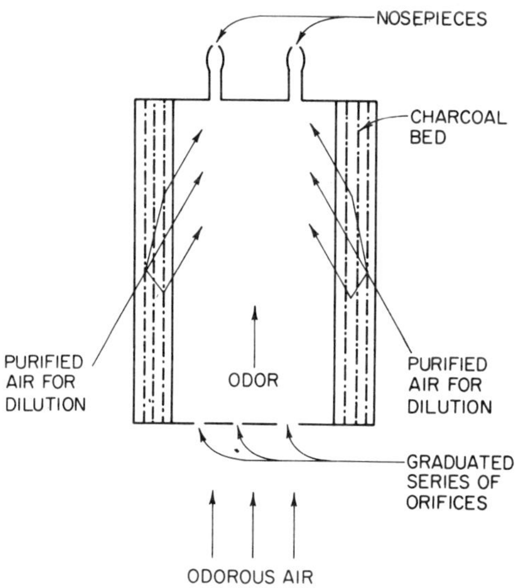

Figure 10. Schematic of scentometer. Odorous air passes through graduated orifices and is mixed with air from the same source, which is purified by passing through charcoal beds. Dilution rates are fixed by the orifice selection.

holes which can be opened or closed by the fingers in the same manner that one fingers a flute. Various finger combinations result in odorous air dilutions of from 1 to 32 by odor-free air which passes through the activated charcoal during inhalation.

While the scentometer has a limited dilution range, the variation in strength of ambient odors encountered should not be much greater. One feature of this device is that the observer can, by breathing only air filtered by activated carbon, combat the olfactory fatigue which could occur in driving through an odorous area, and can return to zero odor reference before trying a different dilution to determine odor concentration. The primary advantage of this device is that it is small, inexpensive, and extremely portable. Huey describes the use of the scentometer in the first United States Interstate Abatement action between Delaware and Maryland, which involved odorous emissions from a rendering plant (*33, 34*).

Lindvall (*18*) compared ambient odor occurrence at specified distances downwind from a pulp mill with predicted occurrences by dispersion models. Concentration of odorous compounds at various distances downwind were computed using the bivariate Gaussian equation (see Chapter 9, Volume I). The probability of the odor occurring at any location was determined by including a factor for the frequency that the wind blew in that direction. "Actual" odor occurrences were determined by panelists. Early attempts to use people who lived in the neighborhood gave erroneous data, since odor observations were made when the wind was not in the direction which would have carried the odor to the panelist. Panelists were used who had some experience in identifying the smell of this particular effluent. The results of one set of tests is shown in Table I. While the correlation between predicted odor occurrences and observed ones seems quite good relatively close to the plant, the prediction method used seems to underestimate at greater distances.

Clarenburg (*35*) has developed mathematical models which he calls penalization functions, for both point sources and area sources, which predict the result of a public opinion poll or the probability of telephone complaints downwind from an odorous source. Correlation between pre-

Table I Comparison of Predicted vs Observed Odor Frequency

Distance from the source (km)	2	5	10	20
Total number of observations	6426	7490	5528	6976
No. of positive observations	696	736	470	360
Observed odor frequency (%)	10.8	9.8	8.5	5.1
Predicted odor frequency (%)	9.1	5.7	3.2	1.7

dicted estimates of complaint frequency and observed values was quite good. The approach offers the possibility of predicting the effect on a community—in the sense of complaint response—resulting from expanding the production of an existing chemical complex, increasing the number of producing units in a complex, or constructing a new industrial complex.

An interesting example of what can be done in a community study with an experienced survey group with no instrumentation was reported by Horstman, Wromble, and Heller (*36*). High school students were asked to make odor observations in their neighborhoods at three different times of the day. Odors were reported as faint or strong, and classified by one of ten descriptors such as rotten egg, wood smoke, and gasoline. Such a survey can at low cost give a good estimate of the geographical extent of an odor problem, the frequency distribution of various responses, and their seasonal variation.

Sableski reports a similar inexpensive technique utilizing firemen (*37*). Firemen were selected as observers because of their 24-hour availability. Observations could be made at predetermined times throughout the day and night. At a given time, all available firemen in the fire stations in the community were instructed to go outside the firehouse, take a few sniffs of air and record in their own words what was smelled. For fire stations where significant data were accumulated, the odor observations and their definition (diesel fumes, smoke, oil, sewer, etc.) were plotted on a wind rose which defined the probable sources of odor, or at least the directions of particularly well-defined sources and their frequency in the community.

VII. Examples of Application of Odor Measurement Results

A. Single Source

Let us assume that we have determined that the odor threshold value of dimethylamine (DMA), $NH_3(CH_3)_2$, a gaseous compound which has a fishlike odor in dilute concentrations, is 0.5 ppm by volume. This means that when 1 ft^3 of DMA in a gaseous form is mixed with 2,000,000 ft^3 of odor-free air, the resultant mixture will just barely have a fishy odor at the odor threshold for the average nose used in the odor panel. In other terms, 1 ft^3/hour of DMA emitted is equal to 2 million odor units per hour.

Downwind odor concentration is directly proportional to the source emission. Therefore, odor response for Y pounds/hour emission can be obtained by direct proportionality with the odor response for 10 pounds/hour emission. Ten pounds per hour of dimethylamine, by the following calculation, is equivalent to 85 ft^3 of DMA per hour at standard temperature and pressure (STP):

$$\frac{10 \text{ pounds DMA/hour}}{45 \text{ pounds DMA/mole}} \times 379 \text{ ft}^3 \text{ STP/mole} = 85 \text{ ft}^3 \text{ (STP) DMA/hour}$$

Since we have have assumed that 1 ft^3/hour of DMA emission is equal to 2 million odor units per hour, then an emission of 85 ft^3/hour is equivalent to 170,000,000 odor units per hour, or 2.8 million odor units per minute. Thus, an emission of 10 pounds/hour of DMA will contaminate to the odor threshold 170 million cubic feet of air per hour, or 2.8 million cubic feet per minute.

The emission of 2.8 million odor units per minute can be plugged into the various stack diffusion equations which can be found in Chapter 9, Vol. I. For mean height of release, we might assume 25 ft above ground as the average level of odor emission from various process vessels and columns, or we can assume ground-level release where this is more reasonable. An equation which gives good approximation with relatively simple mathematics is Halitsky's (*38*) vent equation:

$$\text{Dilution} = M(3.16 + 0.1R)^2$$

where X is distance in vent diameters, and M is a turbulence coefficient. If the figure of 2.8 million odor units per minute is cranked into the standard diffusion equations (using 0.4 for C_y, 0.2 for C_z, and $n = 0.25$ in Sutton's equation; 0.05 for the ratio p/q in the Bosanquet–Pearson equation; and $R = 12$ in. and $M = 3$ for Halitsky's equation) the chart shown on Figure 11 can be produced. This chart shows that if 10 pounds/hour of DMA are emitted to the atmosphere from pressure relief valves, spills, pump leaks, valve leaks, etc., according to the Sutton equation the odor will be below the threshold level at any distance greater than 500 feet; and according to the Bosanquet–Pearson equation it will be below the threshold level at any distance greater than 1600 ft. The Halitsky equation gives results between the two. The agreement is good considering the time factors implicit in the several equations. From this chart, it is possible to check the distance to the nearest residence or neighborhood of concern, or to the plant lot line; and by simple ratios determine what maximum emission could be allowed for such a plant. As an example, if 20 pounds/hour were to be emitted, the odor responses on the figure would all be doubled and an odor threshold response might occur up to 1000 ft (Sutton) or 2800 ft (Bosanquet–Pearson).

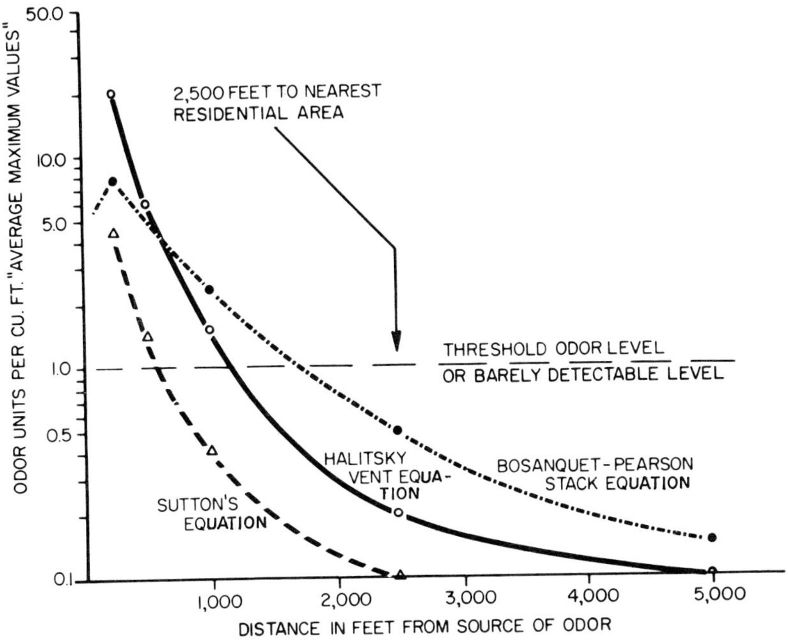

Figure 11. Downwind odor concentrations by three different formulas. Odor threshold concentration is 0.5 ppm, which is, by definition, equal to 1.0 odor unit per ft^3. (1) Odor emission: 2,800,000 odor units/minute from 10 pound/hour assuming odor threshold of 0.5 ppm. (2) Wind velocity 8 mph. (3) "Average maximum values" 1–5 minute duration every $\frac{1}{2}$ hour.

B. Entire Plant

An application of odor measurement in improving neighborhood odors would be to survey all the operations in a plant and determine the odor units emitted from each (*31*). A summary of the various odor units being emitted can be listed showing magnitude of emission and cost to control (Table II). This will help management pick out the largest odor sources rather than the largest stacks or largest volumes and concentrate efforts initially on those which are likely to provide the greatest improvement per dollar of expenditure.

Table II also shows that, if a crash program had to be instituted to improve this odor problem immediately, items 1, 5, and 7 might all be done initially even though some of them are very costly. These steps would reduce three-quarters of the odor emissions immediately. On the other hand, a stepwise approach taking the least costly items first, in order to make marginal improvement to the area until a community problem has been solved, might be another approach, in which case items

Table II Relative Cost of Various Measures for Odor Reduction

Action that can be taken	*Odor reduction (units/minute)*	*Total cost*	*Odor unit reduction cost in dollars per thousand*
1. Throttle agitation steam	2,500,000	\$1500 (slower heat-up)	\$ 0.60
2. Closed heating coils, 10 fat traps	500,000	\$3000 (equipment purchases)	6.00
3. Tank car cleaning techniques	300,000	\$2000 (increased labor)	6.06
4. Cover, 10 fat traps	400,000	\$4000 (equipment purchases)	10.00
5. Control equipment, tallow bleach	2,225,000	\$35,000 (equipment purchases)	15.70
6. Control equipment, still house	1,500,000	\$30,000 (equipment purchases)	20.00
7. Process change	2,500,000	\$90,000	36.00
Complete program	9,925,000	\$165,000	\$16.60

1, 2, and 3 might be tried first. In any case, such a tabulation allows management to pick any level of odor reduction desired and know exactly what it costs. There would be occasions where the character of the odor might have to be considered, but in most cases, one smell is usually as bad as another, and rarely must a particular emission be controlled because it is more annoying than others.

C. Pilot Plant

Another important application is evaluating the effectiveness of odor removal by various devices on a pilot plant scale. Since the pilot plant device is intended to remove odor, the best method of gauging its effectiveness is by measuring the odor into and out of the unit (*39*).

VIII. Odor Threshold Data

Literature on odor thresholds for pure compounds is usually vague as to the techniques of determination, the evaluation of the odor panel, and the reliability or spread of the observed results. Even where pure compound threshold data are available, it is wise to take samples of actual discharges and experimentally evaluate their threshold.

In 1951, the Manufacturing Chemists' Association (MCA) prepared manuals on air pollution and in one section covered physiological effects (*40*). In this section, a long list of odor threshold values was given. Approximately 200 compounds were listed with threshold values in parts per million. Because of widely varying thresholds given in many cases for the same compound, the MCA withdrew this manual from publication. Droege (*41*) studied the various odor threshold tables and attempted to analyze the cause of the variation in data given. He concluded that many of the sources derived their values from Bureau of Mines Report No. 480 (*21*) and that variations between different tables may be traced to some authors using No. 1 intensity, others using the No. 2 intensity from that report.

Table III Odor Thresholds in Air (ppm Volume)

Compound	*ppm*	*Compound*	*ppm*
Acetaldehyde	0.21	Ethyl acrylate	0.00047
Acetic acid	1.0	Ethyl mercaptan	0.001
Acetone	100.0	Formaldehyde	1.0
Acrolein	0.21	Hydrochloric acid gas	10.0
Acrylonitrile	21.4	Methanol	100.0
Allyl chloride	0.47	Methyl chloride	(above 10 ppm)
Amine, dimethyl	0.047	Methylene chloride	214.0
Amine, monomethyl	0.021	Methyl ethyl ketone	10.0
Amine, trimethyl	0.00021	Methyl isobutyl ketone	0.47
Ammonia	46.8	Methyl mercaptan	0.0021
Aniline	1.0	Methyl methacrylate	0.21
Benzene	4.68	Monochlorobenzene	0.21
Benzyl chloride	0.047	Nitrobenzene	0.0047
Benzyl sulfide	0.0021	Paracresol	0.001
Bromine	0.047	Paraxylene	0.47
Butyric acid	0.001	Perchloroethylene	4.68
Carbon disulfide	0.21	Phenol	0.047
Carbon tetrachloride (chlorination of CS_2)	21.4	Phosgene	1.0
		Phosphine	0.021
Carbon tetrachloride (chlorination of CH_4)	100.0	Pyridine	0.021
		Styrene (inhibited)	0.1
Chloral	0.047	Styrene (uninhibited)	0.047
Chlorine	0.314	Sulfur dichloride	0.001
Dimethylacetamide	46.8	Sulfur dioxide	0.47
Dimethylformamide	100.0	Toluene (from coke)	4.68
Dimethylsulfide	0.001	Toluene (from petroleum)	2.14
Diphenyl ether (perfume grade)	0.1	Tolylene diisocyanate	2.14
Diphenyl sulfide	0.0047	Trichloroethylene	21.4
Ethanol (synthetic)	10.0		

Table IV Odor Thresholds of Petrochemicals in Air (ppm Volume)

Compound	*ppm*	*Compound*	*ppm*
Acetic anhydride	0.36	Isobutanol	2.05
Acetone	140	Isobutyl acetate	0.50
Acetophenone	0.60	Isobutyl acrylate	0.012
Acrylic acid	1.04	Isobutyl cellosolve	0.191
Amyl acetate, primary (mixed isomers)	0.21	Isobutyraldehyde	0.236
Amyl alcohol	1.0	Isodecanol	0.042
1,3-Butadiene	1.3	Isopentanoic acid, mixed isomers	0.026
n-Butanol	2.0	Isophorone	0.54
2-Butanol	0.56	Isopropanol (anhydrous)	28.2
Butyl acetate	0.037	Isopropyl acetate	0.97
n-Butylamine	0.24	Isopropylamine	0.95
Butyl cellosolve	0.48	Isopropyl ether	0.053
Butyl cellosolve acetate	0.20	Mesityl oxide	0.051
n-Butyl chloride	16.7	Methanol	53.3
n-Butyl ether	0.47	Methyl amyl acetate	0.40
Butylene oxide	0.71	Methyl amyl alcohol	0.52
Butyraldehyde	0.039	2-Methyl butanol	0.23
Carbitol acetate	0.263	Methyl cellosolve	0.40
Carbitol solvent	1.10	Methyl cellosolve acetate	0.64
Cellosolve acetate	0.250	Methylethanolamine	3.4
Cellosolve solvent	1.3	Methyl ethyl ketone	6.0
Cumene	0.047	2-Methyl-5-ethyl pyridine	0.010
Cyclohexanone	0.24	Methyl isoamyl alcohol	0.20
Diacetone alcohol	1.7	Methyl isoamyl ketone	0.070
Di-*N*-butylamine	0.48	Methyl isobutyl ketone	0.28
Dicyclopentadiene	0.020	Methyl methacrylate	0.34
Diethylamine	0.06	2-Methylpentaldehyde	0.136
Diethyl ethanolamine	0.04	2-Methyl-1-pentanol	0.082
Diisobutyl carbinol	0.160	α-Methyl styrene	0.156
Diisobutyl ketone	0.31	Morpholine	0.14
Diisopropylamine	0.85	2,4-Pentanedione	0.024
Dimethyl ethanolamine	0.045	*n*-Pentanol	0.31
1–4,Dioxane	5.7	2-Picoline	0.046
1–3,Dioxolane	128.0	*n*-Propanol	0.13
Di-*N*-propylamine	0.10	Propionic acid	0.034
Ethyl acetate	13.2	*n*-Propyl acetate	0.15
Ethyl acrylate	0.00036	Propylene	67.6
Ethylamine (70–72% in water)	0.83	Popylenediamrine	0.067
Ethylene	700	Propylene dichloride	0.60
Ethylenediamine	11.2	Propylene oxide	35.0
Ethylene dichloride	40.0	Propionaldehyde	0.080
Ethylene oxide	500	Styrene	0.15
2-Ethylbutanol	0.77	Styrene oxide	0.40
2-Ethylhexanol	0.138	Tetraethyl *o*-silicate	7.2
Ethylhexyl acetate	0.21	Toluene	1.74
2-Ethylhexyl acrylate	0.18	Triethylamine	0.28
Ethylidene norbornene	0.073	Ucon-11 (trichloromonofluoromethane)	209.0
2-Ethoxy-3,4-dihydro-1,2-pyran	0.60	Ucon-113 solvent (trichlorotrifluoroethane)	135.0
N-Ethyl morpholine	0.25	Vinyl acetate	0.55
Glycol diacetate	0.321	Xylene	0.27
1-Hexanol	0.09		

In 1967, A. D. Little, Inc. made the study referred to in Section V, D. The samples tested were supplied by MCA member companies and were of commercially available purity. The threshold level determined by this study is the recognition threshold for 100% of the observers. Because of the low background odor level in the test chamber, this is considered to be a conservative estimate of the threshold level which would be found in ambient air. The data found in this study are presented in Table III. MCA plans to continue this study and extend the list of chemicals for which threshold levels will be obtained.

Hellman and Small (*42*) report the odor properties of 101 petrochemicals. The authors describe the sample presentation device as an "odor fountain." It is similar in concept to sniffing techniques (other than the syringe technique) described in Section V,B except that they claim that the use of a venturi followed by a dwell chamber improves mixing and eliminates the possibility of an odor gradient at the "fountain." The data found in their study are presented in Table IV and the threshold level given is the recognition threshold for 100% of the observers, as with the A. D. Little study. For eight compounds common to both tables, the results compare quite well.

Work is also being done in Japan on the subject of odor measurement. Many papers on the subject—in Japanese—may be found in the *Journal of the Japan Air Cleaning Association* and the *Japanese Odor Research Journal.*

A critical review of the use of odor measurement techniques applied to regulations for the control of odors has been made by Leonardos (*43*). He concludes that odor measurement must include not only threshold values, but also intensity and quality. At present, odor regulations are quite varied and seem to be relatively unsatisfactory, but so far the various aspects of the problem identified by the Stockholm (*3*) and Cambridge (*1*) conferences have not been adequately investigated.

REFERENCES

1. D. A. Kendall and T. Lindvall, eds., "Evaluation of Community Odor Exposure" (Report of a symposium sponsored by the U.S. Environmental Protection Agency). Arthur D. Little, Cambridge, Massachusetts, 1971.
2. A. Dravnieks and P. Laffort, *Proc. Int. Sym. Olfaction Taste, 4th, 1971* (1971).
3. The 3rd Karolinska Institute Symposium on Environmental Health, "Methods for Measuring and Evaluating Odorous Air Pollutants at the Source and in the Ambient Air," Department of Environmental Hygiene. Karolinska Institute, Stockholm, Sweden, 1970.

4. J. E. Yocom and R. A. Duffee, *Chem. Eng. (London)*, 161 (1970).
5. S. H. Katz and V. C. Allison, *U.S., Bur. Mines, Tech Pap.* **267,** 1–22, (1920).
6. S. S. Stevens, *Psychol. Rev.* **64,** 153–180 (1957).
7. A. Dravnieks, *Tappi* **55,** No. 5 (1972).
8. R. A. Duffee, *Proc. 60th Annu. Meet. Air Pollut. Contr. Ass.* **18**(17), 472 (1967).
9. A. Dravnieks, R. Stromberg, and B. Ellis, 154th Nat. Meet. American Chemical Society, Washington, D.C., 1967.
10. H. Zwaardemaker, *Arch. Anat. Physiol., Physiol. Abt.* **31,** Suppl. 59 (1907).
11. J. F. Byrd, H. A. Mills, C. H. Schellhase, and H. E. Stokes, *J. Air Pollut. Contr. Ass.* **14,** 509–516 (1964).
12. J. F. Byrd, *J. Air Pollut. Contr. Ass.* **7,** 59–59 (1957).
13. V. E. Gex and E. A. Fox, *J. Air Pollu. Contr. Ass.* **7,** 60–61 (1957).
14. W. C. L. Hemeon, *J. Air Pollu. Contr. Ass.* **18,** No. 3 (1968).
15. D. M. Benforado, W. J. Rotella, and D. L. Horton, *J. Air Pollu. Contr. Ass.* **19,** No. 2 (1969).
16. National Council of the Paper Industry for Air and Stream Improvement, Inc., "Evaluation of the Use of Humans in Measuring the Effectiveness of Odor Control Technology at the Source," Tech. Bull. No. 56. NCASI, New York, New York, 1971.
17. R. G. H. Prince and H. H. Ince, *J. Appl. Chem.* pp. 314–321 (1958).
18. T. Lindvall, "On Sensory Evaluation of Odorous Air Pollutant Intensities." Uno S. Andersons Tryckeri AB, Stockholm, Sweden, 1970.
19. J. L. Mills, R. T. Walsh, K. D. Luedtke, and L. K. Smith, *Proc. 56th Annu. Meet. Air Pollut. Contr. Ass.* p. 467 (1963).
20. H. Stome, *J. Food Sci.* **28,** 719–725 (1963).
21. S. H. Katz and E. J. Talbert, *U.S., Bur. Mines, Tech. Pap.* **480** (1930).
22. C. P. McCord and W. N. Witheridge, "Odors: Physiology and Control," pp. 45–56. McGraw-Hill, New York, New York, 1949.
23. J. P. Snyder and V. E. Gex, *Chem. Eng.* **59,** 200–201 and 374 (1952).
24. "Standard Method for Measurement of Odor in Atmospheres (Dilution Method)," ASTM D-1391-57, pp. 185–188. Amer. Soc. Test. Mater., Philadelphia, Pennsylvania, 1959.
25. A. Dravnieks and W. H. Prokop, Paper, 66th Annu. Meet. Air Pollution Control Association, Pittsburgh, Pennsylania (1973).
26. R. Cederlof, M. Edfors, L. Friberg, and T. Lindvall, *Proc. 57th Annu. Meet. Air Pollut. Contr. Ass.* pp. 65–107 (1965).
27. F. V. Wilby, *Proc. A.G.A. Operating Sect. Prod. Conf.* American Gas Association (1964).
28. A. D. Little, Inc., "Research on Chemical Odors." Mfg. Chem. Ass., Washington, D.C., 1968.
29. A. Dravnieks, B. K. Krotoszynski, J. Whitfield, A. O'Donnell, and T. Burgwald, *Environ. Sci. Technol.* **5,** 1220–1222 (1971).
30. A. Dravnieks, *Proc. Int. Pollut. Eng. Congr., 1972* (1972).
31. J. F. Byrd, *Eng. Bull. Purdue Univ. Eng. Ext. Ser.* **91,** 114–128 (1956).
32. C. W. Gruber, G. A. Jutze, and N. A. Huey, *J. Air Pollut. Contr. Ass.* **10,** 327–330 (1960).
33. "Report on Interstate Air Pollution in the Selbyville, Delaware-Bishop, Maryland Area," pp. 13–15. Abatement Branch, Div. Air Pollut., United States Department of Health, Education, and Welfare, Washington, D.C., 1965.

34. N. A. Huey, Paper, 61st Annu. Meet. Air Pollution Control Association, Pittsburgh, Pensylvania, 1968.
35. L. A. Clarenburg, *Atmos. Environ.* pp. 333–351 (1973).
36. S. W. Hortsman, R. F. Wromble, and A. N. Heller, *J. Air Pollut. Contr. Ass.* **15**(6), 2101 (1960).
37. J. J. Sableski, *Air Eng.* pp. 16–18 (1968).
38. J. Halitsky, *J. Air Pollut. Contr. Ass.* **12**, 74–80 (1962).
39. J. F. Byrd and E. L. Dewey, *Chem. Eng. Progr.* **53**, 447–451 (1957).
40. "Air Pollution Abatement Manual, Physiological Effects," Chapter 5. Mfg. Chem. Ass., Washington, D.C., 1951 (out of print).
41. H. F. Droege, Paper, 60th Annu. Meet. Air Pollution Control Association, Pittsburgh, Pennsylvania, 1967.
42. T. M. Hellman and F. H. Small, *J. Air Pollut. Contr. Ass.* **24**(10), 979–982, (1974).
43. G. Leonardos, *J. Air Pollut. Contr. Ass.* **24**, 456–468 (1974).

GENERAL REFERENCE

A. Turk, J. W. Johnston, Jr., and D. G. Moulton, eds., "Human Responses to Environmental Odors." Academic Press, New York, 1974.

Part B

AMBIENT AIR SURVEILLANCE

9

Ambient Air Quality Surveillance

Robert J. Bryan

I. Introduction

Air quality monitoring has increasingly come to be interpreted as air sampling and analysis accomplished with the use of continuous automatic analyzers. In the broadest sense, however, the term monitoring can be defined as any systematic, planned means of determining the quality of the air using whatever methods are consistent with the obtaining of data over a reporting interval and at a frequency of collection required by the monitoring objectives involved. Therefore, the use of manually operated batch samplers or automated, sequential batch sampling devices may very well serve the purposes required for air quality monitoring when method specificity, time resolution, and frequency of reporting requirement are met.

Significant developments continue to take place in all phases of air quality monitoring, including more objectively defined sampling strategies, utilization of a wide variety of detection techniques, better definition of performance requirements, availability of improved calibration procedures, and the increased utilization of digital and computer-based data acquisition systems.

One significant problem, in fact, is related in part to the proliferation of analytic methods. This has to do with the determination of equivalency when legalized air quality standards contain a specified method of sampling and analysis to determine the concentration of the contaminant covered by the standard. This subject is treated in more detail later in the chapter, but it seems reasonable to state that insufficient data are available at the present time upon which to make an exclusive selection of one method for each major air contaminant at the expense of rejecting all others.

Together with the development of new and/or refined monitoring methods increasing attention is being devoted to sample site selection and to determination of the number of sites required to give reasonable confidence limits in the definition of pollutant exposure to various receptors and to meet other monitoring objectives. Air quality monitoring has not reached the point where a combination of personal dosimeters and long-open-path sensors have eliminated conventional point monitoring instruments.

II. Basic Considerations

A. Objectives

1. Standard Evaluation

Air quality is increasingly being judged against legally adopted standards. These standards may be designed to provide a guide for defining acceptable air quality or may set forth values beyond which specific acute effects are expected to result. In either case these standards always involve an ambient concentration term. In most cases a time averaging term is also specified, such as an hourly, 8-hour, or annual mean. A reference method is usually specified with any substitute requiring evaluation to determine equivalency.

In the case of emergency standards designed to protect against acute incidents there is a further implied requirement that data be available on a real-time basis. As a result of this requirement, and in part stimulated by the advantage of decreasing manual data manipulation, digital data acquisition systems and telemetry are increasingly being utilized for networks requiring real-time data reporting.

2. Source Detection

Monitoring to detect or determine the effect of a particular source is distinguished by the siting strategy requirements (*1*). There must be sufficient resolution in time and space to distinguish the effect on air quality of the source over advected or background levels. Supplementary information on source strength (emission rate), point of discharge (particularly height above ground), and meteorological influences (wind speed, direction, atmospheric stability) is necessary to design such networks. Diffusion modeling is often carried out using source and meteorological data.

3. Effects Surveillance

Monitoring to determine effects of air pollution requires coordination with the effects study in terms of number and distribution of receptors, suspected nature and severity of effect, time-dosage aspects of expected effect, and presence of confounding factors. A typical example of this type of monitoring is that conducted in conjunction with prospective epidemiological studies (*2*).

4. Background and Trend Data

Monitoring may be conducted to establish current geographical and cyclic patterns or to evaluate trends in air quality. This objective is particularly important in the gathering of data to estimate the impact of future actions possibly affecting pollutant emissions, to determine whether there are any natural or induced cyclical patterns in contaminant levels, and to evaluate the effect of air pollution control strategies.

For this type of monitoring prime consideration must be given to the areal distribution of monitors, and to the frequency and duration of sampling necessary to produce data of the desired statistical reliability.

5. Evaluation of Atmospheric Processes

Many air contaminants undergo changes during their residence time in the atmosphere due to chemical or physical processes. These may include formation of ozone through the photochemical reactions of reactive organic species and nitrogen oxides, the conversion of sulfur dioxide to sulfuric acid or other sulfates, and formation or transformation of aerosols by chemical and condensation mechanisms. In order to understand more fully these processes and indeed the fate of air contaminants, monitoring for highly specific species is often required. Significant attention must be paid to temporal and spatial resolution of measurements made for these purposes in order that the dynamics of the processes being examined can be evaluated on an equivalent basis. Increasing attention is being paid to the vertical as well as horizontal distribution of contaminants in this type of study.

B. Monitoring Strategy

1. Time Resolution

All monitoring methods involve some departure from real ambient behavior insofar as time resolution of data is concerned. That is, no sampling or monitoring device can faithfully represent all the variability that exists in ambient air quality. Saltzman (*3*) has examined the effect of sample averaging time on the extent to which actual contaminant variability is represented in the displayed or reported results. In his analysis, the ambient variability of a given air contaminant is represented by a sine wave of cycle period, t_p, and the effective sample period (or averaging time of the analyzer) by t_s. The analytical expression for the variance transmittance factor is

$$\frac{S^2}{\sigma^2} = \frac{1}{2\pi^2}\left(\frac{t_p}{t_s}\right)^2\left[1 - \cos\left(\frac{2\pi}{t_p/t_s}\right)\right] \tag{1}$$

where t_p = cycle period and t_s = sample time. The graphic results of this analysis (Fig. 1) plot the variance transmittance factor as a function of the ratio of contaminant cycle period to sample period. An examination of this plot shows that at ratios of cycle period to sample period <1 almost none of the actual fluctuations in contaminant concentration about the mean are represented in the results. At ratios above 5.6, however, over 90% of the variability is transmitted.

Thus it becomes important to evaluate the need for time resolution

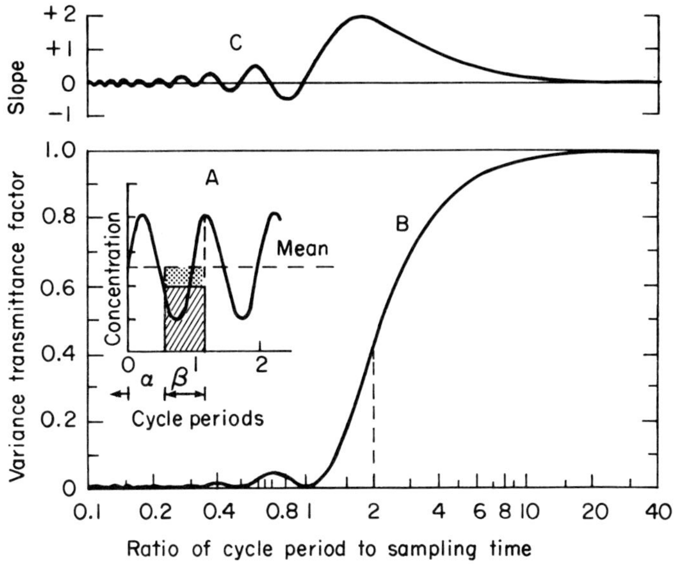

Figure 1. Variance transmittance plot. A: Simulated concentration plot (sine curve); B: variance transmittance plot applied to curve A; C: slope of curve B.

in relationship to the sampling objective during the design of monitoring networks. For cumulative effects, such as corrosion, or chronic effects, such as lead toxicity, the time resolution afforded by continuous analyzers would not be required, whereas for sensory irritants, such as sulfur dioxide or ozone, resolution to time periods less than an hour could be necessary.

Consideration of time resolution requirements is necessary also in the evaluation of source behavior or in atmospheric reaction studies utilizing simulation modeling. In these cases significant cycle periods may well be as short as an hour. If one desires to resolve the structure so that 90% of the variability is detected, the criteria given by Saltzman's expression indicates that an effective sampling period of approximately 10 minutes (60 minutes/5.6) is necessary.

For evaluation of air quality against standards having as their shortest averaging time a period of 1 hour, and for resolution of diurnal pollution patterns, current practice in data acquisition for population-oriented monitoring networks of providing hourly means seems to be quite adequate.

2. Space Resolution

Most air quality data originate from a specific point having some positive means of fixing the location by use of an address or a set of coordinates such as latitude and longitude or those developed from the Universal Transverse Mercator (UTM) system. When obtained by use of a device which samples the air from a fixed probe, those data, in the strictest sense, can be associated only with that specific point. It is common, however, to associate a much larger geographical area as being represented by that point. In the case of a representative site, located in an area of well-distributed air pollutant sources, and having variable meteorological conditions, this assumption may be reasonable. The reasonableness of the assumption improves as the time base for averaging results increases (*4*).

In the case of fixed point sampling, the only way of improving space resolution of air quality data is to increase the number of samplers used. From a theoretical point of view the number of points could be expanded to an extent that the resolution required by any objective, such as determining the maximum exposure of the population or measuring the impact on air quality from a specific source, could be easily achieved. In reality, resource limitations require some compromise in the space resolution of air quality data. This compromise may be expressed in terms of the probability that certain confidence limits can be applied to the data when used to characterize the air quality of an area being represented by a fixed sampling point.

Other means of improving space resolution have been applied through the use of long-path monitoring. Most of the techniques used to date rely on the sensing of absorption or scattering of electromagnetic radiation. Both active and passive techniques have been used. In the case of active devices, energy is actually transmitted by the system into the atmosphere; while passive devices rely on radiation already present in the atmosphere.

In the case of absorption techniques, whether active or passive, an integrated concentration value of the contaminant being monitored over the path is obtained as an output from the monitoring system. If, for example an active system having a path length of 1 km is being used, the output would be in terms of the mean concentration over the path. This would be equivalent to having data from an infinite number of point sensors located along that path from which to calculate a similar average value.

For validating the predictions of air quality models, the open-path approach may have significant advantages as compared to the use of point sampling devices. First of all, many models use area source data as an input which when coupled with the uncertainties in model output make reliance upon a relatively few point sampling monitors for model validation a rather risky approach unless data are collected over a very long time period.

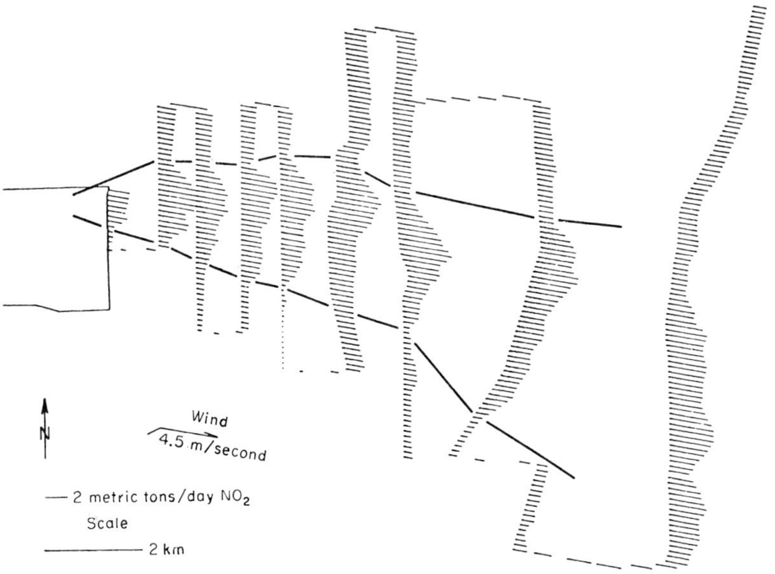

Figure 2. Correlation spectrometer traverse plot—nitrogen dioxide emissions from point source.

One example of the type of space resolution which may be obtained with open-path sensors is shown in Figure 2. In this the sun was used as a source of radiation in conjunction with a van-mounted correlation spectrometer, thus producing integrated data over the nearly vertical path between the sun and the van.

Another approach to open path monitoring involves the use of a LIDAR (an acronym for Light Detecting and Ranging). It is similar to RADAR (an acronym for Radio Detection and Ranging) but uses a pulsed laser system as a source and an optical telescope as the receiver. The laser propagates a pulse of energy into the atmosphere which interacts with aerosols and molecules, which, in turn, scatter light in all directions. A small portion of radiation is backscattered toward the receiver. Monitoring the backscattered energy in time and wavelength provides information on both the range and identity of the scattering species. Means of comparing open-path and fixed-monitoring data have been proposed (*5*).

3. Mobility

The use of mobile or portable air monitoring facilities increases the flexibility of geographical placement of such equipment. It does not increase the total amount of data available. In essence a choice must be made between obtaining greater time coverage at a fewer number of points as compared to reduced time coverage at a greater number of points. Some of the situations in which mobility would seem to be a clear advantage include: (a) emergency episodes—deployment may be called for of additional monitoring equipment at points of suspected major impact. In these cases both mobility and fast response capability are important; (b) environmental impact assessment—contaminant data may be required in areas that might possibly be impacted by existing facilities or planned developments. This type of monitoring is generally highly localized and is commonly collected over time periods of a year or less; (c) monitoring site selection—preliminary data obtained by the use of temporary monitoring facilities are an extremely useful supplement to source, population, and meteorological data in planning the location of fixed monitoring sites; (d) background monitoring—periodic surveys of areas not known to have significant air pollution levels can provide valuable data on possible air quality degradation or on pollutant transport and reactions; and (e) plume dispersal studies—mobile monitoring, whether by point sampling technique or by open-path sensor, can aid in plume transport and dispersal studies. This approach should enable a better determination of lateral and possibly vertical spread of the contaminant mass away from the plume axis with less expenditure of resources, at least in some cases. In the conventional approach to validation of plume dispersion models a fixed grid of ground-based monitors or sampling devices is generally utilized. The strategy for placing these devices is usually set by allowing for some variation in meteorological conditions, and thus may not be optimal for any given condition. Also there may be a problem with elevated sources in evaluating the effect of the source under consideration as compared to other sources contributing to the measurement made some distance from the source. Mobile equipment can be used to help locate the plume by conducting a systematic series of traverses and longitudinal movements which are based upon existent meteorological conditions. An aircraft-based open-path or point sensor can aid in plume definition when the plume is still aloft.

C. Specifications and Definitions

Major problems have been encountered in defining monitoring instrument performance in terms that have widespread acceptance and that are clear enough in meaning so that an objective determination can be made of physical values associated with the terms. A number of such terms have now come into widespread use and have a degree of official acceptance in cases where certain measurement methods are specified as part of air quality standards. These terms and their generally accepted definitions are

1. *Accuracy.* The degree of agreement between a measured value and the true value; usually expressed as ± percent of full scale.
2. *Precision.* The degree of agreement between repeated measurements of the same concentration, expressed as the average deviation of the single results from the mean.
3. *Noise.* Spontaneous deviations from a mean output not caused by input concentration changes.
4. *Lower detectable limit.* The smallest amount of input concentration that can be detected as the concentration approaches zero; usually two times the noise level at zero concentration.
5. *Lag time.* The time interval from a step change in the input concentration at the instrument inlet to a reading of 90% of the ultimate recorded concentration.
6. *Rise time (90%).* The interval between initial response and time to 90% response after a step increase in the inlet concentration.
7. *Fall time (90%).* The interval between initial response and time to 90% response after a step decrease in concentration.
8. *Zero drift.* The change in instrument output over a stated time period, usually 24 hours, of unadjusted continuous operation when the input concentration is zero; usually expressed as percent full scale.
9. *Span drift.* The change in instrument output over a stated time period, usually 24 hours, of unadjusted continuous operation when the input concentration is a stated up-scale value; usually expressed as percent of full scale.
10. *Interference.* An undesired positive or negative output caused by a substance other than the one being measured.
11. *Linearity.* The maximum deviation between an actual instrument reading and the reading predicted by a straight line drawn between upper and lower calibration points.
12. *Full scale.* The maximum measuring limit for a given operational range.

Given the existence of these definitions, the numerical values associated with them need determining from actual field trials and laboratory experiments. One approach is the use of collaborative studies whereby a number of participants carry out analyses of standard samples and in some cases actually come together to conduct sampling and analysis operations on a common sample, such as a dynamically prepared gas sample introduced into a manifold with a number of sampling ports. In this way a sufficient number of measurements can be made that statistical techniques can be applied to the data, resulting in calculated mean values and confidence limits for the parameter measured. Such studies have been conducted on suspended particulates, sulfur dioxide, carbon monoxide, and ozone measurements (*6, 7*). Similar information has been reported from longitudinal studies conducted by single parties (*8, 9*).

Closely related to the problem of method specifications and definitions is that of methods equivalency. This has become particularly important because of the specification of "reference methods" by regulatory agencies. In order to provide for the use of new and possibly improved measurement techniques a means of systematically comparing new techniques with reference procedures must be used.

Where formal determination of methods equivalency is required before use in specified circumstances, such as to judge the level of air quality against a standard, laboratory procedures have been developed which cover two broad subject areas:

a. Performance characteristics. These concern the performance of candidate methods with regard to such specifications as defined in items 1–12 earlier in this subsection. Particular importance is attached to response to interference. In order to be preliminarily acceptable, performance should not deviate from defined limits.

b. Consistent relationships to reference method. Using a test atmosphere, a candidate method is run along side the reference method. The test atmosphere normally would be ambient air or ambient air "spiked" with a known additional amount of the contaminant to be measured. Specified allowable deviations from the reference method can be stated for different levels of the contaminant.

In the final analysis, the actual determination of methods equivalency should arise from field trial over extended periods of time under a variety of conditions. Classical comparison methods such as analysis of variance, and tests of differences between means should be used for evaluation purposes.

III. System Design

A. Station Siting

1. Placement

The placement of air monitoring stations in a network depends upon the nature of the monitoring objectives such as those described in Section

II. Independent of the purely physical aspect of siting, the criteria for placement should be established so that answers to one or more of the following questions can be obtained:

a. Where are the points of maximum contaminant concentrations?
b. What is the exposure of the population to air contaminants?
c. What is the effect of major sources of pollutants on the distributions of levels of contaminants found?
d. What is the effect of control programs on contaminant levels?
e. Are air contaminants transported into the area of concern?
f. At what points are data most needed to evaluate the possible effects of complex sources and to provide information useful in connection with nondegradation requirements?

There are three major categories of influences that must be considered. These include: (a) source influences, (b) demographic influences, and (c) meteorological influences. The most sophisticated joint use of information concerning these influences involves simulation modeling to predict pollution patterns (*10, 11*).

A variety of information inputs are necessary to develop a siting plan which will provide answers to these questions. Among the more important of these information categories are the following:

a. Stationary sources: present and proposed sites for major sources together with emissions data
b. Mobile sources: present and projected daily vehicle miles assigned to geographical analysis zones
c. Population distribution: present and projected population densities by area
d. Meteorological aspects, particularly with respect to prevalent wind regimes and synoptic conditions giving rise to various pollutant transport modes and stagnation
e. Topography
f. Existing and projected land use patterns
g. Federal and state guideline criteria
h. Location of known air pollution effects
i. Information from past or present air sampling operations

Ideally, the information developed should serve as the input to appropriate air quality predictive models which would produce data to develop plots of pollutant patterns expected under various assumed conditions. This has been done in a number of situations and, even though there are serious limitations to the use of models, their use for this purpose is to be encouraged.

Among the limitations of formal predictive air quality models are

a. There is often insufficient meteorological data available to permit achievement of the maximum accuracies available from the models

b. For primary contaminants such as carbon monoxide and sulfur dioxide, existing models for predicting long-term average concentrations are more reliable than those used for predicting short-term or episode conditions

c. Certain meteorological conditions such as those involving stability breakup or fumigation are not easily modeled

d. Models for the prediction of contaminants formed by atmospheric reaction are in an earlier developmental stage than are those used for nonreactive primary contaminants

Therefore, the applications of formal models must necessarily be complemented by the use of professional judgment in such areas as defining meteorologically similar areas, in assessing the impact of point sources under episode conditions, in identifying areas of maximum potential for development of photochemical air contaminants, and in the evaluation of micrometeorological factors and other aspects of specific site location.

With regard to the problem of evaluating a specific location for its suitability as a monitoring site, a major problem exists in determining whether it is representative of the surrounding area. A typical problem is carbon monoxide monitoring in heavily *trafficked* areas. As Ott has shown, 8-hour carbon monoxide concentrations can vary by a factor of 3 between sites less than 3 km apart (*12*). In fact, along a major roadway or in a traffic *stagnation* area significant differences can be expected over distances of tens of meters. It seems reasonable to suggest that curbside probes are inappropriate, but that locations within 15–30 m may be reasonable if a significant number of people are exposed at those distances. Other highly localized sources which are not representative of the surrounding area should also be avoided, such as exhausts from incinerators and space heaters.

Insofar as physical site criteria are concerned, sample lines or manifold inlets should be 3–6 m above the ground to minimize the effect of reentrained surface dust and to reduce likelihood of damage. Inlets also should not be in the lee of taller buildings located closer than two building heights to minimize the chance of sampling stagnant air.

2. Network Density Factors

One of the most vexing problems in the design of air monitoring networks has to do with the required or desirable spacing of individual sites.

It is clear that no two spots are identical in air quality, but differences among sites may range from "not detectable within experimental error" to order of magnitude. It is also generally accepted that the closer two stations are to one another the more similar their air quality data will be. This, of course, is dependent upon the nature of pollutants being measured (direct or secondary), and other factors. The question with which network designers must cope is—when are these differences great enough to justify another station? In other terms, the question might be phrased: "With what confidence and tolerance limits do we wish to know the air quality for any given point?"

A number of investigators have examined this problem and provided some insight as to the nature of the problem (*13–15*). Stalker (*14*) reported the results of analysis of data from up to 119 sampling sites for sulfur dioxide in Nashville, Tennessee.

One of his conclusions (which should be fairly obvious) is that the number of stations required to define air quality within certain confidence limits was directly proportional to the square of the coefficient of variation in sample values and thus indirectly proportional to the averaging time. His analysis, for example, showed that 245 stations or four stations per square mile would be required to estimate the daily mean concentration of SO_2 at any given point with only 5% chance that the estimate differed more than ±20% from the true value. On the other hand he stated that only two stations would be required to estimate seasonal means with the same confidence and accuracy. Ingels (*16*) found significant differences in peak values for several contaminants in pairs of stations where the distances between the two stations making up the pair ranged from 1 to 3 km. The greatest variation was in sulfur dioxide and the smallest in oxidant. Differences ranged from 5 to 107%. Only sulfur dioxide values differed by more than about 50%. Holland *et al.* (*17*) found rather small differences in oxidant and carbon monoxide when comparing mobile station data to fixed station data when the mobile station was located at four different points, all being about 1 mile from the fixed station. Gordon and Bryan (*18*) found coronene, suspended particulate matter, lead, and benzene-soluble organic matter to be highly correlated with traffic density determined from 2-mi^2 grids. The lowest linear correlation coefficient found was 0.964.

It seems apparent then that there is no perfect station siting plan, but that a plan can be optimized with respect to available resources. A formal operations research approach might be used in which values are assigned to such possible benefits as location of a majority of sites at points at which maximum concentrations might be found, minimizing distance of the population to the nearest air monitoring station, the identification of air parcels originating outside the area of jurisdiction, assessment of impact of point and area sources, and the obtaining of background data in areas suitable for development.

The values associated with various plans would be compared to the cost of the air monitoring networks associated with each plan and an optimization technique used to determine the plan showing the maximum benefit-to-cost ratio. To use the foregoing approach in a completely rigorous manner, a major effort would be necessary to quantify the benefits of providing accurate air quality data.

Pending the development of such rigorous techniques together with the requisite quantified cost and benefit data, an approach involving the use of one or more air quality predictive models and other judgmental efforts to identify locations of probable high concentrations, together with population density maps to determine the location of population with relation to these high exposure locations, can be used. The criteria of acceptability could include: (a) degree of coverage of predicted high contaminant concentration areas, and (b) proportion of the population that resides or works within some stipulated distance from an air monitoring station. Furthermore, regions with poor air quality should have greater numbers of stations per unit area than those with better air quality.

Empirical guidelines for determining density of air monitoring networks have been published both in the literature and in governmental regulations. The United States Environmental Protection Agency (USEPA) (*19*) gives a general guide for number of stations based upon population (Fig. 3). An equation for estimating the number of station N based upon degree of pollution and the land area of the regions, considers three distinct areas; X, where the pollution levels are higher than the ambient air quality standards; Y, where the pollution concentrations are above background level but lower than the standards; and the area Z, where the existing concentrations are at background level. The total number of stations N required for the whole region is obtained by summing the number for each of the three subareas:

$$N = N_X + N_Y + N_Z \qquad (2)$$

The number of stations required for each subarea, N_X, N_Y, and N_Z, are

$$N_X = 0.0965 \frac{C_m - C_s}{C_s} X \qquad (3)$$

$$N_Y = 0.0096 \frac{C_s - C_b}{C_s} Y \qquad (4)$$

$$N_Z = 0.0004\, Z \qquad (5)$$

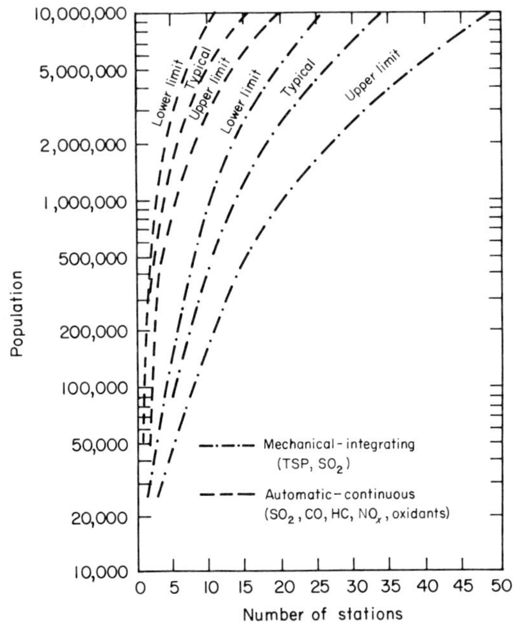

Figure 3. Region population vs number of stations. TSP: total suspended particulate.

where

C_m = Value of maximum isopleth (with a contour interval of 10), $\mu g/m^3$

C_s = Ambient air quality standard, $\mu g/m^3$

C_b = Value of minimum isopleth (i.e., background concentration), $\mu g/m^3$

X = Area wherein concentrations are higher than ambient air quality standard, km^2

Y = Area wherein concentrations are above background level but less than the ambient air quality standard, km^2

Z = Area wherein concentrations are at background level, km^2

The constants used in the above expression have no theoretical basis. Instead, they were developed by an examination of such factors for a number of existing networks which appeared to be reasonably adequate.

Criteria for the minimum number of air monitoring sites based upon a priority classification have been published as USEPA (*20*) regulations. The three priority levels used do not match the definitions given for areas *X*, *Y*, and *Z* in the preceding paragraph. Once the priority of the region has been determined the number of stations required can be calculated by use of the multiplying factors which are given in Table I.

Table I Recommended Number of Air Quality Monitoring Sites

Regional classification	*Pollutant*	*Measurement method*	*Region population*	*Minimum number of air quality monitoring sites*	*Minimum frequency of sampling*
I	Suspended particulates	High volume sampler	Less than 100,000	4	One 24-hour sample every 6 days
			100,000–1,000,000	4 + 0.6 per 100,000	
			1,000,000–5,000,000	7.5 + 0.25 per 100,000	
			Above 5,000,000	12 + 0.16 per 100,000	
		Tape sampler	—	One per 250,000 up to 8 sites	One sample every 2 hours
	Sulfur dioxide	Pararosaniline or equivalent	Less than 100,000	2	One 24-hour sample every 6 days (gas bubbler)
			100,000–1,000,000	2.5 + 0.5 per 100,000	
			1,000,000–5,000,000	6 + 0.15 per 100,000	
			Above 5,000,000	11 + 0.05 per 100,000	
			Less than 100,000	1	Continuous
			100,000–5,000,000	1 + 0.15 per 100,000	
			Above 5,000,000	6 + 0.05 per 100,000	
	Carbon monoxide	Nondispersive infrared or equivalent	Less than 100,000	1	Continuous
			100,000–5,000,000	1 + 0.15 per 100,000	
			Above 5,000,000	6 + 0.05 per 100,000	
	Photochemical oxidants	Gas-phase chemiluminescence or equivalent	Less than 100,000	1	Continuous
			100,000–5,000,000	1 + 0.15 per 100,000	
			Above 5,000,000	6 + 0.05 per 100,000	
	Nitrogen dioxide	24-hour sampling method (Jacobs–Hochheiser method)	Less than 100,000	3	One 24-hour sample every 14 days (gas bubbler)
			100,000–1,000,000	4 + 0.6 per 100,000	
			Above 1,000,000	10	
II	Suspended particulates	High volume sampler	—	3	One 24-hour sample every 6 days
		Tape sampler	—	1	One sample every 2 hours
	Sulfur dioxide	Pararosaniline	—	3	One 24-hour sample every 6 days (gas bubbler
			—	1	Continuous
III	Suspended particulates	High volume sampler	—	1	One 24-hour sample every 6 days
	Sulfur dioxide	Pararosaniline	—	1	One 24-hour sample every 6 days (gas bubbler)

B. *Physical Environment*

1. Shelter

As with any type of scientific equipment, air monitoring instruments will perform more reliably and produce more accurate data when operated under proper environmental conditions. The noise from pumps and data acquisition equipment may disturb workers. Therefore, it is desirable to locate an instrumented multipurpose air monitoring installation in an enclosed, sound-insulated, work space. This may be a room in a building, where exposure and sample line length are acceptable, or it may be in a free standing structure. Increasing use is being made of prefabricated portable buildings which can be transported to the site by truck or tractor. Some of these are mobile in the sense that they have wheels, which can be removed if the station is to remain at the site for some length of time.

The shelter used should be air conditioned to maintain temperature and relative humidity within the range specified for the most sensitive instrument. In specifying the air conditioning load, consideration must be given to the heat load introduced by the operation of the air monitoring equipment. A design temperature of 21°C (70°F) is probably most acceptable for both personnel comfort and instrument needs. Abrupt deviations of more than 2°F or 3°F are undesirable, since some analyzers give spurious response to such rapid changes. A nominal relative humidity of 50% is satisfactory. Normally, humidity is not directly controlled but rather is allowed to "float" with the temperature control. In unusually humid climates, control of humidity may require reheat after refrigeration.

Security and safety considerations include providing safe ventilated storage for gas cylinders, proper fire extinguishers, grounded electrical service, and electrolytically generated hydrogen, where possible.

2. Space

The space provided should be sufficient to enable access to all portions of each instrument requiring service at the site. Particular attention should be paid to providing space for pull-out instrument chassis, for removal and replacement of component parts or entire units, for setup and use of calibration and maintenance equipment, and for cylinder and other storage. Space normally should also be provided for report preparation, and toilet facilities. It is very desirable to have a sink even if there is no connected water or sewer service. The USEPA recommends

Table II Space Allocation—Air Monitoring Stations

	ft^2	m^2
Instrumentation and laboratory area	150	14
Work area (desk and sink)	30	3
Compressed gas cylinder storage	10	1
Toilet facilities	15	1.5
Miscellaneous storage	10	1
	220	20.5

the minimum space allocation for a complete remote air monitoring station to be that shown in Table II. One possible layout of such a station is shown in Figure 4. Roof space and easy access thereto is desirable for locating meteorological equipment, high volume particulate samplers, and other equipment requiring direct exposure.

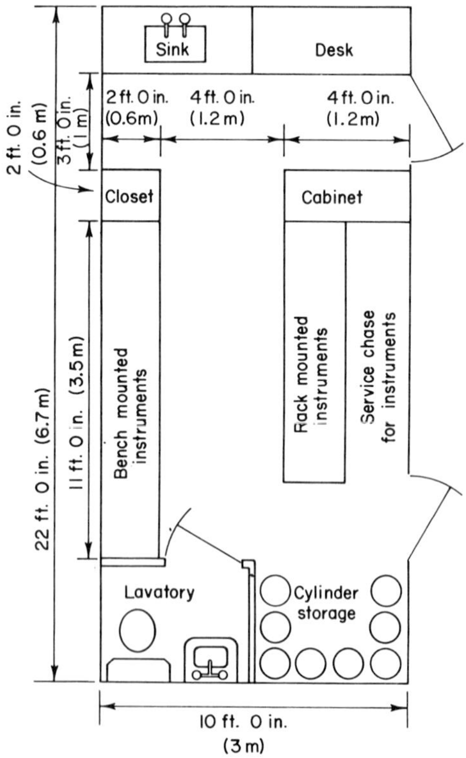

Figure 4. Recommended air monitoring station plan.

3. Utilities

The major utilities' requirement is for electrical power, although both hot and cold running water and a sewer connection are desirable. The electrical service provided should be 230 V, single phase, with amperage sufficient for air conditioning, electrical heating, lighting, air monitoring instruments, sampling manifold blower, meteorological equipment, data acquisition equipment, office equipment, bench equipment, refrigerator, hydrogen generator, water still (if required), and the occasionally used servicing and calibration equipment.

Separate voltage regulated circuits should be provided for each continuous analyzer, and for the data acquisition equipment. Additional convenience outlets should be provided for the intermittantly used maintenance, calibration, bench, and office equipment. Weatherproof receptacles should be located near the rooftop sampling area. Under normal circumstances, all of the equipment operated in an air monitoring station, with the exception of air conditioning units and possibly electric strip heaters and water stills, operate on 115-V, single-phase electrical power. All but the smaller air conditioning units require 230-V single-phase service.

4. Service and Support

As contrasted with the servicing and maintenance requirements for the air sampling instruments themseleves, the support requirements for air monitoring stations are relatively simple. These would include normal building maintenance, custodial service, and care of air conditioning and heating equipment. In cases where reagents are used it may be necessary to remove spent reagents for disposal at another site, depending upon the facilities available at the site. Further, sample probes and manifolds will have to be cleaned at regular intervals.

5. Influence Quantities

Performance specifications for air monitoring instruments are usually expressed so as to be valid only under certain restrictions as to the conditions of the physical environment. These conditions should be understood and considered when designing or establishing air monitoring sites and shelters.

In order that performance specifications for air quality analyzers be more meaningful the International Electrotechnical Commission (*21*) has described a number of "influence quantities" which should be spelled out when performance specifications are given. In general these influences are defined as being "any quantity, generally external to an apparatus, which may affect the performance of the apparatus." The more important of these influences are (a) ambient temperature; (b) relative humidity; (c) barometric pressure; (d) solar radiation; (e) sand and dust content of air; (f) salt content of air; (g) presence of gaseous contaminants; (h) operating position; (i) ventilation; (j) vibration; (k) mechanical shock; (l) sound pressure; (m) power supply voltage; (n) power supply frequency; (o) distortion of a.c. power supply; and (p) presence of electric fields. It should be noted that with the exception of barometric pressure these influence quantities can be controlled by properly designing the facilities necessary for the housing and operation of air sampling instruments. Care should be taken to eliminate or reduce the effect of these influences as much as is possible.

C. Sampling Lines and Manifolds

When an air monitoring instrument is located within a shelter or building, the air sample will have to be brought to the analyzer through some sort of sampling line. It is important that the sample not be degraded or diluted as a result of this passage. Furthermore, there should be a minimum time delay between entry of the air sample to the sampling system and its delivery to the analyzer.

Sample degradation can result from reaction between the contaminant being sampled and the sample line material, e.g., decomposition, adsorption, and diffusion. Atmospheric particulates deposited in the sample line can contribute to these losses. The more reactive air contaminants such as ozone, nitrogen dioxide, and sulfur dioxide are most susceptible to degradation-type losses. All joints and seals in the sample system must be tightly sealed to prevent infiltration of room air into the sample system, thereby diluting the desired outdoor air sample.

A number of generalizations can be made about sampling systems: (a) The manifold design should minimize wall losses. (b) Flow should substantially exceed the total withdrawn by the analyzers withdrawing samples from the manifold. (c) Pressure inside the manifold should be nearly atmospheric. (d) Construction should be such as to completely avoid leaks. (e) Material and construction should facilitate cleaning. (f) The entry should be designed to avoid direct entrance of precipitation.

The most common approach to achieving these goals has been to provide a glass manifold which consists of an entry section which extends through the roof or wall of the shelter, a moisture trap, and a series of modular sections each having several sample connections (*22*). A small integrally constructed squirrel cage exhaust fan is attached to the interior

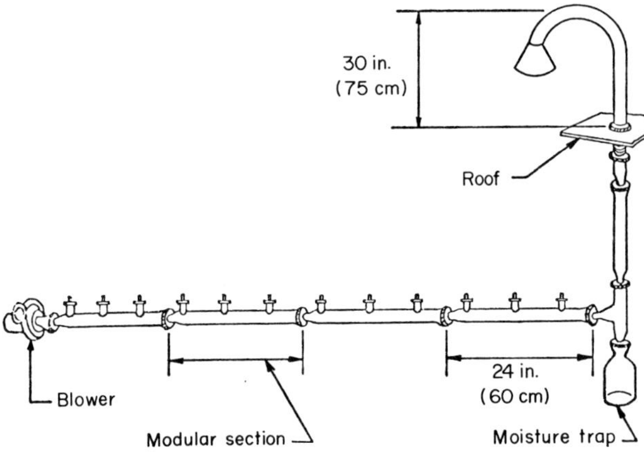

Figure 5. Conventional air sampling manifold.

terminal point of the manifold to provide a continous flow of ambient air through the manifold. The volumetric flow rate should be three to five times the total sampling requirements of all the monitors attached to the manifold. The diameter of the manifold is generally about 50 mm. In any case the manifold should be sized so that a minimal pressure drop exists between the manifold entrance and the exhaust blower. This is important to reduce the chance for leaks, to minimize wall losses which are enhanced by turbulent flow, and to provide a nearly ambient pressure sample at the analyzer. Sample lines from the manifold to the monitor should be glass or a nonreactive material such as Teflon. A typical manifold is shown in Figure 5.

In some cases individual sample lines are used with the sample pump provided in the analyzer being used to draw in the sample. This is satisfactory where short runs (generally not exceeding 2–3 m) of sample line will reach an acceptable outside sample entry point. It should be remembered that as the ratio of the sample system volume to sample rate increases, so will the mixing of the sample increase. Deliberate additions to sample system volume in the form of mixing chambers or surge volumes have sometimes been used to achieve a partial integration of the sample. A specific example is that of carbon monoxide monitoring where extremely rapid variations of concentrations are smoothed to facilitate data reduction.

D. Data Acquisition

Data acquisition is that portion of the data processing function which deals with the transformation of what is usually a continuous analog signal from a monitor to an intelligible format suitable for further examination or reduction to conventional formats: The data reduction and subsequent storage and retrieval operations are treated in Chapter 12, this volume.

1. Display

In continuous automatic air monitoring instruments data are nearly always displayed in real-time form at the point of initial acquisition. This may be in the form of an analog or digital meter output, or in the form of an analog or digital permanent record. Sensitive galvanometers are commonly utilized for analog meter output, while digital voltmeters are most often used for digital display. Display is usually in engineering units, e.g., parts per million by volume. In the case of instruments designed to be compatible with telemetry systems, it has become increasingly common for manufacturers to provide meter output as standard equipment, with on-site recording as an extra cost option. The argument for this approach is that, since the data are being recorded at a central point, a remote (instrument mounted) display device is needed only for such occasional purposes as calibration, servicing, and operation observation. The counter arguments are that a permanent real-time record is valuable during servicing and calibration operations to avoid reliance on observer memory and that during performance testing and calibration an excessive amount of manual data recording would be necessary to obtain the required time resolution of the data. Furthermore, it is common in many telemetry systems to provide data recording only at intervals of a few minutes. While this may be adequate for data reduction purposes, the time resolution necessary to determine signal-to-noise ratios can usually be obtained only with a fast response analog recorder.

2. Recording

The most common type of on-site data recording device is the null-balance recording potentiometer. This is a servo-operated device in which the input from the sensor in the form of a continuously varying analog voltage signal is automatically compared to a reference signal which is varied by the servo system. At the point where the input signal is equal to the reference signal, the system is said to be "nulled." That is, there is zero difference between the input and reference signals and no current flows. A scale pointer and recording device (usually a pen) are mechanically linked to the balancing system, thus producing a visual and permanent record. These recorders must be matched to the input system for input voltage range and should have a chart and pen speed to provide clarity of record and a response time fast enough to represent faithfully the variations in input signal. Where telemetry systems are used to obtain the data actually input into the data processing system, it may be more economical to use one or more multipoint recorders at each site to obtain the field data necessary during calibration and trouble shooting operations.

Several other types of remote-site recording systems have been used or suggested. These include digital tape printers, magnetic tape cassette units, and punched paper tape units. The latter two produce records which are not directly readable at the site, but which are physically transported to a central site where they are read into a computer system. Using this approach the automated data reduction aspects, but not the data transmission features, of a full telemetry system are employed. Where real-time data are not required at a central point and where there is no particular advantage in being able to transmit automated calibration system commands to remote sites, these computer compatible recording systems may offer some advantages in cost. In some cases cassette recording devices are incorporated into telemetry systems. Commands for remote station readout can be issued from the central site at any convenient time after recording.

3. Data Transmission

Data produced at remote air monitoring sites must be transmitted to the point at which immediate use of the data is to be made or at which the data are further processed and stored. Much data are transmitted by telephone, messenger, or mail, particularly when there is no immediate or real-time use for the data and when travel distances and servicing schedules are such that there is no inconvenience or threat to data integrity when such means of communication and transmission are used.

There are circumstances, however, that argue in favor of the use of telemetry systems for transmission of data from remote sites to a central site for display, action, or input to the data reduction system (Fig. 6). Some of these circumstances include: (a) real-time need for data to activate emergency action systems and for use in forecasting or simulation modeling operations, (b) need for early warning of monitoring system failure when sites are visited relatively infrequently, and (c) use of the telemetry system in reverse mode to activate remote calibration systems when stations are serviced infrequently.

In the simplest terms, telemetry can be viewed as a means of extending

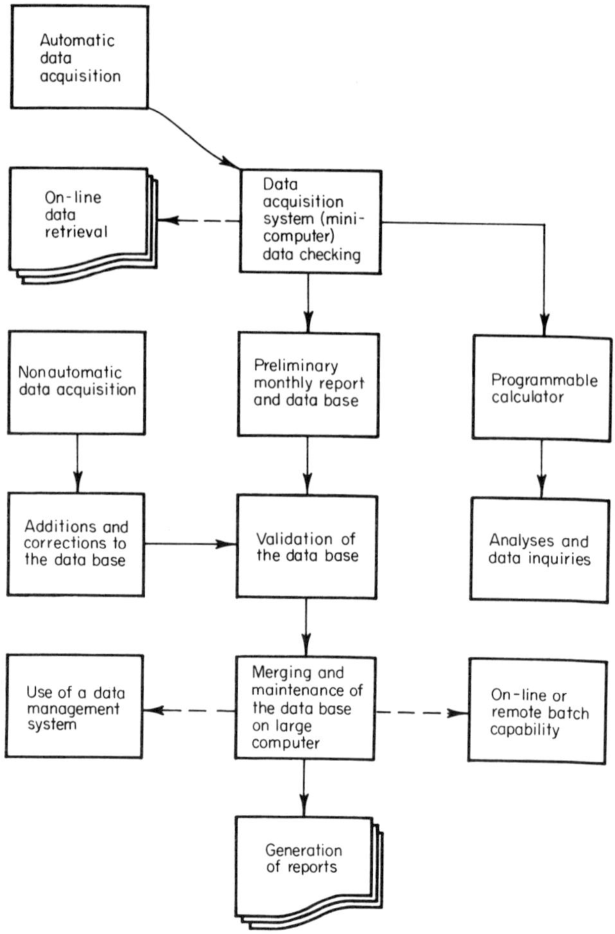

Figure 6. Air quality information management system data flow. Dashed line indicates optional features.

the distance between the sensor and the data recording or display device used. Earliest systems used analog transmission of data over land lines or radio links. In these systems an analog signal was transmitted which varied in proportion to the signal from the sensor. Today, the telemetry, or communication link, is merely one element of an automated data acquisition system, and essentially all such systems operate in the digital mode. Analog-to-digital conversion takes place at the remote site and transmission is in digital format.

A variety of data acquisition and telemetry systems is in use, ranging from rather simple transmission and display of raw data to very sophisticated systems which perform nearly all the data processing operation, including the generation of routine statistical summaries.

These systems are of three basic types. In the first type, raw data are transmitted in a logical sequence to a central site where data are displayed in hard-copy format such as a teletypewriter output and/or a cathode ray tube (CRT). In this type system, the telemetry is used solely for real-time purposes such as for input to an air pollution emergency control plan or for the forecasting of air pollution potential. These forecasts may be used for general information to the public, but most often have the function of determining whether certain operational air pollution control measures such as curtailment of agricultural burning, fuel switching, or industrial production modifications should be invoked.

In the second type of system, data recording in some computer compatible form such as magnetic or punched paper tape is added. There is no further processing of the data, however, to produce averages or for error checking, and there is no ability to recall data from storage. Instead, the interim forms of data storage must be entered into a computer system programmed to perform these latter steps.

In the third, and increasingly popular, form of automated data acquisition system, a minicomputer (sometimes termed as a central processor) is used at the central data receiving point to receive and process data and to otherwise interface with the system. Some of the operations which can be performed include (a) linearize and scale raw data; (b) average instantaneous data points to produce averages over selected time bases; (c) perform edit routines such as checking for "out-of-limits" concentrations and "excessive" rate-of-climb; (d) flag and print out values above preset limits such as those above the emergency standard; (e) produce hourly and daily summaries; (f) adjust concentrations for zero and span drift as new values for these parameters are fed into the system; and (g) handle command functions such as continuous interrogation of a single remote station and activation of remote calibration equipment.

Some of these systems are designed to produce longer term statistical summaries such as monthly reports and to perform other data analysis functions. In order to maximize the performance of the primary data acquisition functions, it is common to perform most longer term or retrospective data processing on an off-line larger general purpose computer. In the United States, it is the usual practice, at least in the case of air pollution control agencies, to record data in the format specified for the SAROAD (Storage and Retrieval of Air Data) system of the USEPA (*23*). Figure 7 is a schematic diagram of an automated data acquisition system of the third type described above. This diagram is highly simplified and does not attempt to indicate the specific means for accom-

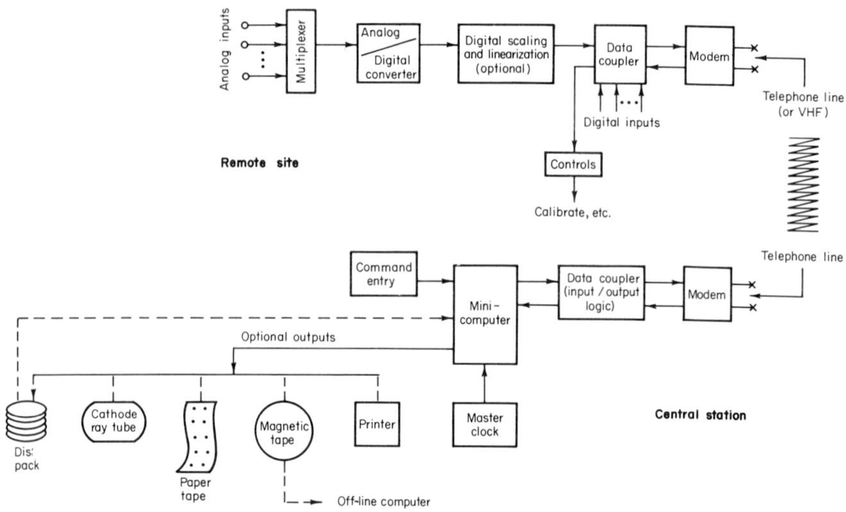

Figure 7. Automated data acquisition system.

plishing the functions shown. For example, some data acquisition systems are packaged while others are modularized. There are a variety of types of multiplexing devices, analog-to-digital converters, and input/output controllers. Further, there are a substantial number of manufacturers producing minicomputers, with most having a variety of options in core memory capacity. Furthermore, it should not be assumed that all of the output devices shown would be procured for any one system.

Another important aspect of these data acquisition systems has to do with the type of software (computer programming) specified. Those using lower-level language such as FORTRAN IV offer greater flexibility of use for engineers and chemists not specifically trained in computer programming but do require greater computer memory than do those systems programmed in assembly language.

IV. Measurement Principles

Methods used in air monitoring can be classified as those primarily involving chemical and those primarily involving physical principles of measurement. However, these terms are not used with clear-cut precision. A number of methods which have been described as physical also involve chemical reactions, e.g., the chemiluminescent techniques in which gas-phase chemical reactions occur, and the flame excitation techniques such as flame photometry and hydrogen flame ionization in which combustion reactions occur. Those methods classed as chemical are in most cases the so-called wet chemical methods which involve interaction between a liquid reagent and the contaminant of interest.

There is a growing shift away from these wet chemical techniques, because, in general, they suffer from problems of interferences, reagent stability, need to control reagent flow rate, complex plumbing, air–reagent contactor efficiency, and relatively slow response time. On the other hand many of these methods are well characterized, sensitive enough as not to require sophisticated electronic amplification, and exhibit little signal drift. Also, initial costs for these instruments are in many cases lower. Thus, some wet chemical analyzers will likely see continued use.

Of the purely physical instruments, principally those involving spectrometric principles, much of the development has taken place in those methods designed for long, open-path measurements. These include use of tuneable laser sources, Raman backscattering, and correlation spectrometry.

A. Gaseous Air Contaminants

1. Colorimetric Analyzers (Spectrophotometry)

a. Principle of Operation. Analyzers based upon colorimetry form a major group of wet chemical instruments. They operate by absorption of the air contaminant in a liquid reagent with subsequent formation of a colored reaction product which is determined by measurement of light absorbance in a flow colorimeter.

Typically, the sample air passes into a scrubber where it comes into contact with the liquid absorbent (Fig. 8). Here the gas reacts with the

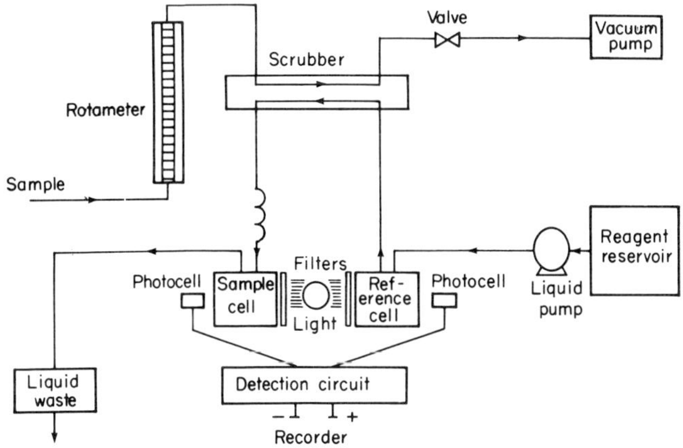

Figure 8. Colorimetric analyzer.

absorbent to form the measurable colored reaction product. There are many different configurations for the scrubber chamber including continuous bubblers, countercurrent flow columns, co-current flow columns, and spray chambers. Some columns use packing to enhance air–reagent contact.

The absorbent then moves to the colorimeter cell where the intensity of the color is measured. The measurement is usually based upon the comparison of the light transmittance through the sample cell to that through a reference cell. In a complete calibration, several standard gas samples containing known concentrations of the specified contaminant are serially introduced so as to develop the relationship between signal output and concentration. The detector portion of the analyzer can be checked by introducing standard solutions of the final reaction product in the absorbing solution to the colorimeter cell. If the overall efficiency of the absorption and reaction is well characterized, relative air and reagent flow rates can be applied to reagent calibration data to produce an overall calibration plot.

The color intensity of the absorption solution can be sensitive to temperature, pH, color development time, purity of reagents, age of solutions, and atmospheric interferences.

b. Uses and Limitations. Colorimetric techniques have been adapted for use in the sampling of almost all pollutants. This adaptability and generally good sensitivity are some of the advantages of colorimetry. The United States Federal Reference Method for manual determination of sulfur dioxide and two of the proposed Federal reference methods for nitrogen dioxide are based upon colorimetric analysis techniques.

There are, of course, disadvantages to the use of colorimetric techniques. Extensive plumbing for both the liquid reagent and the sample air flow are required. There is also an inherent delay in the instrument to allow for color development. Also reagent preparation, transport, and storage can cause significant problems.

2. Electrometric Analyzers

a. Principle of Operation. There are a variety of instruments that are based upon the use of electrochemical reactions (*24–29*). For purposes of description these are subclassified into four categories:

i. *Galvanic coulometric analyzer.* The galvanic coulometric analyzer requires no applied electrical current for the cell. The sample air containing the contaminant flows through an electrolytic solution at one electrode. A balancing reaction occurs at the other electrode causing a small current to flow across the cell. The concentration is determined by the relationship of Faraday's law which states that 1 gm equivalent weight is reduced by 1 Faraday of electricity. By measuring the amount of current across the cell, the concentration of the sample can be calculated by the following relationship:

$$C = \frac{(A)(T)(W_{\mathrm{EQ}})10^6}{96{,}493(\mathrm{m}^3)} \tag{6}$$

where

C = concentration (μg/m^3)
A = current (A)
T = time current passes (second)
W_{EQ} = equivalent weight of pollutant gas (1 gm equivalent weight = atomic weight/valence charge)
m^3 = cubic meters of gas passed during the sample period

ii. *Amperometric coulometric analyzer* (*ozone analysis*). In the principal example of this type of instrument a constant potential is applied across two platinum electrodes with the reagent serving as the electrolyte. The cathode becomes polarized and no current flows. When ozone is present, iodine is released which depolarizes the electrode, thus allowing an amount of current to flow which is proportional to the amount of ozone.

iii. *Bromo-coulometric analyzer.* In this type of instrument an electrolyte is used in which a specific ratio of bromine to bromide ion is maintained by a dual electrode system. One set of electrodes detects this ratio potentiometrically while the other set is used to generate any bromine necessary to maintain the proper balance. The sample air is bubbled through the electrolyte at a fixed rate. Reducing agents in the sample air (primarily sulfur dioxide or hydrogen sulfide) will reduce the bromine to the bromide ion. To maintain the proper balance, the generating electrodes will generate bromine in the amount equivalent to that reduced by the sulfur dioxide or hydrogen sulfide. This current flow can be related to the amount of contaminant through appropriate calibration factors.

iv. *Electrochemical transducers.* These devices depend on selective electrocatalytic oxidation (or reduction) of the pollutant in a cell quite similar to a fuel cell. The sampled air passes by a semipermeable membrane designed to be as selective as is possible for the diffusion of the pollutant of interest. Inside the cell the sample diffuses through a thin liquid film to an electrocatalytic electrode where a charge transfer reaction occurs.

In operation the cell produces an electrical potential that is directly proportional to the concentration of the pollutant and which, within limits, is independent of sample flow rate. Thus unlike the three previous Faraday-law-type electrometric analyzers, these devices are concentration rather than mass dependent.

b. Uses and Limitations. Electrometric analyzers have been most widely used for sulfur dioxide sampling, but have also been used for ozone, carbon monoxide, and nitrogen dioxide. An advantage of these monitors is the relatively low level of maintenance required. In some, the reagent is regenerated by electrolysis so that reagent replacement is relatively infrequent. A limitation of this method is that other reducing or oxidizing agents can react in the electrolyte and cause a direct interference. This requires an efficient scrubbing system to be applied to the inlet to clean the air of unwanted contaminants. By using the correct scrubbing apparatus and electrolyte, the method can be very specific to the desired pollutant.

3. Conductimetric Analyzers

a. Principle of Operation. Conductimetric analyzers are based upon the principle that a solution of a given ionic strength will have a specific measurable resistance. The sample gas passes through the solution and ionizes within it, decreasing the electrical resistance (increasing the conductance). This increase in conductance is measurable and proportional to the concentration of the ionizable gas in the atmosphere (Fig. 9). This

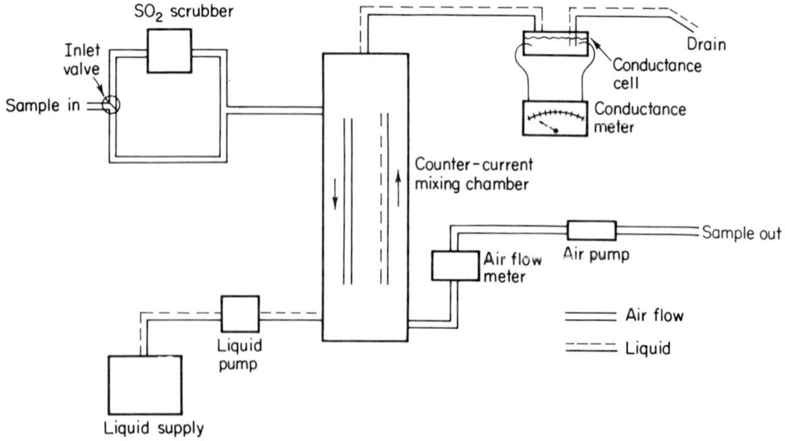

Figure 9. Counter-current type sulfur dioxide conductimetric analyzer.

method is applicable to continuous monitoring, when the conductance is constantly being measured, or to sequential batch-type analysis, when the conductance is measured once during a sampling period and an average concentration is thus derived.

b. Uses and Limitation. Conductimetric analyzers are most widely used for analysis of atmospheric sulfur dioxide. For this application the absorbing solution is either a dilute mixture of hydrogen peroxide in weak sulfuric acid, or deionized distilled water.

In the case of the hydrogen peroxide–sulfuric acid absorbing solution, the following reaction takes place:

$$SO_2 + H_2O_2 \rightarrow H_2SO_4 \quad (7)$$

When plain distilled water is used the reaction takes place according to the following equation:

$$SO_2 + H_2O \rightarrow H_2SO_3 \quad (8)$$

While the technique is simple in concept and operation, it is subject to substantial interference. Acidic gases and salt spray give positive errors, while ammonia interferes negatively.

When sampling sulfur dioxide in distilled water, carbon dioxide will readily ionize and, therefore, provides a direct positive interference. This can be overcome by scrubbing the system periodically of sulfur dioxide and reading a base carbon dioxide concentration. The carbon dioxide reading can then be subtracted from the total reading to yield the sulfur dioxide concentration.

4. Chemiluminescence Analyzers

a. Principle of Operation. Chemiluminescent reactions are those in which light is emitted as the result of the reaction between two chemical species (*30–34*). Both gas–solid-phase and gas-phase reactions have been used—but only the latter forms the basis for current commercial analyzers. In gas-phase systems, the sample air is introduced to a chamber along with another reactant gas. The light emitted is detected with a photomultiplier tube and amplified (Fig. 10). The amount of light that is emitted is proportional to the quanta of the excited species formed. The reactant gas is supplied to the instrument in excess so that the reaction is dependent only upon the concentration of the contaminant being measured. Factors that must be maintained to have an effective chemiluminescent analyzer are (a) constant air flow rate, (b) an adequate source of reactant gas, (c) a properly designed flow reactor chamber, and (d) a stable and sensitive photomultiplier tube.

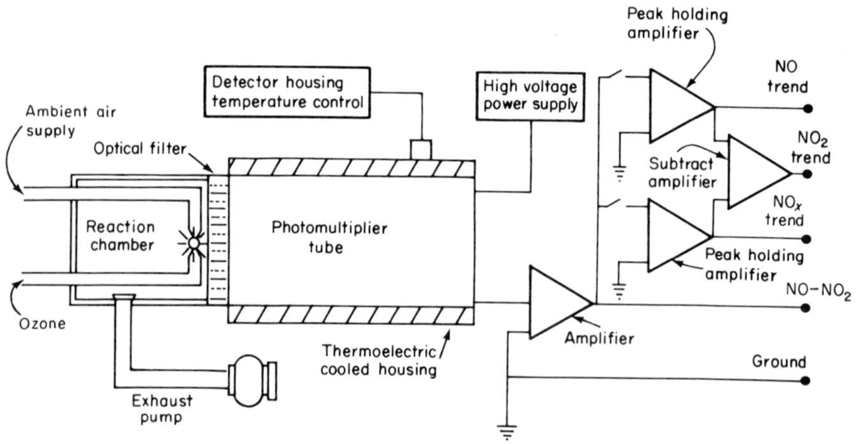

Figure 10. Chemiluminescent nitric oxide (NO)–nitrogen dioxide (NO_2) analyzer.

b. Uses and Limitations. The chemiluminescent reference method for the determination of ozone in the atmosphere is based upon the reaction of ozone with ethylene. Chemiluminescence has also become widely used for the analysis of oxides of nitrogen and nitric oxide, based upon the reaction of nitric oxide with ozone. To measure a concentration of total oxides of nitrogen (NO_x), the nitrogen dioxide is converted to nitric oxide and fed to the reaction chamber.

The advantages of chemiluminescent techniques are (a) specificity of the reactions and emitted light, (b) sensitivity, (c) rapid response time, and (d) absence of liquid reagents. The reactions are not in all cases stoichiometric so that careful and regular calibrations are required.

5. Nondispersive Infrared Analyzers

a. Principle of Operation. The nondispersive infrared analyzer is based on the absorption of infrared energy by the contaminant gas (*35*). The simplest nondispersive infrared (NDIR) analyzer consists of a sample cell and a reference cell, two infrared sources, and a detector cell. The reference cell is usually sealed and contains a gas that is transparent to infrared wavelengths. The sample cell is continuously flushed with the sample (Fig. 11).

An optical chopper alternately interrupts the infrared source for each cell to provide a pulse of energy through each cell. The reference cell passes almost all of the infrared energy on to the detector cell, while the air sample in the sample cell will absorb some of the energy.

The detector consists of two compartments separated by a highly flex-

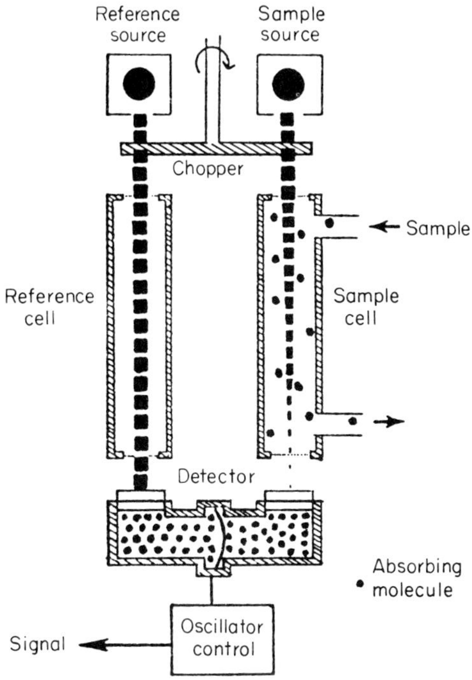

Figure 11. Nondispersive infrared (NDIR) analyzer—analytical section.

ible metal diaphragm. Each compartment is filled (at low pressure) with the gas that is to be detected. Absorption of the infrared energy by the gas in the detector results in heating. Because the infrared energy reaching the reference detector cell is essentially unattenuated, there is more heating, and thus expansion in the reference detector cell as compared to the sample detector cell. This unequal expansion causes the metal diaphragm between the cells to pulse back and forth in phase with the chopper action, with the magnitude of diaphragm distention being a function of the absorbing gas concentration in the sample cell. The pulsing motion of the diaphragm changes the electrical capacitance of an external circuit which ultimately results in a varying output signal.

b. Uses and Limitations. The principal application of nondispersive infrared analysis in ambient air monitoring is for carbon monoxide analysis. However, in the process industries it is widely used for other compounds. Water vapor is an interference in carbon monoxide analysis and must have the same concentration in the sample, the span, and the zero gases. When using dry span and zero gases, it is best to dry the sample by refrigeration or the use of drying agents.

6. Hydrogen Flame Ionization Analyzers

a. Principle of Operation. An air sample is passed through a hydrogen flame which results in the production of ions when hydrocarbons are present in the air sample (*36*). The ionization potential of the hydrogen flame is too low to ionize other species, thus providing selectivity. More specifically carbon atoms having C—H bonds, but not entire molecules, are detected by this technique. A set of electrodes in the flame chamber are connected to an external electrometer circuit to detect and amplify the ionization current, thus providing a signal for the data recording or other acquisition system.

The method is very sensitive to flame conditions and sample flow rate making careful control of hydrogen, combustion air, and sample air flow rate a necessity. Because of the varying response of different hydrocarbon or other organic compounds, it is necessary to report the concentration values obtained in terms of the calibration gas used. Most frequently this is done with methane or propane.

b. Uses and Limitations. Earliest use of flame ionization detectors was with full-scale gas chromatographs. In this case calibration for each compound appearing as a separate peak is possible. In its earliest use as a continuous analyzer, only total hydrocarbon as methane equivalent was measured. Later a substantial number of measurements were made using pretreatment or abstractor columns designed to hold back or retain certain hydrocarbons. The most common approach involves the use of an activated carbon column pretreated with methane so that for a period of time methane would be passed, but other hydrocarbons held back. By use of a sequential sampling valve arrangement, alternate measurements of total hydrocarbons and methane can be obtained. Thus some indication of the photochemically more reactive nonmethane hydrocarbons can be obtained.

Automated special purpose gas chromatographs have been developed which determine carbon monoxide in addition to methane and total hydrocarbons (*37, 38*). It is possible to produce discrete values for each of these groups on a semicontinuous cycle of about 5-minute duration. The system consists of a timed automatic multiport sampling valve, a stripper column, an analytical column, a reactor to reduce carbon monoxide to methane in the presence of hydrogen, a flame ionization detector, an electrometer, and a data acquisition system (Fig. 12). There are advantages, of course, in being able to do multicomponent analysis on a single instrument, but competent operators and servicemen are required to provide a high degree of reliable data output.

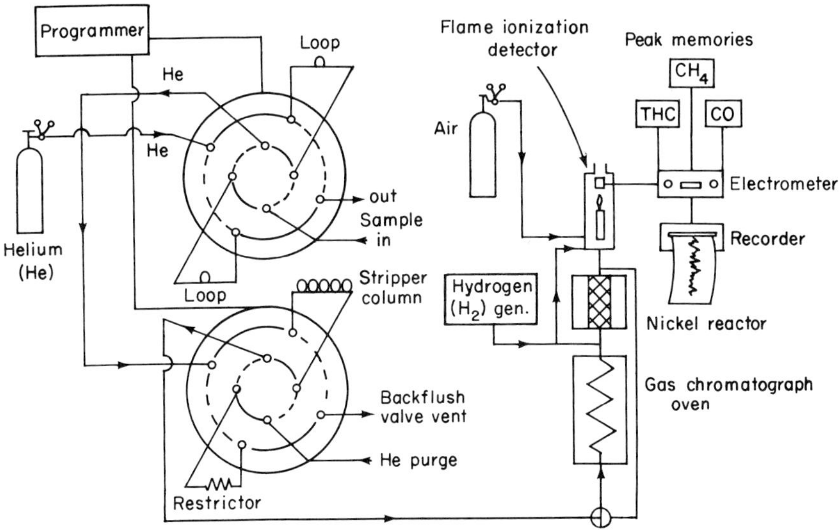

Figure 12. Gas chromatographic analyzer for total hydrocarbons (THC), methane (CH_4), and carbon monoxide (CO).

7. Ultraviolet Absorption

a. PRINCIPLE OF OPERATION. This method is based upon the direct gas-phase absorption of ultraviolet (UV) light. It is not to be confused with colorimetric analysis based upon the absorption of visible or near-UV light by reaction products in liquid solution. Optical components such as cell windows must be transparent to UV light. For this purpose quartz or fused silica are the most widely used materials. Ambient point sampling utilizing ultraviolet absorption has been attempted for relatively few substances, although it is used in source monitoring for several gaseous contaminants. Open-path monitoring using ultraviolet correlation spectrometry has been successful.

b. USES AND LIMITATIONS. Ultraviolet absorption has been used for monitoring ozone because it absorbs UV light over a narrow band (250–260 nm) (*39*). Sulfur dioxide and mercury vapors will interfere with this technique, although mercury vapor interference is small because it is usually present in the air in only low concentrations. Where the concentration of mercury cannot be ignored, a gold surface absorber can be placed at the sample inlet to eliminate the mercury from the system. At normal concentrations any interference from sulfur dioxide or other UV absorbing compounds is eliminated by cycling the air sample through a selective gas filter. The sample stream first passes through the selective removal system for ozone, then enters the absorption cell and absorbs a certain amount of the UV light. This reference value is stored in the electronics of the instrument. In the next cycle a sample enters the cell, bypassing the scrubber. This sample will absorb more of the UV energy if ozone is present. This value is then compared to the stored value with the difference being proportional to the concentration of ozone (Fig. 13). The instrument is based upon Beer's law which can be expressed,

$$\ln (I_0/I_t) = A = abc \tag{9}$$

where

I_0 = intensity of incident UV
I_t = intensity of UV transmitted through cell
A = the absorbance
a = absorptivity
b = path length
c = concentration

The analyzer illustrated is unique in that both I_0 and I_t are measured with the integration time for determining I_t being controlled by the parallel measurement of I_0. This results in a more stable baseline than in earlier UV ozone analyzers before such solid-state circuits were available.

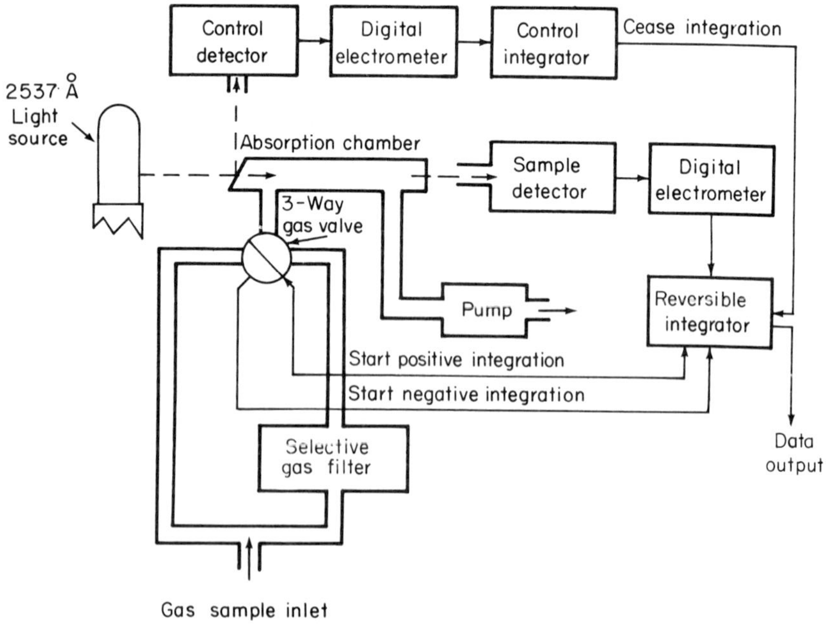

Figure 13. Ultraviolet ozone monitor.

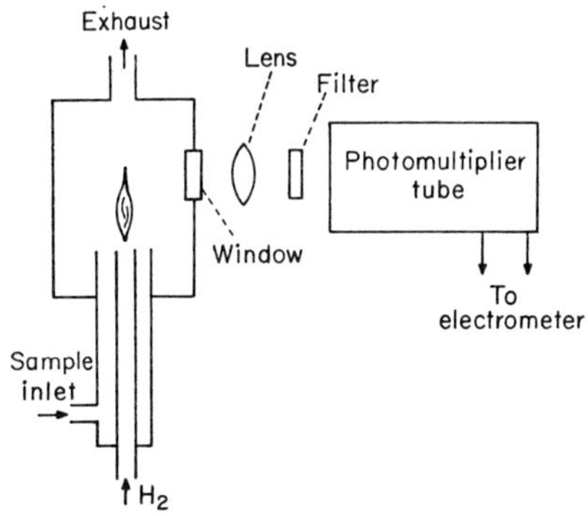

Figure 14. Flame photometric detector.

8. Flame Photometric Analyzers

a. Principle of Operation. The flame photometric detector operates on the measurement of light emitted from an element when burned in a hydrogen flame. This principle has been applied mainly for phosphorous and sulfur analysis (*40, 41*).

The detector consists of a hydrogen burner and a photomultiplier tube. The sample gas enters the detector and is burned in the hydrogen flame. When phosphorus or sulfur are present characteristic flame emission spectra are produced. After a passage through a narrow band-pass filter the emitted light is detected and amplified by the photomultiplier tube. The concentration of the emitting species is proportional to the amount of light emitted (Fig. 14).

b. Uses and Limitations. As applied to ambient air monitoring, the flame photometric detector has been used almost exclusively for sulfur compounds.

The main limitation of such a detector is that, being a sulfur detector, it will not distinguish among sulfur compounds (such as, sulfur dioxide, hydrogen sulfide, methyl mercaptan, etc.) To eliminate this problem and make this detector useful for the measurement of one specific gas, selective chemical filters or gas chromatographic techniques are commercially available.

B. Continuous Analyzers for Particulate Matter

Only a few types of continuous automatic air monitoring devices for suspended atmospheric particulates are available. This situation may be due in part to the relatively greater technological problems in development of such equipment for particulate matter as compared to gaseous contaminants. The essentially nonspecific nature of particulates and the uncertainty of what property to measure has contributed to the relatively low level of developmental activity. Those instruments which have been developed have been based upon measurement of mass or particle count by indirect techniques.

1. Piezoelectric Mass Sensor

The piezoelectric analysis technique involves deposition of particulates from a sample airstream onto an oscillating quartz crystal which is coated with a thin-film substrate to enhance collection efficiency and particle retention (*42*). The resonant frequency of such a crystal decreases linearly with the mass of deposited material. This shift can be detected in a tuned circuit and used to provide a signal output proportional to the mass present.

While the method is very sensitive there are some problems and possible interferences. Perhaps the major problem is that any change in mass is detected including any water that might be deposited. Also the particulates must be captured to be detected, therefore implying knowledge of particulate collection efficiency. There is also a limit to the mass that can be sensed without cleaning the crystal of the deposited material.

Because of its compactness and rapid response characteristics, the instrument has been used to detect particulate plumes while mounted in an aircraft.

2. Nephelometry

Light-scattering techniques have been long used for particle detecting and counting. In nephelometry, the overall scattering resulting from a cloud of particles occupying a fixed volume subjected to an incident source of light is measured (*43*).

In operation air is continuously moved through a detection chamber illuminated by a light source which flashes intermittently. An optical detector is placed in a position at an angle to the direct line of the light path and measures the amount of light scattered during each flash (Fig. 15). This measured amount of scattered light is proportional to the concentration of the particulate matter in the sample. The wavelength of the scattered light; the angle of the detector; and the size, shape, and index of refraction of the aerosol particles all influence the relationship between number of particles and the amount of scattered light. Studies

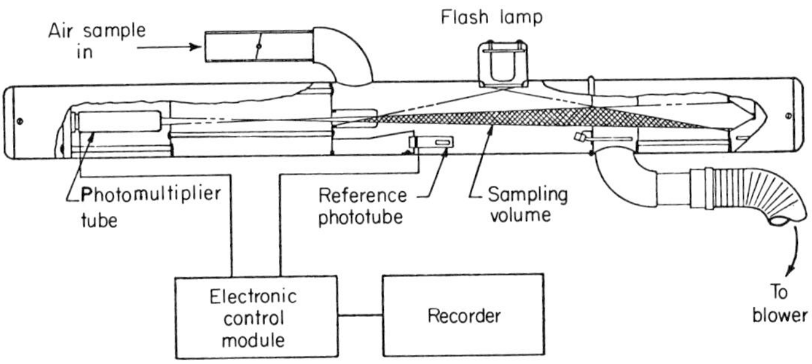

Figure 15. Nephelometer.

have shown a reasonably good correlation between nephelometer measurements and visibility as determined by the human eye.

3. Tape Samplers

Tape samplers operate by passing an air sample through a filter which is in tape form and which is advanced in discrete time intervals from a supply reel to a take-up reel. During the sampling period, which can be varied, the tape is stationary and is sealed from both sides. Traditionally, these tapes have been examined by either light transmission or by light reflectance. This can be done either on the sampling device with direct recording or by manual evaluation after removal from the sampler. The degree of filter staining has been reported as Coefficient of Haze (COH units) in the case of transmittance or Reflectance Units of Dirt Shade (RUDS) when reflectance is used. Correlation of these units with mass loading has been only partially successful.

A more recent type of deposit evaluation which has been shown to have a better correlation with deposit mass is that of β gauging (*44, 45*). In this type device a β-radiation source is located on one side of the exposed tape and a detector on the other (Fig. 16). The attenuation of the β radiation is a function of mass on the tape.

4. Charged Particle Mobility Analyzer

One of the problems with particulate measurement is that of sensitivity over a wide range of particle sizes. Thus techniques which are based upon mass detection will be influenced more heavily by larger particles than by smaller ones. For many reasons, including the ability of submicrometer-sized particles to penetrate deeply within the lung, there is interest in measuring these smaller particles. The charged particle mobility tech-

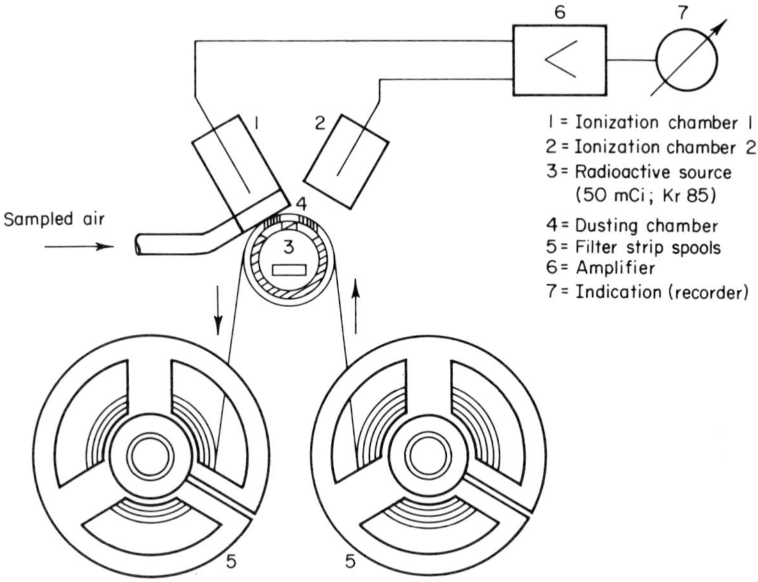

Figure 16. β-radiation particulate mass monitor.

nique is especially suited for submicrometer-sized particles, because, under conditions closely approximating that of diffusion charging, the electrical mobility of an aerosol is a monotonically decreasing function of particle size for particles smaller than approximately 1 μm in diameter (*46*). That is to say the mobility of particles increases as size decreases.

In use, a mobility analyzer functions by introducing the air sample first to a charging section and then passing it through an electrical precipitator where particles with a sufficiently high mobility are precipitated while those with lower mobilities escape collection and are sensed by an electrometer sensor. By changing the voltage on the collector rod, the threshold mobility for collection is changed thus making possible a particle-size determination mode of operation.

C. Open-Path Sensors and Other Developing Techniques

A major portion of the research and development being conducted on air monitoring analyzers has been concentrated on physical methods, principally involving some form of spectrophotometry (*47, 48*). The goal, of course, has been to improve the specificity and sensitivity of monitoring methods with equipment that can be operated reliably under field conditions.

1. Correlation Spectrometry

Correlation spectrometry is well advanced in application (*49, 50*). Specificity is obtained by use of a correlation mask instead of dispersive optics. These masks, which are unique for each component of interest, pass selected portions of the source spectrum. For air monitoring purposes, such instruments are most frequently used in the long-path passive mode using sky light as the light source. They have been applied to both sulfur dioxide and nitrogen dioxide monitoring.

2. Tunable Infrared Lasers

Substantial efforts are underway to develop tunable infrared lasers capable of high output over selected narrow spectral intervals. Both semiconductor diodes and gas lasers are under study (*51–56*). Devices now available require fairly long path lengths (1 to 2 km) to achieve necessary sensitivity. The actual path may be shorter if multiple reflections are used. Sensitivity can be limited by scattering from particulates and turbulence, although dual wavelength techniques may help in reducing these effects. Optoacoustic detectors such as those used in nondispersive infrared spectrometry have been used with some success. The technique is potentially useful for carbon monoxide, ozone, nitric oxide, sulfur dioxide, and hydrocarbons, if engineering problems such as the present requirement of providing cryogenic cooling for diode lasers can be solved.

3. Ultraviolet Lasers

A remote Raman technique using an ultraviolet laser to stimulate Raman emission has been proposed and is in the experimental stage (*57*). One system uses a doubled ruby laser operating at 3472 Å and a 1-m diameter collecting telescope. Present systems are large, complex and costly, but the potential exists to develop the technique for a wide range of contaminants.

4. Light Detection and Ranging (LIDAR)

LIDAR systems for remote sensing of suspended particulates in the atmosphere have certain features in common with Raman systems in that a pulsed laser is used as a source and an optical telescope as a receiver. LIDAR, however, detects backscatter from particulates as compared to the molecular scattering induced in Raman systems (*58*). Because range information is obtained, the technique has been of particular interest in conjunction with estimation of meteorological mixing height, in determining the presence of stratified layers of particulates, and in locating plumes.

5. Microwave Spectroscopy

Microwave spectroscopy has been proposed because of the high degree of specificity which can be theoretically achieved *(59)*. In this technique, small polar molecules convert microwave energy to rotational energy at discrete frequencies which are unique for each pollutant. Because there is very little overlap in these frequencies, a high degree of specificity is possible. The procedure suffers at present from poorer sensitivity than is available with other methods. Work to increase sensitivity is being carried out.

6. Second Derivative Spectroscopy

Second derivative spectroscopy is based upon measurement of the amplitude of the second derivative of the absorption spectra at some selected wavelength *(60)*. The second derivative is a measure of curvature and it can be shown that for any compound which obeys Beer's law the second derivative is a linear function of concentration. The technique is generally limited to those compounds which have sharp narrow band absorption spectra. Because interference can result only from those compounds which exhibit narrow band absorption at exactly the wavelength selected for analysis, the method can be very specific. Development of the technique for ambient air monitoring has centered upon sulfur dioxide, nitric oxide, nitrogen dioxide, and ammonia. Figure 17 shows a schematic of an analyzer designed for continuous monitoring. The second derivative measurement is obtained through the technique of using a wobbler to sinusoidally vary the source wavelength about the selected analysis peak.

7. Ultraviolet Fluorescence

Another optical technique, which has been developed for sulfur dioxide monitoring, is based upon the ultraviolet fluorescence of this compound produced by absorption of the Zn 2138 Å or Cd 2288 Å line *(61)*. By careful selection of excitation and emission filters good sensitivity has been achieved. Sensitivity does not yet compare to some other methods such as the flame photometric procedure. Water vapor decreases the signal and must be removed or a correction applied.

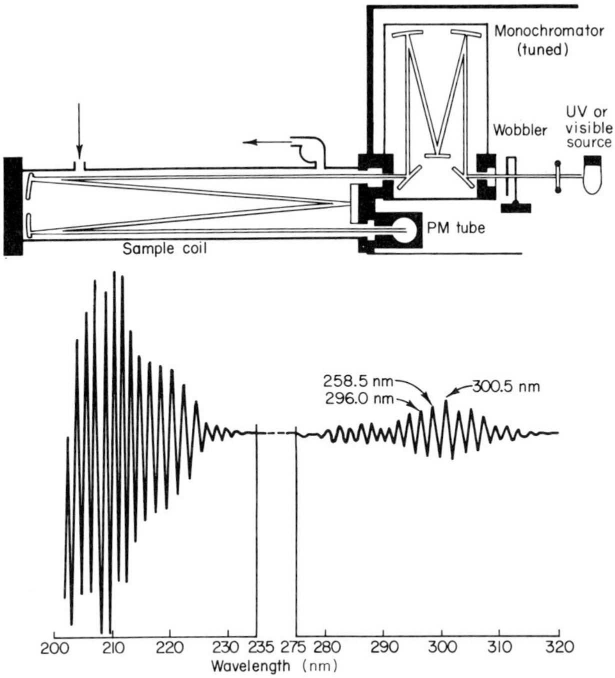

Figure 17. Second derivative spectrometer. PM: photomultiplier tube.

8. Dual-Isotope Fluorescence

An interesting approach to improvement of specificity of the nondispersive infrared (NDIR) technique for carbon monoxide analysis involves the use of the dual-isotope technique (*62*). A broadband infrared source is used to stimulate fluorescence of $C^{16}O$ and $C^{18}O$ contained in a sealed cell. The stimulated radiation is passed first through a filter that alternately passes radiation only from one of the isotopic sources and then is passed through the sample cell to a solid-state detector. Because almost all atmospheric carbon monoxide consists of the $C^{16}O$ isotope, only radiation produced by the $C^{16}O$ source will be absorbed by the carbon monoxide in the sample. The radiation produced by the $C^{18}O$ source is used as a reference beam to account for attenuation by water vapor or other interferences (Fig. 18).

9. Semiconductors

Considerable research interest has also been expressed in the use of semiconductors as direct sensing devices. The electrical properties of these

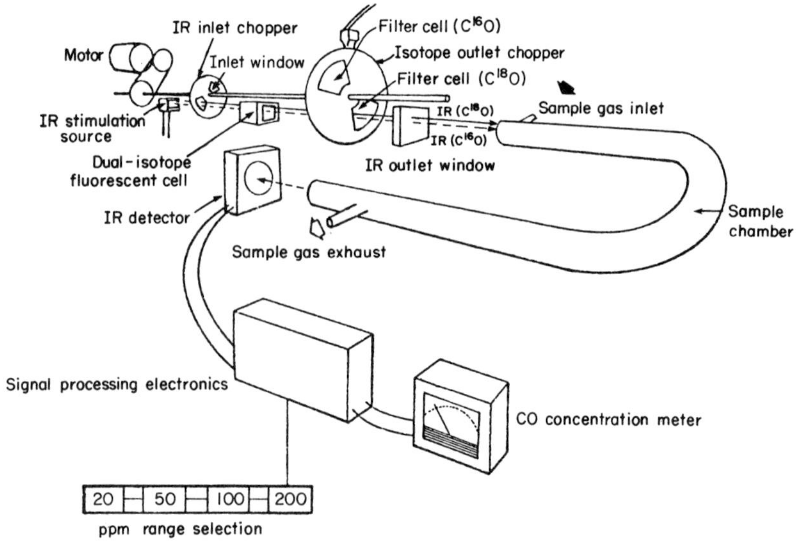

Figure 18. Dual-isotope fluorescence nondispersive infrared (NDIR) carbon monoxide (CO) analyzer.

devices are affected by surface contamination which results from their contact with pollutants. By appropriate selection of permeable membranes, bias voltage polarity and magnitude, and electrode material, a whole family of quasi-specific detectors might be developed.

V. System Operation

The obtaining of valid air quality data over a high proportion of operating time is very dependent upon the manner in which system operations are conducted. The major factors involved are proper installation of equipment, careful calibration of the system, close adherence to established operating procedures, and provision of adequate servicing and maintenance. Guidelines and instructions should be provided which cover in detail all aspects of these procedures, and adequate documentation should be maintained to support the accuracy and other measures of validity of the data.

A. Calibration

Calibration determines the relationship between the observed and true values of the variable being measured. In the case of air quality monitoring systems, it is generally necessary to calibrate both the sensor portion of the analyzer and any flow metering devices that are incorporated.

Chapter 4, this volume covers the technical aspects of calibration. A brief discussion of conceptual principles is presented here.

The terms "dynamic" and "static" sensor calibration have been used rather widely although neither term is completely descriptive of the generally understood meaning. "Dynamic" calibration means introduction of a known sample of the contaminant of interest to the analyzer in the way that an ambient air sample is introduced during normal operation. Thus, not only the detector portion of the instrument but the flow metering components as well are involved and influence the results observed on the data acquisition system. In instruments where signal output is sample flow rate dependent—a majority of cases except for direct spectrophotometric sensors—the accurate determination of this flow rate during calibration is an essential element.

"Static" calibration means the calibration of only the sensor portion of the instrument and involves neither the sample nor reagent flow systems. In most cases, "static" calibration involves a pseudocalibration of a wet chemical analyzer in which a series of reagent samples containing known amounts of material producing the same reaction product in the reagent as does the actual contaminant of interest are introduced to the measuring cell. Examples of such an approach include the use of sulfuric acid for calibration of conductimetric sulfur dioxide analyzers, use of sulfites in the colorimetric sulfur dioxide instrument calibration, nitrites in colorimetric nitrogen dioxide analyzers, and iodine in potassium iodide type oxidant analyzers. When the absorption efficiency and stoichiometry of the reaction are well understood and reproducible, these static techniques can be quite useful because of the relative ease with which they can be conducted. With purely physical instruments, however, it is necessary to provide or generate actual gaseous samples of the contaminant later to be measured. In the case of some instruments, physical or electronic tests may be conducted using optical filters or imposed electrical signals to check the detector and signal processing components of the system.

B. Servicing and Maintenance

Regular schedules of routine preventative maintenance should be established and adhered to in the operation of air quality analyzers. In most cases manufacturer's service manuals and operational guidelines developed by professional societies and governmental agencies will serve, at least initially, for this purpose.

Included among the more commonly required service functions are (a) checking of instrument settings, (b) making flow rate checks, (c) cleaning glassware and sample lines, (d) leak testing, (e) replacement of filters,

(f) replacement of expendable reagents and calibration gases, (g) performance of span and zero checks, (h) replacement of recorder charts, (i) cleaning of recorder slidewire, (j) performance of recorder linearity checks, and (k) checking for recorder dead band (signal range necessary to initiate recorder response). Beyond this level of servicing, specially trained service personnel who are specialists in mechanical and electronic servicing will be needed. Special tools and equipment will be required to support this work. Otherwise service from the manufacturer or other contract source can be utilized.

C. Quality Assurance

The application of formal quality assurance practices in air monitoring is a fairly recent development. Those elements of a program which formerly contributed to the collection of valid data generally were covered under the terms—proper calibration, good operating practice, servicing and maintenance, data editing, etc. While the procedures conducted under these terms play an important role in quality assurance, not all of the elements of a formal quality assurance program are necessarily covered. For this to be the case, quantifiable statements about accuracy, precision, and reliability must be able to be made. Furthermore, means to identify and correct deficiencies in a timely way must be incorporated.

In a recent series of documents produced under the direction of the USEPA covering the establishment of specific guidelines for sampling and monitoring procedures (*63*), a set of objectives and program elements were set forth. These are repeated below:

1. Objectives
   a. Provide routine indications for operating purposes of unsatisfactory performance of personnel and/or equipment
   b. Provide prompt detection and correction of conditions which contribute to the collection of poor data
   c. Collect and supply information necessary to describe the quality of data
2. Program elements
   a. Routine training and/or evaluation of operators
   b. Routine monitoring of the variables and/or parameters which may have a significant effect on data quality
   c. Development of statements and evidence to qualify data and detect defects
   d. Action strategies to increase the level of precision in the reported data and/or to detect equipment defects or degradation and to correct them

Such a program should cover all the aspects of monitoring activities including equipment pretest, reagent preparation (where required), sampling operations, calibrations, obtaining of standard reference materials, servicing and maintenance, and data processing.

In order to implement such a program, information must be developed on the impact of controlled and noncontrolled variables on such data quality measures as accuracy and precision. These variables should include quality of calibration materials, flow rates, span and zero drift, signal-to-noise ratio, temperature, and presence of interferences. In some cases this information may be available from standard reference methods, the results of collaborative studies, or the manufacturer of the instrument. Otherwise it will have to be developed.

At this point a decision must be reached as to the quality of data desired or required, as this has an impact on the costs of administering a quality assurance program. In essence, a decision must be made as to the cost and risk involved in accepting bad data versus the cost of rejecting good data. Once a decision has been made on acceptance or rejection criteria, procedures must be established to control the monitoring operation and to develop an auditing scheme to determine whether or not the criteria for acceptability are being met.

Such auditing and the use of objective acceptance or rejection criteria for data are the two aspects of a formal quality assurance program which set it apart from the "good practices" approach of securing valid data.

There are differences between continuous automatic monitoring and conventional laboratory analysis of discrete samples which make modification of standard auditing techniques necessary. For example the inclusion of standard or reference samples in a batch of actual samples to be run in the laboratory is not possible in continuous air monitoring. At best an automated sequential calibration procedure can be used, but the calibration results will be separated in time from the ambient air concentrations. Thus concepts such as allowable span and/or zero drift between stated intervals must be used instead. For example, if span drift does exceed a predetermined value, data between the last previous acceptable span value and the unacceptable span value must be rejected. Also statistical techniques for detecting likely spurious values by screening for excessive "rate of climb" between two sequential data bits and for identifying "out-of-limits" values must be developed. Criteria must also be developed for flow rate, slope, and linearity of calibration curves, and interference checks.

REFERENCES

1. J. J. Paulus and A. T. Rossano, *Proc. Int. Clean Air Congr. 3rd, 1973* p. C133.
2. W. B. Rigga, D. I. Hammer, J. F. Finklea, F. Hasselbad, C. R. Sharp, R. M. Burton, and C. M. Shy, "CHESS—A Community Health and Environmental

Surveillance System" (presented at Sixth Berkeley Symposium on Mathematical Statistics and Probability, Biology), Health Sect., Part III. Univ. of Calif., Nat. Inst. of Health, Atomic Energy Comm., Berkeley, California (1971).
3. B. E. Saltzman, *J. Air Pollut. Contr. Ass.* **20,** 660 (1970).
4. W. W. Stalker and R. C. Dickerson, *J. Air Pollut. Contr. Ass.* **12,** 111 (1962).
5. W. A. McClenny, R. E. Baumgardner, Jr., F. W. Baity, Jr., and R. A. Gray, *J. Air Pollut. Contr. Ass.* **24,** 1044 (1974).
6. H. C. McKee and R. E. Childers, "Collaborative Study of Reference Method for the Continuous Measurement of Carbon Monoxide in the Atmosphere (Non-dispersive Infrared Spectrometry)," Southwest Research Institute Project No. 01-02811, Contract CPA 70-40. U.S. Environmental Protection Agency, Research Triangle Park, North Carolina, 1972.
7. H. C. McKee, R. E. Childers, and O. Saenz, Jr., "Collaborative Study of Reference Method for Determination of SO_2 in the Atmosphere (Pararosaniline Method)," Southwest Research Institute Project No. 01-02811, Contract CPA 70–40. U.S. Environmental Protection Agency, Research Triangle Park, North Carolina, 1971.
8. H. A. Clements, T. B. McMullen, R. J. Thompson, and G. G. Akland, *J. Air Pollut. Contr. Ass.* **22,** 955 (1972).
9. J. H. Blacker, R. G. Confer, and R. S. Brief, *J. Air Pollut. Contr. Ass.* **23,** 525 (1973).
10. F. C. Hamburg, *J. Air Pollut. Contr. Ass.* **21,** 609 (1971).
11. J. H. Seinfeld, *Atmos. Environ.* **6,** 847 (1972).
12. W. Ott and R. Eliassen, *J. Air Pollut. Contr. Ass.* **23,** 685 (1973).
13. M. Clifton, D. Kerridge, W. Moulds, J. Pemberton, and J. K. Donoghue, *Int. J. Air Pollut.* **2,** 188 (1959).
14. W. W. Stalker, R. C. Dickerson, and G. D. Kramer, *J. Air Pollut. Contr. Ass.* **12,** 361 (1962).
15. J. M. Leavitt, F. Pooler, Jr., and R. C. Wanta, *J. Air Pollut. Contr. Ass.* **7,** 211 (1957).
16. R. M. Ingels, "Statistical Analysis of Difference Between Air Monitoring Data from Adjacent Sampling Stations," Evaluation Report. Los Angeles County Air Pollution Control District, California, 1962.
17. W. D. Holland, E. L. Fisher, M. F. Brunelle, and R. J. Bryan, Pap. No. 68-41, 61st Annu. Meet. Air Pollution Control Association, Pittsburgh, Pennsylvania, 1968.
18. R. J. Gordon and R. J. Bryan, *Environ. Sci. Technol.* **7,** 1050 (1973).
19. United States Environmental Protection Agency, "Guidelines: Air Quality Surveillance Networks," No. AP-98. Research Triangle Park, North Carolina, 1971.
20. United States Environmental Protection Agency, *Fed. Regist.* **36,** No. 158 (1971).
21. International Electrotechnical Commission, "Expression of Performance of Air Quality Infrared Analyzers." IEC, Geneva, Switzerland, 1974.
22. United States Environmental Protection Agency, "Field Operations Guide for Automatic Air Monitoring Equipment," No. APTD-0736. Research Triangle Park, North Carolina, 1971.
23. United States Environmental Protection Agency, "SAROAD Users Manual," No. APTD-0663. Research Triangle Park, North Carolina, 1971.
24. M. Shaw, "Plug-In Sensors for Air Monitoring," Proc. 12th Conf. Methods Air Pollut. Ind. Hyg. Stud. California Department of Public Health, Berkeley, California, 1971.
25. R. W. Garber and C. E. Wilson, *Anal. Chem.* **44,** 1357 (1972).

26. L. B. Potter and M. Feldstein, "A Description of the Philips SO_2 Monitor," Joint Conf. Sensing Environ. Pollut., Pap. No. AIAA 71-1113. Amer. Inst. of Aeronautics and Astronautics, Palo Alto, California, 1971.
27. Intersociety Committee, "Methods of Air Sampling and Analysis," Method 44101-01-69T. Intersociety Committee, Amer. Pub. Health Ass., Washington, D.C., 1972.
28. A. W. Brewer and J. R. Milford, *Proc. Roy. Soc., Ser. A* **256**, 470 (1960).
29. P. Hersch and R. Deuringer, *Anal. Chem.* **35**, 897 (1963).
30. J. A. Hodgeson, K. A. Rehme, B. E. Martin, and R. K. Stevens, Pap. No. 72-12, 65th Annu. Meet. Air Pollution Control Association, Pittsburgh, Pennsylvania, 1972.
31. D. H. Stedman, E. E. Darby, F. Stuhl, and H. Niki, *J. Air Pollut. Contr. Ass.* **22**, 260 (1972).
32. "Ozone Analyzer," *Wasser, Luft Betr.* **16**, 130 (1972).
33. S. W. Benson, G. R. Haugen, and R. S. Jackson, U.S. Patent 3,647,387 (1972).
34. A. Fontijn, A. J. Sabadell, and R. J. Ronco, *Anal. Chem.* **42**, 575 (1970).
35. P. C. Wolf, *Environ. Sci. Technol.* **5**, 212 (1971).
36. A. P. Altshuller, S. L. Kopzynski, W. A. Lonneman, T. L. Becker, and R. Slater, *Environ. Sci. Technol.* **1**, 89 (1967).
37. R. K. Stevens, A. E. O'Keefe, and G. C. Ortman, "A Gas Chromatographic Approach to the Semi-Continuous Monitoring of Atmospheric CO and CH_4," Proc. 11th Conf. Methods Air Pollut. Ind. Hyg. Stud. California Department of Public Health, Berkeley, California, 1970.
38. Anonymous, *Wasser, Luft Betr.* **16**, 131 (1972).
39. B. A. Behl, Pap. No. 72-7, 65th Annu. Meet. Air Pollution Control Association, Pittsburgh, Pennsylvania, 1972.
40. R. K. Stevens, A. E. O'Keefe, J. D. Mulik, and K. J. Krost, *Anal. Chem.* **43**, 827 (1971).
41. D. G. Greer and T. J. Bydalek, *Environ. Sci. Technol.* **7**, 153 (1973).
42. J. G. Olin and G. J. Sem, Pap. No. 71-1, 64th Annu. Meet. Air Pollution Control Association, Pittsburgh, Pennsylvania, 1971.
43. R. J. Charlson, N. C. Ahlquist, H. Selvidge, and P. B. MacCready, Jr., *J. Air Pollut. Contr. Ass.* **19**, 937 (1969).
44. G. J. Sem, J. A. Borgos, and J. G. Olin, *Chem. Eng. Progr.* **67**, 83 (1971).
45. R. B. Husar and S. L. Heisler, "Atmospheric Particulate Mass Monitoring with a Beta Radiation Detector," Proc. 13th Conf. Methods Air Pollut. Ind. Hyg. Stud. California Department of Public Health, Berkeley, California, 1972.
46. K. T. Whitby and W. W. Clark, *Tellus* **18**, 573 (1967).
47. R. K. Stevens and A. E. O'Keefe, *Anal. Chem.* **42**, 143A (1970).
48. "Air Pollution Instrumentation: A Trend Toward Physical Methods," *Science* **177**, 685 (1972).
49. H. W. Goldstein, M. H. Bortner, R. N. Grenda, A. M. Karger, R. Dick, F. David, and P. J. LeBel, "The Remote Sensing of Trace Atmospheric Species by Correlation Interferometry," Proc. 2nd Joint Conf. Sensing Environ. Pollut. Amer. Inst. of Aeronautics and Astronautics, Washington, D.C., 1973.
50. "Remote Sensing Correlation Spectrometry for Pollution Measurement," Proc. 9th Conf. Methods Air Poll. Ind. Hyg. Stud. California Department of Public Health, Berkeley, California, 1968.
51. E. D. Hinckley and H. A. Pike, Pap. No. 73-170, 68th Annu. Meet. Air Pollution Control Association, Pittsburgh, Pennsylvania, 1973.
52. E. H. Cristy and K. H. Fallen, "A Dual-Beam Laser Absorption Spectrometer

for Air Pollution Monitoring," 2nd Joint Conf. Sensing Environ. Pollut. Amer. Inst. of Aeronautics and Astronautics, Washington, D.C., 1973.

53. L. B. Kreuzer, "Laser Optoacoustic Spectroscopy—A New Technique for Measuring Gaseous Pollutants," 2nd Joint Conf. Sensing Environ. Pollut. Amer. Inst. Aeronautics and Astronautics, Washington, D.C., 1973.
54. L. R. Snowman and R. J. Gillmeister, "Infrared Laser System for Extended Area Monitoring of Air Pollution," Joint Conference on Sensing of Environmental Pollutants, Pap. No. AIAA 71-1059. Amer. Inst. of Aeronautics and Astronautics, Washington, D.C., 1971.
55. A. Kaldor, W. B. Olson, and A. G. Maki, *Science* **176,** 508 (1972).
56. R. M. Measures, "A Comparative Study of Laser Methods of Air Pollution Mapping," Rep. No. UTIAS-174. Inst. Aerosp. Stud., University of Ontario, Toronto, Ontario, 1971.
57. H. Tannenbaum, D. Tannenbaum, and H. DeLong, Pap. No. 72-11, 65th Annu. Meet. Air Pollution Control Association, Pittsburgh, Pennsylvania, 1972.
58. S. H. Melfi, "Standard Methods for Analysis and Interpretation of LIDAR Data for Environmental Monitoring," 2nd Joint Conf. Sensing Environ. Pollut. Amer. Inst. of Aeronautics and Astronautics, Washington, D.C., 1973.
59. L. W. Hrubesch, "Microwave Rotational Spectroscopy: A Physical Technique for Specific Pollutant Monitoring," NRSA-CR-136475, UCRL-74603. University of California, Lawrence Livermore Laboratory, Livermore, California, 1973.
60. R. Hager and R. Anderson, *J. Opt. Soc. Amer.* **60,** 1444 (1970).
61. H. Okabe, J. J. Ball, and P. L. Splitstone, *J. Air Pollut. Contr. Ass.* **23,** 514 (1973).
62. Andros 7000, "Carbon Monoxide Analyzer Product Brochure." Andros, Inc., Berkeley, California, 1972.
63. U.S. Environmental Protection Agency, "Guidelines for Development of a Quality Assurance Program," EPA-R4-73-028a, -028b, -028c, -028d. Research Triangle Park, North Carolina, 1973.

10

Air Pollution Effects Surveillance

Charles W. Gruber and George Jutze

I. Introduction

In the main, the data from monitoring the effects of pollutants on materials and growing things is qualitative and will not lead to direct quantitative measurement of air quality in grams per cubic meter or ppm. Instrumental air quality monitoring is expensive and requires large capital outlay. Therefore the number of sampling stations is usually far short of adequate to cover an air pollution control region or other area of concern. Monitoring by effects is low cost and therefore the multiplicity of stations needed for proper coverage can be set up by organizations with limited resources to augment instrumental monitoring.

Instrumental surveillance is usually limited to the pollutants for which ambient air quality or emission standards have been set. Monitoring by effects has the capability of indicating effects from many pollutants which are not covered by such standards.

A. Areas of Concern

Effects monitoring can be utilized in any of the following situations: (a) routine or continuing assessment of air quality across a nation, a state, or province or within a region or jurisdiction of a state or province; (b) surveillance of a single source; (c) locating an unknown source; (d) screening an area prior to instrumental monitoring; and (e) assessing the severity of episodal emissions.

B. Survey Methodology

Effects monitoring can involve as indicators: (a) things which are naturally present in an area such as vegetation, exterior building materials, furnishings, structural and industrial metals, paints and coatings; and (b) selected materials placed in the area of study.

System studies to establish economic costs (*1–3*) are excellent references in planning and conducting an effects study. From them baseline data on effects on materials can be obtained which will be helpful in selecting materials for effects surveillance.

II. Material Effects Indicators

A large number of materials react or interact with specific air pollutants yielding a noticeable or recognizable change in their appearance or properties. Table I (*4*) presents a summarization of these parameters. By inspection of the table, it is obvious that many of the pollutant effects cited may be visually detected. Materials soiling, fading, or darkening; tarnishing; corrosion and etching; and discoloration are all in this category.

Additionally, the flaking of building stone and mortar, malfunction of electrical components, loss of elasticity in rubber, brittleness of paper or leather, and weakening of fabrics are all changes in physical composition or properties which are easily discerned by the user of these materials.

Table I Material Effect—Pollutant Relationships (2)

Pollutant	*Material affected*	*Effect*
Particulates	Metals	Abrasion, chemical reaction
	Building materials	Soiling
	Painted surfaces	Abrasion
	Textiles	Chemical reaction
Ozone	Rubber	Embrittlement
	Dyes	Fading, discoloration
Sulfur oxides	Metals	Corrosion
	Building materials	Chemical reaction
	Paper	Embrittlement
	Leather	
	Textiles	Weakening, disintegration
Oxides of nitrogen	Building materials	Chemical reaction
	Dyes	Fading, discoloration
Hydrogen sulfide	Lead-base paints	Darkening
	Metals	Tarnish
Carbon dioxide	Building stone	Formation of soluble carbonates
Hydrofluoric acid	Metals	Corrosion
	Glass	Etching
	Metals	
	Enamel	

Therefore, once the presence of a specific pollutant or group of pollutants is known or suspected to exist in concentrations capable of producing any or all of the noted effects, it becomes a simple matter to conduct routine surveillance of a region using the following means. (a) Scheduled inspection trips throughout a designated area keeping a log of the visual appearance of selected buildings, metal bridges, painted surfaces, regularly parked cars, and the like. For comparison purposes, the trips should cover areas of known or probable high levels of air pollution as well as a "control" area, within the same climatic region, where air quality is good. (b) Regular survey of selected users of materials which are susceptible to change in their properties or composition, such as, telephone company stations (maintenance of electric contacts/relays), automotive tire supply warehouses, property managers, department store "return" departments (faded or disintegrated fabrics or materials), and librarians.

An example of the selectivity of air pollution effects on materials is shown in Table II (*1*). Those which have poor or fair chemical resistance

Table II Chemical Resistance of Materials to Pollutants *(1)*

Material	*Organic*	NO_x	SO_x	O_3	CO/CO_2, NH_3, HCl, H_2S	*Particulates*
Ferrous metals						
Gray iron	F	E	P	E	F	P
Malleable iron	G	E	F	E	F	F
Alloy steel	G	E	P	E	P	P
Carbon steel	G	E	P	E	P	P
Stainless steel	E	E	E	—	F	F
Nonferrous metals						
Aluminum	G	E	F	E	G	G
Brass and bronze	E	E	F	E	P	P
Chromium	E	E	G	E	G	G
Copper	G	E	P	E	P	P
Gold	E	E	E	E	G	G
Magnesium	G	E	F	E	G	G
Molybdenum	E	E	G	E	G	G
Lead	G	E	E	E	G	F
Nickel	E	E	F	E	F	E
Tin	E	E	P	E	F	F
Silver	E	E	P	E	F	F
Zinc	F	E	P	E	F	P
Stone, brick, cement						
Building stone	G	E	P	E	G	F
Building brick	G	E	G	E	G	F
Cement and concrete	G	E	P	E	G	G
Glass, carbon, graphite						
Glass	E	E	E	E	G	E
Carbon and graphite	E	E	E	E	E	E
Plastic and rubber						
Synthetic rubber	P	G	G	G	E	P
Natural rubber	P	G	F	P	E	P
Polyethylene	G	G	G	F	E	G
Polystyrene	P	E	E	F	E	P
PVC	P	G	E	G	E	P
Phenolics	F	G	E	G	E	P
Polypropylene	G	G	E	F	E	F
Urea and melamine	E	G	G	G	E	F
Polyesters	F	G	G	G	E	P
Acrylics	P	G	G	G	E	P
ABS	F	G	P	G	E	P
Cellulosics	F	G	G	G	G	P
Epoxies	E	G	E	G	E	P

Column heads: *Pollutant*[a]

Table II (*Continued*)

Material	*Pollutant*[a]					
	Organic	NO_x	SO_x	O_3	CO/CO_2, NH_3, HCl, H_2S	*Particulates*
Acetate	G	G	F	F	F	P
Nylons	G	G	P	F	F	G
Fibers						
Cotton	G	G	P	F	F	P
Wool	E	G	F	G	E	F
Rayon	F	G	P	P	F	P
Acetate	P	G	P	G	G	P
Acrylics	P	G	E	G	E	P
Nylons	G	F	P	P	G	P
Polyester	P	G	G	G	G	P
Polyolefins	P	F	E	F	G	P
Finishes and coatings						
Paint	G	G	F	F	P	P
Other materials						
Wood	G	G	F	F	F	F
Paper	G	E	P	G	G	F
Leather	G	G	P	G	G	F
Bituminous materials	G	G	E	G	G	G

[a] Letter designations are defined as follows: E: excellent; G: good; F: fair; P: poor.

would make good materials for effects monitoring. Exterior painted surfaces can be used for monitoring particulates, H_2S (if paint contains lead), and to a lesser degree SO_2 particulates (*3*).

A number of materials demonstrate pollutant effects which are not nominally expressed in related damage units. However, this does not preclude their use as indicator tools in monitoring air contamination nor the development of their relative damage indices. Several approaches which have been used in this regard are presented in the following sections.

A. Stone and Masonry Surfaces

Two pollutant effects on buildings, statues, and masonry in general are easily recognized: (a) physical erosion of surfaces by particulates with abrasive properties and/or chemical deterioration of the stone or mortar due to the acidic action of sulfur dioxide or carbon dioxide; and (b) soiling of surfaces by deposits of oily or sticky particulate matter.

Verification of the pollution impact on buildings and masonry materials is generally established by photographic evidence, if soiling is the problem. It is usually a simple matter to find a building which is being cleaned by sandblasting, and photographs exhibiting both the cleaned and uncleaned surfaces adjoining one another offer graphic proof of the extent that soiling had occurred. Naturally, these results are qualitative and cannot be measured in definitive units. However, records indicating how often cleaning was required in areas of high suspended particulate concentrations versus that done in less polluted areas can establish the relative severity of the problem.

In the case of chemical attack, the building or statuary stone which generally contains large amounts of insoluble calcium, reacts with either SO_2 or CO_2 to form calcium sulfate or calcium bicarbonate, respectively. Being considerably more soluble forms of combined calcium, these compounds develop on the surface and flake off. The extent of attack can be documented by comparing a recent photograph with an earlier photograph, drawing, painting, or written description. Verification that deterioration is due to gaseous air contaminants may be shown by microscopic comparison of the flaked materials with chips of unattacked stone or, by chemical analysis of the flakes for unusually high sulfate content.

B. Lead-Based Paint

Where hydrogen sulfide is present as a pollutant, lead pigments in paint darken due to the formation of lead sulfide. The exposed white painted surface acquires a splotchy, stained appearance, which varies in color from a grayish-brown to black.

Before the widespread use of titanium/zinc-base paints, painted residences throughout entire communities would signal the presence of hydrogen sulfide. However, since some molds and fungus which can grow on painted surfaces (especially in recessed areas) resemble the effect of hydrogen sulfide, care has to be taken to further ensure identification of the pollutant. If staining occurs only on lead-base painted areas, hydrogen sulfide is the most probable cause. Thus, side-by-side exposure of two painted buildings, one coated with lead-base and the other with titanium/zinc-base paint, can serve as surveillance monitors to detect the presence of hydrogen sulfide.

III. Vegetation Effects Indicators

Vegetation effects can monitor a wide range of gases which can be grouped as oxidants (ozone, PAN, and NO_2), sulfur oxides, ethylene and associated olefins, chlorine, and ammonia (*5*). Vegetation effects from specific source gases as opposed to atmospheric reaction products are excellent indicators of the dose rate during short-time high concentration exposure such as would occur following an accidental release of the pollutant.

Two approaches, field surveys and biological monitors, may be used as vegetation indicators of air quality. Because of the complex nature of vegetation effects evaluation, Weinstein and McCune (*6*) recommend that the person responsible for conduct of a survey should be a skilled observer and be able to recognize symptoms induced by pollutants, nutrient stresses, or plant diseases, as well as being conversant with agronomy, horticulture, ecology, plant identification, the characteristics of emissions and their dispersion, air monitoring technology, and statistics.

A. Field Surveys

The spatial and temporal distribution of air contaminants must first be described. Then the possible receptors must be identified. Cole (*7*) has suggested five guidelines for the selection of indicator species:

(a) The species should be sensitive to the pollutant at a level below the sensitivity of vegetation of economic or aesthetic importance

(b) It should be widely distributed

(c) The pollutant-induced markings should be characteristic and easily observed

(d) The species should be present throughout the growing season

(e) The species should grow from a terminal shoot throughout the growing season

Lichens and mosses can serve as suitable indicators of industrial (*8, 9*) or urban (*10*) pollution based on field observations of the relative abundance of certain species.

Following selection of the indicator species and determination of the locations where pollutant-induced effects are expected, the surveyor must select specific sites of predetermined size sufficient to be representative of the area, and, must have cataloged the vegetation symptoms anticipated.

Weinstein and McCune (*6*) recommend that the field examination report should include, in addition to site identification and description, the following items: (a) species, (b) type of marking, (c) age of foliage affected, (d) distribution of the symptoms on the leaf, and (e) distribution of the symptoms over the entire plant. In addition, photographic records should be made.

Figure 1 shows a survey form used in assessing vegetation effects which can be adapted to surveillance of gaseous pollutants by effects on specific vegetation (*11*). The form includes both number and severity, i.e., percent of each plant affected and percent of all plants affected. In utilizing growing things consideration must be given to the variability of resistance

AIR POLLUTION INJURY REPORT

Request Assistance ______ County Bucks

Grower Identification H. M. Phillips - Philadelphia

Location of Observation 340 Willow Grove Avenue

Suspected Pollutant:
SO_2, O_x, P, Other (Specify) Ethylene

Crop & Variety Orchid

Acreage Affected 3 Greenhouses or No. of Plants 1000

Date of Incidence Week of June 8, 1969

Avg. % of Each Plant Affected: 0-10, 11-30, 31-60, 61-80, 81-90, 91-99, 100

Avg. % of Plants Affected: 0-10, 11-30, 61-80, 81-90, 91-99, 100

Loss in (check one): Quality X Quantity ______

Estimated Crop Loss: 100 % or $ 1000.00

Remarks Owner supplied figure of $1.00 per orchid. All were grown in three different greenhouses at the same location.

Date 6/14/69 Reporter's Name John Jones

Figure 1. Completed report of injury (*11*).

among species. A table showing several hundred species of crops, flowering plants, trees, and native shrubs with their sensitivity to six classes of gaseous pollutants (ozone, NO_2, PAN, oxidant, SO_2, and hydrogen fluoride) developed by Benedict *et al.* (*11*) is helpful in determining the presence or absence of these gases.

B. Vegetation Monitors

If field investigation data are not sufficiently specific, biological monitoring may be employed. For this purpose replicate potted plants are located in specially designed chambers where one set is exposed to filtered air while the other is exposed to ambient air (*12*).

In one such experiment (*12*), vegetation was cultured hydroponically in vermiculite in exposure chambers for 1 month at three locations and in a control chamber ventilated with clean air. The exposure chambers were ventilated with ambient air at a rate of about one air change per minute. The control chamber was ventilated with filtered air, which was passed through activated charcoal. Specimen plants were also exposed in a greenhouse during the last 6 weeks of the investigation. Tobacco W_3, pinto bean, geranium, petunia, begonia and gladiolus (Snow Princess) plants were exposed in each chamber. The pots were kept in shallow trays with Hoagland nutrient solution; distilled water was added as needed. Each week the vermiculite was flushed with distilled water, and the plants were placed in fresh nutrient solution. Exposures were arranged so that damage could be determined during the entire exposure period and during each of the 3-week increments during the exposure period. Vegetation was also examined for short-term damage each week when the plants were tended. Table III summarizes the leaf damage reported on the vegetation.

IV. Effects Surveillance by Static Indicators

A. Painted Panels

Freshly painted glass panels can prove useful in detecting the presence of acid droplets and localized sources of airborne materials from paint-spraying, asphalt batching, and the like.

The panels should be given a single coat of enamel (automobile finishes are especially suitable) and exposed for periods of up to 72 hours immediately after drying. At this point, the painted surface has not had time to cure or build up its surface oxidation layer, and is especially suscepti-

Table III Summary of Leaf Damage[a,b] (*12*)

Vegetation	*Exposure period*	*Washington, D.C.*	*Alexandria, Virginia*	*Riverdale, Maryland*	*Bethesda, Maryland*
Tobacco	1	$O_3 + SO_2$ (35%)	$O_3 + SO_2$ (50%)	O_3 (25%)	—
	2	O_3 (10%)	O_3 (10%)	O_3 (25%)	O_3 (20%)
	3	O_3 (10%)	O_3 (10%)	O_3 (45%)	O_3 (30%)
Pinto bean	1	PAN (T)	PAN (L)	PAN (T)	—
	2	PAN (S)	PAN (S)	OX and PAN (M)	PAN (M)
	3	PAN (M)	PAN (S)	OX and PAN (L)	PAN (M)
Begonia	1	None	None	None	—
	2	None	None	OX (M)	OX (L)
	3	None	None	OX (M)	OX (L)
Petunia	4	Chlorosis (L)	Chlorosis (L)	Chlorosis (M)	Chlorosis (M)
Geranium	4	Chlorosis (L)	Chlorosis (L)	Chlorosis (L)	Chlorosis (L)

[a] Number in parentheses means percent leaf damage.
[b] Letter designations are defined as follows: T: trace (0–5%); L: little (5–25%); M: moderate (25–50%); S: severe (50–100%); PAN: peroxyacetyl nitrate; OX: oxidant.

ble to pitting from acidic droplets or particulates as well as offering a blank surface for the collection of over-sprays, colloidal mixtures, and asphaltic or sticky particles.

Painted microscopic slides are most practical for use in this type of survey since microscopic examination is used to detect the presence of pitting, discoloration, or adhesion of particles, as well as to characterize and identify the captured emission. "The Particle Atlas" (*13*) is especially useful for identification of the collected materials or their resultant effect on the paint surface. Naturally, "blank" panels must be retained in the laboratory to substantiate the observations and exposure results.

Glass jars, 8-oz size, can be prepared in the manner described above and are mounted, bottom side up on suitable holders. The jar lid is fastened to a bracket or plate for easy sample changing. Deposition on the side of the jar is directional and thus a suspected source can be located or confirmed.

B. The Effects Package

An effects-sampling system designed to be used in estimating the nature and extent of air pollution on a long-term basis is illustrated in Figure 2 (*14*). Its individual components were selected because of their simplicity and relatively low cost of preparation, operation, and analysis. In keeping with the concepts of simplicity and low cost, the system is designed to be a static testing device requiring no electricity. As a result, this type of monitoring unit can be easily sited in remote locations.

1. Exposure Stand

The effects package has a stand and body to which the various exposure components can be attached or mounted (Fig. 2). The body has an exterior frontal dimension of 18 in. $\times$ 17 in. which slopes to $13\frac{1}{2}$ in. at the rear. The roof measures 30 in. $\times$ 30 in. Each vertical wall contains a lightproof framed aluminum louver which permits the free movement of air through the unit but does not permit the entry of sunlight into the interior compartment. In the front, a door containing a lightproof louver is hinged at the top with keepers in open and shut positions; it permits access to the interior compartment and still maintains the light-tight characteristic of a closed door.

Components such as rubber strips and fabric panels, which may be deteriorated by sunlight even in the absence of air pollution, are exposed inside the interior compartment. Other components which are deteriorated

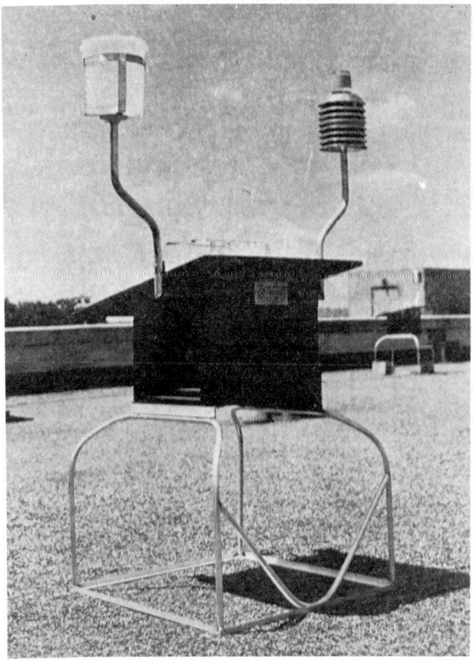

Figure 2. Exposure stand.

by air pollution but not by sunlight are attached to the stand or the roof: metal panels are attached at the rear in such a manner as to permit a 30-deg angle of exposure to the south. An insulated silver plate is mounted under the roof eave. A horizontal aluminum slide holder, 4–6 in. from the roof surface, permits attachments of nylon panels and petri dishes. Two aluminum extension arms, attached to the sides of the body and to the roof, extend to a height of 62 in. for attachment of a dustfall bucket holder on one arm, and a lead candle shelter plus a sticky-paper jar on the other arm.

2. Components

Table IV lists the various components of the effects package, their exposure periods, and effects demonstrated by each item in measurement units.

Table IV Effects Monitoring Techniques

Components	*Exposure time, days*	*Effects of pollutants*
Dustfall bucket	30	Soiling by settlable particulates, in gm/m^2-month
"Sticky" paper	7	Soiling by windblown particulates, in particles mm^2/week
Bleached fabric panel (no dye)	7	Soiling by settleable particulates in NBS[b] units
Steel or zinc panel	90; 365	Metal corrosion by sulfur dioxide, in μm/year
Sulfation plate	30	Accumulation of sulfur dioxide, in μg SO_2/cm^2-day
Sulfation candle	30	Accumulation of sulfur dioxide, in mg $SO_3/100$ cm^2-day
Silver plate	30	Tarnishing by hydrogen sulfide, as loss of reflectance
Nylon panel	30; 90	Materials deterioration by acid aerosols, in fiber breaks/exposure time
Dyed fabric panel[a]	90	Color fading by ozone or nitrogen dioxide, in National Bureau of Standards (NBS) units
Rubber strip	7	Cracking by ozone, as depth in mm/year

[a] Specifications for fabric and dye per pollutant: NO_2, acetate dyed with Disperse Blue 3; O_3, acetate dyed with Disperse Blue 27, Disperse Red 35, and Disperse Yellow 37; O_3, triacetate dyed with same three as before.
[b] National Bureau of Standards.

The techniques used to monitor effects generally have a dual objective: to indicate the occurance of individual air pollutants and to demonstrate material damage caused by pollution.

To indicate the occurence of individual pollutants, the effects package uses a dustfall bucket to collect settleable particulates, and adhesive ("sticky") impaction paper and undyed bleached muslin fabric to collect windblown particulates. Zinc and steel panels, and sulfation plates or candles are used to detect sulfur dioxide pollution. A silver plate is used to indicate the presence of hydrogen sulfide. Nylon panels are used in detecting acidic particulates. Nitrogen dioxide is detected by the use of a dyed fabric. Rubber panels and dyed fabrics are used as an indicator of photochemical oxidants and ozone.

Material damage, caused by air pollution in general can be measured by use of the same components. The degree of damage to zinc and steel panels can be evaluated to establish metal corrosion rates. Damage to the dyed fabric combinations can be used to assess dye destruction of colorfast fabrics. Damage to nylon, silver, and rubber panels are indicative of deterioration of these three products. The sulfation candle and plate can assess potential building damage in SO_2 environments, and the degree of general soiling of materials can be gauged by the dustfall bucket and the "sticky" paper.

3. Evaluating Soiling*

a. Dustfall Bucket (see also Chapter 3, Vol. I; Chapters 1 and 2, this volume). Particulates settled out of the atmosphere by gravity (*16*) are collected by an open-top 5-quart plastic canister with an open-top area of 0.029 m^2.

After exposure of 30 days, the canister is returned to the laboratory where its contents are transferred to a glass beaker, the water is evaporated, and the remaining solids are weighed. The results are reported as the equivalent weight of material (in grams) that settled over a surface of 1 m^2 during the month. The factor for converting grams per square meter to tons per square mile per month is 2.86.

b. Sticky Paper. Particulates in the size range of 20 to 100 μm are collected on sticky paper (*17*). These particulates, which indicate the soiling potential of the atmosphere, are carried by wind from a source of pollution and captured by adhesive-coated paper mounted on a cylinder perpendicular to the earth's surface.

* See Ref. (*15*).

The sampler is called an ambient adhesive impactor or sticky-paper sampler. It is prepared by wrapping a 2-in. strip of contact paper (Fasson's Pli-a-Print, R-135) around a glass jar (diameter, $2\frac{3}{4}$ in.; height $3\frac{1}{4}$ in.) with the adhesive side out. Only the north direction is marked on the paper.

After an exposure of 7 days, the sampler is sprayed with clear enamel to fix the particulates and to eliminate entrapment of others.

In the laboratory, the particulate loading (particles/square millimeter) at eight points, representing eight directions of the compass, is estimated by comparing the sample to a series of standard photographs of samples previously counted microscopically which range from 15 to 100 particles/mm^2. Figure 3 shows a particle comparator available for this purpose.

As an adjunct to the fixed sampler, the rotary ambient adhesive impactor (*17a*) can be used for mobile surveys and would be especially useful in evaluating fugitive dusts of >20 μm. An uncomplicated hand-powered rotary sling, the sampler uses a round $\frac{1}{4}$ in. o.d. $\times$ 2-in.-long target, at the end of a 30-in.-long flexible wire attached to a suitable handle. The target is whirled through a 60-in.-diameter circular path at 200 rpm for approximately 1 minute. The resulting approach velocity will give a collection efficiency of about 87% for particles 20 μm in size and specific gravity of 2.5. Efficiency increases as particle size increases. The resulting catch is evaluated by microscopic counting or by matching to a precounted standard. Results are reported as "number of particles per unit of area" directly from the count or as "particles per cubic meter" when

Figure 3. Gruber particle comparator. (Courtesy of Research Appliance Company.)

the volume of air sampled is determined. The volume sampled is frontal area of the target (cm^2) $\times$ circumference of path (cm) $\times$ rpm $\times$ duration of sampling (minutes) $\times$ 10^{-6} with the volume sampled being cubic meters.

4. Measuring Metal Corrosion*

a. ZINC PANELS. Before exposure, new 4-in. $\times$ 3-in. $\times$ 0.065-in. zinc panels are scrubbed twice in warm detergent solution, rinsed with water, and then rinsed with acetone. The panels are then air dried, weighed to the nearest 10 mg, and placed in a kraft envelope for transport to the exposure location.

At the exposure site, the zinc panel is mounted between plastic insulators at an angle of 30° from the horizontal facing south.

After exposure, generally of a year, but, in special circumstances, 3 months, the panel is scrubbed and rinsed in water, immersed for 5 minutes in a warm (160° to 180°F) solution of ammonium chloride, and then scrubbed and rerinsed with water. If corrosion still remains on the panel, it is dipped for 20 seconds in a hot (200° to 212°F) 5% chromic acid solution, rerinsed with water, and then rinsed in acetone to facilitate air drying before weighing.

Corrosion is measured as weight loss. Because weight loss depends on the size and shape of the panel, it is more meaningful to report results as units of depth of penetration of the corrosion. The weight loss expressed in grams is converted to penetration depth in micrometers per year (μm/year). The measurement is corrected for time exposure expressed in days.

$$\mu\text{m/year} = \frac{3300 \text{ (weight loss, gm)}}{\text{days exposed}} \tag{1}$$

b. STEEL PANELS. Before exposure, new 4-in. $\times$ 3-in. $\times$ 0.03-in. steel panels† are scrubbed with detergent, rinsed several times with water, rinsed with acetone, and then air dried. They are then dipped in concentrated hydrochloric acid to remove rust spots, rinsed several times with water, rinsed with acetone, and air dried over a hot plate. Finally, each panel is weighed to the nearest milligram and placed in a kraft envelope for shipment.

* See Ref. (*15*).

† Low carbon (0.019%), low copper (0.028%) steel; Armco enameling iron DQ plate.

At the exposure site, the panels are mounted between plastic insulators at an angle of 30° from the horizontal, facing south.

After exposure, generally of a year, but, under special circumstances, 3 months, the plates are wire brushed to remove loose rust, rinsed in an inhibited acid solution to remove the rest of the corrosion products, and rinsed with water, then acetone, before air drying over a hot plate. Each cleaned panel is then weighed to the nearest 10 mg.

The corrosion penetration rate in microns per year is calculated from the following relationship:

$$\mu\text{m/year} = \frac{3300 \text{ (weight loss, gm)}}{\text{days exposed}} \qquad (2)$$

5. Measuring Sulfation*

The measurement of sulfation was introduced in the early 1930's in England to give a direct index of the activity of airborne sulfur oxides and sulfates, such as in the attack of polluted air upon building stone, exposed metal, and the like.

a. SULFATION CANDLES (see also Chapter 6, this volume). A sulfation candle is exposed for a month. The candle surface—prepared by coating it with lead dioxide, which is sensitive to sulfur dioxide—is a cylinder 10 cm in height with a total reactive area of 100 cm². When atmospheric sulfur dioxide comes in contact with the lead dioxide on the candle, it is converted to lead sulfate.

After exposure, the lead sulfate is removed by adding sodium carbonate to the surface. Then the amount of soluble sodium sulfate formed is measured turbidimetrically as barium sulfate (*18*).

The amount of sulfur dioxide collected per 100 cm² of surface area per day—measured as barium sulfate—is converted to and reported as the equivalent amount of sulfur trioxide (as mg SO_3/100 cm²-day). Past measurements by similar methods have yielded an empirical relationship between sulfation and sulfur dioxide concentrations: sulfur dioxide in ppm can be estimated from milligrams of sulfur trioxide per square centimeter per day by multiplying the sulfation rate by 0.035 (*19*).

b. SULFATION PLATE. In the sulfation plate method, sulfation is measured by exposing an inverted petri dish (48 mm i.d.), coated on the inside with a layer of lead dioxide paste. Both the plate and the candle

* See Ref. (*15*).

use the same materials and have identical reactions, but the plate has a much smaller surface, only 18 cm^2 compared to 100 cm^2 for the candle. The plate is more stable, easier to fabricate and analyze, and less expensive to expose than the candle. Plates are usually exposed for 30 days. A comparative study of the plate and sulfation candle methods of detection has been made (*19*).

The amount of sulfate, determined the same way as for the candle, is converted to and reported as the amount of sulfur dioxide per square centimeter per day (as $\mu g\ SO_2/cm^2$-day). The factor used to convert the micrograms of sulfur dioxide to milligrams of sulfur trioxide is 0.125, which allows a direct comparison with data collected by the candle.

6. Measuring Materials Deterioration*

Silver plate, nylon panels, dyed fabric panels, and rubber strips are used as indicators of materials deterioration caused by air pollution.

a. Silver Plate. Before exposure, the reflectance of each silver-plated plate (William A. Rogers No. 1120, 6 in. diameter) is measured with a photovolt reflection meter (Model No. 610 with a 610-D search unit) with a green tristimulus filter.

At the exposure site, a plate is exposed for 30 days in an upside down position under an overhanging roof shelter to protect the silver film from particulates and precipitation. After exposure, the reflectance is again measured, and the reflectance loss is calculated as percent loss per month of exposure.

b. Nylon Panels. A sample of good quality (15 denier) nylon hose is stretched and glued between the halves of a standard Polaroid slide mount. The $3\frac{1}{4}$ in. $\times$ $2\frac{3}{8}$ in. panel has an exposure area of 7.7 in.^{2}. The panels are exposed in a horizontal position usually for 30 days but, in special circumstances, for 90 days. After exposure, the samples are examined for breaks in the fiber. An image of the sample can be projected onto a screen and examined or, with the aid of a low-power objective, it can be examined microscopically. The measurement is reported as total number of defects (such as broken fibers causing runs or perforations) per exposed panel.

c. Dyed Fabric Panels (*15*). Samples of dyed fabrics are glued to $2\frac{3}{4}$-in.-square pieces of cardboard, with 2-in. square cutouts, by using silicone contact adhesive.

* See Ref. (*15*).

Analysis of the fabric panels (*20*) is made by using a color-difference meter.* This is a photoelectric color instrument which measures incandescent light reflected at 45 deg from the cloth sample. The different fabric panels are measured for light reflectance through red, green, yellow, and blue filters before and after exposure to obtain a total color difference or fading reading. These differences in "before" and "after" appearances are calculated and reported in National Bureau of Standards color-difference units (ΔE).

d. Rubber Strips. Natural rubber, ozone-sensitive, test slabs measuring 9 in. × 9 in. × $\frac{1}{6}$ in. are cut into 8-mm-wide strips. (The length is not critical; about 50 mm is sufficient.) These strips are exposed for 1 week under a constant stress caused by the gravitational pull of a 360-gm lead weight.

Specifications for ozone-sensitive rubber, as supplied to the Los Angeles County, California, Air Pollution Control District, are as follows:

Plantation rubber	100 parts
Tire reclaim–xylose 5484 or 6008	125
SRF black	33
Stearic acid	1.5
Pine tar	8.0
Zinc oxide	4.0
Captax	0.7
Sulfur	5.0

Cure 40 minutes at 45 psi steam pressure. Well-vulcanized compounds are desirable, since undercures with bloom make reading of the cracking difficult. It is important that the mold used have highly polished surfaces and that the thickness of the stock be uniform and approximately $\frac{1}{6}$ in. Prior to use, the plaques are exposed to the air for several days to reduce tack, increase sensitivity to ozone, and to render the initial cracking more easily visible. This is conveniently done by hanging them on hooks suspended from a cord or wire and allowing sufficient room between plaques for free circulation of air. Properly cut stock should present a smooth rectangular cut when viewed endwise. The cut pieces can be stored indefinitely in clean jars with tight-fitting covers.

After exposure for 7 days, the rubber strips are prepared for microscopic examination. A small corner is cut off to identify the edge of the strip, and the strip is cut in half lengthwise. The half with the cut corner is placed under a 50× lens so that the freshly cut edge may be observed. The cut strip is stretched slightly (about 0.15-in.-length increase) to make crack depth easier to measure; only the cracks within $\frac{1}{2}$ in. from the center of the strip length are measured since the cracking of the strip

* Color and Glass Meter (photovolt 670), equipped with tristimulus filters.

may be inconsistent at the ends where the strips were fastened. The depths of nine consecutive cracks are measured by using a microscope fitted with a Whipple disk calibrated with a stage micrometer 2 mm in length with subdivisions of 0.01 mm. The average of the nine is calculated and reported.

According to Vega and Seymour (*21*), if the exposure is 7 days, the average ozone concentration can be approximated in parts per hundred million by multiplying the average crack depth in millimeters by six.

C. Performance Evaluation

In 1971 the United States Environmental Protection Agency issued a detailed analytical statistical report (*15*) which evaluated the performance of the effects package. The conclusion advanced by the investigators can generally be applied to all effects monitoring, whether utilized in a long-term system configuration or in single or specific combination for short-term problem definition. The report states that:

(a) Static indicators are capable of detecting the existence of major pollutants or potential pollution problems. These are sulfur oxides, particulates, nitrogen oxides, hydrogen sulfide, acid aerosols, and oxidation potential. The conclusion is based upon either direct correlation of sampler data with air quality or emissions data, or on strong evidence of source-receptor relationships. However, evidence of these pollutants does not necessarily imply knowledge of specific concentration levels.

(b) Effects samplers can be used to detect gross differences in pollution levels. This conclusion was substantiated by the differences observed in the frequency distribution of pollutants at various classes of stations, by comparison of pollutant levels with emission levels, and by correlations with air quality data. By computing the frequency distribution of annual averages and comparing individual stations and/or areas to the nationwide distribution, an indication of relative severity can be obtained.

(c) Components can provide evidence of the extent to which materials are damaged by air pollution. The most notable effects are related to corrosion; however, damage to nylon and rubber was also noted. It was shown that corrosion was enhanced by air pollutants, specifically sulfur dioxide.

REFERENCES

1. R. L. Salmon, "System Analysis of the Effects of Air Pollution on Materials." Midwest Res. Inst., Kansas City, Missouri, 1970.

2. "A Survey and Economic Assessment of the Effects of Air Pollutants on Electrical Components," Project Report 179, Vol. 1, Sect. 1–9. ITT Electro-Physics Laboratories Inc., Columbia, Maryland, 1962.
3. J. W. Spence and F. H. Haynie, "Paint Technology and Air Pollution: A Survey and Economic Assessment," Office of Air Program Publ. No. AP-103. United States Enviromental Protection Agency, Research Triangle Park, North Carolina, 1972.
4. R. A. Simon, "Effects on Materials" (presented as lecture in Air Pollution Training Program). United States Environmental Protection Agency, Research Triangle Park, North Carolina, 1970.
5. N. L. Lacasse and W. J. Moroz, "Handbook of Effects Assessment Vegetation Damage," Center for Air Environment Studies, Pennsylvania State University, University Park, Pennsylvania, 1969.
6. L. H. Weinstein and D. C. McCune, "Recognition of Air Pollution Injury to Vegetation," Inform. Rep No. 1. TR-7. Agr. Comm., Air Pollution Control Association, Pittsburgh, Pennsylvania, 1970.
7. G. A. Cole, *Agron. J.* **50,** 553–555 (1958).
8. O. L. Gilbert, *New Phytol.* **67,** 15–30 (1968).
9. D. N. Rao and F. LeBlanc, *Bryologist* **70,** 141–157 (1967).
10. E. Skye, *Acta Phytogeogr. Suecica* **52,** 1–123 (1967).
11. H. M. Benedict, C. J. Miller, and R. E. Olson, "Economic Impact of Air Pollution on Plants in the United States," Final Report. Stanford Res. Inst., Menlo Park, California, 1971.
12. "Washington, D.C., Metropolitan Area Air Pollution Abatement Activity," Project Report. United States Department of Health, Education, and Welfare, Pub. Health Serv., Cincinnati, Ohio, 1967.
13. W. C. McCrone, R. G. Draftz, and J. G. Delly, "The Particle Atlas; A Photomicrographic Reference for the Microscopical Identification of Particulate Substance." Ann Arbor Sci. Publ., Ann Arbor, Michigan, 1967.
14. G. A. Jutze, R. L. Harris, Jr., and M. Georgevich, *J. Air Pollut. Contr. Ass.* **17,** 291 (1967).
15. J. H. Cavender, W. M. Cox, M. Georgevich, N. Huey, G. A. Jutze, and C. Zimmer, "Interstate Surveillance Project," Air Polution Control Office Publ. No. APTD-0666. United States Environmental Protection Agency, Research Triangle Park, North Carolina, 1971.
16. American Society for Testing and Materials, "ASTM Standard Method D-1739," Part 23, pp. 505–508. ASTM, Philadelphia, Pennsylvania, 1968.
17. C. W. Gruber and G. A. Jutze, *J. Air Pollut. Contr. Ass.* **7,** 115–117 (1957).
17a. C. W. Gruber and G. Harms, "Fugitive Dust Measurement by the Rotary Ambient Adhesive Impactor," Pap. No. 74-85. Air Pollution Control Association, Pittsburgh, Pennsylvania, 1974.
18. N. A. Huey, *J. Air Pollut. Contr. Ass.* **18,** 610–611 (1968).
19. N. A. Huey and W. Waller, Pap. No. 69–133. 62nd Annu. Meet. Air Pollution Control Association, Pittsburgh, Pennsylvania (1969).
20. American Association of Textile Chemists and Colorists, *Amer. Dyest. Rep.* **2,** 450–451 (1958).
21. T. Vega and C. J. Seymour, *J. Air Pollut. Contr. Ass.* **11,** 28–44 (1961).

11

Radionuclide Surveillance

Harry F. Schulte

I. Introduction

The concern over radioactive air pollution dates back to the first atomic bomb test. At the Trinity Test in southern New Mexico in July 1945, a network of monitoring stations and mobile monitors was established to look for radiation and radioactive materials released as a result of the test. By present day standards, their equipment was extremely crude, and consisted largely of instruments for detecting the presence of radioactive material on the ground. No actual air sampling equipment was used. Nevertheless, radioactive contamination on the ground was detected miles from the site of detonation, and it could have gotten there only through the air. Since that day, there have been new concerns expressed each time a major new application of radiation or radioactivity is introduced.

Radiation, more specifically ionizing radiation, is a form of energy, largely localized at a site, whose intensity diminishes rapidly as one moves away from the source. Radiation, itself, is not an air pollutant and most forms of radiation are incapable of producing anything that can become an air pollutant. Thus, processes which use intense sources of γ radiation or hospitals using such radiation for intensive cancer therapy are not matters of concern from the standpoint of air pollution. An exception to this is an intense source of neutrons, such as that produced by an accelerator, nuclear reactor, or explosion of an atomic bomb, which can induce radioactivity in surrounding materials. In the case of accelerators, this is usually not significant since the materials activated are usually fixed parts of the equipment. Neutrons from air-cooled reactors can activate components of the cooling air and atomic bombs can activate the debris which they disperse into the air. Our real concern is with radioactive materials in the air. (See also Chapter 5, Vol. I.)

When testing of atomic bombs intensified in the 1950's "fallout," which consists of products of the detonated device and activated dirt and debris, was readily detectable in air samples collected many miles from the detonation. Concern in the United States began to increase when testing was started in Nevada, and networks of sampling stations were established, first around the test site, then throughout the United States, and finally throughout the world. These stations were equipped with air sampling devices, usually filter paper, and airborne radioactive material could be detected anywhere. While these tests were going on almost anyone could obtain radioactive material for experimental purposes by filtering it from the air even with relatively crude equipment.

With the ratification of the Test Ban Treaty this source of concern has diminished, although French and Chinese tests continue to add small amounts of radioactive material to the atmosphere. As nuclear power reactors were successfully developed and built, there was increasing concern over these as sources of radioactive air pollution. Although such reactors, during normal operations, create small amounts of radioactive materials which can become airborne, it is relatively easy to limit such discharges. The major concerns are the possibility of an accident that might cause serious releases of radioactive material to the atmosphere from the fuel elements and the gradual buildup of large quantities of highly radioactive materials in the form of wastes. The latter materials accumulate at processing plants where fuel elements are treated to separate out the wastes and recover the useful fuel materials for reuse.

In recent years, there has been an intensified effort to develop a successful breeder reactor which would produce plutonium in sufficient quantities that it could economically serve as a fuel in the same or other reactors. Such reactors operate at very high temperatures and use liquid metals as coolants and heat transfer media. Partly as a result of this increased interest in plutonium, and partly because of a serious fire in a major plutonium facility in 1966, there has been considerable concern expressed regarding possible hazards to the public from airborne plutonium. The use of plutonium-238 as an isotope power source, especially in space missions, has increased this concern.

Another area of public interest in this field has been that involving the use of atomic explosives for operations, such as stimulating natural gas production by underground explosions. Such explosions, especially in the presence of underground water, produce considerable quantities of tritium, the hydrogen isotope of atomic weight three, which is then present both as the oxide or water and as hydrogen gas. This may leak to the atmosphere from the underground source or it may become part of the natural gas and be released when the gas is used.

Finally, krypton-85, a noble gas, is produced by the fission process and some is released to the atmosphere by nuclear power reactors. The relatively long half-life (10 years) of this gas, plus its chemical inertness, which makes it difficult to remove from air, indicates there may be a gradual buildup of it in the atmosphere as more power is produced by nuclear reactors. It is possible that the concentration of krypton-85 in the atmosphere may be the factor which establishes the ultimate limit on the number of power reactors in operation (*1*). All of these concerns have led to a renewal interest in radioactive substances as air pollutants.

II. Physical and Chemical Forms of Radioactive Pollutants

Radioactive air pollutants may exist as solid or liquid aerosols or they may be gaseous. Radioactive liquid aerosols are rare as are most other liquid air pollutants. Solid aerosols are most frequently metals or their oxides, although their actual composition depends on the operation by which they are formed. Where their release occurs as a result of a fire or an explosion, they are most commonly oxides. A nuclear explosion produces a wide range of products that are initially in elemental form, but are in a high state of activation that causes rapid and intricate chemical reactions. Some products decay rapidly changing from gas to solid or vice versa. The same changes take place under other less dramatic conditions in reactors. Radon, emanating from the decay of radium, is a gas and diffuses rapidly. As it in turn decays, solid radioactive products consisting of single atoms are formed which deposit quickly on dust particles or surfaces.

There is no method of sampling that is common to all radioactive materials. The property of radioactivity implies no common physical or chemical property that enables them to be collected from the air. Methods of collecting particulates are applicable to most solid aerosols, whether radioactive or not, but the measurement of radioactive gases in air requires a variety of techniques depending on the properties of each gas.

III. Purposes of Monitoring Air for Radioactive Substances

Air is monitored for its content and concentration of radioactive material and the reasons for making such measurements may affect the choice of the sampling and measuring technique.

Air may be monitored to determine its general contamination content in order to document changes that occur over a period of time. Thus, before a reactor or nuclear material processing plant is constructed, it has been customary to conduct an extensive program of sampling air, soil, water, and vegetation in the area surrounding the proposed plant. This preoperational survey serves as a baseline to which future measurements may be compared. Such surveys are consistent with the procedures stated under the United States National Environmental Quality Act and predate it by many years. Some radioactive materials occur in all natural environments and their abundance varies from place to place. For these reasons such background studies are essential. Since background radioactive air concentrations are low, high volume sampling is necessary.

Another reason for air sampling is to evaluate a potential localized hazard, such as that which might possibly be produced by a power reactor or a fuel processing plant. Data, here, are to be related both to the results of the background study and to ambient standards. Such sampling may be conducted intensively following the startup of a new plant as one means for determining whether all control systems are functioning properly. Intense sampling is done at the effluent source, i.e., stacks and similar release points. However, sampling in the plant environs, especially in early phases of plant operation, also is necessary to detect releases which may not have been anticipated.

After plant operations have entered a routine phase, sampling is still continued on a somewhat reduced scale. The emphasis, now, is on sampling to determine whether compliance with ambient air quality standards is being achieved. Sampling stations may now be extended farther afield from the plant, and some may be located in nearby communities. Community sampling stations are the most common type of sampling for radioactive air pollutants and are essentially in permanent operation during the life of the plant. This type of sampling is done by plant personnel. Additionally, public agencies may operate similar networks of sampling stations to assure compliance with regulations. Individual samples may cover relatively long time spans, and equipment requirements are for moderate sampling rates and a high degree of equipment reliability.

Air sampling may be done as part of a specific study of limited geographic extent or of global extent. The presence of radioactive materials in the air affords a means of studying large-scale air movements, evaluating meteorological parameters, and checking diffusion models. The United States maintains sampling networks operated variously by the Public Health Service, the Environmental Protection Agency, the Energy Research and Development Administration, the Nuclear Regulatory Commission, and state agencies. In some cases, air sampling is done to evaluate the radiation exposure of special population groups. Certain areas of the world have high concentrations of radioactive materials in soil which can be transferred to air by wind and diffusion forces. Valuable studies have been made on such populations to determine whether any biological effects can be related to the elevated radiation dosage of the population. Air concentrations must be measured along with measurements on soil, water, food, and vegetation. Air is also sampled regularly at high altitudes to detect tests of nuclear explosives that may be in violation of the Test Ban Treaty. This is a very special application using specialized equipment and will not be discussed in this chapter.

Air sampling for radioactive materials is usually part of a more general type of surveillance which also includes measurements of surface contamination, soil and vegetation contamination, and general ambient radiation levels. Results of such studies may be quite complex involving the interaction of many ecosystems.

IV. Radioactive Particulate Sampling and Quantitation

Many operations involving the processing of radioactive materials lead to the production of solid aerosols, such as dusts or fumes. Other operations may produce aerosols that are initially liquid droplets, such as sprays or mists, but that quickly dry to form solid particulate aerosols. Accidents involving high temperatures may volatilize materials which quickly condense from the vapor phase to form solid particulates. Sampling methods for these depend on the separation of the solid phase from the gaseous medium, air, and not on the specific chemical properties of the radioactive elements.

*A. Filters**

Filtration is the most widely used method for separating a solid aerosol from the air in which it is suspended. Its advantages include simplicity, low cost, versatility, and the fact that the quantity of collected material usually can be measured without removing it from the paper. A wide assortment of papers of various sizes and properties are available. In addition to papers originally developed for chemical manipulations, there are now many papers available which were specifically developed for air sampling purposes. Each type of filter paper has advantages and disadvantages which must be weighed carefully.

Use of filters for sampling for radioactive materials imposes requirements that may not be significant in sampling for nonradioactive pollutants. In fact, even the type of radioactivity to be measured is important in selecting the proper paper. The most exacting requirements are imposed when direct counting of α radioactivity is to be used to quantitate the collected material. In this case, it is desirable to have collection take place as nearly as possible on the surface of the paper with very little penetration of particles into the paper. Alpha particles are readily stopped by even very thin materials with the result that materials which penetrate into the paper will not be included in the amount measured (*1a*). Uranium, plutonium, polonium, and radon decay products are the most important α emitters that may be determined in this manner.

Beta radiation is absorbed less by thin layers of filter paper, with the result that surface collection is somewhat less important for β emitters,

* See also Chapter 1, this volume; Chapter 4, Vol. IV.

such as strontium-90. For the γ emitters, penetration into the paper is not a significant factor in filter selection. However, if the collected material is to be removed from the paper and can be removed quantitatively, the matter of surface collection can be ignored.

The analytical grades of chemical filter paper have been used for air sampling for a long time and are available in a variety of grades and sizes. The fiber diameters of these filters are relatively large and efficiency increases significantly as the filtration velocity increases. With the widely used Whatman No. 41 paper, efficiency is very low at 15-ft/minute filtration velocity but rises to 90% at 80 ft/minute for 0.3-μm-diameter particles, as shown in Figure 1 (*2a*). The denser or harder papers are more efficient, but their flow resistance is high, requiring larger pumps or lower flow rates.

Filter efficiency is also a function of the particle diameter of the aerosol being filtered, or more accurately, of the aerodynamic diameter which, for spherical particles is the diameter multiplied by the square root of the density. The efficiency of a given filter, then, can best be expressed as a function of aerodynamic diameter and filtration velocity. An example of this dependence is shown in Figure 1, which shows experimental data on Whatman No. 41 filter paper. It should be noted that for sizes above 1 μm or for filtration velocities above 80 ft/minute efficiencies are all high and essentially invariant. For filters with smaller fiber diameters than Whatman No. 41, variations of efficiency with particle size and filtration velocity are less and efficiencies are higher. Analytical grade papers are comparatively low in cost and are easily wet- or dry-ashed for subsequent radiochemical analysis. Aerosol particles do penetrate into the paper so there may be some losses on direct α counting.

In addition to the cellulose fibers which make up the analytical grade papers, other filters which can be used in air sampling include those made

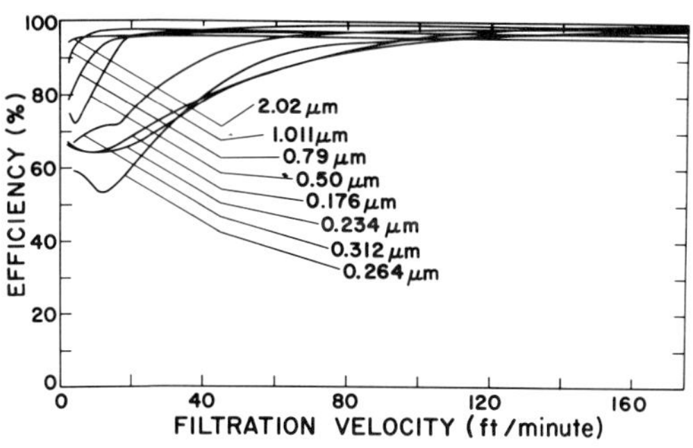

Figure 1. Efficiency of Whatman No. 41 filter paper as a function of particle size and filtration velocity for unit density spheres (*2a*).

of glass fibers, mixtures of glass fibers and cellulose, cellulose and asbestos, synthetic organic fibers, and controlled pore size plastics and metals.

Glass fibers are available in very small diameters and can be made into very efficient papers that collect aerosols near the surface. Their efficiencies remain high over a broad range of particle sizes and filtration velocities. Such papers are more expensive than cellulose fiber filters, have a moderately high pressure drop at most sampling flow rates, and are somewhat fragile to handle. These papers can be leached with strong reagents to remove the deposit if radiochemical analysis is required.

Papers made from mixtures of cellulose fibers and either asbestos or glass have characteristics similar to those of all-glass fibers but are somewhat easier to handle because they are less fragile. The cellulose allows somewhat more penetration into the filter material, causing some loss with direct α counting. Because of the presence of the cellulose, these mixed papers cannot be leached, and because of the asbestos or glass, they cannot be ashed. As a result, radiochemical analysis of the collected material is almost impossible. For this reason, also, it is difficult to assess the magnitude of the error due to penetration into the filter. These papers are useful where the deposit is measured by direct counting and where trends or variations in concentrations are sought using the same sampling conditions day after day.

Membrane filters are materials having pores of controlled diameters and are usually made of plastic materials, although metallic membranes also are available. Their pore size is determined in the manufacturing process and filters of widely varying pore size are available. Originally, they were considered essentially sieves, but detailed study has shown that the pores are really curving paths; hence, the collection mechanisms are similar to those of fibrous filters. Particles smaller than the pore diameter can be collected, and very fine particles are collected with high efficiency. Collection efficiency is high for these very thin filters and most material is collected very close to the upstream surface, resulting in very low losses with direct radioactivity counting.

The original membrane filters were made of cellulose esters and were very fragile. Membranes incorporating fibers to give flexibility and strength and using more durable plastics such as vinyl polymers are now available. Because of their organic composition, these membranes can be dissolved readily by certain organic solvents, or they can be ashed easily. By these means, the deposit can be separated readily from the filter for subsequent analysis. Another useful property of these filters is the fact that they can be rendered completely transparent by adding a drop of immersion oil of the proper refractive index. The deposit then can be examined under the light microscope. Techniques also have been developed to transfer the deposit to an electron microscope grid for examination by that instrument (*3, 4*).

Membranes made of consolidated metal particles also are available and are useful in sampling corrosive or hot gas streams. Silver membranes have been used widely for this purpose. Membrane filters are generally more expensive than fibrous filters and require relatively high pressure drops to achieve useful flow rates. They are widely used because of their high efficiencies and special properties, and are especially valuable as standards against which other sampling methods can be rated.

True sieves made of plastics with controlled hole sizes are made by subjecting thin films of nonporous plastics to bombardment by energetic particles such as neutrons. The film is then soaked in a developing solution, such as sodium hydroxide, which attacks the neutron-damaged plastic, producing straight holes through the plastic where the neutrons have gone through. Hole diameter is controlled by regulating the duration of development or etching. Even with these films used as filters, some particles smaller than the hole are collected by inertial and electrostatic forces. These filters are mainly used for studying the size characteristics of aerosols, although their further development may reveal other useful properties (*5*).

Filters made of fine organic fibers, such as polyamide plastic, have the desirable properties of high collection efficiency, strength, and ease of ashing. They have not been widely used in the United States and are not readily available there.

Air sampling efficiency should be high, but it need not be 100%, particularly if the efficiency is known. However, if the efficiency is less than 90%, it is likely to be affected considerably by variations in flow rate and particle size. If it is important to ensure that all particle sizes are collected with high efficiency, then a filter having these properties must be selected, and it must be operated at a filtration velocity giving uniformly high collection efficiency. Information on filter efficiency over a broad range of filtration velocities and particle sizes for available filters are still scarce and, to some extent, conflicting (*6, 7*). Additional data on this topic probably would reveal filter media having properties more desirable than those most used at present. It is fortunate that most sampling filters in common use are fairly efficient in all particle size ranges when operated at moderately high filtration velocities (100 ft/minute) and, hence, moderate pressure drops.

The actual instruments for sampling air for the determination of radioactive material content by filtration are basically the same as those used for sampling nonradioactive particulate pollutants. They consist of a

Figure 2. An air sampler used by the United States Environmental Protection Agency.

source of suction (usually a pump), a flow rate or total flow indicator, and a filter in a suitable holder (Figs. 2–4).

B. *Impactors**

Impaction devices have been used in air sampling for radioactive materials, but mostly for industrial hygiene rather than for air pollution studies. In these devices, the airstream is speeded up by passing it through a jet and then allowing it to impinge or impact on a surface that may or may not be coated with a sticky substance to retain the dust. The material is collected in a small area immediately in front of the jet, and the size range collected is a function of jet velocity and the system dimensions. Since air pollution studies usually involve long collection times under outdoor conditions where natural dust exists, impactors usually are unsatisfactory. Under such conditions, natural dust piles up at the

* See also Chapter 1, this volume.

Figure 3. Another type of air sampler also used by the United States Environmental Protection Agency.

collection point, and the collection efficiency diminishes rapidly. Moving slides and tapes have been used for this purpose but are satisfactory only for relatively short-period sampling. Impactors are useful for studying short-term emissions or releases in the field. Specialized forms of impactors are used where particle-size data are required.

*C. Impingers**

Impingers use impaction under a liquid surface and are little used in radioactivity surveillance. They occasionally have been used in combination with other collectors for sampling hot, wet emissions, such as those from an incinerator for radioactive combustible wastes. Here, the sampling method is somewhat similar to that used for stationary source sampling for nonradioactive substances except that one or more impingers *precede* the filter. The impingers are immersed in ice water and both trap

* See also Chapter 1, this volume.

Figure 4. Trailer equipped for air sampling. A gasoline powered generator on the trailer furnishes electric power. On the left is a high volume sampler; a sampler for collecting a series of samples is in the center and a cascade impactor for particle-size analysis is on the right.

the larger particulates and cool the gases. An entrainment trap must be placed between the impingers and the filter.

D. Precipitators*

The thermal precipitator, though highly efficient as a collector, is not used for sampling air pollutants because of its extremely low sampling rate. It may be used, under exceptional circumstances, to collect a sample directly on electron microscope grids for high magnification observation.

The electrostatic precipitator is rarely used in studying radioactive air pollution because of the difficulty of adequate maintenance and unstable operation when lengthy sampling periods are required. The high voltage required is a hazard if the sampler must be left unattended.

E. Settling Trays†

Settled dust collection devices have been used widely for measuring dust fall. For radionuclide surveillance, "radioactive fallout" is still a

* See also Chapter 1, this volume; Chapter 5, Vol. IV.

† See also Chapter 1, this volume.

very important measurement. Where the radioactive air pollutant is a γ or an energetic β emitter, the concentration per unit area on the ground is a more realistic measure of the hazard than is the air concentration. Ground-deposited material becomes a source of hazardous radiation whose intensity increases steadily during the period of deposition. After deposition ceases the radiation continues, decreasing with dispersal and penetration of the radioactive material below the surface and with radioactive decay. Contamination of soil or vegetation is usually measured in terms of contamination per unit area. Thus, the "fallout tray" is a standard instrument in the assessment of the more harmful aspects of radioactive air contamination. Because of their simplicity, low cost, and minimum maintenance or service requirements, such trays may be used in large numbers and in places where aspirating air samplers cannot be used.

The usual tray for detecting radioactive material is a simple metal sheet coated with a sticky substance or with a sheet of gummed paper. After a fixed period of exposure, the metal sheet may be placed in a counter having a specially designed probe for direct counting of radioactivity, or the material may be removed from the tray and the radioactivity determined in various ways. For coating the metal sheet, it is convenient to use a mixture of a nondrying alkyd resin containing a small amount of a plasticizer, such as castor oil. This mixture is thinned with toluene and applied to the metal surface with a clean paint brush. The toluene evaporates leaving the resin-plasticizer coating. After exposure, the collected material can be washed off with toluene and the material wet- or dry-ashed for analysis. Care must be taken in the ashing process not to lose volatile radioactive materials, which may be present. Paper having a gummed surface on both sides may be applied directly to the metal plate instead of the resin. After exposure, the gummed paper may be stripped off and ashed. Radiochemical analysis for various elements may be performed on the washed-off material.

If the particles falling out are of fairly uniform size, and if turbulent atmospheric conditions are not present, a rough estimate of the average air concentration of the fallout material may be made from the tray results.

$$C = R/V \tag{1}$$

where

C = Air concentration in mass (or activity) per cubic meter
R = Fallout rate in mass (or activity) per square meter per minute
V = Particle fall (or deposition) velocity in meters per minute

For particles larger than about 20 μm, V may be taken as the settling velocity calculated from Stokes' law. Deposition velocities for smaller particles are strongly affected by air currents and eddy diffusion. Such particles, as well as vapors, may also collect on the underside of a tray, if the coated surface is exposed in a manner to permit this. Interpretation of settling tray results is difficult in quantitative terms. Trays are useful when relatively large quantities of radioactive materials may be present, particularly where particles of fairly large size are moving downward through the atmosphere. For strong γ emitters, the same information as is obtainable from trays can be obtained by measuring the radiation from material directly deposited on the ground or even by measurements from aircraft.

F. Quantitation of Collected Samples

Where particulate radionuclides are sampled from the atmosphere, in many circumstances, the total quantity of radioactive material can be determined directly by electronic counting of the activity on the filter paper, impactor plate, settling tray, or other collector. Strictly speaking, this is true only if a single radionuclide is present but the range of usefulness is much wider than this. Thus, if air around a plant primarily devoted to processing fuel elements is sampled, the filter paper may contain decay products of radon and thoron in addition to various isotopes of uranium and plutonium and a large variety of radioactive fission products. If several days are allowed to elapse after sampling, before the paper is counted directly in an α counter, the total count will reflect only the α emitters, uranium and plutonium. This is known as the "gross α count." If the activity concentration represented by the gross α count is less than that which is acceptable for the most hazardous α emitter, plutonium, then no further refinement of the quantitation technique need be undertaken. Similarly, a gross β count can be determined which may be compared with the standard for the β emitter of most concern, e.g., strontium-90. There are other ways of estimating concentrations where a mixture of isotopes are present that do not require removal of the material from the filter and some of these will be noted later (*8*).

The electronic equipment required for the determination of radioactive material on a filter paper is complex and cannot be described in detail here. For direct counting of air samples, the size of the probe or detection device should match the size of the sampling filter paper or be some simple fraction of the paper size. Special probes can be built to handle most filter sizes and shapes used with standard scalers and count-rate meters.

Thus, the same basic equipment can be used for counting filter papers and for other radioactivity measurements.

Proportional counters have been used widely for counting filter papers for α and β activity. These frequently give erratic results, especially at high levels of activity, because of the distortion of the electric field inside the counting chamber by the nonconducting filter paper. Scintillation counting is more widely used for counting all types of air samples. Since α particles have very short ranges in air, filter papers must be placed very close to the scintillation material for α counting. Excellent results with low-level samples can be obtained using small disks of scintillating material in or on clear plastic placed in actual contact with the deposited material on the filter paper. The counting device is then a photomultiplier which "sees" the light flashes inside a dark chamber containing the sample and the scintillating disk (*9*). More recently, semiconductor devices have come into use as counter detectors. Their principal advantage is their ability to discriminate between the energy levels of the various particles they detect.

Counting of γ activity usually is done with a crystal as a scintillation source, although Geiger tubes with end windows have been used. Beta counting of air samples may be done with scintillating crystals or powders on plastic films or with thin-window proportional counters. Multichannel analyzers may be used, particularly with γ emitters, to give qualitative information on the isotopes present; discriminators or single-channel analyzers may be used to eliminate all but a specific energy band in the counting. For α emitters, such methods are less satisfactory but special techniques are available as noted below.

If the gross α, β, or γ count of the material on a sampling filter exceeds that acceptable for the most hazardous isotope, or that acceptable for specific mixtures, then it may be necessary to dissolve the material from the filter and chemically separate the particular elements of concern (*10*). It should be noted that chemical treatment will not separate the individual isotopes of an element. Thus, no chemical treatment will separate plutonium-238 from plutonium-239, or strontium-89 from strontium-90. If it is necessary to determine the concentration of each of these individual isotopes in the presence of the other, it usually is done by the use of energy discriminating counters or, for short half-life material, by serial counting and interpretation of results based on differing decay rates or differing rates of in-growth of daughter products.

The problem of absorption losses is always present in α counting of filter paper samples. As previously noted, some airborne material penetrates slightly into the looser, thicker papers, and the α particles emitted from it may not be detected by direct counting. If a large amount of nonradioactive dust is collected along with the radioactive material, it may prevent the escape of the α particles to the counting area. This is particularly serious in air pollution work since large quantities of inert dust usually are present. Studies may have to be made to evaluate such losses under various conditions.

The air usually contains appreciable quantities of radioactive particulates formed by the decay of radon and thoron escaping from rocks and soil. Such naturally radioactive particles also are collected on filter paper when other radioactive contamination is being measured. If the samples are counted immediately after the end of the sampling period, the results may be very high due to the presence of these short-lived, natural radioactive materials. This is particularly serious in the case of α counting for plutonium, uranium, radium, and polonium, because most of the natural materials are also α emitters and levels of concern for the long-lived material are very low.

Counting can be delayed for several days to permit the decay of the natural products or several counts can be made and a correction calculated. The daughter products of radon decay quickly and may be assumed to have disappeared after several hours. If appreciable quantities of thoron daughters are present, these will decay with the half-life of ^{212}Pb (ThB), which is 10.6 hours. Several days must elapse before such material decays below significant levels. If results are required sooner, two counts should be made at least 12 hours apart. The activity of the long-lived α emitters in the sample is then given by

$$C = (C_2 - C_1 e^{-\lambda \Delta t})/(1 - e^{-\lambda \Delta t}) \tag{2}$$

where

C = activity of long-lived material
C_1 = activity on filter at time t_1
C_2 = activity on filter at time t_2
Δt = elapsed time between counts = $t_2 - t_1$ (hours)
λ = decay constant for ^{212}Pb = 0.693/10.6 = 0.0653 (hr^{-1})

If counts are made exactly 12 hours apart, then $e^{-\lambda \Delta t}$ is 0.456 and the equation becomes

$$C = (C_2 - 0.456C_1)/0.544 \tag{3}$$

or for a 24-hour time lapse

$$C = (C_2 - 0.208C_1)/0.792 \tag{4}$$

Samplers having discriminator circuits with solid-state detectors have been used to screen out the unwanted counts. Such systems are complex and are seldom warranted for air pollution samples. These and other correction methods have been employed where continuous recording devices are used in fixed locations and are discussed later in this chapter.

When the material being measured has a short or medium length half-life, multiple counts over a period of time are necessary. These results are then plotted and the curve extrapolated back to the time of sampling to get the correct activity at that time. If a single nuclide is being measured, a plot of the activity on a log scale against elapsed time on a linear scale will give a straight line that can be extrapolated readily. Such plots for sodium-24 (half-life 15 hours) and for molybdenum-99 (half-life 67 hours) are shown in Figure 5. Thus, if a sample containing only sodium-24 is counted 2 days after it is collected, only about 10% of the activity that was originally on the paper is still there. In the case of molybdenum-99, about 69% is still there. Decay rate tables for many of the common nuclides are available to facilitate such calculations (*11*).

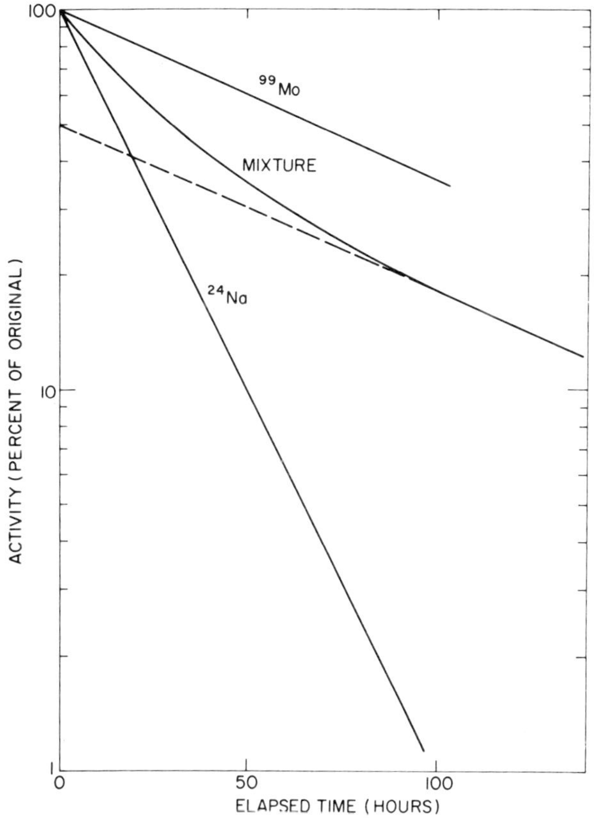

Figure 5. Decay curves for ^{24}Na and ^{99}Mo and for a mixture initially containing equal activities of each.

If the sample consists of a mixture of two nuclides, the decay curve initially follows that of the short half-life material and, at later times, parallels that of the longer half-life material. In Figure 5, the decay curve is shown for a mixture which initially contained equal amounts of activity from sodium-24 and molybdenum-99. If the latter part of this curve, where it is parallel to that of molybdenum-99, is extrapolated back to $t = 0$, the percentage of molybdenum in the original mixture can be determined. Such simple mixtures rarely occur and, in actual practice, mixtures of nuclides usually are analyzed by α- or γ-ray spectroscopy.

If the contaminant being measured is "mixed fission products" (MFP), the result of a nuclear bomb explosion, the decay curve will be a straight line when log scales are used for both activity and time. The time, in such a plot, is the elapsed time since the explosion. Such a plot is shown in Figure 6, with the decay curve for sodium-24 drawn on these coordinates for comparison. The MFP decay curve has a slope of —1.2. Such a straight line can be obtained only if the time of the explosion is known. Indeed, by a process of trial and error, the time of the explosion can be obtained since the correct time is the only one that will yield a straight line in this type of plot.

When very short half-life materials are collected, appreciable decay

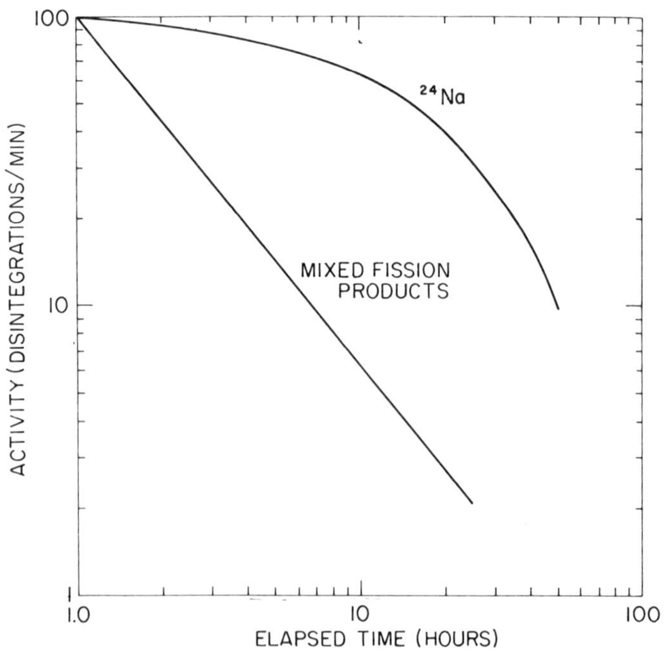

Figure 6. Decay curves for ^{24}Na and for mixed fission products.

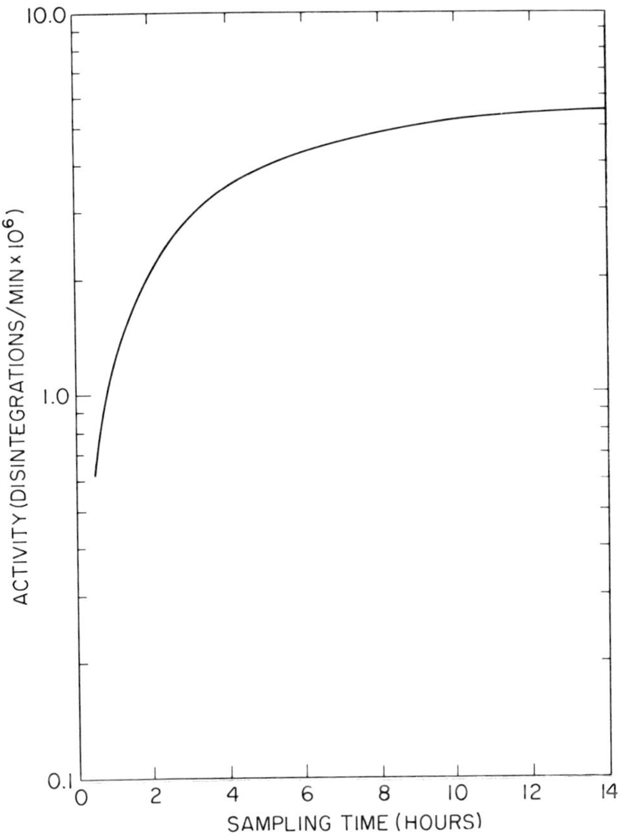

Figure 7. Buildup of activity on filter paper during sampling.

occurs during the collection. After some time, the rate of decay of the material on the filter during collection becomes equal to the rate of buildup of new material on the filter. After this, there is no further increase of activity on the filter as sampling continues. This buildup to equilibrium is illustrated in Figure 7. Here it is assumed that the material being collected has a half-life ($T_{1/2}$) of 3 hours ($\lambda = 0.231$ hr^{-1}), the sampling rate (F) is 100 liters/minute, and the air concentration being sampled (C) is 10^{-7} μCi/cm^3 or 220 disintegrations/minute/liter of air. The activity (A) on the filter reaches an equilibrium value of 5.7×10^6 disintegrations/minute after about 10 hours of sampling. After equilibrium, the relation between activity on the filter and the air concentration is given by Equation (5).

$$C = A/F = 0.693A/FT_{1/2} \tag{5}$$

Spectrometry or the ability to identify and measure the quantities of various radionuclides present in a complex mixture has advanced rapidly in recent years. It is especially useful for γ emitters since the high penetrating quality of γ radiation makes it unnecessary to separate the collected material from the filter paper. Crystals are most widely used as energy discriminating detectors for γ radiation. These function best at low temperatures that greatly reduces the amount of random background counts or "noise." Hence, such crystals usually are mounted on a holder bathed in liquid nitrogen. For many applications, the semiconducting detector is replacing the crystal for both γ and α counting. This is usually a material surface, such as germanium, into which a trace of another material, lithium, has been "drifted." The advantage of this type of detector is the very high degree of energy resolution that can be obtained. This gives valuable information if a large number of nuclides are present or if the composition of the collected material is unknown.

For accurate α spectrometry, the collected material must be separated from the filter and from as much extraneous inert matter as possible. The radioactive residue is then plated in some manner on as small a metal disk as possible. It is this surface that is counted. Since α particles have very short ranges, the source should be as nearly a point source as possible. Alpha particles escaping from portions of the deposit embedded in the filter paper lose part of their energy in escaping and, thus, are "degraded" before reaching the detector. This also happens if material is deposited on the metal disk along with a deposit of inert dust. These are the reasons for separating the radioactive material from the filter and the inert dust. Fortunately, the number of nuclides that are α emitters are limited, so that chemical separation is practicable. If the spectra are "degraded," very poor resolution of energies is obtained. The energies of most α particles are quite close together, and high resolution is necessary. Since β particles are emitted in a broad range of energies, β spectrometry is seldom useful for identification or quantitation.

Many α emitters and β emitters also emit low-energy γ rays and X rays, and these sometimes are useful for identification and quantitation. Thus, plutonium-239 is primarily an α emitter but it also emits an X ray with an energy of only 17 keV. A layer of moisture or a film of oil will completely absorb the α particles but will have a negligible effect on this weak X ray.

To achieve the high resolution spectrometry that is now possible, elaborate and expensive equipment is required. Such equipment requires a great deal of maintenance by highly trained specialists. This fact should be carefully considered before deciding upon spectrometry as the method of choice for quantitation.

*G. Particle-Size Determination**

In the field of industrial hygiene or health physics, the particle size of an aerosol breathed by a worker is a very important parameter in assessing the health hazard from breathing it. Frequently, the particle-size distribution is such that a high percentage of the suspended material does not enter the lung because it is too large. This usually is not the case for air pollutants because most of the large particles have settled out or have been removed before the material is released to the atmosphere. As a result, most of the material usually may be assumed to be of a size that can reach the lungs. However, particle-size measurements may be important for other reasons. The specific activity of a radioactive particle usually is proportional to its mass and, hence, to the cube of its diameter. The extent of biological hazard may be related to this specific activity for a given particle. The same is true, in general, for other toxic materials, except possibly mineral dusts where the number of particles in a particular size range may be more significant. A single large particle of radioactive material can give intense radiation to a small volume of tissue. This ability to do damage simply by its presence is unique to radioactive particles.

The determination of particle size for radioactive dusts presents more difficult problems than for other dusts. It is seldom possible to make a microscopic measurement or a light-scattering determination of particle diameters. The mass concentration of a radioactive contaminant in air usually is so minute, even at concentrations far above permissible levels, that it cannot be seen on the collection medium. The concentration of ordinary dust, even in very clean areas, is far in excess of that of radioactive dust. In addition, there is seldom any visible characteristic of the radioactive particles that render them distinguishable under the microscope. For these reasons, indirect sizing methods usually are used.

The cascade impactor is the instrument most widely used for sizing radioactive dusts (*12, 13*). In this sampling instrument, the air is drawn through a series of progressively smaller holes or jets and, after each jet, allowed to impact on a plate coated with an adhesive or dust-retaining material. Since the jet velocities increase as the jet size decreases, progressively smaller particles are impacted and retained. Thus, the original suspended material is separated into a series of size classifications similar to that obtained in a screen analysis of a bulk solid sample. If the impactor has been properly calibrated, the characteristic sizes deposited on each stage will be known. Thus, it is sufficient to measure the

* See also Chapter 1, this volume.

quantity of radioactive material on each stage in order to reconstruct the original size distribution.

The determination of particle size with the cascade impactor is fairly accurate when the impactor is calibrated by optically sizing particles on each stage, using an aerosol as similar as possible to the one to be sampled. Particle shape and density, as well as size, affects the stage constants. A reasonably good correction for density differences can be made but shape factors usually are not known. Other errors may be introduced by leakage of air into various parts of the impactor, by deposition inside the instrument body, and by resuspension of deposited aerosol from heavily loaded stages. Calibration of the instrument is a difficult task and must be done carefully (*14*). A large number of particles must be sized on each stage by a well-trained microscopist. A somewhat simpler method of calibration, in which each stage is sized separately, has been reported (*15*). The cascade centripeter, which also separates an aerosol into a series of size ranges, meets some of the objections to the cascade impactor (*16*). There has been little use made of this instrument in air pollution work. It deserves further investigation since it is less limited than the cascade impactor in the total quantity of material that it can collect.

Several aerosol spectrometers, that separate an aerosol by size in a complete spectrum, can be adapted for use with radioactive materials. In Timbrell's aerosol spectrometer, a commercial version of which is available from Fleming Radio, Ltd., England, the air passes horizontally in a thin layer above a long surface, and the particles settle on the surface (*17*). Large particles settle nearest the entrance with the rest strung out in order of decreasing size. Distance from the entrance then becomes a measure of size, and the amount settled out at various distances can be measured to give the size distribution. Since settling velocities are low, this system is satisfactory only for relatively large particles, particularly for those larger than 10 μm. In the conifuge, centrifugal force is used to speed up the settling (*18*). The aerosol is fed into the tip of a hollow space between two cones rotating rapidly. The air emerges at the base of the cone. During passage down the annular space, the centrifugal force increases rapidly with the increasing linear velocity. Particles are driven to the outside wall where they are deposited on an adhesive-coated surface. Distance down the wall from the entrance is again a measure of size and can be suitably calibrated. This instrument is expensive and difficult to build. It is principally a laboratory instrument.

The Goetz aerosol spectrometer is similar to the conifuge but the air traverses a spiral track down the annular space between the cones (*19*). Unlike the Timbrell spectrometer and the conifuge, the air is not guided into the deposition space in a thin layer. As a result, distance from the entrance is not a measure of the size of all particles deposited at that point but only of the maximum size particle deposited there. Interpretation of the resulting data is quite complex. It, too, is principally a laboratory instrument. A much more practical device which combines the features of the conifuge and the Goetz instrument is the spiral centrifuge (*20, 21*). In this, the air enters in a thin film along with filtered air as in the conifuge and traverses a spiral channel in a rapidly revolving cylinder. A metal collecting strip forms the outer wall of the spiral channel. After operation, the strip is removed and cut into sections which can be placed in a counting chamber for radioactivity quantitation. The distance down the channel is a measure of aerodynamic size while the counter measures the quantity of material collected at that point. From this, a complete size spectrum of the radioactive material can be constructed. The instrument is relatively simple to calibrate and yields accurate results. A lighter and more portable version has been constructed and could be used in air pollution field studies, especially near a potential source (*22*). The concentric aerosol spectrometer is a similar instrument (*23*).

Radioactive particles can be identified under the microscope by first placing the collected aerosol in contact with photographic film for a period of time and then developing the film. In some cases, the particles are left fixed in position on the film during development and, in others, the film is developed separately and then placed back in contact with the particles in the exact original position. When examined under the microscope, the radioactive particles are identified by the darkened spots under the particles on the film and can then be sized optically.

With α-emitting aerosols, a nuclear track system is useful. In this, the collected aerosol is placed in contact with a nuclear track photographic film for a fixed period of time. The film is then developed and examined under the microscope. Tracks can be seen where α particles were emitted and penetrated the film emulsion. The number of tracks emanating from a single point is a measure of the amount of radioactive material in a single particle at that point. From the calculated mass of material, the size can be estimated (*24*). This assumes that there is no nonradioactive material in the particle. Both the nuclear track and the film darkening method are extremely tedious and time consuming. For these reasons they are seldom used for routine work.

V. Radioactive Gas Sampling and Quantitation

Unlike the case of radioactive particulates, separation of radioactive gases from air is not simple and, in some cases, may not even be possible.

Table I Properties of Common Radioactive Gases

Gas	*Emission*	*Max. β energy (keV)*	*Half-life*	*Principal source*
$^{131}I_2$	β,γ	608	8.0 days	Reactors, bombs, chemical fuel processing
T_2[a] HTO	β	18	12.5 years	Reactors, accelerators
^{41}Ar	β,γ	1250	1.8 hours	Reactors
^{133}Xe	β,γ	330	5.2 days	Reactors, fuel processing
^{135}Xe	β,γ	930	9.2 hours	Reactors, fuel processing
^{85}Kr	β,γ	740	10.0 years	Reactors, fuel processing
$^{14}CO_2$	β	155	5700.0 years	Laboratories, bombs
$^{35}SO_2$	β	167	87.0 days	Laboratories
$^{13}N_2$	Positron	1190	10.0 minutes	Accelerators
$^{15}O_2$	Positron	1680	2.0 minutes	Accelerators
$^{106}RuO_4$	β,γ	39	1.0 years	Fuel processing
^{222}Rn	α		3.8 days	Mines and mills, refinery

[a] Tritium.

As a result, each radioactive gas poses a distinctive sampling problem. Fortunately, the number of radioactive gases encountered in air pollution work is relatively small. A list of such gases and their usual sources are given in Table I.

It is possible to postulate the existence of many other radioactive gases formed by chemical combinations containing radioactive isotopes of carbon, sulfur, chlorine, iodine, and others. Under special circumstances, such materials might occur, as in the vicinity of a plant producing such compounds or in a laboratory handling them. Air pollution from such a source would be present only as a result of a clear failure of the control system which would be included in such an establishment.

A. Iodine

Radioactive isotopes of iodine are, by far, the most common radioactive gaseous air pollutants. The principal isotope encountered is iodine-131 but other isotopes may also be present. Iodine is formed in the fission process and by the decay of other fission products, mainly tellurium. Since iodine is readily volatile in elemental form, it is released continuously in some types of reactors and is released in large quantities during processing of spent reactor fuels and in explosions of nuclear devices. Because iodine concentrates in the thyroid gland, the ingestion or inhalation of this element in small quantities can cause appreciable local radiation to this gland; hence concentrations must be kept low. Since, thyroid irradiation from this source may be more harmful for children than adults, iodine is a particularly serious air pollutant.

The most widely used method for the collection of iodine is adsorption on activated charcoal. Special activated carbons, containing elements which increase its holding power for iodine, are available. Respirator cartridges for protection against organic vapors are a convenient form for use of charcoal as an air sampling medium. Holders similar to filter paper holders can be built easily for these cartridges and then combined with a flowmeter and pump to form a sampling system (*25*) (Fig. 8). Refillable cartridges can also be made and are much cheaper for use where large numbers of samples must be collected (*26, 27*). Samples can be quantitated most conveniently by placing the adsorber cartridge directly on a scintillation crystal or in a well counter for γ counting. Using discriminator circuits in a gated single-channel analyzer, a high degree of sensitivity can be obtained.

In contrast to the use of filters for particulates, when an adsorbent becomes overloaded, the contaminant begins to penetrate the collector. The capacity of activated charcoal for iodine is relatively high, and the mass concentration of even highly radioactive concentrations of radioiodine is very minute. Thus, the charcoal rarely becomes saturated, and radioiodine rarely penetrates the adsorbent when it occurs in elemental form.

Figure 8. Chemical cartridge in holder for sampling for iodine.

While this may sound relatively simple, iodine monitoring in the atmosphere is very complex and much remains to be learned about it. Where the iodine is completely gaseous and entirely in elemental form, the charcoal adsorption method gives reliable results. The problem lies in the fact that one can never be entirely certain that the iodine meets these two conditions. At ordinary temperatures, some iodine may be present as agglomerated solid particles or iodine atoms may attach themselves to other particles in the atmosphere. Such materials may penetrate the loosely packed adsorbent but can be collected on a good filter paper. For this reason, a filter paper is usually placed before or after the collection cartridge during air sampling. Both should be counted in measuring the iodine concentration. The relative amounts of iodine on the filter paper and the cartridge depend on the method of formation of the iodine and its subsequent history as a vapor before collection. Even with filter paper and a charcoal adsorbent, some iodine is difficult to collect. A considerable fraction of the iodine may be present as methyl iodide since this compound appears to be formed readily if any carbon is present. Other organic iodine compounds, as well as hydrogen iodide and hypoiodous acid, may also be present.

Materials other than charcoal have been used for sampling for iodine. Copper and silver mesh have been used as absorbents and their efficiency seems to be dependent on humidity among other factors (*28*). Silver-coated filter papers and charcoal-loaded filter papers have also been used. Both of these give high efficiencies with iodine formed by chemical reactions in the laboratory, but give varying efficiencies, sometimes very low, with iodine produced by reactors or industrial fuel processing operations (*23, 29*). Chemical absorbers have been used successfully (*30*) and silver-treated zeolites are being used rather widely (*31*). The latter are also effective in collecting methyl iodide and other organic iodides. Several sampling methods should be tried in any air sampling situation—the one yielding the highest concentration assumed to be the most accurate. However, much important work remains to be done in this field.

Since the limiting concentration for radioiodine exposure is based on the dose to the thyroid of a child who drinks milk from a cow on a pasture contaminated by airborne deposition of iodine, it is often simpler and more significant to measure the amount of iodine on the pasture grass rather than the concentration in the air (*32*).

B. Tritium

Tritium is a radioactive isotope of hydrogen having an atomic weight of 3. It is produced in many types of reactors and in very small quantities whenever hydrogenous materials are exposed to extremely high radiation fluxes. Cooling water for accelerator targets is an example of such exposure. Thus, tritium conceivably can be found near reactors, high energy physics laboratories, tritium production plants, and laboratories where tritium is used for experimental purposes. While it may be incorporated in many chemical compounds, it usually is present in the form of gaseous molecular hydrogen or as water vapor. It is much more hazardous as water vapor than as hydrogen gas because, as water, it is easily and rapidly incorporated into the body water of an individual by inhalation, ingestion, or direct passage through the skin. Tritium, as molecular hydrogen, when dispersed in air, gradually oxidizes to tritium oxide or water vapor, and standards are based on this assumption, unless there is good reason to be assured it is in the form of molecular hydrogen.

Tritium decays with a half-life of 12 years by the emission of a β particle of very low energy. Because of this low energy, tritium usually is monitored by introducing the air sample directly into an ionization chamber. In the device originally known as the "sniffer," the air to be sampled is drawn continuously through an ion chamber of 1-liter capacity. The amplified ion current actuates an indicating meter, a recorder, or an alarm. This relatively simple device, in modified and improved form, is available from several manufacturers (*33*). The original sniffer, which is still in use, is shown in Figure 9, along with a more modern version.

It should be noted that the simple sniffer measures the concentration of ions already present in the air as well as those formed during passage through the chamber. Ions may be present in the air from flames, fixed radioactive sources, or other ion generators, as well as from tritium in the air. The presence of ions formed by these other methods of generation must be ruled out if the sniffer device is to be used for measuring tritium concentration. Since tritium begins to generate ions in air as soon as it is mixed with air, and since these have an appreciable lifetime, the indication made by the chamber is only semiquantitative and depends on the length of time between tritium release to air and passage through the chamber. For air pollution work, this time usually is sufficiently long that the equilibrium number of ions will be associated with the tritium.

The instrument can be made more quantitative by placing ahead of the chamber an ion trap for collecting existing ions. Thus, it measures only ions produced by the tritium during passage through the chamber and a quantitative measure of concentration can be obtained. This makes the instrument much less sensitive since the equilibrium concentration of ions is much greater than the concentration formed during the brief passage through the chamber. To obtain adequate sensitivity a vibrating

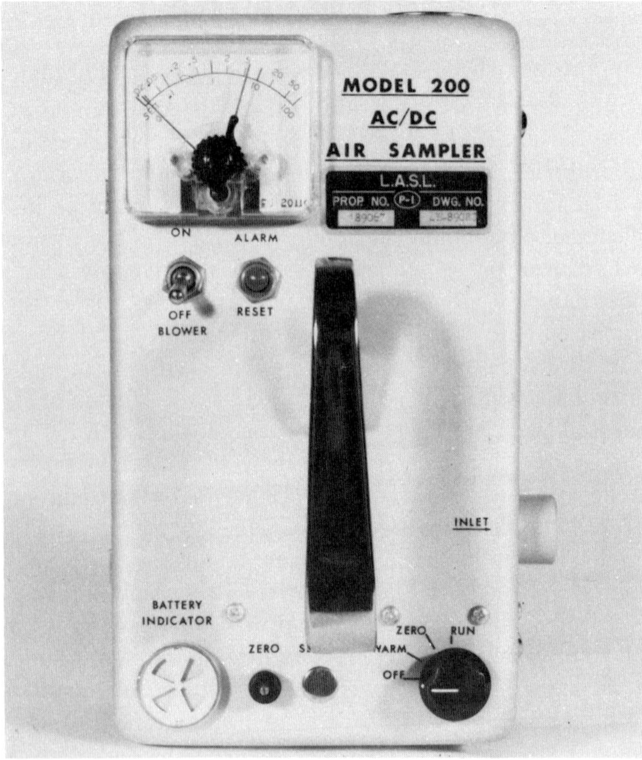
MODEL 200
AC/DC
AIR SAMPLER
L.A.S.L.
PROP. NO.
P-1
DWG. NO.
ON
ALARM
OFF
BLOWER
RESET
INLET
BATTERY
INDICATOR
ZERO
ZERO
RUN
OFF

reed electrometer may be used for measuring the ion current in the trapped ion device. A filter ahead of the chamber removes dust and further improves its performance. A device incorporating all these elements can read and record the tritium concentration continuously.

Larger ion chambers provide greater sensitivity and are used in devices which serve as fixed monitoring stations. A frequently used unit of this type is known as the Kanné chamber (*34*). Such chambers also respond to other radioactive gases, usually with a higher degree of response. Thus, the presence of such gases must be ruled out if tritium is to be measured. Indeed, flow-through ion chambers can be used for monitoring most radioactive gases. Iodine, however, readily attaches itself to metal surfaces and cannot be measured in this manner.

Since tritium in the form of water vapor is the principal hazard, a useful method for its determination in air pollution work is the adsorption of water vapor from the atmosphere onto silica gel, and determination of the tritium content of the collected moisture. A simultaneous determination of the humidity must be made. Tritium determination on the silica gel is made by distilling off the water, mixing all or a portion of the distillate with a scintillating liquid and counting the mixture in a liquid scintillation β counter. This method is particularly useful in atmospheric studies since large volumes of air can be passed through the silica gel over a considerable period of time. About 200 gm of silica gel will serve for sampling at 100 cm^3/minute for a period of 2 weeks (*35*).

There is no air-cleaning method for removing tritium from air other than removing all the water vapor. Even this only leads to a problem of disposal of contaminated liquid water. Therefore, tritium must be controlled at its source before it has a chance to mix with large volumes of air. The tritium problem, including experience in monitoring for it, has been the subject of a critical review (*36*).

C. Noble Gases

Noble gases, particularly argon, krypton, and xenon, may occur as atmospheric radioactive pollutants. Argon-41 is formed whenever air, which always contains argon, is exposed to a high neutron flux. Thus, the air used to cool the interior of a reactor always contains argon-41. Even though the initial quantity of argon in air is small and the half-life of argon-41 is relatively short, considerable argon-41 may be present since

---

Figure 9. Original tritium sniffer and a commercial model. Air passes through the ion chamber at the front. The instrument can be set to alarm at a specific concentration.

the reaction forming it has a very high cross section. Krypton and xenon, particularly krypton-85, xenon-133, and xenon-135, are products of the fission process and, hence, may be emitted from certain reactors having unclad fuel elements. These gases also are emitted from fuel processing operations where the fuel element is dissolved during chemical processing. Such plants include measures for trapping these materials and holding them until they decay.

The usual method of monitoring for these noble gases is by means of a simple radiation detector suspended in the atmosphere or in a large container (*37*). The radiation emitted is energetic enough for this measurement, and the permissible concentration is based on the dose delivered to the whole body immersed in the atmosphere containing the contaminant. A simple thin-window Geiger counter may be used. The Kanné chamber or other flow-through ion chambers discussed in the preceding section may be used (*38*), or a sealed ion chamber inside a large volume container can be used to measure the γ-ray component. The latter may be a more significant measure since meeting the standard for γ exposure will usually ensure that the β standard also is met (*39*). Methods for krypton-85 have been reviewed by Budnitz (*40*) and by Jaquish and Moghissi (*41*).

Charcoal has a limited adsorption capacity for krypton and xenon, but this capacity can be increased considerably by cooling the charcoal. Hence, a combination charcoal-freeze out trap can be used for collecting a sample. The trapped gases can then be released into an ion chamber or a chamber containing a Geiger counter for measurement. This permits the measurement of very low concentrations of krypton and xenon, when necessary. Since permissible air concentrations of these gases are relatively high, such techniques are almost never used except in special studies. Indeed, as air pollutants, these noble gases have been considered significant mostly because their presence interferes with low background counting systems.

D. Radon and Thoron

Radon and thoron also are noble gases but because they are energetic α emitters, they have considerable health significance. They occur naturally in air practically everywhere, since they escape from rocks and soil, both of which contain minute amounts of radium and thorium. Their concentration is higher in areas where there are substantial amounts of uranium and thorium ores. These gases may occur as air pollutants in the vicinity of uranium mines, mills, and refineries, or where radium, its ores and by-products, are processed.

Radon and thoron can be monitored in an ion chamber like the other noble gases but the chamber must be preceded by a filter to remove solid decay products and by an ion trap to remove existing ions. They can also be monitored by passing the air through a long tube or a chamber having a filter at each end. The first filter removes the decay products already present while the second filter collects decay products formed during passage through the tube. After measuring the amount of α-active material on the second filter, the concentration of radon or thoron can be calculated from this and the volume rate of flow through the tube or chamber (*42*). For either gas, air can be allowed to flow into an evacuated flask, the inner surface of which is coated with a scintillation powder, except on its flat base. The flask is then placed in a light-tight chamber on a photomultiplier tube which records the light flashes emitted (*43*). Figure 10 shows a metal flask with a clear glass base on the front shelf of a counter for counting the light flashes. The flask is placed inside one of the black chambers. An integrated environmental sample can be obtained by adsorption on charcoal in a method described by Sill (*44*).

Neither radon or thoron alone are significant environmental health hazards. However, they decay to solid daughter products, most of which are also α emitters and can be inhaled. In the case of radon, which is the

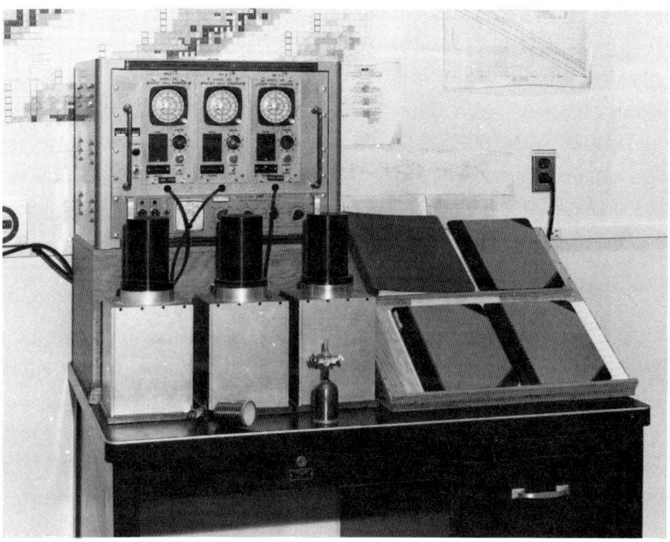

Figure 10. Scintillation counter used for radon determination. The gas is collected in the cylindrical vessels in the front which are lined with a scintillator. These are placed inside the black cylinders directly on the photomultiplier tubes.

more important case, the immediate decay scheme includes three α emitters. If these are all in equilibrium, the radiation dose delivered to the respiratory tract tissue is 80% from the daughters and only 20% from the radon since the daughter products remain deposited on the tissue. Thus, in evaluating radiation dosage in areas where radon is present, as in uranium mines, the daughter products are sampled by means of filter paper, as described in Section IV,A. Because of the rapid decay of these materials and the fact that several radioactive materials are present simultaneously, the results are evaluated empirically in terms of a concentration unit called a "working level" (*45*). Exposures in terms of this unit have been correlated epidemiologically with the incidence of lung cancer in uranium miners.

Tailings from uranium mills have been incorporated in the fill material used in some housing construction. The exposure of people in such buildings to radon and its progeny emanating from this material requires the measurement of very low levels of radioactivity. One technique uses a long-term sampler drawing air through a small filter paper mounted close to a piece of thermoluminescent crystalline material. The decay of the collected α-active material produces changes in the crystalline material. When the latter is heated, it emits a quantity of light proportional to the amount of ionizing radiation absorbed and the quantity of light can be measured. This instrument is calibrated by taking a series of regular filter paper samples at intervals during the instrument sampling period and determining the working level concentration in the usual manner. Another novel approach to this problem is the use of plastic track-etch films hung in areas where radon is presumed present. Such films, when etched in alkaline material produce holes or pits where they have been struck by α particles. This system is difficult to calibrate and is not yet proven in practice for this purpose.

E. Carbon-14 and Sulfur-35 Dioxides

Carbon-14 dioxide and sulfur-35 dioxide may be formed as a result of operations in an isotope laboratory or the incineration of wastes from such a laboratory. They are readily sampled by means of liquid scrubbers containing sodium hydroxide or barium chloride with an oxidant. The determination of collected radioactive material is done most simply by liquid scintillation counting. The precipitated barium carbonate or barium sulfate also can be filtered off and the filter paper counted in a suitable instrument. These gases also can be determined directly in a calibrated sniffer similar to that used for tritium.

Minute quantities of carbon-14 dioxide are formed in the atmosphere by bomb testing and by cosmic radiation. Highly specialized sampling equipment is required for detection of such minute amounts.

F. Other Gases

The normal air components, nitrogen and oxygen, may become radioactive as a result of exposure to intense radiation fields, such as are present where the beam of an accelerator emerges from the vacuum chamber. The half-lives of these materials are so short that they are unlikely to be hazardous outside the target room. Unshielded Geiger counters or other detectors can be used for their direct measurement.

When reactor fuel elements are dissolved in highly oxidizing solutions, ruthenium, which is formed by the fission process, may be oxidized to the volatile tetroxide and released. Ruthenium-106 is the most hazardous isotope of this element. The air can be sampled by passing it through an absorber containing a dry organic material, such as polyethylene pellets. The ruthenium is held tenaciously on the pellets. The amount of ruthenium on the sample is determined by γ counting (*46*).

Freeze-out traps can also be used for collecting air samples of most radioactive gases. The difficulty of using such systems in the field eliminates them as practical devices for this purpose in air pollution work.

VI. Automated, Direct-Reading, and Specialized Samplers

Combined sampler-counter units have been developed for particulates using a scintillation counter probe placed near the filter paper during the sampling period. The counter usually is a count-rate instrument, and the output is connected to a recorder which then measures the buildup of activity on the filter paper. The slope of the activity versus time curve is a measure of the concentration at any one time, while the count rate at the end of the sampling period is a measure of the average air concentration during the sampling period. Natural activity seriously interferes with this system. Since α particles have very short ranges and cannot penetrate paper, it is necessary to have the detector on the upstream side of the filter and very close to it for this type of measurement. This is a disadvantage since it limits the area for airflow and the counter probe easily becomes contaminated. Hence, for air pollution work such devices are not usually used for α emitters, although they can be used for β and γ emitters where the detector can be placed on the downstream side. Instruments also have been built using filter tape, moving intermittently or continuously, as a collector with the detector so located as to count one sample while another is being collected. These instruments are very useful in air pollution studies, particularly where β- and γ-active isotopes are involved, since they show diurnal fluctuations which can be associated with pollution processes (*47*).

Reactors and fuel processing plants may emit a variety of pollutants, necessitating complex sampling systems (*48*). Thus, around a reactor, it may be desirable to use a sampler consisting of filter paper, followed by a chemical cartridge, followed by a chamber. The latter may be an ion chamber or a container housing a Geiger counter or a scintillation detector.

There is little that is unique to stack sampling for radioactive materials. Isokinetic sampling is desirable in most cases, although it is not always a requirement since radioactive particles frequently are in very small size ranges. Because many pollutants, such as radioiodine, occur in elemental and highly reactive forms, sampling lines should be kept as short as possible. These materials have a strong tendency to "plate out" on surfaces of all sorts. In particular, exposure to metal surfaces should be avoided. Where solid pollutants are formed by the decay of radioactive gases, their unique physical state as individual atoms of solids makes them very easily attracted and adherent to surfaces.

VII. Surveys, Networks, and Reporting

There is very little that is unique about radioactive materials with respect to survey techniques as compared with those used for nonradioactive materials described elsewhere in this book. The high sensitivity of radioactivity measurements, and the fact that the radioactive effect is readily converted into electronic pulses, makes continuous recording devices possible in most cases. Thus, radioactive air pollution is readily documented. In some cases, sampling times must be limited because of the short half-life of the pollutant being measured.

Originating with the concern over radioactive fallout, networks of sampling stations have been set up throughout the United States and other countries. These networks include unusual sampling systems involving the use of airplanes, rockets, and high altitude balloons but the sampling equipment uses the same principles as ground-based samplers. Exact sampling locations are not critical since such pollutants are so widespread. Data from these and other networks are collected by the agencies operating them but most such information collected by United States organizations is published regularly by the United States Environmental Protection Agency in the periodical *Radiation Data and Reports*. Along with air sampling results, data are published on the analyses of milk, water, soil, vegetation, and other materials.

In addition to such routine studies, reports are also prepared on planned new nuclear installations. These include preconstruction and preoperational studies. These reports are usually filed with the United States Nuclear Regulatory Commission and are available for public perusal.

VIII. Standards and Standard Methods*

Although maximum permissible concentrations of radioactive materials for exposure of the public were formulated long before similar standards—now called ambient air quality standards—were available for nonradioactive materials, there has been little additional progress in this field. These standards were first put forth by the International Commission on Radiological Protection (ICRP) and in the United States, by the National Council on Radiation Protection (NCRP), and were widely adopted by various governmental agencies. For prime contractors of the United States Energy Research and Development Administration (ERDA), these standards are substantially incorporated in an ERDA manual and administered by its Division of Operational Safety. For licensees of the Nuclear Regulatory Commission (NRC), similar values are incorporated in the United States Code of Federal Regulations, Title 10, Part 20. All of these are ambient air quality standards; no emission standards have been developed in this field by either ERDA or the Office of Radiation Programs of the United States Environmental Protection Agency (USEPA). These ambient standards are derived from occupational standards, being, in general, one-three-hundredth of the occupational standard for elements having a genetic effect, and one-ninetieth for other elements. Some deviations from this generality have occurred for a few specific nuclides but these are the basic principles. These same values also have been used as standards by most states within the United States. Both ICRP and NCRP have committees actively reviewing these standards, but extensive revisions require a long period of time (*49, 50*).

Some progress has been made in standardizing methods of sampling and analysis, although much of this has concerned occupational exposure. In the field of air pollution, the publications of the Intersociety Committee on Methods of Air Sampling and Analysis has included tentative methods for the determination of gross α and gross β particulates, iodine-

* See Chapter 11, Vol. V.

131, lead-210, plutonium, radon, strontium-89, strontium-90, and tritium in the atmosphere (*8*). The American Society for Testing and Materials Committee D-22 also has a subcommittee on "Sampling and Analysis of the Atmosphere for Radioactivity." Both ERDA and USEPA have been assisting efforts at developing standards and some problems may arise from conflicting "standards."

IX. Perspective on Radioactive Materials as Air Pollutants

The fact that exposure to radioactive material can produce long-delayed effects or genetic changes is properly of concern and leaves room for possibilities of harm which some feel are not given adequate attention. These problems have been extensively investigated including studies of the various ecological pathways by which radioactive materials can reach and be incorporated into the human body. The fact that there seems to be no safe threshold of exposure to radiation has been a source of concern but, as air quality standards have been adopted in recent years for nonradioactive materials, it is apparent that there are no wholly safe thresholds for these either. Air pollution is never considered as an isolated method of exposure but attempts are made to estimate the additive effects through various food chains, water, and surface contamination, as well as air. The concept of dose—whether wholly valid or not—has provided a common unit in which additive effects can be expressed.

Most plants handling or processing radioactive materials emit very little airborne radioactive pollution. Indeed, there is a tendency to use ambient air quality standards as emission standards for effluents. The possibility of accidental discharges of large quantities of radioactive material as a result of natural disasters or internal process failures, fires, or explosions is quite properly the most serious concern. As a result, these plants devote much more than ordinary attention to providing protection against the occurrence of such accidents or to the containment of potential pollutants should they occur.

X. Radioactivity Measurements as Tools in Air Pollution Studies

Radioactivity has been a valuable tool in the study and control of air pollution. Because of the high sensitivity of radiation detection devices, minute quantities of short half-life isotopes have been used to trace localized air movements in polluted areas. By this means, potential and actual sources of pollution can be detected. Since most radiation detectors are direct reading, an entire air flow area can be mapped quickly.

Where it is inadvisable to introduce even a slight amount of radioactive material into the air, the method of neutron activation may be used. In this method, a small amount of an uncommon element is introduced in the air or up a stack. Air samples are collected around the area even though concentrations are far below measurable levels for this element. The collected sample is exposed to an intense neutron flux in a nuclear reactor or similar source. As a result of a nuclear reaction, the trace of the uncommon element in the sample is made radioactive, or it is converted into another element which is radioactive and can be detected. By this means, very minute concentrations of the original element can be measured and areas of high concentration plotted. Other methods of activation—by α radiation, protons, and γ rays—extend the range of this technique. X-ray spectrometry and fluorescence also offer tools for use in this field. Even stable isotopes, such as carbon-13 measured by nuclear magnetic resonance, may be useful in air pollution studies. These tools are spin-offs from developments in the nuclear field. It must be admitted, however, that the direct application of these techniques in air pollution studies has been very limited so far.

Testing of nuclear weapons in the atmosphere, although undesirable from the pollution standpoint, did introduce tracer materials into the atmosphere which could be detected in minute concentrations. From such concentration measurements, directions and velocities of worldwide air movements could be traced. Interchanges of air between various layers of the atmosphere also have been studied by this means (*51*). All of this has added greatly to our store of meteorological knowledge and has benefitted us in the application of such knowledge to air pollution problems.

REFERENCES

1. E. C. Oscarson, W. H. Ellett, and N. S. Nelson, *Health Phys.* **28**, 302 (1975).
1a. D. H. Denham, *in* "Health Physics Operational Monitoring" (C. A. Willis and J. S. Handloser, eds.), Vol. 2, p. 1093. Gordon & Breach, New York, New York, 1972.
2. R. G. Stafford and H. J. Ettinger, Rep. No. LA-4650. Los Alamos Scientific Lab., Los Alamos, New Mexico, 1971.
2a. R. G. Stafford and H. J. Ettinger, *Atmos. Environ.* **6**, 353 (1972).
3. F. Kalmus, *J. Appl. Phys.* **25**, 87 (1954).
4. B. C. Isom, Rep. No. LA-PR-5359. Los Alamos Sci. Lab., Los Alamos, New Mexico, 1973.
5. K. R. Spurny, J. P. Lodge, E. R. Frank, and D. C. Sheesley, *Environ. Sci. Technol.* **3**, 453 (1969).

6. W. J. Smith and N. F. Surprenant, *Amer. Soc. Test. Mater., Proc.* **53,** 1122 (1953).
7. L. B. Lockhart, R. L. Paterson, and W. L. Anderson, Report No. 6054. U.S. Naval Research Lab., Washington, D.C., 1954.
8. Intersociety Committee, "Methods of Air Sampling and Analysis," p. 376. Amer. Pub. Health Ass., Washington, D.C., 1972.
9. N. A. Hallden and J. H. Harley, *Anal. Chem.* **23,** 1861 (1960).
10. Intersociety Committee, "Methods of Air Sampling and Analysis," p. 390. Amer. Pub. Health Ass., Washington, D.C., 1972.
11. K. H. Falter, "Radioactive Decay Correction Factors," Pub. Health Serv. Publ. No. 999-RH-12. U.S. Govt. Printing Office, Washington, D.C., 1965.
12. K. R. May, *J. Sci. Instrum.* **22,** 187 (1945).
13. A. A. Anderson, *Amer. Ind. Hyg. Ass., J.* **27,** 160 (1966).
14. T. T. Mercer, *Ann. Occup. Hyg.* **7,** 115 (1964).
15. J. C. Couchman and H. M. Moseley, *Amer. Ind. Hyg. Ass., J.* **28,** 62 (1967).
16. R. F. Hounam and R. J. Sherwood, *Amer. Ind. Hyg. Ass., J.* **26,** 122 (1965).
17. V. Timbrell, *Brit. J. Appl. Phys., Suppl.* **3,** S86 (1954).
18. K. F. Sawyer and W. H. Walton, *J. Sci. Instrum.* **27,** 272 (1950).
19. A. Goetz, H. J. R. Stevenson, and O. Preining, *J. Air Pollut. Contr. Ass.* **10,** 378 (1960).
20. W. Stöber and H. Flachbart, *Environ. Sci. Technol.* **3,** 1280 (1969).
21. O. R. Moss, H. J. Ettinger, and J. R. Coulter, *Environ. Sci. Technol.* **6,** 614 (1972).
22. P. Kotrappa and M. E. Light, *Rev. Sci. Instrum.* **43,** 1106 (1972).
23. M. I. Tillery, *Amer. Ind. Hyg. Ass., J.* **35,** 62 (1974).
24. J. A. Leary, Rep. No. LAMS-905. Los Alamos Sci. Lab., Los Alamos, New Mexico, 1949.
25. H. J. Ettinger, Rep. No. LA-3363. Los Alamos Sci. Lab., Los Alamos, New Mexico, 1965.
26. C. W. Sill and J. K. Flygare, *Health Phys.* **2,** 261 (1960).
27. Intersociety Committee, "Methods of Air Sampling and Analysis," p. 382. Amer. Pub. Health Ass., Washington, D.C., 1972.
28. P. J. Barry, *Health Phys.* **15,** 243 (1968).
29. H. J. Ettinger, *Health Phys.* **12,** 305 (1966).
30. J. K. Soldat, *J. Air. Pollut. Contr. Ass.* **10,** 265 (1960).
31. J. H. Keller, T. R. Thomas, D. T. Pence, and W. L. Maeck, *Proc. At. Energy Comm. Air Clean Conf., 12th, 1972* p. 322 (1973).
32. U.S. Atomic Energy Committee, "Regulatory Guide 1.42." U.S. Nuclear Regulatory Comm., Washington, D.C. 1974.
33. B. C. Eutsler, G. L. Evans, R. D. Hiebert, R. N. Mitchell, M. C. Robbins, and R. J. Watts, *Nucleonics* **14,** 114 (1956).
34. J. J. Fitzgerald and B. W. Borelli, Knolls Atomic Power Lab. Rep. No. KAPL-1231. General Electric Co., Schnectady, New York, 1954.
35. Intersociety Committee, "Methods of Air Sampling and Analysis," p. 422. Amer. Pub. Health Ass., Washington, D. C., 1972.
36. D. G. Jacobs, U.S. At. Energy Comm., Rep. **TID 24635** U.S. Atomic Energy Commission, Washington, D.C., 1968.
37. J. D. Jones, *in* "Health Physics Operational Monitoring" (C. A. Willis and J. S. Handloser, eds.), Vol. 2, p. 1067. Gordon & Breach, New York, New York, 1972.
38. R. A. Meck, *in* "Health Physics Operational Monitoring" (C. A. Willis and J. S. Handloser, eds.), Vol. 2, p. 1029. Gordon & Breach, New York, New York, 1972.

39. R. A. Meck, *in* "Health Physics Operational Monitoring" (C. A. Willis and J. S. Handloser, eds.), Vol. 2, p. 1067. Gordon & Breach, New York, New York, 1972.
40. R. J. Budnitz, *in* "Noble Gases" (A. A. Moghissi and R. E. Stanley, eds.) pp. 192–198, CONF-730915. U.S. Govt. Printing Office, Washington, D.C., 1975.
41. R. E. Jaquish and A. A. Moghissi, *in* "Noble Gases" (A. A. Moghissi and R. E. Stanley, eds.), pp. 169–174, CONF-730915. U.S. Govt. Printing Office, Washington, D.C., 1975.
42. J. W. Thomas and P. C. LeClare, *Health Phys.* **18**, 113 (1970).
43. H. F. Lucas, *Rev. Sci. Instrum.* **28**, 680 (1957).
44. C. W. Sill, *Health Phys.* **16**, 371 (1969).
45. D. A. Holaday, D. E. Rushing, R. D. Coleman, P. F. Rushing, H. L. Kuznetz, and W. F. Bale "Control of Radon and Daughters," Pub. Health Serv. Publ. No. 494. U.S. Dept. of Health, Education, and Welfare. U.S. Govt. Printing Office, Washington, D.C., 1957.
46. D. K. Jenson, *Health Phys.* **12**, 923 (1966).
47. G. L. Helgeson, *Health Phys.* **9**, 931 (1963).
48. "Environmental Radioactivity Surveillance Guide," ORP/SID 72-2. United States Environmental Protection Agency, Washington, D.C., 1972.
49. International Committee on Radiological Protection, *Health Phys.* **3** (1960).
50. National Council on Radiation Protection, "Maximum Permissible Body Burdens and Maximum Permissible Concentrations of Radionuclides in Air and Water for Occupational Exposure," NCRP Report No. 22. NCRP Publications, Washington, D.C., 1959.
51. H. E. Moore, S. E. Poet, and E. A. Martell, *J. Geophys. Res.* **78**(30), 7065 (1963).

12

Air Quality Data Handling and Analysis

Charles E. Zimmer

I. Introduction

The purpose for operating air quality monitoring systems is to provide environmental managers and researchers with pertinent information for decision making and planning. To provide this information in the required format and at the appropriate time, it is necessary to develop specific procedures for the handling and analysis of data. These procedures should encompass all aspects of data recording, validation, storage and retrieval, presentation, and statistical methods of analysis. Depending upon the volume of data generated by the air quality monitoring system and the complexity of data presentation and statistical analyses, the data handling system may be either manual or computer oriented. Furthermore, if there are requirements for immediate access to field sampling data, it may be necessary to make use of real-time telemetry systems.

II. Data Recording

Air quality measurement instrumentation can be classified into two general categories: (a) intermittent samplers, and (b) continuous analyzers. With the former, a sample is collected in the field, as with a high volume sampler or a bubbler, and the sampling medium is returned to a laboratory for determination of the concentration of the pollutant(s) involved. The latter system involves a sensor which makes continuous measurements of a pollutant and requires some type of data logging device at the field station. Such a data logging device may be a continuous strip chart recorder, a more sophisticated system in which data may be recorded on magnetic tape, or a telemetry system by which data are transmitted in real-time to a central computer facility. Each type of system presents a different problem insofar as data recording is concerned.

A. Manual Procedures

Procedures for manually recording intermittent and continuous sampling system data are quite different.

1. Results Derived in the Laboratory

Information necessary to determine the concentration of an air contaminant is derived from technicians in a field sampling location and from an analyst in a central laboratory. The field technician provides information regarding sampling time, flow rate, and sensor calibration. The analyst in the laboratory uses this information, along with the results of the chemical or physical analysis, to determine pollutant concentration.

A typical format for recording intermittent data is that used by the United States Environmental Protection Agency (USEPA) for the sub-

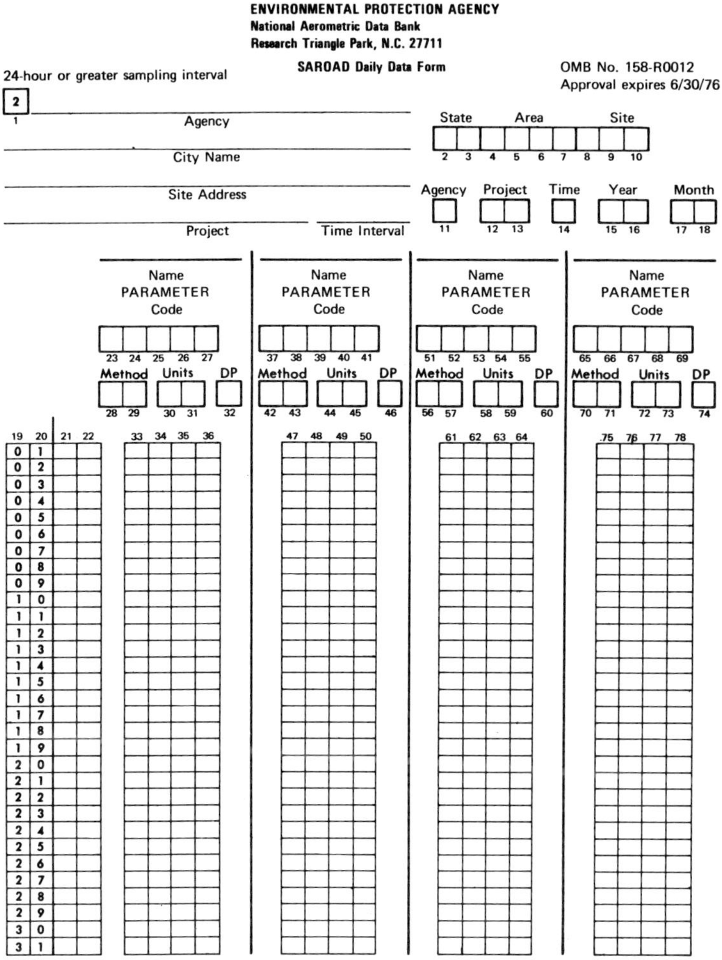

ENVIRONMENTAL PROTECTION AGENCY
National Aerometric Data Bank
Research Triangle Park, N.C. 27711

SAROAD Daily Data Form

OMB No. 158-R0012
Approval expires 6/30/76

24-hour or greater sampling interval

2 (1)

Agency — State (2 3) Area (4 5 6 7) Site (8 9 10)

City Name

Site Address — Agency (11) Project (12 13) Time (14) Year (15 16) Month (17 18)

Project — Time Interval

Name PARAMETER Code (23 24 25 26 27) Method (28 29) Units (30 31) DP (32)

Name PARAMETER Code (37 38 39 40 41) Method (42 43) Units (44 45) DP (46)

Name PARAMETER Code (51 52 53 54 55) Method (56 57) Units (58 59) DP (60)

Name PARAMETER Code (65 66 67 68 69) Method (70 71) Units (72 73) DP (74)

19 20 | 21 22 | 33 34 35 36 | 47 48 49 50 | 61 62 63 64 | 75 76 77 78

01, 02, 03, 04, 05, 06, 07, 08, 09, 10, 11, 12, 13, 14, 15, 16, 17, 18, 19, 20, 21, 22, 23, 24, 25, 26, 27, 28, 29, 30, 31

Figure 1. SAROAD daily data form.

mission of data to its National Aerometric Data Bank (Fig. 1). This data record form is designed for use with a punched card oriented computer storage and retrieval system. Because a month's data for as many as four pollutants can be recorded on a single page, this form can be used for the storage of data with a completely manual system.

2. Reduction from Strip Charts

Measurements of pollutant concentrations made with a continuous monitoring device may be recorded on a strip chart. In order to make use of the strip charts, it is necessary to reduce the analog record to a digital value and then convert the digitized value to units of concentration. Data reduction can be in the form of either instantaneous or time-averaged pollutant concentrations. The frequency of recording instantaneous values, or the length of time over which the average value is read, should be based upon the requirements for management decision making. Generally, hourly average concentrations of gaseous pollutants are about the minimum averaging times necessary.

If an hourly average concentration is to be calculated from instantaneous values, the frequency of reading the instantaneous values could range from once each minute to once each hour. The frequency of reading directly affects the precision of the calculated hourly average. Statistically, the precision of an average, calculated from a sample of n observations selected from a population of N possible observations, can be expressed in terms of a confidence interval. The equations for the lower (X_L) and the upper (X_U) limits of the $(1 - \alpha)$ percent confidence interval are given by

$$X_L = \bar{X} - \frac{t_{\alpha,n-1}s}{\sqrt{n}} \tag{1}$$

$$X_U = \bar{X} + \frac{t_{\alpha,n-1}s}{\sqrt{n}} \tag{2}$$

where

$\bar{X}$ = sample average
$t_{\alpha,n-1}$ = student "t" for a probability α and a sample size n
s = standard deviation calculated from the sample of n observations

It is obvious from Equations (1) and (2) that the width of the confidence interval is directly related to the sample size n (i.e., as n increases the width of the interval decreases). Thus, it seems desirable to read a large number of instantaneous data values each hour.

From a practical standpoint, the cost of data reduction and subsequent calculation to determine the hourly averages is related to the number of data points to be read. In the end then, the decision as to how frequently to read instantaneous values requires a compromise between the desired precision and available resources. It is recommended that instantaneous values be read every 5 to 10 minutes depending, of course, upon whether the trace is noisy or smooth.

More commonly, the reduction of data from strip charts is in the form of hourly average concentrations. The time-average concentration is determined by visually averaging the trace on the chart. This process (often referred to as eyeball integration) is very subjective, and its precision is dependent upon the nature of the trace. If the trace is very noisy, that is, the sensor is responding rapidly to changes in concentration of the pollutant, it may be very difficult for an individual to make a reliable determination. One solution to this problem is to put a mixing (surge) chamber in the air intake line of the sensor to obtain an average over several minutes. The volume of the mixing chamber can be determined on the basis of the sensor response time and the air flow rate.

Data which are reduced from strip charts are usually recorded on forms designed for keypunching. An example of a typical form is presented in Figure 2. This form generates two punched cards for each day.

One way to increase productivity in reducing data from strip charts is through the use of a chart reading device. Basically such a device still requires the technician to visually read the chart, however the chart reader automatically transfers the data values to punched cards, thus eliminating the need to manually record and then keypunch the data.

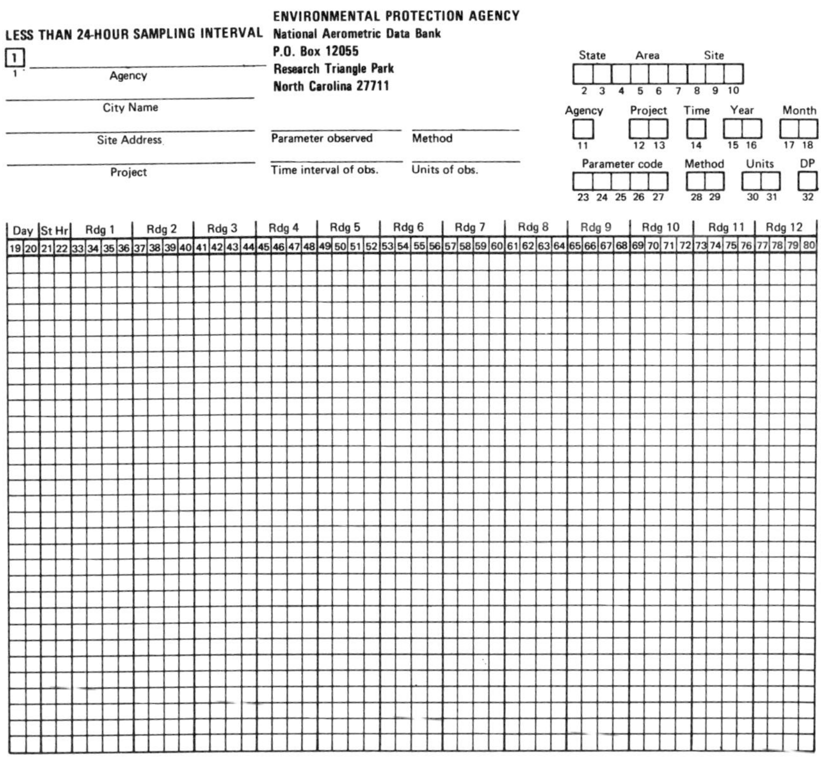

LESS THAN 24-HOUR SAMPLING INTERVAL

ENVIRONMENTAL PROTECTION AGENCY
National Aerometric Data Bank
P.O. Box 12055
Research Triangle Park
North Carolina 27711

1

Agency

City Name

Site Address

Project

Parameter observed

Method

Time interval of obs.

Units of obs.

State (2 3) Area (4 5 6) Site (7 8 9 10)

Agency (11) Project (12 13) Time (14) Year (15 16) Month (17 18)

Parameter code (23 24 25 26 27) Method (28 29) Units (30 31) DP (32)

Day		St Hr		Rdg 1				Rdg 2				Rdg 3				Rdg 4				Rdg 5				Rdg 6				Rdg 7				Rdg 8				Rdg 9				Rdg 10				Rdg 11				Rdg 12			
19	20	21	22	33	34	35	36	37	38	39	40	41	42	43	44	45	46	47	48	49	50	51	52	53	54	55	56	57	58	59	60	61	62	63	64	65	66	67	68	69	70	71	72	73	74	75	76	77	78	79	80

Figure 2. SAROAD hourly data form.

Because the task of reducing data from strip charts is highly repetitive, the technician is prone to making errors. A program of quality control should be developed to assure the reliability of data. An acceptance sampling procedure can be set up in which some portion of the data read by a technician is read by another person. Criteria can then be developed to determine whether or not the data reduced by a technician should be accepted or read again by another individual.

B. Automated Procedures

The volume of data generated by air quality monitoring activities can very quickly exceed that which can be adequately handled by a manual system. In addition, a manual system may not provide pertinent data in the time frame required for immediate action during an air pollution episode. The addition of data acquisition hardware provides a means for minimizing the manpower necessary to handle large quantities of data. Finally the use of telecommunications provides immediate access to field sampling data for action by enforcement groups during an air pollution episode.

1. Data Loggers

To eliminate the necessity for reducing data from strip charts, the output from a continuous sensor can be written on a medium suitable for direct computer input. The hardware to effect this transition is referred to as a data logger. Basically the data logger converts the analog signal from the sensor to digital form. The digitized value can be either an instantaneous impulse from the sensor, or, through the use of an electronic integrator, it can be a time-averaged value. As with manual reading of strip charts, the frequency of reading instantaneous impulses, or the length of time for integration, is determined by the way in which the data will finally be used.

Current practice is generally to have the data logger read instantaneous values and use the computer to calculate time-averaged values. Systems now in use employ frequencies of from once each minute to once each 15 minutes. The frequency long used with the United States Environmental Protection Agency's Continuous Air Monitoring Program is one each 5 minutes. Again, because, even with the use of a computer, the cost of data handling is directly related to the amount of data being handled, some compromise may be in order. Generally a recording time of once each 5 minutes is adequate for the precision required of air quality data.

2. Telecommunications

Fully automated air quality monitoring systems include the use of telecommunications to provide a direct link between the field monitoring stations and the central office of a control agency. Data transmission may be handled either in real-time or in a delayed batch mode. Justification for a fully automated system may be based upon cost or the necessity to have immediate access to field measurement data. Hickey *et al.* (*1*) concluded that fully automated systems may not be justified on the basis of cost alone if there are fewer than 20–25 continuous sensors in the monitoring network. Further, they concluded that for larger networks fully automated systems are economically more desirable when costs are figured over several years.

Cost alone may be a secondary consideration in justifying the added cost of an on-line system for an area with a high potential for air pollution episodes. In this situation, failure to initiate an emission reduction program at the first indication of an impending period of high air pollution, may result in adverse effects which far outweigh the cost of the on-line system.

a. Delayed Batch Mode Operation. In most areas of the United States, air pollution episodes occur relatively infrequently. Thus, for a major portion of the time it is not necessary that data from field stations be instantaneously transmitted to the central station. Generally, it is acceptable to provide data to the central station once or twice a day. Even with the use of standard voice grade telephone lines, the amount of data generated at a station for a 24-hour period can be transmitted in a matter of a few minutes. During an air pollution episode the frequency of transmission can be increased to hourly if necessary. The cost of data transmission over standard telephone lines includes the normal monthly service charge and normal long distance charges when the line is being used.

A flow diagram for a typical delayed batch mode, on-line system is shown in Figure 3. The output from a number of sensors is digitized, converted to engineering units, organized into a suitable record format and written on a magnetic tape. Periodically the data generated during the interval since the last transmission is transmitted to the central station. A minicomputer at the central station corrects the data for zero and span drift, computes time-averaged values, prepares a preliminary output tabulation and writes the data on a magnetic tape in a format compatible with that required by a large general purpose computer. Periodically the general purpose computer reads the data tapes and updates the master file. Typically the terminal hardware at a sampling station

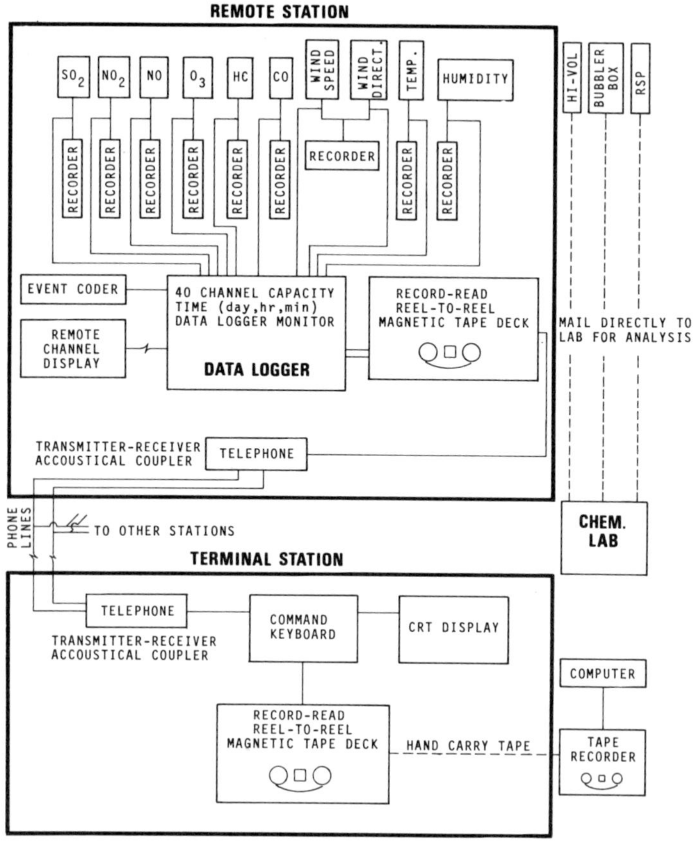

Figure 3. Delayed batch mode system flow diagram.

is designed for data input only. It is possible, however, with appropriate terminal hardware to provide retrieval access to the remote facility. When such capability is available, a user can access the data base at the central computer facility and retrieve data for that (or other) sampling station(s).

b. REAL-TIME SYSTEM. The ultimate in a fully automated data handling system is transmission of data from the field sampling station to the central computer facility as the measurements are being made (i.e., in real-time). Such a telecommunication system requires the use of a dedicated telephone line. Naturally the cost of such a line far exceeds that for delayed batch mode operation. The overall configuration of a system of this type is similar to that shown in Figure 4, the difference being that the data are transmitted instantly, as well as being written on magnetic tape for use as backup in the event of equipment malfunction. Consider-

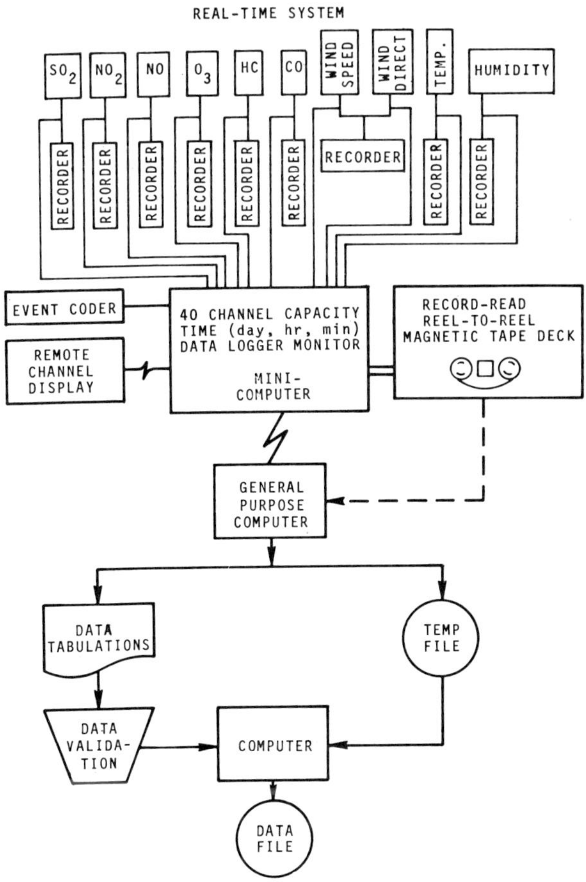

Figure 4. Real-time system.

ing the cost of the communication tie and the use of the data in management decision making, a real-time system is probably not required by most air pollution control agencies.

C. Preparation for Computer Input

There exists a variety of media on which data may be recorded in preparation for computer input. To a large extent the medium is determined by the volume of data being generated and, naturally, the input hardware of the computer itself.

1. Punched Cards

Historically, punched cards have been the primary input medium for computer systems. Data entered on punched cards must be uniquely identified as to (a) the site where the sample was collected, (b) the date and time of collection, (c) the duration of the sample, (d) the parameter being measured, (e) the method of sampling and analysis, and finally (f) the datum value itself.

For an example of a typical punched card format see Figure 1. This format was designed to handle data for sampling times of 24 hours or greater. Each element on the form which is to be keypunched is identified by the card column into which the data are to be punched. In the upper left hand corner of the form the preprinted "2" for card column 1 is used by the computer to identify the fact that the data are for sampling durations of 24 hours or more.

The sampling site is uniquely identified by numerical codes for state, area (county or city), and site within the area. This identification requires nine digits. Next there is a code for the agency responsible for collecting the data, the type of project (i.e., long-term surveillance, special sampling, etc.), and time (i.e., duration of sample, 24 hours, 48 hours, monthly, etc.). Next the year and month are recorded.

Each pollutant to be recorded on the form is identified by a parameter code, the method of sampling and analysis, the units of measurement, and the position of the decimal point in the reported datum value. Each actual datum value is recorded for the appropriate calendar day. Note that this form requires that the sampling interval for all parameters begins (and ends) at the same time.

Another way of designing an input form is to make it specific for selected pollutants. In this fashion the computer is directed to read certain fields that contain the information for specified pollutants (Fig. 5). With a form of this type, it is still necessary to uniquely identify the sampling site where and the date when the data were collected. With this form, designed for use by a local agency, the city and state codes are really unnecessary. The computer is instructed that the data in columns 10–13 will always be total suspended particulates in micrograms per cubic meter, that in columns 14–17 is sulfur dioxide in micrograms per cubic meter, etc.

2. Magnetic Tape

With continuous analyzers, data generated at the field sampling station can be recorded directly on magnetic tape which can subsequently be read directly into a computer. Again it is necessary to uniquely identify the sampling site where the data were collected, the date and time of sample collection, and the parameter being measured. In addition, it is

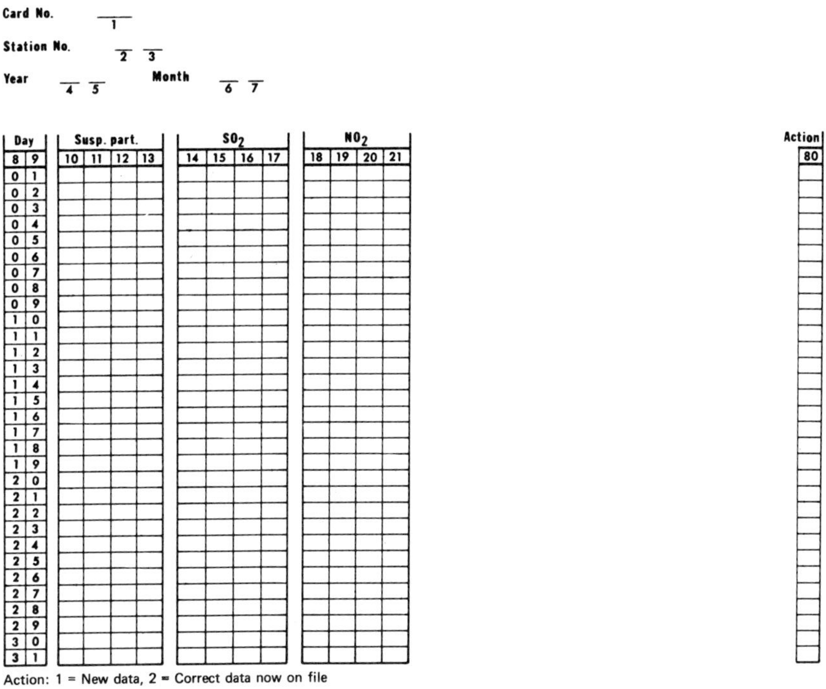

Figure 5. Cincinnati, Ohio daily data form.

desirable to have built-in edit checks which can be used by the computer to determine the validity of the data.

A typical record format used with a magnetic tape recorder is shown in Figure 6. For each 5-minute interval, the instantaneous digitized values from a maximum of 16 sensors (i.e., channels 02 to 17) are written in a 156-character record. In addition to station, date, and time, a code is included for Station Calibrate Status. This provides for an electronic check on the output from each sensor. The first two channels (00 and 01) are reserved for a zero and span check on the electronics. This information is used by the computer to determine if the data logger was properly recording on the magnetic tape.

A total of eight characters are recorded for each sensor: channel status (1), channel number (2), polarity (1), and data (4). After the last channel has been recorded, an interrecord gap signals the end of record.

3. Optical Character Recognition

Computer technologists are constantly striving to improve the process of getting information from its source of generation into the computer.

RECORDING FORMAT – RAW DATA

IICCDDDHHMMA BYYPXXXX IRG

II	= Station Number	ØØ – 99
CC	= Year	72, 73, etc.
DDD	= Day of Year	ØØ1 – 366
HH	= Hour of Day	ØØ – 23
MM	= Minuter	ØØ – 59
A	= Status Status	Ø = Normal Operation
		1 = Low Cal (10%)
		2 = High Cal (70%)
B	= Channel Status	Ø = Normal Operation
		1 = Zero Calibrate
		2 = Span Calibrate
		3 = Out of Service
YY	= Channel	ØØ – 17
P	= Polarity	– or +
XXXX	= Value	
IRG	= Interrecord Gap	

Figure 6. Recording format for magnetic tape data handling system.

With a punched card oriented system data must be entered on a suitable data record form at the place of origin. Next the data must be keypunched and then verified to ensure minimum errors. This process is both costly and time consuming. Methods are now available to record data on a source document and then have the source document processed by an optical reader directly into the computer. Because of the cost of the optical reading device, a very high volume of data is necessary to justify its use. In planning a data handling system, the potential for optical character recognition (OCR) hardware should be investigated with the management of the computer center.

The design of data forms for OCR requires consideration of the same factors as discussed for punched card systems. In addition, however, technicians responsible for transcribing data must be given special training as to the forms of handwriting acceptable to the optical reader.

III. Data Validation

A very important final step in data handling is that of data validation. The purpose of data validation is to detect and then correct any data values which may not actually represent air quality conditions at a given sampling station. To be effective, the data validation procedures should be handled completely independent of the procedures used in initial data collection. Moreover, it is advisable that the individuals responsible for data validation not be involved directly with initial data collection.

A. Manual

Whether data handling is carried out manually or with a computer-oriented system, it is important to have an individual review tabulations of all data. When the volume of data is quite large, it may be advisable to use the procedures described in the following section to assist with the manual inspection.

As an individual scans tabulations there is no absolute way that it can be determined that all values are, or are not, valid. Typically, in manual inspection the purpose is to spot unusually high (or low) values which might be indicative of a gross error somewhere in the data collection system. Obviously to recognize a pollutant concentration as being extreme, the individual must be quite knowledgeable of air quality conditions which generally prevail at that station. Data values which are considered questionable are not summarily thrown out, but rather are referred back to the data collection group for verification. It is apparent that this procedure is sensitive to spurious extreme values but not to intermediate values which could also be grossly in error.

Through manual data validation it may also be possible to detect uncorrected drift in the zero baseline of a continuous sensor. A zero drift may be indicated when the daily minimum concentration shows a tendency to increase or decrease over a period of several days. For example, at most sampling stations the early morning (3:00 to 4:00 a.m.) concentrations of carbon monoxide generally tend to reach a minimum of 2 to 4 ppm. If the minimum concentration differs very much from this value there may be reason to suspect a zero drift. The existence of a zero drift will be obvious when the original strip chart is reviewed.

B. Computer Assisted

Procedures for data validation can easily be incorporated into the basic software system used with automatic data processing. The computer can be programmed to scan each datum value prior to preparing an output tabulation of new data. Questionable data values are then flagged on the data tabulation to indicate the type of error condition which may exist.

The inspection for extreme values can be handled quite easily by the computer. Tables of high and low values for each pollutant at each sampling station can be built into the program. The check for extreme values can be further refined to include different values depending upon the time of day, weekends, and week days, etc.

Because of the computer's ability to perform computations and make comparisons extremely rapidly, it is possible to make some determination concerning the validity of data values which are not necessarily high or low. One way to do this is to test the difference between successive data values. Ordinarily one would not expect very rapid changes in the concentration of some pollutants from one 5-minute value to the next, or from one hourly value to the next. Whenever the difference between two successive values is greater than a predetermined value, the tabulation can be flagged with an appropriate symbol. For example, with hourly sulfur dioxide concentrations if two values differ by more than 0.05 ppm (113 $\mu g/m^3$) it may be considered desirable to check the data values against the original strip chart to assure validity.

IV. Computer Storage Media

A. Types of Storage Media

1. Magnetic Tape

Magnetic tape is widely used as a medium for the permanent storage of data. The physical space requirements for maintaining a data base on magnetic tape is less than that of most other storage media. Data that would require several hundred thousand punched cards for storage can be written on one 2400-ft reel of 1600 bits per inch (BPI) magnetic tape. Next, because of the high speed tape drives now available, data can be read into the computer very rapidly. Likewise, magnetic tapes can be read repeatedly with a very low probability of damage. The sequential nature of air quality data is ideally suited to storage on magnetic tape.

Some care in the storage and handling of magnetic tape is necessary. Ideally magnetic tapes should be stored in a fireproof vault. Lacking a suitable vault, a room with controlled temperature and humidity can be used. In this situation it is advisable to maintain a second copy of each tape in a separate room, so that a fire or other calamity will not destroy the data base. Another reason for maintaining two copies of each tape is the ever present problem of human error. Even though there are means of protecting a tape once information has been written on it, all computer centers have experienced mistakes in tape handling. The use of more than one tape is especially important with file updating, in which data must be transferred from one tape to another as new data become available.

2. Direct Access Storage

The ultimate in computer storage media are the direct access devices such as magnetic disks, magnetic drums, and data cells. Devices such as these are ideally suited for use with a data base which is constantly being accessed. The speeds at which data are transferred into and out of direct access storage devices provide for more efficient operation from the standpoint of the use of the central processor of the computer system. The cost of a direct access storage device is much greater than that of a magnetic tape device. Because of the differential cost, direct access storage must be justified on the basis of savings in computer processing costs.

B. Selecting the Storage Medium

The use of data record forms, paper tape, and punched cards as permanent storage media for air quality data is to be avoided in all but very specialized situations. Determining whether to use magnetic tape or direct access storage requires some additional considerations.

A magnetic tape can be purchased in the United States for about $20.00. A disk, capable of storing about twice the quantity of data that can be written on a magnetic tape, costs about $500. The differential cost can be very quickly offset through the savings in computer processing cost if the data are being frequently accessed. Since air quality data are being added continually, the data base may be accessed on a daily or weekly basis for file updating. Additionally, as the new data are validated, the file must be accessed to make changes in data already on the file. After data have been on the file for a period of time, the frequency of access very quickly decreases.

The selection of the storage medium is, of course, dependent upon the hardware configuration of the computer system. If the system has both tape and disk drives, some combination of the two should probably be considered. The system adopted by the city of Cincinnati, Ohio utilizes both magnetic disk and magnetic tape. New data are stored on a disk file for a period of 3 months, and afterward are transferred to magnetic tape for permanent storage. This procedure takes advantage of the high speed associated with direct access storage and the lower cost of sequential storage for historical files.

C. File Structure

A unit of data is termed a field, which consists of a fixed or floating point number, a packed decimal number, or a series of one or more bytes. Related fields, when grouped together, form a logical record (e.g., a logical record might be the necessary identification information and the 5-minute data values for 1 hour for a specific pollutant). The interchange of information between core storage and secondary storage (i.e., magnetic tape or disk) is by blocks composed of one or more logical records. A block is referred to as a physical record in secondary storage.

The effectiveness of a data management system is determined by the way in which data files are structured to make maximum utilization of data storage facilities. For example, on a magnetic tape there is a fixed length gap of $\frac{3}{4}$ in. between each physical record. The use of long physical records minimizes the portion of the total tape which is taken up by record gaps.

Logical records are grouped to form a file. The individual records can be constructed according to one of four formats; fixed length records, variable length records, undefined length records, and spanned records. The first two are most commonly used and are discussed below.

The size (i.e., number of characters or bytes) of a fixed length record is constant for all records in a file. They may be stored blocked or unblocked. Blocking of logical records is often used to create large physical records to minimize the number of record gaps in the file. Computer programmers tend toward the use of fixed length records because they are more easily handled in sorting and computation.

Variable length records provide for more flexible storage of data in secondary storage. It is necessary to include the record length as part of the information in the logical record. Variable length records tend to maximize the efficiency of secondary storage.

In a situation where missing data may occur, space must be provided for the missing data in a fixed length record. The use of variable length records permits the use of a smaller record size and usually results in a more efficient use of secondary storage.

V. Statistical Parameters

A large body of data, such as that resulting from the operation of air quality monitoring equipment, must be reduced in volume when it is presented to management for decision making. Single values determined from a body of data are referred to as statistics. The more commonly used statistics can be grouped into two classes: (a) those which describe location, and (b) those which describe dispersion.

A. Location Statistics

Typically, series of observations have a tendency to cluster at some particular location on the scale of measurement of the variable in question. In many instances knowledge of the value of the variable at the point of cluster may be more informative than the actual observations which were made. Statistics which are used to identify the point of cluster include the arithmetic mean, the median, and in the case of some air quality variables, the geometric mean.

1. Arithmetic Mean

The equation for the arithmetic mean ($\overline{X}$) is

$$\overline{X} = \frac{1}{n} \sum_{i=1}^{n} X_i \tag{3}$$

where $X_i = X_1, X_2, \ldots, X_n$, the individual measurements and n is equal to the number of measurements.

2. Median

The median is the middle value when the observations are arranged in order of magnitude. Thus, if there are five observations and they are arranged such that $X_1 < X_2 < X_3 < X_4 < X_5$, the median would correspond to the value of X_3. With four observations, $X_1 < X_2 < X_3 < X_4$ the median is the value midway between X_2 and X_3. In general if n (the number of observations) is odd, the median is the value of the $\frac{1}{2}(n+1)$ observation. If n is even, the median is midway between the $\frac{1}{2}n$ and the $\frac{1}{2}(n+2)$ observation.

3. Geometric Mean

The equation for the geometric mean ($\overline{X}_g$) is

$$\overline{X}_g = \exp\left(\frac{1}{n} \sum_{i=1}^{n} \ln X_i\right) \tag{4}$$

Selection of a location statistic is dependent upon the variation in the observed data. The location statistic, which is selected, is that which best characterizes the actual concentrations. For example, consider the following measurements of ambient suspended particulate concentrations; 75, 60, 80, 90, and 475. The arithmetic mean ($\overline{X}$) for this data is 156, the median is 80, and the geometric mean is 109. Clearly; the arithmetic mean does not represent the set of data. The median of 80 probably best characterizes the data. More is said about the rationale for selecting a location statistic in Section VI.

B. Measures of Variability

A location statistic, in and of itself, is not sufficient to characterize a set of measurement data. It is also necessary to indicate to what extent the data varied about the value selected to represent the data. Measures of variability frequently used with aerometric data are the standard deviation (s) and the standard geometric deviation (s_g).

1. Standard Deviation

The most widely used statistic to express the variability of data is the variance (s^2) given by

$$s^2 = \frac{\sum_{i=1}^{n} (X_i - \overline{X})^2}{n - 1} \tag{5}$$

The variance is, in essence, the average square deviation of the individual measured values from the arithmetic mean. The use of squared deviations is necessary from a mathematical standpoint because the sum of the deviations about the mean

$$\sum_{i=1}^{n} (X_i - \overline{X}) = 0 \tag{6}$$

The standard deviation (s) is given by

$$s = \sqrt{s^2} \tag{7}$$

The standard deviation is expressed in the same units as the data from which it was determined. Further discussion of the standard deviation and its relationship to the normal probability distribution is presented in Section VI.

2. Standard Geometric Deviation

Because of the nature of the underlying probability distribution for some parameters, the standard geometric deviation is often used as a measure of variability for air quality data. Its use is based upon studies which have shown that the logarithm of some air quality measurement data tend to follow the normal probability distribution (*2*). The standard geometric deviation (s_g) is given by

$$s_g = \exp\left(\sum_{i=1}^{n} \frac{(\ln X_i - \overline{\ln X})^2}{n-1}\right)^{1/2} \tag{8}$$

VI. Methods of Statistical Analysis

A. Frequency Distributions

A convenient way to present a large body of data is in the form of a frequency distribution. The frequency distribution shows the number of data values falling within selected intervals over the scale of measurement of the data. An example of a frequency distribution of 60 measurements of the ambient concentration of suspended particulates is shown in Figure 7.

Another way to present these data is in the form of a cumulative frequency distribution (Fig. 8). The cumulative frequency distribution shows the number (or percent of the total number) of data values which are less than, or equal to, specified concentrations.

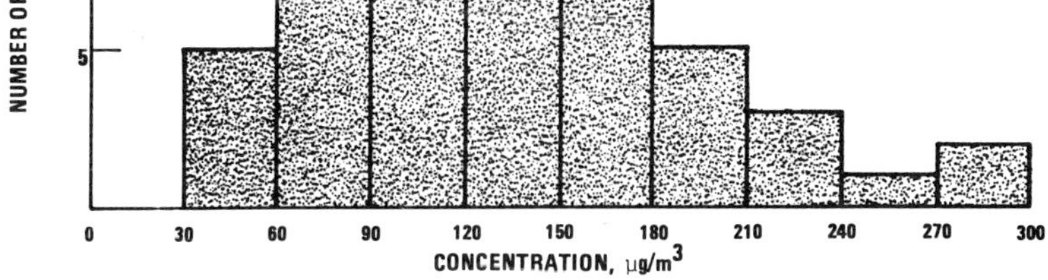

Figure 7. Frequency distribution of concentrations of total suspended particulate matter.

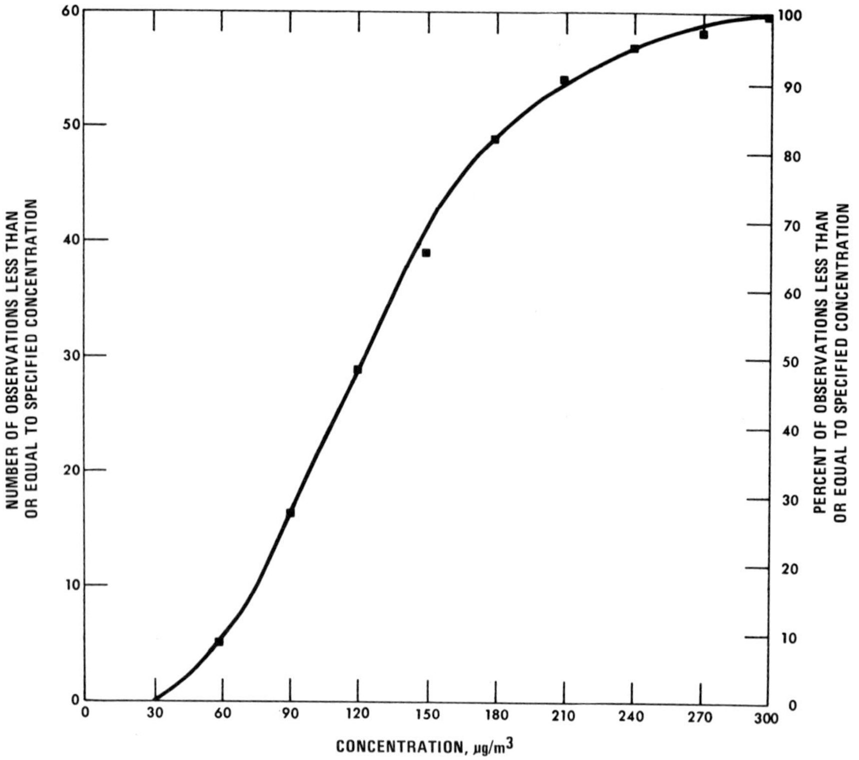

Figure 8. Cumulative frequency distribution of concentrations of suspended particulate matter.

1. Probability Distributions

The histogram of the frequency distribution shown in Figure 7 is for a sample of 60 observations selected from the population of all possible observations. It may be possible to postulate the nature of the underlying theoretical distribution of the population. It may also be possible to determine a mathematical model, $y = f(x)$ which best describes this theoretical distribution. If the function $f(x)$ is constructed to have the properties

$$f(x) \geq 0 \tag{9}$$

and

$$\int_{-\infty}^{+\infty} f(x)\, dx = 1 \tag{10}$$

it is referred to as a probability density function. The probability that a value x, selected at random from the population, will lie in the interval

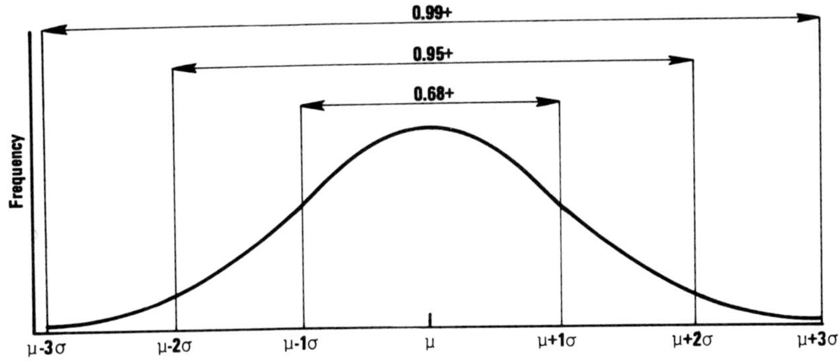

Figure 9. Area under normal curve between specified intervals.

between a and b is given by

$$\Pr(a \leq x \leq b) = \int_a^b f(x)\,dx \tag{11}$$

The normal (or Gaussian) probability density function (Pd f) is widely used in statistical methods of analysis. It is referred to as a two-parameter distribution in that it is uniquely described by the mean (μ) and the standard deviation (σ). The Pd f is given by

$$y = (1/\sqrt{2\pi}\,\sigma)\exp[-(X - \mu)^2/2\sigma^2] \tag{12}$$

A graph of the normal distribution, including probabilities within selected intervals measured about the population mean, is shown in Figure 9.

In contrast to the normal distribution, which is symmetrical about the mean (μ), distribution of air contaminants exhibits a positive skewness (Fig. 7). The upper end of such a distribution extends much farther above the mean than does the lower end. Distributions of air contaminant concentrations have been studied quite extensively (*2–4*). To date there is no general agreement among investigators as to which is the correct probability density function. There is general agreement that a logarthmic transformation of the data closely approximates the normal distribution. This distribution, referred to as the log-normal distribution, like the normal is also a two-parameter distribution. The log-normal distribution is specified by the geometric mean ($\overline{X}_g$) and the standard geometric deviation (s_g).

2. Extrapolation with Intermittent Data

Because of the tendency for measured concentrations of many air contaminants to approach log normality, such data are often presented in the form of cumulative frequency distributions plotted on logarithm-probability graph paper (Fig. 10). The horizontal (probability) axis of logarithm-probability paper is constructed such that the cumulative frequency distribution, of a log-normally distributed variable, will be linear.

To prepare a graph of the cumulative frequency distribution it is necessary to order the measurement data from the lowest to the highest concentration. Next select about five or six concentration values, over the entire range of concentrations, and determine the percent of all measured concentrations which are less than or equal to the selected concentrations. The plotting positions of the five or six points are then determined by the concentration and the corresponding percentile value.

Another way to plot the graph of the cumulative frequency distribution is to begin by preparing a frequency distribution of the data. Next determine the percentiles associated with the upper boundary of each interval in the distribution. The data presented in Figure 9 were plotted in this latter fashion.

The geometric mean ($\overline{X}_g$) and the standard geometric deviation (s_g) can be estimated from the graph of the cumulative frequency distribution. The value of $\overline{X}_g$ corresponds to the 50th percentile = 118 $\mu g/m^3$

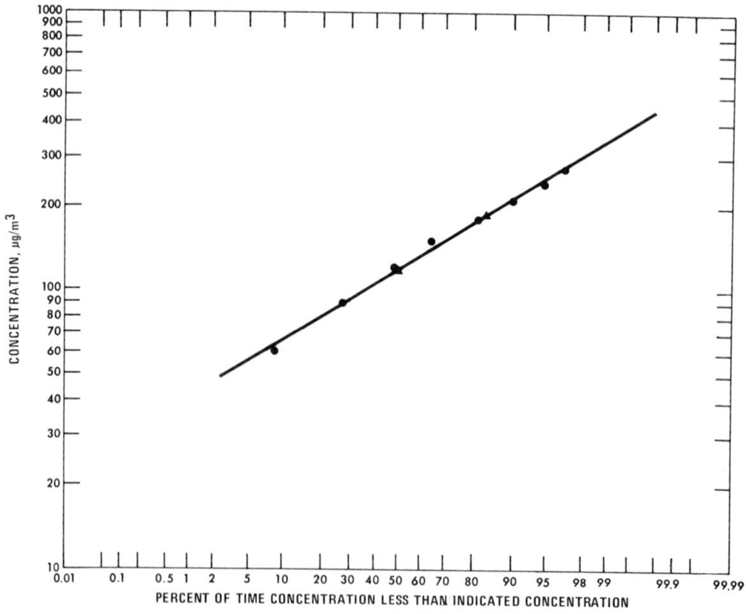

Figure 10. Cumulative frequency distribution of ambient concentrations of particulate matter.

for the data in Figure 10. Similarly the value for s_g is

$$s_g = \frac{\text{concentration at 84th percentile}}{\text{concentration at 50th percentile}} = \frac{188}{118} = 1.59 \tag{13}$$

B. Pollution Concentration Maps

National ambient air quality standards, such as have been adopted by the United States for selected air contaminants, specify sample averaging time, frequency of occurrence, and concentration. Sample averaging times are 1, 3, 8, and 24 hours, and 1 year. Standards which specify maximum concentrations for 24 hours or less also state that the maximum shall not be exceeded more than once per year. It is further specified that the standard must not be exceeded anywhere in the air quality control region. Thus, it is important to be able to characterize pollution levels throughout a geographical area.

The geographical distribution of air pollution can be presented in the form of contour maps quite similar to ones used to show elevation on a topographic map. Maps such as these are usually referred to as isopleth maps. The isolines shown on such maps can be used to depict changes in average concentration or changes in the frequency of occurrence of concentrations above a specified threshold value (such as a maximum 24-hour concentration of suspended particulates of 260 $\mu g/m^3$).

1. Average Concentrations

In order to prepare an isopleth map to show changes in annual average concentration throughout a geographical area, it is necessary to know the annual average pollutant concentrations at a number of points. The range of average concentrations to be shown should be separated into from four to six groups. To the extent practicable the boundaries of the groups should be in round numbers, as for example 50, 60, 70 or 55, 65, 75, etc. Finally points of equal concentration are connected by isoconcentration lines. The fitting of isoconcentration lines can be done manually; however, it requires considerable skill in performing visual interpolation. Where a computer system with an XY plotter is available, computer programs are available to fit the contour lines on a more consistent mathematical basis.

For many areas, the number of locations, for which measured pollutant concentration data are available, is quite limited. An isopleth map based upon a small number of data points will show very little detail. It is possible to augment existing air quality measurement data with estimates made by means of a mathematical dispersion model. One such model, which has been used extensively in recent years is the Air Quality Display Model (AQDM) (*5*) made available by the United States Environmental Protection Agency. The application of this model requires detailed meteorological and source emission data.

In practice AQDM is fitted to an area by using ambient data from

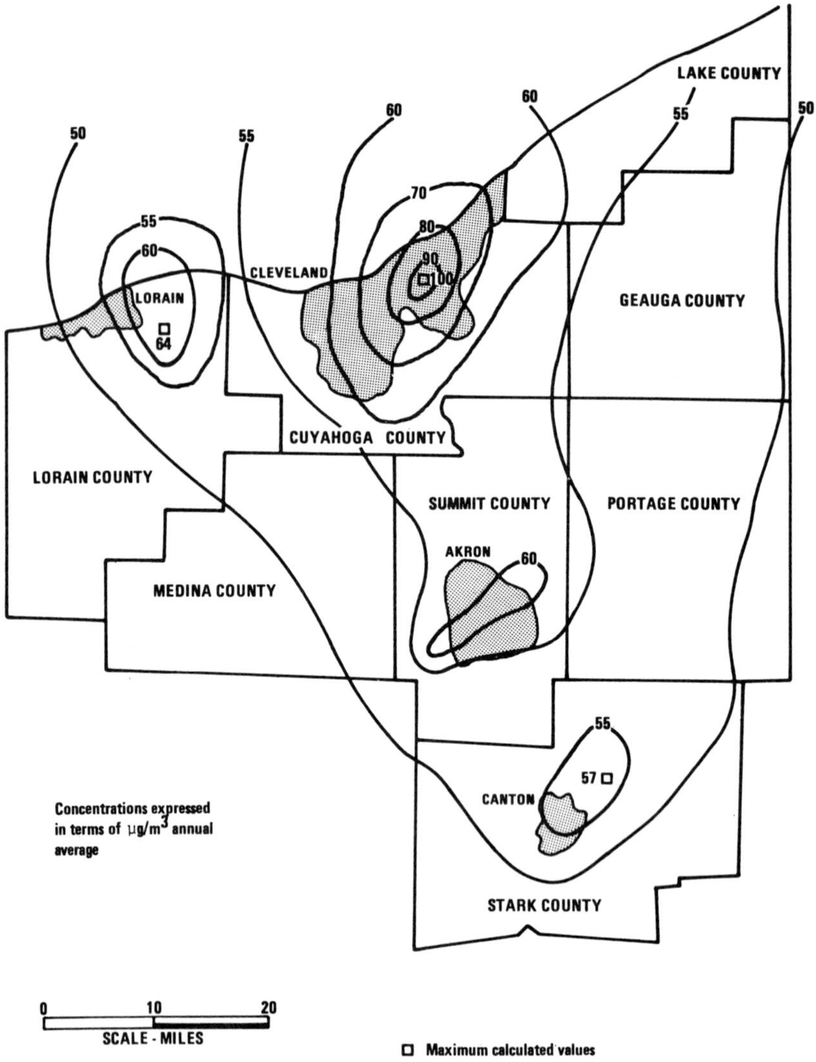

Figure 11. Geographical distribution of ambient particulate matter concentration as determined by the air quality display model. Cleveland, Ohio, Air Quality Control Region. Concentrations expressed in terms of $\mu g/m^3$ annual average. □: Maximum calculated values.

a number of sampling locations. Next the model is used to estimate ground-level pollutant concentrations at the points of intersection on a receptor grid (a maximum of 237 points for the AQDM). An example of an isopleth map prepared with the aid of AQDM is shown in Figure 11.

2. Percent of Time above Selected Concentrations

Where measurements of ambient pollutant concentrations are made, it is a simple matter to determine the percent of time a specified concentration is exceeded. As with the annual average concentrations, from four to six categories of percent of time are selected. Contour lines are then fitted according to these categories. The AQDM model, previously discussed, is not capable of providing the frequency of exceeding specified values at each receptor point and so cannot be used to augment the available air quality data.

Smith and Jeffrey (*6*) presented pollution maps showing the spatial distribution of SO_2 for a portion of London, England. These maps were developed from measurements made at 100 sampling locations. In addition to presenting isopleths of average concentration, they also included contour maps showing the percent of times when the daily average SO_2 concentration exceeded specified concentrations. Figure 12 shows the number of times when the daily average SO_2 concentration exceeded 500 $\mu g/m^3$.

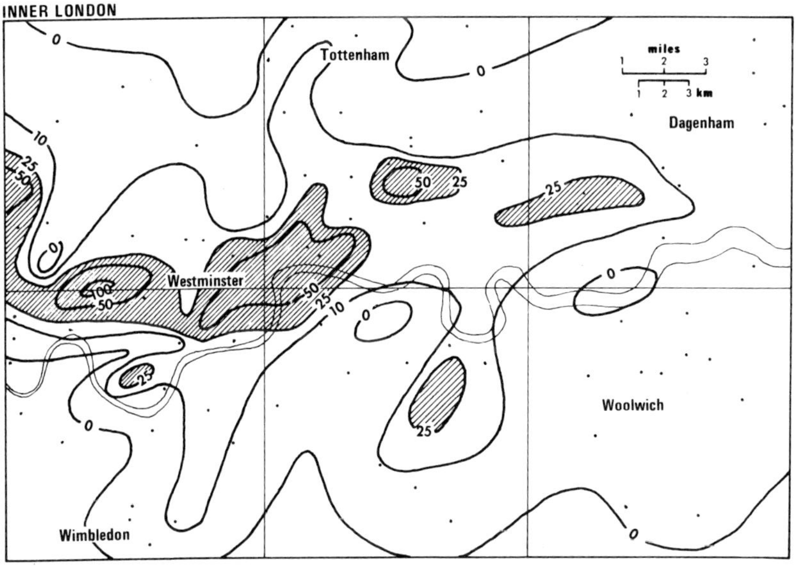

Figure 12. Number of days when the daily average sulfur dioxide concentration exceeded 500 $\mu g/m^3$, London, England (*6*).

Maps showing the spatial distribution of pollutant concentration can be quite useful in determining compliance with air quality standards. Pollution maps such as Figures 11 and 12 can provide an indication of where annual and short-term standards are being exceeded.

C. Concentration versus Sample Averaging Time

Data from the Continuous Air Monitoring Program (CAMP) were used by Larsen (7) to study the relationship between pollutant concentration and sample averaging time. Larsen began by calculating the percent cumulative frequency distribution for the basic data which were 5-minute instantaneous concentrations. The 5-minute data values were then used to calculate concentrations for averaging times from 10 minutes to 1 year. Percent cumulative frequency distributions were then calculated from the data for each averaging time. An example of the cumulative frequency distributions for sulfur dioxide measured at the CAMP stations in four cities are presented in Table I. Larsen then plotted the distributions for the various averaging times on log–log graph paper. An example of such a plot for Washington, D.C. is presented in Figure 13.

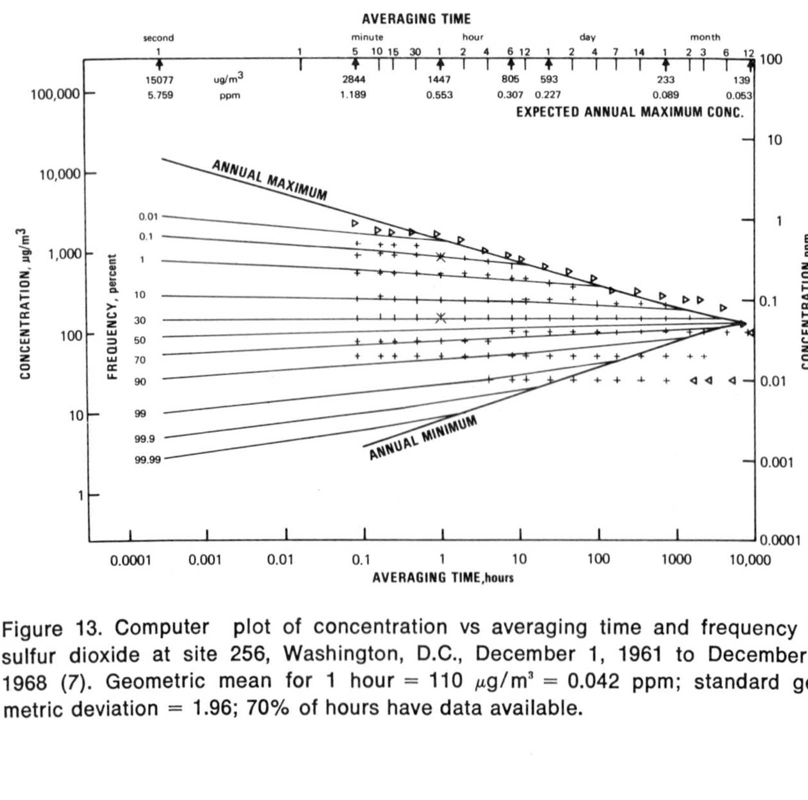

Figure 13. Computer plot of concentration vs averaging time and frequency for sulfur dioxide at site 256, Washington, D.C., December 1, 1961 to December 1, 1968 (7). Geometric mean for 1 hour = 110 $\mu g/m^3$ = 0.042 ppm; standard geometric deviation = 1.96; 70% of hours have data available.

Table I **Sulfur Dioxide Concentration (ppm), at United States Environmental Protection Agency Continuous Air Monitoring Program Sites, by Averaging Time and Frequency, 1962–1968** (*7*).

City and averaging time	*Annual maxi-mum*	*1962 through 1968*				*Maximum for year*							*Percent data available*	*Percent of time concentration is exceeded*							
		Geom. mean	*SGD*[a]	*Max*	*Min*	*1962*	*1963*	*1964*	*1965*	*1966*	*1967*	*1968*		*0.01*	*0.1*	*1*	*10*	*30*	*50*	*70*	*90*
Philadelphia																					
5 minutes	2.67	0.055	2.41	1.25	0.79	1.25	1.05	1.00	1.11	0.79	1.12	1.10	77	0.96	0.72	0.48	0.20	0.09	0.05	0.03	0.01
1 hour	1.19	0.060	2.18	1.03	0.66	1.03	0.85	0.84	0.94	0.66	0.77	0.88	78	0.85	0.67	0.45	0.20	0.09	0.05	0.03	0.01
8 hours	0.61	0.064	1.99	0.71	0.43	0.56	0.58	0.54	0.71	0.43	0.56	0.55	79		0.52	0.37	0.19	0.10	0.06	0.04	0.01
1 day	0.43	0.067	1.88	0.46	0.35	0.35	0.46	0.43	0.35	0.36	0.35	0.36	81			0.31	0.17	0.10	0.07	0.04	0.02
1 month	0.15	0.075	1.50	0.16	0.12	0.13	0.12	0.15	0.13	0.13	0.13	0.16	88				0.12	0.10	0.08	0.07	0.04
1 year	0.08	0.081	1.00	0.10	0.06	0.09	0.06	0.09	0.08	0.09	0.10	0.08	100						0.08		
St. Louis																					
5 minutes	2.32	0.028	2.72	1.42	0.85			1.16	1.42	1.25	0.85	1.33	83	0.94	0.56	0.28	0.10	0.04	0.02	0.01	0.00
1 hour	0.94	0.032	2.43	0.96	0.55			0.73	0.96	0.84	0.55	0.68	83	0.73	0.49	0.25	0.10	0.05	0.02	0.01	0.00
8 hours	0.44	0.034	2.18	0.36	0.23			0.31	0.36	0.33	0.24	0.23	84		0.29	0.18	0.09	0.05	0.03	0.02	0.00
1 day	0.30	0.036	2.05	0.26	0.16			0.26	0.19	0.18	0.21	0.16	86			0.14	0.08	0.05	0.03	0.02	0.01
1 month	0.09	0.042	1.59	0.08	0.05			0.08	0.06	0.06	0.05	0.06	90				0.06	0.05	0.04	0.03	0.01
1 year	0.05	0.047	1.00	0.06	0.03			0.06	0.05	0.04	0.03	0.03	100						0.03		
San Francisco																					
5 minutes	0.56	0.005	2.87	0.33	0.16	0.16	0.33	0.26					64	0.18	0.13	0.07	0.03	0.01	0.01	0.00	0.00
1 hour	0.22	0.006	2.54	0.26	0.11	0.11	0.26	0.16					64	0.16	0.11	0.07	0.03	0.01	0.01	0.00	0.00
8 hours	0.10	0.007	2.27	0.10	0.06	0.06	0.07	0.10					65		0.09	0.06	0.03	0.02	0.01	0.00	0.00
1 day	0.07	0.007	2.12	0.08	0.05	0.05	0.05	0.08					65			0.05	0.03	0.02	0.01	0.00	0.00
1 month	0.02	0.008	1.63	0.03	0.01	0.01	0.02	0.03					75				0.02	0.01	0.01	0.01	0.00
1 year	0.01	0.010	1.00	0.02	0.01		0.01	0.02					67						0.01		
Washington																					
5 minutes	1.14	0.040	2.14	0.87	0.42	0.42	0.56	0.87	0.44	0.47	0.44	0.53	70	0.49	0.35	0.21	0.10	0.06	0.03	0.02	0.00
1 hour	0.56	0.042	1.96	0.62	0.35	0.38	0.48	0.62	0.35	0.45	0.37	0.42	70	0.45	0.34	0.21	0.10	0.06	0.03	0.02	0.00
8 hours	0.31	0.045	1.81	0.35	0.22	0.23	0.35	0.32	0.27	0.34	0.22	0.22	71		0.26	0.18	0.10	0.06	0.04	0.02	0.01
1 day	0.23	0.046	1.72	0.25	0.15	0.18	0.25	0.22	0.20	0.25	0.15	0.18	72			0.16	0.10	0.06	0.04	0.02	0.01
1 month	0.09	0.050	1.42	0.11	0.07	0.10	0.11	0.09	0.08	0.10	0.07	0.10	76				0.09	0.06	0.04	0.02	0.01
1 year	0.05	0.053	1.00	0.05	0.04	0.05	0.05	0.04	0.05	0.04		0.04	86						0.04		

[a] Standard geometric deviation.

Based upon his study of cumulative frequency distribution of measured concentrations of gaseous pollutants in eight major cities, Larsen found a number of empirical relationships.

1. The maximum concentration (C_{max}) for a given sample averaging time is given by

$$C_{max} = \overline{X}_g s_g^z \tag{14}$$

where

$\overline{X}_g$ = geometric mean concentration
s_g = standard geometric deviation
z = number of standard deviations from geometric mean concentration to maximum concentration

Values of z corresponding to maximum concentrations for selected averaging times are presented in the tabulation below.

Averaging time (hours)	z
1	3.81
3	3.53
8	3.26
24	2.94

For example, the values of $\overline{X}_g$ and s_g for hourly average SO_2 at the Washington, D.C. CAMP station were 0.042 $\mu g/m^3$ and 1.96, respectively. The estimated value for the maximum hourly concentration is

$$C_{max\ hr} = (0.042)(1.96)^{3.81} = 0.55 \text{ ppm}$$

2. If the standard geometric deviation is known for an averaging time a, then the standard geometric deviation for an averaging time b is given by

$$s_{gb} = s_{ga}{}^{v^{0.5}} \tag{15}$$

where

$$v = \frac{\ln(t_{tot}/t_b)}{\ln(t_{tot}/t_a)} \tag{16}$$

and

t_a = averaging time a
t_b = averaging time b
t_{tot} = the total averaging time (usually 1 year)

Referring again to the Washington, D.C. CAMP data (Table I), suppose measurements had been made of the 24-hour concentrations of SO_2 and the standard geometric deviation was determined to be 1.72. Estimate the value of the standard geometric deviations for 1-hour measurements:

$$v = \frac{\ln(8760/1)}{\ln(8760/24)}$$

$$v = 1.5386$$

$$s_{gb} = (1.72)^{(1.5386)^{0.5}}$$

$$s_{gb} = 1.96$$

These empirical relationships provide a means of extrapolation from the sample averaging time used with the ambient sampling program to any other sample averaging time.

D. Correlation and Regression Analysis

Situations often occur wherein measurements of several aerometric parameters are made at the same time—either at the same location or at different locations. There are two statistical methods which can be used to determine whether or not there is a relationship between two (or more) variables, i.e., correlation analysis and regression analysis. The former is used to provide a measure of the degree of the dependence between the variables. The latter is used to determine the mathematical function which characterizes the relationship between the variables.

1. Correlation Analysis

The degree of dependence between two variables is expressed in terms of the population correlation coefficient ρ. In practice an estimate of ρ is obtained by determining the sample correlation coefficient r from a sample of observations selected from the population of all possible observations. Two properties of the correlation coefficient which are important are (1) r is a pure number without units or dimension, and (2) r always lies in the interval -1 to $+1$. Positive values indicate a tendency for both variables to increase (or decrease) together, negative values indicate that large values of one variable are associated with small values of the other variable.

The equation for calculating the correlation coefficient from a sample of simultaneous observations for two variables is

$$r = \frac{\sum_{i=1}^{n} (X_i - \bar{X})(Y_i - \bar{Y})}{\left[\sum_{i=1}^{n} (X_i - \bar{X})^2 \sum_{i=1}^{n} (Y_i - \bar{Y})^2\right]^{1/2}} \tag{17}$$

For computational purposes Equation (17) can be transformed to

$$r = \frac{n \sum_{i=1}^{n} X_i Y_i - \left(\sum_{i=1}^{n} X_i\right)\left(\sum_{i=1}^{n} Y_i\right)}{\left(\left[n \sum_{i=1}^{n} X_i^2 - \left(\sum_{i=1}^{n} X_i\right)^2\right]\left[n \sum_{i=1}^{n} Y_i^2 - \left(\sum_{i=1}^{n} Y_i\right)^2\right]\right)^{1/2}} \tag{18}$$

To illustrate the use of the correlation coefficient, consider the situation where measurements of the 24-hour concentration of suspended particulates are being made at two nearby sampling sites. If it can be established that there is a dependence in the pollutant concentrations at the two sites, the decision may be made to discontinue one of the sites. Measurements of suspended particulates on 10 randomly selected days are presented in the tabulation below.

Day	*Site A* (X)	*Site B* (Y)
1	90	50
2	80	50
3	100	95
4	150	185
5	105	120
6	95	82
7	195	165
8	167	175
9	65	55
10	120	80

The sample correlation coefficient is $r = 0.89$ suggesting a high degree of dependence between measurements of particulates made at the two sites. Because of the rather small sample size, the possibility exists that a sample correlation of 0.89 could have occurred by chance, when in fact there is no relationship between the two variables. Snedecor and Cochran (*8*) provide a table of critical values of correlation coefficients for selected sample sizes and both the 5 and 1% levels of significance. For a sample of 10 observations these critical values are 0.63 and 0.76, respectively. Because the calculated value, $r = 0.89$ exceeds the critical value 0.76, it can be concluded that there is, in fact, a relationship between measurements of suspended particulates at the two sites.

2. Regression Analysis

The statistical procedure used to determine the mathematical relationship between a dependent variable and one or more independent variables is referred to as regression analysis. The linear model which describes the relationship between a dependent and one independent variable is given by

$$Y = \alpha + \beta X + \epsilon \tag{19}$$

where

α = intercept
β = slope
ϵ = random error, assumed to be normally distributed with mean = 0 and variance = σ^2

In practice the parameters α and β and the variance σ^2 are estimated by a, b, and $s_{x \cdot y}^2$, respectively, calculated from a sample of observations by

$$b = \frac{n \sum_{i=1}^{n} X_i Y_i - \sum_{i=1}^{n} X_i \sum_{i=1}^{n} Y_i}{n \sum_{i=1}^{n} X_i^2 - \left(\sum_{i=1}^{n} X_i\right)^2} \tag{20}$$

$$a = \frac{1}{n}\left(\sum_{i=1}^{n} Y_i - b \sum_{i=1}^{n} X_i\right) \tag{21}$$

$$s_{x \cdot y}^2 = \frac{\sum_{i=1}^{n} (Y_i - a - bX_i)^2}{(n - 2)} \tag{22}$$

For the example previously cited, the linear relationship between measured values made at site A and predicted values at site B is

$$\begin{aligned} Y &= -26 + 1.13X \\ s_{y \cdot x}^2 &= 662 \\ s_{y \cdot x} &= 25.73 \end{aligned} \tag{23}$$

Thus, if a value of 120 μg/m^3 was measured on a given day at site A, the predicted value at site B is

$$Y = -26 + 1.13\ (120) = 110\ \mu\text{g/m}^3$$

The reliability of a predicted value (Y) can be expressed in the form of a confidence interval. The equation for determining the upper and lower limits of the confidence interval is

$$Y \pm t_{n-2,\alpha} s_{y \cdot x} \left(1 + \frac{1}{n} + \sum_{i=1}^{n} \frac{(X_0 - \overline{X})^2}{(X_i - \overline{X})^2}\right) \tag{24}$$

where

$t_{n-2,\alpha}$ = student "t" value for $(100 - \alpha)\%$ level of confidence and $n - 2$ degrees of freedom

$t_{8,0.05}$ = 2.306 for 95% level of confidence when $n = 10$

X_0 = measured value at site A

$\bar{X}$ = mean value of the $n = 10$ observations at site A

Solving (19) yields

$$110 \pm (2.306)(25.73)(1.1006)$$
$$110 \pm 65$$

Thus, while the predicted concentration at site B is 110 $\mu g/m^3$, the 95% confidence interval for the predicted value extends from 45 to 175 $\mu g/m^3$.

REFERENCES

1. H. R. Hickey, W. D. Rowe, and F. Skinner, *J. Air Pollut. Contr. Ass.* **21,** 11 (1971).
2. R. I. Larsen, C. E. Zimmer, D. A. Lynn, and K. G. Blemel, *J. Air Pollut. Contr. Ass.* **17,** 2 (1967).
3. D. A. Lynn, "Fitting Curves to Urban Suspended Particulate Data," Proceedings of the Symposium on Statistical Aspects of Air Quality Data (L. D. Kornreich, ed.), EPA-6J0/4-74-038. U.S. Environmental Protection Agency, Research Triangle Park, North Carolina, 1974.
4. N. D. Singpurwalla, *Technometrics* **14,** 703–712 (1972)
5. "Air Quality Display Model," TRW Systems Group, U.S. Dept. of Health, Education, and Welfare Contract No. PM22-68-60. Office of Technical Information and Publications, U.S. Environmental Protection Agency, Research Triangle Park, North Carolina, 1969.
6. F. B. Smith, and G. H. Jeffrey, "The Prediction of High Concentrations of Sulfur Dioxide in London and Manchester Air," Proceedings of the Symposium on Statistical Aspects of Air Quality Data (L. D. Kornreich, ed.), EPA-6J0/4-74-038. U.S. Environmental Protection Agency, Research Triangle Park, North Carolina, 1974.
7. R. I. Larsen, "A Mathematical Model for Relating Air Quality Measurements to Air Quality Standards," Publ. No. AP-89. Office of Technical Information and Publications, U.S. Environmental Protection Agency, Research Triangle Park, North Carolina, 1971.
8. G. W. Snedecor and W. G. Cochran "Statisitcal Methods," 6th ed., p. 185 and Table A 11, p. 557. Iowa State Univ. Press, Ames, Iowa, 1967.

13

Global Monitoring and Surveillance of Air Pollution

Richard D. Cadle

I. Introduction

A. Man as a Source of Global Trace Atmospheric Constituents

Only in recent years has mankind paid much attention to the possibility that he is polluting his atmosphere on a global scale. Perhaps this was because the atmosphere has seemed so vast and the cleansing mechanisms so effective that significant worldwide contamination seemed unlikely. Furthermore, arousing sufficient interest to establish effective action even against urban air pollution has, until recently, been extremely difficult, and when this difficulty is combined with a lack of satisfactory analytical techniques for determining many trace constituents in relatively pure air, the lack of concern with the possibility that there may be important global pollution is understandable. Of course, as for so many generalizations, there has been at least one exception. For a number of years the carbon dioxide (CO_2) concentration has been known to be increasing (Figs. 1 and 2) presumably as a result of man's activities (*1–4*). At the time this increase was recognized, the temperature of the earth's atmosphere was also increasing slightly. The increasing temperature was attributed to an increase of CO_2 (the "greenhouse effect") (*5*). However, since about 1940 the mean annual temperature has been decreasing (*4*) (Fig. 3), and this decrease has also been attributed by some to man's activities, in this case to particulate pollution. It is very unlikely, however, that both the increase and decrease in the mean annual temperature could have been anthropogenic.

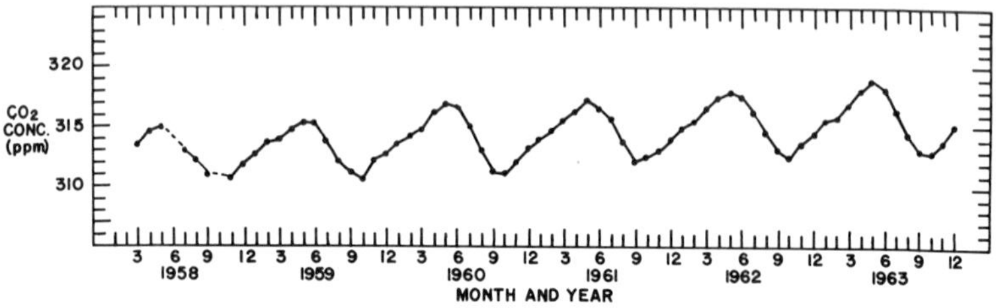

Figure 1. Monthly average concentration of atmospheric CO_2 at Mauna Loa Observatory vs time (*1*).

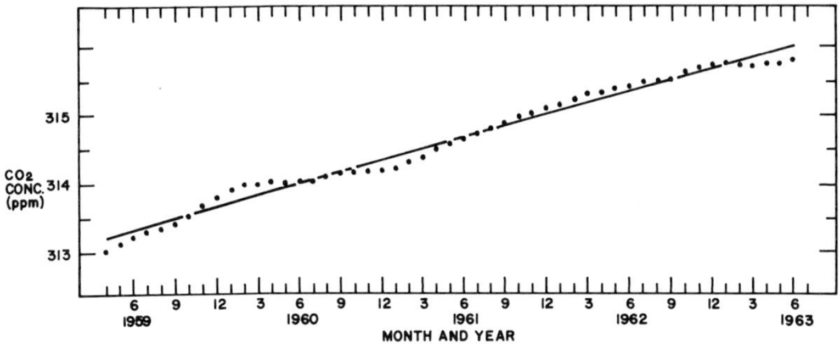

Figure 2. Twelve-month running mean of the concentration of atmospheric CO_2 at Mauna Loa Observatory. Means are plotted vs the sixth month of the appropriate 12-month interval. The straight line indicates a rate of increase of 0.68 ppm/year (*1*).

Since about 1965, there has been a great increase in concern about the various undesirable effects of man on his environment. Part of this concern is a realization that even though man may not now be contaminating the earth's atmosphere to an important extent globally (and this depends somewhat on what is meant by globally), global pollution could become an important problem in a few decades if controls are not implemented before serious global pollution develops.

The global sources, sinks, and transport of air pollutants are reviewed in Chapter 7, Vol. I. Numerous estimates have been made of the contribu-

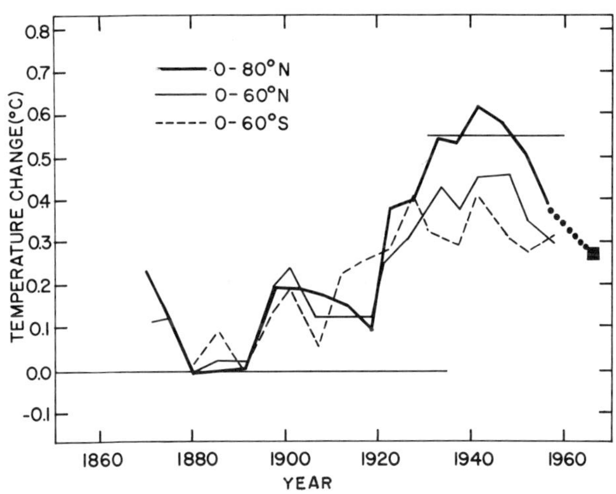

Figure 3. Mean annual temperature for various latitude bands, 1870–1967. The horizontal bar shows the mean value of temperature in the 0° to 80° N band for 1931 to 1960 (*4*).

Table I Estimates of the Annual Contribution of Man and Nature to Several Atmospheric Trace Constituents

Constituent	*Contribution (10^6 metric tons per year)*		*Reference*
	Man	*Nature*	
Sulfur calculated as sulfate	150	400	*6*
	220	425	*7*
$NO + NO_2$, calculated as NO_2	45	700	*8*
Carbon monoxide	245	70	*8*
	200	5000	*9*
Particles smaller than 20 μm radius	185–415	770–2200	*4*
Carbon dioxide	1.2×10^4	8×10^5	*8*

tions of man and nature to trace constituents in the atmosphere (*4, 6–9*); a few of these are summarized in Table I. By far the greatest uncertainty in these estimates is the extent of natural emissions; nonetheless, such estimates demonstrate that the relative contributions by man are significant. Predictions based on the rapidly growing ("exploding") world population and power consumption indicate that man's contributions to pollution will greatly increase during the next few decades unless drastic control measures are established (*4, 10*).

Problems of Defining Global Pollution

A difficulty in attempting to determine man's and nature's relative effects on the atmosphere from estimates such as those of Table I is that the sources, whether anthropogenic or natural, are often strong but local. Thus, unless the mean lifetimes of the pollutants in the atmosphere are long, any effects of the pollutants may be local but not global. On the other hand, many local sources, distributed over much of the earth, may pollute much of the atmosphere of the earth even though the effect from each individual source is almost local. An example is provided by cities in the northern hemisphere; and it is also important to note that because the rate of removal from the atmosphere of most pollutants decreases exponentially with time, small amounts of a pollutant may remain in the atmosphere long after its mean lifetime has elapsed. Thus, any attempt to define precisely the extent of local, regional, or global pollution is at best arbitrary and, perhaps, rather useless; we can, however, hope to monitor trace constituents in the atmosphere at numerous more or less remote locations to determine whether, and to what extent, man is contaminating his atmosphere and whether this contamination is increasing.

B. Importance of Global Air Pollution

There are a number of reasons for monitoring global air pollution. Especially compelling are the facts that particulate and gaseous trace constituents of the atmosphere affect the climate in many ways, and that changes in the concentrations of such constituents could produce marked changes in the climate. Another reason is that the atmosphere transports toxic pollutants large distances from their sources to locations where they can enter the biosphere. An example might be the transport of DDT and of metallic contaminants to rivers and lakes.

Aircraft operating in the middle stratosphere emit pollutants which are believed to decrease the ozone concentration, thereby increasing the intensity of harmful solar ultraviolet radiation that reaches the earth's surface. Chlorofluorocarbons used as "aerosol" can propellants and for refrigeration are believed to reach the stratosphere essentially unchanged but are decomposed in the stratosphere, liberating chlorine which likewise decreases the ozone concentration. Thus, the monitoring of stratospheric air will become increasingly desirable.

C. Community Monitoring versus Global Monitoring

Both similarities and differences exist between community monitoring and global (background) monitoring. The atmospheres of cities, and in fact, of entire regions, are now so highly polluted that the concentrations of the pollutants are often orders of magnitude greater than the concentrations of the same substances in the natural atmosphere. We know we are already in trouble in such highly polluted places, and monitoring is undertaken not only to establish the severity of the problem but also to determine the effectiveness of pollution controls. Globally, we are mainly attempting to determine whether man's activities are increasing the concentrations of potentially objectionable substances. Thus, chemical analytical techniques for global monitoring must often be orders of magnitude more sensitive than those for community monitoring. Although techniques for the two types of monitoring for the same substances are often based on quite different scientific principles, nonetheless, much the same list of substances may be monitored both in communities and globally.

Another difference is that for monitoring to be truly global, it must be international. Influential atmospheric chemists in different countries often prefer different analytical techniques; since the development of analytical techniques is a rapidly advancing field, standardization is probably not yet desirable, but direct comparison of the methods, using the same air mass, preferably in the field as well as in the laboratory, is imperative if the results obtained by different countries are to be comparable. The fact that global monitoring should be international implies the need for the existence of an international body, such as the World Meteorological Organization (WMO), to be responsible for the collection, storage, analysis, and dissemination of the data. Thus, the entire data handling and analyzing system is much more complicated for truly global than for community monitoring.

II. Aims of Global Monitoring

A. Baseline, Benchmark, and Background Data

The terms "baseline data" and "benchmark data" are often considered to be synonymous, although there is a slight difference in emphasis. With regard to atmospheric pollution, they refer to information concerning the concentrations of pollutants at some given time with which the concentrations at some other time, usually later, can be compared. Probably the term baseline is most properly used to refer to concentration data which are obtained when the atmosphere is uncontaminated or nearly uncontaminated and which are used to determine the extent of contamination at some later date; the term benchmark refers to concentration data which are obtained when the air is already contaminated and which are used to determine the effectiveness of control measures. For simplicity, however, in this chapter baseline and benchmark will be considered to be synonymous terms.

Background data relating to trace atmospheric constituents are data for regions remote from man's polluting activities. Thus, unlike baseline and benchmark data, the emphasis is on location rather than on time. Background data, of course, can serve as baseline or benchmark data, but can also be used in other ways. For example, they establish the lower limits of concentrations of trace constituents that, in principle, can be approached in a community by air pollution control measures.

Since baseline data are to be used for making comparisons, great care must be taken when obtaining the data to ensure that meaningful comparisons will be possible. An important consideration is whether different techniques for determining the concentrations of a given pollutant give comparable results. Of course, if all the techniques had as great precision and accuracy as might be desired there would be no problem. However, this is usually not the case. As mentioned above, different techniques used more or less simultaneously for global monitoring should be experimentally compared. It is essential that detailed descriptions of the sampling and analytical procedures be included with the published data so that the procedures can be reproduced by future generations of scientists.

An excellent demonstration of the importance of making precise, accurate baseline measurements is the finding that the concentrations of oxygen in the troposphere between 1910 and 1970 have changed with time over the globe no more than in the third decimal place (*11*). Demonstrating this for oxygen, though, is much simpler than would be demonstrating such long-term invariance for many trace atmospheric constituents which, unlike oxygen, exhibit marked concentration fluctuations during short intervals of time and space. The spacial fluctuations are in three dimensions and on various scales. To take an extreme example, mixing ratios of fine particles usually decrease with increasing altitude in the troposphere, and are orders of magnitude higher in a dust storm than over the oceans. Thus, sampling networks operated over long time periods are essential to obtain satisfactory baseline information concerning variable atmospheric constituents.

Another interesting example of the importance of ensuring that measurement techniques during monitoring can be reproduced is the "La Porte anomaly" (*12*). An examination of appropriate records suggests that a notable increase in precipitation, moderate rain days, thunderstorm days, and hail days has occurred since 1925 at La Porte, Indiana, a city 30 miles east of the industrial area of Chicago, Illinois. These increases have been ascribed to inadvertent weather modification resulting from increasing industrial air pollution from Chicago. As Changnon (*12*) points out, if the change at La Porte is real, it is indicative of the sizable precipitation increases that could be purposely achieved, assuming also the cause and effect relationship between pollution and precipitation. But the validity of the weather records has been questioned because of changes in the observers and the location of the rain gauge. Without taking sides in the dispute, the fact that the dispute exists emphasizes the importance of obtaining data by comparable techniques if trends are to be established. (See also Chapter 12, Vol. I and Chapter 1, Vol. II.)

All else being equal, the better the coverage by the network, the better will be the baseline information on which to base air pollution control decisions. In general, the simplest system to operate is a network of ground-based monitoring units. However, such a system alone is apt to produce misleading data, if for no other reason than that the land and oceans are sources and sinks for many atmospheric trace constituents. Therefore, a ground-based network should be supplemented by measurements taken from aircraft, balloons, and space vehicles.

Scientists and engineers often ignore the fact that our environment is not neatly compartmentalized, and fail to consider the relationships among air, water, and solid-waste pollution. Air pollutants often eventually become water pollutants; ridding ourselves of solid waste by burning usually produces air pollution. Furthermore, baseline measurements of concentrations are most useful if we have considerable knowledge of the geophysical and geochemical cycle of each contaminant considered, including the amounts of emissions of nature and of mankind. To be most effective, air pollution monitoring should be part of a broad program of environmental quality monitoring. Obviously, global monitoring is an expensive operation and compromises must be made.

Occasionally, baseline measurements are made because of the possibility of contamination from a specific source. An example is the 1973–1975 investigation of the composition of the stratosphere intended to provide baseline information to aid in predicting the effects of vehicles such as supersonic transports operating in the stratosphere. Table II (*13*) shows some of the results obtained by personnel at the National Center for Atmospheric Research as part of this study.

B. Detection of Increased Concentrations

Use of the terms baseline and benchmark imply that sometime after obtaining such information, additional measurements will be made to detect any changes which may have occurred in the meantime. Occasionally, these data can also be used to look backward in time if reliable early data exist. The additional measurements can be made in a more or less continuous manner or they can be postponed until there is reason to believe or fear that changes in concentration may have occurred. Each approach has advantages and disadvantages. An obvious advantage of continuous monitoring is that the amount of "noise" or lack of precision in the data is determined. Also, short-term "real" variations may become evident. Such features can be seen in Figures 1 and 2. The obvious seasonal variations are probably produced principally by land plants, especially forests of North America and Asia, which withdraw CO_2 from the atmosphere in summer and release it to the atmosphere during autumn, winter, and spring (*1*). Another reason for continuous monitoring is that it guarantees the detection of sudden, important concentration changes that might otherwise be overlooked. The disadvantages of continuous monitoring are its expense and the large amount of data that must be processed, stored, and made available.

The advantages and disadvantages of intermittent monitoring are implicit in the above discussion of continuous monitoring. Intermittent

Table II Concentrations of Atmospheric Trace Constituents Collected from the Midlatitude Stratosphere in 1970[a] (13)

Date, 1970	Latitude, N	Longitude, W	Altitude, ft. × 10^{-3}	Concentrations, $\mu g/m^3$ [b,c] SO_4^{2-}	Si	Na	Cl	K	Mn	Br
Oct. 21	33°00′	103°45′	55	0.032	0.030	0.026	0.0086	0.034	1.3×10^{-4}	5.1×10^{-4}
Oct. 21	31°00′	96°50′	55	0.037	0.0095	0.0022	—	0.0068	0.4×10^{-4}	3.9×10^{-4}
Oct. 21	39°15′	95°00′	60	0.032	0.0074	0.0084	0.026	0.0062	1.1×10^{-4}	4.0×10^{-4}
Oct. 29–30	34°15′	105°00′	50	0.024	—	—	—	—	—	2.2×10^{-4}
Oct. 29–30	30°27′	97°15′	50	0.021	—	—	—	—	—	2.8×10^{-4}
Oct. 29–30	25°45′	90°40′	50	0.011	—	0.0017	0.0028	—	—	2.0×10^{-4}
Oct. 29–30	20°30′	86°15′	50	0.022	0.16	0.012	0.012	0.016	5.2×10^{-4}	5.7×10^{-4}
Oct. 30	33°00′	102°00′	60	0.004	—	—	0.011	—	—	0.9×10^{-4}
Oct. 30	29°20′	95°00′	60	0.052	—	—	0.0082	—	—	6.9×10^{-4}
Oct. 30	24°45′	89°25′	60	0.020	—	—	0.0020	—	—	—
Oct. 30	19°20′	85°30′	60	0.040	0.0012	—	—	—	—	3.1×10^{-4}

[a] Sample times, 1 hour.
[b] All polystyrene filters.
[c] — indicates not detectable over blank.

monitoring is usually, but not always, cheaper; it has the disadvantages of overlooking minor variations and possibly a sudden major change. A potential problem is that if there is a long time lapse between measurements the data may not be strictly comparable because of variations in the analytical equipment and techniques.

One way to overcome some of the disadvantages of intermittent monitoring is to monitor continuously for a long enough time to reveal minor variations, such as the seasonal CO_2 variations and gradual trends. The measurements are then discontinued, to be resumed later. Another way to avoid the expense of continuous thorough monitoring is to develop some type of alert system which can lead to activation of the complete monitoring system when an alarm is given.

C. Basic Research

In addition to obtaining benchmark information, global monitoring of trace constituents yields fundamental information concerning the sources, concentrations, and sinks of the trace constituents measured. For example, measurements of sulfur dioxide concentrations over the North Atlantic Ocean by Georgii (*14*) yielded values between 0 and 3 $\mu g\ SO_2/m^3$, the highest values being between 30° and 45° north latitude. These results strongly suggest that the air over the North Atlantic is highly contaminated by effluents from the urban complexes of the eastern United States, but similar data would have to be obtained for the South Atlantic to confirm or refute this interpretation.

Removal of atmospheric trace constituents by precipitation is the main sink for many such substances. We can distinguish between two processes for removal by precipitation. Rainout involves association of the trace constituents with cloud droplets which become precipitation (rain, snow, etc.), while washout is removal of gases or particles by the falling precipitation below the clouds. Since precipitation is such an effective cleansing mechanism for tropospheric air, data obtained by monitoring the trace constituents in precipitation are essential for establishing cycles of trace constituents, for example, the work of Kellogg *et al.* (*6*) on the sulfur cycle (Fig. 1 in Chapter 7, Vol. I).

Of course, estimating cycles is not the only reason for monitoring trace constituents in precipitation. For example, rain and snow supply substances, beneficial and harmful, to plants. Precipitation contamination provides a means for studying contamination of the atmosphere. Although the relationships between the concentrations of constituents of air and their concentrations in precipitation are not known quantitatively, there is reason to believe that concentration patterns of contaminants in precipitation reflect their patterns in air. Furthermore, the chemical analysis of precipitation provides a crude means of monitoring changes in air contamination with time (*15*).

III. Substances to Be Monitored

A. Need to Monitor Culprit Substances

The minimum amount of monitoring required to determine whether some property of our atmosphere is changing as a result of some activity of mankind is to monitor only that property. However, such minimum monitoring will often (perhaps usually) not provide the data needed to take effective action if important changes in the atmosphere occur, or indeed to determine whether mankind was responsible for the changes. For example, if for many years we monitor atmospheric turbidity, which is determined by the particle loading of the atmosphere, we may find that the turbidity is increasing dangerously and that the increase may be attributed to man's activities. Physical and chemical analysis of the particulate matter might lead us to the sources and thus to effective remedial action. But the particles may have been produced indirectly, by chemical reaction, from gases emitted by man's actions, and if such gases had also been monitored, much more effective control measures might have been taken. However, we should not monitor a long list of substances just because we know how to measure their concentrations. Monitoring is expensive and obviously considerable careful judgment is required.

B. Trace Constituents That Affect the Climate

Two types of trace atmospheric constituents can more or less directly affect the climate. One type includes the gases that absorb radiation from the sun, the earth's surface or both; the other consists of the particles in the air that scatter and absorb such radiation. Absorbing gases include ozone, carbon dioxide, and water vapor (Figs. 4–7) (*16–18*). All airborne particles scatter radiation, at least to some extent, and most or all absorb radiation from the sun and earth in some wavelength ranges. Furthermore, many particles serve as condensation or freezing nuclei, or both.

In addition to the trace constituents that have a direct effect on climate, numerous trace gases in air undergo chemical reactions which produce particulate matter. The chemical reactions which produce such particles in city smog are discussed in Chapter 6, Vol. I. Although it is hard to demonstrate, such reactions must occur at the lower concentrations

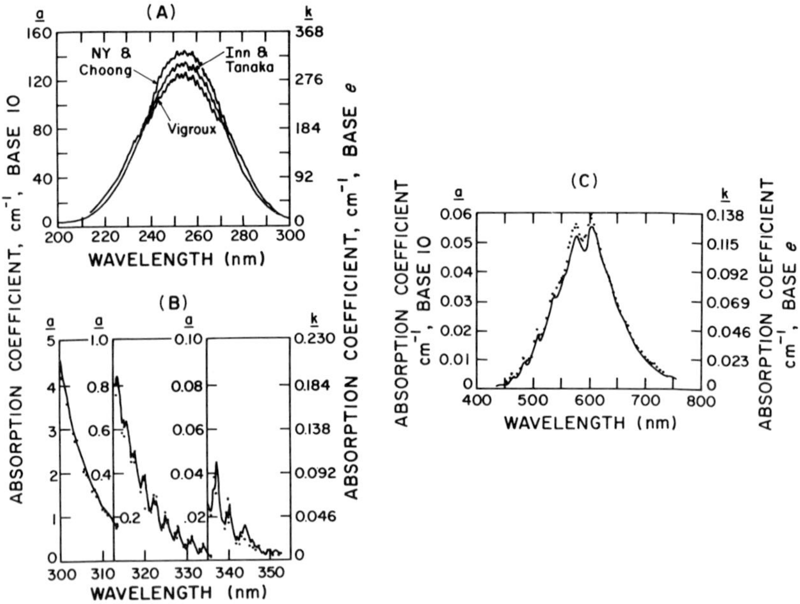

Figure 4. Absorption coefficients of ozone: (A) 200–300 nm, (B) 300–350 nm, (C) 400–800 nm (*16*).

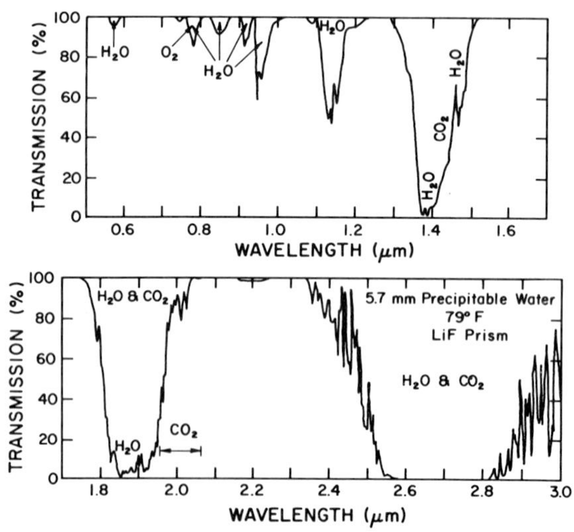

Figure 5. Atmospheric transmission at sea level over a 0.3-km path, 0.5 to 3.0 μm (*17, 18*).

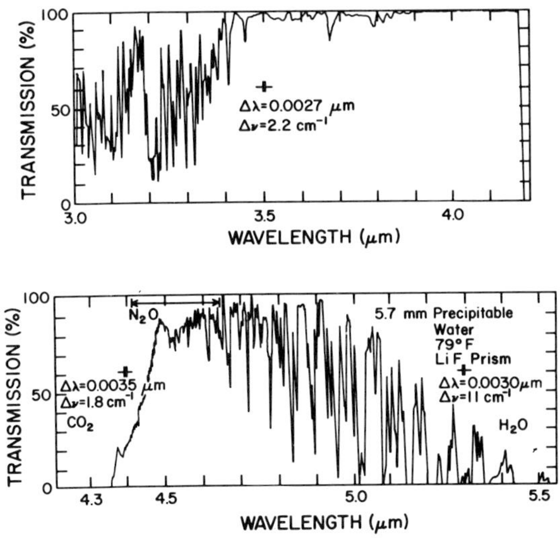

Figure 6. Atmospheric transmission at sea level over a 0.3-km path, 3.0 to 5.5 μm (*17, 18*).

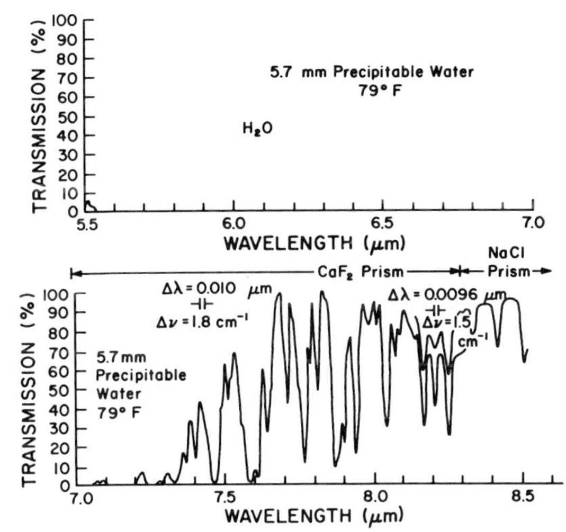

Figure 7. Atmospheric transmission at sea level over a 0.3-km path, 5.5 to 8.5 μm (*17, 18*).

outside of urban areas and contribute to the particulate loading of the atmosphere. Table III gives a list of gases and vapors emitted by man's activities with typical concentrations of these substances occurring at present in the ambient atmosphere. All of these should be monitored if we are to learn the possible impact of man on the climate of the earth.

Table III Some Atmospheric Gases and Vapors Involved in Chemical Reactions Leading to Particle Formation

Constituent	*Typical concentration in the ambient atmosphere, parts per billion by volume*
NO	0.5
NO_2	1
H_2S	0.05
NH_3	4
Organic vapor	?
SO_2	1.2
CO	190
O_3	40

C. Biologically Active Substances

Close to their sources (which may be factories, agricultural spraying, or accidental releases) gases and particles often harm and sometimes kill men, animals, and plants. Farther away from such sources, the toxic materials are often so diluted with air that they seem to be nearly or completely harmless since they have no apparent direct effect. However, even after such dilution, toxic substances may be extremely harmful, especially after they enter a biological cycle. For example, DDT [dichlorodiphenyl trichloroethane or 1,1,1,-trichloro-2,2-bis(*p*-chlorophenyl)-ethane] is removed from air by rain and directly or indirectly reaches the oceans. According to the SCEP (Study of Critical Environmental Problems) report (*10*), marine fish are almost universally contaminated with DDT residues, and there is "presumptive evidence" that this contamination causes reproductive failure in certain kinds of fish such as sea trout. The declining catches of Dungeness crabs in California seem to be associated with observed DDT residues in the developing larvas. Fish-eating birds have undergone reproduction failure and decreases in population. Numerous other examples relating to DDT alone could be cited. Thus, the air can serve as a route whereby toxic substances ultimately enter a biological cycle.

The numbers of different toxic substances emitted into the atmosphere by man is immense, and obviously not all can or should be monitored, at least on a global basis. Many can be eliminated as candidates for monitoring, for example, because their toxicity is low or because they are rapidly destroyed in the atmosphere. But many substances previously believed to be unimportant on a global basis have recently been found to be very important indeed. Another way to keep the amount of monitoring within reason is to monitor groups of substances rather than all of the individuals in the groups. For example, many insecticides and toxic industrial chemicals are organic halides, such as chlorinated hydrocarbons, which includes DDT and the polychlorinated biphenyls. The atmosphere can be analyzed for organic halides as a group. However, many of the organic halides being used have an appreciable vapor pressure. For the most part, the presently used analytical methods are based on the assumption that these substances are associated with airborne particles. These methods start by collecting particles on a filter. When the concentrations are very low, as in the ambient atmosphere, this assumption can produce large errors since the organic halides may be largely in the gas phase. Thus, both the gas and the particulate phases of the atmosphere should be monitored for these substances.

Another group of toxic substances that can at least be collected simultaneously consists of most toxic metals and their cations, which are associated with airborne particles. Such metals include mercury, cadmium, copper, and lead. Perhaps 10–15 metals should be monitored (*19*). Of course, many of these substances occur naturally. For example, Cadle *et al.* (*20*) estimated the following concentrations in Kilauea (Hawaii) volcano fume normalized to fume undiluted with air and associated with the particulate matter in parts per million by weight, ppmw: Hg, 0.14; Cd, 0.13; Cu, 3.2; Pb, 0.60; As, 0.084; Se, 0.14; Cr, 0.70; Sb, 0.70; Co, 0.07. The main component was sulfate, 14×10^4 ppmw. Using the method of calculation of Kellogg *et al.* (*6*) we find that the concentration of mercury corresponds to about 400 kg of mercury emitted in the particulate phase of the fume during a typical Hawaiian eruption, and to an annual emission rate of 2 metric tons of mercury from all the world's volcanoes.* Although these figures should not be taken very seriously because of all the assumptions, they demonstrate the great extent to which nature introduces such minor constituents into the atmosphere.

Radioactive nuclides from various sources need to be monitored and are being monitored globally. (See also Chapter 5, Vol. I and Chapter 11, Vol. III.) Global atmospheric monitoring of biologically active substances will be most useful if undertaken in conjunction with monitoring other parts of our environment (the oceans, for example) for such materials.

* The annual industrial consumption of mercury is about three orders of magnitude greater than this.

D. Measurements for Basic Knowledge

As mentioned above, atmospheric monitoring is undertaken to provide basic knowledge concerning the atmosphere as well as to determine man's impact on his environment. The number of hypotheses that might be tested in this way is almost limitless, but a couple of examples may suggest how this use of monitoring operates. In each case, many more than a few isolated measurements are required.

Judging from carbon isotope ratios, carbon monoxide is produced in nature by the oxidation of methane, with an atmospheric residence time of 0.1 years (*9*). Since most methane is believed to be produced in the northern hemisphere, the southern hemisphere should have a much lower methane concentration than the northern, if these theories concerning methane and carbon monoxide formation are correct. A very few measurements suggest that concentrations in the two hemispheres are about equal, but much more information is needed. Another suggestion for global-scale monitoring is by Junge (*21*) who states that tropospheric ozone may be a very powerful tool for studying the exchange mechanism between the stratosphere and the troposphere. He suggests establishing a meridional network of surface ozone recording stations and operating it continuously for several years. Precipitation collection and analysis were mentioned earlier in connection with establishing cycles of contaminants.

E. Meteorological Measurements

Meteorological measurements should be made at the same time and in the same places as the measurements of the concentrations of trace constituents. Wind velocity is probably the most important, but temperature, pressure, and relative humidity should also be measured if it is reasonable to do so.

IV. Monitoring Locations

A. Ground Level

Our atmosphere is three-dimensional and a reasonably complete global monitoring system must also be three-dimensional. Nonetheless, man spends most of his life at or near the earth's surface, and operating air monitoring systems at the earth's surface is, at least for the present, the most uncomplicated method of monitoring. Because of the cost of establishing and operating even ground-based monitoring stations, it is tempting to state that the minimum number needed to obtain the desired information should be determined. But such a statement would be an almost meaningless platitude. In principle, the network should be representative of the whole globe, and this could imply a need for hundreds or thousands of stations. In practice, however, a single, simple, well-run monitoring station can supply exceedingly meaningful data, one of the best examples being the carbon dioxide monitoring equipment on Mauna Loa, Hawaii.

The Commission for Air Chemistry and Radioactivity of the International Association of Meteorology and Atmospheric Physics (CACR) (*22*) has emphasized that the number of stations should not be the same for each constituent since each has its own characteristic variability in time and space which in turn depends on the distribution of its sources and sinks and on its residence time in the atmosphere. They propose a spacing of about 1000 km for the short-lived constituents, such as SO_2, corresponding to a total number of stations on the order of 100. They suggest that these include some oceanic stations, especially weather ships, in the northern Pacific and Atlantic, and a few in the heavily polluted regions of the eastern United States and Europe.

For long-lived species such as CO_2, the CACR suggests that the minimum number of stations be about 10, more than one-half to be placed on a north–south line near the center of the Pacific Ocean (for example, the north coast of Alaska, the Aleutian Islands, Hawaii, the Line Islands, a South Pacific Island, and the South Pole). Other stations to assess longitudinal differences might be placed on or near the north and south Atlantic Oceans, the Indian Ocean, and on a desert. A network suggested by the World Meteorological Organization (WMO) is described in Section VII, A.

Ground-based global monitoring stations should preferably be located where the air is essentially free of local pollution, and will probably remain free of such pollution for many years. However, some stations may be located at spots where local pollution may reach the station when the wind is from certain directions, and mechanisms must be provided in the station system so that data obtained under such conditions can be identified and discarded.

A number of international environmental monitoring networks have already been established for rather specialized purposes. For the most part, these stations were established where facilities for other purposes already existed. An example is the isotopes-in-precipitation network (deuterium, tritium, and oxygen-18) (Fig. 8) (*23*). Clearly, the requirements for operating a station must be taken into account in choosing its location, but this is largely a matter of logistics.

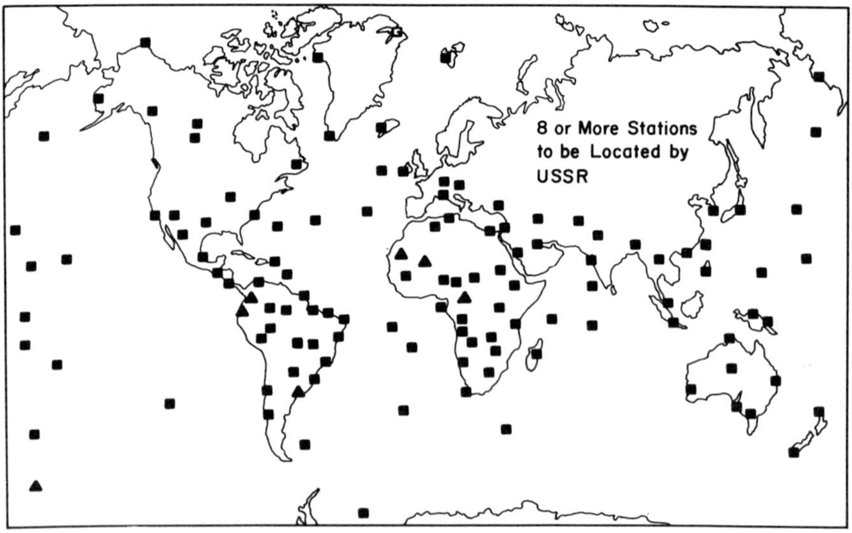

Figure 8. Stations cooperating in the IAEA/WMO isotopes-in-water network, including new stations and those recommended for addition (*23*). ▲: Stations recommended for addition; ■: present network stations.

B. Aircraft and Balloons

Monitoring trace atmospheric constituents from aircraft and balloons will probably largely supplement monitoring from ground stations for many years to come because of the relative costs. Of course, airborne monitoring of radioactivity has been conducted for many years, and balloon determinations of ozone profiles are almost routine. But unless great sums of money are to be spent, monitoring locations and frequency will largely remain controlled by the availability of appropriate airports, airstrips, balloon launching areas, aircraft, and trained crews.

An effective way to partially decrease the cost is to "piggyback" on flights made for other purposes. An example is a project undertaken by the United States National Aeronautics and Space Agency (NASA) to analyze for a number of trace atmospheric constituents from commercial airliners. The monitoring must, of course, be restricted to air routes, but such equipment could also be placed and operated on certain types of military aircraft. Rockets could be used to carry monitoring equipment but, at present, rockets appear to be impractical for routine monitoring for several fairly obvious reasons.

C. Satellites

Satellites have many unique advantages for studying global environmental problems (*24*). Since they are situated a great distance from the earth and orbit it, they can quickly survey the entire globe. Hence, location is not a problem and, in principle, satellites are ideal monitoring platforms. However, at present there are serious limitations which are discussed in Section V,C.

V. Measurement Techniques

A. *Ground Level*

1. Turbidity*

Atmospheric turbidity has been defined as any condition of the atmosphere which reduces its transparency to radiation, especially visible radiation (*25*). The subject has been reviewed by many authors (*26–29*). The measurements usually use the sun as the light source, and a host of instruments of varying degrees of complexity have been used. Probably the most commonly employed turbidity-measuring instruments are the pyrheliometer and the Volz sun photometer. The latter is inexpensive, simple to use, and has furnished the basis of a network of turbidity-measuring stations in the United States (*29*). The Volz sun photometer is shaped like a small box, 5.5 × 2.2 in. with a diopter on the side that measures the solar elevation angle and is calibrated in terms of the optical air mass. One end of the box is an aperture plate that provides a viewing angle of 3.75°. Light passes through a Kodak Wratten No. 65 filter and falls onto a selenium photocell, the output of which is indicated by a micrometer mounted in the top of the box. Every Volz sun photometer is calibrated against a standard photometer before use. The standard has a No. 1P28 phototube, which has a linear response, in place of the selenium photocell, the output of which is nonlinear. Volz has also developed an 11-wavelength photometer.

Turbidity measurements are usually reported as turbidity coefficients (B) which may be defined in different ways. Volz adopted the definition of Schüepp (*30*): $\log(J/J_0) = -(\tau_r + \tau_x + B)m$, where J is the observed solar radiation at the wavelength 0.5 μm, adjusted to the mean sun–earth distance; J_0 is the solar radiation at this wavelength outside the atmosphere at the mean sun–earth distance; τ_r and τ_x are the scattering coefficient for air molecules and the absorption coefficient for ozone, respectively; and m is the optical air mass. Other methods for expressing turbidity that are often used are those of Linke and Boda (*31*) and of Ångström (*28, 32*).

Robinson (*26*) has pointed out that at a few stations measurements

* See also Chapter 12, Vol. I and Chapter 1, Vol. II.

are being made of total and diffuse radiation in broad spectral bands defined by filters and that measurements of diffuse radiation in the solar infrared can be used to indicate the extent of particulate loading. A difficulty in determining the particulate loading of the atmosphere from turbidity measurements is that undetected tenuous cirrus contributes to the turbidity. Robinson (*26*) suggests that such cirrus is perhaps a common phenomenon.

2. Trace Gases

A number of methods are available for analyzing the air for each trace gas of interest, ranging from very simple "wet chemical" methods to very sophisticated instrumental techniques. A compendium of methods of air sampling and analysis has recently been assembled by the Intersociety Committee and published by the American Public Health Association (*33*). The methods briefly described below are listed as examples and are selected largely as a result of experience with monitoring by the United States National Center for Atmospheric Research (*34*).

For wet-chemical methods, the sampling apparatus shown in Figure 9 is very convenient and consists of a prefilter to remove particles, a glass and Teflon bubbler, a critical orifice, and a vacuum pump. The bubbler (*35*) contains a solution of reagent that extracts the trace constituent as the air sample is drawn through the dispersion tube and bubbles up through the solution. The tube has a fritted glass disperser at its lower end. The critical orifice, which controls the flow rate, is a hypodermic syringe needle which is inserted through the wall of a plastic or rubber

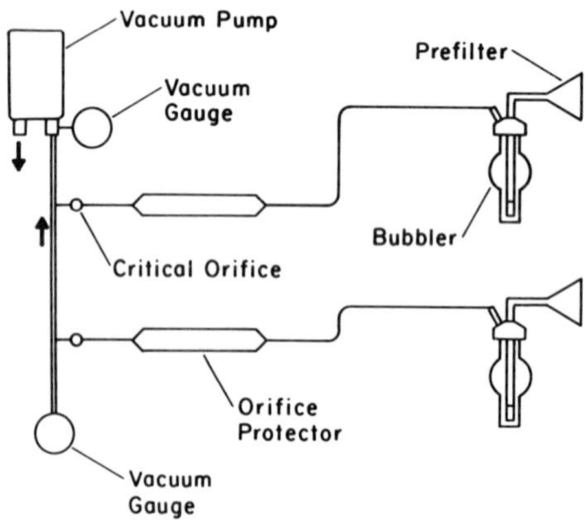

Figure 9. Gas-sampling apparatus, schematic (*34*). (Reprinted by permission of the MIT Press, Cambridge, Massachusetts. Copyright 1971 by the Massachusetts Institute of Technology.)

tube that serves as a vacuum manifold (*36*). So long as the pressure drop across the needle exceeds 0.5 atm, the flow rate will be nearly constant. The orifices are protected by a bed of silica gel which removes spray from the solution that might otherwise clog the needles. Carbon vane pumps can be used, but oil-lubricated pumps should be avoided to prevent hydrocarbon-vapor contamination of the sampling site.

a. SULFUR DIOXIDE. (See also Chapter 6, Vol. III.) Sulfur dioxide (SO_2) is often measured with conductimetric techniques which involve oxidizing the sulfurous acid formed when SO_2 dissolves in water to form sulfuric acid (*33*). Soluble electrolytes interfere, and many versions of the method are not sufficiently sensitive for global monitoring.

A better approach is a modified West–Gaeke method using the apparatus of Figure 9 (*33, 37, 38*). Sulfur dioxide is removed from the air sample with an aqueous solution of sodium tetrachloromercurate. The resulting complex is then reacted with formaldehyde and pararosanaline to form a colored product. The intensity of the color is measured with a colorimeter. Another method which may ultimately prove to be the most sensitive, involves gas chromatography using a flame photometric detector.

b. OZONE. (See also Chapter 7, Vol. III.) Many instruments are available for measuring ozone (O_3) in the atmosphere, although not all have the sensitivity required to measure it in unpolluted air (*33*). The Komhyr ozone meter is especially useful and, like many ozone meters, is based on the oxidation of iodide to iodine by the ozone (*39*). All of the instruments based on this reaction respond to some extent to strong oxidants such as nitrogen dioxide in addition to ozone. However, ozone is the major strong oxidant in the unpolluted atmosphere. Another method which is gaining in popularity involves the chemiluminescent reaction of ozone with ethylene.

c. NITROGEN DIOXIDE. (See also Chapter 7, Vol. III.) The wet-chemical method for nitrogen dioxide (NO_2) involves its collection from air with an aqueous solution of *N*-naphthylethylenediamine hydrochloride, acetic acid, and sulfanilic acid (*33, 40, 41*). The NO_2 forms nitrous oxide which diazotizes the sulfanilic acid, and the resulting diazonium compound is coupled with *N*-naphthylethylenediamine to form a colored product which is measured colorimetrically.

d. NITRIC OXIDE. (See also Chapter 7, Vol. III.) Nitric oxide (NO) can be measured by oxidizing it to NO_2 and then applying the above method for the latter (*33*). The sampling apparatus (Fig. 9) is modified by substituting a tube containing triethanolamine on firebrick and a chromate–sulfuric acid filter in place of the prefilter assembly (Figure 10). Nitrogen dioxide is removed from the air by the triethanolamine. The NO is then oxidized to NO_2 by the chromate–sulfuric acid filter. Nitric oxide can also be determined with instrumentation based on the chemiluminescent reaction of NO with O_3.

e. HYDROCARBONS. (See also Chapter 5, Vol. III.) Unless very expensive equipment is to be operated in the field, air samples for hydrocarbons, or the hydrocarbons adsorbed on some surface, are returned to a central laboratory for analysis. For example, air samples can be collected in evacuated stainless-steel canisters or in gas-tight syringes made of stainless steel, glass, and Teflon which can be locked once the samples are collected. The samples can then be analyzed by gas chromatographic or mass spectrometric techniques. A typical chromatogram of the gaseous hydrocarbons collected from air in the Panama jungle is shown in Figure 11. However, sample integrity can possibly be compromised by long-term storage. (See Chapter 4, Vol. III for additional information.)

f. CARBON MONOXIDE. (See also Chapter 5, Vol. III.) No suitable wet-chemical methods are available for carbon monoxide (CO) but there are several instrumental methods (*33*). An automated chromatograph is commercially available which has a sensitivity of about 0.05 ppmv, adequate for global monitoring. It also determines methane (CH_4). Another instrumental method involves the reaction of CO with hot mercuric oxide to

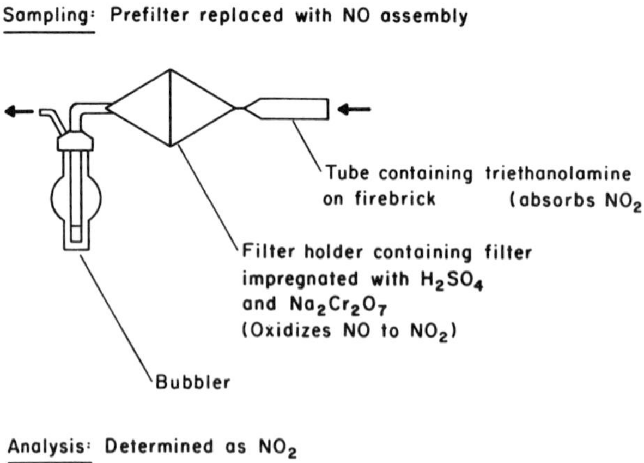

Figure 10. Analysis for nitric oxide (*34*). (Reprinted by permission of the MIT Press, Cambridge, Massachusetts. Copyright 1971 by the Massachusetts Institute of Technology.)

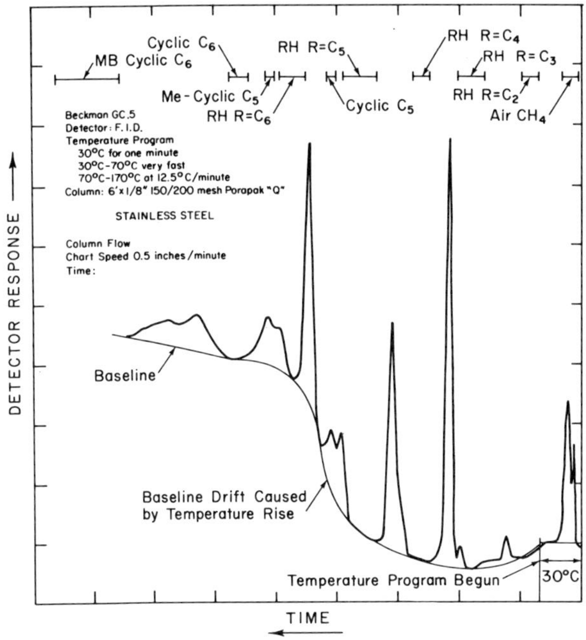

Figure 11. Typical gas chromatogram of an air sample from Albrook Forest, Panama (*34*). (Reprinted by permission of the MIT Press, Cambridge, Massachusetts. Copyright 1971 by the Massachusetts Institute of Technology.)

liberate mercury vapor which is determined photometrically. However, various organic vapors, such as olefins, interfere.

g. Carbon Dioxide. Carbon dioxide (CO_2) can be continuously and accurately measured with a nondispersive infrared analyzer (*3*). The instrument measures the loss of energy of a beam of infrared radiation passing through a sample of the gas, responding only to those wavelengths where carbon dioxide absorbs. Such an instrument has been successfully operated for many years on Mauna Loa, Hawaii.

h. Hydrogen Sulfide. (See also Chapter 6, Vol. III.) Although a number of methods have been used for analyzing for hydrogen sulfide (H_2S) in the ambient atmosphere, the most reliable seems to be that developed by Natusch *et al.* (*42*). It is a fluorescence method in which the H_2S is collected from the air by reaction with an $AgNO_3$ impregnated filter. The extent to which an extract of the filter quenches the fluorescence of a reagent containing fluorescein mercuric acetate is then determined.

i. AMMONIA. (See also Chapter 7, Vol. III.) Ammonia (NH_3) can be analyzed wet-chemically by scrubbing it from the air. The ammonium ion is then reacted with sodium phenate and hypochlorous acid, yielding the colored indophenol (*43*) which is determined with a colorimeter.

3. Particulate Matter*

Numerous methods exist for collecting particles from air. These include the use of filters, impactors, thermal precipitators, diffusion batteries, centrifugal separators, and sedimentation chambers (*44*). However, filters will usually be most appropriate for monitoring because of the relatively large amount of material that can be collected and the high collection efficiency for an extremely wide range of particle size. Fiber filters with fibers averaging less than 1 μm in diameter are especially appropriate because of the low-pressure drop across the filters and their high collection efficiency for "Aitken" particles. The National Air Sampling Network of the United States Environmental Protection Agency (USEPA) uses glass fiber filters. All the samples are analyzed for weight concentrations of particles, benzene-soluble organic matter, and gross β radioactivity. Some samples are also analyzed for NO_3^-, SO_4^{2-}, Be, Mn, Pb, Sn, Fe, Cu, Ti, Zn, Cr, Ni, Mo, Bi, Cd, and Sb (*45*).

Scientists at the National Center for Atmospheric Research have been preparing their own filters from polystyrene, using the method of Cadle and Thuman (*46*), because of the difficulty of obtaining commercially available filters that are sufficiently free of trace elements to permit neutron activation determination of such elements in airborne particles. The preparation involves dissolving polystyrene in an organic solvent to produce a rather viscous solution. This solution is sprayed under pressure through a nozzle with an annular jet of air. The solution is drawn out into numerous fine fibers which would collapse into droplets before evaporation of the solvent except for the high viscocity of the solution. The resulting fibers are drawn by suction against Dacron netting to form a mat. Numerous polymers other than polystyrene can be used. All operations should be undertaken in a clean room. Table IV shows the collection efficiencies of polystyrene filters as a function of fiber diameter and particle size, for an aerosol with a linear flow rate of 14 cm sec^{-1}.

Numerous techniques can be used to analyze the material collected on the filters. As already mentioned, the filters can be weighed to determine the total amount of particulate material collected, although if this is done, weighings should be made at the same relative humidities because of the absorption of water vapor both by the filter material and by the

* See also Chapters 1–3, Vol. III.

Table IV Collection Efficiency for a Linear Aerosol Flow Rate of 14 cm sec^{-1} as a Function of Fiber and Particle Size *(46)*
[Reprinted with permission from R. D. Cadle and W. C. Thuman, *Ind. Eng. Chem.* **52,** 315 (1960). Copyright by the American Chemical Society.]

Median fiber diameter, μm	*Filter density, g/cm²*	*Δp,[a] mm H_2O*	*Efficiency for indicated particle diameter, %*			*F[d] for indicated particle diameter, mm^{-1}*		
			0.01 μm[b]	*0.1 μm[c]*	*1.0 μm[c]*	*0.01 μm[b]*	*0.1 μm[c]*	*1.0 μm[c]*
2.0[e]	0.0022	10	68	8	90	11	0.83	23
2.0	0.0105	17	97.2	57	99.0	21	5.0	27
0.4	0.0027	120	97.2	85.7	>99.99	3.0	1.6	—
0.4	0.0050	408	99.9	98.3	>99.99	1.7	1.0	—
0.3	0.0032	680	99.2	99.2	>99.99	0.71	0.71	—

[a] Δp is the pressure drop across the filter.
[b] LiF aerosols.
[c] $(NH_4)_2SO_4$ aerosols.
[d] F is defined by the equation (C_1/C_0) exp $(-F\,\Delta p/100)$, where C_0 is the number concentration of particles in the aerosol to be filtered and C_1 is the number concentration of particles leaving the filter.
[e] Mats compressed with a weighted roller to give them extra strength. All other mats listed were not compressed.

collected particulate matter. Neutron activation, wet-chemical analysis, X-ray fluorescence, emission spectroscopy, and atomic absorption spectroscopy are among the methods employed to analyze the collected particulate material.

Measurements can also be made of the radioactivity of such material. The methods used for analyzing air for organic halogen-containing compounds, such as DDT, usually are based on the assumption that the compounds are in particulate form, and involve collection on a filter. As mentioned earlier, most such compounds that are toxic have an appreciable vapor pressure, and at large distances from the places of emission they may exist largely or entirely in the gas phase. Therefore, methods need to be developed that take account of the fact that such substances can exist in two phases.

The removal for examination of individual particles from filters is usually an unsatisfactory operation at best, although methods for doing this have been described by McCrone, Draftz, and Delly (*47*). If individual particles are to be examined, collection on an available surface, by impaction, for example, is usually preferable to collection by any type of filter.

4. Precipitation*

Automated precipitation collectors have been developed that remain open only during periods of precipitation, thereby excluding most dry particles (*15*). The lid is activated by a moisture-sensing grid which is thermostatically controlled so that snow and sleet as well as rain are collected. The grid temperature maintained is sufficiently high that moisture is evaporated from the grid after precipitation ceases, causing the lid to close. The collected precipitation is mailed to a central laboratory where it is analyzed. A small amount of toluene is added to the polyethylene mailing bottles to minimize biological activity in the sample. Many of the analytical methods described above are appropriate for analyzing the samples.

Automated methods of analysis are always appropriate when feasible, and especially so for precipitation samples. Automatic wet-chemical, colorimetric analyzers, such as the "Auto-Analyzer" of Technicon, Inc., can greatly reduce the time spent on analysis.

5. Meteorological Measurements*

Meteorological sensors can be mounted on a small tower and the data recorded on a digital tape recorder. Numerous commercial sensors are

* See also Chapter 10, Vol. I.

available that are satisfactory and can be battery operated. Convenient meter readout and selector switches enable the field observer to ascertain whether the sensors are functioning properly.

B. Aircraft and Balloons

As previously noted, monitoring of the atmosphere should be three-dimensional. For some purposes, such as the impact of supersonic transport aircraft on the environment, ground-level monitoring, except by remote techniques such as laser radar, is completely useless. Balloons and aircraft are useful platforms for three-dimensional monitoring and each has obvious advantages and disadvantages. In general, balloons can reach much higher altitudes than aircraft. Single flights with small balloons are quite inexpensive and do not require airports, and they can readily be used to obtain "vertical" profiles of concentrations. Aircraft, on the other hand, can carry operators, are much more independent of weather conditions, and, within wide limits, go wherever desired.

1. *In Situ* Techniques

In principle, the same types of analytical techniques that are used at ground level can be used on airborne platforms. Since an airplane can carry an operator, simple, wet-chemical, nonautomated techniques can be used, although pressurization and heating of the air may be required at higher altitudes. The pressurizing equipment of the airplane is usually not satisfactory because of the likelihood of contamination. The Komhyr ozone analyzer (*39*) was developed for use on small balloons, is very useful for near-surface monitoring, and has been modified at the National Center for Atmospheric Research for use on WB 57F aircraft flying in the stratosphere.

The analytical equipment used for monitoring from aircraft can often be much more elaborate than is practical for ground-based monitoring, the equipment for which must often be located in very remote spots. For example, WB 57F aircraft are equipped with 12 filter holders in what would normally be the bomb bay. These holders can be sequentially lowered into a sampling tube beneath the aircraft and raised after sampling, in the fashion of a record changer, by controls in the navigator's cockpit.

Balloon sampling places much more stringent restrictions on the sampling equipment than does aircraft sampling, for very obvious reasons. There is no operator, so the equipment must be automated and is practically always battery operated. Since small balloons are much cheaper to fly than large ones, miniaturizing the equipment can be very important. When large numbers of flights are to be made over a considerable period of time, simple, expendable equipment may be necessary because of the low balloon recovery rate. In this regard, monitoring with balloons is very different from atmospheric research with balloons, during which only a few flights are made and which may involve the use of very expensive equipment.

2. Remote Techniques

Remote techniques are those methods that determine concentrations at some distance from or integrated over a considerable distance from the measuring device. LIDAR (LIGHT DETECTING and RANGING) has already been mentioned. Searchlights, when used to study clouds and dust layers, are remote sensors. Most of the remote methods appropriate for aircraft use are based on spectroscopy, especially infrared spectroscopy. Instrumentation for the latter takes many forms, such as far-infrared spectroscopy, correlation spectroscopy, interferometer spectroscopy, and second-derivative spectroscopy. Such methods are usually too expensive for balloon monitoring although they may be essential for certain types of atmospheric research using balloons.

C. Satellites

Satellites have a great potential for the monitoring of air pollution, especially on a regional or global scale. However, they are limited to remote sensing over large distances. Their requirements for instrument sensitivity, low "noise," and fine resolution result in major problems. At the time of writing this chapter, essentially no air pollution studies had been made from satellites in spite of a number of detailed studies that had been made of the potentialities of such studies. For example, a project was undertaken at the Convair Corporation, under NASA sponsorship, to develop and test a remote sensing instrument for air pollution monitoring from a satellite. Several possible pollution detection techniques were evaluated: dispersive absorption spectroscopy, matched filter absorption spectroscopy, filter cell nondispersive techniques, interferometer techniques, and microwave methods. A nondispersive infrared technique was finally selected (*48*) for continued study.

Gille *et al.* (*49*) have proposed a device for determining the concentrations of trace atmospheric gases by measuring the thermal emission of the atmosphere seen at the limb (the outer edge of the apparent disk of a celestial body, in this case, of the earth). Their calculations indicate that H_2O, HNO_3, N_2O, CH_4, and NO_2 can be detected and their vertical distributions obtained by scanning the earth's limb with instrumentation within the present state of the art. The calculations also indicate that interference should not be a serious problem.

VI. Data Handling, Analysis, Storage, and Retrieval

One of the more horrendous problems that results when large amounts of data are collected over a long period of time is what to do with it. The problem is especially difficult when a number of nations are involved. A worldwide monitoring network should probably be coordinated by a single international organization to be effective. Even so, it is probably too much to expect that the instrumentation or even the general analytical techniques would be the same for all stations. Different countries might very well insist on different approaches. Comparisons of methods of analyzing for the same atmospheric constituent would, of course, be essential but the immediate outputs might be quite different in form. Thus, it would be necessary to reduce the data to a common form for ultimate analysis, storage, and retrieval.

The portion of a network operated by a given country would usually be run by a single agency of that country. It should then be the responsibility of that agency in each country to process the data so that it could be incorporated into the data handling system of the international organization. Ultimately, the data should be stored on magnetic tape, by a microphotographic technique, or both. However it is stored, a method of rapid, flexible data retrieval is essential. The retrieval system will be especially useful if it includes an analysis capability. Obviously, a computer-based system is required. Numerous books have been written on the general subject of information retrieval, examples being Schecter (*50*), Kent (*51*), and Kent and Taulbee (*52*).

Repositories of data not unlike that which would be obtained from a global air pollution network already exist. Examples are the records of the United States Public Health Service Radiation Surveillance Network and the International Geophysical Year Nuclear Radiation Program. Data from more than 10 years of daily observations from a number of sites are recorded.

Various statistical techniques exist, some quite sophisticated, for condensing and ordering such data. An example of one of the more sophisticated methods is factor analysis, which was developed to interpret intelligence test scores and has been used extensively in the social and biological sciences. It condenses large amounts of data, expressing them in linearly independent sets of coefficients.

The data can also be processed and regurgitated in the form of graphs and charts. A sort of animation is also possible and instructive. For example, a method was developed at the National Center for Atmospheric Research to portray cinematographically the time and space variations in concentrations of fission products in near-surface air. The data were those of the United States Public Health Service Radiation Surveillance Network, and the method used a Control Data 6600 computer with dd80 microfilm output. The film also showed meteorological parameters such as the 500-mbar pressure surface (*53*).

A bulletin, annual report, or other publication summarizing the results obtained should be issued periodically by the international organization responsible for the network. Such a publication is apt to be the most useful and most often referred-to product of a global network. Furthermore, analysis of the data, when of more than casual interest, should be published in the well-known technical and scientific journals.

From the above discussion, it should be apparent that data handling for global monitoring involves much the same problems as that for community or nationwide monitoring, but is complicated by its international character. A more detailed discussion of handling the data produced by monitoring is given in Chapters 9 and 12, this volume.

VII. Existing and Planned Global Monitoring and Surveillance Systems

A. The WMO System

In May 1970 a committee of the World Meteorological Organization (WMO) suggested the creation of a global network of monitoring stations consisting of two types of stations with distinctly different objectives, as follows:

(a) *Regional air pollution stations*—intended to reveal long-term changes in atmospheric composition which may be related to changes in regional land use practices or other activities. Parameters to be measured at these stations are turbidity and chemical composition of precipitation. Regional stations are encouraged to make other measurements as well.

(b) *Baseline air pollution stations*—designed to document long-term changes in atmospheric environmental parameters of particular significance to weather or climate, with a minimum influence from local or regional factors. A list of constituents to be measured at these stations has been developed by a group of experts in atmospheric chemistry who met in July 1971 at the invitation of WMO, as follows:

Priority Group I

Carbon dioxide (CO_2)
Turbidity
Constituents of precipitation: SO_4—S; Cl; NH_4—N, NO_3—N; Ca; Mg; Na; K; heavy metals: Hg, Pb, Cd, Zn, V, Cu, Ni; alkalinity (or acidity); electrical conductivity; and pH

Priority Group II

Sulfur dioxide (SO_2) } or total sulfur
Hydrogen sulfide (H_2S) } or total sulfur
Oxides of nitrogen (NO and NO_2)
Ammonia (NH_3)
Total ozone (Dobson instrument)
Total precipitable water
Particle composition (CL^-, SO_4^{2-}, NO_3^-, NH_4^+, Ca^{2+}, Mg^{2+}, Na^+, K^+)

Priority Group III

Carbon monoxide (CO)
Methane (CH_4)
Nitrous oxide (N_2O)
Ozone (O_3) at ground level
Condensation nuclei (Aitken nuclei)
Particle size distribution
Heavy metals (Pb, Hg, Cd) in air
Total light hydrocarbon compounds (excluding methane)

A number of "regional stations" are now participating in the network. The WMO hopes that eventually there will be 100–150 baseline stations. Data processing and reporting for the precipitation data are being undertaken by the United States Environmental Protection Agency; the turbidity data are being processed and reported by the United States National Oceanic and Atmospheric Agency (NOAA).

B. The NCAR System

As an outgrowth of several years experience in analyzing the atmosphere in remote locations for trace gases, the National Center for Atmospheric Research (NCAR) has developed a remote monitoring system that can be used in sparsely inhabited areas to measure various trace constituents. The initial aim of the program has been to develop a suitable technology rather than to focus on analysis for particular chemical constituents. The program has been given the acronym TRACE (Trace Remote Atmospheric Chemical Evaluation). Three of the systems have now been built: one is operating on the roof of NCAR in Boulder, Colorado; one is located on Mauna Loa volcano in Hawaii, and, at the time of writing this chapter, one has been set up in American Samoa on Cape Matatula. The stations on Mauna Loa and Cape Matatula are operated by NOAA personnel; since NCAR is not a monitoring organization, the project will probably be taken over by NOAA or some other United States government agency once the stations have been demonstrated to be effective.

The TRACE system consists essentially of field sites with sampling and meteorological equipment, local field laboratories, and a central laboratory serving all of the stations. The system was designed to have the following characteristics: (1) battery operation, except for heating in cold climates; (2) automatic sampling (but not automatic analysis); (3) self-sufficiency, with only a local field laboratory for servicing the station; (4) relatively low cost; and (5) simplicity in performing station servicing and laboratory preparations.

The sampling equipment consists of four units, as shown in the block diagram in Figure 12. The units, as deployed on Mauna Loa, are shown in Figure 13. All except box No. 1 can remain indefinitely at the sampling site. The sampling components of boxes 2 and 3 must be transported to

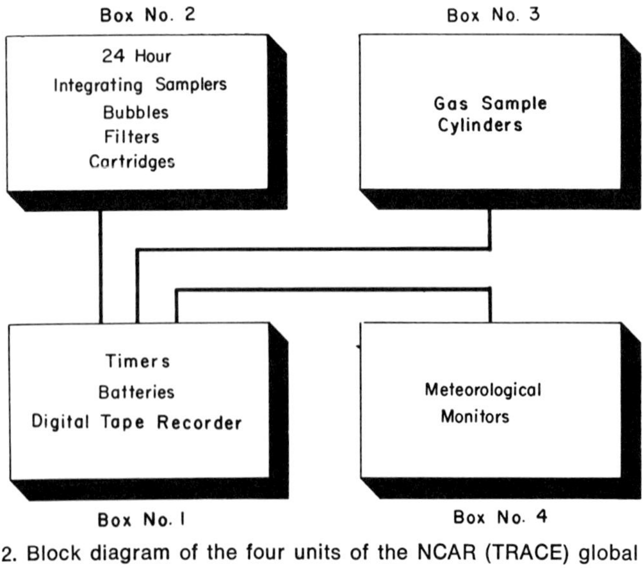

Figure 12. Block diagram of the four units of the NCAR (TRACE) global monitoring system.

Figure 13. The NCAR (TRACE) global monitoring system set up at the observatory at the 11,000 ft level on Mauna Loa Volcano, Hawaii.

and from the sampling site, and the batteries in box No. 1 must be recharged repeatedly.

C. Other Global and International Monitoring

A large number of international and global monitoring systems have been established for various purposes. Their activities have been summarized in a document entitled *National and International Environmental Monitoring Activities,* published by the Smithsonian Institution (*23*). For example, 53 network maps are shown, and the monitoring includes such diverse events or materials as cosmic rays, solar flares, radioactive fallout, milk, flash floods, and human bones.

VIII. Use of the Data in International Decision Making

A. Data Interpretation

Although we certainly know how to use the results of urban and regional monitoring systems, we do not as yet know how to use the results of global monitoring except to the extent that the results may show that the atmosphere is not being polluted on a global scale. This is because we know so little about the effects, on climate for example, of changes in the concentrations of atmospheric trace constituents. Research into such effects should be of very high priority. Nonetheless, we should not wait for the successful outcome of such research to undertake global monitoring. We need baseline data concerning the trace constituent composition of the atmosphere on a global scale while much of it remains relatively unpolluted. We also need to know which trace substances are increasing in concentration so that when appropriate theory has been developed we can make meaningful predictions concerning man's effect on important aspects of air quality.

B. International Response

A possibility which could prevent the need for ever having to develop an international response to the threat of imminent disaster from global pollution is that local and regional pollution would become so bad that it would be controlled at the national level before global pollution becomes important. Lacking adequate theory, it would be a mistake to depend on this possibility since pollutants that are relatively unimportant locally might have important global effects. As discussed earlier, carbon dioxide is an example.

The chief polluters of the atmosphere are the developed countries, and these are the ones that would have the incentives and resources to control the offending effluents. For the most part, underdeveloped countries are scarcely interested in controlling pollution on a local level, let alone globally. Industrial and agricultural development is much more important to them. However, achieving cooperation even among the developed countries to control pollution, when such control would be expensive and highly restrictive (almost certainly the case), would be very difficult. The existence of an international monitoring network prior to the need for control would help to establish cooperation with regard to control among the nations participating in the monitoring network, but would not guarantee it.

Historically, international cooperation has been achieved only when a highly compelling advantage existed for participation by each nation involved. This situation was especially evident during the United Nations Conference on the Environment held in Stockholm in June 1972. The conference members agreed that a global atmospheric monitoring system should be established, to be coordinated by the World Meteorological Organization along the lines already worked out by WMO. Perhaps the most important outcome of the conference was to set up the United Nations Environmental Programme to coordinate the environmental activities of the several United Nations agencies. The new organization located its headquarters in Nairobi, Kenya.

Excellent discussions of the problems of taking action based on the results of global monitoring have been written by Spafford (*54*) and Matthews (*55*).

REFERENCES

1. J. C. Poles and C. D. Keeling, *J. Geophys. Res.* **70,** 6053 (1965).
2. M. A. Goldman, *J. Geophys. Res.* **79,** 4550 (1974).
3. C. D. Keeling, *Tellus* **12,** 200 (1960).
4. Massachusetts Institute of Technology, "Inadvertent Climate Modification (SMIC)." MIT Press, Cambridge, Massachusetts, 1971; J. M. Mitchell, Jr., *in* "Symposium on the Global Effects of Environmental Pollution" (S. F. Singer, ed.), pp. 139–155. Reidel, Dortrecht, Holland, 1970.
5. J. S. Sawyer, *Nature* (*London*) **239,** 23 (1972).
6. W. W. Kellogg, R. D. Cadle, E. R. Allen, A. L. Lazrus, and E. A. Martell, *Science* **175,** 587 (1972).
7. E. Robinson and R. C. Robbins, *J. Air Pollut. Contr. Ass.* **20,** 233 (1970).
8. E. Robinson and R. C. Robbins, "Sources, Abundance, and Fate of Gaseous Atmospheric Pollutants," Suppl. Rep., Project No. PR6775. Stanford Research Institute, Menlo Park, California, 1969.
9. B. Weinstock and H. Niki, *Science* **176,** 290 (1972).
10. C. L. Wilson and W. H. Matthews, eds., "Man's Impact on the Global Environment." MIT Press, Cambridge, Massachusetts, 1970.
11. L. Machta and E. Hughes, *in* "Man's Impact on the Climate" (W. H. Matthews, W. W. Kellogg, and G. D. Robinson, eds.), p. 449. MIT Press, Cambridge, Massachusetts, 1971.
12. S. A. Changnon, Jr., *Bull. Amer. Meteorol. Soc.* **49,** 4 (1958).
13. R. D. Cadle, *EOS, Trans. Amer. Geophys. Union* **53,** 812 (1972).
14. H. W. Georgii, *J. Geophys. Res.* **75,** 2365 (1970).
15. J. P. Lodge, Jr., J. B. Pate, W. Basbergill, G. S. Swanson, K. C. Hill, E. Lerange, and A. L. Lazrus, "Chemistry of United States Precipitation," Report. Nat. Cent. Atmos. Res., Boulder, Colorado, 1968.
16. United States Air Force, "Handbook of Geophysics," rev. ed. Macmillan, New York, New York, 1961.
17. W. L. Wolfe, ed., "Handbook of Military Infrared Technology." Office of Naval Research, Washington, D.C., 1965.
18. H. W. Yates and J. H. Taylor, "Infrared Transmissions of the Atmosphere," NRL Report 5453, ASTIA AD 240188. U.S. Nav. Res. Lab., Washington, D.C., 1960.
19. A. P. Altshuller and R. D. Cadle, *in* "Man's Impact on the Climate" (W. H. Matthews, W. W. Kellogg, and G. D. Robinson, eds.), p. 375. MIT Press, Cambridge, Massachusetts, 1971.
20. R. D. Cadle, A. F. Wartburg, W. H. Pollock, and B. W. Gandrud, 54th Annu. Meet. Amer. Geophys. Union, Paper V162, 1973; *EOS, Trans. Am. Geophys. Union* **54,** 511 (1973).
21. C. E. Junge, *Tellus* **14,** 363 (1962).

22. Commission for Air Chemistry and Radioactivity, *in* "Man's Impact on the Climate" (W. H. Matthews, W. W. Kellogg, and G. D. Robinson, eds.), p. 239. MIT Press, Cambridge, Massachusetts, 1971.
23. Smithsonian Institution, "National and International Environmental Monitoring Activities, A Directory." Smithsonian Institution, Washington, D.C., 1970.
24. P. Hanst, J. Lehmann, H. Reichle, and M. Lepper, *in* "Man's Impact on the Climate" (W. H. Matthews, W. W. Kellogg, and G. D. Robinson, eds.), p. 502. MIT Press, Cambridge, Massachusetts, 1971.
25. R. E. Huschke, ed., "Glossary of Meteorology." Amer. Meteorol. Soc., Boston, Massachusetts, 1959.
26. G. D. Robinson, *in* "Man's Impact on the Climate" (W. H. Matthews, W. W. Kellogg, and G. D. Robinson, eds.), p. 360. MIT Press, Cambridge, Massachusetts, 1971.
27. K. Y. Kondratyev, "Radiation in the Atmosphere." Academic Press, New York, New York, 1969.
28. A. J. Drummond, ed., "Precision Radiometry." Academic Press, New York, New York, 1970.
29. E. C. Flowers, R. A. McCormick, and K. R. Kurfis, *J. Appl. Meteorol.* **8,** 955 (1969).
30. W. Schüepp, *Meteorol. Geophys., Bioklim., Ser. B* **1,** 257 (1949).
31. F. Linke and K. Boda, *Meteorol. Z.* **39,** 161 (1922).
32. A. Ångström, *Geogr. Ann.* **11,** 156 (1929).
33. Intersociety Committee, "Methods of Air Sampling and Analysis." Amer. Pub. Health Ass., Washington, D.C., 1972.
34. J. B. Pate and R. D. Cadle, *in* "Man's Impact on the Climate" (W. H. Matthews, W. W. Kellogg, and G. D. Robinson, eds.), p. 462. MIT Press, Cambridge, Massachusetts, 1971.
35. A. L. Wartburg, J. B. Pate, and J. P. Lodge, Jr., *Environ. Sci. Technol.* **3,** 767 (1969).
36. J. P. Lodge, Jr., J. B. Pate, B. E. Ammons, and G. A. Swanson, *J. Air Pollut. Contr. Ass.* **16,** 197 (1966).
37. J. B. Pate, B. E. Ammons, G. A. Swanson, and J. P. Lodge, Jr., *Anal. Chem.* **37,** 942 (1965).
38. P. W. West and G. C. Gaeke, *Anal. Chem.* **28,** 1816 (1956).
39. W. D. Komhyr, *Ann. Geophys.* **25,** 203 (1969).
40. B. E. Saltzman, *Anal. Chem.* **26,** 1949 (1964).
41. I. C. Huygen, *Anal. Chem.* **42,** 407 (1970).
42. D. F. S. Natusch, H. B. Klonis, H. D. Axelrod, R. J. Teck, and J. P. Lodge, Jr., *Anal. Chem.* **44,** 2067 (1972).
43. J. W. Tetlow and A. L. Wilson, *Analyst* **89,** 453 (1964); J. E. Harwood and A. L. Kuhn, *Water Res.* **4,** 805 (1970); J. E. Harwood and D. J. Huyser, *ibid.* pp. 501 and 695 (1970).
44. R. D. Cadle, "The Measurement of Airborne Particles." Wiley (Interscience), New York, New York, 1975.
45. Anonymous, "Air Pollution Measurements of the National Air Sampling Network," Pub. Health Serv. Publ. No. 637. Superintendent of Documents, U.S. Govt. Printing Office, Washington, D.C., 1958.
46. R. D. Cadle and W. C. Thuman, *Ind. Eng. Chem.* **52,** 315 (1960).
47. W. C. McCrone, R. G. Draftz, and J. G. Delly, "The Particle Atlas." Ann Arbor Sci. Publ., Ann Arbor., Michigan, 1967.

48. C. B. Ludwig, R. Bartel, and M. Griggs, "A Study of Air Pollutant Detection by Remote Sensors," NASA CR 1380. Nat. Aeron. Space Admin., Washington, D.C., 1969.
49. J. C. Gille, P. L. Bailey, J. M. Russell, III, and F. B. House, *Bull. Amer. Meteorol. Soc.* **53,** 1052 (1972).
50. G. Schecter, ed., "Information Retrieval." Thompson Book Co., Washington, D.C., 1967.
51. A. Kent, "Textbook on Mechanized Information Retrieval." Wiley (Interscience), New York, New York, 1962.
52. A. Kent and O. E. Taulbee, "Electronic Information Handling." Spartan Books, Inc., Washington, D.C., 1965.
53. National Center for Atmospheric Research, Annual Report for 1967. NCAR, Boulder, Colorado, 1968.
54. W. O. Spofford, Jr., *in* "Man's Impact on the Climate" (W. H. Matthews, W. W. Kellogg, and G. D. Robinson, eds.), p. 537. MIT Press, Cambridge, Massachusetts, 1971.
55. W. H. Matthews, *in* "Man's Impact on the Climate" (W. H. Matthews, W. W. Kellogg, and G. D. Robinson, eds.), p. 564. MIT Press, Cambridge, Massachusetts, 1971.

Part C

SOURCE SURVEILLANCE

14

Stack Sampling

Harold J. Paulus and Raymond W. Thron

I. Introduction

A. General Considerations

The evaluation of an industrial plant or a group of plants, or even of an entire city, with respect to air pollution resolves itself into a study of specific pollution sources. The exhaust stack, the roof monitor, or the open window thus becomes important, because it is here that the investigators can determine the composition and quantity of the material being emitted.

Stack sampling is usually done where there is acknowledged or suspected pollution or where it is desired to measure the efficiency of collection equipment. In some instances an apparently innocuous individual source is sampled to acquire data on the cumulative effect of many such sources or to determine the effect of process changes on the effluent from the process. Regardless of the study purpose, the sampling procedure utilized should enable a reasonably accurate evaluation of the source expressed in terms of the rate of emission of specific pollutants. Even the simplest stack study requires the determination of gas flow, temperature, and humidity in the stack and, in some instances, will indicate process imbalance by virtue of such conditions as unusually high or low stack temperatures, low moisture levels, or erratic gas flow rates.

The reasons for performing stack sampling are not always the same. Samples may be necessary for any one, or more, of the following reasons: (a) To obtain data concerning the emissions for an emission inventory or to identify a predominant source in the area.—An example of this would be to determine the hydrocarbon release from a new type of organic solvent used in a degreasing tank. (b) To determine compliance with regulations.—If authorization is obtained to construct an incinerator and the permit states that the maximum allowable particulate emission is 230 mg/standard m^3 corrected to 12% CO_2, a source test must be made to determine compliance with the permit. (c) To gather information which will enable selection of appropriate control equipment.—If a source test determines that the emission is 3000 mg of particulate matter/m^3 and that it has a weight mean size of 5 μm, a control device must be chosen which will collect enough particulate matter to meet a required standard, e.g., 200 mg/m^3. (d) To determine the efficiency of control equipment installed to reduce emissions.—If a manufacturer supplies a device guaranteed to be 95% efficient for removal of particulate matter with a weight mean size of 5 μm, the effluent stream must be sampled at the inlet and outlet of the device to determine if the guarantee has been met. (e) To evaluate emission changes as a result of process or equipment modification—If, for example, a plastics manufacturing company installs additional in-plant venting of vinyl chloride gases to a central stack. (f) To establish data as legal evidence.—A private company or a governmental regulatory body may initiate a special, probably more intensive, program of stack sampling to establish findings of fact.

B. Planning for Stack Sampling

1. Primary Considerations

Testing and monitoring requires a considerable amount of planning of equipment and procedures for sampling and subsequent analysis of samples in the laboratory. Timing must also be planned, both because the industrial processes involved frequently undergo cyclic changes, and because actual sampling time is but a small fraction of the total time involved in assembling equipment, calibration, and travel. Such details as power supply, scaffolding, and access to the sampling source must be arranged for in advance.

The amount of sample required for suitable laboratory analysis dictates the sampling procedure to be employed. For example, in sampling ammonium chloride in a stack, a membrane filter may plug within a few seconds. Such a sample would be too light to weigh and insufficient for other laboratory procedures. Despite the wide range of equipment available, it sometimes appears ironic that what should be a simple stack sampling procedure often requires a detailed literature search and the development of specilized sampling equipment. Complete familiarity with the process involved is therefore necessary before undertaking a source sampling study. Operations directly related to the source should be carefully reviewed to determine the peculiarities of the processing equipment used, the time cycles involved, and the peak loading which might cause effluent and temperature variations. A review of this type can uncover a wide variety of significant process idiosyncrasies which, if not known, could completely invalidate results. Errors can result because peak loading periods were missed, an insufficient number of samples were taken, or temperature variations and composition changes are not accounted for, so that the results are not representative of the process. For example, for a few minutes in every hour the gas flow characteristics and the particulate matter loading may double in the electric furnace melting of brass scrap when scrap containing large amounts of oil is loaded into the furnace, thereby producing high stack gas temperatures and emission of incompletely burned oil. Similarly, in both gray iron cupolas and open hearths, stack loading varies, depending upon the charging, melting, and pouring cycle. The sampling procedure must account for these variations.

Gases are usually collected by absorption, adsorption, freeze-out, and grab sampling. Particulate material sampling is more difficult than gas sampling because of particle size, charge, and shape. A satisfactory sample is one that is representative of the flow pattern in the duct or stack, is free of contamination, and has undergone no chemical or physical change.

2. Roof Monitor and Vent Sampling

Sampling of roof monitors and general purpose, natural-draft vents is frequently necessary and often difficult. Roof monitors are used for low velocity displacement of large air masses, and accessibility is usually difficult. Since high contaminant dilution is typical, the sample volumes required may be large.

The airflow rate through a continuous roof monitor can be estimated on the basis of manufacturer's design data, indoor and outdoor temperatures, and meteorological conditions. A more suitable procedure is actual measurement of the exhaust flow rate at the monitor opening. Zoning of roof monitor sections for both airflow measurement and sampling may be advantageous, depending upon building use, plant layout, and indoor contaminant sources.

Building vents, either of the natural draft or mechanical type, present no special problems. Low velocities are common to natural-draft vents, whereas any range of discharge velocities can be encountered where mechanically driven equipment is used. The stack sampling procedures discussed later are applicable. However, it is frequently necessary to have a duct type of outlet installed at the vent exhaust so that relatively uniform and measurable flow patterns can be obtained and representative samples collected.

Smoke bombs of the 2- or 5-minute-duration type are sometimes helpful in establishing airflow patterns in buildings where roof monitors and vents are used. Industrial buildings are frequently under negative pressure and the skillful use of smoke bombs can quickly demonstrate the effect of an unbalanced supply and exhaust condition and indicate areas of difficulty related to air pollution sampling.

3. Stack Sampling Statistics

When one considers that he may be taking a sample at the rate of 0.5 liter/minute from a stack discharging 2000 m^3/minute to the atmosphere, the chances for error become quite large. A truly representative sample will be accurate and unbiased. If it is not representative, it may be biased, because of some consistent phenomenon, such as sampling technique, or in error, because of some uncontrolled variation such as the sensitivity of the analytical method employed.

Source testing can be considered as simple random sampling (*1*). A simple random sample of size n from a population of size N is a sample selected in such a way that every group of n different units has the same probability of being selected as the sample.

For the sample, determine the sample mean, $\bar{y}$:

$$\bar{y} = \frac{y_1 + y_2 + \cdots + y_n}{n} \tag{1}$$

If the sample is unbiased, estimate the source mean $\bar{Y}$ as being equal to the sample mean so that $\bar{Y} = \bar{y}$.

For example, take eight samples of carbon dioxide from an incinerator stack during a stack sampling test as shown in Table I. The sample mean is

$$\bar{y} = \frac{6.0 + 7.2 + 4.8 + 5.3 + 5.7 + 5.5 + 6.3 + 5.8}{8} = 5.825$$

The source mean is the same because we assume that the sample is unbiased as seen by

$$\bar{Y} = \bar{y} = 5.825$$

The variation of the sample and the population (source) may also be assumed to be equal if the sample is unbiased. The variance s^2 is defined as

$$s^2 = \frac{\Sigma_1^n (y_i - \bar{y})^2}{(n - 1)} \tag{2}$$

The variance of the source is usually calculated by Equation (3).

$$s^2 = \frac{1}{n - 1}\left[\Sigma y_i^2 - \frac{(\Sigma y)^2}{n}\right] \tag{3}$$

For the example of Table I:

$$\Sigma y_i^2 = 275.04 \qquad \Sigma y = 46.6 \qquad n = 8$$
$$s^2 = (8 - 1)^{-1}[275.04 - (46.6)^2/8] = 0.514$$

Table I Stack Waste Gas, CO_2 Percentages

Test no.	*CO_2, %*
1	6.0
2	7.2
3	4.8
4	5.3
5	5.7
6	5.5
7	6.3
8	5.8

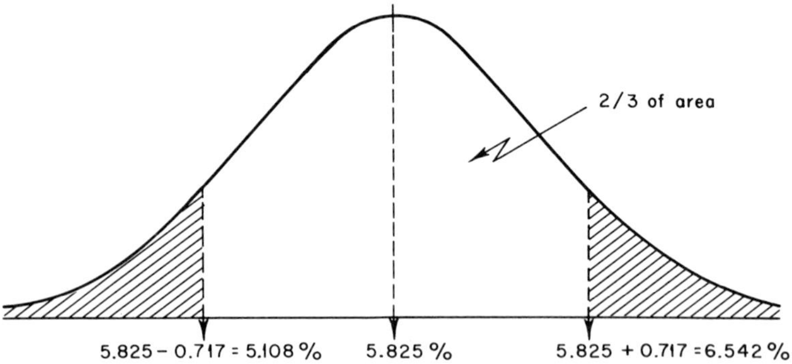

Figure 1. Distribution of carbon dioxide from an incinerator source.

The standard deviation of the sample is defined as the square root of the variance.

For the example of Table I:

$$s = (s^2)^{1/2} = (0.514)^{1/2} = 0.717$$

The sample thus represents a population (source) which, if normally distributed, has a mean of 5.825% and a standard deviation of 0.717%. The normal distribution of the sample is represented graphically in Figure 1. The curve is bell-shaped and has a single peak or mode at the mean value (u). It is symmetrical about the mean in the sense that the value of the distribution is identical at $u + c$ and $u - c$ for any value of c.

The inference from this calculation is that the true mean value for the concentration of carbon dioxide in the incinerator stack stands a 66.7% chance of being between 5.108 and 6.542%. The best single number for the carbon dioxide emission concentration is 5.825%, the mean value.

The confidence limits of these results may be calculated by applying further statistical procedures. Generally, only the values for the mean and standard deviation need to be reported. The reader is referred to any good statistical text to expand upon this brief analysis (*2*).

II. Gas Flow Measurements

A. Velocity of Waste Gas

The first step in stack sampling is always the measurement of the rate of flow of gas or air through the duct, stack, roof monitor, window, or other cross section at which pollutant concentrations are to be sampled. In particulate matter sampling, the approximate range of stack or duct velocity must be known before sampling starts to enable the use of most isokinetic sampling procedures, i.e., sampling in which the flow of gas into the sampling device has the same velocity and direction as the ambient atmosphere being sampled. A wide variety of instruments for measuring air velocity are available for this purpose (*3–6*). The standard pitot tube combination with a differential pressure gauge is the most widely used and the most satisfactory velocity-measuring instrument. As shown in Figure 2, the pitot tube consists of two concentric tubes, one

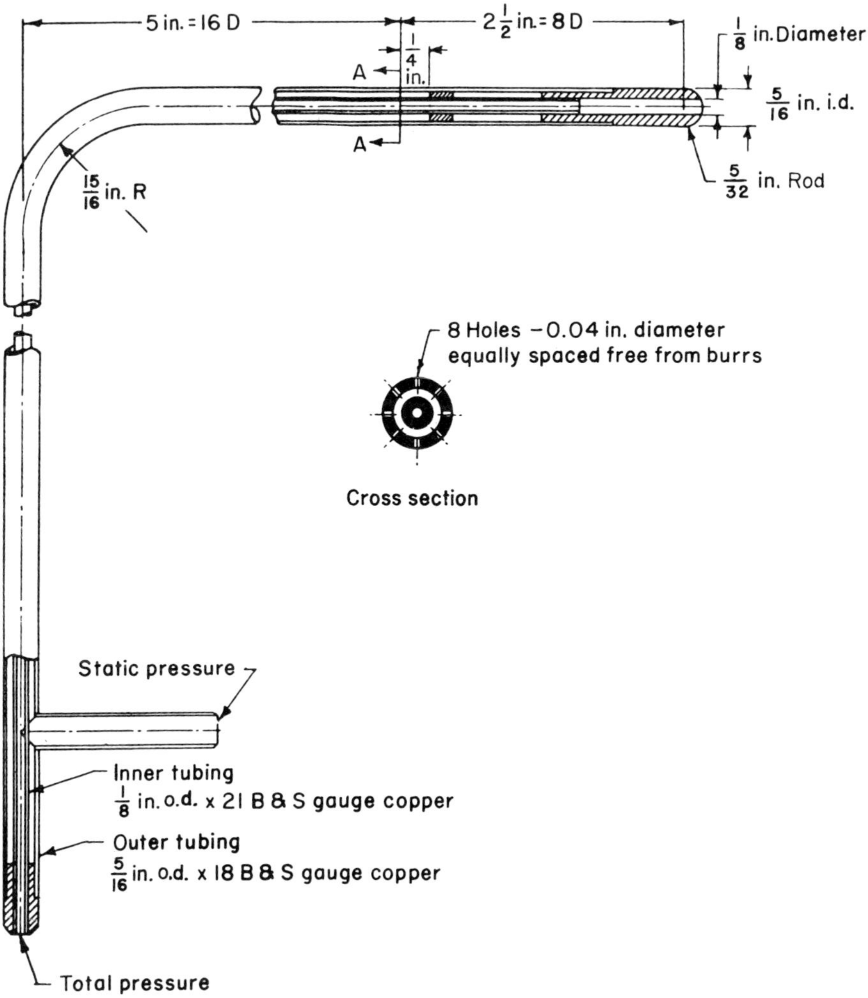

Figure 2. Standard pitot tube (*6*).

serving to measure impact pressure, and the other to measure static pressure. Since both taps are connected across a differential manometer, the pressure reading indicates velocity pressure. Velocity can be determined by referring to a velocity pressure table or by calculating by use of the formula:

English units

$$V = 1096.5\sqrt{p_v/S} \tag{4}$$
$$V = 953\sqrt{p_v(1/G)(T/P)}$$
$$V = 4005\sqrt{p_v} \qquad \text{for air at 70°F and 29.92 in. Hg}$$

SI units

$$V = 265.4\sqrt{p_v/S}$$
$$V = 389.8\sqrt{p_v(1/G)(T/P)}$$
$$V = 242.2\sqrt{p_v} \qquad \text{for air at 21°C and 760 mm Hg}$$

where

V = velocity, feet per minute (meters per minute) at duct conditions
p_v = velocity pressure, inches (millimeters) of water
S = gas density at duct conditions, pounds per cubic foot (kilograms per cubic meter)
G = specific gravity of gas (air = 1)
T = absolute temperature of gas in duct, °R (°K)
P = absolute static pressure of gas in duct, inches (millimeters) Hg

Jamison *et al.* (7) describe an electrically controlled pneumatic system designed for purging a standard pitot tube when flow rate is being determined in a duct heavily loaded with liquid droplets and/or solids. A three-position manual switch actuates solenoid valves so that, in the "read" position, the pitot tube is connected to the manometer; in the "vent" position to the atmosphere; and, in the "purge" position, to a compressed air supply.

Figure 3 illustrates a type of double pitot tube (type-S) which consists of two opposing openings, one made to face upstream and the other facing downstream during velocity measurement. The instrument design serves to minimize difficulties of plugging of the fine static pressure holes of the standard pitot tube when used in dirty gas streams and, by virtue of design, gives a higher manometer reading than the standard pitot tube. Its main disadvantages are greater sensitivity to nonparallelism with flow and a variable calibration coefficient depending upon flow conditions. It is necessary that double pitot tubes and similar specially designed instruments be calibrated under flow conditions which are substantially similar to those at test locations.

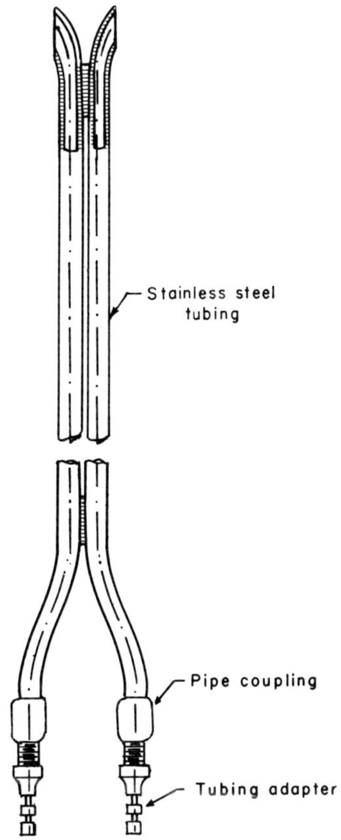

Figure 3. Double pitot tube. (Courtesy Western Precipitation Division, Joy Manufacturing Co., Los Angeles, California.)

A pitot tube traverse for determining average gas velocity in a duct is made whenever possible at a section 5–10 diameters downstream from any major gas stream disturbance such as an elbow or branch entry, 3–5 diameters upstream from similar turbulent conditions, and, in circular ducts, should consist of two sets of readings, 90° apart. Each set of readings should be spaced so as to be at the center of annular rings of equal area in a circular duct, and at the centers of a minimum of nine hypothetical squares on at least three lines in a rectangular duct (Fig. 4). For approximate results in circular ducts, a centerline reading can be used provided the measurement location is preceded by at least 10 diameters of straight duct. It is necessary that the centerline velocity pressure be multiplied by 0.81, or the centerline velocity by 0.90, to obtain average duct velocity. Since accessible duct cross sections preceded by $7\frac{1}{2}$ diame-

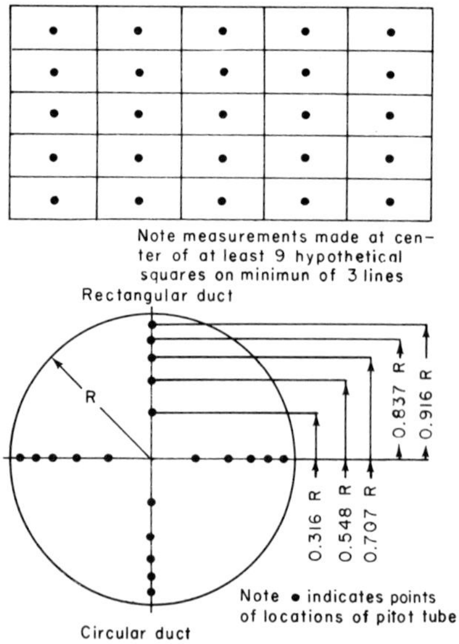

Figure 4. Location of pitot traverse points.

ters of duct are frequently not available (Fig. 5), it is generally necessary to select a measuring cross section which is a compromise between accessibility and aerodynamic conditions optimum for measurement. Further, since the ideal sampling location is often nonexistent, a series of pitot traverses are required to "feel out" the most suitable sampling location. Where necessary, in excessively turbulent ducts, the number of pitot traverse points should be increased. Conditions of flow also serve to govern the number of sampling points and stack samples in order to obtain representative results. Pitot traverse points are usually also used as sampling points. The differential manometer used with the pitot tube limits the minimum velocity measurement to about 800 ft/minute (243.8 m/minute) equivalent to about 0.04 in. (1.02 mm) of water. The Whalen gauge (*8*) is more sensitive than the inclined manometer and can be used for measuring velocities as low as 200 ft/minute (61 m/minute). The gauge utilizes two liquids of different specific gravities and is a null method instrument which must be carefully leveled and handled. Its delicate nature does not suit it to the usual field study.

Hama and Curley (*9*) have described an electrical sensing manometer which can be used with the standard pitot tube to measure velocities as low as 300 ft/minute (91.4 m/minute). The instrument is essentially a vertical U-tube containing an electrolyte (water with a few grains of salt), and measurement is achieved through micrometer depth gauges.

Figure 5. Ideal pitot traverse locations are seldom available.

A small battery is used as a source of current. It is necessary, in use, to isolate the instrument from vibrations and to maintain a constant liquid temperature.

Use of the pitot tube to evaluate intermittent nonsteady flow in a duct system is often difficult or impossible. Harding *et al.* (*10*) overcame this problem by devising a rack holding six pitot tubes at appropriate traverse point positions within the stack, thus obtaining an instantaneous traverse across the stack (Fig. 6). The pitot tubes were connected to dial-typc differential pressure gauges, and a single frame movie camera and flood-

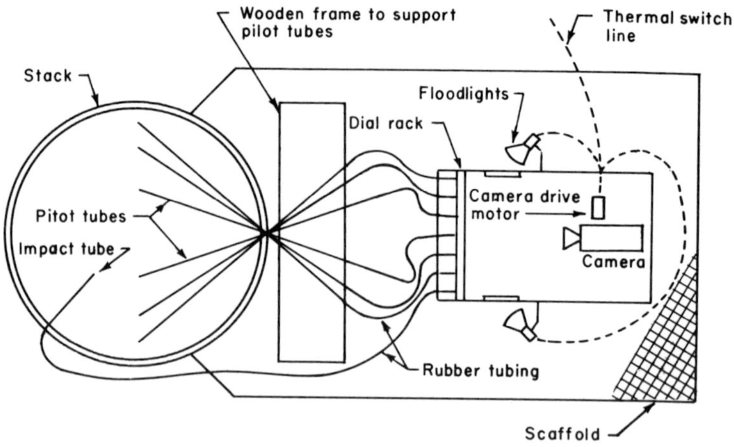

Figure 6. Plan view of photo—pitot equipment positioned for measuring vent discharge (*10*).

lights were employed to photograph the dial array and a clock, once every 20 seconds.

In using the pitot tube, it is important to recognize the fact that variations in temperature, pressure, and humidity affect density and, accordingly, enter into the determination of velocity pressure. The procedure for determining the flow rate of gas with a pitot tube traverse consists essentially of measuring the velocity pressure readings where required, converting them to equivalent velocities, averaging these velocity values (one does not average the velocity pressures), and finally, multiplying average velocity by the cross-sectional area of the duct. This procedure will give the actual flow in cubic feet (cubic meters) of gas per minute at the temperature, pressure, and humidity of the gas being measured.

Instruments such as the rotating vane anemometer, the swinging vane anemometer, the heated thermometer anemometer, and the thermal anemometer are also used for measuring velocity, but are not well suited to stack-sampling procedures. They do find use in measuring gas flow through roof monitors, windows, and other building openings.

The operation and performance of the exhaust fan during the sampling period can influence the airflow, and consequently, the accuracy of the results. Variable airflow in the stack may result from variations in the fan operation caused by ordinary wear and tear, a defective fan belt, an unbalanced blower, or variable electrical power flow. Fan operation and maintainance should be thoroughly checked before a test is made. If the fan is found to need repair, this work should be completed before the test is made.

Fans, similar to other gaseous and fluid acceleration machines, operate with respect to their characteristic curves. These curves define the performance of the fan. The type of fan being utilized and its performance characteristics should be recorded so that these data can be considered in the waste gas flow calculations. Variable airflow can also be caused by pressure head changes, differences in waste gas densities, specific volume, and relative humidity. Information on fans and characteristic fan curves may be found in mechanical engineering manuals.

B. Gas Conditions

Along with the determination of airflow in the stack, waste gas conditions must be carefully measured and evaluated before a proper test can be made. It is also necessary in many cases to make condition measurements of the waste gas while a sample is being taken. Waste gas conditions can change during the sampling period, influenced by changes in an industrial process, and in the charging rate of a combustion process. Gas condition measurements involve the determination of waste gas temperature, variations in temperature, moisture content, waste gas composition, and pressure. Accurate measurements of these factors are necessary for an accurate test of the stack. These values are also necessary in making calculations of dust loading and in reporting results.

Most United States regulations on stack sampling require that the results be reported in grains per standard cubic foot of dry waste gas. However "standard" in this case may refer to 60°F (15.6°C) or 70°F (21.1°C) depending on the regulation in question. Pressure is usually based on 29.92 in. (760 mm) Hg. There are also requirements in which the results must be corrected to 12% CO_2 or 50% excess air. A thorough review of gas laws and stack volume calculations may be found in Cooper and Rossano (*11*).

1. Temperature

The temperature of the waste gas in the stack may be measured by various means including mercury thermometers, bimetallic thermometers, thermocouples, thermistors, and optical pyrometers. The choice among these instruments depends on the temperature involved and the situation. Thermometers are based on the expansion of mercury in a sealed tube as a function of temperature, and are useful in the range from —30°F (—34.4°C) to 500°F (260°C). Bimetallic thermometers may be used up to 500°F (260°C) and operate on the principle of thermal expansion differences of two dissimilar metal strips. They are rugged in construction and their dial-type readout make them easy to read. Thermocouples operate in the wide temperature range from —200°F (—129°C) to 2500°F (1371°C), depending on the combination of metals used. Their principle of operation is that the union of the two different metal conductors produces a voltage difference which is a function of temperature. The readings of thermistors are based on their change in electrical resistance with temperature and may be used in the temperature range from —200°F (—129°C) to 1000°F (538°C). For higher temperature above 3000°F (1649°C), measurement of radiant energy from a body or surface may be made by an optical pyrometer.

2. Pressure

For many stack sampling tests, knowledge of the barometric pressure at the test elevation is sufficient. The barometric pressure for a given location may be obtained from the local weather bureau. Its value will vary from standard atmospheric pressure, which is defined as 29.92 in. (760 mm) Hg at sea level. The static pressure of the waste gas in the stack may be measured by comparison with barometric pressure using a straight tube inserted into the stack, connected to a pressure gauge or manometer. Perry (*12*) has described a micromanometer which can be used to make very precise pressure measurements.

3. Moisture Content

The amount of water in the waste gas stream has an important bearing on accurate stack sampling tests. The evaluation of moisture content in the stack gas is influenced by the temperature of the gas. Cooper and Rossano review the subject in detail (*11*). Among the methods for measuring moisture content in waste gas are (a) the cooled wet bulb–dry bulb method and (b) silica gel adsorption. In the wet bulb–dry bulb

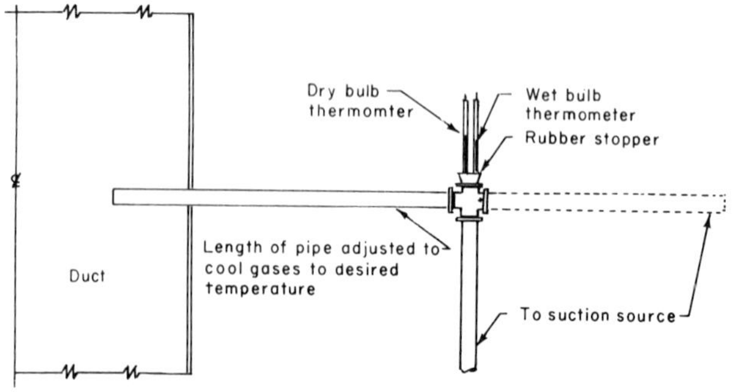

Figure 7. Gas cooler for reading dry and wet bulb temperatures.

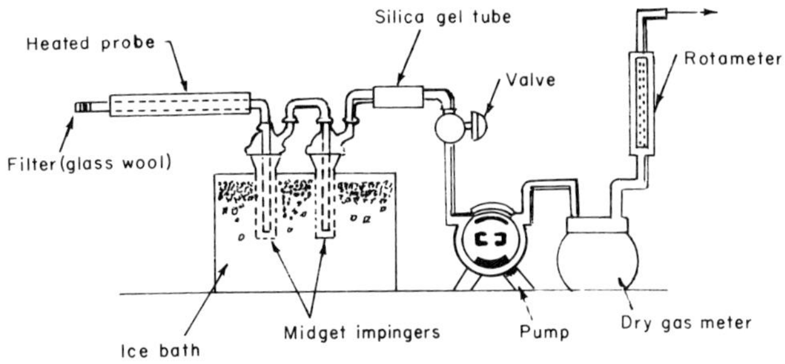

Figure 8. Moisture-sampling train.

method, the waste gas is cooled to approximately 212°F (100°C) by drawing it through a pipe past two thermometers (Fig. 7). It is sometimes necessary to insulate the pipe connected to the stack in order to obtain accurate measurements. This method has proved successful for waste gas temperatures from 212° to 270°F (100–132°C) particularly if the gas stream is near saturation.

The adsorption method recommended by the United States Environmental Protection Agency (USEPA) involves drawing the waste gas through a silica gel drying tube as shown in Figure 8. The water content is determined by the difference in weight of the silica gel before and after sampling.

4. Gas Composition and Density

It is necessary to determine the composition and density of stack gases in order to calculate velocities, gas concentrations related to specified conditions, and emission rates expressed in terms of weight of pollutant per unit weight of waste gas. The sample may be collected for analysis by either grab sampling as shown in Figure 9 or by integrated gas-sampling train as shown in Figure 10. For emissions from combustion, and certain other processes, gas composition may be determined by an Orsat

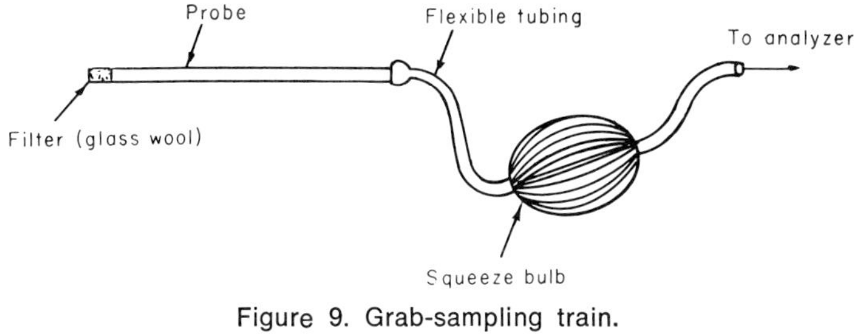

Figure 9. Grab-sampling train.

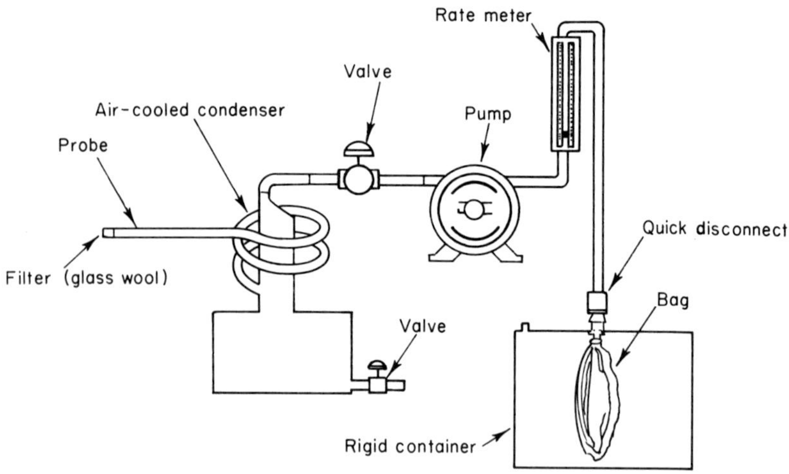

Figure 10. Integrated gas-sampling train.

Analyzer. The quantities of carbon dioxide, oxygen, and carbon monoxide are determined, with the percentage of nitrogen being the sum difference from 100%. With this information, the average molecular weight of the flue gas may be computed from the relative amounts and respective molecular weights of each constituent present.

C. Data Sheets

Typical data sheets for the field tabulation of data pertaining to gas flow conditions including velocity traverse, waste gas temperature, barometric and static pressure, moisture content, and waste gas composition are shown in Figures 11 and 12 (*13*). Calculations using these data are presented later in this chapter.

III. Particulate Matter Sampling

A. Isokinetic Sampling

Particulate matter, especially in the range of 3 μm or greater, presents a problem in that inertial effects of the particles can result in selective or erroneous samples, if the sampling velocity (velocity in the sampling nozzle) is not the same as the velocity of the gas stream at the sampling point. As illustrated in Figure 13, when the velocity of the gas within the sampling nozzle is less than the gas velocity in the duct, portions

PLANT ______________________________

DATE ______________________________

RUN NO. ______________________________

STACK DIAMETER, in. (cm) ______________________________

BAROMETRIC PRESSURE, in. (mm) Hg ______________________________

STATIC PRESSURE IN STACK (P_g), in. (mm) Hg ______________________________

OPERATORS ______________________________

SCHEMATIC OF STACK CROSS SECTION

Traverse point number	Velocity head, in. (mm) H_2O	$\sqrt{\Delta p}$	Stack Temperature (T_s), °F (°C)
	AVERAGE		

Figure 11. Velocity traverse data sheet.

of the gas stream approaching at a higher velocity are deflected. The result is that the light particles in the low micrometer size range follow the deflected gas stream and do not enter the probe. The heavier particles, by virtue of their inertia, continue into the probe, with the result that a nonrepresentative high concentration of coarse particles is collected, and the sample weight is in error on the high side. Conversely, when

Moisture content (silica gel tubes)

	Tube 1		Tube 2	
Final weight	______	grams	______	grams
Initial weight	______	grams	______	grams
Weight gain	______	grams	______	grams
Sample volume	______	liters	______	liters
A =	______		______	
Sample volume	______		______	
B =	______		______	
A/B (100%)	______	%	______	%

$$\text{A.} = \left(\frac{\text{Wt. gain}}{\text{18 gm mol.}}\right)\left(\frac{\text{24.1 liter}}{\text{mol.}}\right)$$

B. = A + Sample volume

Average moisture content ______ %

Approximate specific gravity

(% Moisture) (18) = ______

(% Dry gas) (28.95) = ______

$$\text{Sum}\ \frac{\_\_\_\_\_\_}{28.95} = \_\_\_\_\_\_\ \text{S.G.}$$

Orsat

CO_2 ______ %

O_2 ______ %

CO ______ %

N_2 ______ %

Figure 12. Basic gas condition data sheet (*13*).

the velocity within the probe is higher than that of the gas stream being sampled, a convergent airstream will develop at the nozzle face, with an excessive amount of the lighter particles entering the probe. The result is a sample indicating an excessively high concentration of light particles and a sample weight that is in error on the low side. The latter condition is also due to the inertial effect of the heavier particles that instead of converging into the nozzle's face with the gas stream travel past the edge of the nozzle and are not collected.

Watson (*14*) developed the following equation as an expression of the relationship of sample validity to nozzle and duct flow conditions.

$$\frac{C}{C_0} = \frac{U_0}{U}\left\{1 + f(p)\left[\left(\frac{U}{U_0}\right)^{1/2} - 1\right]\right\}^2 \tag{5}$$

where

C = concentration measured

C_0 = true concentration

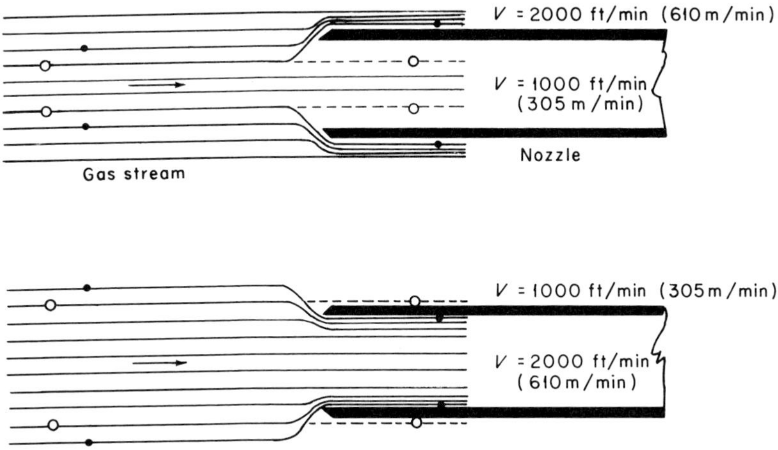

Figure 13. Selective sampling due to nonisokinetic conditions.

U_0 = stream velocity
U = mean air velocity at sampling orifice
$p = \mu^2 \rho U_0 / 18 \eta D$
D = diameter of orifice
ρ = specific gravity of particles
μ = diameter of particles
η = viscosity of gas

Figure 14 illustrates the relationship of $f(p)$ and p based on wind tunnel data. Watson (*14*) conducted his experiments in a wind tunnel of low turbulence, utilizing spherical spore particles of two basic varieties, one, 32 μm in diameter and the other 4 μm in diameter, in order to deter-

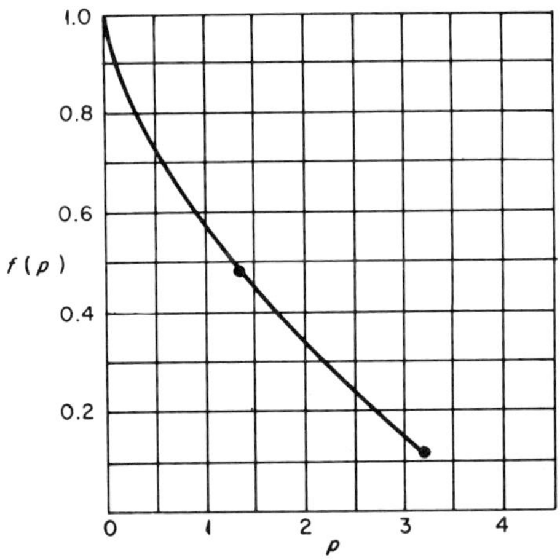

Figure 14. Relationship between $f(p)$ and p (*14*).

mine $f(p)$. In his experimental runs, the wind tunnel velocity was maintained at about 900 ft/minute (274 m/minute), which is on the low side in terms of actual processing equipment or pollution sources.

Badzioch (*15*) concluded that over a range or velocity ratios U_0/U from about 0.23 to 2:

$$C/C_0 = U_0/U + (1 - \alpha) \tag{6}$$

where α is an inertial parameter representing the fraction of particles present in the deflected (due to anisokinesis) volume. The value of α approaches unity for very coarse particles and approaches zero for very fine particles.

Badzioch further concluded that the use of blunt-edged nozzles only negligibly affects the inertial parameter α. The implication is that data obtained with blunt-edged nozzles are not likely to differ significantly from data obtained with sharp-edged nozzles.

The effect of the velocity at which tests of this phenomenon are conducted is illustrated in Figure 15, derived from data by Hemeon and Haines (*16*). Dennis *et al.* (*17*) investigated the aerodynamic characteristics of five specially designed probes in gas streams having velocities ranging from 1000 to 6000 ft/minute (305 to 1829 m/minute). Their laboratory work verified the findings of these and other investigators.

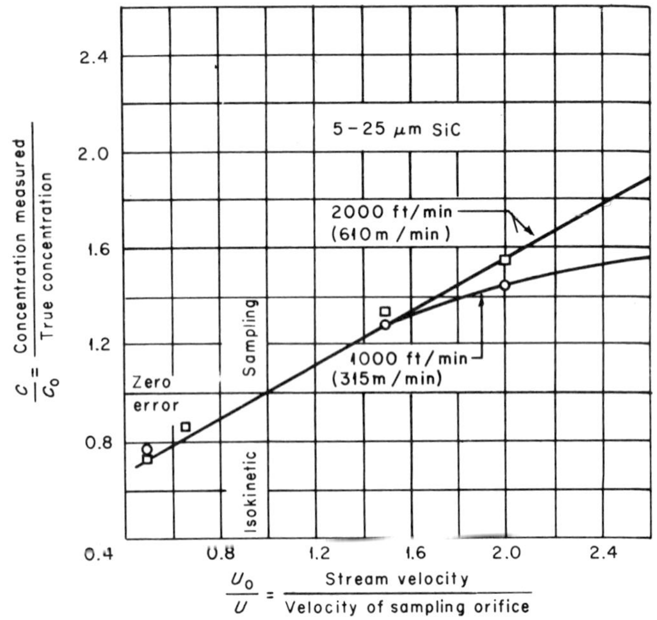

Figue 15. Errors due to departure from isokinetic sampling (*16*).

Vitols (*18*) applied computer analysis techniques to obtain plots of particle trajectories with respect to the inlet of a sampling nozzle in a mathematical study of errors associated with anisokinetic sampling of particulate matter. His theoretically predicted concentration ratios (C/C_0) demonstrate good overall agreement with the experimental results of Badzioch (*15*).

Lundgren and Calvert (*19*) investigated the bias resulting from anisokinetic sampling of aerosols in the 0.5–10 μm range utilizing both side-port and conventional-type probes. A side-port probe is a straight section of tubing positioned at right angles to the airflow direction. The aerosol sample is obtained by drawing in air through a side inlet or an end opening. Graphs were developed to illustrate sampling biases for the conventional (straight inlet) and side-port probes, based on the study findings that such biases were functions of two dimensionless parameters, an inertial parameter K, and velocity ratio U_0/U. The impaction parameter is a function of particle density, duct velocity, particle diameter, gas velocity, and probe inlet diameter. It was concluded that the side-port probe might be used to obtain representative values under conditions of low impaction parameter values.

The conclusions reached from these studies are that for precise work the correction chart for deviations from isokinesis must be determined experimentally for each combination of dust sizes, dust densities, and gas velocity that may be met in practice, but that, where precision does not demand or time does not allow such experimental determination, the following rules may be used: (a) Isokinetic sampling is unnecessary for smoke and fumes which are not admixed with particulate matter of over 5 μm in size; and (b) for the size distribution of dusts encountered in practice, Table II may be used.

In the absence of specific knowledge of the size distribution of the ma-

Table II Effect of Departure from Isokinetic Conditions on Sample Concentrations

	C/C_0	
U_0/U	*Range*	*Typical value*
0.6	0.75–0.90	0.85
0.8	0.85–0.95	0.90
1.2	1.05–1.20	1.10
1.4	1.10–1.40	1.20
1.6	1.15–1.60	1.30
1.8	1.20–1.80	1.40

terial to be sampled, it is necessary that isokinetic sampling procedures be used to assure sampling reliability. However, once systematic sampling has proved the particulate matter all to be below the size (approximately 5 μm) for which such sampling procedure is necessary, future sampling need not be isokinetic. Hemeon and Haines (*16*) investigated the effect of nozzle size on the collection of both coarse and fine particulate material and found that with particles in the 5–23 μm range there was little deviation in sample weights collected, with each of three nozzles, $\frac{1}{8}$, $\frac{1}{4}$, and $\frac{3}{8}$ in. (3.175, 6.35, and 9.525 mm) diameter. Where the particle size ranged from 420 to 500 μm, a condition atypical of stack sampling, it was found that use of the small diameter nozzle resulted in a lower weight of sample being collected.

The American Society of Mechanical Engineers Power Test Code No. 27 (1957) for Determining Dust Concentration in a Gas Stream (*20*) mentions the importance of a nozzle of not less than $\frac{1}{4}$ in. (6.35 mm) diameter in which "the tapering, if any, shall be on the outside, not on the inside." There is an advantage in using nozzles larger than $\frac{1}{4}$ in. (6.35 mm) because they enable the sampling of a proportionately greater volume and the obtaining of a larger sample.

It is possible to accomplish isokinetic sampling of a number of sampling points with constant nozzle diameter by varying the flow rate at each sampling point so that the velocity in the nozzle is equal to the velocity at the sampling point. This can be accomplished by computing from the velocity profile the nozzle flow necessary at each sampling point and by adjusting flow to that value as the nozzle is moved from point to point. Since rapid on-the-spot computations are necessary, it is desirable to have aids to computations wherever possible. These may be in the form of graphical charts or a programmed minicomputer. Figure 16 illustrates a graphic aid (*21*) for conversion of pitot and temperature readings to the orifice flowmeter setting required. Derivation of the graphical chart is as follows for given chart values and a $\frac{3}{16}$ in. (4.76 mm) diameter orifice:

$$Q_0 \text{ (orifice ft}^3\text{/minute)} = \left(\frac{\pi(D_2')^2}{4}\right)\frac{C_0}{\sqrt{1-\beta^4}}\sqrt{2gh}$$

$$= 5.21\,\frac{C_0 D_2{}^2}{\sqrt{1-\beta^4}}\sqrt{\frac{\Delta p\, T_2}{P_2 G}} \tag{7}$$

$$Q_n \text{ (nozzle cfm)} = 0.121\sqrt{\frac{\Delta p\, T_2}{P_2 G}} \quad \text{for orifice meter specified}$$

$$= A_n\left(953\sqrt{\frac{p_v T}{GP}}\right) \tag{8}$$

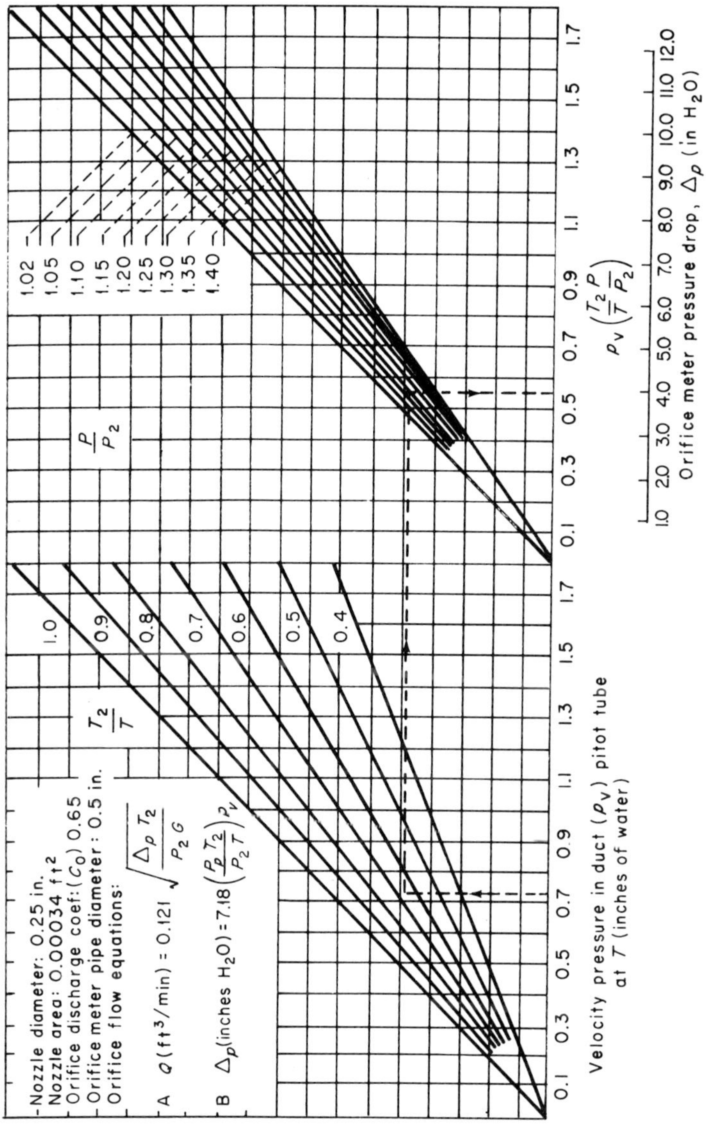

Figure 16. Graphic calculations for differential pressure across orifice (*21*).

Since $Q_n = Q_0$

$$A_n 953 \sqrt{\frac{p_v T}{GP}} \quad \Delta p = 0.121 \sqrt{\frac{\Delta p \; T_2}{GP_2}} \left(\frac{TP_2}{T_2 P}\right)$$

$$= \left(\frac{953 A_n}{0.121}\right)^2 \left(\frac{P}{P_2} \frac{T_2}{T}\right) p_v = 7.18 \left(\frac{P}{P_2} \frac{T_2}{T}\right) p_v \tag{9}$$

Derivation for the above formula using SI units is as follows:

$$Q_0 \text{ (orifice liter/minute)} = 1000 \left(\frac{\pi D_2'^2}{4}\right) \frac{C_0}{\sqrt{1 - \beta^4}} \sqrt{2gh}$$

$$= 12.97 \frac{C_0 D_2^2}{\sqrt{1 - \beta^1}} \sqrt{\frac{\Delta p \; T_2}{P_2 G}}$$

$$= 1.93 \sqrt{\frac{\Delta p \; T_2}{P_2 G}} \qquad \text{for orifice meter specified}$$

$$Q_n \text{ (nozzle liter/minute)} = 1000 A_n \left(389.8 \sqrt{\frac{p_v T}{GP}}\right)$$

Since $Q_n = Q_0$

$$1000 A_n \left(389.8 \sqrt{\frac{p_v T}{GP}}\right) = 1.93 \sqrt{\frac{\Delta p \; T_2}{P_2 G}} \left(\frac{TP_2}{T_2 P}\right)$$

$$\Delta p = \left(\frac{389.8(1000 A_n)}{1.93}\right)^2 \left(\frac{P}{P_2} \frac{T_2}{T}\right) p_v = 40.7 \left(\frac{P}{P_2} \frac{T_2}{T}\right) p_v$$

where

D_2' = orifice diameter, feet (meters)
D_2 = orifice diameter, inches (centimeters)
C_0 = orifice coefficient = ratio of orifice diameter to pipe diameter, dimensionless
g = acceleration due to gravity, 32.17 ft (9.8 m) per second per second
h = orifice pressure differential in feet (meters) of fluid flowing at orifice outlet conditions
Δp = differential pressure across orifice, inches (millimeters) of water
P_2 = absolute static pressure at orifice outlet, inches (millimeters) of Hg
T_2 = absolute temperature at orifice outlet, °R (°K)
G = specific gravity (air = 1), dimensionless

A_n = nozzle area, square feet (square meters)
P = absolute static pressure in duct, inches (millimeters) of Hg
T = absolute temperature in duct, °R (°K)
P_r = velocity pressure, inches (millimeters) of water

Pressure and temperature corrections are made automatically, and, in cases where moisture condensation is to be corrected for, it is simply a matter of applying a new abscissa scale to the chart, or correcting the Δp value to be held on the orifice manometer by multiplying by the moisture correction factor squared.

Figures 16 and 17 and the above derivation of the graphical chart indicate temperature and pressure measurements made at the flowmeter outlet. Some investigators take measurements at the flowmeter inlet, and either location is satisfactory just so long as one is consistent on the meter calibration and the mathematics involved.

Isokinesis can also be attained with the use of a "null"-type nozzle. The usual null-type nozzle (several variations of which are shown in Fig. 18) incorporates hollow chambers surrounding a nozzle whose diameter is the same or larger than the sampling tube and uses static pressure tubes which are connected to both the interior and exterior sides of the sampling nozzle, thereby indicating static pressure in the main gas stream as well as in the sampling nozzle itself. Each static pressure tap is con-

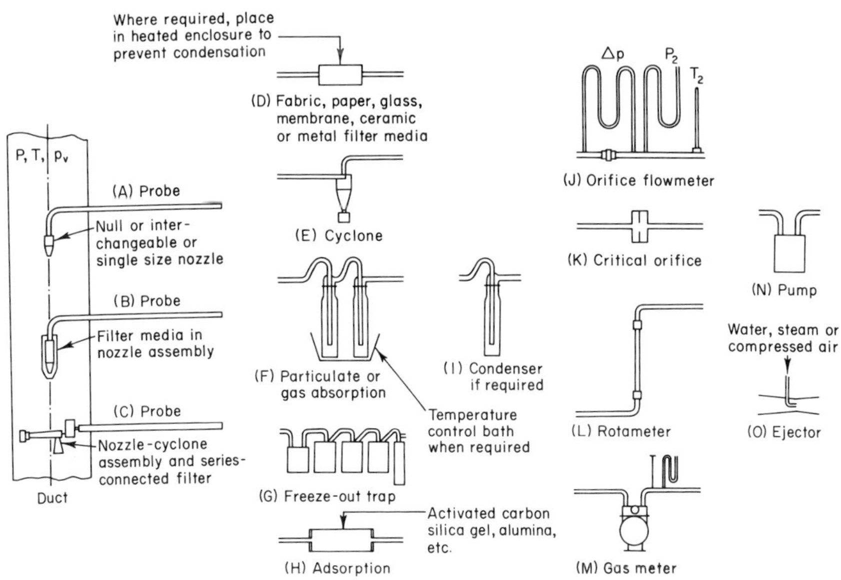

Figure 17. Sampling system components.

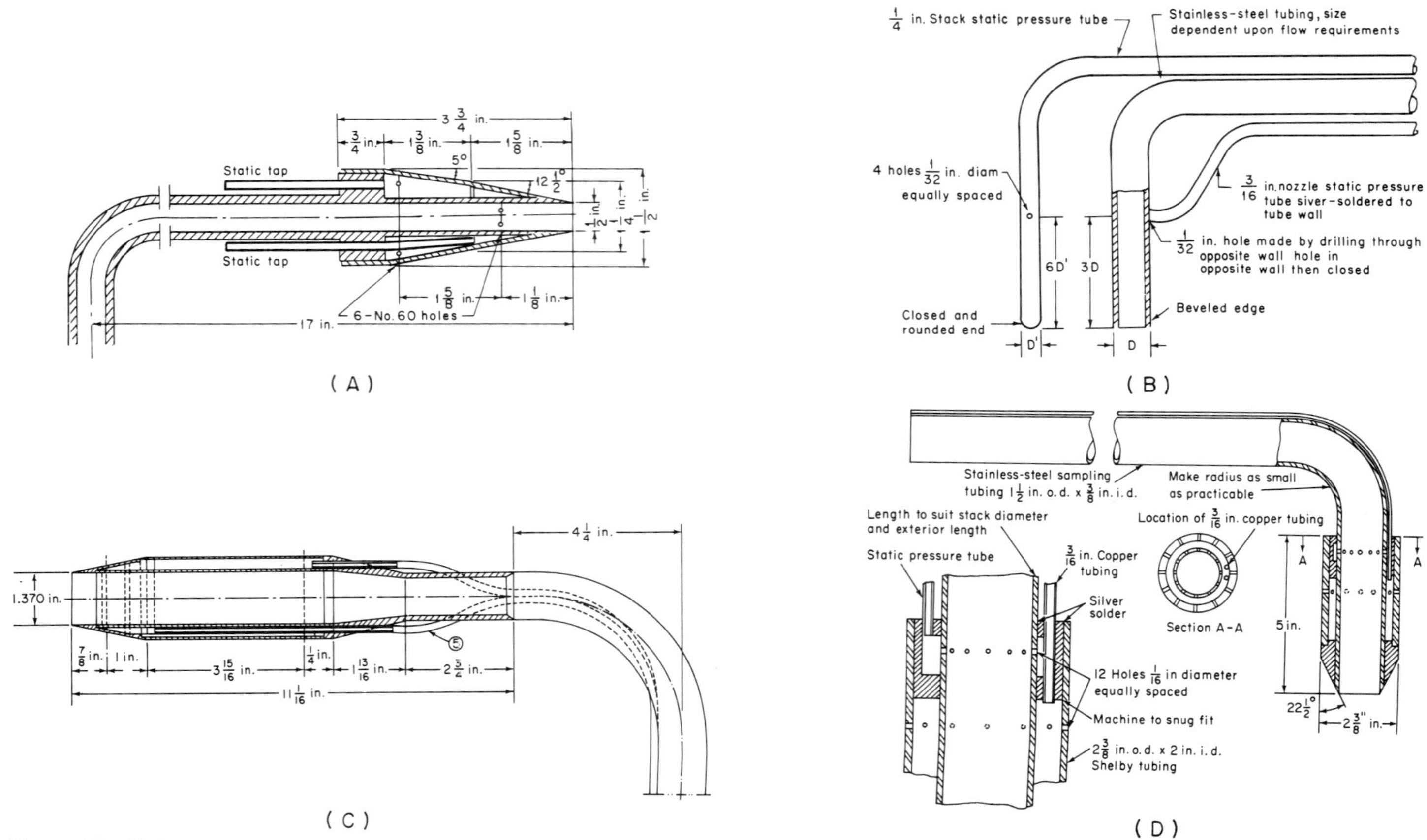

Figure 18. (A) Null-type stack sampling nozzle. [Courtesy of Western Precipitation Division, Joy Manufacturing Co., Los Angeles, California (*5*).] (B) Null-type sampling nozzle. (*16*). (C) Null-type stack sampling nozzle (*17*). [Reprinted with permission from R. Dennis, W. R. Samples, D. M. Anderson, and R. Silverman, *Ind. Eng. Chem.* **49,** 294 (1957). Copyright by the American Chemical Society.] (D) Null-type stack sampling nozzle (*16*).

nected to one of the two legs of a manometer. In operation, flow through the sampling tube is adjusted to continuously maintain no pressure differential across the manometer, i.e., a null condition. Figure 19 illustrates the sampling errors due to departure from null condition and the lack of reliability when using the null nozzle in low velocity (1000 ft/minute) (305 m/minute) gas streams.

The reason that null sampling nozzles of the type described do not guarantee isokinetic sampling conditions is that a static balance between pressures measured on the outer and inner probe walls is not necessarily indicative of a matched velocity condition. Thus, the null condition may not actually exist, because of deviations caused by differences in turbulence for duct and probe flow, the shape of the entry nozzle, its degree of surface roughness, and the location of the static holes in the interior and exterior of the nozzles.

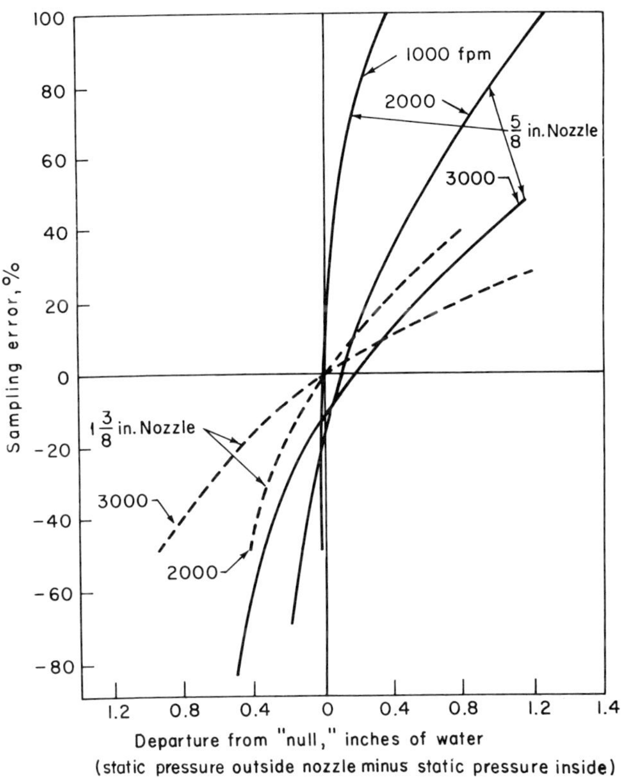

Figure 19. Sampling error caused by departure from null conditions when using null-type probes. Dashed line from Dennis *et al.* (*17*); solid line from Hemeon and Haines (*16*).

Toynbee and Parkes (*22*) claim to have overcome these problems. They postulate that the loss of pressure on the inside of the usual null nozzle design is of such magnitude as to prevent successful operation. In their nozzle design the internal bore of the rear section is greater than that of the nozzle tip, with the resulting slight expansion of gases producing a reduction in frictional loss of pressure inside the nozzle. Figure 20 illustrates their nozzle, which expands to a 1 in. (25.4 mm) internal diameter. Their experimental nozzles which ranged in internal tip diameter from 0.88 in. (22.35 mm) to 1.00 in. (25.4 mm) were tested in a velocity range of from 600 to 1800 ft/minute (183 to 549 m/minute) at temperatures up to 500°F (260°C). They developed the following semiempirical correlation relating gas flow at isokinetic conditions to the required nozzle tip diameter D for polished stainless-steel nozzles:

$$D = [0.33 \log \mathrm{Re} - 0.491]^{1/4} \tag{10}$$

where Re is the Reynolds number based on the nozzle diameter, a gas density of 0.077 lb/ft^3 (1234 gm/m^3) and a gas viscosity of 0.0175 cP. A series of curves enable selection of a nozzle for any predetermined set of average gas velocity and average temperature conditions (Fig. 21).

The alternate procedure for maintaining isokinetic conditions requires the use of a series of nozzles of different diameters. The sampling rate is kept constant, and the nozzle is selected on the basis of its inlet diameter, so as to match the duct velocity. This technique is particularly advantageous when the collection device exhibits a constant pressure drop, such as the impinger or the cyclone (without after-filter), and when only one pump calibration point, predetermined in the laboratory, is used in the field for maintaining a constant flow rate. The nozzles are changed to

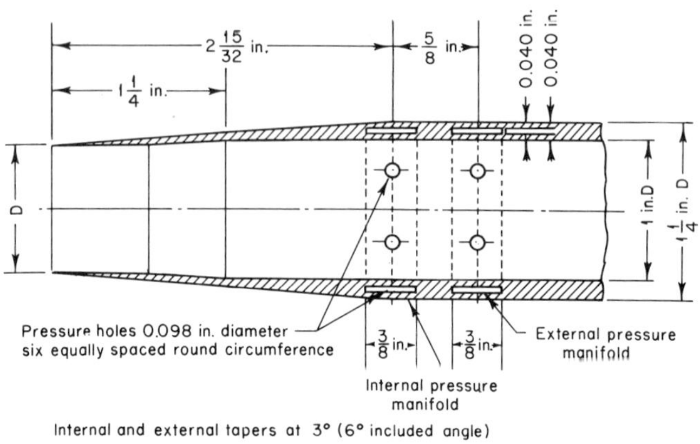

Figure 20. Sampling probe (pressure leads from manifold not shown) (*22*).

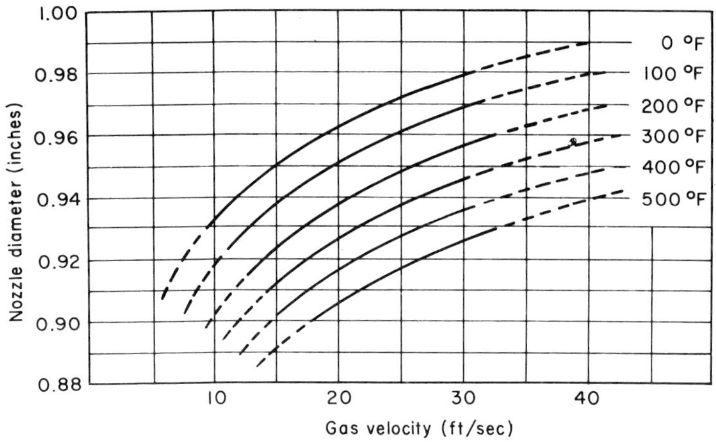

Figure 21. Design of nozzles for stainless-steel probes (*22*).

suit the velocity at the sampling point, nozzle size being a function of gas densities in the stack and at the flowmetering device. Therefore, a means for rapid choice of nozzle diameter in the field should be provided in the form of a chart, nomograph or programmed minicomputer usable at the sampling site. Table III lists a nozzle series.

B. Particulate Sample Collection

1. General Comments

The approach to stack sampling has changed considerably over the past 10–15 years. This change has been brought about by equipment changes as well as the types of pollutants considered as particulate matter. In general, the changes have improved the accuracy of tests and reduced the errors inherent in an exercise with so many variable factors.

Stack sampling equipment has been greatly improved in regard to durability, ease of operation, manipulation, and dependability. However, more improvement is necessary to reduce human errors and to improve the accuracy of the tests. In this section the discussion of equipment will include sampling trains used in the past as well as those now considered standard. The official USEPA stack sampling procedures are reviewed later in this chapter.

In stack sampling, one of the basic concerns is the type of particulate matter to be sampled. Particulate matter types include dry solids, wet solids, and condensibles. If dry solid particulate matter is to be collected, an ordinary filter system can be used with concern being paid to the

Table III Nozzle Series Isokinetic Source Sampling

Nozzle number	Letter or number of drill	Nozzle size: Diameter in.	mm	Area ft^2	mm^2	Velocity: ft/min at 1 ft^3/min	m/min at 1 m^3/min
1	27	0.1440	3.658	0.00011	10.22	8853	95,300
2	25	0.1495	3.797	0.00012	11.15	8144	87,700
3	22	0.1570	3.988	0.00013	12.08	7450	80,200
4	20	0.1610	4.089	0.00014	13.01	7080	76,200
5	18	0.1695	4.305	0.00016	14.86	6390	68,800
6	16	0.1770	4.496	0.00017	15.79	5830	62,700
7	14	0.1820	4.623	0.00018	16.72	5540	59,600
8	12	0.1890	4.801	0.00019	17.65	5140	55,300
9	10	0.1935	4.915	0.00020	18.58	4900	52,700
10	8	0.1900	5.055	0.00021	19.51	4640	49,900
11	6	0.2040	5.182	0.00023	21.37	4420	47,600
12	4	0.2090	5.309	0.00024	22.30	4210	45,300
13	3	0.2130	5.410	0.00025	23.23	4040	43,500
14	2	0.2210	5.613	0.00027	25.08	3760	40,500
15	1	0.2280	5.791	0.00028	26.01	3530	38,000
16	B	0.2380	6.045	0.00031	28.80	3240	34,900
17	D	0.2460	6.248	0.00033	30.66	3030	32,600
18	F	0.2570	6.528	0.00036	33.44	2780	29,900
19	H	0.2660	6.756	0.00039	36.23	2590	27,900
20	J	0.2770	7.036	0.00042	39.02	2390	25,700
21	L	0.2900	7.366	0.00046	42.73	2180	23,500
22	N	0.3020	7.671	0.00050	46.45	2010	21,600
23	P	0.3230	8.204	0.00057	52.95	1760	18,900
24	U	0.3390	9.611	0.00062	57.60	1600	17,200
25	T	0.3580	9.093	0.00070	65.03	1430	15,400
26	V	0.3770	9.576	0.00078	72.46	1290	13,900
27	X	0.3970	10.084	0.00086	79.89	1160	12,500
28	—	0.4375	11.113	0.00104	96.62	960	10,300

temperature of water on the filter. When a wet solid or high humidity waste gas stream is to be sampled, the stack gas should be conditioned by removal of water vapor and the solid impinged on the collecting surface. Recent discussions on stack sampling techniques have been primarily on the matter of condensibles. These are mainly organic gases that condense into solid particles at temperatures below the stack gas temperature and close to ambient temperature. In the USEPA sampling train, the condensible solids are collected in wet impingers following cyclone separation of the larger particles and filtration of the other solids.

2. Stack Sampling Equipment

A series of stack sampling system components is illustrated in Figure 17, which also indicates those points where temperature and pressure measurements are to be made. Methods for determining temperature, relative humidity, and specific gravity of gases are discussed elsewhere.

The stack sampling components illustrated in Figure 17 are indicative of the wide variety of sampling systems that can be used in a stack study, depending upon the source to be sampled, the contaminants involved, and the data desired. All incorporate combinations of probes to go into the stack and collection devices (Table IV) for removing the contaminants being measured from the gas that is being sampled through the probe. Much of the probe equipment shown is not commerically available, so that this type of equipment must be constructed as needed. The major disadvantage in using probe type (A), in Figure 17, is the length of pipe between the nozzle and the collecting device, in which material deposition can occur. Such probes are frequently jacketed and heated in order to prevent condensation. The error introduced by deposition of materials in the probe is eliminated with the use of nozzle assemblies (B) and (C) (Fig. 17) which are placed directly in the gas stream and which, by attaining gas stream temperature, also eliminate the condensation problem. One variation of nozzle assembly type (B) (Fig. 17) is described by Dennis *et al.* (*17*) and is illustrated in detail in Figure 22. Nozzle assembly type (C) (Fig. 17) includes a miniature cyclone and, when required, a fabric filter system in series, all placed directly in the duct as shown. This arrangement has been developed by Hawksley *et al.* (*23*). The assembly is shown in detail in Figure 23.

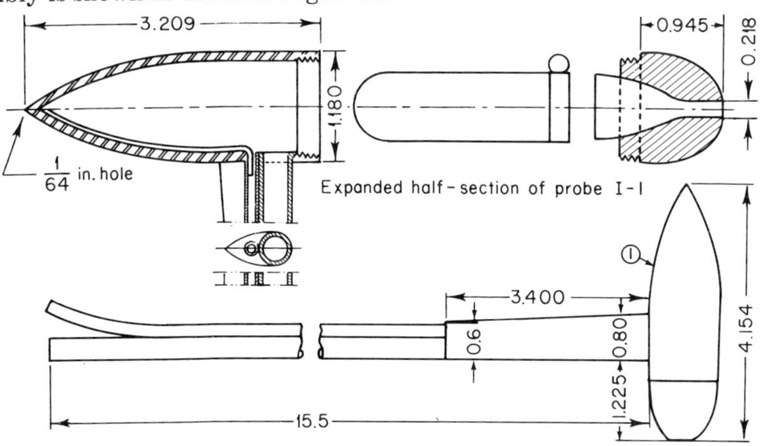

Figure 22. Null-type stack sampling nozzle (*17*) [Reprinted with permission from R. Dennis, W. R. Samples, D. M . Anderson, and R. Silverman, *Ind. Eng. Chem.* **49,** 294 (1957). Copyright by the American Chemical Society.]

Table IV Collection Devices Used in Source Sampling

Principle	*Instrument or media*	*Pressure drop*[a]	*Remarks*[b]
Absorption	Fritted bubbler (medium frit) stem in impinger flask	2.4 in. Hg at 0.25 ft^3/min to 5.8 in. Hg at 1 ft^3/min	Constant pressure drop device utilizing various collection liquids with high efficiency range for absorbable materials. Temperature control required with hot gases.
Adsorption	Cartridge or tube containing activated carbon, silica gel, or alumina	Varies with type of material and method of use	Efficiency of collection (activity and retention) varies with sorbate. Generally with carbon, any gas or vapor whose molecular structure contains fewer than 3 atoms, independent of hydrogen, is not practically adsorbable. The use of carbon is limited to gas temperatures of 120°F or less, depending upon boiling point.
Centrifugal action	Cyclone	Varies with cyclone design. Typical range: 3–36 in. H_2O	Suitable for high temperature (1500°F) if constructed of stainless steel. Can be designed for efficient collection of particles ≥ 10 μm in size. Constant pressure drop devices.
Filtration	Alundum thimbles RA 84	2.2 in. Hg at 0.25 ft^3/min to 6.9 in. Hg at 0.75 ft^3/min	Suitable for high temperature (1000°F) use. Sampling holders commercially available. Utility improved by lining with glass fiber mat.
	RA360	0.8 in. Hg at 0.25 ft^3/min to 3.2 in. Hg at 1 ft^3/min	
	Paper thimble (Whatman)	0.4 in. Hg at 0.74 ft^3/min to 1.2 in. at 2 ft^3/min	95–99% efficiency (count) for particles ≥ 1 μm. For low temperature useage (180°F max).
	Glass thimble (Schleicher and Schuell)	0.45 in. H_2O at 1 ft^3/min to 4.6 in. H_2O at 4 ft^3/min	Efficiency data unavailable. Suitable for high (600°F) gas streams.
	Glass paper (Mine Safety Appliance Co.	4.4 in. H_2O at 28 ft/min	DOP[c] efficiency of 99.93% and 99.97% at 28 ft/min. No. 1106 suitable for high

Table IV *(Continued)*

Principle	*Instrument or media*	*Pressure drop*[a]	*Remarks*[b]	
	No. 1106 and 1106-B)		(600°F) temperature gas streams.	
	Fiberfrax refractory fiber sheet stock	3.1 in. H_2O at 28 ft/min	DOP efficiency 99% at 28 ft/min	
	Membrane	See Fig. 21, Chapter 4, Vol. IV	For low temperature use (180°F max).	
	Chemical filter paper (Whatman)	See Fig. 21, Chapter 4, Vol. IV	For low temperature use (180°F max).	
	Fabrics			
	Cotton (Wellington)	$\frac{1}{2}$ in. H_2O at 20 ft/min	Max temp. 180°F	95–99% eff. (count) for particles ≥1 μm
	Wool; white (Albany)	$\frac{1}{2}$ in. H_2O at 185 ft/min	Max temp. 215°F	95–99% eff. (count) for particles ≥1 μm
	Nylon (Stevens)	$\frac{1}{2}$ in. H_2O at 7.5 ft/min	Max temp. 225°F	95–99% eff. (count) for particles ≥1 μm
	Asbestos (U.S. Rubber)	$\frac{1}{2}$ in. H_2O at 50 ft/min	Max temp. 275°F	95–99% eff. (count) for particles ≥1 μm
	Orlon (Stevens)	$\frac{1}{2}$ in. H_2O at 42 ft/min	Max temp. 275°F	95–99% eff. (count) for particles ≥1 μm
Impingement	Greenburg–Smith impinger	2.9–3.2 in. Hg at 1 ft³/min	Constant pressure drop device utilizing various collection liquids and exhibiting collection efficiency range of 90–99% (count) for particles ≥1 μm. Temperature control required with hot gases.	

[a] Pressure data based on new filtration media.
[b] Max: maximum; temp.: temperature; eff.: efficiency.
[c] Using 0.3 μm dioctyl phthalate test aerosol.

A wide variety of filtration type media [Figs. 17 (B) and 17 (D)] have been used for sampling, with the choice usually dependent upon the source to be sampled, sampling equipment available, analytical requirements, test specifications written into air cleaning equipment purchase contracts, air pollution control regulatory requirements, and individual preference. The system illustrated in Figure 24, consisting of a filter media holder which can be heated, is commonly used with glass paper. Where an equipment assembly is dependent upon thimbles (Fig. 25) for particulate separation, the choice of paper, ceramic, or fiberglass

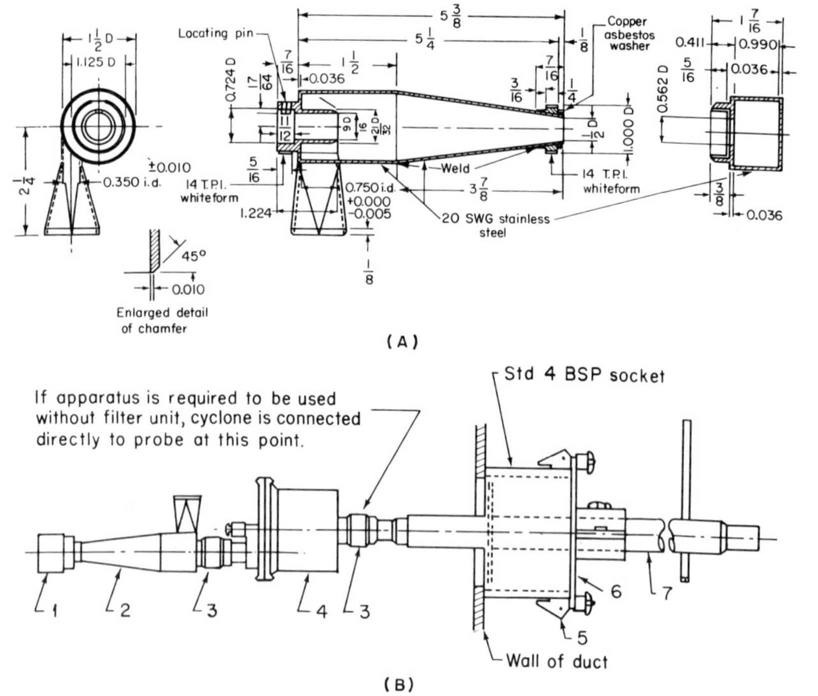

Figure 23. (A) Dimensions of cyclone-type stack sampling nozzle (*23*). (B) General arrangement of cyclone-filter probe (*23*). 1. hopper; 2. cyclone; 3. union connection; 4. filter house; 5. hinged clamping screw; 6. probe plug; 7. probe tube.

exists, each with its peculiar characteristics. Paper has a temperature limitation; ceramic is expensive and heavy, with a high initial pressure drop; and fiberglass is fragile and may shed.

Any of the many filter fabrics can be sewn into tubular (sock) shapes or can be used as flat pads. Membrane filters (*24, 25, 25a*), a woven metal cloth, and porous metal shapes have also been used. Sampling time for all of the filtration-type devices is limited by increased resistance resulting from the buildup of collected material. Where an external dry cyclone [Fig. 17(E)] or a wet scrubber such as the impinger [Fig. 17(F)] is used as the collection device, without any filtration media in the sampling system, resistance buildup does not commonly occur, since these are essentially constant pressure drop (with constant flow) devices. The cyclone, once calibrated, can also be used as a flowmeter (*26*).

The collection devices to be used for the great many aerosols of interest are dependent upon the physical and chemical properties of the materials to be collected and, in many instances, are selected as the result of detailed study of the contaminant characteristics. For example, sulfuric acid mist in the fine droplet form may or may not be satisfactorily collected in a bubbler or an impinger using water, depending upon the droplet size. Ammonium chloride, which is soluble in water, is not successfully collected with conventional wet collection devices when present in the submicrometer size range. The latter is best handled with a fabric filtration-type device.

3. Size Selection

A stack sample collected on filter paper may be weighed to determine dust loading and compliance with regulations. However, a filter paper sample cannot give an accurate particle-size distribution of the solids being emitted. Information on size distribution is necessary for: (a) selection of control equipment, (b) determination of control equipment efficiency, and (c) evaluation of the amount of respirable dust being emitted. The major emphasis in size selection has been with cyclone separation before filtration and the use of cascade impactors.

Cyclone separators in stack sampling trains are similar to the personal samplers used in industrial hygiene practice to evaluate the occupational exposure of workmen (*27*). The cyclone collector in personal samplers is designed to collect nonrespirable particles while the smaller respirable particles pass through the cyclone to be later collected on a membrane filter for weight determination. The principle purpose for using a cyclone collector in a stack sampling train is to remove the larger and heavier particles so that in a 3-hour test the subsequent filter will not fill up too rapidly. The cyclone collector could separate out the nonrespirable dust if the proper flow rate is used. However, when a variable flow rate is used for isokinetic sampling it prevents a true cut size of the particulate matter in the waste gas stream.

Several investigators have utilized the cascade impactor as a means of sample collection and particle-size separation in the stack sampling train. Gussman and Gordon (*28*) redesigned a stack probe and developed a method to prevent deposition of aerosol on the first stage of the impactor prior to the actual sampling period. Pilat *et al.* (*29*) designed a sampling probe using the cascade impactor principle for determining size distribution of stack gas particulate matter. The impactor is operated inside the stack to achieve true isokinetic sampling with a minimum of wall loss and condensation. The impactor includes seven stages, a single inlet jet and six multijet stages, followed by a filter. The single jet of the inlet nozzle eliminates the problem of particle loss on the top of the first multijet stage. Additional work has been done by other investigators to improve the cascade impactor method for size distribution of stack particles (*30, 31*).

Figure 24A. Collection device for flat filter media. (Courtesy Ford Motor Co., Dearborn, Michigan).

4. Flow Rate Control

Flow rate control is best accomplished by the use of a simple bypass arrangement around the suction device and either a calibrated orifice meter or, where a constant pressure drop collection device is used, calibration of the pressure drop across the device in terms of flow. Volume meters of the wet or dry type and Rotameters have been used successfully by some investigators (Fig. 26) but necessitate firm fastening in rugged carrying cases. They are both fragile, and the dry meter is fairly large and sometimes heavy. Effects of dirt, moisture, and positioning must also be considered. Where Rotameters, or wet or dry meters, are used for calibrating equipment, consideration should be given to manufacturers' use instruction, to pressure drop through the metering instruments, and to periodic calibration with a primary air displacement device such as a spirometer. The critical pressure (constant flow rate) type of orifice meter (*12, 32*) can be used in source sampling when the temperature, pressure, and resistance ranges of the sampling system are known, and the suction source can produce critical pressure across the orifice.

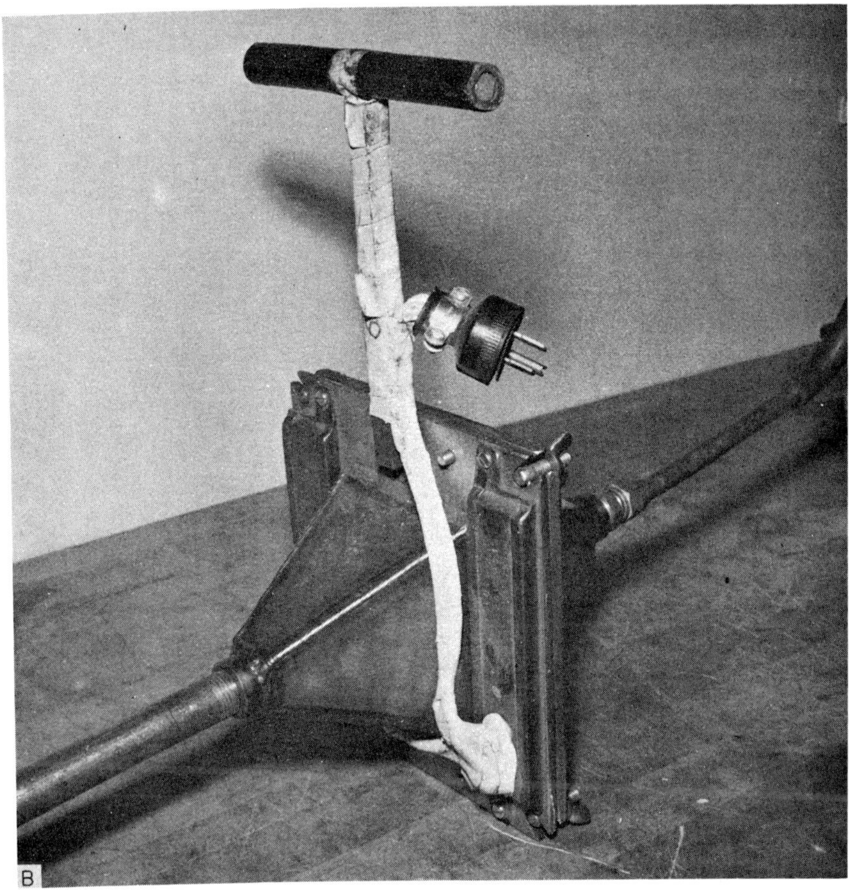

Figure 24B. Same collection device as in Figure 24A but in closed position. Note heating elements (Courtesy Ford Motor Co., Dearborn, Michigan).

5. Manipulation of Equipment

It is good source sampling practice to insert the probe, turn on the suction device, and carefully keep the probe at an exact sampling point for a predetermined time period. If the duct is under suction, the open air end of the suction device should be plugged when it is shut off to eliminate the possibility of reverse flow through the sampling system. Where the probe is to be washed free of settled matter and condensate, it should be handled with extreme care, to enable collection and subsequent evaluation of the washed material, which may be appreciable in terms of sample weight.

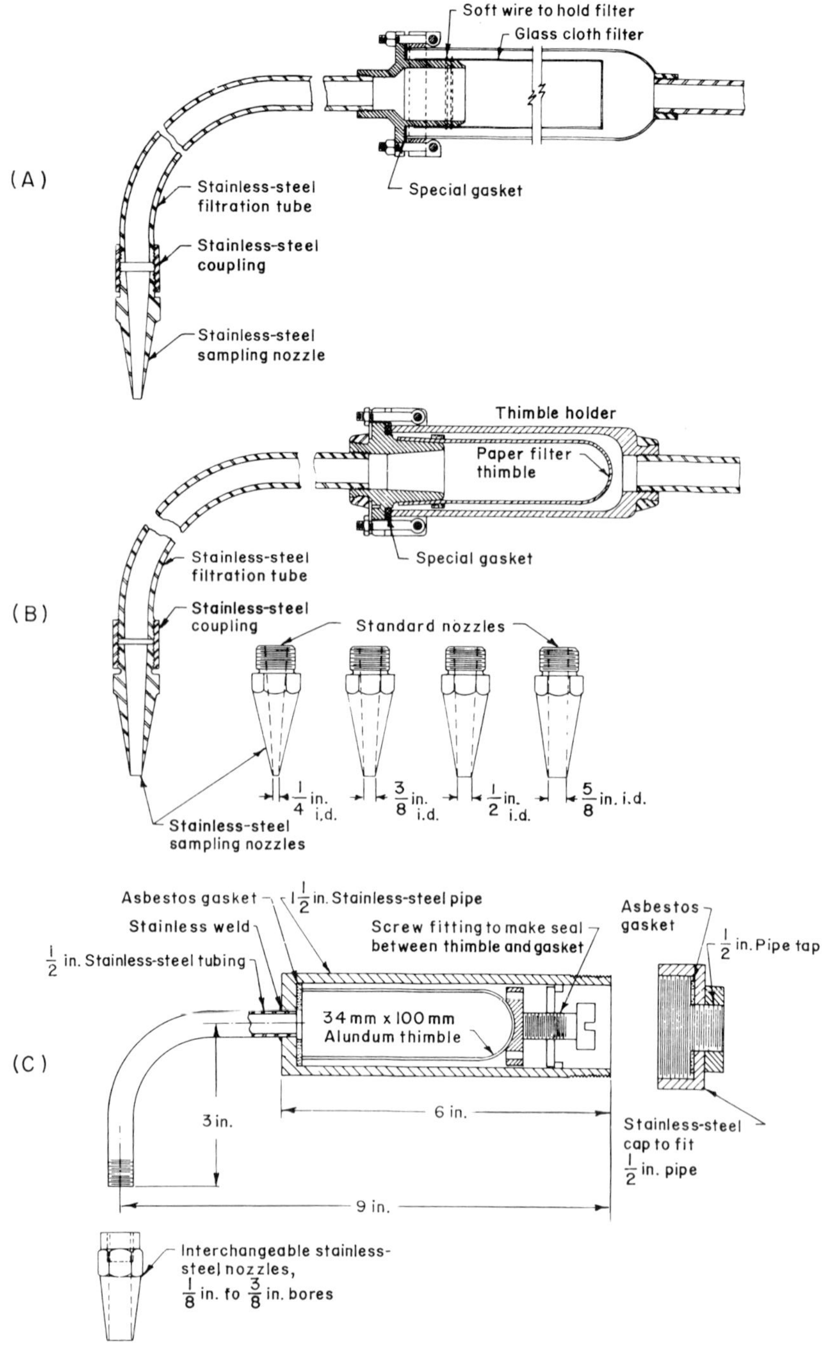

Figure 25. (A)–(C). Nozzle, short probe, and collector device assemblies. (Courtesy Western Precipitation Division, Joy Manufacturing Co., Los Angeles, California.)

Figure 26. Volume meters mounted in carrying case. (Courtesy Ajem Laboratories, Chicago, Illinois.)

6. Sampling Duration

For any stack study, equipment selection and sampling time will depend upon accuracy of results required. Where such results determine the acceptance or rejection of air-cleaning equipment involving large capital outlays or the payment of performance penalties or bonuses, precise sampling procedures are necessary, and accurate results are required. On this basis a large number of points must be sampled for an extended period of time at several cross sections of duct. The total sampling period will be lengthy. For example, in a rectangular 5 ft × 5 ft (152.4 cm × 152.4 cm) duct, precise sampling procedures call for a velocity profile determined at 25 centerline points, with subsequent sampling for equal time periods at each of these points. Where particulate matter is involved, sampling procedures must be isokinetic and the metering of gas flow accurate. Assuming the use of alundum thimbles with a 10-minute sampling period per point, the minimum sampling time would be 250 minutes. Allowing time for relocating probes and changing thimbles, a minimum sampling period of about 5 hours per cross section would be required. Two or more sampling trains may be used simultaneously to minimize the overall sampling time requirement. Sampling time is not necessarily the most significant time factor, since the setting up of equipment, preliminary testing, and the tearing down of equipment, as well as travel time, frequently far exceed it. A complete test involving duplicate runs and cross sections before and after a collector can thus easily require from 3 days to a week and a crew of two to four men.

For other than equipment acceptance testing, precision is usually sacrificed in the interest of shortening test duration and decreasing the number of people required to run a test. Where tests are to be routinely conducted by plant personnel, several days spent in initially establishing iso-duct-concentration contours, by taking individual samples at each sample point for various flow rates by the precise methods discussed above, will allow the selection of either a single sampling point or a traverse line, the use of which can cut subsequent routine sampling time to a minimum. Where a single or traverse line sampling point is chosen for routine sampling, a relatively high flow rate sampling device [50 ft^3/min (1416 liters/minute)] such as a miniature cyclone or a cyclone-filter system can be used. Correction factors accounting for deviation from 100% collection efficiency and from isokinetic sampling can be used and the sampling device may serve as the metering device. Calibration curves can be prepared in advance on the basis of gas temperature and density, eliminating the need for any additional flowmetering equipment. Under these conditions, the sampling procedure is reduced to (a) making the pitot traverse and determining duct temperature and pressure, (b) selecting a sampling nozzle diameter to conform to the above criteria, and (c) traversing the duct through one or more sampling holes with a probe–cyclone–fan combination with neither flowmeter nor flow regulator. At a 50 ft^3/minute (1416 liters/minute) flow rate, a cyclone or cyclone-filter combination will collect as much sample in 30 seconds as an alundum thimble will collect in 50 minutes. The entire sampling operation using such devices can be made independent of laboratory facilities because the large sample-to-tare ratio will minimize the weighing error and enable the use of a portable field balance in place of the laboratory type of balance. Stern (*26*) describes the combination of this technique with the use of a two-stage electrostatic precipitator, the latter to sample an aliquot of the cyclone outlet gas, to allow computation of the cyclone efficiency correction factor during the test.

Where a sampling device of less than 100% collection efficiency is used without application of an efficiency correction factor to test for compliance for air pollution control regulations, the net effect is to apply a less stringent standard, as much less stringent as is the amount of deviation of the collection efficiency of the sampling device from 100%.

The British Standard Institute has drafted a set of standards (*33*) for the simplified measurement of dust emission from chimneys. Five different apparatus assemblies were considered and judged suitable to their specified sampling requirements. An accuracy of ±25% satisfies the simplified technique, and it is further stipulated that "the apparatus used shall be such that the whole operation of sampling can be completed within one hour by not more than two operators, who need not be necessarily skilled technicians but who shall be familiar with the handling and reading of instruments such as manometers and chemical balances."

The importance of considering steady and unsteady flow rates and uniform and nonuniform concentrations of pollutants within a cross section being sampled is discussed by Achinger and Shigehara (*34*). Suggestions concerning basic sampling approaches for each combination of time variation and cross-sectional variation are made. Statistical sampling approaches are suggested for sampling under unsteady and nonuniform source conditions.

7. Source of Errors

Anyone who has made a stack sampling test will agree that the procedure is not exact. There are many sources of error in the evaluation of dust loading in the stack. Shigehara *et al.* (*35*) reviewed two of the major sources of error: (a) velocity measurement using the pitot tube, and (b) the use of a sampling train which utilizes a pitot tube and an orifice meter to attain isokinetic conditions. They made their error evaluation by partial differentiation of the pitot tube equation and the isokinetic sampling train equation. In this fashion the relative order of magnitude of error was determined. In the case of gas velocity measurements using the pitot tube, the greatest source of error was found to be the velocity head measurement. For the isokinetic sampling train equation, the errors were shown to be in the measurements of the velocity head, the moisture content of the stack gas, and the nozzle diameter of the sampling train. Larsson (*36*) indicates that a minimum gas flow velocity of 5 m/second is necessary to get reliable values in stack measurement. He emphasizes that, with lower flows, discrepancies are large and that single values are not representative.

Gnyp *et al.* (*37, 38*) have reviewed the factors affecting the accuracy of stack sampling tests. In their study they stressed the special consideration of sampling site location, number of sampling points, sampling time at each point, total sampling time, and sampling method. Among their recommendations for accurate testing of stack emissions is the requirement of at least a 10-minute collection at a sampling point when a relatively clean gas stream is encountered. They suggest that the overall test should be as long as possible (2–6 hours) to cover one or more cycles for cyclic processes.

Methods of computation, analysis, and the significance of bias and random errors in source sampling were further reviewed by the USEPA (*38a*). On the basis of their error analysis, the determination of emission rates by isokinetic stack sampling can be expected to be within 10.4% of the true mean 99.6% of the time. Most of the sources of error contribute only in a very small way to the total error in the calculated emission rate. The most significant errors resulted from the measurement of differential pressure (Δp) with the pitot tube.

C. Data Sheets and Calculations

A typical data sheet for recording all pertinent data involving a stack sampling test is shown in Figure 27. Care should be taken to record accurately all data necessary to make proper calculations.

Using the data gathered for the evaluation of waste gas conditions and during a stack sampling test, typical calculations are presented for isokinetic sampling of an airstream containing particulate matter.

1. Sampling System Nozzle-Probe Apparatus

Fiberglass filter media, condenser, orifice meter, and suction pump in series. Single nozzle size, adjustable suction source to achieve isokinetic sampling conditions.

Plant ________
Location ________
Operator ________
Date ________
Run no. ________
Sample box no. ________
Meter box no. ________
Meter $\Delta H_@$ ________
C Factor ________

Schematic of stack cross section

Ambient temperature ________
Barometric pressure ________
Assumed moisture, % ________
Heater box setting ________
Probe length, m. ________
Nozzle diameter, in. (mm) ________
Probe heater setting ________

Traverse point number	Sampling time (e). min.	Static pressure (P_s), in. (mm Hg)	Stack temperature (T_s), (°C)	Velocity head (ΔP_s)	Pressure differential across orifice meter (ΔH), in. (mm Hg)	Gas sample volume (V_m), m^3	Gas sample temperature at dry gas meter		Sample box temperature, (°C)	Impinger temperature (°C)
							Inlet (Tm_{in}), (°C)	Outlet (Tm_{out}), (°C)		
Total							Avg.	Avg.		
Average							Avg.			

Figure 27. Field data sheet.

2. Sampling Conditions

Item	*Symbol*	*Units**	*Value†*
(1) Barometric pressure	P_b	inches (mm) Hg	29.80 (757)
(2) Duct diameter	D	inches (cm)	40 (101.6)
(3) Duct temperature	t	°F (°C)	350 (177)
(4) Duct static pressure	p	inches (mm) H_2O	3 (76.2)
(5) Cooled gas temperature dry bulb	t_d	°F (°C)	190 (88)
Cooled gas temperature wet bulb	t_w	°F (°C)	120 (49)
(6) Meter temperature	t_2	°F (°C)	80 (27)
(7) Meter pressure at outlet	P_2	inches (mm) Hg	3.50 (88.9)
(8) Pitot traverse data			

Reading number	*Velocity pressure in duct, p_v [inches (mm) H_2O]*	*Velocity pressure adjusted to t_{std} and p_{std} {[70°F (21°C) and 29.92 in. (760 mm) Hg] allowing the use of Eq. (4)}**	*Velocity in duct, V [ft/min (m/min)]*
(1)	0.66 (16.8)	1.02 (26.0)	4045 (1233)
(2)	0.68 (17.3)	1.05 (26.7)	4103 (1251)
(3)	0.68 (17.3)	1.05 (26.7)	4103 (1251)
(4)	0.75 (19.0)	1.16 (29.4)	4314 (1315)
(5)	0.78 (19.8)	1.20 (30.6)	4386 (1337)
(6)	0.83 (21.1)	1.28 (32.6)	4531 (1381)
(7)	0.73 (18.5)	1.12 (28.6)	4238 (1292)
(8)	0.71 (18.0)	1.09 (27.8)	4181 (1275)
(9)	0.70 (17.8)	1.08 (27.5)	4162 (1260)
(10)	0.70 (17.8)	1.08 (27.5)	4162 (1260)

(9) Velocity in duct (average)	V	ft/min (m/min)	4223 (1287)
(10) Sampling nozzle, inside diameter	d	inches (mm)	1/4 (6.35)
(11) Sampling nozzle area	A_n	ft^2 (m^2)	0.00034 (0.000032)
(12) Sampling rate at nozzle	Q_n	ft^3/min (m^3/min)	1.44 (0.0412)

$$Q_n = VA_n = 4223 \times 0.00034 = 1.44 \text{ ft}^3/\text{min}$$
$$(Q_n = VA_n = 1287 \times 0.000032 = 0.0412 \text{ m}^3/\text{min})$$

* SI units indicated in parentheses.
† Values obtained using SI units indicated in parentheses.

Item	*Symbol*	*Units**	*Value†*
(13) Moisture content of air in duct	H	lb/lb dry gas (kg/kg dry gas)	0.0635 (0.0635) (from psychrometric chart using t_d, t_w)
(14) Moisture content of air in meter outlet	H_2	lb/lb dry gas (kg/kg dry gas)	0.0223 (0.0223) (from psychrometric chart, t_2, saturated)
(15) Moisture not measured by meter	H_3	lb/lb dry gas (kg/kg dry gas)	0.0412 (0.0412)

$$H_3 = H - H_2 = 0.0635 - 0.0223 = 0.0412$$
$$(H_3 = H - H_2 = 0.0635 - 0.0223 = 0.0412)$$

(16) Specific volume of water vapor condensed at meter outlet conditions	W	ft³/lb dry gas (m³/kg dry gas)	1.02 (0.029)

$$W = \frac{H_3(t_2 + 460)}{P_b - P_2} \times \frac{p_{std}}{(t_{std} + 460)} \times \frac{(\text{ft}^3 \text{ gas/mole at } p_{std} \text{ and } t_{std})}{\text{mol wt of water}}$$
$$= \frac{0.0412(80 + 460)}{(29.80 - 3.50)} \times \frac{29.92}{(70 + 460)} \times \frac{387}{18} = 1.02$$
$$\left(W = \frac{H_3(t_2 + 273)}{P_b - P_2} \times \frac{p_{std}}{(t_{std} + 273)} \times \frac{(\text{m}^3 \text{ gas/mole at } p_{std} \text{ and } t_{std})}{\text{mol wt. of water}} = \frac{0.0412(27 + 273)}{(757 - 88.9)} \times \frac{760}{(21 + 273)} \times \frac{10.96}{18} = 0.029\right)$$

(17) Specific volume of gas at meter outlet conditions	W_g	ft³/lb dry gas (m³/kg dry gas)	16.0 (0.454)

$$W_g = \left[\frac{\text{ft}^3/\text{mole STP}}{\text{mol wt of gas}} + H_2\left(\frac{\text{ft}^3/\text{mole STP}}{\text{mol wt of H}_2\text{O}}\right)\right]\left(\frac{p_{std}}{t_{std} + 460}\right)\left(\frac{t_2 + 460}{P_b - P_2}\right)$$
$$= \left[\frac{387}{29} + 0.0223\left(\frac{387}{18}\right)\right]\left(\frac{29.92}{70 + 460}\right)\left(\frac{80 + 460}{29.80 - 3.50}\right) = 16.0$$
$$\left(W_g = \left[\frac{\text{m}^3/\text{mole STP}}{\text{mole wt of gas}} + \text{H}_2\left(\frac{\text{m}^3/\text{mole STP}}{\text{mol wt of H}_2\text{O}}\right)\right]\left(\frac{p_{std}}{t_{std} + 273}\right)\left(\frac{t_2 + 273}{P_b - P_2}\right) = \left[\frac{10.96}{29} + 0.0223\left(\frac{10.96}{18}\right)\right]\left(\frac{760}{21 + 273}\right)\left(\frac{27 + 273}{757 - 88.9}\right) = 0.454\right)$$

Item	*Symbol*	*Units**	*Value†*
(18) Sample flow rate at meter	Q_m	ft^3/min (m^3/min)	1.00 (0.026)

where P = duct static pressure (abs) − 3 inches (mm) H_2O = 0.23 inches (mm) Hg.

$$Q_m = Q_n \times \frac{W_g}{W_g + W} \times \frac{t_2 + 460}{t + 460} \times \frac{P_b - P}{P_b - P_2}$$

$$= 1.44 \times \frac{16}{17.02} \times \frac{540}{810} \times \frac{29.57}{26.30} = 1.00$$

$$\left(Q_m = Q_n \times \frac{W_g}{W_g + W} \times \frac{t_2 + 273}{t + 273} \times \frac{P_b - P}{P_b - P_2} = 0.0412 \times \frac{0.454}{0.483} \times \frac{300}{450} \times \frac{680.8}{668.1} = 0.026\right)$$

(19) Sampling time	S	minutes	10
(20) Sample volume at nozzle	M	ft^3 (m^3)	14.4 (0.412)

$$M = Q_n S = 1.44 \times 10 = 14.4$$

$$(M = Q_n S = 0.0412 \times 10 = 0.412)$$

(21) Sample weight	Y	grams	1.4
(22) Stack loading	Z	grains/ft^3 (mg/m^3)	1.50 (3398)

$$Z = \frac{Y}{M} \times \frac{15.43 \text{ grains}}{\text{gram}} = \frac{1.4}{14.4} \times 15.43 = 1.50$$

$$\left(Z = \frac{1000Y}{M} = \frac{(1000)(1.4)}{0.412} = 3398\right)$$

(23) Duct area	A	ft^2 (m^2)	8.73 (0.811)
(24) Flow rate in duct at duct conditions	Q	ft^3/min (m^3/min)	36,867 (1043.8)

$$Q = VA = 4223 \times 8.73 = 36{,}867$$

$$(Q = VA = 1287 \times 0.811 = 1043.8)$$

(25) Duct loading	E	lb/hour (kg/hour)	474 (213)

$$E = \frac{Z \times Q \times 60}{7000 \text{ grains/lb}} = \frac{1.50 \times 36{,}867 \times 60}{7000} = 474$$

$$\left(E = \frac{Z \times Q \times 60}{10^6} = \frac{3398 \times 1043.8 \times 60}{10^6} = 213\right)$$

* SI units indicated in parentheses.

† Values obtained using SI units indicated in parentheses.

IV. Gaseous Sampling

A. General Considerations

The procedure used for sampling a gas or mixture of gases in a duct or stack differs little from that employed in the sampling of particulate matter insofar as basic routine is concerned. Isokinetic sampling procedures are not required. There is nevertheless the problem of selecting the proper collecting device which will enable reliable qualitative and quantitative analysis. Flow rates for gaseous sampling are generally much lower than with particulate sampling, ranging approximately from 1 to 5 liters/minute.

Several different methods are available for collecting the gaseous constituents from waste gas streams. The method which is used has its special application, depending upon temperature, humidity, and the gaseous constituent being analyzed.

The USEPA (*39*), the Los Angeles, California, Air Pollution Control District (*40*), The American Society for Testing and Materials (*41*), and the British Standards Institution (*33*) are but a few of the many sources of information on recommended sampling procedures.

Industrial hygienists have for many years utilized air sampling procedures and equipment suited to the measurement of concentrations of contaminants which are in the same range as must be measured for air pollution source testing (Table V). The sampling time–sampling rate ranges shown have been based in most cases on an analytical sensitivity to a contaminant concentration of one-half the American Conference of Governmental Industrial Hygienists' threshold limit value (*42*) for 8 hours/day, 5 days/week adult worker exposure. While many of these procedures can serve as a starting point in assembling a source sampling system, consideration must be given to the presence in the gas stream of interfering substances which could upset the sampling and/or subsequent analytical procedure.

Obviously, since different gases, mists, and vapors all exhibit a wide variety of physical and chemical characteristics, no all-purpose collection device exists. For example, in the sampling of gasoline vapors, activated carbon serves as an effective adsorption medium, whereas its collection efficiency falls off sharply for butane or propane, and it is totally unsatisfactory for methane.

A condenser [Fig. 17 (I)] may be required in the sampling system, depending upon the humidity of the stack gas being sampled, its purpose being to enable calculation of the humidity of the stack gas and/or to condense the moisture in the sampling system before it enters the flowmeter and suction equipment. The orifice [Fig. 17 (J)] is an excellent means of measuring flow rate by virtue of its size and operating characteristics. The critical orifice [Fig. 17 (H)] is sometimes used as a metering device, depending upon the sampling system and the flow rate desired; it requires a suction source sufficient to create critical orifice conditions. The Rotameter [Fig. 17 (L)], usually constructed of glass, and the gas meter [Fig. 17 (M)] are generally less desirable than the orifice meter because of their fragility, weight, and size. A pump [Fig. 17 (N)] is the usual choice of equipment for providing the required suction, although in some instances an ejector [Fig. 17 (O)] utilizing steam, compressed air, or water is used.

Data collection sheets and calculations are similar to those discussed in particulate mater sampling, Section III,C of this chapter. Figure 27 can be used for recording the waste gas data.

B. Sampling Methods

1. Liquid Absorption

Absorption, in which the gas sample is bubbled or drawn through an absorbing liquid, is the most common collecting procedure. A wide variety of equipment designed for gas washing purposes has been described and used (*39–41*). The Greenburg–Smith impinger (Fig. 28), which is commonly used at a sampling rate of 1 ft^3/minute (28.31 liters/minute) is an effective absorption device for certain gases and can be effectively used in a series train to increase efficiency of absorption. The gas sample is forced through a single orifice submerged beneath the absorbing medium. Various types of bubblers (Fig. 28) for absorption have been used; they are commonly of the type having a submerged sintered tip through which the gas sample is dispersed in a collecting liquid in fine bubble form or they may be made simply of a glass container having a glass inlet tube submerged in glass beads and filled with the collecting medium. Another variety of absorption apparatus is the nebulizer or atomizer through which a stream of gas is drawn. A typical example is a venturi shape in which a fine spray is formed in the venturi throat, the gas and droplets make contact, absorption takes place, and the fine droplets are separated from the gas stream by impaction in a cyclone or similar separator. The absorbing liquid thus collects and is recycled for a predetermined time. Basically, absorption efficiency is inversely proportional to bubble or droplet size and is thereby a direct function of surface contact area.

Table V Sampling Procedure Outline for Some Specific Contaminants[a]

Pollutant	*Collecting device*	*Collecting medium*[b]	*Sampling rate (ft^3/min)*	*Minimum sampling volume (ft^3)*	*Analytical method*	*Minimum concentration that can be analyzed in minimum sampling volume*
Acetone	Mylar bag	None			Gas chromatograph	
	Combustible gas indicator	None				
Acid vapors and mists	Impinger or fritted bubbler	Aqueous alkali or water	Up to 1	Sulfuric acid–60 ft^3		0.5 mg/m^3
				Nitric acid–0.1 ft^3		2.5 mg/m^3
				Hydrochloric acid–5 ft^3		3.5 mg/m
Aldehydes	Impinger or fritted bubbler	1% $NaHSO_3$ solution	0.1	1	Titration	Formaldehyde–1 ppm
Alkali mists	Impinger or fritted bubbler	Water	Up to 1	20		
	Impinger or fritted bubbler	Dilute H_2SO_4	Up to 1	20	Titration	Hydroxide– mg/m^3
Ammonia	Fritted bubbler	0.1 *N* H_2SO_4	0.1	1	Colorimetric	12.5 ppm
Aniline	Fritted bubbler	Dilute H_2SO_4	0.05	1	Colorimetric	2.5 ppm
Antimony	Impinger	Water	1	10	Colorimetric or spectrograph	
Arsenic	Impinger	Water	1	3	Colorimetric	0.25 mg/m^3
Arsine	Midget impinger	Silver diethyldithiocarbonate in pyridine	0.1	3 or Color change from yellow to red	Colorimetric	0.025 ppm
Benzene (Benzol)	Bubbler	Isooctane	0.25	3	UV Spectrophotometer	
	Mylar bag	None			Gas chromatograph	
Beryllium	Millipore filter	None	3	10	Spectrograph	0.001 mg/m^3
Cadmium	Impinger	Water	1	10	Polarograph	0.025 mg/m^3
Carbon dioxide	Mylar bag	None			Infrared	
Carbon disulfide	Fritted bubbler	Alcoholic KOH	0.05	0.1	Colorimetric	10 ppm
	Fritted bubbler	Diethylamine–Cu acetate	0.05	0.1		
Chlorinated hydrocarbons	Combustion tube with fritted bubbler	Na_2CO_3–arsenite reagent	0.1	1	Titration	
	Mylar bag	None			Gas chromatograph	
Chromic acid	Impinger	Water or aqueous alkali	Up to 1	Sample until yellow color is detected	Titration or colorimetric	0.05 mg/m^3
Cobalt	Millipore filter	None	3	15	Spectrograph	0.05 mg/m^3
Copper	Millipore filter	None	3	15	Spectrograph	0.1 mg/m^3
						0.5 mg/m^3
Cresol	Fritted bubbler	Aqueous alkali	0.05	1	Colorimetric	2.5 ppm
Cyanides	Fritted bubbler	Aqueous alkali	0.1	2.5	Colorimetric	2.5 mg/m^3 as CN
Dimethylaniline	Fritted bubbler	Alcohol	0.1	1	Colorimetric	2.5 ppm
Dinitrotoluene	Fritted bubbler	Alcohol	0.1	5	Colorimetric	0.75 mg/m^3
Ethane	Mylar bag	None			Gas chromatograph	
	Ethane detector	None				
Ethyl alcohol	Fritted bubbler	H_2SO_4–K_2CrO_7	0.05	0.5	Titration	500 ppm
	Mylar bag	None			Gas chromatograph	

Fluorides (inorganic salts)	Electrostatic precipitator	None	3	15		1.25 mg/m^3
Halogenated compounds	See chlorinated hydrocarbons					
Hydrogen cyanide	Fritted bubbler	Aqueous alkali	0.1	2.5	Colorimetric	5 ppm
Hydrogen fluoride	Fritted bubbler	Aqueous alkali	Up to 1	15	Colorimetric	1.5 ppm
Hydrogen sulfide	Bubbler column	1% $CdCl_2$	0.05	3 or Until yellow precipitate appears	Gravimetric	5 ppm
Lead	Impinger	Dilute HNO_3	1	10	Polarograph	0.75 mg/m^3
	Millipore filter	None	3	10		
Manganese	Impinger	Water	1	10	Colorimetric	2.5 mg/m^3
	Millipore filter		3	10		
Mercury	Impinger	Dilute HNO_3	1	10	Colorimetric	0.025 mg/m^3
	Mercury vapor detector	None				
Methane	Mylar bag	None			Gas chromatograph	
	Ethane detector	None				
	Combustible gas indicator	None				
Methanol	Combustible gas indicator					
Nickel carbonyl	Mylar bag	None			Gas chromatograph	
Nitrobenzene	Grab sample flask	None		500 ml	Colorimetric	0.5 ppm
Nitrogen dioxide	Grab sample flask	None		500 ml	Colorimetric	2.5 ppm
Nitrogen oxides	Grab sample flask	None		500 ml	Colorimetric	2.5 ppm
Oil mist	Impinger	Perchlorethylene	1	20	Infrared	2.5 mg/m^3
Ozone	Fritted bubbler	0.1 *N* NaOH + 1% KI solution	0.5	10	Colorimetric	0.05 ppm
Phenols	Fritted bubbler or impinger	Aqueous alkali	0.1	10	Colorimetric	2.5 ppm
Phosgene	Fritted bubbler	Sat. aniline + sat. diphenylurea	0.1	1	Gravimetric	0.05 ppm
Phosphoric acid	Impinger or fritted bubbler	Aqueous alkali	Up to 1	10	Colorimetric	0.5 mg/m^3
Phosphine	Midget impinger	Silver diethyldithiocarbonate in pyridine	1	3 or color change from yellow to red	Colorimetric	0.15 ppm
Propane	Mylar bag	None			Gas chromatograph	
	Combustible gas indicator	None				
Selenium	Impinger	Water	1	10	Gravimetric	0.1 mg/m^3
Stibine	Bubbler	6% $HgCl_2$ + 6 *N* HCl	0.1	10	Colorimetric	0.05 ppm
Sulfur dioxide	Fritted bubbler	2% Glycerol in 0.05 *N* NaOH	Up to 1	10	Colorimetric	2.5 ppm
Tellurium	Impinger	Water	1	20	Spectrophotometer	0.05 mg/m^3
Thallium	Fritted bubbler	10% HNO_3	0.1	3	Spectrograph or polarograph	0.05 mg/m^3
Toluene	Mylar bag	None			Gas chromatograph	
Tolulene diisocyanate	Fritted bubbler	HCl + acetic acid	0.1	3	Colorimetric	0.01 ppm
Xylene	Mylar bag	None			Gas chromatograph	

[a] Adapted from the Michigan Department of Public Health.
[b] Sat. = Saturated.

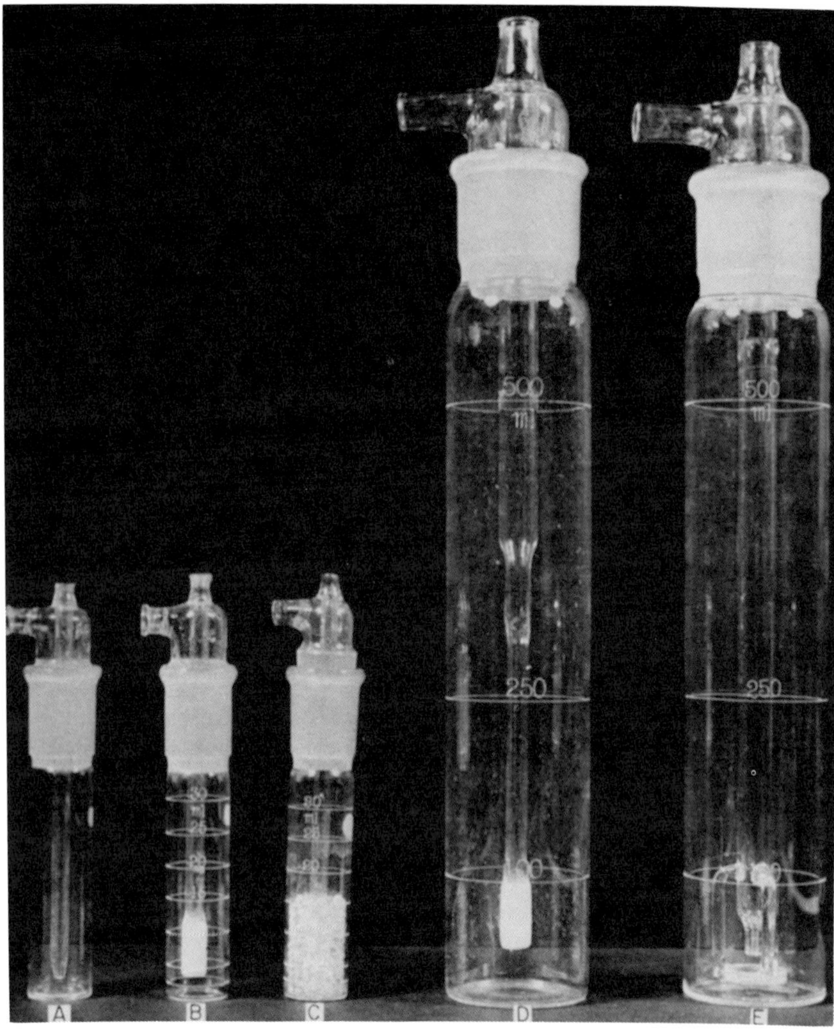

Figure 28. Wet scrubbing equipment. (A) midget impinger (1-mm stem opening); (B) midget impinger flask (fritted bubbler stem); (C) midget impinger (glass beads); (D) Greenburg–Smith impinger flask (fritted bubbler); (E) Greenburg–Smith impinger.

It is important, when sampling for any gas, that collection efficiency of the equipment be known for the contaminant concerned. Sexton (*43*) reported on the absorption efficiency of single gas washing bottles (500 ml tall-form) containing coarse porosity fritted disks for a variety of gases and mists. The work was done in conjunction with the development of a field test kit suited to field testing and on-the-spot titration-type chemical analysis (Table VI).

2. Solid Adsorption

Adsorption is another commonly used procedure for collecting a specific gas or combination of gases. Usually, a container is filled with an adsorbent such as activated carbon, alumina, or silica gel through which the gas stream is passed. Feldstein *et al.* (*44*) discuss the use of silica gel in the collection of organic solvents such as ketones, hydrocarbons, and chlorinated organic compounds. Adsorption efficiency can be affected by the type of material used, total surface area per unit weight of solid, the amount of adsorption which previously took place, and losses due to chemical reactions. Also, there may be a competition for the adsorption areas, particularly if multiple gases are present in the waste gas stream. The contaminant gases are bound by molecular forces to the adsorbent

Table VI Collection Efficiency of Gas Washing Bottle[a] for Various Contaminants (*43*)

	Collection medium		*Sampling rate*			
Chemical	*Description*	*Amount (ml)*	*ft³/min*	*liters/min*	*Test concentration (ppm)*	*Absorption efficiency (%)*
Acetic acid	Water	200	0.12	3.40	520	99+
Ammonia	Water	200	0.12	3.40	162	84
Chlorine	Water	200	0.12	3.40	43	Very low
Chlorine	2% Na_2CO_3		0.12	3.40	43	77
Ethylenediamine	Water	200	0.12	3.40	103	99+
Hydrochloric acid	Water	200	1.00	28.32	47	97
Hydrochloric acid	Water	200	0.12	3.40	47	100
Nitric acid	Water	200	0.12	3.40	85	99+
Nitrogen dioxide	Saltzman reagent	200	0.01	0.28	4–35	95
Nitrogen dioxide	Saltzman reagent	500	0.01	0.28	21–27	100
Nitrogen dioxide	Water	400	0.01	0.28	1000	76
Nitrogen dioxide	Water	400	0.01	0.28	3000	89
Perchloric acid	Water	200	0.12	3.40	12	99+
Sulfuric acid	Water	200	0.12	3.40	22	99+

[a] Single gas washing bottle, 500 ml tall-form, coarse porosity fritted disk.

surfaces and, if condensation in the adsorbent does not occur, the gases remain physically and chemically unchanged. A variety of other adsorbents are available for specific gases. Following collection, the contaminant gases are removed either by heat or eluted by an inert carrier gas.

3. Freeze-Out

Freeze-out or condensation consists essentially of a trap immersed in a low temperature bath, such as dry ice and acetone or liquid nitrogen, through which the gas sample is drawn and condensable contaminants collected. Bath temperature is necessarily selectively controlled according to the contaminant to be collected. Atmospheric water collects easily. Various combinations of freeze-out traps, desiccants, and adsorbents have been used for collecting specific materials.

Freeze-out is a useful method for trapping large numbers of pollutants. This method is dependent upon the concentration of the contaminant, the boiling points of the respective materials, temperatures of the cold traps, equipment design, capacity, and arrangement, flow rate through the sampling train, and the moisture content of the waste gas. Figure 29 illustrates a five-component freeze-out train utilizing three different cooling systems connected so as to progressively cool the gas sample. The first trap contains ice and salt at a temperature of —80°C and the fifth and final trap contains liquid nitrogen at a temperature of —195°C. Flow rate is a function of relative humidity, ranging from 0.1 to 0.2 ft³/minute (2.83 to 5.66 liters/minute).

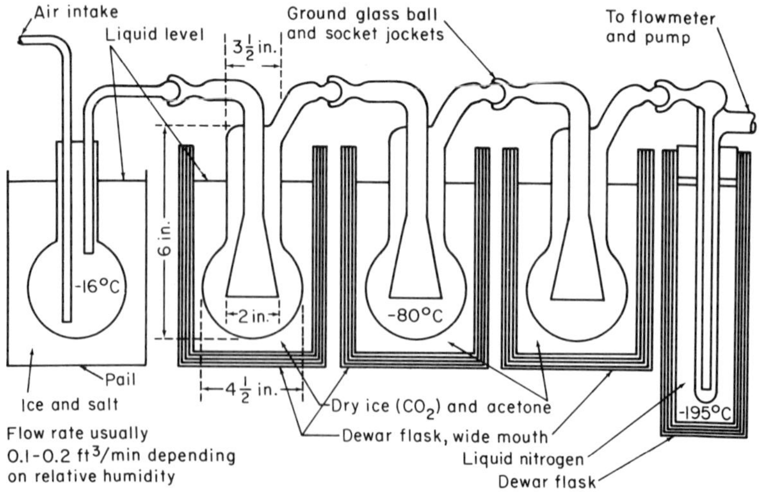

Figure 29. Freeze-out trap for collecting gaseous contaminants.

4. Grab Sampling

Grab sampling involves withdrawing a source sample into a container and subsequently taking the container to the laboratory where the contents are analyzed. In the laboratory the pollutant gases may be analyzed by absorption into a liquid medium or by instrumental methods. Losses of the gaseous constituents can result from inadequate container sealing which allows seepage, moisture condensation, and adsorption on, or reaction with, the container wall.

Containers for the grab sample are of the rigid wall and flexible bag types. Glass and stainless steel are most often used for rigid wall containers since they generally are inert to most gaseous constituents. Heating the containers by heating tapes or wires may be useful in preventing condensation during sampling (*45*).

Where analytical equipment of extreme sensitivity, such as the mass spectrometer or gas chromatograph, is available, it is sometimes possible to take a grab sample from the stack into an evacuated bottle or plastic bag without any attempt to concentrate the contaminant in an absorbent, adsorbent, or freeze-out trap. A technique for collection of a sample of this type at a sampling rate controlled to proportionately sample a cyclic process, so that one integrated sample represents the contaminant discharged to the atmosphere over a long period of time, has been described by Brief and Drinker (*46*). (Figure 10 shows a typical integrated gas sampling train.)

Collection into flexible bags is usually accomplished by pumping in the waste gas from the duct. However, the waste may require cooling to prevent bag damage. Flexible bags may be made of mylar, polyethylene, and other similar materials. The walls of the bags should be inert to the waste gases. As with rigid wall containers, collection in flexible bags is subject to losses caused by moisture condensation or adsorption on the walls of the bags. Baker and Doerr (*47*) and Altshuler *et al.* (*48*) have documented the storage properties of several gases in different bag materials. VanderKolk (*49*) found good agreement between the sampling techniques of mylar bag and gas washing bottle collection. The advantages and disadvantages of these two sampling techniques with respect to organic solvent emissions are discussed by VanderKolk.

5. Continuous Sampling

Continuous sampling affords the opportunity for avoiding reliance on single or limited batch samples. Where continuous sampling can be accomplished it will provide a much better representation of actual waste gas emissions. It will permit the observation of maximum and minimum values, and provide a better basis for calculating average emissions.

Continuous sampling oftentimes may be coupled with continuous instrumental monitoring. The information obtained from continuous sampling and subsequent analysis can be used to better ascertain compliance with air pollution control regulations, to indicate sudden changes in emission levels which may signal equipment malfunction, and to indicate potential process changes that would affect air pollution emissions.

Care must be exercised that significant losses do not occur in the sample line between the waste gas duct and the monitoring instruments. The tubing and connections should be made of inert materials so that there can be no reaction between the gas and the tubing. Removing particulate matter immediately after it enters the sample line lessens the chances for line plugging and monitoring instrument malfunction. Water vapor in the waste gas may create losses of other gaseous constituents by solubility in the condensed water. Selectively removing the water or heating the entire sample line may overcome the water vapor problem. A means for pressure reduction must be used, if the sample is taken from a high pressure source. A pressure reducing valve in the sampling line generally can reduce the pressure to near-ambient conditions (*11*). Lahmann *et al.* (*50*) describe equipment for automatic gas sampling which is coupled with apparatus automatically recording volume sampled per unit of time and measuring temperature and pressure. Neuberger and Junker (*51*) have developed steam eductors for use in removing continuous or chronologically ordered samples of stack gas. Their system is particularly useful in continuous sampling. Brown (*52*) further discusses gaseous sampling system design.

V. Special Sampling Methods

A. United States Environmental Protection Agency (USEPA) Procedures

The several Standards of Performance for New Stationary Sources, published by the USEPA in the *Federal Register* (*39*), include recommended methods and procedures for collecting and analyzing emissions. Several states have adopted these methods and require that all stack emission tests follow these procedures.

The USEPA procedure specifies the necessary units in the sampling trains for different pollutants. The USEPA sampling train for particulate matter stack emissions is shown in Figure 30, and the sampling train for sulfur dioxide in Figure 31. The important features of the particulate

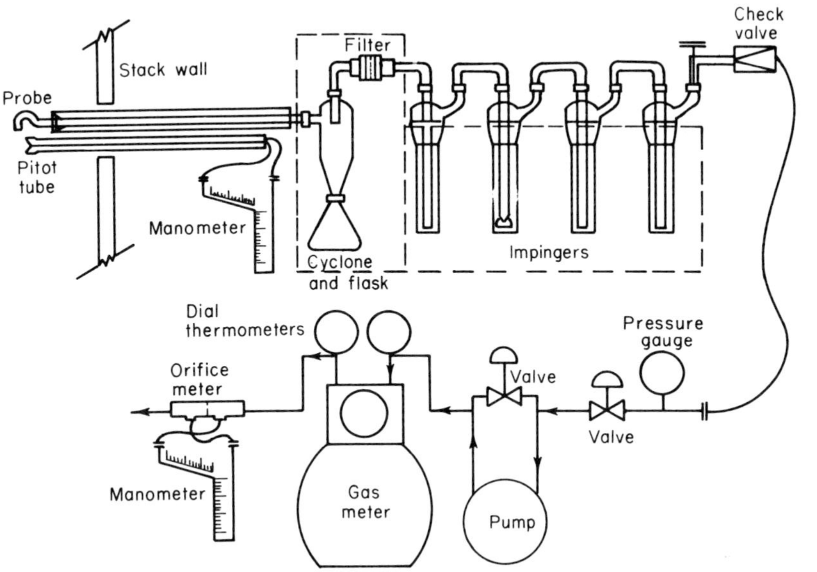

Figure 30. USEPA particulate sampling train.

matter train include the heated probe, the cyclone for size selection, the filter in the heated chamber, and the series of impingers for the collection of condensable gases, for trapping water from the impingers and for drying the airstream by means of silica gel. Isokinetic sampling is achieved by balancing the velocity pressure of the waste gas stream in the stack

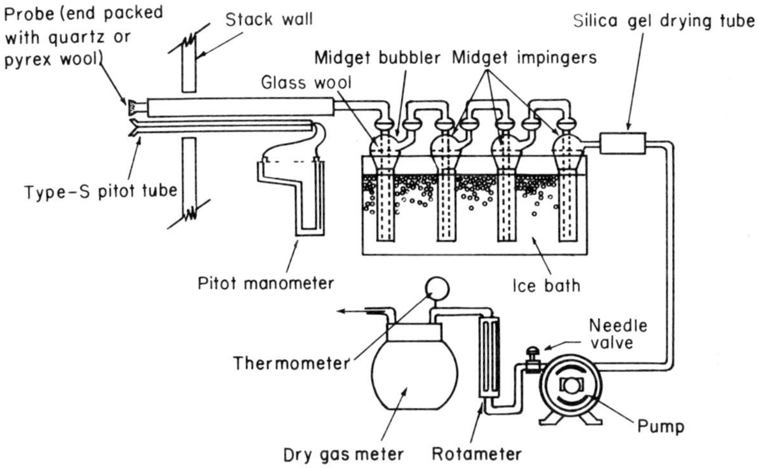

Figure 31. USEPA SO_2 sampling train.

measured by the type-S pitot with the velocity pressure entering the sample probe. The function of the cyclone system is mainly to collect the larger particles to prevent rapid buildup of the particles on the filter and thereby to allow a longer sampling period before the filter must be changed. Condensable particulates are collected in the first two impingers and, in part, represent organic compounds present as gases at stack temperature, but changed to particles by condensation at the lower temperatures existing in the impingers. All of the solids collected in the cyclone, the filter, and recovered from the rest of the train are included in the weight to be used in the emission determination. The condensable gases collected in the impingers are not included in the total weight.

Comparative stack sampling tests have been made using the USEPA method and other methods. Selle and Gronhovd (*53*) reported on simultaneous determinations of particulate matter concentration using the stack sampling methods of USEPA and the American Society of Mechanical Engineers (ASME) on a test furnace burning 75 lb/hour (34 kg/hour) of low-sulfur North Dakota lignite at dust concentrations between 0.01 and 0.5 grains/standard ft^3 (23 to 1144 mg/m^3). The two methods checked within about 5% at the higher concentration, at which the condensed impinger residues were not significant. However, at the lower dust concentration, the USEPA method gave values from 50 to 200% higher than the ASME method, the impinger residues providing a substantial part of the total weight. The authors stress the need for extremely careful handling of the USEPA method equipment and samples to prevent contamination of the total samples with small quantities of extraneous material which can significantly affect the calculated dust loading. Other comparative studies on stack sampling measurements from fuel-fired boilers were conducted by the National Council of the Paper Industry for Air and Stream Improvement (*53a*).

In the sulfur dioxide sampling train midget bubblers and impingers are utilized with the appropriate collecting medium. A silica gel drying tube placed after the wet collection of the gas serves to measure the moisture content of the waste gas and also to dry the air before it enters the pump and the dry test meter.

B. Special Industry Procedures

A wide variety of source sampling procedures have been developed and utilized by governmental agencies, consultants, gas cleaning equipment suppliers, and manufacturers. The following examples serve to illustrate that sampling techniques and subsequent analytical procedures are, of necessity, specific to the contaminants and the processes from which the contaminants are evolved, and require strict attention to detail in their development and use.

1. Power Plants

A procedure (*54*) is described in which isokinetic sampling is maintained with a pitot tube attached to a probe during particulate matter sampling. Temperatures are recorded at all sampling positions utilizing chromel-alumel thermocouples, with a high velocity thermocouple probe used at higher gas temperature locations.

The particulate matter sample is brushed and washed from all train components and added to the filter sample for total weight. Oxides of sulfur are sampled with the Shell Development–Emeryville method (*55*) utilizing a glass train, made up of a glass probe, two isopropyl alcohol adsorbers for SO_3 and a hydrogen peroxide water absorber for collection of SO_2. A glass wool filter is used to remove particulate matter which may contain sulfates and cause interference during analysis. The oxides of nitrogen are sampled with a 2-liter evacuated flask containing sulfuric acid and hydrogen peroxide and analyzed by the phenoldisulfonic acid method. Polynuclear hydrocarbons are collected on filters, after cooling and analyzed by ultraviolet visible spectroscopy; total gaseous hydrocarbons are collected in Mylar bags and analyzed with a flame ionization meter, formaldehyde is collected in midget impingers; and organic acids are collected by bubbling the flue gas through two 500-ml impingers in series, each containing 5% NaOH solution.

A utility company has developed a source sampling procedure based on the use of a "null"-type sampling nozzle series ($\frac{3}{4}$ to $1\frac{1}{4}$ in.) (19.05 to 31.75 mm) diameter. Selection of a nozzle is based on a predetermined velocity for the sampling location in the duct and a cyclone collection device flow range of from 10 to 20 ft^3/minute (4.72 to 9.44 liters/second). The collection device is a stainless-steel single inlet cyclone equipped with four paper secondary filters [$3\frac{1}{2}$-in. (88.9-mm) diameter disks] in parallel, and a double outlet. A sampling time of about 10–12 minutes per point, over 20–24 points in a duct, is typical; and heated hose [electrical wrap around wire at 300 W/25 ft (300 W/7.62 m)] is used for connecting train components. The particulate sample is collected in the cyclone filter device, and any material depositing in the probe-tubing system is blown into a preweighed thimble after each sampling cycle and the weight added to the total. Gas analysis is done with an Orsat-type device by direct aspiration (*56*).

Smith *et al.* (*57*) discuss analytical technique and difficulties encountered in sampling waste gas emissions from large coal-fired steam generators which have interfering acidic components in the stack gases. Shaw (*58*) further discusses the state of the art of sampling from large coal-fired boilers. He recommends that the flue gas should be sampled at the lowest possible temperature by using water injection at the mouth of the sampling probe.

2. Incinerators

Large diameter sampling nozzles [$\frac{3}{4}$ to 2 in. (19.0 to 50.8 mm) inside diameter] enable capture of the flake and char materials peculiar to incinerator effluents, and the water cooling of the probe reduces loss of sampled particles due to continued combustion and enables use of lower temperature range filtration media (*59*). A "null"-type nozzle facilitates sampling in that it accomodates the rapid and frequent gas flow rate changes typical of the incineration process. Sampling should be done in a vertical section of stack to avoid stratification which occurs in horizontal breechings and ducts. A calibrated $1\frac{1}{2}$ in. (38.1 mm) diameter stainless-steel cyclone also serves as flowmeter and is followed by a fabric filter. Sampling is done at a rate of from 3 to 20 ft^3/minute (1.42 to 9.44 liters/second).

Boubel and Wise (*60*) collect source samples from "tepee"-type incinerators utilizing a portable tilt-up column to operate and retrieve a sampling train from ground level. This arrangement was necessitated by the hazardous temperature and exhaust gas conditions at the emission point. The sampling train consists of a 100-mesh screen (100 openings per 2.54 cm), a modified midget impinger containing distilled water, and a moisture trap, and is operated at a 4 cfm (1.89 liters/second) sampling rate. Fife (*61*) provides a comparison of older and newer methods of incinerator sampling.

3. Foundry Cupolas

A typical collection system used for foundry cupola sampling consists of two Greenburg–Smith impingers, a paper thimble, and a vapor trap (containing a thermometer) in series. The impingers and vapor trap flask are kept in an ice bath, and the thimble is placed after the impingers to protect it against the high temperature, its main purpose being to collect the particulate matter passing through the impingers. In foundry cupola studies, this has been found to be 3 to 75% of the total weight of particulate, depending upon the flow rate through the impingers. The sampling rate is at or below 1 ft^3/minute (0.472 liters/second), which is optimum for the impingers. It is important to sample when blast air is not being bypassed. This can best be judged by continuous temperature measurement. Cupola operational data, including blast volume, blast temperature, cupola jacket temperature, blast pressure, and the character of the charge should be recorded so as to relate findings to operational characteristics.

Tamori *et al.* (*62, 63*) discuss the problem associated with samplers when extracting high temperature cupola gas. Holders and filter papers made of quartz glass can effectively endure temperature up to 1000°C. Asbestos and quartz glass fiber have been successfully used as packing in the sampler.

4. Industrial Ovens

A typical sampling train for industrial ovens uses two impingers in series followed by a Whatman paper thimble. Organic acids and oxides of sulfur samples are collected with the impinger train containing 5% NaOH solution. Aldehyde samples are collected in an impinger or fritted bubbler. Carbon dioxide is determined with an Orsat-type apparatus, and carbon monoxide is approximated with a portable analyzer. Oxides of nitrogen samples are collected in two 500-ml bottles and one 2-liter flask by means of a common manifold. The 2-liter flask contains 25 ml of dilute solution of hydrogen peroxide in 0.1 *N* sulfuric acid and is used for the determination of total oxides of nitrogen by the phenoldisulfonic acid method. The 500-ml bottles should each contain 50 ml of Saltzman solution, which is a selective absorbent for nitrogen dioxide. One of these two bottles is shaken for 1 minute after sampling to remove the nitrogen dioxide by absorption in the reagent and then is evacuated to discard the remaining nitric oxide gas. Both bottles are allowed to stand for 1 day before being analyzed to permit oxidation of nitric oxide to nitrogen dioxide in the bottle that was not evacuated. Assuming that all the nitrogen dioxide has been absorbed during the minute of shaking and that the nitric oxide would convert to the dioxide and be absorbed on standing for 24 hours, the difference in NO_2 assay of the two bottles will indicate the amount of nitric oxide present initially in the gas samples. Hydrocarbon samples are collected in mylar bags, preferably with a membrane or fiberglass filter in the sampling line to remove particulates. The samples are analyzed as soon as possible thereafter for the determination of total hydrocarbons as hexane, total carbon, or for individual organic components. A hydrogen flame ionization detector can be used for a direct measurement of total hydrocarbons expressed either as hexane or "total carbon."

Bolduc and Severs (*64*) developed a modified chromatograph combustion infrared analyzer unit which could be used in conjunction with a source testing study for on-site analysis. Their analytical objective was to determine CO, CO_2, and total hydrocarbons, although they demonstrated that with the use of the chromatograph and given sufficient time for the analytical procedure, the individual organic components could be identified and quantitated. A somewhat similar procedure is described by Goldstein and Elliott (*65*). A chromatograph is used for reasonably complete separation and quantitative determination of the individual compounds in the hydrocarbon sample, and a hydrogen flame ionization detector is used for the analysis.

5. Aluminum Reduction

Aluminum reduction presents special sampling problems. Rooth *et al.* (*66*) found that fluoride sampling methods were unsatisfactory because the sampling trains are not equipped to collect submicron particulate fluorides. They describe a sampling train which includes a replaceable probe nozzle, a heating and cooling device, cyclones for the collection of coarse particulates, a cellulose nitrate filter for collection of fine particulates, and absorption bottles for hydrofluoric acid. Their improved sampling train allows the separate determination of gaseous fluorides, coarse particulate fluorides, and fine particulate fluorides. Ivie (*67*) gives further discussion on gaseous fluoride sampling from industrial stacks.

6. Radioactive Gases and Particles

Radioactive gases and particles discharged from hot cells have been monitored at the source by a flat plate counter with a sensitivity of 10^{-7} $\mu Ci/cm^3$ (*68*). The counter was designed to count β radiation attendant to gaseous γ emitters, and the particulates were collected with a unique electrostatic precipitator which enables counting with high geometry while the particulates are deposited on a ring of corona points. The sampled gas, at a flow rate of 0.9 ft^3/minute (0.425 liter/second), passes through an annulus between a ring of corona points and a coaxially mounted Geiger–Mueller tube on which the particulates deposit.

7. Basic Oxygen Furnace

McShane and Bulba (*69*) describe the performance of a sampling and monitoring instrument which responds to variations in the fume emission from a basic oxygen process. Samples are collected on a filter tape and the density of the collected sample measured by means of the attenuation of a β radiation source located under the filter. The attenuation signal is measured with a Geiger–Mueller detector tube and relayed to a remote ratemeter.

REFERENCES

1. W. G. Cochran, "Sampling Techniques," p. 11. Wiley, New York, New York, 1963.
2. E. Parzen, "Modern Probability Theory and its Applications." Wiley, New York, New York, 1964.
3. American Conference of Governmental Industrial Hygienists, "Industrial Ventilation," 13th ed. ACGIH, Lansing, Michigan, 1974.
4. American Conference of Governmental Industrial Hygienists, "Air Sampling Instruments," 4th ed. ACGIH, Cincinnati, Ohio, 1972.
5. "Methods for Determination of Velocity, Volume, Dust and Mist Content of Gases," Bull. WP50. Western Precipitation Corp., Los Angeles, California, 1968.
6. "Guide and Data Book 1975–Fundamentals and Equipment" (Annual). Amer. Soc. Heat., Refrig., and Air Cond. Eng., New York, New York, 1975.
7. R. M. Jamison, V. W. Hanson, and O. M. Arnold, *Air Eng.* **7**, 26 (1965).
8. A. C. Willard, A. P. Kratz, and V. S. Day, *Ill., Univ., Eng. Exp. Sta., Bull.* **120** (1921).
9. G. M. Hama and L. C. Curley, *Amer. Ind. Hyg. Ass., J.* **28**, 204 (1967).
10. C. I. Harding, E. R. Hendrickson, and R. S. Sholtes, *J. Air Pollut. Contr. Ass.* **16**, 12 (1966).
11. H. B. H. Cooper, Jr. and A. T. Rossano, "Source Testing for Air Pollution Control." Environ. Sci. Serv., Wilton, Connecticut, 1971.
12. J. H. Perry, "Chemical Engineers Handbook," 5th ed. McGraw-Hill, New York, New York, 1973.
13. "Air Pollution Source Sampling Manual." Nalco Chem. Co., Chicago, Illinois, 1972.
14. H. H. Watson, *Amer. Ind. Hyg. Ass., Quart.* **15**, 21 (1954).
15. S. Badzioch, *Brit. J. Phys.* **10**, 26 (1959).
16. W. C. L. Hemeon and G. F. Haines, Jr., *J. Air Pollut. Contr. Ass.* **4**, 159 (1954).
17. R. Dennis, W. R. Samples, D. M. Anderson, and L. Silverman, *Ind. Eng. Chem.* **49**, 294 (1957).
18. V. Vitols, *J. Air Pollut. Contr. Ass.* **16**, 79 (1966).
19. D. Lundgren and S. Calvert, *Amer. Ind. Hyg. Ass. J.* **28**, 208 (1967).
20. "Determining Dust Concentration in a Gas Stream," Power Test Code No. 27. Amer. Soc. Mech. Eng., New York, New York, 1957; "Test Code for Dust Separating Apparatus," Power Test Code No. 21. Amer. Soc. Mech. Eng., New York, New York, 1941.
21. C. A. Gallaer, *Power* **101**, 88 (1957).
22. P. A. Toynbee and W. J. S. Parkes, *Int. J. Air Water Pollut.* **6**, 13 (1962).
23. P. G. W. Hawksley, S. Badzioch, and J. H. Blackett, *J. Inst. Fuel* **31**, 147 (1958).
24. M. W. First and L. Silverman, *Arch. Ind. Hyg. Occup. Med.* **7**, 1 (1953).
25. A. Goetz, *Amer. J. Pub. Health* **43**, 150 (1953).

25a. S. P. Belyaev and L. M. Levin, *J. Aerosol Sci.* **5**, 325 (1974).

26. A. C. Stern, *Combustion* **4–5**, 35 (1933).
27. M. Lippmann, *Amer. Ind. Hyg. Ass., J.* **31**, 138 (1970).
28. R. A. Gussman and D. Gordan, *Amer. Ind. Hyg. Ass., J.* **27**, 252, (1966).
29. M. J. Pilat, D. S. Ensor, and J. C. Bosch, *Atmos. Environ.* **4**, 671 (1970).
30. I. Tamori, T. Shirasawa, T. Tajiri, and I. T. Horda, *J. Jap. Soc. Air Pollut.* **5**, 53 (1970).
31. J. F. Carr and R. T. Jarman, *J. Inst. Fuel* **44**, 33 (1971).
32. T. Merriman and T. H. Wiggin, "American Civil Engineers Handbook," 5th ed. Wiley, New York, New York, 1944.
33. "Simplified Methods for Measurement of Grit and Dust Emissions from Chimneys," Tech. Commun. SFE/28. Brit. Stand. Inst., London, England, 1959.
34. W. C. Achinger and R. T. Shigehara, *J. Amer. Pollut. Contr. Ass.* **18**, 605 (1968).
35. R. T. Shigehara, W. F. Todd, and W. S. Smith, Paper 70-35, 63rd Annu. Meet. Air Pollution Control Association, Pittsburgh, Pennsylvania, 1970.
36. O. Larsson, *J. Ass. Heat., Vent. Sanit. Eng. (Stockholm)* **60**, 509 (1969).
37. A. W. Gnyp, S. J. W. Price, C. C. St. Pierre, and J. Steiner, *Water Pollut. Contr.* **111** (7), 40 (1973).
38. A. W. Gnyp, S. J. W. Price, C. C. St. Pierre, and J. Steiner, *Water Pollut. Contr.*, **111** (8), 40 (1973).
38a. "Administrative and Technical Aspects of Source Sampling for Particulates," Publ. No. EPA 450/3-74-047. United States Environmental Protection Agency, Research Triangle Park, North Carolina, 1974.
39. United States Environmental Protection Agency, *Fed. Regist.* **36**, 24876 (1971).
40. Los Angeles County Air Pollution Control District, "Source Sampling Manual." Los Angeles, California, 1965.
41. American Society for Testing and Materials, "ASTM Standards of Methods of Atmospheric Sampling and Analysis," Committee D-22. ASTM, Philadelphia, Pennsylvania, 1959.
42. American Conference of Governmental Industrial Hygienists, "Threshold Limit Values for 1975." ACGIH, Cincinnati, Ohio, 1975.
43. R. W. Sexton, *Amer. Ind. Hyg. Ass., J.* **25**, 346 (1964).
44. M. Feldstein, S. Balestrieri, and D. Levaggi, *Amer. Ind. Hyg. Ass., J.* **28**, 381 (1967).
45. R. O. Blosser and H. B. Cooper, "Analytical Equipment and Monitoring Devices for Gases and Particulates," Tech. Bull. No. 35. National Council of the Paper Industry for Air and Stream Improvement, Inc., New York, New York, 1968.
46. R. S. Brief and P. A. Drinker, *AMA Arch. Ind. Health* **17**, 1 (1958).
47. R. A. Baker and R. C. Doerr, *Int. J. Air Water Pollut.* **2**, 142 (1959).
48. A. P. Altshuler, A. F. Wartburg, I. R. Cohen, and S. F. Sleva, *Int. J. Air Water Pollut.* **6**, 75 (1962).
49. A. L. VanderKolk, *Amer. Ind. Hyg. Ass., J.* **28**, 588 (1967).
50. E. Lahmann, G. Menk, and M. Moeller, *Wasser, Luft Betr.* **13**, 467 (1969).
51. E. D. Neuberger and T. J. Jurker, "Sampling System for Stack Gas," U.S. Patent 3,641,821. Calgon Corp., Pittsburgh, Pennsylvania, 1972.
52. J. E. Brown, *Chem Eng. (New York)* **75**, 164 (1968).
53. S. J. Selle and G. H. Gronhovd, *J. Eng. Power* **95**, 145 (1973).
53a. National Council of the Paper Industry for Air and Stream Improvement, Inc., *Air Qual. Inform. Tech. Bull.* **75** (1974).
54. R. W. Gerstle, S. T. Cuffe, A. A. Orning, and C. H. Schwartz, *J. Air Pollut. Contr. Ass.* **14**, 353 (1964); **15**, 59 (1965).

55. Emeryville Method Ser. 4S16/59a. Anal. Dept., Shell Develop. Co., Emeryville, California, 1959.
56. J. E. Delange, *J. Air Pollut. Contr. Ass.* **18,** 95 (1968).
57. J. F. Smith, J. A. Hultz, and A. A. Orning, *U.S., Bur. Mines, Rep. Invest.* **RI-7108** (1968).
58. J. T. Shaw, *Brit. Coal Util. Res. Ass., Mon. Bull.* **34,** 252 (1970).
59. F. R. Rehm, *J. Air Pollut. Contr. Ass.* **15,** 127 (1965).
60. R. W. Boubel and K. R. Wise, *J. Air Pollut. Contr. Ass.,* **18,** 84 (1968).
61. J. A. Fife, *Pub. Works* **102,** 56 (1971).
62. I. Tamori, K. Nabuyuki, and I. Kazunari, *J. Pollut. Contr.* **6,** 398 (1971).
63. I. Tamori, K. Nabuyuki, and I. Kazunari, *J. Pollut. Contr.* **6,** 87 (1971).
64. M. J. Bolduc and R. K. Severs, *Air Eng.* **7,** 26 (1965).
65. R. Goldstein and J. H. Elliott, "Experimental Program for the Control of Organic Emissions from Protective Coating Operations," Interim Rep. No. 5. Los Angeles County Air Pollution Control District, Los Angeles, California, 1960.
66. R. Rooth, O. K. Bockman, K. O. Hagen, and A. Takerud, "Measurement of Fluorine Emission from Aluminum Smelters—a New and More Correct Method," Paper No. A73-23. Amer. Inst. Mining, Met. Petrol. Eng., New York, New York, 1973.
67. J. O. Ivie, Paper No. 73–171, 66th Annu. Meet. Air Pollution Control Association, Pittsburgh, Pnnsylvania, 1973.
68. W. H. Ray, *Amer. Ind. Hyg. Ass., J.* **23,** 496 (1962).
69. W. P. McShane and E. Bulba, *J. Air Pollut. Contr. Ass.* **18,** 214 (1968).

15

Source Monitoring

John S. Nader

I. Introduction

The monitoring of pollutant concentration or of total mass flow of pollutant emissions from stationary sources is a major problem facing plant owners, control agencies, and equipment and instrument manufacturers. Industry needs such measurements to keep a record of process operations and emissions in response to regulatory requirements. Control officials need the information for enforcement. Equipment manufacturers require effective methods to evaluate equipment performance. Instrument manufacturers need guidelines on performance requirements for appropriate instrumentation. Instrument systems that automatically and continuously record emissions are increasingly important as tools to achieve these objectives.

In this chapter, source monitoring means measurement of emissions on a continuous and automatic basis. Automatic means that the measurement system is capable of unattended operation for extended periods of time, at least 24 hours, but preferably more than a week. Continuous means that the measurements are made either instantaneously as a function of time or periodically with a frequency great enough to meet the monitoring requirements. Thus, a measurement every 10 minutes is probably sufficient for most emission control regulations whereas instantaneous measurements may be necessary for the performance evaluation of some pollution control systems. Trade-offs need to be made between the rate at which data are obtained and their accuracy. Ideally, a monitoring system is selected to meet specific needs and is tailored to the more or less unique properties of the emissions from a particular process within a particular industry. Under these circumstances, it is necessary to take into account the specific process, the nature of the control devices, the peculiarities of the source, and how the data will be utilized.

A. Systems Concept

Source monitoring can best be treated as a system concept ideally consisting of six unit operations: sampling site selection; sample transport;

Table I System Concept of Stationary Source Measurements

Operation	*Objective*
Sampling site selection	*Representative* sampling consistent with intended interpretation of measurement
Sample transport (when applicable)	Spatial and temporal transfer of sample extract with *minimum* and/or *known effects* on sample integrity
Sample treatment (when applicable)	Physical and/or chemical conditioning of sample consistent with analytical operation *with controlled* and/or *known effects* on sample integrity
Sample analysis	Generation of qualitative and quantitative data on pollutant or parameter of interest
Data reduction and display	*Calibration* and *processing* of analog data and display of final data in format consistent with measurement objectives
Data interpretation	*Validly* relating the measurement data to the source environment within the constraints and limitations of the sampling and analytical operations

sample treatment; sample analysis; data reduction and display; and data interpretation (Table I).

Sampling site selection must provide a sample representative of the environment being measured and consistent with the intended use of the data. For example, in the measurement of particulate matter mass emission rate, it is necessary, in most cases, to traverse the stack diameter in two directions at right angles to each other to average out effects of stratification.

Sample transport and sample treatment are not always required. *In situ* methods do not extract a sample. The analysis is done on the gas stream in the stack, and there is no explicit sample transport. Extractive methods, however, withdraw a sample from the stack and transport it through probes and sampling lines. Where ambient air-type analyzers are used, sample treatment, i.e., physical and/or chemical conditioning of the sample to make it compatible with the analytical equipment may be necessary. Where sample treatment is required, it generally includes reduction in water vapor, temperature, or particulate matter content.

Sample analysis has as its immediate objective the generation of quantitative data on the pollutant or parameter of interest. Sensitivity of the method is generally not critical because of the relatively high concentrations involved in source emissions. Of more importance is its discriminatory capability, i.e., its ability to identify and quantify one pollutant in the presence of several others at about the same level of concentration.

Data reduction and display and data interpretation often go hand-in-hand. Considerable data interpretation occurs when reduced data are displayed in tabular or graphic form. Monitoring systems are available with minicomputers to simultaneously reduce concentration and flow data to emission rate data for direct display on a continuous basis.

B. Monitoring Approaches

On-site monitoring provides continuous data on one source to demonstrate its operation and maintenance, whereas remote monitoring from a central point can provide discontinuous data on several sources with adequate accuracy for such an application. Remote monitoring also can provide data without the need of access to plant facilities or interaction with plant operation.

C. Source Categories and Pollutant Monitoring Needs

In the United States installation and operation of monitoring systems have been prescribed for a number of industries (Table II) (*1–4*). Performance specifications for monitors for gaseous pollutants and visible emissions are summarized in Tables III and IV, respectively (*5, 6*).

Table II Source Emissions Requiring Continuous Monitoring by United States New Source Performance Standards (*2*)

	Pollutant				
Source	*Sulfur dioxide*	*Nitrogen oxides*	*Visible emissions*	*Carbon monoxide*	*Hydrogen sulfide*
Power plants	X	X	X		
Sulfuric acid plants	X				
Nitric acid plants		X			
Petroleum refineries	X		X	X	X
Iron and steel mills (BOF)[a]			X		
Primary copper smelters	X		X		
Primary zinc smelters	X		X		
Primary lead smelters	X		X		
Steel plants (electric arc furnaces)			X		
Ferroalloy production facilities			X		
Coal preparation plants			X		

[a] BOF: Basic oxygen furnaces.

Table III Typical Performance Specifications for Monitoring System for Pollutant Emission from a Stationary Source *(5, 6)*

Parameter[a]	*Specification*
Accuracy	≤20% of reference mean value[b]
Calibration error	≤5% of test gas value[b]
Zero drift (2 hours)	≤2% of span[b]
Zero drift (24 hours)	≤4% of span[b]
Calibration drift (2 hours)	≤2% of span[b]
Calibration drift (24 hours)	≤2½% of span[b]
Response time	≤15 minutes maximum
Operational period	168 hours[c]

[a] Parameter definitions are:

Accuracy (relative)—The degree of correctness with which the measurement system yields the value of concentration of a sample relative to the value given by a defined reference method. This accuracy is expressed in terms of error which is the difference between the paired concentration measurements. The error is expressed as a percentage of full scale span of the measurement system.

Calibration error—The difference between the pollutant concentration indicated by the measurement system and the known concentration of the test mixture.

Zero drift—The change in measurement system output over a stated period of time of normal continuous operation when the pollutant concentration at the time of the measurements is zero.

Calibration drift—The change in measurement system output over a stated period of time of normal continuous operation when the pollutant concentration at the time of the measurements is the same known up-scale value.

Response time—The time interval from a step change in pollutant concentration at the input to the measurement system to the time at which 95% of the corresponding final value is reached as displayed on the measurement system data presentation device.

Operational period—A minimum period of time over which a measurement system is expected to operate within certain performance specifications without unscheduled maintenance, repair, or adjustment.

[b] Expressed as sum of absolute mean value plus 95% confidence interval of a series of tests.

[c] The operational period value is a minimum.

In Germany, monitoring systems are prescribed for mass concentration of particulate matter emissions and for sulfur dioxide emissions *(7, 8)*, and guidelines are given for testing and selection of appropriate instrument systems *(9, 10)*. Monitoring needs for additional sources and pollutant emissions are evolving.

Table IV Typical Performance Specification for Transmissometer System for Monitoring System for Visible Emission from Stationary Source (5, 6)

Parameter[a]	*Specification*
Performance	
Calibration error	≤3% opacity[b]
Zero drift (24 hours)	≤2% opacity[b]
Calibration drift (24 hours)	≤2% opacity[b]
Response time	≤10 seconds
Operational period	168 hours[c]
Design	
Spectral response	Peak and mean response within 500 to 600 nm; less than 10% of peak response outside of 400 to 700 nm
Angle of view	Five degrees maximum (total angle)
Angle of projection	Five degrees maximum (total angle)

[a] See definitions in Table III. Other parameter definitions are:

Spectral response—The relative response of a transmissometer to radiation of a different wavelength.

Angle of view—The maximum (total) angle of radiation that is seen by the photodetector assembly of an optical transmissometer.

Angle of projection—The maximum (total) angle of radiation that is projected by the lamp assembly of an optical transmissometer.

[b] Expressed as sum of absolute mean value plus 95% confidence interval of series of tests.

[c] The operational period value is a minimum.

II. Sampling and Analysis Overview

The first source monitors were *in situ* smoke meters which measured the attenuation by the particulate matter in the stack of a light beam projected across the stack. The first monitors to measure the gaseous components of stack gases were of the extractive type utilizing ambient air-type monitors to measure the concentration of pollutant in a sample extracted from the stack. Electrooptical analytical techniques are common to all the monitoring approaches; however, the distinguishing feature of the various monitoring approaches is the location in the measuring system where the analysis occurs (Fig. 1). In "extractive" monitoring it occurs in a sample in an instrument outside the gas stream. In "*in situ*" monitoring, it occurs in the gas stream being monitored. In "long path" or "extended source" monitoring, a form of "*in situ*" monitoring, it looks at

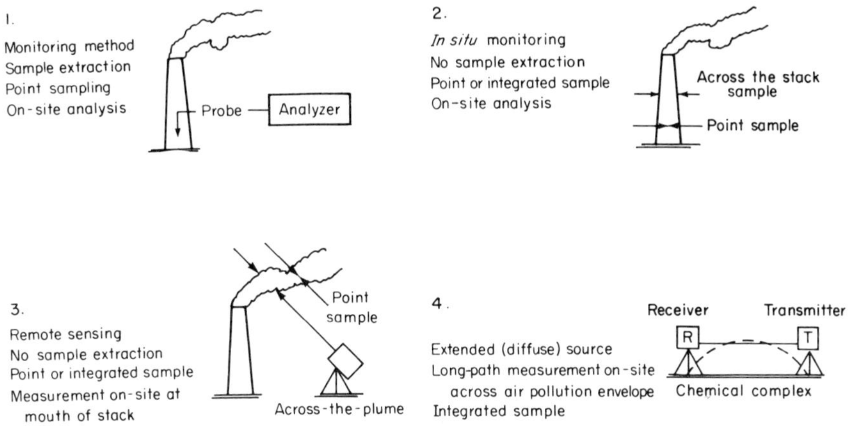

Figure 1. Sampling approaches to monitoring source emissions.

a dimensional extension of the sample stream. In "remote" monitoring, the optical effect being measured occurs in the atmosphere or in a plume at a point distant from the measuring instrument.

A. Extractive Monitoring

Extractive monitoring offers the advantage of utilizing available ambient air-type monitors by appropriate coupling to stack sources, in which case, sampling is the primary area of concern. Extractive sampling is needed where traverses across the stack must be made to obtain concentration gradient data.

The coupling of the pollutant analyzer with the stack makes the sampling interface a significant component of an extractive monitoring system. Some interface requirements are given in Table V.

Sampling errors due to probe losses can be significant (as much as 60%) as reported in data collected on emissions from a coal-fired power plant (*11*) when sampling precautions are not taken in particulate matter measurements. Preliminary experiments on incinerator stacks (*12*) comparing isokinetic sampling with fixed-flow sampling (at average stack velocities) showed surprisingly good correlation between measurements using both approaches to sampling. Further studies of various process operations will be required to delineate the capability of fixed-flow sampling as a representative sampling approach. Avoidance of the need for automatic isokinetic sampling for extractive monitoring systems will have significant impact on the simplicity, reliability, and cost of commercially available systems.

Table V Stationary Source Sampling Interface Requirements

Objective	*Method of attainment*
Maintain gas temperatures above water dew point to avoid water condensation	Heated lines, dilution
Remove water vapor	Freeze-out traps, semipermeable membranes, refrigeration
Remove particulates	Particulate collection systems, e.g., filters and cyclones
Dilute sample stream to reduce concentration of pollutants, to bring temperature down to ambient conditions, and/or to reduce water dew point temperature	Mechanical dilution systems, membrane diffusion systems
Maintain integrity of mass concentration, particulate size distribution and chemical composition	Isokinetic sampling, minimize probe and transport losses and in transit chemical transformations

B. In Situ Monitoring

In situ monitoring is usually in the gas stream in the duct leading to the stack or somewhere in the stack between its base and top. *In situ* measurement systems generally involve electrooptical techniques of analysis and utilize an optical path that can range from the total width of the stack to a small fraction of its width. The concentration measured within the optical path length is spatially averaged in the process of analysis. Therefore, where spatial integration across the concentration gradient within the stack is acceptable, the average over the optical path length will give such a value. As the optical path length is reduced to a very small percentage of the stack dimension, *in situ* monitoring approaches point sampling in extractive monitoring.

There is a subtle point of difference between spatial integration by traversing and that by *in situ* optical techniques. This involves the question of time. Traversing consists of sampling a sequence of a limited number of points in time. *In situ* methods sample all the points along the optical path length simultaneously. Concentration gradients can exist both in time and in space (stratification). The *in situ* optical technique will adequately integrate spatial gradients occurring within short time cycles. Traversing done sequentially must have a time cycle shorter than that of the concentration variation to measure as adequately on a comparable basis. However, to overcome the question of temporal gradients, traversing can be done by a multiple point sampling probe.

In situ measurements offer a decided advantage over extractive techniques in that sample transport and sample treatment are eliminated, although sample transport may be retained in those *in situ* systems that extract a sample within the stack. Many problems associated with the sampling operation are avoided by *in situ* techniques. In principle, elimination of the sampling operation potentially leads to simpler systems with more reliability and lower costs. In the case of gases, water condensation, particulate filtration, and gas adsorption in sample lines that occur in extractive sampling are avoided.

Although similar advantages apply to sampling for particulate matter, thus far only *in situ* opacity and mass concentration measurement techniques for particulate matter have been developed. This parallels the state-of-the-art in development of continuous monitors for particulate matter in ambient air which, in general, is well behind that for gases. Nonetheless, several potential advantages for *in situ* monitoring of particulate matter in stacks can be pointed out. Isokinetic sampling ceases to be of concern. There is no distortion of size distribution data, mass data, or composition data associated with size, since nonrepresentative sampling and particulate impaction no longer occur.

Stacks that are several hundreds of meters in height are in a sense similar to sample lines and can provide the opportunity for particulates to agglomerate, impact on, or diffuse to, the walls, or be reentrained in the gas stream and thus provide differences in particulate matter data depending upon the point of measurement. Stratification can also cause a difference unless an effort is made to traverse the stack width in an extractive measurement to give a spatial integration equivalent to an *in situ* across-the-stack measurement.

C. Remote Monitoring

Remote monitoring is usually of the emission exiting from the stack and generally at a point in the atmosphere near the mouth of the stack. Depending upon the industry and the nature of the emissions, it is possible that the pollutant concentrations are not the same at a sampling location in the stack and a remote location beyond the mouth of the stack.

Stacks vary in height from tens to hundreds of meters in height. The residence time and temperature difference can provide conditions for changes. Pollutant emissions include volatile substances that can exist as a gas at the higher temperatures within the stack but which may condense at the lower temperatures that exist upon exiting from the stack. Therefore, remote measurements cannot be said to be equivalent to *in situ* or *in-stack* measurements until it has been demonstrated for a given industry–pollutant combination that no physical or chemical transformation has taken place in the space and time interval between the two sampling locations.

Remote monitoring techniques are for the most part extensions of *in situ* techniques. They also involve, primarily, electrooptical analytical methods. The distinction is that in remote monitoring the instrumentation is physically distant and detached from the stack from which the emissions are being measured. Remote monitoring can range from monitoring a distant point to spatial integration over the optical path length across a plume. In remote monitoring, as in *in situ* monitoring, sample transport and sample treatment are avoided.

D. Long Path (Extended Source) Monitoring

Source emissions often involve more than a single stack and in many cases unconfined and uncontrolled points of emission. A chemical complex such as an oil refinery is an example of a multipoint source having many stacks, leaking valves, etc. Such a source can be classified as an extended source that involves an area as much as several hundred meters or more in diameter. Extended sources also include solid or liquid waste piles and open burning areas. In all these cases, the emissions constitute an air pollution envelope over the extended source. To monitor such sources an approach similar to the *in situ* across-the-stack technique can be used by scale-up to accommodate the larger dimensions of the air pollution envelope about the extended source. Thus, the *in situ* across-the-stack technique applied to the extended source with its much larger path length defines the long path technique.

The earlier discussion of the features of *in situ* measurements also apply to long path techniques. If one considers that *in situ* techniques are a first step in advancing the state-of-the-art of source emission measurement beyond extractive techniques, then long path techniques are a step beyond *in situ* techniques.

E. Electrooptical Schemes

An electrooptical system basically consists of a transmitter and a receiver. The transmitter provides a source of electromagnetic radiation that is directed through an optical sample path to the receiver. The receiver detects and analyzes the electromagnetic radiation incident upon its sensing element.

1. Single-Ended and Double-Ended Systems

When the transmitter and receiver are together on the same end of the optical sample path, they constitute a single-ended or monostatic system. When the transmitter and receiver are on opposite ends of the optical sample path, they constitute a double-ended or bistatic system. Single-ended systems are essentially more practical to service because everything is accessible at one location. Double-ended systems may present alignment problems.

2. Active and Passive Systems

Systems are designated as active or passive mainly with respect to the ultimate source of electromagnetic radiation that is detected. An active system provides the source of energy in its transmitter. In a passive system the source of energy is independent of the measurement system. The energy source may be in the surrounding environment such as scattered sunlight or the sun itself or it may be in the sampled environment such as the hot gases in the stack effluent.

Passive systems are limited in their capabilities and/or applicability by the amount and nature of the energy available. Systems utilizing sunlight are not applicable at night. Systems utilizing hot gas temperatures are not applicable to emissions at ambient temperatures. Active systems can be designed to provide the source with energy compatible with the required sensitivity and application. Passive systems are inherently simpler and thereby less expensive. In effect, the transmitter is eliminated and only a receiver is required. A passive system is by definition a single-ended system. An active system may be either a double-ended or a single-ended system.

F. Environmental Interferences

In comparing *in situ* methods with extractive techniques, the point was made that the former avoided sample transport and sample treatment. Although this reduces the complexity of the system, it imposes more burdens on the analytical phase of the operation. Since the gas stream is optically sampled and analyzed at the environmental conditions existing within the stack, the potential interferences are significant and can vary greatly for different pollutant–industry combinations. The predominant potential interferences within the stack environment common to many operations are high temperature, high concentrations of water vapor in gas or liquid phase, relatively high particulate loading, and high concentrations of pollutants other than the one being measured. Extending the *in situ* technique to remote methods introduces added potential for interferences from the ambient atmosphere. The environmental interferences in the plume for all practical purposes will be quite similar to those within the stack. However, the ambient air between the remote instrumentation and the optical path length being sampled within the plume must be accounted for. For example, in a stack plume at a remote distance of 500 m, the concentration of sulfur dioxide in the intervening air between the monitor and the stack emission is a potential interference for remote measurement of sulfur dioxide in the plume. In the situation in which atmospheric visibility is reduced to 100 m by a combination of particulate matter and water vapor loading, ambient environmental conditions present an interference problem that can conceivably render most, if not all, remote optical transmittance techniques inoperative.

G. Direct and Indirect Analysis

A direct measurement involves an analysis in which the parameter of interest is sensed directly or is uniquely related to the parameter that is sensed. For example, a gravimetric measurement of mass would be a direct measurement. An indirect measurement involves an analysis of a parameter that is not necessarily unique to the parameter of interest. For example, the optical attenuation of light by particulate matter is a factor that is relatable to the mass of particulate matter present, but additional factors enter into the relationship between them. Accordingly, an opacity measurement is potentially an indirect measurement of mass. The validity of an indirect measurement is predicated upon the uniqueness of the relationship to the measured parameter of interest. This unique relationship must be established for the particular circumstances under which the measurement is made by verification that the other factors are constant or vary in a known and predictable manner.

Verification is particularly necessary for indirect methods of measurement applied to various industrial emissions. Demonstration of correlation, for example, between opacity and mass in one industry does not necessarily translate to another industry, nor to another operation within the same industry. In each instance, correlation must be established on a case by case basis.

H. Performance Requirements and Data Utilization

The performance requirements of a monitoring system must be considered in the context of the pollutant–industry combination to which the system is to be applied and the data utilized. Accuracy of monitoring systems is determined relative to a reference method. It is possible that the absolute accuracy exceeds the relative accuracy, particularly in those cases where the reference method is a manual method lacking the measurement capability of the instrumental method.

The relative accuracy is determined by simultaneous measurements with the reference and monitoring methods for a specific pollutant emission from a specific industrial operation. The accuracy of the monitoring system is not critical in the generation of data to observe long-term trends, such as deterioration of performance of control equipment, since data on a relative basis are adequate. Start-up. shutdown, and malfunction operations generate emission levels that are factors of 2 to 10 times higher than normal emission levels.

On the other hand, when emissions are being measured to determine compliance with emission standards and other legal regulations, the accuracy of the monitoring data needs to be comparable to that of the compliance test methods themselves.

III. Particulate Matter Emissions

Particulate matter emissions are monitored for their visibility, mass concentration, size distribution, and chemical composition. Monitoring requirements have been prescribed for visible emissions (*2*) and mass concentration (*7*). Various research studies, especially those on health effects and efficiency of control systems, indicate the need for monitoring systems for size distribution and chemical composition. In some applications a monitoring system may be capable of measuring more than one parameter. A notable example is the opacity monitor which gives a direct measure of visible emissions (opacity) and may also provide an indirect measure of mass concentration.

A. Visible Emissions (Opacity)

Opacity is defined as the ratio of the light attenuated by the particulate matter emission to the incident light. The light transmittance of the emission is the ratio of light transmitted to the incident light. Emissions which transmit all incident light are invisible, have a transmittance of 100%, and an opacity of zero. Emissions which attenuate all incident light are totally opaque, have an opacity of 100% and a transmittance of zero. Monitoring systems for opacity or transmittance are referred to as transmissometers or more commonly "smoke meters." Many versions of transmissometers are commercially available as stack monitors of visible emissions (Table VI) (*13–19*).

Table VI Status of Visible Emissions (Opacity) Monitoring Techniques

Analytical scheme	*Sampling approach*[a]	*Development status*[b]	*References*
Light transmission	I	C	(*13–16*)
Light transmission	I, Impaction	C	(*17*)
Light scatter	I	P	(*18*)
Lidar	R	F	(*19*)

[a] I ≡ *in situ;* R ≡ remote.
[b] C ≡ commercial; P ≡ prototype; F ≡ feasibility.

1. In-Stack Opacity

In an in-stack transmissometer a light source transmits a collimated beam of light directly across the stack contents. The opacity measured is a function of both aerosol concentration and optical path length. When the required measurement is the opacity of the emissions at the exit of the stack, a measurement of any other location in the stack, such as at the base of the stack or in the duct leading to the stack, has to have its optical path length adjusted to the exit diameter. Figure 2 gives the relationship of effluent transmittance at the stack exit as a function of in-stack transmittance for various ratios of stack exit diameter to transmissometer optical path length (*5*).

To obtain true transmittance data the light collimation angle must be limited at the transmitter and receiver to reduce stray light scatter. In practice, a compromise is necessary between the ideal collimation angle of zero degrees and a finite angle which introduces a systematic error to the transmittance measurement. As zero degree collimation is approached, instrument construction costs, operating stability, and optical alignment problems increase. A transmissometer having a 5-deg collimating angle applied to the emissions of a pulverized-coal-fired steam generator (*20*) gave an opacity measurement about 5% low relative to the true (0-deg) value. This error would decrease for the same transmissometer applied to emissions with a particle size distribution shifted toward a smaller mass mean diameter.

When transmissometer data are used to determine compliance with a regulation based on the visibility of emissions, it is necessary to measure the opacity of the emissions in the visible range of the radiation spectrum. For other uses of the data it may be possible to operate at nonvisible wavelengths. In-stack transmittance measurements of emissions from a pulverized-coal-fired power plant were shown to vary as a function of wavelength of monochromatic light in the visible range, there being an

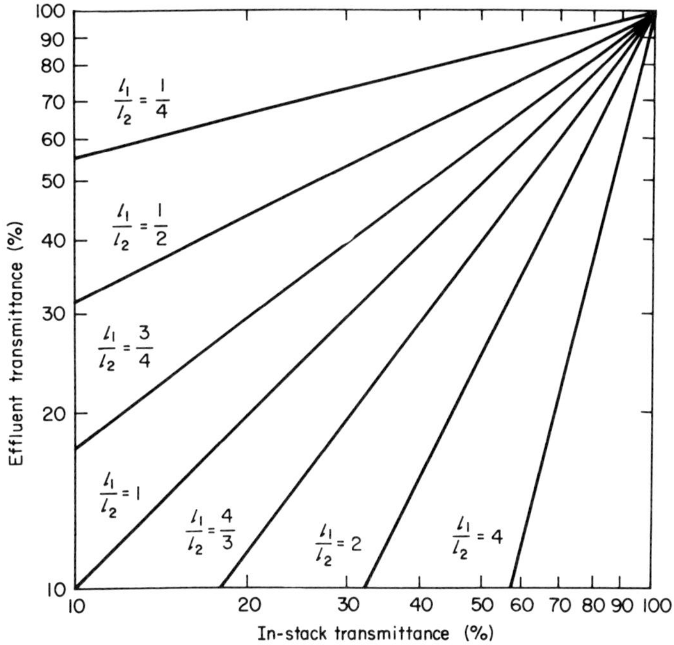

Figure 2. Effluent transmittance as a function of in-stack transmittance for various ratios of stack exit diameter (l_1) to in-stack path length (l_2).

increase in opacity of about 7% in going from red to blue light (*20*). In a study on experimental white (oil) and black (carbon) plumes, this effect was more pronounced with the transparent white plume as compared to the absorbing black plume (*21*).

Good operational stability requires maintenance of optical alignment and clean optical surfaces. Therefore, installation and design must minimize the effects of temperature, vibration, and other conditions that can cause optical misalignment and soiling. The design shown schematically in Figure 3 employs a chopped, dual-beam, optical system that automatically compensates for the effects of temperature and voltage changes and component aging (*16*). The same source is optically divided into a measuring beam and a reference beam. The measuring beam is reflected back by a corner-cube retroreflector which tolerates small alignment changes, ±0.4 deg off the optical axis (*22*). The reference beam provides automatic gain control to compensate for any changes in source intensity or detector response. Zero calibration checks can be made by inserting, in the light path outside the optical window of the source detector (transceiver), a test reflector for zero opacity, adjusted during instrument setup to simulate the retroreflector under "clean stack" conditions. The zero

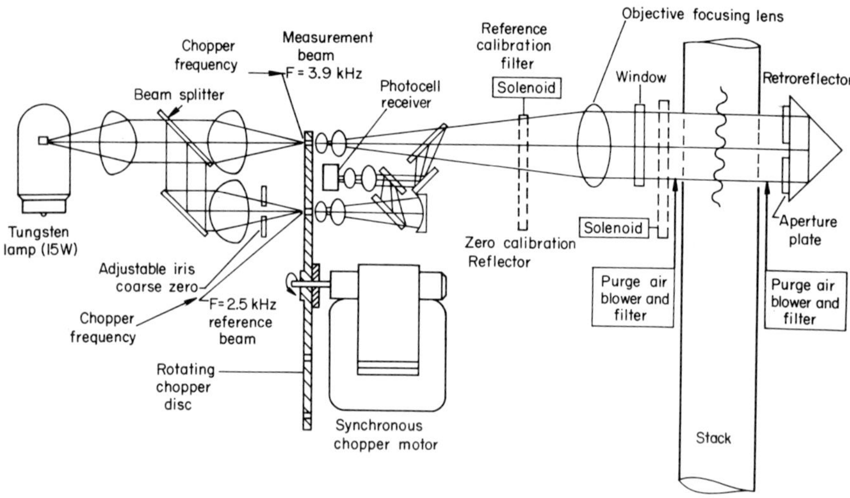

Figure 3. A single-ended opacity monitoring system.

calibration check indicates any drift due to the instrument measurement circuit or the accumulation of dust on the optical windows.

Both the transceiver and reflector units are equipped with specially constructed air purging attachments to keep the optical windows free of dirt deposits. Drift in the zero calibration check is usually an indication of the need to clean the optical window surfaces. The window purge system can provide 3 or more months of unattended operation with adequately clean windows. Frequent replacement of purge air filters may be necessary to maintain the effectiveness of the purge air system when the atmosphere surrounding the stack is very heavily laden with particulate matter.

2. Out-of-Stack Opacity

In situ measurement of solid particulate matter cannot be made when water droplets are present, such as when scrubbers are used for emission control. Water vapor does not cause trouble but water droplets do. In such cases, it is possible to use an external transmissometer. The sample stream is heated to vaporize the water droplets.

3. Plume Opacity

The *lidar* (light detection and ranging) technique (*19, 21, 23*) gives a true measure of plume opacity. Conventional lidar systems are large, complicated, expensive, and require a professionally trained operator and a large van. A relatively inexpensive lidar which lends itself to a portable and relatively simple stack monitor uses a low powered laser (argon ion) operating in a continuous wave, frequency modulated system in contrast to high energy pulsed ruby laser systems (*24*).

4. Selective Opacity

Opacity measurements can be selectively biased toward a portion of the particulate matter emission size spectrum by introducing a size fractionating step prior to the detection step (*17*). Such instruments are designed to assess the nuisance aspect of dust that is emitted from stacks and settles out nearby, causing complaints. The large particles involved are selectively sampled by an inertial technique and settle on a glass plate where they are measured by light attenuation.

Incident light that is attenuated by particulate emissions is treated as two components, scattered light and absorbed light. If the absorption coefficient is zero for the dust being measured, the scattered light is a measure of the incident light that has been attenuated. Light scatter instrumentation has been developed for ambient air monitoring of visibility and is being adapted to source monitoring of visible emissions. A prototype system has been developed as an extractive monitor (*18*). The prototype instrument has had preliminary tests in a diatomaceous earth processing plant. The sensitivity of the system is adequate to monitor a 1% opacity during most of the operation of the baghouse control system and to detect a 5% increase in opacity when a failing bag is shaken.

The extractive approach allows the monitoring of sources which otherwise present interference problems from condensed water vapor. Nevertheless, the extractive technique must overcome problems associated with the interface subsystem. These potential problems include: secondary light scatter from variations in reflections of the sampling chamber walls caused by deposits of the sampled particulates; and the need to verify the assumption that the absorption coefficient is zero for the dust being measured if the scattered light is to be equated with the attenuated light.

B. Mass Concentration

Techniques for monitoring the mass concentration of particulate matter are categorized in Table VII on the basis of the general physical principles employed (*11–16, 18, 18a, 25–40*). This shows that few mass monitoring instruments were commercially available in mid-1975, and that these primarily employed indirect methods, requiring correlation
a true measure of plume opacity. Conventional *lidar* systems are large,

Table VII Status of Mass Monitoring Techniques

Analytical scheme	*Sampling approach*[a]	*Development status*[b]	*References*
Electromechanical			
Gravimetric	E, Electrostatic precipitation	P	(*25*)
Frequency change (piezoelectric)	E, Electrostatic precipitation or impaction	P	(*26, 27*)
Impact momentum	I	F	(*28, 29*)
Optical			
Light attenuation	I	C	(*13–16*)
Light scatter	I	P,C	(*18, 18a*)
Electrical			
Ion attenuation	E	P	(*13, 30, 31*)
Charge transfer	E	C	(*32*)
Charge transfer	E, I	P	(*13, 33–36*)
Nuclear			
β Radiation attenuation	E, Filter	C	(*11, 37*)
β Radiation attenuation	E, Filter	P	(*12, 38–40*)

[a] E ≡ extractive; I ≡ *in situ*.
[b] C ≡ commercial; P ≡ prototype; F ≡ feasibility.

with direct mass measurements to validate their applicability to specific stack emissions.

1. Electromechanical

Electromechanical techniques try to directly measure the mass of the particulate matter.

a. Gravimetric. This method consists of intermittent sampling and weighing. Weighing is based on a balance of opposing torques about a pivot point. The torque due to the collected mass is opposed by mass or equivalent forces generated electromagnetically. A prototype system using electrostatic precipitation for collection and an electromagnet for the restoring torque has been developed and tested in a stack environment (*25*). The measurement must be conducted at a high enough temperature to avoid condensation. Sensitivity is limited to about 1 to 10 mg by vibration in the vicinity of the stack. The particle collection surface requires cleaning or replacement. As an extractive technique, the probe introduces errors of particulate losses in sample transport.

b. PIEZOELECTRIC. When a crystal is made part of an electronic oscillating circuit, the oscillating frequency of the circuit is determined by the natural vibrational frequency of the crystal. The deposition and adhesion of particulate matter on the crystal changes its vibrational frequency. The magnitude of this frequency change is directly proportional to the mass of the adhering material. Frequency changes can be measured electronically to one part in 10^7.

Electrostatic precipitation and impaction have been used to deposit particulate matter onto the crystal surface. Application of this method to a stack source involves extractive sampling and is therefore subject to particulate matter entrainment in the sampling probe. Although electrostatic precipitation has the potential for collecting particles of all sizes, in practice there is loss of efficiency for particles which are inadequately charged for the sampling conditions. When particle collection is by impaction, collection efficiency drops off significantly for particle sizes much below the impactor cutoff size.

An error is introduced in the weighing by the particles which fall off the vibrating crystal surface. Adhesion of particles to the crystal surface involves an interaction of molecular attraction forces, electrostatic forces, surface tension forces, and inertial forces. Particle size becomes a critical parameter in this interplay of forces such that particles above a critical size do not adhere to the crystal and thereby affect its vibrational frequency.

Commercial devices of this type have been developed for ambient air monitoring (*26, 27*), but their applicability to emissions from stacks remains to be evaluated.

c. IMPACT MOMENTUM. In this method, particles are impacted onto a thin lamellar plate (lamella) that is part of a capacitance in an electronic circuit. The momentum of impaction causes a movement of the lamella and a change in capacitance that is sensed and measured. Since the airborne particles are initially accelerated to a constant velocity, the measured capacitance change is independent of the particle velocity and dependent on mass only. Prototype instruments have been built to demonstrate the feasibility of the method (*28, 29*) to monitor mass concentrations above 15 mg/m^3 of particles in the size range above 1 μm. Parameters requiring further study include aerodynamic stability of the lamella, correlation with mass, minimum mass detection sensitivity and sensitivity for particles in the submicrometer range, operation at high stack temperature, sampling probe losses, and disruptive accumulation of particulate matter on the lamella. Results of the study of aerodynamic stability indicate that the technique does not lend itself to a practical field device (*27*).

2. Optical

The light attenuation or transmittance measurement instruments commercially available as monitors of visible emissions (Table VI) can be used to monitor indirectly the mass concentration of particulate emissions. The mass concentration of the particulate matter must be shown to correlate with the measured optical property and the calibration of the instrument system must be shown to hold constant for the emissions of a stack source under all operating conditions of the plant. The physical characteristics of the particulate emissions affecting the calibration include particle shape, size, density, index of refraction, and density. These physical properties must not change if the calibration is to remain constant. In many source emissions these physical properties may not change significantly under normal operating conditions.

a. LIGHT ATTENUATION. If, for a specific application, a valid correlation can be made between light attenuation and mass concentration, a light attenuation instrument can be calibrated in terms of mass concentration. Studies have been conducted to evaluate this correlation for emissions of particulate matter from a variety of industries. The sources evaluated included coal-fired power plants (*13, 27, 30, 41*), lignite-fired power plants (*25*), cement plants (*42*), and pulp and paper plants (*43*). Other industries under study are petroleum refineries, asphalt-concrete plants, and sewage sludge incinerators (*15*). Generally, in such evaluations mass concentration is determined gravimetrically from filter samples taken over a time interval concurrent with light attenuation measurements.

Data reported on emissions from a normally operating coal-fired power plant showed correlation coefficients for two different instruments of 0.63 and 0.87 for about 300 measurements; deviations at a concentration of 150 mg/m^3 were ±68 and ±39%, respectively, at the 95% confidence interval (*30*). The poorer correlation was attributed to window contamination. Additional measurements were made under minimum load operation, and operation with and without soot blowing. The results showed different correlation curves for each operating condition; one instrument performed relatively better than the other. Other data (*13*) indicated trouble-free continuous operation throughout a test period of 6 to

10 weeks without the need to clean the optical surfaces. Linear correlation was found for a mass concentration range from 300 to 1000 mg/m^3. Correlation and performance evaluations were conducted over one 3-month interval and two 2-month intervals (January–March, July–August, October–November) representing different seasons of power plant operation (*14*). Emissions were increased at various times by cutting off one or more electrostatic precipitator stages. Correlation curves were essentially the same for the three different time periods with coefficients of 0.93, 0.98, and 0.99. Size distribution measurements were made with full electrostatic precipitator controls on, and at times when controls were partially cut back. No significant size differences were observed. Mass concentration ranged from 55 to 360 mg/m^3. The composite correlation curve for the data for all three time intervals has little scatter of points and a correlation coefficient of 0.97 (Fig. 4). The composite data included mass concentrations from 0.05 to 0.35 g/m^3 coresponding to an opacity range from 3 to about 25%, respecively. No problems with window contamination were encountered with continuous operation spanning the 1-year period. Size distribution measurements were made on the emissions at various operating conditions including cutback of the electrostatic precipitator stages, and no significant changes were observed in the mean size or the size distribution.

Comparative measurements at lignite-fired power plants (*25*) gave results initially similar to those at coal-fired plants. Extended operation,

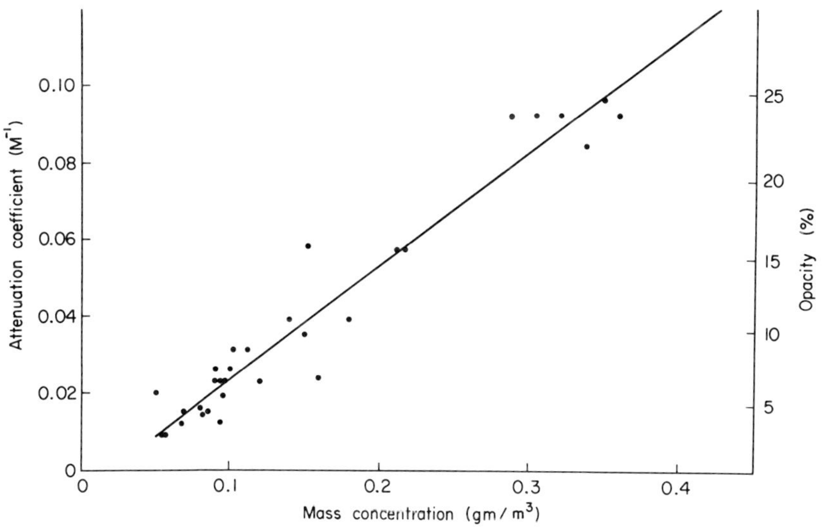

Figure 4. Correlation data between opacity and mass measurements of particulate matter in emissions from a coal-burning power plant.

however, showed that calibration at the beginning and end of a 3-month test period yielded regression lines of differing slope and parallel off-set. The sampling approach had an effect on the scatter of points about the regression line. Very little scatter was shown for mass concentration measured along the axis of the optical beam. Scatter for mass concentration measured on a grid pattern to represent the entire duct cross section was very great. The conclusion on reliability of performance was that monitors on lignite-fired plants were subject to dirtying of the optical surfaces within a short period of time; weekly maintenance was deemed necessary if calibration was expected to hold.

Comparative measurements were conducted on emissions from the clean gas duct of a rotary cement kiln with a suspension gas preheater and a subsequent rotary dryer (*42*). Correlation between light attenuation and mass concentration was studied for operation conditions that included normal operations, shutdown and start-up, and other operations with deviations. A total of 57 comparative values for all the operations gave a linear correlation with a coefficient of 0.98 for a dust loading that ranged from 10 to 450 mg/m^3 and a deviation of correlation of $\pm 11\%$ at a mass concentration of 150 mg/m^3. The correlations for the individual operations were not significantly different from the correlation for the data from the combined operations, ranging from 99 to 100%. After 3 months of continuous operation with servicing according to operating instructions, the monitors were removed and inspected. Cleaning of the main filter was not yet necessary. The optical boundary surfaces and the purging air attachments of transceiver and reflector (Fig. 3) were also practically clean and free of dust. Performance results indicated that the zero point and the sensitivity remained constant during the 3-month interval.

An interesting study (*43*) was conducted on the relationship between plume opacity and particle mass concentration measured downstream from electrostatic precipitators on a kraft pulp mill recovery furnace. The results of 18 tests of simultaneous measurements of particle mass concentration, size distribution, and light transmittance were analyzed. In the 10-day period of the tests random sampling of various operating conditions included conditions with both electrostatic precipitators on, with one precipitator on, and with both precipitators off. The range of light transmittance was from 83 to 41% corresponding to mass concentrations of about 0.7 to 5 gm/m^3, respectively. The correlation coefficient was 0.99. The experimental correlation was also found to be in good agreement with a theoretical prediction based on Mie light scattering theory and measured particle-size distribution. A linear multiple regression analysis of the measured data showed that variation in particle-size distribution from 0.34- to 9.6-μm mass mean diameter had little effect on the correlation. This result was in agreement with predictions based on theoretical calculations. It is important to note, however, that the transmittance was measured using a device with a light source having an average wavelength of 1.2 μm. Theoretical calculations show that the operating wavelength has an important role, relative to the geometric mass mean particle size and the geometric standard deviation of the size distribution, in whether or not a good correlation can be expected. Thus, it is possible, assuming other conditions are favorable, that the proper choice of operating wavelength can optimize the correlation for a given industry having particulate emissions with adequately characterized physical properties.

b. Light Scattering. Instrumentation has been developed to measure only the light scattered by the particulate matter in an airstream (*18, 18a*). This approach is claimed to be a more sensitive indirect measure of mass of particulate matter than the light attenuation approach. Nevertheless, it suffers the same constraints on its validity as the light attenuation technique. Experimental data on correlation between mass and light scatter on selected source emissions need to be obtained to determine the applicability and limitations of the method.

3. Electrical

The ability of particles to capture ions and to transfer accumulated charges is a physical characteristic that is measurable with relatively simple electronic instrumentation. The electrical characteristics of particles depend on their size, shape, resistivity, dielectric constant, and composition. Instruments utilizing these characteristics (*30*) must be calibrated for each specific application, and the constancy of this calibration must be validated periodically. Their theory of operation has not been adequately developed, but limited empirical data are available to judge their performance.

a. Ion Attenuation. Ion attenuation consists of reduction in ion concentration as particles pass through a stream of ions. Assuming constant particle-size distribution and composition, the attenuation of ion concentration is a function of particle concentration and particle mean size. Quartz dust with a mean particle size of 0.7 μm shows a linear relationship between the attenuation ratio and the mass concentration above 1 mg/m^3 (*31*). Extrapolation of these limited data indicate almost complete attenuation of the ion current at 100 mg/m^3. This indicates the need for dilution with ion-free air for application to source emission levels.

Ion current appears to be very sensitive to particles below 0.1 μm. The technique appears to have a response that is a function of surface area. Consequently, the method has more sensitivity to a mass of particulate matter consisting of small particles than to the same mass consisting of large particles.

b. Ion Charge Transfer. This technique consists of the transfer of electrical charge between the particles in the sample stream and a sensing element when they come into physical contact with each other. The transfer of charge by physical contact of dissimilar materials is identified with the "triboelectric effect." Charge transfer is measured as flow of electrical current from the sensor element by appropriate electronic circuiting. The composition and surface properties of the particles and the sensor, and the flow characteristics of the sample stream, determine the probability and nature of physical contact between the particles and the sensor element, hence the electrical signal that is generated and the correlation of this signal with particulate mass flow. These interdependencies are poorly understood.

A German instrument (Konitest, no longer manufactured) sampled the particles from a stack and introduced a swirling motion to the airstream in the sensor by means of a tangential inlet duct (*13*). Centrifugal force threw the particles against the sensor tube wall where the electrical charge is generated. A later Konitest design used a venturi nozzle instead of a cyclone as the sensing element (*33*). This instrument gave good correlation with mass concentration when applied to a power plant stack. Over a range of 0.1 to 2.2 gm/m^3, its response was linear, with a signal output of 7.4×10^{-9} A per 100 mg/m^3 of dust concentration. A test program evaluating the Konitest on the emission of a coal-fired power plant (*34*), consisting of 312 measurements, gave a correlation coefficient of 92% and a deviation, at the 95% confidence level, of ±33% at a mass concentration of 150 mg/m^3. Potential problems of the application of the Konitest to emissions from coal- or oil-fired combustion facilities are: long-term contamination of the sensing tube; sensitivity to submicrometer particles which do not contribute proportionately to mass; change in calibration due to changes in particle composition.

A contact charge transfer instrument which senses particles passing through a venturi nozzle with a conical metal probe in its throat (*44*) has been developed. A transfer of charge takes place when particles hit the probe at velocities above 100 m/second. A permanent installation featuring automatic isokinetic sampling is available as well as a portable unit (*32*). The portable system has been utilized in the laboratory and in industrial facilities including coal, oil, and incinerator flyash, metal oxides and auto and jet engine exhausts. The monitor has a range of five decades (10^5 to 1). In a coal-burning combustion source, this range corresponds to a mass concentration range of 0.001 to 100 gm/m^3. Its lower limit on sensitivity has not been fully explored. Recalibration of the instrument is necessary if substantial changes in particle size distribution, particularly in the submicrometer size, occurs. Its response to changes in particulate matter composition can vary sufficiently to require recalibration.

Another instrument (*35*) utilizes two stages of fluidized beds (*45*) as a means of minimizing variations in the charge transfer process. This instrument can be located in the gas stream as an *in situ* device. An air injector pulls the sample air through the device and ejects it back into the stack. The electronic detection circuit is outside the stack and connected to the sensor element by an electrical cable. The sensor element is the container wall of the second-stage fluidized bed. The action of the fluidized bed makes the device self-cleaning. Preliminary data on its operation in the stack of an oil-fired boiler indicated no contamination problem for continuous operation in excess of 6 weeks (*36*). Sensitivity is of the order of 1 mg/m^3 at which the signal output approaches the system's electronic background noise. Further development is needed to delineate the capability of the method in specific applications and the correlation of output signal with mass concentration. Application of the initial prototype design to a unique source which produced highly charged particles showed that a significantly high initial charge on the particles can affect a calibration that may have been obtained under normal charging conditions.

4. Nuclear

When β particles impinge on matter, some are absorbed, some are scattered, and some are transmitted. The reduction in the incident beam intensity is known as β radiation attenuation. The attenuation is primarily a function of the β particle energy, the mass concentration of the attenuating matter in the radiation path, and the electron density of the matter. The electron density is the ratio of the atomic number (number of electrons per molecule) to the atomic weight (mass of the molecule). This ratio is essentially the same (between 0.45 and 0.50) for most of the elements below the very heavy ones. (Hydrogen is an exception but its presence does not normally cause a significant error.) Consequently, β radiation attenuation is an indirect method of mass measurement which is less subject to interferences by particle size, composition, shape, or other physical properties than any other indirect measurement methods.

Both gases and particulate matter attenuate β radiation. Therefore, *in situ* application of β attenuation as a particulate matter measuring technique is not feasible because the mass of gas molecules in a gas stream far outweighs the mass of the particulate matter, and β attenuation methods cannot discriminate between the two. Consequently, implementation of this technique requires an extractive system in which a particulate sample is collected by one of the several available particulate matter concentration methods, and subsequently analyzed.

The feasibility of β radiation attenuation as an analytical technique has been demonstrated for both ambient air (*38*) and stationary source particulate matter (*12*, *37*). Beta radiation attenuation instruments are commercially available. Only a few are complete systems with sampling interface components for application to stack emissions. All use a filter medium for particle collection. Potential sources of sampling and analytical errors are loss of sample in the probe, varying filter thicknesses, uncertain flow measurement, losses in filter efficiency, statistics of radiation counting, and deviation from the Beer–Lambert law for attenuation.

Results reported on applications of commercially available β-gauge monitors to various source emissions present conflicting conclusions on the accuracy of measurement.

Data have been reported (*37*) for accuracy errors of about $\pm 10\%$ in the analytical part of the system (exclusive of sampling errors, i.e., probe losses, etc.) These errors would include varying filter thicknesses, uncertain flow measurement, losses in filter efficiency, statistics of radiation counting, and deviation from the Beer–Lambert law for attenuation.

Two first-generation commercial β radiation attenuation instruments were evaluated in a coal-fired power plant in comparison with manual gravimetric measurements (*30*). Results showed correlation coefficients of 0.94 and 0.81 for 29 and 13 measurements, respectively. The deviations at the 95% confidence level were ± 22 and $\pm 31\%$ for a mass concentration value of 150 mg/m^3.

Data have been reported on experience with the β-gauge monitor on a coal-fired and lignite-fired power plant, a cement plant, and a ferroalloy plant (*38a*). Accuracy on coal-fired emissions was stated to be within 30%. The report stated that no dust deposits could be detected in the probes in the measurement of the emissions from the lignite-fired power plant. In this case the accuracy was within 16%. No information was given on how the gravimetric measurements were made. It is conceivable that the gravimetric measurements did not include cleaning but were limited to a filter catch. Under such circumstances, a favorable comparison is not surprising.

Other data on application of β-gauge monitors to a coal-fired power plant (*11*), a cement plant, and a ferro-alloy plant, and an oil-fired power plant (*38b*) have indicated β-guage probe losses ranging from 30 to 86% of the total catch determined by the EPA reference method 5 for particulate matter emissions (*2*). The data showed an unpredictable and variable loss that tended to invalidate the assumption that the buildup of particulate in the probe came to an equilibrium condition or that the probes remained clean. Correlation between the EPA reference method and the β-gauge monitor data varied from 0.14 to 0.80.

The various collection techniques which have been used in prototype systems include filters, electrostatic precipitators (*39*), impactors (*40*), cyclones (*37*), and cyclone–filter combinations (*12*). All these methods appear to operate adequately with the β radiation sensor in their respective system.

Two prototype β radiation attenuation instruments designed and developed for both auto exhaust and stack emissions were briefly evaluated on a coal-burning power plant (*40*). The reference measurement was a parallel filter run simultaneously with the monitor and gravimetrically analyzed. The average ratio of the monitor data to the corresponding filter data was 0.98 for one monitor and 0.51 for the other monitor.

Only one stationary source instrument system has been developed to interface with a stack and sample isokinetically on an automatic basis. This was a prototype system designed specifically for application to incinerator flyash emissions (*12*). Field test data on several municipal incinerators showed the need for improved material components in the sampling system to minimize corrosion and increase long-term reliability.

The poor accuracy as a result of probe losses and its impact on correlation indicate the need for further development of an improved sampling interface to overcome the problems associated with transport of particulate matter in extractive monitors. A boundary layer interface system for extractive sampling of particulate emissions has been developed and will be evaluated with the β-gauge monitor applied to various source emissions (*38c*).

C. Size Distribution

Particles in source emissions encompass a broad range of sizes from less than 0.001 to 100 μm or greater—at least five size decades. No single method is foreseen that will be capable of measuring the entire size range. No commercial system is available to continuously monitor size distribution of particulate matter emissions from source stacks. Monitors have been developed for sizing ambient air particulate matter but these do not as yet have interfaces designed to allow their use for stack emission measurements (*46–52*). A prototype system is being developed with capability of both ambient air and source emissions measurement (*53*). A feasibility study of an electrooptical particle sizing technique is being conducted to execute a measurement by optically viewing the gas stream through a window without sampling (*54*).

1. Size Distribution Parameters

There are three distinct particle parameters with which the size distribution is to be associated, namely, number (N), surface (S), and mass concentration (M). Each of these particle parameters can be applied to measurement of all particles independent of their composition or of only particles having a specified elemental or molecular composition. For example, one can wish to measure the number concentration versus size distribution of all particles in a sample; the mass concentration versus size distribution of only the particles containing sulfate; or the surface area concentration versus size distribution of only the particles containing lead.

The size of the particle must also be clearly identified because several definitions exist. Monitoring systems are usually designed to measure the equivalent aerodynamic size of a particle (d_e), which is assumed in this discussion unless otherwise specified.

2. Sequential Size Classifying and Sensing

Many potential size distribution measurement systems for particulate matter consist of two separate operations executed by two independent system components: a classifying sampler that separates the particles into various size fractions and a concentration sensor that measures the amount of particles in each size fraction. Size classifying techniques and concentration sensors are discussed in other chapters. It is possible to develop a number of systems based on various combinations of these size classifiers and particle sensors. Table VIII identifies a number of potential size distribution monitoring techniques. Operable size ranges (Table VIII) are associated with the size-fraction sampling technique. No one technique spans more than about two decades of size. All three techniques are required to cover the size range from 0.001 to 100 μm. The referenced monitoring techniques have been developed as prototype or commercial systems for ambient air measurements, but are potentially capable of measuring source emissions if properly interfaced as extractive systems. The critical performance requirements of the interface are adequate dilution to avoid overloading the analyzer and maintenance of size distribu-

Table VIII Operable Size Ranges of Potential Size Distribution Monitoring Techniques with Sequential Sampling and Detection Operations

Concentration detection scheme	*Size-fraction sampling range*[a]			*References*
	Inertial	*Electrical*	*Diffusional*	
Electromechanical	0.2–20 M, P[b]			*(48)*
Optical	0.2–100 N, P[b]	0.005–0.6	0.001–0.05	*(49)*
Electrical	0.2–100	0.005–0.6 S, C[b]	0.005–0.05	*(50)*
Nuclear	0.2–100 M, P[b]			*(53)*

[a] Equivalent aerodynamic particle diameter in micrometers and densities of 1 gm/cm^3.
[b] M is mass concentration; N is number concentration; S is surface concentration; C is commercially available; P refers to prototype development.

tion in the process of extraction and transport. Probe losses that may be insignificant for gross mass measurement can be significant in biasing size distribution data, i.e., a probe that fails to transport most of the very small particles will have little effect on gross mass measurement.

3. Concurrent Size Classifying and Sensing

Electrooptical techniques in which electromagnetic radiation interacts with particles generate data simultaneously on concentration and size in one system. The size and concentration information inherent in the signal initially sensed in the interaction is sorted out by data reduction electronics in the system (Table IX) *(47, 51, 52, 54, 55)*.

A commercial size analyzer using laser light scattering has been developed for application to *in situ* monitoring of stack emissions, primarily for aqueous aerosol particle size distribution of emissions from large industrial cooling towers *(54)*. The system is claimed to be adjustable to cover a wide range of particle sizes from submicrometer to about 1000 μm. As with all the potential techniques using concurrent size-classifying and sensing operations, considerable research and development needs to be conducted to assess the applicability of each specifically executed system to each selected industry.

D. Chemical Composition

There are no commercial monitors available for measuring the composition of the particulate matter in stack emissions. X-ray fluorescence

Table IX Potential Electrooptical Size Monitoring Techniques with Simultaneous Sampling and Detection Operations[a]

Technique	*Size range*[b]	*Status*[c]	*Remarks or reference*
Lightscatter pulse analysis	0.2–10	C	(*47, 51*)
Lightscatter pulse analysis	>1	F	(*55*), *In situ*, no sampling
Angular lightscatter pattern analysis	0.2–2.	P	(*52*)
Angular lightscatter pattern analysis	0.5–1000	C	(*54*), *In situ*, no sampling

[a] Unless indicated otherwise the status of these techniques is given for ambient air measurement. The methods will require appropriate interface systems for extractive sampling from stack emissions.
[b] Physical diameter in micrometers equivalent to physical diameter of calibration particles.
[c] See definitions, Table VII, footnote *b*.

of particulate matter in source emission samples has been demonstrated to be a viable analytical method (*56*). Since no sample preparation is necessary for particulate matter collected on a suitable filtering medium, the development of a continuous X-ray fluorescence monitor for elemental composition of stack emissions extracted from a stack and collected on a filter has great promise.

In situ measurement of molecular composition has been evaluated in a feasibility study of Raman scattering and fluorescence techniques (*57*). It was concluded from calculations and preliminary laboratory experiments that both fluorescence and Raman *in situ* monitoring systems can yield useful information about the composition of a particulate stream.

IV. Gas Emissions

The development of gas monitoring systems and their commercial availability has been far more advanced than particulate matter monitoring systems. Process gas monitors have been used for years by the various process industries, e.g., iron and steel producers, oil refineries, power plants, and chemical manufacturers. These monitors have been of the extractive type. The first approach for many instrument manufacturers has been to try to adapt their process gas or ambient air pollutant monitors for the monitoring of waste gas emissions. In most cases, this has been unsuccessful because these monitors are not equipped to handle the hostile environment of stack sources. Consequently, process gas or ambient air monitors require a practical gas sampling interface to couple the monitor to the stack source. This requirement has received little or no attention on the part of instrument manufacturers. Furthermore, advanced electro-optical methods using the *in situ* approach or across-the-stack techniques have been developed only recently and do not have a historical background of performance capability.

In the transfer of measurement technology from process gas and ambient air monitoring to waste gas emissions, the attention and experience of instrument manufacturers and developers has been focused on those pollutants and industries identified by emission regulations (*2, 10*). Since the development of these regulations has been an evolutionary process, initial efforts have been directed at those pollutants whose emissions have the most significant impact on ambient air quality. This explains why monitors for emissions of oxides of sulfur, nitrogen, and carbon are in a more advanced state of development than those for other pollutants.

Gas monitoring in the process industry and the ambient air has resulted in the development of many analytical methods. These are well documented and have been discussed extensively with regard to their applicability to, and utilization in, monitors for stationary source emissions (*58–75*) (Table X). There are, for example, as many as ten manufacturers of sulfur dioxide stationary source monitors using the nondispersive absorption (spectroscopic) technique. Some of them do not provide a complete monitoring system. Many require the purchaser to provide his own interface sampling system adapted to his particular source.

A. Sulfur Compounds

Instrument manufacturers concentrated their initial efforts on the development of sulfur dioxide monitoring systems for application to power plants, acid plants, and smelters. Additional sulfur compounds requiring monitoring include: sulfur trioxide, sulfuric acid mist, hydrogen sulfide, mercaptans, total reduced sulfur, and total sulfur.

1. Sulfur Dioxide

There are prototype sulfur dioxide monitoring instruments operating in all modes, extractive, *in situ,* and remote sensing. Combustion sources fired with fossil fuel have received the most attention in the development of commercially available monitoring systems.

Southern Services had developed a manifold interface system for use with commercially available analyzers, and have evaluated several inter-

Table X Commercially Available Gas Emissions Monitoring Systems

Analytical scheme	*Sampling approach*[a]	*Pollutant capability*[b]	*Remarks or reference*
Chemielectromagnetic			
Colorimetry	E	SO_2, NO_x, H_2S, TS	*(65)* Wet chemical method
Chemiluminescent	E	NO_x	*(70)*
Electromagnetic (electrooptical)			*(71)* Potentially applicable to E, I, and R monitoring systems
Flame photometry	E	SO_2, H_2S, TRS, TS	Specificity by sample conditioning *(59)*
Nondispersive infrared	E	SO_2, NO, CO, HC, CO_2	*(71)*
Nondispersive, ultraviolet and visible	E	SO_2, NO_x, NH_3, H_2S	*(71)*
Dispersive, infrared and ultraviolet	I	SO_2, NO, CO, CO_2, HC, H_2S	*(71)*
Dispersive, infrared and ultraviolet	E	SO_2, CO, CO_2, HC, NO	*(71)*
Correlation, ultraviolet	I	SO_2	*(71, 72)*
Correlation, ultraviolet and visible	R	SO_2, NO_2	*(73)*
Derivative, ultraviolet	I	SO_2, NO_2, NO, O_2, NH_3, CO	*(71, 74)*
Fluorescence, ultraviolet	E	SO_2	*(75)*
Electrical			
Conductivity	E	SO_2, NH_3, HCl	*(65)* Wet chemical method, specificity by sample conditioning
Coulometry	E	SO_2, H_2S, TRS	*(65)* Wet chemical method, specificity by sample conditioning
Electrochemical	E, I	SO_2, NO_x, CO, H_2S, O_2	*(71, 116a, 116b)*
Flame ionization	E	HC	*(71)*
Thermal			
Oxidation	E	CO	*(71)*
Conductivity	E	SO_2, NH_3, CO_2	*(59)*
Hybrid			
Gas chromatography, flame ionization	E	CO, HC	*(75a)*
Gas chromatography, flame photometry	E	Sulfur compounds	*(75b)*

[a] E ≡ extractive; I ≡ *in situ;* R ≡ remote.

[b] HC ≡ hydrocarbons; TRS ≡ total reduced sulfur; TS ≡ total sulfur; NO_x ≡ total oxides of nitrogen, but system may be specific for NO, NO_2, or both.

face units furnished by various manufacturers. Their conclusion was that commercial sample conditioning units were inadequate but that the unit of their own design has given satisfactory performance (*76*).

The United States Environmental Protection Agency (*77*) has shown that neither the *in situ* nor the extractive sampling approach offers a distinct advantage in reliability and accuracy and that particulate matter is an analytical interference for some *in situ* systems.

a. EXTRACTIVE. Continuous monitoring requires long-term operation with minimum maintenance, short response time, low zero and calibration drift. An evaluation of sulfur dioxide monitors commercially available in 1970, on emissions from a coal-fired power plant, included nondispersive infrared (NDIR), nondispersive ultraviolet (NDUV), conductometric, coulometric, and electrochemical monitors, concluded that the instruments then available were capable of meeting emission monitoring needs (*78*). At that time, no performance criteria had been prescribed by the United States Environmental Protection Agency (USEPA). The data generated helped provide the technical basis for the development of such USEPA criteria (*5*). The tests showed the need for improved particulate and water vapor removal interfaces, and improved zero and calibration drift stability. No particular measurement technique was conclusively shown to be superior to any other. Continual instrument modifications required during the study indicated the need for more field experience. Systems having a prior history of operating experience of 4 or more years on stack emissions of sulfur dioxide gave longer term reliability.

The conclusion of the adequacy of commercially available sulfur dioxide monitors for power plant emissions supported the findings of an earlier German study by Dr. Thoenes of TUV ("Technische Überwachunge-Vereine") at Essen (*58*) who evaluated two commercially available monitors, an NDIR and a conductometric. These monitors were found to meet performance requirements prescribed by German regulations (*79*).

A sampling interface must be specific for the source/analyzer combination to which it will be applied and the performance requirements of the monitoring system (Fig. 5), and must have the following characteristics (*80*):

1. The systems must be able to be successfully calibrated by injection of zero and span gases at the probe tip.

2. There must be negligible sulfur dioxide absorption in sample lines. Teflon, 316ss, or polypropylene are recommended. There is small absorption in nylon lines. Polyethylene and Tygon lines are to be avoided because of their large absorption. Sample lines of $\frac{1}{4}$ in. (0.64 cm) o.d. tubing

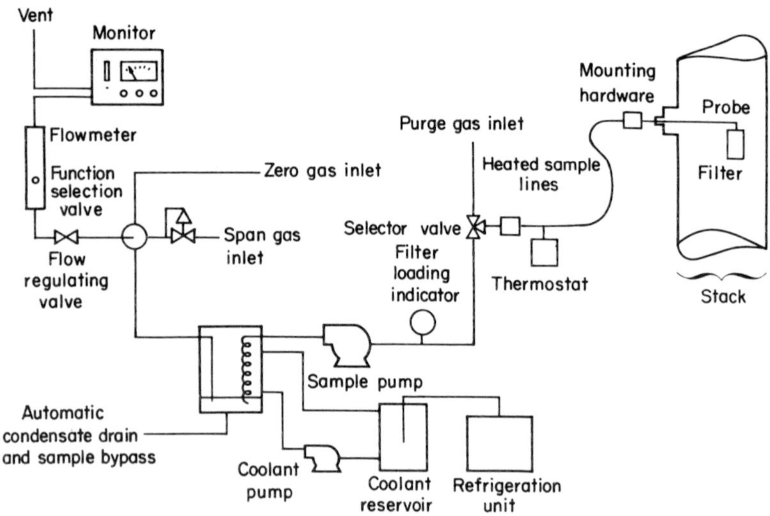

Figure 5. A typical interface system for an extractive monitor.

are recommended for typical analyzer flow rates of the order of 1 standard liter per minute. This produces a response lag of about $\frac{1}{2}$ minute for a 100 ft (30 m) length.

3. Coarse filtration of particulate at the stack must be provided. Either external or internal filters can be used. An internal filter is recommended when calibration gases are to be injected at the probe tip.

4. A refrigerated condenser must be provided for moisture removal or control for NDIR instruments; a condensate trap for the other monitors. Loss of sulfur dioxide by absorption in condensate is negligible, both by calculation and experiment.

Dilution is used as a sample preconditioning procedure in interface systems for one or more of the following reasons: to reduce the high concentration of the pollutant emission sample for compatibility with the detection range of the analyzer, especially ambient air types; to cool the hot gases but maintain their temperature above the water vapor dew point so that condensation does not take place; and to reduce high concentrations of particulate matter.

Dilution systems are usually external to the stack and require filters to clean the diluting air. One *in situ* dilution system uses a specially constructed sampling probe as a controlled-flow/orifice combination to give dilution ratios which can be varied from 2 to 20 (*81*). The novel feature of this probe as contrasted with conventional external dilution methods (*82*) is the elimination of particulate filters, heated sample lines, and

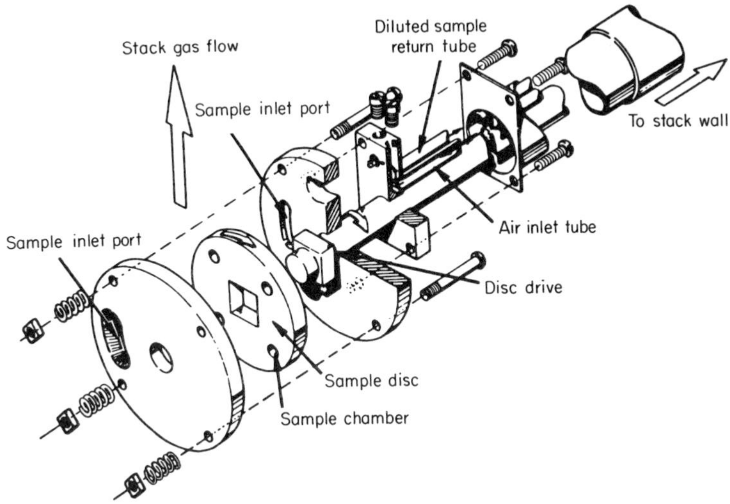

Figure 6. An *in situ* dilution system for interfacing with an extractive monitor (*83*). [Reprinted with permission from J. B. Homolya, *Chem. Technol.* **4**, 426(1974). Copyright by the American Chemical Society.]

condensation devices. Another *in situ* dilution device, which extends the possible dilution ratio to 1800 (Fig. 6), incorporates a rotating perforated disc placed in the stack gas atmosphere (*83*). Small volumes of gas extracted by the disc are mixed with a metered stream of diluent air. The range of attainable dilution ratios is adjustable by changing either the rotational speed or the sampling disc or the number and volume of disc perforations. Its reliability under stack environmental conditions has been demonstrated on a coal-fired power plant (*84*).

Permeation sampling devices (*85, 86*) utilize a specially chosen membrane that allows the selective diffusion of the gas of interest through the membrane while restricting the flow of other, potentially interfering, gases. Dilution is achieved by a stream of clean air or nitrogen that sweeps the diffused gas to the analyzer. The membrane provides complete particulate matter removal.

The emission control system operating upstream of the monitoring system may affect the operation of the monitoring system. For example, a wet scrubber can appreciably lower stack gas temperatures at the monitoring location and the dimethylamine (DMA) used in some scrubbers, because it absorbs in the same region of the ultraviolet as sulfur dioxide, can be an interference in some UV spectroscopic analytical techniques.

b. *In Situ.* Electrooptical systems for *in situ* measurements are relatively new as commercially available monitoring systems (Table X).

Two different *in situ* system, each measuring sulfur dioxide by UV absorption, had significantly different responses to particulate fly ash as an interference (*77*). One was unaffected. The other showed a negative interference with increased particulate concentration (increased opacity). This result further demonstrates that measurement systems employing similar sampling approaches and similar analytical principles can perform differently although sampling the same emissions.

The tunable diode infrared laser (*87, 88*) offers excellent discrimination against potential interferences. However, because the laser and detector require cooling at liquid helium temperatures, it is an impractical instrument for routine field monitoring. Future development to permit operation at thermoelectric cooling temperature and detection at ambient temperature would provide a compact, solid-state monitoring system at relatively low cost.

c. Remote Sensing. Commercially available remote sensing systems (Table XI) (*72, 89–103*) have limited applicability in their present state of development (*89, 90*). One commercial system, using the dispersive correlation technique (Fig. 7), depends upon available sunlight as its

Table XI Development Status of Electrooptical Techniques for Remote Monitoring of Gas Emissions

Analytical scheme	*Pollutants*	*Development status*[a]	*Remark or reference*
Dispersive correlation, ultraviolet and visible	SO_2, NO_2	C	(*72*), Field evaluation studies (*89–91*)
Dispersive, infrared emission	SO_2, NO, CO, CO_2, HF, HCl	RP	Includes all pollutants with infrared spectra (*89, 92, 93*)
Gas filter correlation, infrared emission	SO_2	FP	(*94*)
Diode laser, infrared absorption	SO_2, CO	RP	Extended source applicability (*95*)
Gas filter correlation, infrared emission	NO, H_2S	F	(*96*)
Vibrational raman, ultraviolet	SO_2, NO	RP	Source applicability (*97, 98*)
Laser heterodyne, infrared	SO_2, CO_2	F	(*99*)
Rotational raman, ultraviolet	NO	F	(*100*)
Fluorescence, ultraviolet	SO_2	F	(*101*)
Differential absorption lidar, ultraviolet	SO_2, NO	F	(*102, 103*)

[a] C = commercially available; FP = field prototype; RP = research prototype; F = feasibility.

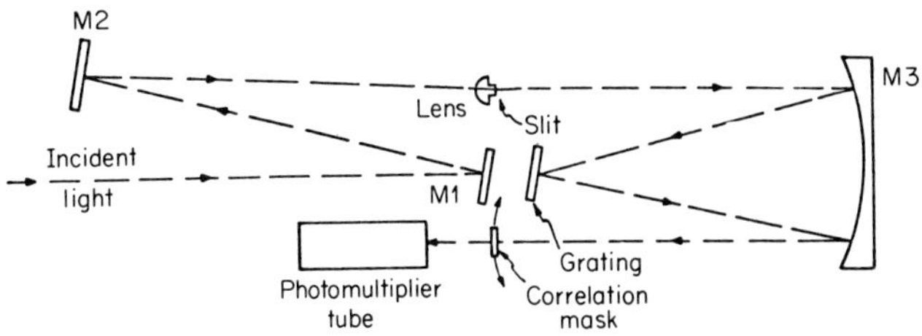

Figure 7. Remote sensing by the dispersive correlation technique. M1, M2, and M3 are mirrors.

source of UV energy for absorption measurement of sulfur dioxide (*72*). As a result, its applicability is limited to daylight hours. Furthermore, its operation in the plume sensing mode (*89, 90*) is affected by the turbidity of the atmosphere as the intervening light scatter medium between the plume and the sensor. This interference increases with increasing distance between the plume and the monitoring system.

The infrared emission filter correlation technique (Fig. 8) depends upon the temperature of the emitted gas as its source of energy for operation as a passive system (*94*). It showed a sensitivity of 1000 ppm meters on emissions from a coal-fired power plant for a 2-m diameter stack and a stack temperature of 150°C. Concentration determinations were within 10% of concurrent reference measurements. Changes in particulate loading in the emissions, in turbidity of the atmosphere between the plume and the instrument, and in the distances of observation up

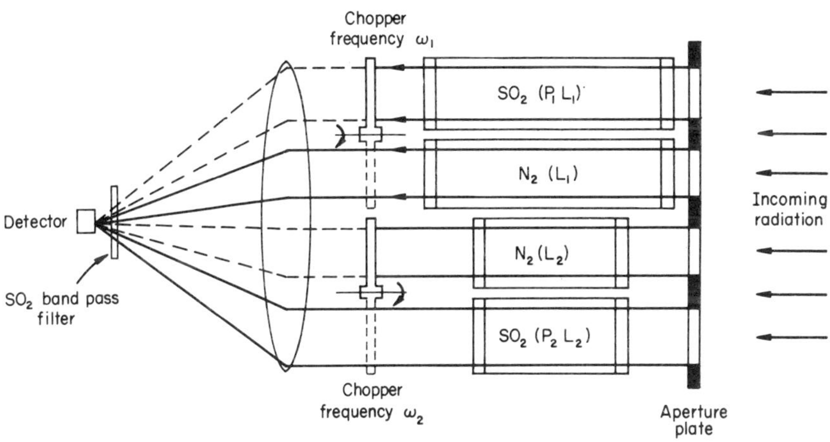

Figure 8. Remote sensing by the gas filter correlation technique. P: SO_2 partial pressure; L: cell path length; SO_2 (P_1L_1) and SO_2 (P_2L_2): correlation cells; N_2 (L_1) and N_2 (L_2): reference cells.

to 450 m had no observable effect on the measured concentrations. In contrast to the ultraviolet absorption technique discussed above, this method is capable of measuring sulfur dioxide during both daytime and nighttime hours. Its applicability, however, is limited to sources having emissions above ambient temperature. A power plant with wet scrubbers will have effluent gas temperature reduced to nearly ambient conditions, and thus will not be a good candidate for remote monitoring by this technique.

2. Sulfur Trioxide and Sulfuric Acid

In most emissions involving sulfur trioxide (SO_3), water vapor and sulfur dioxide are also present. With extractive sampling procedures, the sulfur trioxide is hydrated with the water vapor present to form sulfuric acid at temperatures below the acid dew point. Sulfur dioxide can act as a positive interference through catalytic oxidation. The sulfur trioxide concentration can also be decreased through reaction with trace metals to form sulfate salts. At present no commercial monitoring system is available.

Prototype continuous monitors are being developed (*104, 105*) with automated sampling and collection for optimum sulfur trioxide discrimination in the presence of sulfur dioxide. One prototype (*104*) is an automated version of the Shell method (*106*). The analysis of the collected sulfuric acid mist, however, is done by conductivity instead of by titration. The second prototype design attempts selective collection and analysis by impacting sulfuric acid mist on a filter medium, driving off the sulfur trioxide into a water absorber by a heated airstream, and analysis of the absorbing solution by conductivity measurement.

An *in situ* infrared absorption approach (*107*) has the advantage of executing the analysis of sulfur trioxide in the gas phase and avoids the potential discrimination problems associated with the above prototype developments.

3. Hydrogen Sulfide, Total Reduced Sulfur, and Total Sulfur

The obnoxious odors associated with petroleum refineries and kraft pulp mills are largely caused by hydrogen sulfide (H_2S) and mercaptans (RSH). When these sulfur compounds are measured together, they are measured as total reduced sulfur (TRS). TRS plus sulfur dioxide and other organic sulfur compounds are measured as total sulfur (TS). Sulfur dioxide can react as an oxidizing agent toward the reduced sulfur compounds and be a serious interference in TRS measurements.

A state-of-the-art review has been conducted on available monitoring systems for H_2S, TRS, and TS and a field evaluation of selected instruments has been made *(60)*. The review divided the systems into three categories: preferred commercial, the alternate commercial, and the potential methods.

a. PREFERRED COMMERCIAL. These techniques are coulometric, gas chromatographic (GC) with flame photometric detection (FPD), and reagent-impregnated filter tape. The reagent in the tape samplers is selected to react with H_2S to form an insoluble sulfide precipitate which is photometrically analyzed for light transmission. Mercaptans may or may not react. This group of techniques has the capability of measuring several (or all) types of sulfur compounds.

b. ALTERNATE COMMERCIAL. These techniques include ultraviolet absorption spectrometry, colorimetry, and electrochemistry (Table X). Although the principles have been demonstrated, application of these methods continues under development.

c. POTENTIAL METHODS. These are fluorescence and infrared absorption spectrometry.

Of eight instruments utilizing the preferred commercial techniques (four different coulometers, two flame photometric detector (FPD) systems combined with gas chromatography (GC), and two lead acetate tape samplers), three were determined by laboratory tests to be inadequate for a hostile stack environment despite their ambient air capability. The remaining five instruments still encompassed the three commercially preferred measurement techniques. The results of field tests of these five instruments conducted on a refinery Claus plant and on a kraft pulp mill recovery furnace can best be summarized in terms of the following general conclusions.

As initially installed, none of the instruments was able to provide routine monitoring data. Manufacturers's instruction manuals were inadequate, particularly in advising corrective action for malfunctions and in their recognition of potential problem areas. Coulometric titration was the method which had had the most extensive field application. The effectiveness of the chemical filters, on which the method relies for selective analysis, depends upon the source environment being measured. When individual components are of concern, the combination of GC/FPD offers the preferred approach. Correctly measuring total sulfur or total reduced sulfur compounds in the presence of SO_2 and COS or CS_2 proved to be a problem, which may be by GC separation or pretreatment to convert other sulfur compounds to SO_2 or H_2S. The lead acetate tape system requires an adequate preconditioning interface. Pyrolysis and catalytic reduction to hydrogen sulfide appears to be a promising preconditioning method for this purpose.

B. Nitrogen Compounds

Nitric oxide and nitrogen dioxide are often monitored in combination as nitrogen oxides (NO_x). Some techniques are specific to nitrogen dioxide; nitric oxide is oxidized to nitrogen dioxide to obtain NO_x. Other techniques may be specific to nitric oxide. To monitor NO_x in this case, it becomes necessary to reduce the nitrogen dioxide to nitric oxide. The other nitrogen compound of potential interest in emissions is ammonia (NH_3).

1. Nitric Oxide, Nitrogen Dioxide, and Nitrogen Oxides

As in the case of sulfur dioxide emissions, interest in monitoring NO_x emissions is mainly directed at combustion sources and acid plants. A two-part study was conducted in 1971 on instrumentation for monitoring NO_x from stationary sources by evaluation of commercially available systems on a coal-fired power plant (*108–111*).

The field study included seven different commercial instruments; four were based on the nondispersive infrared technique (nitric oxide only), one on the nondispersive visible absorption method (nitrogen dioxide and NO_x), and two on an electrochemical detection scheme (NO_x). Instrument parameters evaluated in the laboratory tests were accuracy, calibration drift, repeatability, zero drift, sensitivity, response time and lag, interferences by other gases, and susceptibility to environmental changes. The parameters evaluated in the field were accuracy, routine maintenance, ruggedness, and repair frequency. The monitors were not complete systems with their own sampling interfaces. For the field tests, a manifold system was set up from which the monitors sampled; a sample pretreatment system was built into the test facility upstream of the manifold. Five instruments were operative for the three series of field tests; two NDIR systems failed to complete the tests. Four of the five systems sampled the flue gas (dry and filtered) from the manifold for all three test series; the nondispersive visible absorption system sampled the manifold for one test series and sampled the untreated emissions upstream of the pretreatment system for two test series. The reference test method for NO_x (*2*), utilizing phenoldisulfonic (PDS) acid as a reagent, and referred to as the PDS reference method, was run on the manifold gas for two of the test series and once on the untreated emissions.

All of the test results were combined for the various instrument parameters to give an overall index of performance. Based on a perfect performance rating of 1.00, the performance index values ranged from 0.60 to 0.70. One can conclude that, overall, these monitors need improvement. The relatively narrow difference between the best and worst performing instrument shows that no one technique or instrument stands out as an obvious choice for performance. The better performing systems simply reflected a longer history of development and application experience. A significant fact that must not be overlooked is that four of the five monitors were not tested as complete systems with their own interface but shared a common pretreatment interface. Therefore, the various instruments can only be judged on a relative but incomplete basis.

In a subsequent but similar evaluation study (*76*) by public utilities, several sample conditioning units were furnished by the instrument manufacturers, but their performance was not as satisfactory as the sample conditioning system designed by the utilities themselves to overcome the many problems encountered.

A complete system evaluation study was performed on commercially available NO_x monitoring systems for a period of about 3 months, September–November 1973, on a large coal-fired power plant (*5*). The objectives of the study were to compare the results of various analytical and sampling approaches to the measurement of NO_x and to evaluate and validate the applicability of performance specifications and test procedures being proposed for monitoring NO_x emissions. The systems included the following techniques: *in situ* ultraviolet absorption (nitric oxide), extractive nondispersive infrared, extractive chemiluminescent (NO_x) with permeation and water condensation interfaces and extractive visible light absorption (nitrogen dioxide and NO_x). The PDS reference method was run concurrently at various times during the 3-month period. A range of particulate loadings was created by cutting off various stages of the electrostatic precipitators. The relative accuracy results (the sum of the absolute mean error and the 95% confidence interval), expressed as a percentage of the mean measurement by the PDS method, ranged from 15 to 35%. The conclusions were (a) the various analytical and sampling approaches, exemplified by commercially available systems, can give comparable results within a relative accuracy of ±20%; (b) performance specifications and test procedures are applicable and valid but the relative accuracy specifications of performance may need to be modified if maximum possible utilization is to be made of available monitoring systems.

2. Ammonia

Commercial monitoring systems with ammonia measurement capability are available (Table X). The techniques include nondispersive ultraviolet absorption, derivative ultraviolet spectroscopy, conductivity, both electrical and thermal. No information is available on the performance of any of these systems on stack emissions.

C. Carbon Oxides and Hydrocarbons

Carbon oxides and hydrocarbons are emitted from a variety of stationary sources that include incinerators, boilers, foundry cupolas, refinery catalyst regenerators, and asphalt batching plants. When their source is incomplete combustion, hydrocarbons are generally accompanied by carbon monoxide. When the source of carbon monoxide is carbon dioxide reduction, carbon monoxide emissions will not contain appreciable hydrocarbons. Carbon dioxide is not treated as a pollutant gas. Its concentration is used as a supporting measurement in the calculation of pollutant emissions under standard conditions to account for the effect of varying amounts of dilution air.

1. Carbon Monoxide and Total Hydrocarbons (THC)

Table X identifies commercially available carbon monoxide and hydrocarbon monitors. There are seven different systems for carbon monoxide. These include two *in situ* and five extractive methods. The five methods available for hydrocarbons are all extractive approaches.

A combined laboratory (pilot plant) study and field evaluation at three selected industrial sources was conducted on some of these monitoring systems (*63*). Three of the five hydrocarbon methods and two of the seven carbon monoxide methods were selected for evaluation. The hydrocarbon monitors selected were one flame ionization detector (FID), one nondispersive infrared monitor (NDIR), and two gas chromatographic monitors (GC) with flame ionization detectors (FID); the carbon monoxide instruments were four nondispersive infrared monitors (NDIR) and one gas chromatographic monitor (GC) with flame ionization detector (FID). Field work was done during the spring and summer of 1972 for about a 2- to 3 week period at each of the following three sites: a municipal incinerator, a gray iron foundry, and a petroleum refinery.

There were scrubbers upstream of the stack sampling point on the incinerator and foundry being tested. The various monitoring systems sampled from a common manifold with the interface conditioning the gas stream upstream of the manifold. The conditioning included coarse particulate filtering, water condensation, fine particulate filtering, and gas mixing in a 30-liter chamber for calibration and zero gas tests.

a. CARBON MONOXIDE. The carbon monoxide concentrations in the incinerator studies ranged from 200 to 500 ppm (dry refuse) and as high as 1000 ppm (wet or difficult-to-burn refuse). The NDIR and GC + FID systems gave comparable results. The exception was that one of the four different manufacturers' NDIR's gave data statistically different than the others; the correlation coefficient between various paired combinations including the GC + FID system was above 0.9 while the one NDIR system showed a correlation coefficient of about 0.5 in comparison with any other instrument.

The foundry carbon monoxide concentration ranged from several thousand ppm to 10–15%. All the instrument systems were optimized for operation at the high concentrations, above 1% carbon monoxide. Only two of the four NDIR systems operated satisfactorily and agreed with the Orsat and wet chemical determinations. The GC + FID system had problems with saturation in spite of instrument modifications consisting of an extremely small sample loop and reduced amplifier gain.

At the refinery, the carbon monoxide boiler was the control equipment for emissions from a catalytic cracking unit. Carbon monoxide concentrations at the inlet of the carbon monoxide boiler were 10 to 12%, and at the outlet <50 ppm of carbon monoxide in the presence of very high concentrations (12 to 16%) of carbon dioxide. The results for high carbon monoxide concentrations at the boiler inlet were the same as in the case of the foundry. The same two out of the four NDIR systems gave satisfactory results. The GC + FID system had difficulty with saturation problems.

At the outlet of the CO boiler, the low CO concentration and high CO_2 concentration put a stringent test on the discriminatory capabilities of the various NDIR systems. Only one of the four NDIR systems gave data consistent with the results indicated by the wet chemical data. Subsequent controlled laboratory tests simulating the relative CO and CO_2 concentrations encountered in the field tests were conducted to test for the interference of CO_2 under these circumstances. The results verified the conclusion that CO_2 concentrations at 12 to 16% could be a significant interference problem (as much as 170 ppm CO equivalence) for some monitoring systems measuring CO concentrations below 50 ppm. This interference problem appeared with the systems with gas filters but not with the one system that had an interference-type filter.

b. HYDROCARBONS. The hydrocarbon concentrations in the incinerator studies ranged from 2 to 20 ppm with occasional values as high as 100 ppm. The ethane sensitized NDIR could not provide useful data due to high sensitivity to water vapor interference and the variable composition of the hydrocarbons. The FID and the GC + FID results were comparable.

The foundry hydrocarbon concentrations varied rapidly in the range from 120 to several thousand ppm and could vary as much as 2 decades in a 30-minute period. The FID system alone gave satisfactory data when compared to manual (laboratory) hydrocarbon analysis. The GC + FID system proved inadequate to cope with highly fluctuating hydrocarbon emissions. The system had a long response time of 45 minutes. This was presumed to be due to adsorption and desorption effects of the inlet tubing. The ethane sensitized NDIR gave inaccurate data compared to the FID and manual data.

Hydrocarbon concentrations at the inlet of the carbon monoxide boiler ranged from 150 to 300 ppm and were essentially all methane (CH_4). The FID and GC + FID systems gave satisfactory data. Data in terms of total hydrocarbon (FID and GC + FID) were comparable to data in terms of methane (GC + FID) since methane was the only constituent. The ethane sensitized NDIR had a water vapor interference and was relatively insensitive to methane (1:10). Therefore, it was deemed to be inapplicable to this pollutant/source combination. No monitoring studies were conducted on the outlet of the carbon monoxide boiler where manual tests showed a hydrocarbon concentration range of 0.5 to 5 ppm.

2. Carbon Dioxide

Carbon dioxide is not identified as a pollutant; the need for monitoring occurs primarily as a supporting measurement to determine emission rate or as an assessment of a process parameter or process. Commercial systems are available (Table X) and include nondispersive and dispersive infrared techniques and thermal conductivity methods.

D. Halogens

Continuous monitoring systems for emissions of the halogen gases, hydrogen fluoride (HF) and hydrogen chloride (HCl), from stationary sources are not commercially available. The electrical conductometric technique (Table X), which is inherently nonspecific to acid gases, can be made applicable to hydrogen chloride or hydrogen fluoride monitoring in those environments where the interfering gases exist at negligible concentrations or by appropriate sample conditioning. Data on the accuracy or reliability of such applications do not appear available. The corrosiveness of these gases, as in the case of sulfuric acid mist, presents a materials problem for sampling such emissions on a continuous basis.

Serious wall adsorption effects and likely chemical reactions may occur when sampling and analyzing emissions for low concentrations of these gases. *In situ* and remote techniques would overcome these problems. An infrared technique called "gas filter correlation" (GFC) based on optical correlation has been successfully demonstrated in the laboratory. The laboratory research prototype (*112*) has shown the following performance characteristics:

Threshold sensitivity	HF <0.1 ppm
	HCl <0.25 ppm
Concentration range	0.3 to 2500 ppm
Accuracy	±10%
Specificity	No interferences
Response time	90% of true value within 10 seconds

A field prototype GFC system has been developed with multipollutant capability (*113*) for application to *in situ* measurements of stationary source emissions. In addition to hydrogen chloride and hydrogen fluoride capability, the system will measure sulfur dioxide, nitrogen oxide, and carbon monoxide by infrared absorption. The multipollutant capability requires a reference gas cell, containing the particular pollutant to be measured, for each pollutant. A potential problem that will require evaluation is the stability of gases such as hydrogen chloride and hydrogen fluoride in a reference cell. Preliminary data have been obtained demonstrating satisfactory applicability of the field prototype to sulfur dioxide emissions from a coal-fired power plant (*114*).

E. Other Gases

Gaseous pollutant emissions include many constituents which may be subject to monitoring upon being identified as hazardous substances or as objectionable emissions subject to emission control. In principle, electrooptical techniques are potentially capable of monitoring any pollutant gas having an absorption spectrum within the spectral range of the method. A technique demonstrated to be applicable to a specific gas may be readily modified instrumentally to measure another gas. On the other hand, the research and development effort required to establish the technical criteria for the necessary changes may be extensive.

Vinyl chloride was recently identified as a pollutant of concern in emissions from the plastics manufacturing industry. Its absorption spectrum and concentration levels were established. The field prototype GFC system for hydrogen chloride and hydrogen fluoride was found to be applicable to vinyl chloride monitoring provided the system is modified to operate at 10 to 11 μm instead of 3 to 4 μm.

In all cases, reasonable effort needs to be made to evaluate and demonstrate the performance of a system at an emission source under actual stack environmental conditions.

V. Data Reduction

Continuous monitors commonly present data on pollutant concentrations as a function of time on a strip chart recorder. This analog form of continuous data presentation has many advantages, the primary one being that a graphic picture of the quantitative nature of the pollutant emission over a period of time is immediately available. The graphic picture allows one to observe from the recording whether or not the monitoring system is malfunctioning. However, there are other needs for quantitative data that require it be reduced into a format more compatible with its end use, i.e., directly for control purposes, or to determine compliance with emission standards.

A. Standard Units

Emission standards are expressed either as concentration units or as emission rates (*1, 2*). The former are used to assess the performance of emission control systems; the latter to determine compliance with emission standards.

Concentration units are expressed in terms of mass of pollutant per unit of gas volume, e.g., grams of sulfur dioxide per cubic meter of gas. Emission rates are expressed in terms of mass of pollutant per unit of time, e.g., kilograms of sulfur dioxide per hour, or mass of pollutant per unit of product output or of raw material input, e.g., kilograms of sulfur dioxide per 10^6 BTU input. Ancillary data are required on volume flow rate in the stack, on raw material input rates or product output rates to reduce concentration data to the units of emission rate standards.

B. Emission Rate Monitoring

Continuous monitoring data on emission rates may be obtained manually or instrumentally, i.e., data on concentration and ancillary parameters may be reduced manually subsequent to the concentration measurement, or instrumentally in real-time concurrent with the concentration measurement. The manual approach is the more common procedure because continuous data on the supporting parameters, i.e., volume flow rate and product output are not normally available.

Where concentration data are reduced manually to determine averages over periods of time ranging from 2 to 6 hours or more, the question is raised on how to reduce the data from continuous strip charts. The desirable approach is to select and average the minimum number of instantaneous values that will provide a reasonably accurate average compared to an average based on complete integration of all the values under the curve. Little information is available on utilization and reduction of raw emissions data typical of different operating characteristics of various industries.

An interesting research study was conducted to determine the optimum approach to manually reducing continuous analog sulfur dioxide monitoring data to 2-hour average concentrations for emissions from a coal-fired plant *(115)*. Three sulfur dioxide monitoring systems, utilizing both *in situ* and extractive measurement techniques, were operated simultaneously for a 3-month period. The data from these systems were reduced and presented as 2-hour average concentrations based on data selected at time intervals ranging from 1 to 60 minutes. Four different plant operating conditions were experienced, in which the maximum time rate of concentration change did not exceed ±15% per minute. Minute-to-minute variations were usually within 10%. The results of this study showed that averaging instantaneous data values at intervals as high as 30 minutes gave a 2-hour average that was not appreciably different from a 2-hour average based on essentially complete signal integration.

Automated emission rate monitoring systems are presently being studied and commercially available systems are beginning to appear on the market *(116)*. Although the state-of-the-art is adequate at present to make available such systems, the reliability and performance capability of these systems remains to be proved. The complexity of an emission rate monitoring system is greater than that of a concentration monitor by virtue of the additional supporting parameter measurements. In principle the added complexity tends to increase system cost and to decrease the probability of reliable performance. It becomes obvious that the accuracy of emission rate data can be no better than that of the concentration data and will very likely be less because it includes the errors introduced by the supporting parameter measurements.

Two systems are available as mass emissions monitors *(116)*. The continuous stack emission monitoring system (CSEM) has sensors in the stack to measure the temperature, concentration, and volume flow rate of the gas pollutant. The outputs are transmitted electrically to the plant control room and processed by an analog computer with the mass emission rate displayed in real-time. Zeroing and calibration of the sensors are done remotely from the control room.

The second system is the ratio emission monitor (REM) in which the mass of pollutant per unit of heat input for power plant emissions is continuously monitored by analytical devices only. The method is based on a material balance in which the premise is that all carbon atoms entering the system in the fuel feed leave in the waste gases. In the ideal case where a power plant is utilizing a fuel of constant composition and all the carbon in the fuel is oxidized to CO_2, the mass emission of pollutant per unit of heat input is shown to be directly proportional to the ratio of the pollutant gas concentration to the carbon dioxide concentration in the emissions. The constant of proportionality is defined in terms of parameters determined from empirical data for the power plant. The emission rate per unit heat input is calculated directly from the sulfur dioxide to carbon dioxide ratio without the need for volume flow measurement in the stack. The advantage of this ratio technique is particularly significant where nonsymmetrical flows exist in stacks or ducts.

C. Supporting Parameter Monitors

The requirement for parameter measurements additional to concentration depends on the emission standard and the particular industry to which the standard applies. For example, sulfur dioxide emissions are required on a dry basis (data reduced to zero moisture content in the gas stream). In a power plant, the moisture content will vary depending among other things upon dilution, and moisture content information will be necessary as supporting data. On the other hand, sulfur dioxide measurements on emissions from a sulfuric acid manufacturing process are directly acceptable on a dry basis.

It is possible that supporting data can be adequately estimated as a constant factor for a given process without introducing significant error into concentration or emission rate data depending upon the operational characteristics of the industrial process and the effect of deviations from the estimate on the accuracy of the final determination. For example, one can assume water vapor in a power plant emission to be 10%, whereas it may actually vary between 8 and 12%. Deviations of $\pm 2\%$ in the water vapor estimate will introduce an error not exceeding $\pm 2\%$ in the sulfur dioxide concentration on a dry basis. Such an error may be acceptable for monitoring purposes. Where simplifying assumptions or estimates of supporting data are not acceptable, a continuous monitor for the supporting parameter is the only recourse. In such cases, commercial monitors are generally available.

1. Oxygen

Two commercially available types of oxygen monitors operate in the extractive mode and have process application experience over a relatively long period of time. One type uses a polaragraphic sensor as the detector, and the other uses the magnetic susceptibility of oxygen (paramagnetism) as its means of analysis. A recent development (*116a*) has been introduced as a commercially available extractive monitor that operates on the electrochemical principle. Two additional monitoring systems operate in the *in situ* mode. One uses derivative ultraviolet spectroscopy and the other uses the electrochemical cell (Table X).

2. Moisture

Resistive and capacitive sensors have been developed for monitoring moisture content of gas streams. Data are not available on the applicability of these methods to stack sources or on the reliability of available instrumentation for continuous monitoring.

Electrooptical techniques in the infrared are theoretically capable of water vapor determination although at present such systems do not exist commercially. A potential problem with such *in situ* techniques is that moisture can exist in both the solid (condensed) and gaseous phase, whereas an *in situ* spectroscopic method will normally measure only the gaseous phase.

3. Velocity

Transducers for gas flow velocity monitoring are commercially available and include thermal (hot wire anemometer), pressure (pitot tube), and mechanical (rotameter) techniques. The thermal type is potentially limited in stack application by high stack gas temperature and possible physical damage to the sensor filament by the particulate content in the gas stream. Limited data on utilization of pitot tubes for continuous monitoring indicates good potential for reliable performances.

An *in situ* electrooptical technique was developed as a prototype system that has potential merit for stack velocity measurement (*117*). It is basically a remote technique in which a laser beam is split and focussed on the particulate matter in the stack gas stream. The two beams create an interference fringe pattern with known spacings between the interference lines. Particulate matter passing by the interference lines scatter light which is detected and analyzed for the time separating the scattered light pulses. The ratio of the fringe spacing to the time necessary for the particle to traverse this distance is a measure of the gas stream velocity.

VI. Special Monitoring Requirements

A. Unconfined Emissions

Although efforts are made in emission control technology to confine emissions for appropriate control, a number of process operations cannot be economically confined. Data on such emissions are necessary to assess their impact on ambient air quality, e.g., roof emissions in the phosphate fertilizer and iron and steel industry, and leaking valves, vented gases, and vapors in the petrochemical industry. The area around an industry having such emissions is treated as an area source.

B. Area Sources

Where there are a large number of confined and unconfined point source emissions, it is impractical to monitor each source individually. Yet these emissions contribute to the air pollution envelope surrounding the plant. The plant complex as a whole effectively constitutes a stack of significantly larger dimensions than the usual confined source. In this case, the effective source diameter may be that of the air pollution envelope, ranging from 30 to 300 m or more. *In situ* electrooptical techniques are applicable to monitoring such area sources and are referred to as long path monitoring.

The *in situ* and remote electrooptical techniques of Table X are potentially capable of adaptation to long path monitoring. The ultraviolet correlation spectrometer (*73*) has been incorporated into a mobile measurement system to obtain emission rate directly for sulfur dioxide emissions from a power plant complex that may include one or more stacks. This technique is called the perimeter measurement approach (*91*). The correlation spectrometer is operated in the upward viewing mode as the perimeter of the area source is traversed. Concurrent supporting measurements of wind velocity and direction are made. These data are used to calculate the emission rate of sulfur dioxide from within the source perimeter into the ambient air mass moving past the source complex. Results in which the emission rate by the perimeter approach was compared to that calculated from mass balance data for a single stack source have indicated agreement within 15% (*118*).

C. Volatile and Intermittent Emissions

Continuous monitoring often needs to be tailored to the nature of the emissions and the immediate obectives of the measurement. Volatile and intermittent emissions require special sampling and measurement methods. Volatile emissions include pollutants which exist in the gaseous phase within the stack at elevated temperature conditions, but which condense to a liquid or solid after the stack effluent cools to ambient temperature.

Processes such as smelting produce emissions that vary significantly with time in a cyclical manner. The time period of the emission cycle necessitates a monitoring system with an appropriate response time. The general guideline is that the monitor response time should be less than the duration of the shortest gas concentration–time variation. For good resolution of concentration changes, the response time should be about one-tenth the smallest time period of concentration change. Emissions may have very rapid fluctuations of no immediate interest, such as pulsed particulate emissions associated with the rappings of an electrostatic precipitator in a power plant. These are seen as fractions of a second pulses on the analog output of an opacity monitor. In such cases, integration of these pulses is accomplished electronically by the recorder response time to smooth out the analog output and to permit better evaluation of the data.

REFERENCES

1. "Requirements for Preparation, Adoption, and Submittal of Implementation Plans," United States Environmental Protection Agency, *Fed. Regist.* **36,** 15486 (1971).
2. "Standards of Performance for New Sources," United States Environmental Protection Agency, *Fed. Regist.* **36,** 24876 (1971); **39,** 9308, 20790, 37470, 37922, 39872 (1974); **40,** 2803, 33152, 43850 (1975); **41,** 2332, 3826 (1976).
3. "Use of Supplementary Control Systems and Implementation of Secondary Standards," United States Environmental Protection Agency, *Fed. Regist.* **38,** 25697 (1973).
4. "Emissions During Startup, Shutdown, and Malfunction," United States Environmental Protection Agency, *Fed. Regist.* **38,** 28564 (1973).
5. J. S. Nader, F. Jaye, and W. Conner, "Performance Specifications for Stationary-Source Monitoring Systems for Gases and Visible Emissions," EPA-650/2-74-013. United States Environmental Protection Agency, Research Triangle Park, North Carolina, 1974.
6. "Promulgated Emission Monitoring and Performance Testing Requirements—Stationary Sources," United States Environmental Protection Agency, *Fed. Regist.* **40,** 46240 (1975).
7. "Technical Introduction to the Maintenance of Air Quality, Trade Regulations,"

Circular of the Federal Ministry of the Interior, GMBI 433. Bonn, Federal Republic of Germany, 1964.
8. "Uniform Federal Practice for the Monitoring of Emissions from Plants According to §16 of the Trade Regulations," Circular of the Federal Ministry of the Interior, UB II-005-05.43-1/72. Bonn, Federal Republic of Germany, 1973.
9. "Guidelines for Continuous Recording Emission Measuring Equipment," Circular of the Federal Ministry of the Interior, U II 2-005-05.43-11/70. Bonn, Federal Republic of Germany, 1971.
10. "Appropriate Measuring Instruments for Continuous Recording of Emissions," Circular of the Federal Ministry of the Interior, U II 4-005-05.43-3/71 (1971); U III 2-005-05.43-8/72. Bonn, Federal Republic of Germany, 1972.
11. J. W. Davis, *in* "Analytical Methods Applied to Air Pollution Measurement" (R. K. Stevens and W. F. Herget, eds.), Sect. III, pp. 285–297. Ann Arbor Sci. Publ., Ann Arbor, Michigan, 1974.
12. W. Romanek, M. R. Jackson, and A. Lieberman, "A Prototype Fly Ash Monitor for Incinerator Stacks," NTIS PB-187-705T. Nat. Tech. Inform. Serv., Springfield, Virginia, 1968.
13. A. Schutz, *Staub-Reinhalt. Luft* **26**, 1 (1966).
14. Communication with W. D. Conner, In-House Research. United States Environmental Protection Agency, National Environmental Research Center, Research Triangle Park, North Carolina, 1975.
15. E. R. Reisman, W. B. Gerber, and N. D. Potter, "In-Stack Transmissometer Measurement of Particulate Opacity and Mass Concentration," EPA-650/2-74-120. United States Environmental Protection Agency, Research Triangle Park, North Carolina, 1975.
16. H. P. Beutner, *J. Air Pollut. Contr. Ass.* **24**, 865 (1974).
17. D. H. Lucas, W. L. Snowsill, and P. A. E. Crosse, *Meas. Contr.* **5**, 9 (1972).
18. D. S. Ensor and L. D. Bevan, "Application of Nephelometry to the Monitoring of Air Pollution Sources," Preprint 73-AP-14, Annu. Meet. Pac. Northwest Int. Sect., Seattle, Washington. Air Pollution Control Association, Pittsburgh, Pennsylvania, 1973.
18a. F. M. Shofner, G. K. Kreikebaum, and H. W. Schmitt, "*In Situ* Continuous Measurement of Particulate Mass Concentrations," Preprint Paper No. 75-41.1, Annu. Meet. Air Pollution Control Association, Pittsburgh, Pennsylvania, 1975.
19. W. B. Johnson, R. J. Allen, and W. E. Evans, "Development of Lidar Stack Effluent Opacity Measurement System," NTIS PB-227-347/2. Nat. Tech. Inform. Serv., Springfield, Virginia, 1973.
20. C. M. Peterson and M. Tomaides, "In-Stack Transmissometer Techniques for Measuring Opacities of Particulate Emissions from Stationary Sources," NTIS PB-212-741/3. Nat. Tech. Inform. Serv., Springfield, Virginia, 1972.
21. W. D. Conner and J. R. Hodkinson, "Optical Properties and Visual Effects of Smoke-Stack Plumes," EPA AP-30. United States Environmental Protection Agency, Washington, D.C., 1972.
22. W. K. Bühne, *Staub-Reinhalt. Luft* **31**, 22 (1971).
23. C. S. Cook, G. W. Bethke, and W. D. Conner, *Appl. Opt.* **11**, 1742 (1972).
24. R. A. Ferguson, "Feasibility of a CW Lidar Technique for Measurement of Plume Opacity," NTIS PB-231-992/AS. Nat. Tech. Inform. Serv., Springfield, Virginia, 1973.
25. L. Duwel, *Staub-Reinhalt. Luft* **28**, 43 (1968).
26. R. L. Chuan, *Aerosol Sci.* **1**, 111 (1970).

27. J. G. Olin and G. J. Sem, *Atmos. Environ.* **5,** 653 (1971).
28. M. Benarie and J. P. Quetier, *Aerosol Sci.* **1,** 77 (1970).
29. M. Benarie and J. Quetier, "Particle Detector by Mechanical Impact Sensing," EPA-600/2-75-025. United States Environmental Protection Agency, Washington, D.C., 1975.
30. G. J. Sem, J. A. Borgos, J. G. Olin, J. P. Pilney, B. Y. H. Liu, N. Barsic, K. T. Whitby, and F. D. Dorman, "State of the Art: 1971 Instrumentation for Measurement of Particulate Emissions from Combustion Sources," Vol. I, Particulate Mass—Summary Report, NTIS PB-202-655, 1971; Vol. II, Particulate Mass—Detail Report, NTIS PB-202-666. Nat. Tech. Inform. Serv., Springfield, Virginia, 1971.
31. W. Coenen, *Staub-Reinhalt. Luft* **24,** 350 (1964).
32. A. H. Gruber and E. K. Bastress, "A Comparison of a New Continuous Electric Charge Particulate Monitor with the EPA Sampling Train," Proc. Tech. Ass. Pulp Pap. Ind. Eng. Conf., Boston, Massachusetts, 1973. Tech. Ass. Pulp Paper Ind., Atlanta, Georgia.
33. R. Prochazka, *Staub-Reinhalt. Luft* **26,** 22 (1966).
34. H. Schnitzler, *Schreihe Ver. Wass.-Boden Lufthyg. Berlin-Dahlem* **33** (1970).
35. J. C. Guichard, *Staub-Reinhalt. Luft* **33,** 169 (1973).
36. Personal communication with the instrument developer, Dr. J. C. Guichard, at Institute for Chemical Research and Applications, Vert-le-Petit, France.
37. H. Dresia, P. Fischotter, and G. Felden, *VDI* (*Ver. Deut. Ing.*) *Z* **106,** 1191 (1964).
38. J. S. Nader and D. R. Allen, *Amer. Ind. Hyg. Ass., J.* **21,** 115 (1960).
38a. H. Dresin and F. Spohr, *Staub-Reinhalt Luft* **31,** 19 (1971).
38b. M. R. Jackson, "Evaluation of Field Prototype Beta Gauges," Final Report on Contract No. 68-02-1216. United States Environmental Protection Agency, Research Triangle Park, North Carolina, in press.
38c. M. B. Ranade, "Sampling Interface for Quantitative Transport of Aerosols," EPA-650/2-74-016. United States Environmental Protection Agency, Washington, D.C., 1973.
39. W. Horn, *Staub-Reinhalt. Luft* **28,** 20 (1968).
40. G. J. Sem and J. A. Borgos, "State of the Art: 1971 Instrumentation for Measurement of Particulate Emissions from Combustion Sources," Vol. IV, NTIS PB-231-919/AS. Nat. Tech. Inform. Serv., Springfield, Virginia, 1973.
41. E. E. Uthe and C. E. Lapple, "Lidar Study of Stack Plumes," NTIS PB-212-530. Nat. Tech. Inform. Serv., Springfield, Virginia, 1972.
42. W. K. Bühne and L. Duwel, *Staub-Reinhalt. Luft* **32,** 19 (1972).
43. S. Larssen, D. S. Ensor, and M. J. Pilat, *Tappi* **55,** 88 (1972).
44. A. Schutz, *Staub-Reinhalt. Luft* **24,** 359 (1964).
45. J. C. Guichard and J. L. Magne, "Applications of Liquified Bed in Gaseous Atmosphere to the Production of Aerosols," Interior Note No. 51-1967. Inst. Chem. Res. Appl., Vert-le-Petit, France, 1967.
46. G. J. Sem, J. A. Borgos, K. T. Whitby, and B. J. H. Liu, "State of the Art: 1971 Instrumentation for Measurement of Particulate Emissions from Combustion Sources," Vol. III, NTIS PB-233-393. Nat. Tech. Inform. Serv., Springfield, Virginia, 1972.
47. W. R. Zinky, *J. Air Pollut. Contr. Ass.* **12,** 578 (1962).
48. F. J. Monig, N. Schwarzer, and W. Stober, "Bestimmung der Aerosol-Massenverteilung in Einer Aerosol-Zentrifuge Mit Hilfe Von Schwingquarzen," p. 58.

Annu. Conf. Soc. Aerosol Res. Bad Soden, Federal Republic of Germany, 1973 (in press).

49. D. W. Cooper and L. A. Spielman, *Atmos. Environ.* **8,** 221 (1974).
50. K. T. Whitby and W. E. Clark, *Tellus* **18,** 573 (1966).
51. D. Sinclair, *J. Air Pollut. Contr. Ass.* **17,** 105 (1967).
52. C. C. Gravatt, Jr., *J. Air Pollut. Contr. Ass.* **23,** 1035 (1973).
53. J. Wagman and C. M. Peterson, "A Continuous Monitor for Size Distribution of Airborne Particulate Emissions," *Proc. Int. Clean Air Congr., 3rd, 1973* pp. 66–68. International Union of Air Pollution Prevention Associations, VDI-Verlag GmbH, Dusseldorf, Federal Republic of Germany. (1973).
54. Technology Trends, *Opt. Spectra* **8,** 52 (1974).
55. D. W. Roberds, W. M. Farmer, and A. E. Lennert, "Interferometric Instrumentation for Particle Size Analysis," EPA-650/2-73-004. United States Environmental Protection Agency, Washington, D.C., 1974.
56. L. S. Birks and J. V. Gilfrich, "Development of X-Ray Fluorescence Spectroscopy in Elemental Analysis of Particulate Matter," NTIS AD-762-061. Nat. Tech. Inform. Serv., Springfield, Virginia, 1973.
57. M. L. Wright and K. S. Krishman, "Feasibility Study of In-Situ Source Monitoring of Particulate Composition by Raman or Fluorescent Scatter," NTIS PB-225-042/1. Nat. Tech. Inform. Serv., Springfield, Virginia, 1973.
58. H. W. Thoenes and W. Guse, *Staub-Reinhalt. Luft* **28,** 53 (1968).
59. J. S. Nader, *in* "Air Sampling Instruments for Evaluation of Atmospheric Contaminants," Part V, Sect. U. Amer. Conf. Govtl. Ind. Hyg., Cincinnati, Ohio, 1972.
60. H. J. Hall and R. H. Salvesen, "Evaluation of Measurement Methods and Instrumentation for Odorous Compounds in Stationary Sources," Vol. I, NTIS PB-212-812/2; Vol. II, NTIS PB-223-654/5. Nat. Tech. Inform. Serv., Springfield, Virginia, 1972.
61. L. Parts, P. L. Sherman, A. D. Snyder, and F. C. Jaye, *ISA AID* **72440.** Analytical Instrument Division 157, Instrument Society of America, Pittsburgh, Pennsylvania, 1972.
62. C. J. Halstead, G. H. Nation, and L. Turner, *Analyst* **97,** 64 (1972).
63. J. Driscoll, J. Becker, J. McCoy, M. Young, and J. Ehrenfeld, "Evaluation of Monitoring Methods and Instrumentation for Hydrocarbons and Carbon Monoxide in Stationary Source Emission," NTIS PB-226-657/5WP. Nat. Tech. Inform. Serv., Springfield, Virginia, 1972.
64. C. D. Hollowell, G. Y. Gee, and R. D. McLaughlin, *Anal. Chem.* **45,** 63A (1973).
65. Environmental Instrumentation Group, Lawrence Berkeley Laboratory, "Instrumentation for Environmental Monitoring: Air," LBL-1. Tech. Inform. Div., Lawrence Berkeley Laboratory, Berkeley, California, 1973.
66. J. S. Nader, *J. Air Pollut. Contr. Ass.* **23,** 589 (1973).
67. R. L. Chapman, "Continuous Stack Monitoring Techniques, Sampling and Standardization," Proc. Int. Symp. Environ. Meas., pp. 143–163. Beckman Instruments Process, Geneva, Switzerland, 1973.
68. C. D. Hollowell and R. D. McLaughlin, *Environ. Sci. Technol.* **7,** 1011 (1973).
69. P. C. Wolf, *Mech. Eng.* **45,** 22 (1973).
70. J. A. Hodgeson, *Toxicol. Environ. Chem. Rev.* **II,** 81 (1974).
71. P. L. Hanst, *Appl. Spectrosc.* **24,** 161 (1970).

72. A. J. Moffat, J. R. Robbins, and A. R. Barringer, *Atmos. Environ.* **5**, 511 (1971).
73. A. J. Moffat and M. M. Millan, *Atmos. Environ.* **5**, 677 (1971).
74. R. N. Hager, Jr. and R. C. Anderson, *J. Opt. Soc. Amer.* **60** 1444 (1970).
75. H. Okabe, P. L. Splitstone, and J. J. Ball, *J. Air Pollut. Contr. Ass.* **23**, 515 (1973).
75a. J. S. Parsons and S. Mitzner, *Environ. Sci. Technol.* **9**, 1053 (1975).
75b. C. D. Hollowell, G. Y. Gee, and R. D. McLaughlin, *Anal. Chem.* **45**, 63a (1973).
76. R. D. McRanie, "A Study of Instrumentation for Monitoring Emissions from Coal-Fired Boilers," Proc. Instrum. Soc. Amer. Conf., Houston, Texas, 1973. Instrum. Soc. Amer., Pittsburgh, Pennsylvania, 1973.
77. J. B. Homolya, *Sci. Total Environ.* **3**, 349 (1975).
78. F. C. Jaye, "Monitoring Instrumentation for the Measurement of Sulfur Dioxide in Stationary Source Emissions," NTIS PB-220-202. Nat. Tech. Inform. Serv., Springfield, Virginia, 1973.
79. Ministerialblatt fur das Land Nordrhein-Westfalen, "Minimum Requirements for Measuring Instruments for SO_2 Concentrations," No. 124, Ed. A, p. 1551. Bonn, Federal Republic of Germany, 1966.
80. K. J. McNulty, "Investigation of Extractive Sampling Interface Parameters," EPA 650/2-74-089. United States Environmental Protection Agency, Research Triangle Park, North Carolina, 1974.
81. C. E. Rodes, *Proc. Anal. Instrum.* **11**, 125 (1973).
82. A. J. Romano, D. W. Huszagh, J. Klein, R. Doering, and L. Newman, *Combustion* **45**, 36 (1974).
83. J. B. Homolya, *Chem. Technol.* **4**, 426 (1974).
84. W. H. Hedley, "Dilution Device for Coupling Monitoring to Source Emissions," EPA 650/2-055. United States Environmental Protection Agency, Washington, D.C., 1974.
85. C. E. Rodes, R. M. Felder, and J. K. Ferrell, *Environ. Sci. Technol.* **7**, 545 (1973).
86. R. L. Chapman, *Environ. Sci. Technol.* **8**, 520 (1974).
87. E. D. Hinkley, *Opto-Electron.* **4**, 69 (1972).
88. E. D. Hinkley, "Development of *In-Situ* Prototype Diode Laser System to Monitor SO_2 Across the Stack," NTIS PB-223-628/9. Nat. Tech. Inform. Serv., Springfield, Virginia, 1973.
89. H. M. Barnes, Jr., W. F. Herget, and R. Rollins, *in* "Analytical Methods Applied to Air Pollution Measurement" (R. K. Stevens and W. F. Herget, eds.), Sect. III, pp. 245–267. Ann Arbor Sci. Publ., Ann Arbor, Michigan, 1974.
90. "A Report to Department of Health, Education, and Welfare of Optical Measurements of SO_2 and NO_2 Pollution Using Barringer Correlation Spectrometers," NTIS PB-193-485. Nat. Tech. Inform. Serv., Springfield, Virginia, 1969.
91. L. Langan and A. J. Moffat, "The Application of the Correlation Spectrometer to Ambient Air Quality and Source Emissions," Proc. 2nd Joint Conf. Sens. Environ. Pollut., p. 117. Instrum. Soc. Amer., Pittsburgh, Pennsylvania, 1973.
92. M. L. Streiff and C. B. Ludwig, "Remote Sensing of Air Pollution in Urban Areas," NTIS PB-230-885/AS. Nat. Tech. Inform. Serv., Springfield, Virginia, 1973.
93. W. F. Herget, "The Application of Electro-Optical Techniques to Sensing of Stationary Source Pollutants," Proc. 2nd Joint Conf. Sens. Environ. Pollut., p. 155, Instrum. Soc. Amer., Pittsburgh, Pennsylvania, 1973.

94. R. Bartle, "Infrared Sensor for the Remote Monitoring of SO_2," NTIS PB-243-478/5BE. Nat. Tech. Inform. Serv., Springfield, Virginia, 1975.
95. Communication with E. D. Hinkley, "Diode Laser Multipollutant Ambient Air Monitoring." United States Environmental Protection Agency Interagency Agreement No. IAG-D4-0465 with the National Science Foundation, Washington, D.C., 1974.
96. Communication with D. E. Burch, "Evaluation of a Remote Passive Technique Based on Gas Cell Correlation Spectroscopy for Measuring Concentrations of NO and H_2S Emissions." United States Environmental Protection Agency Contract No. 68-02-0766 with Philco-Ford Corporation, Newport Beach, California, 1974.
97. S. H. Melfi, M. L. Brumfield, and R. W. Storey, *Appl. Phys. Lett.* **22,** 402 (1973).
98. S. Nakahara, K. Ito, A. Fuke, S. Komatu, H. Inaba, and T. Kobayasi, *Opto-Electron.* **4,** 169 (1972).
99. Communication with R. T. Menzies, "Remote Detection of SO_2 and CO_2 with a Heterodyne Radiometer," Contract No. NAS 7-100. Nat. Aeron. Space Admin., Jet Propulsion Laboratory, Pasedena, California, 1973.
100. W. H. Smith, *Opto-Electron.* **4,** 161 (1972).
101. M. C. Fowler and P. J. Berger, "Feasibility Study of the Use of Resonance Scattering for the Remote Detection of SO_2," EPA-650/2-74-020. United States Environmental Protection Agency, Research Triangle Park, North Carolina, 1974.
102. Communication with S. A. Ahmed, "Remote Monitoring of Air Pollution Produced by Stationary Sources, Area Sources, and Present in Ambient Conditions." United States Environmental Protection Agency, Grant No. R 803109 with City University of New York, New York, New York, 1974.
103. R. Byer and M. Garbuny, *Appl. Opt.* **12,** 1496 (1973).
104. E. Richter and S. Sattelberger, *Bennst.-Warne-Kraft* **24,** 339 (1972).
105. Communication with Walden Research Corporation on United States Environmental Protection Agency Contract No. 68-02-0789 for the Development of an H_2SO_4 Acid Mist Measurement System for Stationary Sources, Boston, Massachusetts, 1973.
106. H. Goksyr and K. Ross, *J. Inst. Fuel* **35,** 177 (1962).
107. Communication with D. E. Burch, Philco-Ford Corporation, Newport Beach, California, 1974.
108. L. P. Parts, P. L. Sherman, and A. D. Snyder, "Instrumentation for the Determination of Nitrogen Oxides Content of Stationary Source Emissions," Vol. I, NTIS PB-204-877. Nat. Tech. Inform. Serv., Springfield, Virginia, 1971.
109. A. D. Snyder, E. C. Eimutis, M. G. Konicek, L. P. Parts, and P. L. Sherman, "Instrumentation for the Determination of Nitrogen Oxides Content of Stationary Source Emissions," Vol. II, NTIS PB-209-190. Nat. Tech. Inform. Serv., Springfield, Virginia, 1971.
110. L. Parts, P. L. Sherman, A. D. Snyder, and F. C. Jaye, *ISA AID* **72440,** 157 (1972).
111. A. D. Snyder, E. C. Eimutis, M. G. Konicek, P. L. Sherman, and F. C. Jaye, *ISA AID* **72441,** 165 (1972).
112. E. R. Bartle, E. A. Meckstroth, and S. Kaye, "Development HC1 and HF Detection System," AFRPL-TR-71-59. Air Force Rocket Propulsion Laboratory, United States Air Force Base, Edwards, California, 1971.

113. D. E. Burch and D. A. Gryvnak, *in* "Analytical Methods Applied to Air Pollution Measurement" (R. K. Stevens and W. F. Herget, eds.), Sect. III, pp. 193–233. Ann Arbor Sci. Publ., Ann Arbor, Michigan, 1974.
114. Communication with W. F. Herget, United States Environmental Protection Agency, National Environmental Research Center, Research Triangle Park, North Carolina, 1974.
115. J. B. Homolya, "Data Output Requirements for Monitoring SO_2 Emissions from a Stationary Source," Paper No. 716, Vol. 3. Proc. Instrum. Soc. Amer. Conf., Houston, Texas, 1973. Instrum. Soc. Amer., Pittsburgh, Pennsylvania, 1973.
116. "Continuous Emission Monitor for Power Plant Applications." Company Brochure. Western Research and Development, Ltd., Calgary, Alberta.
116a. D. T. Ross, R. E. Pocock, and E. T. McGandy, "Electrochemical Oxygen Analyzer with Dry Jet Sampling System; Preprint Paper No. 75-60.4, Annu Meet. Air Pollution Control Association, Pittsburgh, Pennsylvania, 1975.
116b. Lear Siegel, Inc., Englewood, Colorado, Brochure No. CM50/PL/001A/6/75.
117. L. O. Heflinger, B. J. Matthews, and H. Shelton, "Portable Laser Velocimeter for Stack Velocity Measurements," NTIS PB-213-263/7. Nat. Tech. Inform. Serv., Springfield, Virginia, 1972.
118. Communication with A. Jepsen, Environmental Measurement, Inc., Annapolis, Maryland, 1975.

16

Measurement, Testing, and Surveillance of Air Pollution from Mobile Sources

William F. Biller and John Harkins

I. Introduction

Over most of the history of vehicles powered by internal combustion engines, gaseous emissions were measured by the simple Orsat apparatus, which permitted measurement of carbon dioxide, carbon monoxide, and oxygen by volume changes in a 100-cm^3 sample of exhaust gas as the components were absorbed consecutively into chemical reagents. An approximate measure of exhaust hydrocarbons could be obtained with the more complicated combustion Orsat. Interest was primarily on determining air–fuel ratio that could be calculated from the exhaust composition. Interest in the air polluting constituents of exhaust gases from automobiles stems from the work of Haagen-Smit in the mid 1950s, which called attention to hydrocarbons and nitrogen oxides as the major sources of Los Angeles smog. This led to rapid development of new methods for analysis of vehicle emissions.

The sampling, analysis, and characterization of emissions from mobile sources have continued to evolve as emission levels have decreased and as emission standards have been extended from the automobile to heavy-duty vehicles and aircraft. The techniques used for measuring automotive emissions have set the pattern for measuring emissions from all other mobile sources. Since almost all of the methods in current use for other than the automobile have at one time or another been applied to automotive emissions, this chapter will be principally devoted to the measurement of emissions from automobiles. The treatment of the other mobile sources, i.e., trucks, buses, motorcycles, aircraft, locomotives, boats, and ships, in the last section, will cover those measurement methods unique to such sources.

II. Classification and Composition of Emissions from Automobiles

There are four kinds of emissions from automobiles—exhaust gases, crankcase emissions, evaporative emissions, and refueling emissions. The major constituents of the exhaust gas emissions are nitrogen and the nitrogen oxides (nitric oxide and nitrogen dioxide), the products of complete combustion of gasoline (carbon dioxide and water), and the products of incomplete combustion (carbon monoxide, unburned hydrocarbons, and partly oxygenated fuel derivatives such as aldehydes and alcohols). Particulate matter, sulfur oxides, and sulfates are also present. The hydrocarbons in the exhaust include hundreds of individual compounds representing many of the original fuel molecules ranging from C_4 to C_{10} plus their reaction products after being subjected to high temperature and pressure and undergoing extensive cracking to yield lower molecular weight compounds and a large portion of olefinic compounds. Thus, the observed range is C_1 to C_{10}, and the exact composition is a function of the original fuel, the engine, and its specific operating conditions.

The crankcase emissions result from partially burned charge that is pushed down past the piston rings into the crankcase during combustion. This material contains a large component of fuel and air, with minor amounts of some of the products of combustion (carbon monoxide, carbon dioxide, and water). These emissions are fully controlled in modern vehicles and, therefore, will not be discussed further in this chapter.

Evaporative emissions occur from the carburetor bowl and the fuel tank. The carburetor losses occur primarily during the so-called hot soak period, when the engine is shut down after running and the heat of the engine can heat the fuel in the carburetor bowl above its boiling point. This causes distillation of a substantial amount of fuel into the intake manifold and to the atmosphere through the carburetor vents. Evaporative emissions from the fuel tank occur while the vehicle is running and the fuel tank temperature rises. The resulting increase in vapor pressure causes evaporation of fuel into the vapor space of the tank, pushing other vapor-laden gases out of the tank vent into the atmosphere. This same mechanism can be operative when the vehicle is standing still with the engine off because of normal diurnal temperature variation. Evaporative emissions are primarily the lower-boiling fractions of the fuel. However, due to the nature of the evaporation process, all the fuel molecules are represented to some extent in the fraction evaporated.

Refueling emissions are quite similar to evaporative emissions. They occur during a fueling operation when incoming fuel displaces the hydrocarbon-laden vapors in the fuel tank. At the same time, droplets of fuel are entrained in the displaced gases escaping to the atmosphere. Spillage losses during the refueling operation also contribute to the total refueling emissions.

Each of these emission sources presents its own characteristic problems for sampling, analysis, prototype certification, production-line testing, inspection-lane testing, and on-the-road emission surveillance.

III. Measuring Exhaust Gas Emissions from Automobiles

Exhaust gas emissions represent the dominant historical interest in development of emission measurement methods for motor vehicles and have in one way or another fathered most of the techniques employed in the measurement of emissions from other liquid-fuel-burning mobile sources.

A. Analytical Methods

The principal need in the analysis of exhaust gases is for convenient rapid instrumental methods that are specific for the compound measured and have a linear response over the broad range of concentrations encountered. This need has been by and large met for the compounds of primary interest. Wet chemical methods are still required for measurement of the oxygen-containing organic compounds such as aldehydes.

1. Hydrocarbons

a. Flame Ionization Method. The predominant method for measuring exhaust gas hydrocarbons is the flame ionization detector (FID) (Fig. 1.) It is based on the fact that a diffusion flame of pure hydrogen in air is essentially free of ions and is therefore nonconducting. However, when a small amount of sample gas containing hydrocarbons is premixed with the hydrogen, the concentration of ions increases greatly and the flame

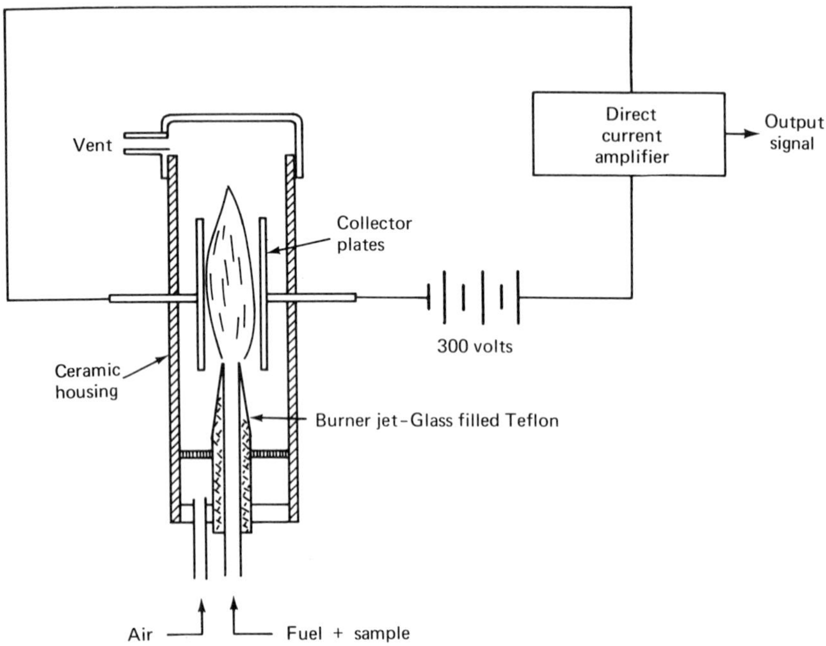

Figure 1. Flame ionization detector hydrocarbon analyzer burner–detector.

becomes conductive. If an electric potential of several hundred volts is established between collector plates surrounding the flame a current can be detected that, under constant gas flow through the detector, is proportional to the concentration of hydrocarbons in the sample gas. FID instruments require accurate regulation of the flow of hydrogen, sample gas, and air to the detector; a well-designed burner; and a high-impedance, low-level, fast-response amplifier (Fig. 2). Commercial instruments are either operated at ambient temperature or are heated. In the latter case, the area in Figure 2 within the dashed lines is maintained at about 350°F (175°C). For application to automotive exhaust gas, the ambient temperature instrument is normally used. The heated FID is used primarily for analysis of diesel engine exhaust to minimize condensation of the high molecular weight exhaust hydrocarbons resulting from the incomplete combustion of diesel fuel.

The flow rates required for optimum operation of the instrument are normally established by the manufacturer. For specific applications, the user may want to modify them to achieve best performance. Typical flow rates are: sample gas, 3–5 ml/minute; fuel mixture, 75 ml/minute; and air, 200 ml/minute (*1*). The sample flow is normally set at the lowest setting consistent with bringing the instrument into the desired sensitivity range after the fuel and air flow rates have been optimized (Fig. 3). The optimum setting will correspond to the flat portions of the curve near the peak response so as to get high response along with minimum sensitivity to changes in flow.

The FID has a linear response to hydrocarbon concentrations over an extremely wide range. The same instrument can measure concentrations both in the percent and the parts per million (ppm) ranges. Normally the instrument is provided with four to five ranges to cover this range in

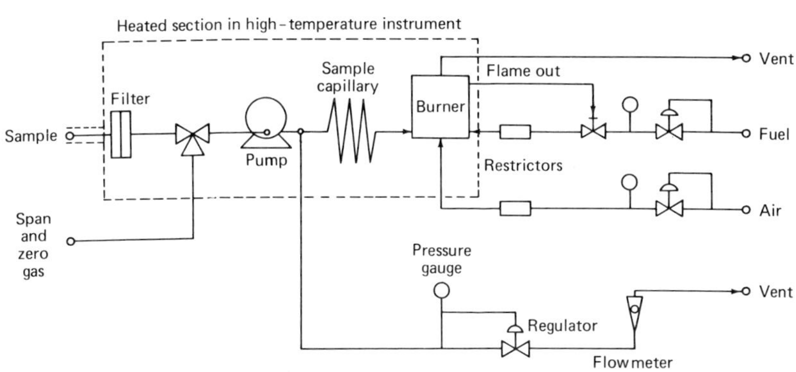

Figure 2. Flame ionization detector hydrocarbon analyzer.

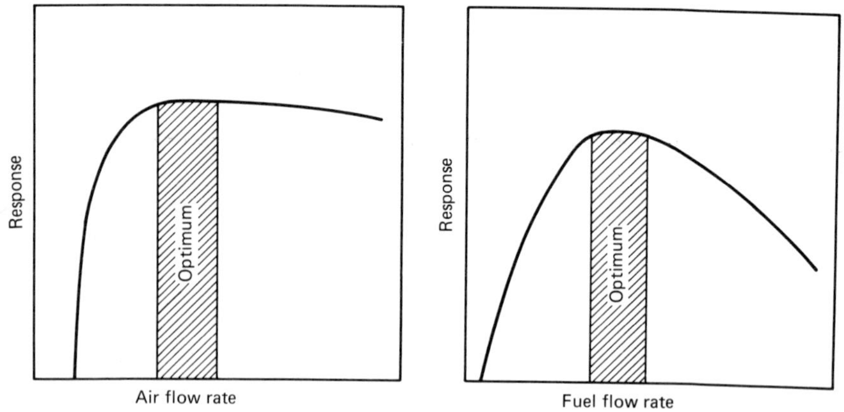

Figure 3. Response of flame ionization detector hydrocarbon analyzer to fuel and air flow rates.

convenient steps. The FID response to hydrocarbons results primarily from carbon atoms within the molecule that are combined with hydrogen. For hydrocarbons, therefore, the response is both proportional to the volumetric concentration of hydrocarbons in the exhaust and to the number of carbon atoms per molecule. This latter characteristic is relatively independent of the structure or type of hydrocarbon molecule. Thus, if a mixture of hydrocarbon compounds is present in the sample, the instrument response is in proportion to the carbon atoms as hydrocarbon per unit volume of sample. The measure commonly used is parts per million carbon (ppmC) by volume. The FID provides, therefore, an unambiguous measure of total hydrocarbons in a sample gas.

The FID does not respond to the carbon oxides, nitrogen oxides, water vapor, oxygen, or nitrogen present in exhaust gases. But it does respond to the partially oxygenated compounds such as aldehydes, ketones, and alcohols, which result from the partial combustion of hydrocarbons. The response is not in direct proportion to the number of carbon atoms per molecule in this case because of carbon–oxygen bonds. The quantity of compounds with these bonds is generally small compared to that of hydrocarbons and causes little error in the measurement of total organics as ppmC. With advances in the control of automotive exhaust emissions, this situation may no longer generally apply.

While the FID measurement of hydrocarbons is free from interferences from other compounds, it does exhibit what has been called "oxygen synergism." That is, the response to hydrocarbons is to some extent a function of the oxygen concentration in the sample gas. It varies with hydrocarbon type, burner design, burner fuel composition, and instrument operating conditions. This effect is reduced by using hydrogen mixed with either nitrogen or helium, usually in the ratio of 40% hydrogen to 60% inert diluent. The operating conditions can be further adjusted to reduce this error to under 2% for up to 5% oxygen in the sample gas (*2*). For those cases in which the oxygen content of the sample is relatively constant, the error can be reduced by calibrating the instrument with a standardized gas containing oxygen in the same concentration range as the sample gas.

b. NONDISPERSIVE INFRARED. Because of the ability of hydrocarbons, carbon oxides, and nitrogen oxides to absorb infrared radiation, various infrared methods were examined at an early stage of the interest in exhaust as an air pollutant. Of the instruments available, the nondispersive infrared gas analyzer (NDIR) was shown to have the necessary sensitivity and selectivity (Fig. 4). It employs a Luft infrared detector, which is a device containing two chambers permanently filled with the same compound or compound type it is desired to measure in the sample gas. In the case of hydrocarbons, the detector is filled at reduced pressure (about 10 Torr) with *n*-hexane or *n*-heptane. A thin metallic diaphram separates the two chambers and contains a small pinhole to allow the gas to slowly equilibrate between the two chambers. However, any momentary differences in pressure caused by unequal absorption of infrared radiation will flex the diaphram. The movement is detected by making the diaphram part of an electrical condenser and sensing the change in capacitance. Infrared radiation from two collimated light sources is passed through a sample cell and a reference cell in parallel and into the two chambers of the detector. The light is modulated with a motorized chopper to give 10 light pulses per second. The reference cell normally contains dry air, which absorbs little infrared radiation. When air or nitrogen is passed through the sample cell, both chambers of the detector receive a substantial influx of infrared radiation. However, the gas in the detector only

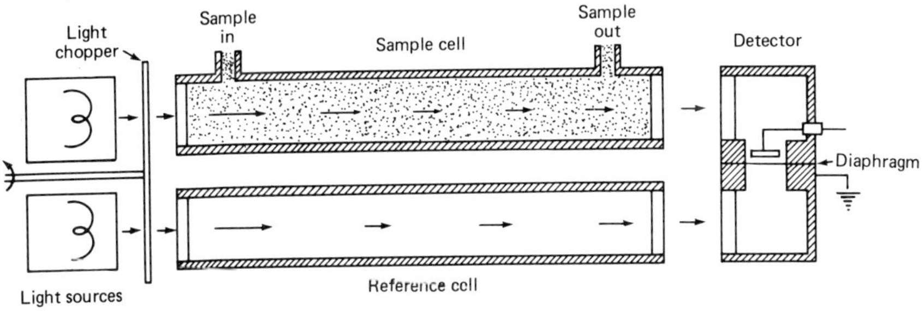

Figure 4. Nondispersive infrared gas analyzer.

absorbs the wavelengths corresponding to its specific infrared absorption bands. The absorbed energy heats the gas as each pulse arrives, causing a momentary increase in pressure. If the light is balanced by mechanically shielding off some of the light so that both halves of the detector see the same amount of energy, the pressure pulses within each chamber of the detector balance each other, and no signal is generated. The instrument is said to be zeroed. (Current instruments use only a coarse mechanical adjustment and complete zeroing electronically.) If a sample of exhaust gases is now passed through the sample cell, the hydrocarbons in it will absorb some of the infrared energy at the same wavelengths as the hydrocarbon in the detector, and a smaller amount of such radiation will reach the sample gas side of the detector. As a result the pressure pulses in the sample gas side of the detector will be weaker; the resulting imbalance of pressure will cause the diaphram to have a net deflection toward the sample side, and a signal will be generated. The higher the concentration of hydrocarbons in the sample cell, the greater the signal. By passing gases containing known concentrations of hydrocarbons through an instrument that has been zeroed on air or nitrogen, a calibration curve of response against concentration can be obtained.

The instrument is nondispersive, because no attempt is made to limit the infrared radiation passing through the sample to a narrow range of frequencies through use of a grating, a prism, or an optical filter. Instead, a broad range of frequencies passes through the sample gas and into the detector. However, the detector can only respond to the specific band of frequencies absorbed by the gas it contains. For example, a hydrocarbon-sensitized instrument will not detect the absorption of infrared energy by carbon monoxide in the sample cell, since the absorption bands of hydrocarbons and carbon monoxide occur at different frequencies. The Luft detector allows greater utilization of the available energy in a compound's absorption band than dispersive instruments and yet is highly selective to a single type or class of compound.

The response of an NDIR instrument is initially nearly linear, but at sufficiently high concentrations, it curves markedly toward the concentration axis. The cell size is chosen small enough to give as nearly linear a response as is consistent with sufficient signal-to-noise ratio to give an accuracy of 1% of full scale in the concentration range of interest. For exhaust hydrocarbons, cell lengths vary from 5 to 35 cm. These general characteristics greatly limit the dynamic range of NDIR instruments relative to the FID. Some improvement in range with a single instrument can be obtained by stacking cells, that is, placing a short and a long sample cell in series.

Despite the excellent selectivity of the NDIR method, the hydrocarbon-sensitized instrument does have a small response to both carbon dioxide and water vapor because of slight overlapping of absorption bands. This problem can be largely controlled by placing a filter cell of about 2 cm in length in line with the sample cell and filling it with pure carbon dioxide gas saturated with water vapor. This procedure greatly reduces the amount of radiation in the region of overlap entering the detector.

An NDIR instrument sensitized with *n*-hexane or *n*-heptane does not have the uniform response of an FID to all hydrocarbons on a parts per million of carbon basis (*3*). The NDIR instrument responds to the vibrational frequencies of the carbon–hydrogen bond (3.4 μm) in saturated or paraffinic hydrocarbon molecules. The presence of double bonds in the molecule will cause shifts in the C—H bond. Consequently, the NDIR instrument will be significantly more responsive to paraffinic than olefinic hydrocarbon compounds and will have a drastically reduced response to aromatic compounds. For the paraffins, there is a rough proportionality of response to molecular size. The response to methane, however, is negligible. Thus, an NDIR sensitized with *n*-hexane or *n*-heptane is sensitive to both concentration and composition and will underestimate the total hydrocarbon on a parts per million of carbon basis. Comparison with the FID shows the NDIR instrument to be on the average about 45% low in estimating automotive exhaust hydrocarbons.

It has been found possible by using a mixture of acetylene, benzene, and ethylene in the detector to obtain a more uniform parts per million of carbon response (*3*). However, there is a significant loss of sensitivity and increased interference from carbon dioxide and water. It also has proved difficult to manufacture a multigas detector with uniform behavior. Therefore, such instruments were in use for only a short time and can no longer be obtained commercially.

As a result of nonuniform response to hydrocarbons and the limited dynamic range of the NDIR, it has been largely replaced by the FID for automotive exhaust hydrocarbon measurements. Its application to light-duty diesel exhaust was always limited because of the general difficulty of heating an NDIR instrument and the low hydrocarbon concentration of diesel exhaust. However, the NDIR is simpler to operate than the FID and needs no hydrogen fuel supply. It therefore continues to be used in state inspection–maintenance programs and is somewhat more practical than the FID for over-the-road testing. Hydrocarbon NDIR instruments are still used in the United States for certification of heavy-duty truck and bus engines, but this is likely to change in time.

c. Optical Filter Infrared. In the last several years, high-quality optical filters for infrared radiation that pass a well-defined and very narrow band of infrared frequencies have become available commercially (4). The frequencies passed can be accurately controlled by the manufacturer. Coupled with solid state detectors such as lead selenide (PbSe) and solid state amplifiers, a compact, rugged instrument can be made that is well suited to state inspection and maintenance programs and use by garages in maintaining and repairing exhaust control systems. Thermopiles and thermistor infrared detectors are also used on commercial instruments. The instruments typically simultaneously measure hydrocarbons and carbon monoxide. Some also measure carbon dioxide. They also contain a sample pump, room temperature water trap, and filter. The three general configurations of this type of instrument are illustrated in Figure 5. The configuration shown in Figure 5a is a dual-beam analyzer with an optical filter and a solid-state detector. The modulated infrared beams coming from the sample and reference cells are alternatedly focused by a

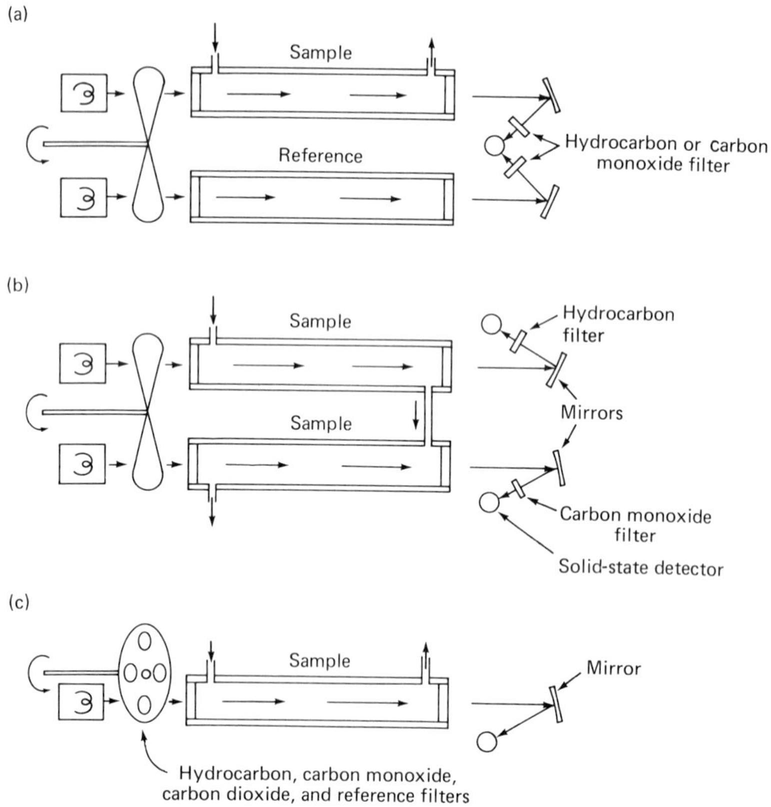

Figure 5. Typical optical filter infrared analyzer configurations: (a) dual beam, (b) single beam, (c) chopper filter.

mirror onto the detector after passing through the optical filter. The configuration shown in Figure 5b has no reference cell. In this case, the zero point is taken as the response of the detector to the total energy of the source passed by the filter. While this method has the advantage of simplicity, its zero stability is strongly dependent upon source temperature and cell thermostating. The system shown in Figure 5c makes the optical filters an integral part of the light chopper. It has filters for hydrocarbon, carbon monoxide, and carbon dioxide. A fourth filter acts as a reference. It passes a narrow band of frequencies close to but not overlapping the hydrocarbon and carbon monoxide bands. The reference frequency reduces sensitivity of the instrument to the temperature effects cited for the Figure 5b type instrument. Only one sample cell need be used in this configuration.

The response characteristic of optical filter instruments to hydrocarbons approximate those of NDIR. The specific behavior is sensitive to the choice of the center band frequency and the bandwidth. Slight shifts in the band center alter the instrument's relative response to equimolar concentrations of propane and hexane. In a Luft-type NDIR, this relative response is close to 0.5 and thus provides good carbon counting characteristics, at least for the paraffins. This characteristic is of value, since it provides more flexibility in the choice of calibration and span gases. As seen in Figure 6 relative response is a function of the frequency band passed by the filter and varies over a large range in commercially available instruments (*4*). A center band frequency of 3.43 μm gives the

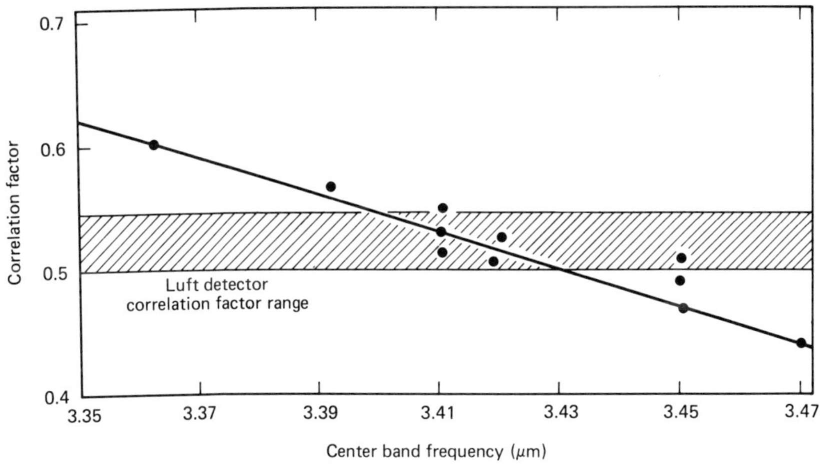

Figure 6. Variation of hexane–propane correlation factor in commercial optical filter infrared hydrocarbon analyzers.

desired 0.5 relative response. As a result, the state of New Jersey requires all instruments approved for their inspection and maintenance program to have a center band frequency in the range 3.41–3.45 μm.

Given an optimum centering of the band, the bandwidth of the optical filter determines the relative response to various hydrocarbon classes. In general, the broader the bandwidth, the greater the response to non-paraffinic hydrocarbons and methane. However, too broad a width can cause increased sensitivity to other exhaust compounds. A half bandwidth of about 0.35 μm appears to be optimum (*4*). The overall correlation of optical filter instruments with the Luft detector NDIR is still poor.

Because such instruments are used in a garage environment, various techniques have been used to provide an internal calibration and thus avoid the use of calibration and span gases. Mechanical wedges to interrupt the light path have been used. Other designs employ solid bars, screens, optical filter blades, or rotating choppers to give a reproducible attenuation of the signal. The trend, however, is in the direction of electronic calibration, which appears to be reasonably satisfactory for garage use. Occasional but periodic checks with calibration gases continue to be essential to assure that mechanical or electronic methods are functioning properly. The use of gases also provides a good overall check of instrument operation and can, for example, be used to check for leaks in the built-in sampling system.

Commercial instruments employ nonlinear meter scales, which give a direct concentration reading and thus avoid the use of calibration curves. If the calibration shifts, an internal adjustment is necessary to realign the response to correspond with the scale.

d. Gas Chromatography. The principle use of gas chromatography in measuring exhaust gas hydrocarbons is the determination of composition. Capillary columns and temperature programming combined with FID separates C_1 through C_{10} hydrocarbons in exhaust from gasoline engines into somewhat over a hundred characterizable peaks (*5*). Many of the peaks represent individual compounds. Others can be identified as to type and molecular size (carbon atoms per molecule). The hydrocarbons in diesel exhaust extend up into the C_{20} range and give rise to broad peaks corresponding to many compounds. As a result, no fully satisfactory gas chromatography method is available for diesel exhaust. The gas chromatography method can be used to measure nonmethane hydrocarbons when such information is needed. While nonmethane hydrocarbons probably give better indication of contribution to the formation of photochemical oxidants and ozone, they are not as convenient to determine. As a result, hydrocarbon mobile emission standards in the United States continue to be expressed as total hydrocarbons, and most measurements are on this basis.

2. Carbon Monoxide

Infrared methods are used almost exclusively for measuring carbon monoxide in exhaust from light-duty vehicles. In research and development and United States federal certification testing, the Luft detector NDIR instrument is commonly used. Optical filter types employing 4.6–4.7 μm filters with bandwidths of 0.1–0.2 μm are predominant in inspection–maintenance applications.

In the range of carbon monoxide concentration of 1% and higher normally found in the exhaust from gasoline engines with no control devices, a 0.65-cm cell is commonly used in a Luft detector NDIR. While there are slight interferences from carbon dioxide, water vapor, and ethylene, discrimination is sufficient so that no steps need be taken to reduce the interference. Below 1% carbon monoxide concentration, the cell sizes increase and reach 25 cm for a 0–100 ppm instrument. In the low ranges, steps are necessary to reduce interference.

The use of a filter cell containing carbon dioxide saturated with water vapor can be effective. Optical filters can be used to narrow the range of frequencies in the carbon monoxide absorption band reaching the detector gas and thus reduce band overlap from interfering compounds. Water can also be removed before entering the instrument by use of refrigerated traps or drying agents. Carbon dioxide can be removed with an alkaline absorbent such as Ascarite.

A very effective approach in the low ppm range is to combine an optical filter with a stacked detector system. The stacked system consists of two detectors in series. The first detector through which the infrared beam passes has a transparent window that allows the beam to pass into a second detector chamber. The signal detected by the first chamber is due primarily to the infrared absorption of carbon monoxide and to a small extent that of interfering components. The second chamber signal is weaker than the first, but its response relative to interfering components is much smaller. The two signals are subtracted in a mixer amplifier to yield a relatively pure signal for carbon monoxide. With instruments of this type, it is possible to obtain discrimination ratios of carbon monoxide to water vapor of greater than 20,000 and of carbon monoxide to carbon dioxide of greater than 40,000.

Because of the nonlinear response of NDIR instruments, several calibration gases covering a range of concentrations must be used. The shape of the curve is reasonably constant, so that having once determined the curve shape, a single calibration gas (called a span gas) can be used to set up the instrument. This is done by first zeroing the instrument on air and then passing span gas through while adjusting the amplifier gain to give the meter or chart deflection indicated by the calibration curve for the concentration of carbon monoxide in the span gas. Since the calibration curve can change slowly during use, it should be checked periodically. The calibration curve can normally be fit by a third-order polynomial, which is useful in those cases in which the output of the instrument is processed by a computer. The signal can also be linearized electronically. However, this procedure does not alter the requirement of checking the calibration periodically and relinearizing it when necessary.

3. Carbon Dioxide

As with carbon monoxide, infrared instruments of either the Luft detector NDIR or optical filter type are the almost exclusive choice for the measurement of carbon dioxide in automobile exhaust emission measurement applications. Carbon dioxide has strong infrared absorption bands. In the concentration ranges normally present, there is no interference from other exhaust components. Short cell lengths of 3–6 mm are adequate for gasoline engine applications, and lengths of up to 12 mm length for diesel engine applications.

The infrared absorbance of carbon dioxide is sufficiently strong that at high concentrations (10–15%) most of the incident energy can be absorbed. As a result, accuracy can be lost in this concentration range, and the shortest feasible cell length is desirable.

4. Nitrogen Oxides

In reference to exhaust emissions, the term "nitrogen oxides" commonly refers to the sum of nitric oxide and nitrogen dioxide, and this is the measurement usually required, because most exhaust samples contain sufficient oxygen to react with some of the nitric oxide formed during combustion to convert it to nitrogen dioxide before analysis. Thus, total nitrogen oxides relate directly to engine output, whereas its individual components may not. The methods available tend to measure either nitric oxide or nitrogen dioxide. These can be measured separately and summed to give total nitrogen oxide or one oxide can be fully converted to the other before the measurement is made.

a. Nondispersive Infrared. Both nitric oxide and nitrogen dioxide have weak infrared absorption bands. The use of long sample cells is sufficient to overcome this deficiency for nitric oxide, and with suitable precautions, measurements down to the 0–100 ppm range can be made. However, despite much effort, a satisfactory NDIR for determining nitrogen dioxide in exhaust has not been developed. The NDIR instrument for nitric oxide has serious interferences from water vapor, carbon monoxide, and carbon dioxide. These can be substantially reduced by the use of optical filters and stacked detector cells. It is generally advisable to further reduce water vapor concentration by condensation of the water vapor in a refrigerated trap. In instruments not using stacked detector cells, passing the exhaust through a drying agent is frequently employed, particularly when concentrations are below a few hundred parts per million. Drierite has proved to be suitable. However, only a small quantity of indicating Drierite should be mixed with it, since the cobalt nitrate, which acts as the indicator, reacts with nitric oxide. Dehydrite (magnesium perchlorate) and molecular sieve drying agents have also been successfully used. In using drying agents, it is advisable to first test for possible reaction with nitric oxide.

b. NONDISPERSIVE ULTRAVIOLET. Nitrogen dioxide does not have a strong absorption band in the infrared to provide the basis for a practical measurement. However, it does have sufficiently strong absorbance in the ultraviolet region to allow measurement down to the 0–50 ppm range. Such instruments have the general configuration of NDIR instruments. The infrared source is replaced by ultraviolet light emitters, and the Luft detector is replaced by a photocell. The method has essentially no interferences from other exhaust components. Nondispersive ultraviolet (NDUV) nitrogen dioxide and NDIR nitric oxide instruments are commonly used together in the same sample train to yield a measurement of total nitrogen oxides.

c. CHEMILUMINESCENT ANALYSIS. The chemiluminescent method reacts ozone with nitric oxide to produce nitrogen dioxide in an excited state (*6*). The excited molecules almost immediately give off their excess energy as photons of light, which are detected by a photomultiplier tube (Fig. 7). The instrument manufactures its own ozone, which is introduced into the reaction chamber, where it mixes with the incoming sample gas containing nitric oxide. As with the FID method, careful controls of flow rates are necessary. Two types of instruments are available, one which maintains the gases in the reaction chamber at 5–10 Torr, and the other in which reaction chamber pressure is close to atmospheric. The low-pressure unit is better suited to the concentration ranges of nitric oxide encountered in light-duty vehicle exhaust. The chemiluminescent method

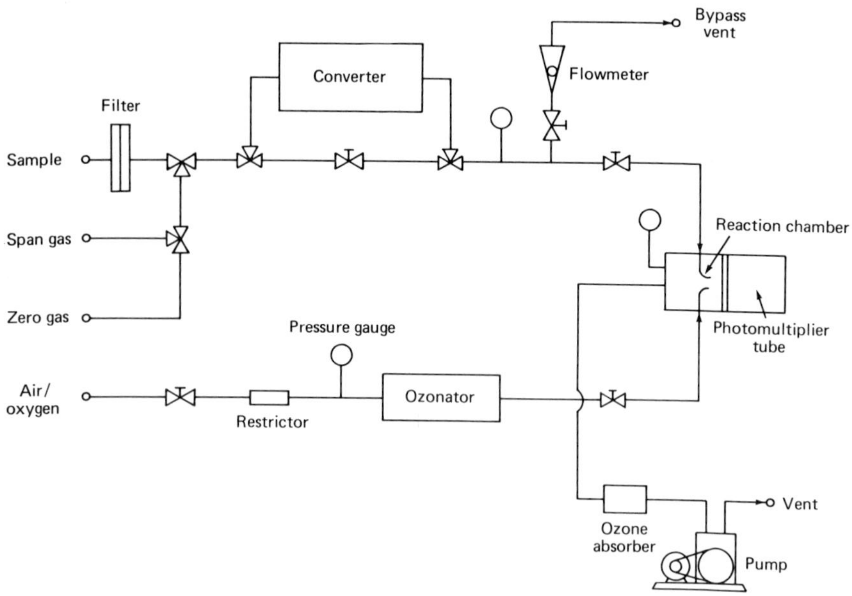

Figure 7. Chemiluminescent analyzer.

has a large dynamic range, its response to nitric oxide is linear over the range, and it is free of interferences. These good features have made it the predominant method for use with light-duty vehicles, and it is rapidly becoming standard for all mobile sources.

The chemiluminescence method can be used to determine total nitrogen oxides by combining the instrument with a converter that reduces any nitrogen dioxide present in the sample gas to nitric oxide prior to entering the instrument's chemiluminescent reactor (Fig. 7). The converter is a coil of stainless steel maintained above 750°C and can be obtained with the instrument. The nitrogen dioxide content of an exhaust sample can be measured by taking the difference in readings when the converter has gas passing through it, and when it is bypassed. It is advisable to check conversion efficiency frequently to insure the proper operation of the converter. Conversion efficiency can be checked by mixing air with a nitric oxide calibrating gas blend in a clean plastic bag and alternately measuring the nitric oxide (i.e., converter bypassed) and nitrogen oxides (i.e., sample gas through converter) until most of the nitric oxide has been converted to nitrogen dioxide. The total nitrogen oxides will remain constant if the conversion is complete (*6*). A more rapid procedure compares the total nitrogen oxides in a dynamically blended mixture of nitric oxide calibration gas and oxygen both before and after the oxygen has been passed through an ozone generator. The ratio of the two measurements after each has been corrected for the nitric oxide remaining after ozonization is a close approximation of the conversion efficiency and is equal to it at 100% conversion (*7*).

The converter works best on an exhaust stream containing a few percent or more of oxygen. If it is operated on a sample gas containing little oxygen and a few percent carbon monoxide, the conversion efficiency will begin to fall off and ultimately fall to near zero. Conversion efficiency can be restored by passing air through the heated converter. This behavior indicates the conversion of nitrogen dioxide to nitric oxide occurs predominately on the stainless-steel surface. The carbon monoxide probably acts by combining with the active sites on the surface and may displace oxygen in the process. Passing air through the converter removes the carbon monoxide (probably by oxidation) and restores the active sites.

d. Wet Chemical Methods. Chemical analysis of exhaust gas employing absorbing reagents and a standard laboratory colorimeter was commonly employed to measure nitrogen oxides in exhaust gases before the advent of instrumentation of adequate sensitivity. The most widely employed technique was the phenoldisulfonic acid procedure (*8*). A modification of the Saltzman method for atmospheric nitrogen oxides was also used (*9*). In both cases, a sample of known volume is brought into contact with oxidizing agents to convert the oxides to nitrates, which are then converted to light-absorbing compounds with a suitable reagent. The absorbance is measured with a laboratory colorimeter that has been calibrated with standard nitrate solutions. These methods can be useful where instrumentation is not available or where an independent check of an instrument reading or calibration gas is desired.

5. Aldehydes

Of the partially oxidized hydrocarbons in exhaust gases, the aldehydes are probably present in highest concentration and have been of greatest interest. Unfortunately, no instrumental methods are available for their analysis, and wet chemical methods must be used. The three methods in use are the chromotropic acid (*10*), the MBTH (*11*), and the DNPH (*12*) methods. All three techniques employ measurement of light absorbance after reaction with a suitable reagent. The chromotropic acid method is used to measure formaldehyde. The MBTH method provides a measure of the lighter aldehydes up to approximately the C_5 compounds. The DNPH method measures total aldehydes but has the disadvantage of being an involved and time-consuming procedure.

6. Oxygen

The measurement of oxygen in exhaust gases is occasionally required. Two instrumental methods are commonly used—the polarographic method and the paramagnetic method (*13*). Both instruments require several seconds or more to respond and therefore are suitable only for measuring emissions under constant operating conditions or in batch samples. The polarographic method relies on the electrochemical reduction of oxygen in a sensor cell consisting of a gold anode and a silver cathode separated by a potassium chloride gel. The cell is separated from the sample gas by a Teflon gas-permeable membrane. Different gases are reduced at different applied potentials. Of the gases commonly in exhaust, oxygen has the lowest reduction potential and is therefore readily measured by this method with little interference. Nitric oxide will reduce to some extent at the oxygen potential, but is normally present at much lower concentrations than oxygen and therefore does not cause a problem.

The paramagnetic oxygen analyzer measures the effect of oxygen on the magnetic susceptibility of the exhaust gas. Oxygen is strongly paramagnetic, while most of the other gases in exhaust are diamagnetic, except for nitric oxide and nitrogen dioxide.

The measurement is made by passing the sample gas through a cell containing a dumbbell-shaped test body suspended on a quartz fiber in a nonuniform magnetic field. The presence of oxygen in the exhaust sample alters the magnetic field and causes the test body to rotate to a new equilibrium position. The rotation is sensed either optically or electronically. While interferences from other gases are small, an accurate measurement may require correction for carbon dioxide, carbon monoxide, and nitric oxide. Their interference can be determined from the tabulation (*13*).

Gas at 100% concentration	*Equivalent percent oxygen*
Carbon dioxide	−0.623
Carbon monoxide	−0.354
Nitric oxide	44.2
Nitrogen dioxide	28.7

7. Analytical Systems for Exhaust Gas Measurement

The measurement of exhaust gases may involve determining the con-

its own instrument. Each instrument will require one or more calibration and span gases and some provisions for sample cleanup to minimize contamination of the instrument. The measurement task is made vastly simpler by incorporating the instruments into an integrated system. Such systems can be assembled by the user. They are also available commercially from instrument manufacturers built into a single cabinet and often can be fabricated to meet individual requirements. The commercial systems normally incorporate many automatic features, such as coordinated operation of valves to establish desired gas flow routes. The system will vary with the nature of the measurements to be made. A system suitable for continuous measurements of exhaust sampled directly from the tailpipe of an automobile (Fig. 8) will require the gases to be passed through a refrigerated trap to remove excess water vapor that would otherwise condense in the sample lines or instruments. It is also necessary to pass the gases through an efficient filter to remove particles down to submicron size that are not removed by the condensation trap. A fiberglass paper is normally used. It has high efficiency into the submicron range with little pressure drop. Care must be taken not to wet these filters, since their efficiency may be strongly impaired. General practice is to use a parallel arrangement with separate lines and separate sample cleanup for each instrument. Stainless-steel metal bellows pumps are commonly

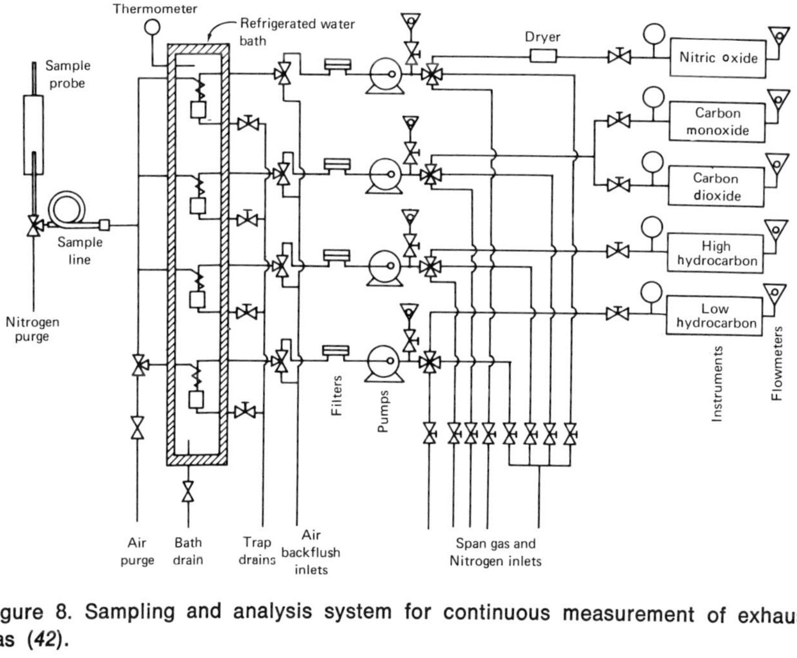

Figure 8. Sampling and analysis system for continuous measurement of exhaust gas (*42*).

used to pull the sample gases through the cleanup section and push it through the instruments. In this way, the pump is protected from contamination, and the instruments are under positive pressure, which reduces leakage. Flow rates are typically 5 ft^3/hour (0.15 m^3/hour). Calibration and span gases are brought into the system through suitable valving. Pressure gauges and flowmeters are provided so that flow conditions through the instruments are the same during calibration as when the sample is flowing. Because of the number of measurements, and the fact that, for the most part, the analyses are conducted simultaneously, it is almost mandatory that recorders be included in the system to provide a permanent record. Zeroing, calibration, and spanning readings are also normally recorded. The outputs of the system can also be fed directly to an on-line computer.

As will be discussed, a common arrangement is to analyze a batch sample of exhaust diluted by a larger volume of air and collected in a plastic bag (Fig. 9). In this case, refrigerated traps are not essential. Althrough traps may still be used to help further reduce water vapor in the carbon monoxide leg, they should not be used in the nitrogen oxide leg, since the dilution will convert a significant portion of the nitric oxide in the sample into water-soluble nitrogen dioxide. While dilution is usually for the purpose of making a mass emission measurement, it also helps reduce loss of hydrocarbon in the sampling lines.

A number of circumstances may require that one or all of the sampling legs be heated (Fig. 10). The analysis of diesel exhaust will require that hydrocarbon be maintained at an elevated temperature from the point of sampling up to entrance of the heated FID instrument. Diesel exhaust hydrocarbon measurements cannot be made on bag samples, since the latter cannot be heated. For continuous measurement of exhaust from a vehicle operating over a cycle of changing conditions, it will also be desirable to use a heated sampling system even with gasoline engine exhaust to prevent hang-up (momentary holdback of hydrocarbons in the sampling lines from condensation or adsorption) during periods of high hydrocarbon concentration.

B. Characterization of Emissions and Mass Emission Measurement

An important question is how to most meaningfully express emissions. The analytical methods in use measure concentration. However, concentrations of pollutants in source emissions are not particularly meaningful and, taken alone, can be misleading. For example, it was originally believed that the high concentrations of hydrocarbons observed during deceleration meant this operating condition was the one most needing

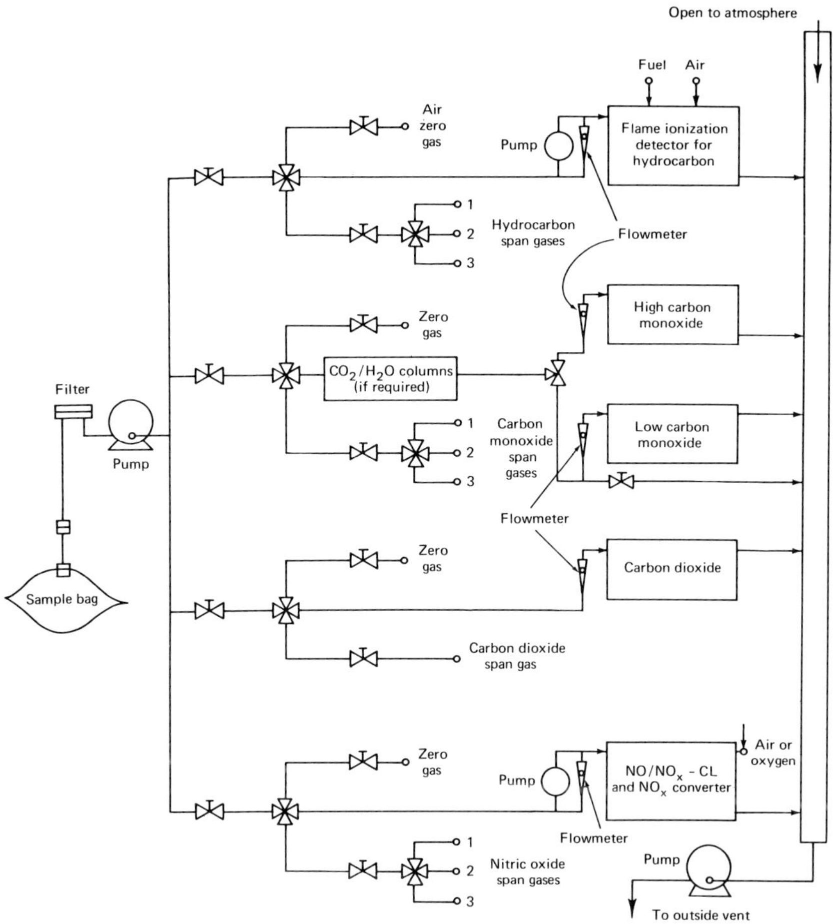

Figure 9. Sampling and analysis system for batch measurement of diluted or dry exhaust gas (7).

control. A substantial effort ensued to reduce emissions during deceleration. It was soon recognized that it was also important to take into account exhaust flow rates. In decelerating conditions, the exhaust flows will be about 5 ft^3/minute, whereas at a cruising condition, for which the hydrocarbon concentrations are generally low, the exhaust flow might easily be 50 ft^3/minute. This tenfold difference can make the overall quantities of hydrocarbons emitted at the two conditions comparable in amount.

While it is desirable to know how the emissions from one vehicle operating mode compare to those from another, it is also of value, particularly

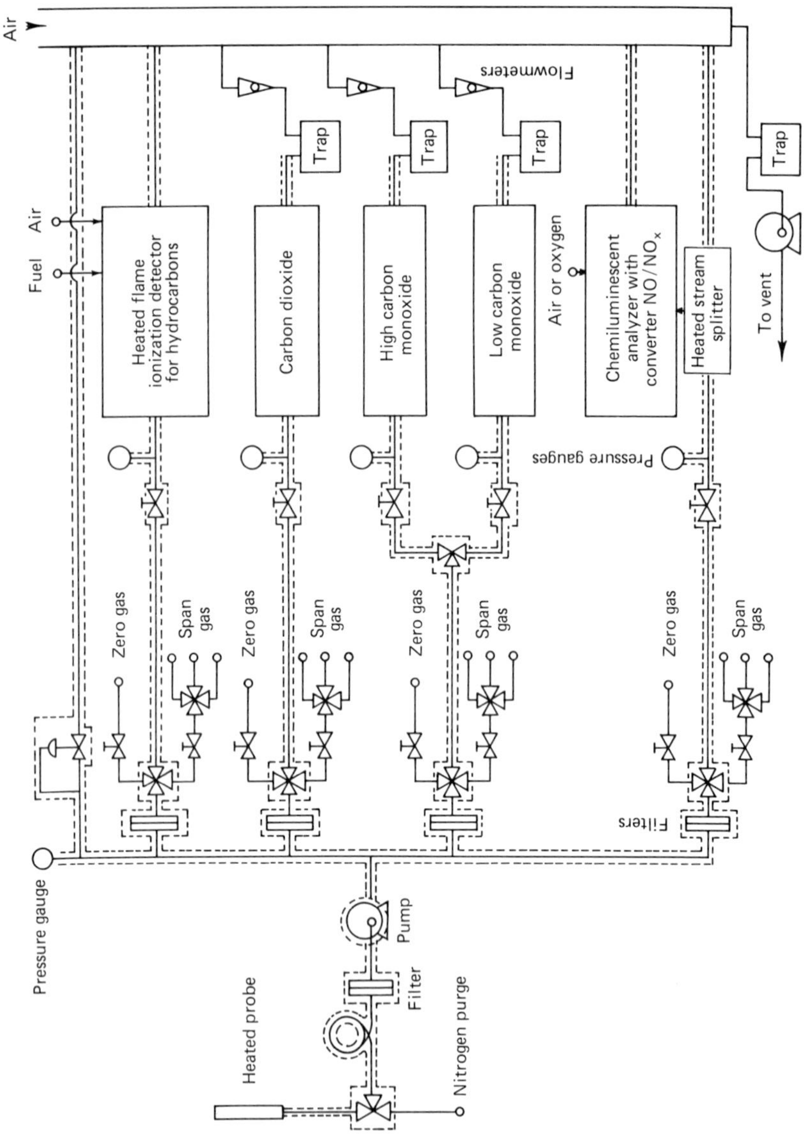

Figure 10. Heated sampling and analysis system. Broken lines indicate heated line or component.

for regulatory purposes, to have a measure of the average or net emissions from a vehicle operating over a typical driving regime. To meet this need, chassis dynamometer test cycles, such as the seven-mode cycle (Table I and Fig. 11), were developed to represent vehicle operation. As originally used by both California and the United States federal government, each of the seven modes was assigned a weight that was proportional to the average exhaust flow rate of passenger cars operating in the mode and the average relative length of time vehicles spend in the mode in urban driving. The measured concentration of a given exhaust component in each mode was multiplied by the appropriate weighting factor, and the resultant products were summed. This procedure yielded a weighted average concentration of the component that was approximately proportional to the weight of the component that would be emitted if the vehicle were operated for the appropriate time in each mode. With this procedure as a basis, the early California and United States federal emission standards were expressed as concentrations.

This method of weighting the concentrations was not fully satisfactory. First, the flow rate used was an average value, which did not reflect individual engine differences. The first attempt at correction was to set the concentration standard for vehicles with small engines at higher values than those for more standard-sized United States vehicles. A further refinement was to apply correction factors, which were dependent on engine size and provided for differences in exhaust flow rates with size to the weighted average concentration. Both these procedures were improve-

Table I Seven-Mode Chassis Dynamometer Test Cycle

Mode no.	*Mode* (*miles/hour*)	(*km/hour*)	*Time in mode* (*seconds*)	*Cumulative time* (*seconds*)	*Weighting factor*[a]
1	Idle	Idle	20.0	20.0	0.042
2	0–25	0–40	11.5	31.5	0.244
	25–30	40–48	2.5	34.0	—
3	30 cruise	48 cruise	15.0	49.0	0.118
4	30–15	48–24	11.0	60.0	0.062
5	15 cruise	24 cruise	15.0	75.0	0.050
6	15–30	24–48	12.5	87.5	0.455
	30–50	48–80	16.5	104.0	—
7	50–20	80–32	25.0	129.0	0.029
	20–0	32–0	8.0	137.0	—

[a] Unnumbered weighting factors are transitional factors used to form a continuous, driveable cycle.

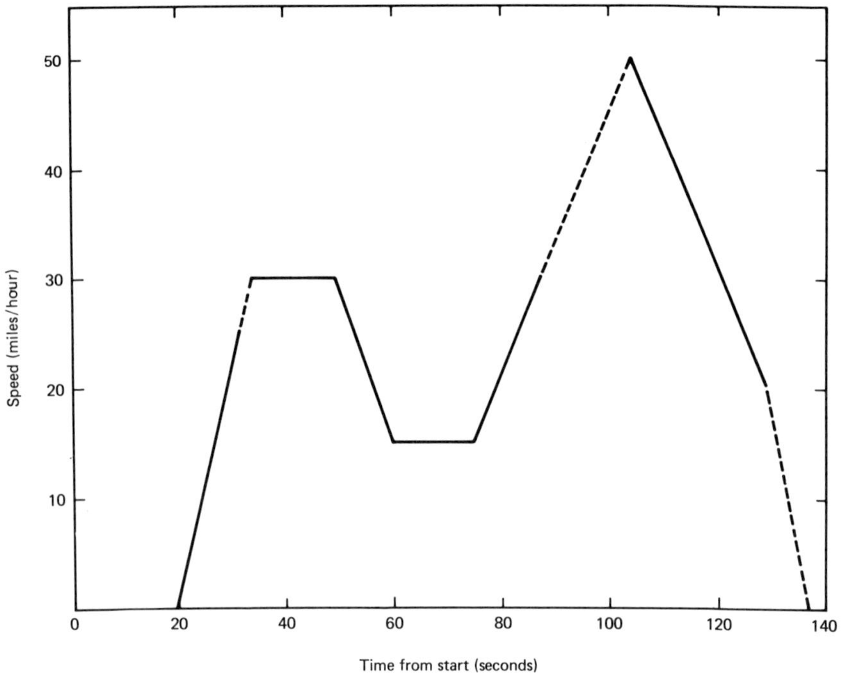

Figure 11. Seven-mode chassis dynamometer test cycle.

ments, but, because they still depended on average behavior, did not fully account for individual differences in flow rates. Second, even with engines of the same size and similar design, there can be differences in flow rates arising from dilution of the exhaust by lean operation or through the use of control systems that add air to the exhaust. To adjust for this effect, a factor was applied to the concentration for each mode to correct back to a stoichiometric condition in which the air supplied is just sufficient to burn the fuel completely to carbon dioxide and water. This was done by assuming that, at the stoichiometric air-to-fuel ratio, the total carbon in the dry exhaust (the sum of the concentrations of carbon dioxide, carbon monoxide, and hydrocarbon, as carbon, all expressed as percent) is close to 15% (*14*). The correction factor is, therefore, $15.0/[\%CO_2 + \%CO + \%HC$ (as carbon)]. Although the method underwent some further refinement (*15*), it was still an approximation and, therefore, not completely satisfactory.

What was needed, therefore, were emission measurement methods that could put emissions from all sizes and types of engines on a truly comparable basis. There was general agreement that the most useful measure would be the actual weight of pollutant emitted by an individual vehicle relative to a convenient operational measure (*16*). For a single operating mode, this could be grams per second in mode, grams per mile traveled in mode, grams per pound of fuel consumed, or grams per brake-horsepower-hour of energy expended. For a test cycle involving a multiple number of modes, the same units could be used but would represent averages over the cycle. The present United States federal and California standards for light-duty vehicles are therefore expressed in grams per vehicle mile (*17*).

A number of mass emission measurements have been proposed and several developed into practical methods. All the methods which have been seriously considered to date involve measurement of the concentration of the component of interest in a sample gas. The mass emission measurement, therefore, involves a calculation of the following general form:

$$W(\mathrm{X}) = D(\mathrm{X})[\mathrm{X}]V \qquad (1)$$

where $W(\mathrm{X})$ = weight of component X emitted; $D(\mathrm{X})$ = density of component X as a pure ideal gas under standard conditions in same weight units as $W(\mathrm{X})$ and volume units as V; $[\mathrm{X}]$ = molecular fraction of component X in the exhaust, or parts of X to parts of exhaust by volume; V = volume of gas at standard conditions containing the weight $W(\mathrm{X})$ of component X. Volume V may be measured directly or indirectly. In the latter situation, its explicit value may not appear in the calculation. In some procedures, concentration is measured after dilution of the exhaust gas by room air. In these cases, V refers to the diluted exhaust. $W(\mathrm{X})$ and V may be expressed as rates such as per unit time, per unit of energy consumption, or per unit distance traveled. Densities for the components of principal interest are shown in the tabulation below.

Component	*Density* *gm/ft*^{3} [*68°F (20°C), 1 atm*]	*gm/m*^{3} *0°C, 1 atm*
Hydrocarbons[a] (as $CH_{1.85}$)	16.33	618.9
Carbon monoxide	32.97	1249
Nitrogen oxides (as NO_2)	54.17	2053

[a] When hydrocarbon concentration is expressed on a carbon basis.

Use of these units requires that the concentration be expressed as mole fraction, or parts component per part by volume of exhaust, and that the volume be expressed as cubic feet at 68°F and 760 Torr. As explicitly stated, Equation (1) is directly applicable to only single-state operating modes. With some modification, it can be applied to open cycles, such as the seven-mode cycle as originally used. The operating modes of an open cycle taken together do not constitute a driveable cycle. In the case of the seven-mode cycle, transitional modes, which are not counted, have been added to make the cycle driveable. Given an open cycle of n modes, Equation (1) becomes:

$$\overline{W(\mathrm{X})} = D(\mathrm{X}) \sum_{i=1}^{n} f_i[\mathrm{X}]_i Q_i \tag{2}$$

where $\overline{W(\mathrm{X})}$ = weighted average emissions per unit time over the n modes; n = number of modes; f_i = relative time spent in the ith mode, $\Sigma f_i = 1$; $[\mathrm{X}]_i$ = concentration of component X during ith mode; Q_i = volumetric flow rate in the ith mode at standard conditions. The term f_i refers to the average time a population of in-use vehicles spends in the ith mode relative to the other modes in the test. It is not the time in mode during the test.

In application of Equation (2), some of the modes may be accelerations or decelerations in which the concentrations and flow rates change over the mode. In this case, the mode can be divided into a number of segments of equal duration and average values for the concentration and flow rate for the mode calculated. The average values are substituted into the equation. A more accurate treatment is to calculate the average of the products of the concentration and flow rates over the segments.

In a closed cycle, the modes do constitute a driveable cycle, and the time in mode during the test is proportional to the relative time a population of in-use vehicles spend in the mode compared to the other modes in the cycle. In this case, the emissions at every point in the cycle count. Since the concentration and flow rate can vary continuously, an integral form of Equation (1) is needed:

$$W(\mathrm{X}) = D(\mathrm{X}) \int_0^{\tau} [\mathrm{X}, t] Q(t)\, dt \tag{3}$$

Where τ is the period of the test. Equation (3) yields total emissions over the test. Dividing by τ yields the weighted average emissions per unit time over the test period. The mass emission measurement methods to be discussed are for the most part concerned with continuous closed cycles and, therefore, are described by Equation (3).

A general classification scheme for mass emission measurement methods is given in Table II. There are two main divisions, batch and continuous

Table II Classification of Mass Emission Measurement Methods

- I. Batch measurement
  - A. Integrating methods
    - 1. Constant-volume sampling
      - a. Positive displacement pump
      - b. Critical-flow venturi
      - c. Centrifugal blower with tube bundle
      - d. Room constant-volume sampler
    - 2. Proportional sampling
      - a. Flow-actuated proportioning valve
      - b. Critical-flow venturi sampling valve
    - 3. Total sampling
  - B. Nonintegrating methods
- II. Continuous measurement
  - A. Exhaust dilution system
    - 1. Constant-volume sampling
    - 2. Temperature-corrected critical-flow venturi
  - B. Exhaust flow measurement
  - C. Intake air flow measurement
  - D. Fuel flow measurement (carbon balance)
  - E. Tracer injection

methods. In batch methods, the sample gas is collected in a container, normally a plastic bag, for later analysis. Various means are used for determining the associated volume measurement. In continuous methods, the concentrations of exhaust components and flow rate are measured continuously over the period of the test.

1. Batch Measurement Methods

Batch measurement methods can be classified into two types, integrating and nonintegrating. In the integrating methods, sample gas is collected continuously during a dynamic closed test cycle in a manner that mechanically integrates the instantaneous product of concentration and flow rate. The concentrations of component gases in the sample bag are weighted averages that when multiplied by the total flow over the test cycle and the respective densities yield the total mass emissions of each component over the test. The nonintegrating methods merely capture sample gas for later analysis, and can be applied only to steady-state operating modes. These latter methods are straightforward and of limited value and therefore will not be discussed further.

a. CONSTANT VOLUME SAMPLER. Of the integrating batch methods, the constant volume sampling technique (CVS) is the most widely used (*7*, *16*). A large air pump, whose capacity is greater than the highest exhaust flow to be encountered during the test, pulls the total exhaust into the sampling system (Fig. 12). Any deficit in exhaust flow is made up by allowing filtered air to flow in and mix with the exhaust close to the point of sampling. A heat exchanger brings the temperature of the diluted exhaust entering the pump to within several degrees of a preset value (about 110°F or 43°C). The pressure of the mixture is also constant and close to atmospheric pressure. Under these conditions, a properly designed system will introduce a constant volume of diluted exhaust to the pump. If a side stream of sample gas is withdrawn at a constant rate, just before entrance to the pump, and collected over the test, its concentration in the bag at the end of the test is a correct weighted average over the test. This is seen from Equation (3). The standard volumetric flow rate $Q(t)$ in this case applies to the diluted exhaust and is constant. It may, therefore, be factored out of the integral to yield:

$$W(\mathrm{X}) = D(\mathrm{X})Q_T \left((1/\tau) \int_0^\tau [\mathrm{X}, t]\, dt\right) \quad (4)$$

The quantity Q_T is the total volume of diluted exhaust at standard conditions passing through the CVS during the test. The quantity within the

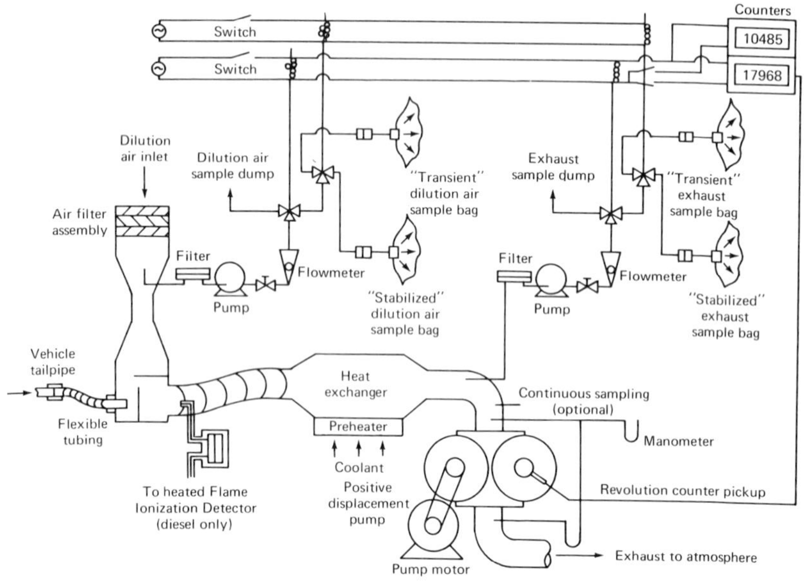

Figure 12. Constant volume sampler (CVS) (*7*).

brackets is the average concentration of X in the diluted exhaust and also the concentration of sample gas in the bag. Thus, Equation (4) may be rewritten as

$$W(\mathrm{X}) = D(\mathrm{X})[\mathrm{X}]_b V \tag{5}$$

where V is the total volume at standard conditions passing through the sampler during the test and $[\mathrm{X}]_b$ is the time averaged concentration in the bag at the completion of the test. The equation has the same form as Equation (1). Typical total flow rates for CVS systems for light-duty vehicles are 300 scfm (standard cubic feet per meter). Exhaust flow from an automobile engine during a typical emission test can vary from 5 to 150 scfm, so that instantaneous dilution ratios range from 2 to 60, with 10 a good average value.

The most commonly used pump for application of the CVS method to light-duty vehicles and engines is a Roots-type positive displacement pump. The positive pumping characteristic of these pumps assures stable flow over the test cycle.

An approach that has gained increasing acceptance employs a critical-flow venturi to produce a constant volume system (*17*). In this system, a high-capacity centrifugal pump is used to pull the diluted exhaust through the venturi. Sonic velocities are produced on the downstream side of the venturi, which produce a constant flow when there is constant pressure and temperature on the upstream side. The system, in its usual form, uses a heat exchanger to condition the gas before entering the venturi. A constant flow sidestream is taken for a bag sample.

A third system employs a centrifugal pump to pull the diluted sample through a bundle of tubes. The resistance offered by the bundle leads to constant flow through the bundle. One of the tubes is used to collect the sample for the bag. The method has not received wide acceptance because of questions concerning ability of the system to maintain constant flow.

Another method that has been proposed is referred to as the "Room CVS." In this method, an engine is placed in a tightly sealed engine test cell. A centrifugal blower forces air into the cell at a fixed rate in excess of the highest exhaust flow expected from the engine during the test. A single duct is provided to allow gas to escape from the cell. The combusted gases from the engine are exhausted directly into the room and mix with the room air. Because the engine is also operating on the room air, it adds very little net volume increase over expansion due to temperature increases. The temperature expansion can be corrected for by placing heat exchangers (air conditioners) in the room to hold the temperature constant. It is therefore not necessary to make any additional provisions to assure constant flow through the centrifugal blower bringing air into the room. The air escapes the room through a single venting duct and can be sampled at a constant rate into a plastic bag to provide a weighted concentration. At the time of writing, there has been little experience with this method. However, it may have application to large diesel or gas turbine engines that would require an impractically large conventional CVS system.

A reasonably accurate measure of the absolute flow rates of CVS systems using a constant displacement pump can be obtained from a knowledge of the pump's displacement. In the case of the critical-flow venturi system, the dimensions of the venturi can be used to calculate the flow. It is customary and desirable, however, to calibrate CVS systems by one or both of two methods. The first is to connect to the input of the system a large laminar flowmeter (*18*) that has been carefully calibrated by the manufacturer or a testing laboratory. The laminar flowmeter (Fig. 13) is essentially a bundle of small-diameter passageways. The pressure drop across the meter is directly proportional to the volumetric flow rate. Flowmeters of this kind are commercially available. The second method is to inject propane from a preweighed container into the system inlet while pumping air alone and collect the diluted sample in a plastic bag. The total weight of propane injected is determined by weighing the propane container before and after the test. The concentration of propane in the bag can then be used, along with the weight loss, to determine the total flow rate through the system. The two methods used together provide an overall check of the system. The system calibration should be fairly constant over a reasonable period of time. Detailed instructions for calibrating CVS systems can be found in the United States Code of Federal Regulations for certification testing of light-duty vehicles (*19*).

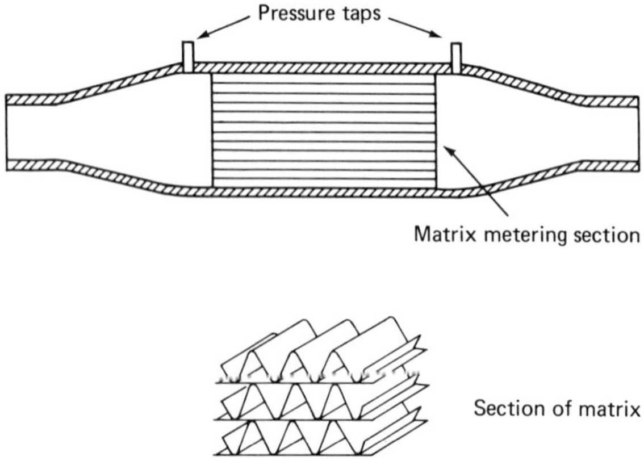

Figure 13. Laminar flowmeter.

In making measurements of low-emission vehicles, it becomes important to correct for background levels in the diluent air. A dilution factor is defined as follows (7):

$$D = (1/13.4)([CO_2]_b^0 + [CO]_b^0 + [HC]_b^0) \tag{6}$$

where the concentrations are the measured mole fractions in the sample bag at the end of the test. The factor D is the approximate volume fraction of exhaust gas in the sample bag. It is essentially the reciprocal of the correction factor used in application of the early California and United States federal standards based on concentrations. The coefficient 13.4 in Equation (6) can be interpreted as typical of the average sum of concentrations in undiluted exhaust from low emission cars over a United States federal certification test (7). The dilution factor is applied as follows to obtain the corrected bag concentration:

$$[X]_b = [X]_b^0 - [X]_d(1 - D) \tag{7}$$

where $[X]_d$ is the concentration of component X in a bag sample of dilution air collected during the test. This sample is obtained by drawing a side stream of the dilution air at a constant rate into a bag just before it mixes with the exhaust (Fig.12). Although they are approximations, both Equations (6) and (7) yield good estimates, and are not likely to lead to significant error.

The dilution of the exhaust in the bag by as much as 10 volumes of air prevents condensation of the water in the exhaust. Thus, an analytical system of the type shown in Figure 9 may be used with CVS systems. The relatively large volume of air in the bag, however, will cause substantial conversion of the nitric oxide in the bag to nitrogen dioxide; and it is therefore important to employ an instrument or instruments that can give a measure of total nitrogen oxide, such as a chemiluminescent analyzer with converter or a combination NDIR instrument for nitric oxide and NDUV instrument for nitrogen dioxide.

The CVS method is a particularly convenient and practical way to obtain an automatically integrated measure of emissions over a continuous, dynamic operating cycle. It has, for this reason, become the dominant means of measuring mass emissions. One of its major drawbacks is that dilution makes extra demands on the sensitivity of the analytical methods. A second drawback is that the undiluted concentration is not normally known. The method measures a concentration proportional to the product of the undiluted exhaust concentration and the exhaust flow rate. In research and development applications, it is desirable to know these parameters separately. If the need is sufficient, this can be done by measuring one of the exhaust constituents before and after dilution and using the concentration ratio to calculate the undiluted exhaust flow rate.

b. PROPORTIONAL SAMPLER. The second type of integrating batch method is proportional sampling (also referred to as variable rate proportional sampling). In this approach, either undiluted or diluted exhaust gases are drawn directly into the sampling bag through a valve that is controlled so as to sample a fixed portion of the total flow under all conditions. In this case, the flow rate of the exhaust (undiluted or diluted) is not constant. Since constant proportionality is maintained, the sample bag performs an automatic integration and produces a properly weighted concentration of exhaust components in the bag. This is seen as follows: The concentration of component X in the bag at the end of the test is given by

$$[\mathrm{X}]_b = \frac{\int_0^T [\mathrm{X}, t]q(t)\,dt}{\int_0^T q(t)\,dt} = \frac{1}{v}\int_0^T [\mathrm{X}, t]q(t)\,dt \tag{8}$$

where $q(t)$ is the flow rate to the bag at standard conditions and v is the volume at standard conditions of exhaust collected in the bag at the end of the test. Because proportionality is maintained,

$$q(t) = kQ(t) \tag{9}$$

Integrating both sides of Equation (9) over the test period yields

$$v = kV \tag{10}$$

Substituting Equations (9) and (10) into Equation (8) gives

$$[\mathrm{X}]_b = (1/V)\int_0^T [\mathrm{X}, t]Q(t)\,dt \tag{11}$$

That is, the bag concentration is the correctly weighted average of the exhaust concentration. Substituting Equation (11) into Equation (3) gives

$$W(\mathrm{X}) = D(\mathrm{X})V[\mathrm{X}]_b$$

which is Equation (5). It is seen from this equation that the system needs to provide a measure of total flow during the test.

Two procedures have been used for obtaining a proportional sample. The first employs a proportionating gas valve, which is controlled by a servo-mechanism activated by a flow measuring device. A laminar flow-type meter connected to the air intake of the engine has been used for this purpose (*20*). Such systems have proved practical for collection of integrated batch samples from vehicles operating over the road. Integration of the flow meter signal gives the total flow over the test. In principle the temperature and pressure at a point of flow measurement must be continuously measured so that the flow rate can be corrected to standard conditions. Since, in the case of air flow measurement, these parameters will show little change over a test, an initial and final reading is sufficient. The method takes undiluted exhaust directly into a bag. To prevent condensation of water and reduce loss of components to the walls, the bag is normally prefilled with nitrogen to a volume of four to six times that of the exhaust volume to be collected. It is important in this application that all the air going to the engine or emission control system pass through the flowmeter. Furthermore, in determining the total exhaust flow, the air flow must be corrected for the increase in the number of moles resulting from combustion. That is, the exhaust occupies a greater volume than the combustion air when the two are compared at standard conditions. The increase will vary with air–fuel ratio and carbon content of the fuel. For an engine operating near the chemically correct air–fuel ratio on a typical fuel, this increase is about 6%. Application of this percentage used as an overall correction, will in most situations probably result in an error of no more than a few percent.

Measuring the exhaust flow would be more direct. The laminar flowmeter, with its narrow passage ways and complex construction, is not well suited to measuring exhaust flow from gasoline engines because of the particulate matter, the rapid and wide temperature fluctuations, and the strongly pulsating flow. Only recently have meters that appear to be practical for exhaust flow measurement become available. The most promising of these is the vortex flowmeter (*21, 22*). In this instrument, a bluff body (nonstreamlined) is placed in the stream to be measured. As the gas moves past the body, vortices are formed that shed from the body at a frequency that is directly proportional to the velocity of the stream. Various means are available to measure the rate of shedding. Given the cross-sectional area, and the temperature and pressure in the flowmeter, the flow rate at standard conditions can be calculated. In commercially available meters, these calculations can be performed by the electronic package supplied with the instrument. The vortex flowmeter has a linear response over a wide range of flows; it offers little resistance to flow and has no narrow passages to clog. It has no moving parts and is relatively small. The meter calibration factor is independent of fluid density, temperature, pressure, composition, and viscosity. For testing engines or vehicles on chassis dynamometers, exhaust measurement is the more convenient and reliable approach. For road measurements, air measurement is the more practical procedure.

The second proportional sampling method is a variation of the critical-flow venturi CVS system (*17*). In this method, a small critical-flow venturi is used to collect the sample just before the diluted exhaust enters the main venturi. If both the main and sample gas venturi are held at critical flow, a fixed split ratio will be maintained. A true proportional sample will be collected in the bag, even though the flow rate of the diluted exhaust to the main venturi varies. Because of this, it is possible to do away with the heat exchanger normally employed when the critical flow venturi system is used as a CVS. A temperature sensor at the split point is all that is needed to make a continuous calculation of the total flow to the venturis at standard conditions. Since the sample gas is already prediluted, no further precautions are needed to prevent condensation. In laboratory testing of automotive engines or vehicles operating on chassis dynamometers, the critical flow method would, in most situations, be preferable to the use of a flow-actuated proportionating valve. The former method is simpler, and the dilution with air leads to much smaller variations in flow and temperature. For road testing, the flow-actuated proportional sampler would need to be used because of its smaller size.

c. TOTAL EXHAUST SAMPLER. Another method that has been used for integrating samples for mass emission measurement is to collect the total exhaust over an operating cycle in one or more large plastic bags. At the end of the test, the total flow into the bag can be determined by pumping the contents of the bag through a dry test meter. Total volume at standard conditions and measured concentrations are substituted into Equation (5) to calculate the mass emissions over the test. Condensation of water in the bag is a problem. Any nitric oxide converted to nitrogen dioxide by reaction with oxygen present in the exhaust is lost by dissolving in the condensed water. Condensation can be reduced by predilution of the bag with nitrogen or air, but this procedure is of limited use in large bags.

None of the batch-integrated methods discussed can be used in the determination of hydrocarbons in the exhaust gases from automotive diesel engines, because the high molecular weight diesel exhaust hydrocarbons would be lost on the surfaces of the bag. One of the continuous methods to be discussed must be used for diesel exhaust.

2. Continuous Measurement Methods

In continuous measurement methods, sample gas is brought directly to the analytical instrumentation and measured continuously over the test. These methods are aplicable to both open cycles, Equation (2), and closed cycles, Equation (3).

a. EXHAUST DILUTION SYSTEMS. The CVS and the temperature corrected critical flow methods, which have been principally used as batch integrating methods, can also be used for continuous measurements. In this case, the bag is not used. Instead, sample gas is passed continuously through the analytical instrumentation. Since with CVS the total flow rate is constant, only the concentrations in the diluted exhaust need be integrated over the test cycle.

$$W(\mathrm{X}) = D(\mathrm{X})V \int_0^{\tau} [\mathrm{X}, t]\, dt \tag{12}$$

The integration of the concentration signal can easily be performed electronically if the calibration curve of the analytical instrument is linear. FID and chemiluminescent instruments are intrinsically linear, while the signals from nondispersive infrared gas analyzers can be linearized electronically. Commercial instruments can be obtained with this feature built in. Since the total flow is not constant with the temperature-corrected critical-flow venturi system, the product of concentration and flow rate must be integrated as required by Equation (3). However, another small correction is necessary to achieve full accuracy with this system. Since it requires time for sample gas to pass from the point of sampling to the analytical instrument, which then takes a small additional time to register its reading, there will be a difference of phase between the flow rate as measured by the temperature of the gas at the split point and the concentration as measured by the instruments. The correct form of the mass emission expressions is

$$W(\mathrm{X}) = D(\mathrm{X}) \int_0^{\tau} [\mathrm{X}(t + \delta)]Q(t)\, dt \tag{13}$$

The term δ is the difference between the time a given exhaust gas segment is sampled at the small venturi and the time the analytical instrument has given in excess of 90% of its response to the segment. This small phase difference is reasonably constant in a properly designed analytical system. The need to continuously form the product of concentration and flow rate after taking the phase difference into account complicates the application of the temperature-corrected venturi system to continuous analysis of closed dynamic operating cycles. The required calculations are most conveniently done by feeding the output information to a computer. The CVS system with heat exchanger, therefore, has a distinct advantage in this application. The dilution system can be conveniently used as a combination batch and continuous mass emission measurement system. For example, in the case of diesel exhaust, a portion of sample can be continuously analyzed for hydrocarbons and bag samples simultaneously collected for determination of carbon monoxide and nitrogen oxides at the end of the test (*23*).

b. Exhaust Flow Measurement. With the advent of practical exhaust-gas flowmeters, such as the vortex flowmeter, undiluted exhaust can be brought directly to the instrument system for analysis. Equation (13) applies, and $Q(t)$ is the actual exhaust flow rate at standard conditions. The major desirable feature of this method is that it gives independent measures of actual exhaust concentrations and flow rate at each operating condition throughout the test and therefore directly yields basic measurements of engine behavior. These measurements can be obtained with dilution systems, but it is necessary, in this case, to measure one of the components before and after dilution in order to develop a correction factor. The other good feature of the direct exhaust-flow measurement method is that there is no dilution of the exhaust, and therefore less demand on the sensitivity of the analytical instrumentation. The principle disadvantage is the complexity of the numerical integration of the correctly phased product of concentration and flow rate.

Since the exhaust flow measurement is on wet exhaust, all concentration measurements made on dry exhaust must be corrected to a wet basis. It is preferable in using this method with either gasoline or diesel exhaust to measure hydrocarbons with a heated FID to minimize condensation or adsorption of hydrocarbons on the walls of the sampling lines. The use of an all-heated system such as shown in Figure 10 will both reduce hydrocarbon losses and allow a wet measurement of all components and avoid the need for a correction.

c. Intake Air Flow Measurement. Measurement of the intake air may be substituted for the exhaust measurement. It has the advantage of a more constant temperature correction in determining standard flow rate but the disadvantage of being an indirect measure of exhaust flow.

d. Fuel Flow Measurement (Carbon Balance). Measurement of fuel flow is another alternative to determining exhaust flow. The technique is commonly referred to as the carbon balance method (*24*). It requires that the concentrations of carbon dioxide, carbon monoxide, and hydrocarbon in the exhaust be measured along with component of interest. The carbon-to-hydrocarbon ratio of the fuel also must be known. Since the carbon monoxide, carbon dioxide, and hydrocarbon account for all the carbon in the exhaust, their emission rate must be equal to the rate of consumption of carbon in the fuel. Consequently,

$$W(\mathrm{X}) = M(\mathrm{X}) \frac{[\mathrm{X}]}{[\mathrm{CO_2}] + [\mathrm{CO}] + [\mathrm{HC}]} \frac{F}{(12.011 + 1.008\alpha)} \tag{14}$$

where $W(\mathrm{X})$ = emission of component X in weight per unit time; $M(\mathrm{X})$ = molecular weight of component X; F = weight of fuel consumed per unit time; α = the number of moles of atomic hydrogen per mole of carbon in the fuel. The quantity $(12.011 + 1.008\ \alpha)$ is the weight of fuel per mole of carbon in the fuel.

The emissions, $W(\mathrm{X})$, are given as a rate, because the principle application of the carbon balance method to date has been to open cycles of steady-state modes. In this case, Equation (14) applies to individual modes and Equation (2) is used to calculate the mass emission for the test. Its future use will probably be limited to open cycles because of the difficulties involved in accurately measuring instantaneous fuel-flow rates under dynamic conditions in either gasoline or diesel engines.

e. TRACER INJECTION. The injection of stable tracer gases into the exhaust at a known fixed rate is a potential continuous mass emission measurement method. Sample gas is taken for analysis at a point sufficiently downstream for thorough mixing of the tracer and the exhaust. The flow rate can be calculated from

$$Q = Q(\mathrm{Tr})/[\mathrm{Tr}] \qquad (15)$$

where $Q(\mathrm{Tr})$ = volumetric injection rate of tracer gas at standard conditions; $[\mathrm{Tr}]$ = mole fraction of tracer gas in exhaust. The tracer is injected through a capillary with sufficient pressure on the upstream side to insure a constant injection rate under all conditions. Because the tracer concentration is measured at a finite time after sampling, phase difference must be taken into account. If the phase differences of the analytical instruments for the components of interest are adjusted so that they all are the same as that for the tracer gas analysis, the phase effect cancels out. As seen from Equation (15), exhaust flow rate is inversely proportional to the concentration of the tracer gas. Thus the tracer method has the undesirable feature of giving a nonlinear measure of flow rate, which measures the highest flow rates at the low end of the tracer concentration scale where the measurement error is largest. Helium, nitrous oxide, and carbon dioxide have been studied as tracers (*21*). A mass spectrograph is required if helium is used. The nitrous oxide and carbon dioxide tracers can be detected with nondispersive infrared analyzers. Because carbon dioxide is already present in the exhaust, a sensitive differential technique is needed to measure the small increment in carbon dioxide concentration resulting from tracer injection.

C. Vehicle Operating Cycles

There is substantial value in measuring exhaust emissions directly from engines mounted on an engine dynamometer in a laboratory because of the wide range of engine loads that can be studied. However, by far the greater number of measurements have been made on the emissions from complete vehicles rather than separate engines. Measurement of emissions from vehicles provides data and conditions more closely approximating the emission of pollutants to the atmosphere. This is particularly true under changing operating conditions, such as acceleration and deceleration, which are not as easily duplicated on an engine dynamometer. Many studies have been performed using portable analytical instrumentation and sampling equipment on vehicles operating on the road. However, they are cumbersome and do not permit the more elaborate measurement that can be made in the laboratory. As a result, the preponderance of measurements made have been on vehicles operating over predetermined driving cycles on chassis dynamometers.

A chassis dynamometer driving cycle for emission measurement attempts to simulate the average driving pattern for a given geographic area during a specified time period such that the measured emissions represent the emissions to the atmosphere that would be observed if the vehicle were driven on the road in the same geographic area and time period. A cycle meeting this description could be referred to as a comprehensive driving cycle. It would be mainly derived from traffic data rather than emission data and should be equally applicable to the measurement of all exhaust components.

A driving cycle can be described as a connected series of driving modes. A driving mode can be characterized by two parameters for a vehicle operating on a level road, the vehicle speed, and its rate of change of speed. From this point of view, a vehicle travels through an infinite number of modes every time it changes its speed from one to another. In practice, driving cycles usually hold the rate of change of speed constant during periods of acceleration and deceleration, so that it is more common to think of the mode as extending over the whole or a portion of the period during which the rate of change of speed is constant. In this case, the mode is designated by the lowest and highest speeds and the rate of change of speed.

Driving cycles also can be described as open or closed. In an open cycle, only some of the modes, which will be called emission modes, are used in the emission measurement. These modes are connected by transitional modes whose purpose is to get the vehicle from one group of emission modes to the next in a way that is driveable. Time spent in each mode is not critical. The seven-mode federal cycle was originally designed as an open cycle (Table 1 and Fig. 11).

The average concentration of a given exhaust component determined over a mode is normally corrected for dilution effects and then weighted by another factor that is proportional to the product of the time spent in the mode in traffic and an average exhaust flow rate during the mode. The corrected and weighted averages per mode are then summed to give the average emission over one cycle. The cycles may be repeated several times with different weightings given to each. Customarily, the first few cycles are considered to represent the warm-up period after a vehicle has been standing idle, while the later cycles are considered representative of a vehicle at operating temperature.

In a closed cycle, the emission modes are connected together in such a way as to form a driveable cycle. There are no transitional modes. The time spent in each mode is proportional to the time vehicles spend in the mode in traffic for a comprehensive cycle. With the absence of transitional modes, flexibility in building the driving cycle is largely lost. Putting together a driveable closed cycle can require considerable ingenuity. It may be necessary, for example, to put in semitransitional modes that do not affect the final answer significantly even though emissions during these modes are counted. The complex LA 4 cycle used in the United States federal certification procedure (*25*) is an example of a comprehensive closed cycle (see also Chapter 12, Volume V). The utility of the closed cycle is that it can be used with CVS or other continuous sampling systems that provide an integrated weighted average emission over the cycle.

Short cycles are used where testing time is at a premium, such as in state inspection of in-use vehicles. A short cycle is often a shortened version of a comprehensive driving cycle. The development of short cycles of this kind are possible because the emissions from different driving modes tend to correlate. Because a short cycle is usually derived on the basis of matching emissions from a given comprehensive cycle rather than from traffic patterns, a short cycle that gives maximum correlation for one exhaust component will not necessarily be optimum for another component. Thus, some compromise will usually have to be made to obtain a reasonable correlation of all the components it is desired to measure. This compromise is more difficult to make in closed than in open cycles.

In the open short cycle, the weighting factors no longer have the significance of being proportional to the product of the average time spent in mode and the average exhaust flow rate. They are, instead, weighting factors that give the best agreement between the short cycle emissions and the corresponding comprehensive cycle emissions. They can be, therefore, the coefficients obtained from a multiple regression analysis, and they generally will be different for each component to be measured. While short cycles have been developed, they have not been much used in practice. The United States seven-mode federal cycle is neither a comprehensive nor a short cycle. It resulted from an arbitrary shortening of the original comprehensive California open cycle developed to simulate Los Angeles driving during peak pollution periods.

Vehicle preconditioning is another important aspect of short-cycle development. The complete United States federal procedure requires that a vehicle stand with ignition off for 12 hours prior to the test (*7*). Since lack of time is often the major factor leading to the need for a short cycle, extensive preconditioning involving cooling of the vehicle to ambient conditions is not practical. Thus, getting good correlation of the full United States federal procedure with an essentially hot short cycle is very difficult. Even the hot cycles of the federal test correlate only moderately well with the cold-start portion of the test.

Historically, the development of driving cycles has been primarily through cut-and-try techniques. They have been customarily derived by comparing trial cycles to extensive traffic data on the way vehicles are driven in a given area. The original California comprehensive cycle was developed in this way. Traffic counts are employed to determine the percentage of traffic on various roads from highways to arterial connecting roads, and vehicles are driven along representative routes to determine times in different modes and general behavior (*26*).

The development of the present United States federal test cycle (the LA 4 cycle) departed from the cut-and-try method. It essentially duplicated a single-speed time record for a vehicle driven in normal traffic over a specified route in California. While this procedure gives a certain level of realism, it is difficult to judge its general applicability. More recently, a mathematical method has been developed for operating on the data obtained from many cars using a chase car, an approach that yields a representative cycle (*27*). The chase car is an excellent technique for obtaining data on how vehicles are driven. For a given period of time an instrumented chase car follows a randomly choosen vehicle on a given route, closely simulating its behavior. If the vehicle moves off the route, another vehicle is followed. In this way a large number of vehicles displaying the driving characteristics of many different drivers can be studied. A skilled operator in a chase car can closely duplicate the behavior of the vehicle he is following (*28*).

The development of a driving cycle from chase car data that represent the way people drive their automobiles is a process of abstraction and simplification. There is, strictly speaking, no average way vehicles are driven. The speed–time traces for many chase runs over each of a number of selected driving patterns are analyzed to develop a matrix characteristic of each pattern. Each element in the matrix represents a transition from one speed, indicated by the row number, to another speed indicated by the column number. The matrix is square and is referred to as a transition matrix. The matrix representation for the seven-mode federal cycle considered as a closed cycle is shown in Figure 14.

The diagonal elements of the matrix are constant speed modes. The off-diagonal elements are acceleration or deceleration modes, depending of whether the element is to the right or left of the diagonal elements. The transition matrix describes a driving mode in the same manner as the United States federal cycle. That is, an acceleration or deceleration is deliniated by the speed at which it starts and the speed at which it ends. The matrix elements of the transition matrix can be used to designate a number of characteristics of the mode. For example, the relative time spent in the mode, the average rate of change of speed for the mode (for diagonal elements, this is zero), and the relative frequency of occurrence of the mode.

Referring back to the transition matrix representation of the seven-mode cycle, the sequence in which the elements occur is given by

$$a_{1,1} \longrightarrow a_{1,4} \longrightarrow a_{4,4} \longrightarrow a_{4,2} \longrightarrow a_{2,2} \longrightarrow a_{2,5} \longrightarrow a_{5,3} \longrightarrow a_{3,1}$$

This path (Fig. 15) would be traced if the following rules were followed:

1. Start at 1,1.
2. Proceed from one element to the next so that the transition is to

Speed at end of mode (miles/hour)

Speed at start of mode (miles/hour)	0	15	20	30	50
0	20	0	0	14	0
15	0	15	0	0	29
20	8	0	0	0	0
30	0	11	0	15	0
50	0	0	25	0	0

Figure 14. Matrix representation of seven-mode cycle. Matrix elements are time in mode (seconds).

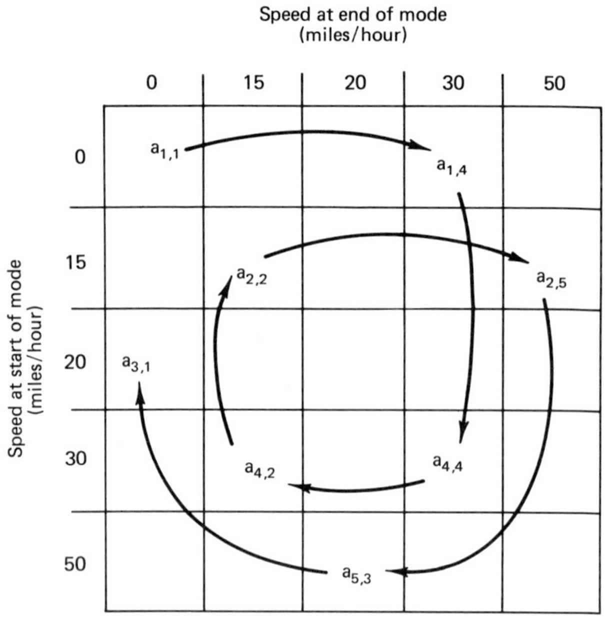

Figure 15. Path through seven-mode cycle transition matrix.

a nonzero element whose row number is equal to the column number of the next preceding occupied element.

3. Occupy all nonzero elements in the matrix once.

The path in this case is uniquely defined by these rules. The arrows in Figure 15 show the path through the matrix. In a complex case, a Monte Carlo technique is used to determine matrix paths that give practical driving cycles representing the overall data (*27*).

IV. Measuring Evaporative and Refueling Emissions from Automobiles

A. Measurement of Fuel Evaporative Emissions

There are two methods currently in use for the measurement of evaporation losses from the fuel systems of light-duty gasoline-powered motor vehicles. The first method employs activated carbon traps connected to the fuel system at locations where vapors are expected to escape. The gain in weight of the traps due to the adsorption of vapors represents the fuel evaporative emissions. This procedure is discussed in detail in SAE Recommended Practice J170a (*29*).

The second method employs an enclosure in which the vehicle is housed during the diurnal and hot-soak phases of the test. Escaping vapors from all openings in the fuel system are retained in the enclosure and the increase in hydrocarbon concentration measured by the flame ionization method represents the evaporative emissions. This method is presented in detail in SAE Recommended Practice J171a (*30*).

1. Activated Carbon Trap Technique

This measurement method provides for connecting activated carbon traps to expected sources of vapor emissions from the vehicle fuel system. Evaporative emissions are determined by measuring the increase in weight of all the traps in the system. The sum of the weight increases represents the total evaporative emission from all the sources trapped.

The trap (Fig. 16) is a cylindrical container of approximately 300 ml capacity. The canisters should be capable of withstanding air pressure of 2 psi (14 kPa) when immersed in water for 30 seconds without leakage. The trap should be filled with 150 ± 10 gm of activated charcoal that has been oven dried just before use and meets the following specifications:

Surface area, minimum: 1000 m^2/gm (N_2, BET method) (*31*)
Adsorption capacity, minimum (carbon tetrachloride): 60% (weight)
Volatile material including adsorbed water vapor: None
Screen size analysis: 1.7–2.4 mm, 90% minimum; 1.4–3.0 mm, 100%.

A typical trap installation is shown in Figure 17 for a fuel system vented to the atmosphere through the air cleaner. Some vehicles may have addi-

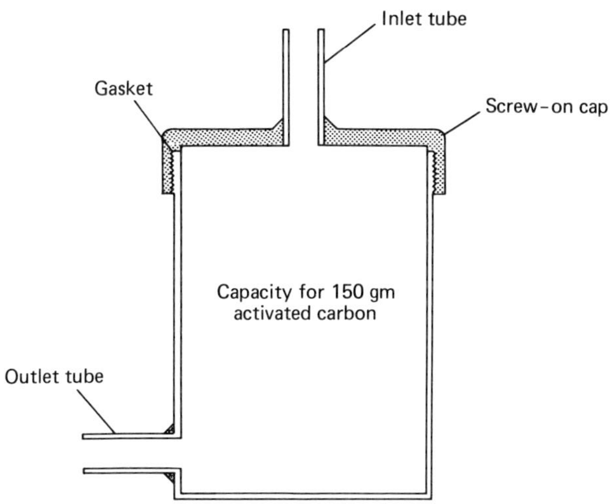

Figure 16. Trap for collection of evaporative emissions (7).

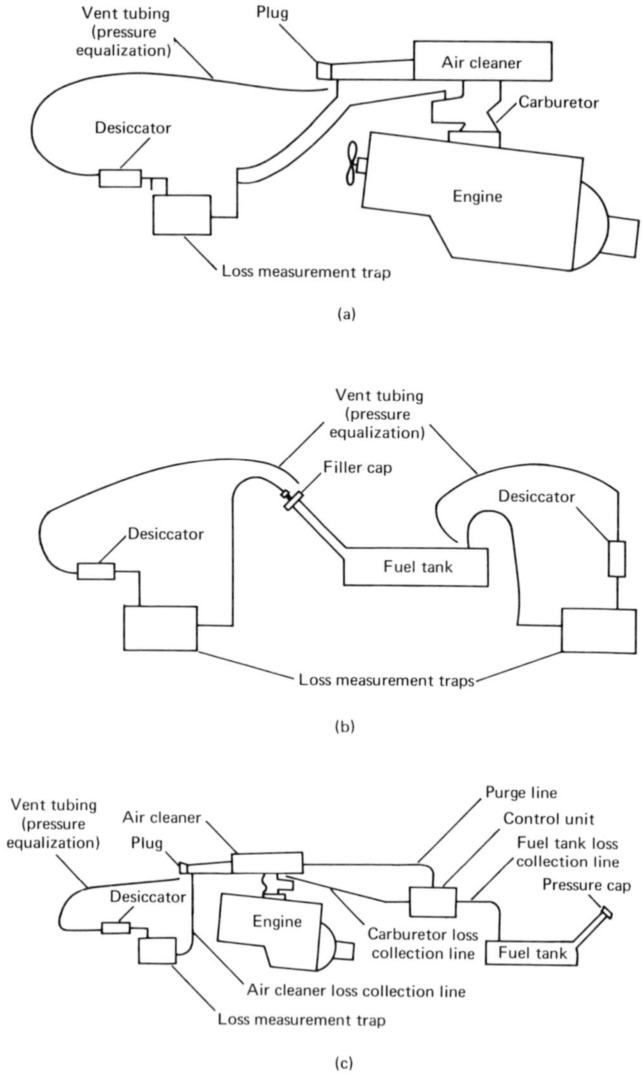

Figure 17. Collection systems for evaporative emissions (7): (a) carburetor evaporative emissions; (b) fuel tank evaporative emissions; (c) collection system for vehicles equipped with evaporative emission control systems.

tional openings where fuel vapor may escape to the atmosphere, and traps must also be connected to them. The traps should be connected to the fuel system vents with Teflon, stainless steel, aluminum, or equivalent $\frac{5}{16}$ inch (7.9 mm) tubing. Joints should be made with heavy-wall fuel-resistant $\frac{5}{16}$ inch (7.9 mm) I.D. flexible rubber vacuum hose.

Traps should be installed as close as possible to the vapor source, but must not be exposed to heat from the engine. They must be placed lower than the vapor source to induce flow of vapors by gravity, but not disturb the function of the fuel system components. The lines and any manifold systems used to connect multiple vents to a single trap should be checked to make sure there are no sharp bends in the connection lines. All lines must have a continuous fall between the fuel system and the trap. The pressure equalization tube must be routed from the trap outlet tube to a location as close as possible to the vapor connection vapor source. In all cases, the modifications and connections to the fuel system, connecting tubes, traps, and other equipment must not disturb the normal functioning of the fuel system components. On vehicles without evaporative emission control systems, multiple traps in series may be necessary to insure that the vapor adsorption capacity of the trap(s) not be exceeded during the test.

The tare weight of the individual traps must be determined within 12 hours of the evaporative emission test. If the weight has changed by more than 0.5 gm from the original weight, the trap must be redried by passing dry nitrogen through the trap. After the evaporative emission tests, the traps are removed, and the inlet and outlet are capped and the trap is allowed to come to room temperature. The traps are then weighed to the nearest 50 mg and the gross weight recorded on the canister.

The test sequence consists of three phases intended to simulate typical usage in a metropolitan area during the summer months:

1. A 1-hour diurnal soak during which time the tank fuel is heated from 60° to 84°F (15.6° to 28.9°C)
2. A 7.5-mile (12.1 km) (approximate) run conducted on a chassis dynamometer from a "cold" start
3. A 1-hour hot soak immediately following the 7.5-mile (12.1 km) run

Evaporative emissions from the vehicle fuel system are measured during each of the three phases. However, measurements during the run phase on a vehicle with an evaporative control system may be omitted if it is apparent that, by system design, all "running losses" will be inducted into the engine.

2. Vehicle Enclosure Technique

The vehicle enclosure method provides for sealing the vehicle in an enclosure during the soak phases of the test. Evaporative emissions are determined from the changes in the hydrocarbon concentrations in the enclosure. This method produces a single measurement of evaporative emissions during a soak, regardless of the number of sources. It can be used to measure evaporative emissions without modifications to the vehicle, and it can also be combined with the trap method to assist in quantitative allocation of emissions to specific sources. It is not applicable to the run phase and therefore is best suited to vehicles that fully control running losses.

The typical enclosure is designed to accommodate vehicles with up to a 130 inch (330 cm) wheelbase. Its interior surfaces must be impermeable to hydrocarbons. These surfaces are normally of polyvinyl fluoride (Teflon or equivalent); a sheet of approximately 0.006 inch (0.15 mm) thickness. A door must be provided to allow entrance of the maximum size automobile. If a flexible door is used, sealing is normally accomplished by a pressure-sealing zipper. Enclosure temperature should be limited to 90°F (32°C). Therefore, during the hot soak phase, it may be necessary to direct the discharge from an internal blower against the enclosure wall to help control peak temperatures.

The measurement of the enclosure hydrocarbons is accomplished by means of a flame ionization detector (FID) analyzer. The analyzer should be capable of measuring hydrocarbons in the range of 10–3000 ppm carbon. A typical sample train for the analyzer is shown in Figure 18. The sample should be withdrawn from the enclosure through a tube located on the geometric center of the wall opposite the door or on one of the sides.

A blower (portable or fixed) of 1500–5600 ft^3/minute (42–158

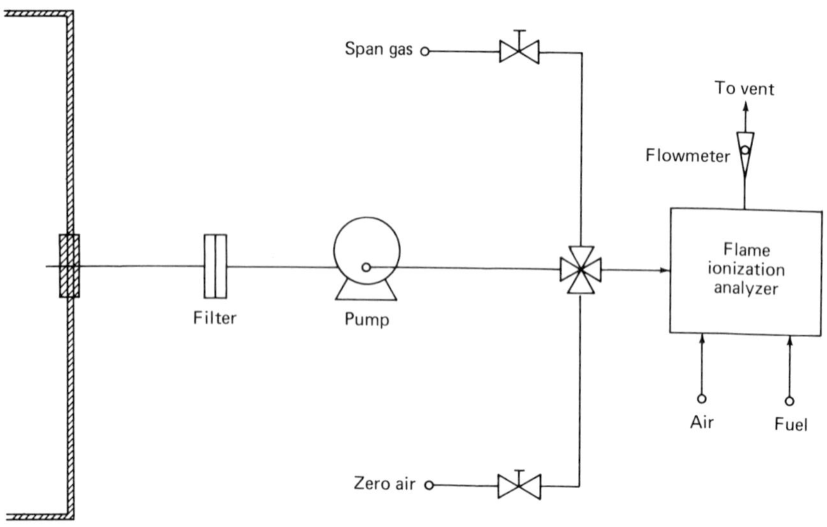

Figure 18. Sample train for measurement of evaporative emissions in an enclosure.

m^3/minute) capacity is required for purging the enclosure between tests, and an explosion-proof blower of 100–200 ft^3/minute (3–6 m^3/minute) capacity is required for mixing the enclosure atmosphere during tests. Two thermocouples for monitoring the enclosure temperature should be installed 30 inches (76 cm) above the floor and 4 inches (10 cm) in from either side, both midway along the length of the enclosure.

A test is conducted by placing a vehicle in the enclosure after the enclosure has been purged for several minutes immediately prior to test. An initial hydrocarbon measurement is made, and both hydrocarbon and temperature (vehicle and enclosure) measurements are made at 5-minute intervals throughout the soak period. The net hydrocarbon concentration is the final hydrocarbon reading minus the initial reading.

Equation (1) can be used to calculate the evaporative loss within the enclosure. In this case, $D(X)$ is the average density of evaporative hydrocarbons at standard conditions assuming ideal gas behavior. The SAE Recommended Practice J171a (*30*) indicates the average ratio of hydrogen to carbon atoms in hydrocarbon evaporative losses as 2.2 for carburetor emissions and 2.33 for fuel tank emissions. These values are higher than the 1.85 ratio commonly used for exhaust gas hydrocarbons. The ratios given for evaportive emissions lead to the densities tabulated below when hydrocarbon is measured on a carbon basis. The average density can be used with less than 1% error.

	gm/cubic foot at 68°F (20°C) 1 atm	*gm/m³ at 0°C, 1 atm*
Carburetor emissions as $CH_{2.2}$	16.75	634.8
Fuel tank emissions as $CH_{2.33}$	16.90	640.5
Average	16.83	637.9

The volume V for use in Equation (1) is the volume occupied by the air–hydrocarbon mixture within the enclosure corrected to standard conditions. The volume units and standard conditions used should be appropriate to the density value taken from the table. The gas volume before correction to standard conditions is the difference between the volume of the enclosure and the vehicle volume for a standard size United States car with windows and trunk open [approximately 50 ft^3 (1.42 m^3)].

B. Measurement of Vehicle Refueling Emissions

Refueling losses may be divided into two kinds, those that occur as a result of the displacement of tank vapor and entrained fuel droplets during the refueling operation and those that occur from spillage and fuel dripping from the refueling nozzle. The displaced and spill losses present different measurement problems.

1. Displaced Refueling Losses

The displaced losses normally account for 90% or more of the total refueling emissions. They are also more reproducible than spill losses, given the same fuel quantity, volatility, and tank temperatures. The enclosure technique for measuring evaporative emissions is also employed for measuring displaced refueling emissions (*32*). To obtain good measurements, a proper fueling facility must be available. Such a system is usually portable and consists of a fuel tank with temperature control equipment and a service-type dispensing pump, meter, hose, and nozzle. During a test run, the unit stands on the outside of the vehicle enclosure with only the hose and nozzle on the inside. The vehicle fuel tank is heated by external application of heating pads prior to the refueling operation in order to bring the fuel to the desired temperature for the run.

In another measurement technique (*33*), a small plastic film enclosure that is large enough to contain only the vehicle fuel tank (Fig. 19) is used. An operator outside the enclosure operates a dispenser nozzle within the enclosure through a plastic glove insert attached to the wall. An arrangement of this kind permits careful control of temperatures and more precise separation of displaced and spill losses.

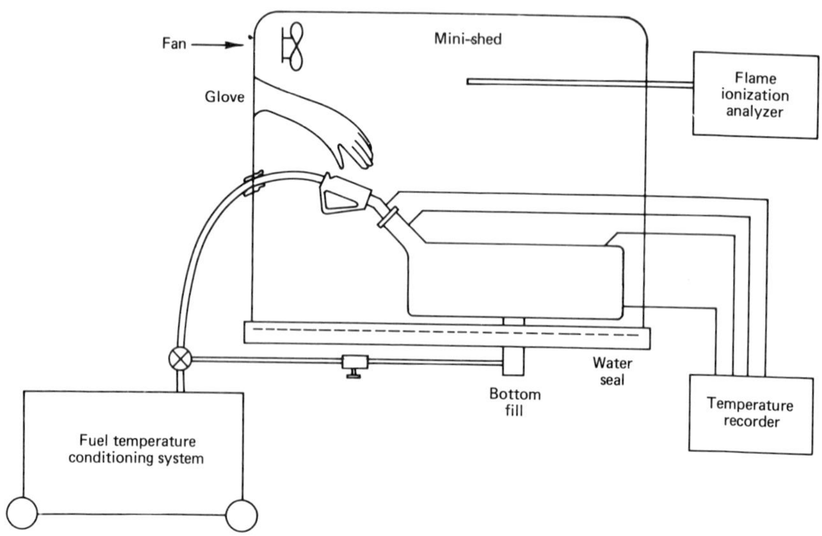

Figure 19. Mini-shed for measurement of refueling emissions.

2. Spill and Nozzle-Drip Losses

Spill and drip losses are largely a matter of the practices of the service station attendant and of the attention he gives the refueling operation. Differences among the tendencies of nozzle–pump systems to drip and the effect of automatic fuel cutoff system and filler pipe designs can be studied in the laboratory using proper enclosures. However, because of the importance of the operator, the proper place to measure spill losses is the service station. It is also necessary that the attendant be unaware he is being observed. Under these circumstances, measurement equipment cannot easily be used. Training techniques have been developed for teaching an observer to gauge the magnitude of such losses (*34*).

In laboratory enclosure measurements, the procedure is to allow gasoline to flow until the nozzle cuts off the flow automatically. The refueling operation is then continued a predetermined number of times (usually three), with the nozzle automatically cutting off each time. This procedure will measure the sum of the displaced and spill losses. The displaced losses are determined by repeating the test, delivering the same volume of gasoline to an automobile, but taking extra precautions to prevent any spills from occurring. The difference in the two total losses is the spill loss. The operator technique must be carefully specified so as to minimize variance from this source.

Nozzle-drip losses can be measured by fashioning a small catch pan or funnel that is fixed to the end of the filler pipe so that as the nozzle is withdrawn, any drippage will fall into the pan and be collected and measured in an accurate graduate (*33*). The manner of withdrawing the nozzle must be carefully specified.

To measure losses under the more realistic conditions of the field, observers are trained to estimate the amount of spill loss by estimating the diameter of the wet area made by the spill on the ground. Experiments in the laboratory on concrete aprons showed that spill diameter correlates reasonably well with quantity spilled and that observers can be trained to make such estimations with reasonable accuracy (*34*). Where part of the spill is on the vehicle, an estimate is made for that part of the loss and a note made on the observation form. The observer also classifies the loss as prefill, spit-back loss, overfill loss, or postfill loss.

V. Testing, Surveillance, and Inspection of Automobile Emissions

Testing, surveillance, and inspection encompass emission certification procedures carried out on vehicle prototypes for future production that enable the manufacturer to produce and market vehicles complying with national emission standards; the testing of new vehicles as they come off the assembly line to provide further assurance that the mass produced vehicles are performing in the same manner as the prototype vehicles that were certified; and the further testing of the vehicles in the hands of the public to assure their continued compliance. The latter tests include: routine inspection by states and provinces, with maintenance required on vehicles not meeting national, state, or provincial requirements; follow-up testing of vehicles that have been properly maintained, to test for continued compliance with emission standards; and the general surveillance of vehicles in the population to determine the overall impact of national state, and provincial programs.

A. Prototype Certification Procedures (United States)

United States federal certification of prototype light-duty vehicles seeks to assure that the mix of vehicle models the manufacturer has selected for production will comply with the federal emissions standards over the useful life of the vehicle, if maintained according to the manufacturer's recommendation. Certification is based on the successful conclusion of the federal procedure by fleets of individually assembled vehicles representative of the planned production for the model year to be certified. This procedure has the feature of permitting regulatory action to be taken in time to prevent the manufacture of noncompliant vehicles. At the same time, it gives the manufacturer lead time to introduce substitutions into his product line to compensate for the vehicles that have failed certification testing. The federal prototype certification procedure has been in operation since the 1968 model year. In its early stages, it was primarily an application of procedures developed by the state of California. However, since that time, it has undergone substantial growth in scope and sophistication. A complete description of the process is contained in the United States Code of Federal Regulations (*7*). The procedure starts when the manufacturer submits to the United States Environmental Protection Agency (EPA) an application in which he describes the product line he plans to market and the procedures he plans to use to obtain certification data. EPA reviews the material and selects the fleets to be tested. This is done by dividing the manufacturer's product line into engine families. These are groupings in which the engines within a given family are expected to have the same emissions characteristics. The choice is made on the basis of critical engine dimensions, location of intake and exhaust valves, method of aspirating air, and the combustion cycle. Two broad fleets are selected, the emission data fleet and the durability fleet. The emission data fleet is selected so that within each engine family there are two to four vehicle/control system combinations chosen on the basis of projected sales along with up to four additional vehicles representing configurations that have features indicating they may have the highest emissions in the family.

The durability data fleet is chosen by classifing all the vehicles into engine family–exhaust emission control–evaporative emission control combinations. The model in each combination with the highest projected sales volume is selected for testing. The selection is done so that each emission control combination is represented twice in the durability fleet. As soon as the durability data fleets have been selected by EPA, the manufacturer begins to assemble the vehicles that will be used in his test fleet. The vehicles are assembled by mechanics from preproduction components.

Although these components are made from the same specifications as those that will be massed produced, they are not normally fabricated on production machinery. There may, therefore, be subtle differences from their mass-produced counterparts. Each of these assembled vehicles is checked to be sure of its calibrations (carburator flow, spark timing, exhaust gas recirculation flow, etc.) are within the specifications the manufacturer has set forth in his application. An initial check of emissions measured by the Federal Emission Test Procedure (7) is made to determine the vehicle's "zero mile emissions." Because the engines are new, the exhaust emissions tend to be somewhat higher than after the vehicle has been thoroughly broken in by running several thousand miles. If the vehicle's emissions are in line with expectations, and if EPA and the manufacturer concur, the testing of the durability fleet begins. The fleet vehicles are run for 50,000 miles on the manufacturer's test track on a fixed driving regime 24 hours a day, 7 days a week, with intermittent shut downs. The use of mileage accumulation dynamometers is becoming more frequent because they require less personnel to operate and reduce the incidence of accidents and other errors that may result in the loss or disqualification of vehicles being tested. Once vehicles are placed on test, the maintenance allowed is carefully prescribed by the regulations. Scheduled maintenance is performed at the mileages specified by the manufacturer's maintenance instructions to the purchaser. Scheduled major tune-ups following manufacturer's specifications may be performed no more frequently than every 12,500 miles of scheduled driving. Unscheduled maintenance can only be conducted when there is a sufficiently clear indication of malfunction that would lead the owner of the vehicle to have the vehicle repaired. Diagnosis of the problem and required repairs and maintenance can be done only with the equipment normally available at a commercial repair facility. Should the repair of a malfunction result in an unrepresentative vehicle, it is removed from the fleet and a new vehicle is started. Emission tests are performed on the durability fleet every 5000 miles up to and including 50,000 miles, at which time the test is terminated. Idle adjustments are allowed before each 5000 mile emission test.

In general, the preparation of the vehicles for the emission data fleet and the zero-mile test requirements is the same as for the durability fleet. In these cases the vehicle is run to 4000 miles by the normal mileage accumulation method used for the durability fleet, and the emission test made at this point. No scheduled maintenance is allowed in the 4000 mile period. Unscheduled maintenance is allowed on the same basis as in the durability fleet. The emission testing is done primarily by the manufacturer. EPA, however, may require that any of the vehicles under test be submitted for emission testing at the EPA facility. In practice, EPA frequently will test vehicles from the emission data fleet.

Should the 4000 mile test on a given data test fleet vehicle fail to meet the federal standards, the manufacturer must drop all models falling within that classification from his product line. The manufacturer may, if there is sufficient time, devise a new family to replace that which has been disqualified and start both new emission data and new durability data vehicles. If a vehicle fails the emission test for the durability fleet, it is sometimes possible with relatively small changes to define a new engine family. If the altered classification is shown to be compliant, the emission data vehicles belonging to the same engine family must then be set up in the new way. These changes generally can be made without greatly affecting the manufacturer's planned product line. There is substantially less flexibility in the case of failure of compliance by emission data vehicles, since all the emission data vehicles corresponding to a given engine family must comply with the regulation in order that the family be certified.

For each durability vehicle, separate plots are made of emissions at 5000–50,000 accumulated miles. Similar plots are made for evaporative emissions. Straight lines are fitted to each plot by the method of least squares. From these, the interpolated emissions at 4000 and 50,000 miles are obtained. For exhaust gas emissions, these values are ratioed to give a deterioration factor for each emission from each durability vehicle. For evaporative emissions, the differences are taken between the interpolated emissions at 4000 miles and at 50,000 miles to obtain a deterioration increment. When decrements are obtained (i.e., when the emissions become lower with age), a zero value is substituted. In the exhaust emission test, results obtained at 4000 miles on each data emission vehicle are multiplied by the appropriate deterioration factors to give projected emissions at 50,000 miles. In the case of evaporative emissions, the appropriate deterioration increment is added to the 4000 mile result. For an engine family to be certified, the projected emissions at 50,000 miles of each emission data vehicle in the family must comply with the federal emission standards for that model year.

There are two basic criticisms of the federal procedure for certification for prototype vehicles. The first, and perhaps more significant, is the assumption that rapid accumulation of 50,000 miles on a 24 hour, 7 days a week basis gives a true picture of the behavior of the emission control system over the useful life of the vehicle. There are no substantiating data at present that deterioration factors determined in this manner correlate with the actual behavior of vehicles over their useful life. On the other hand, a realistic accumulation of mileage would be impractical. It may be possible, however, over the years, to gain sufficient experience with control methods so as to devise laboratory tests that give a higher correlation with actual behavior in the field. The second criticism is that of judging a total vehicle class by a very small number of vehicles. Car manufacturers market a broad variety of vehicles that, because of the large numbers of families, result, under present procedures, in large certification test fleets. If a sufficient number of cars were to be used to develop statistically valid data, the sizes of these fleets would increase substantially. Present thinking is that if a manufacturer can demonstrate attainment of the standard with prototype vehicles, he should be able to produce the same result with production vehicles. The current method of allowing the manufacturer to introduce new families after cars have failed, combined with the testing of a relatively small number of cars seems, for the present at least, to be a practical compromise. Time will tell whether the overall procedures produce vehicles in the hands of the public that meet the United States federal emission standards over their useful life.

The emission test procedure for determining conformity with exhaust emission standards consists of a precisely described (7) sequence of fueling, preconditioning, and testing of both fuel evaporative emissions and exhaust emissions.

The exhaust gas emission test is designed to determine hydrocarbon, carbon monoxide, and oxides of nitrogen mass emissions while the vehicle operating on a chasiss dynamometer simulates an average trip of 7.5 miles in an urban area. The driving schedule, referred to as LA 4, was developed from detailed data on time and speed obtained as a test vehicle was driven over a route in Los Angeles. Exhaust emissions are sampled using the CVS method (Fig. 12).

In preparation for the exhaust gas emission test, the vehicle is preconditioned by standing with key off for no less than 12 hours at an ambient temperature of 76°–86°F in a dry area. The complete test consists of a cold-start drive over the complete driving schedule and a hot-start drive over the schedule. The vehicle is allowed to stand with key off for 10 minutes between the two tests. The cold-start test is divided into two periods. The first is called the cold-start transient phase and ends at the end of deceleration, which is scheduled to occur at 505 seconds into the driving schedule. The second period is the stabilized phase; it is the remainder of the schedule, including the engine shutdown. The hot-start test also has two periods. The first is the hot-start transient phase and ends at the same point in the driving schedule as the first phase of the cold-start test. The second, called the stabilized phase, is assumed to give emissions identical to the stabilized phase of the cold cycle and is therefore not counted. Thus, the hot-start test terminates after the first 505 seconds into the second driving schedule. During these tests, the ambient temperature is held between 68° and 86°F. The sample gas from the CVS is collected into three bags representing, respectively, the cold-start transient phase, the stabilized cold and hot starts, and the transient phase of the hot-start test. The gas samples along with samples of dilution air are analyzed by the analytical system shown in Figure 9 no later than 20 minutes after completion of the tests.

Mass emissions are calculated by Equation (5) after the measured exhaust concentrations are corrected for contamination of the dilution air using Equations (6) and (7). If chemical absorption agents have been used to remove water and carbon dioxide from the diluted exhaust prior to the carbon monoxide analysis, a correction is applied to the carbon monoxide concentration in order to account for the shrinkage in sample gas volume. The nitrogen oxides concentration is corrected to a standard ambient humidity to account for the effect of humidity on production of nitrogen oxides by light-duty engines. These latter corrections are detailed in the United States federal regulations (7). Corrected mass emissions of a given component are calculated in grams for the transient cold- and hot-start tests and the combined stabilized cold–hot-start tests. The overall mass emission in grams per mile is obtained by applying weighting factors to each of the three results, summing them, and dividing by 7.5 miles.

The dynamometer for the test must be set up so that it correctly simulates the vehicle inertial effects and provides power absorption equivalent to road operation. The inertial effect is most commonly simulated through the use of fly wheels. In some dynamometers, it is done electrically. The power absorption is set by adjusting to a given horsepower consumption at 50 miles/hour. Specific values for the inertial weights and the road load power at 50 miles/hour have been established by the United States Environmental Protection Agency for various vehicle weight classes (Table III). These settings have a general utility in vehicle testing and, of course, have the added value of conforming to the United States federal certification procedure. The road load power may be set in an alternative maner by operating the vehicle on a level road at 50 miles/hour and recording the intake manifold vacuum. The vehicle is then run at the same speed on the chassis dynamometer and the power absorption set to attain the same intake manifold vacuum.

The fuel evaporative emission test simulates the fuel hydrocarbon evaporative emissions to the atmosphere as a consequence of urban driving as well as diurnal temperature fluctuations during parking. The test is specifically aimed at collecting and weighing (a) diurnal breathing losses from the fuel tank and other parts of the fuel system when the fuel tank is subjected to a temperature increase representative of the diurnal range, (b) running losses from the fuel tank and carburetor resulting from a simulated trip on a chassis dynamometer, and (c) hot soak losses from the fuel tank and carburetor that result when the vehicle is parked and the hot engine is turned off. An activated carbon trap was used for collection of evaporative emissions through the 1975 model year but was then replaced by shed methods, which give a more complete measure of these emissions.

Table III Chassis Dynamometer Settings for Light-Duty Vehicles[a]

Loaded vehicle weight (pounds)	*Equivalent inertia (pounds)*	*Road load horsepower at 50 miles/hour*
Up to 1125	1000	5.9
1126–1375	1000	6.5
1376–1625	1500	7.1
1626–1875	1750	7.7
1876–2125	2000	8.3
2126–2375	2250	8.8
2376–2625	2500	9.4
2626–2875	2750	9.9
2876–3250	3000	10.3
3251–3750	3500	11.2
3751–4250	4000	12.0
4251–4750	4500	12.7
4751–5250	5000	13.4
5251–5750	5500	13.9
5751 and above	5500	14.4

[a] From (*7*).

In order to assure that no differences in either exhaust or evaporative tests arise due to differences in fuel, specifications for the test fuel have been established (Table IV). This fuel is widely used in emission testing and development work.

The certification test procedures for light-duty vehicles powered by diesel engines is very similar to those for vehicles powered by gasoline engines. Aside from difference in the specified test fuel, the major difference is in the method of measuring hydrocarbon emissions. A heated flame ionization detector is used to give a continuous analysis of a sample taken from the CVS just after dilution but prior to entrance into the heat exchanger. As indicated in an earlier section, the diluted concentration in the CVS is directly proportional to the mass emission rate. The integration of the concentration over the three test periods yields the desired mass emissions for each period.

The United States federal test procedure has undergone continuing change over the several years of its existence. Most of the change has come about from the increasing demands placed on the procedure as the emission standards have become more restrictive.

Table IV Specifications for Test Fuels for Emission Measurements from Light-Duty Vehicles[a]

Item	*ASTM*[b] *designation*	*Leaded fuel*	*Unleaded fuel*
Octane (research minimum)	D1656	100	96
Lead (organic) (gm/U.S. gallon)		1.4	0.00–0.05
Distillation range	D86		
Initial boiling point (°F)		75–95	75–95
10% point (°F)		120–135	120–135
50% point (°F)		200–230	200–230
90% point (°F)		300–325	300–325
Endpoint (°F, maximum)		415	415
Sulfur (wt %, maximum)	D1266	0.10	0.10
Phosphorus (gm/U.S. gallon, maximum)		0.01	0.005
Reid vapor pressure,[c] (pounds/square inch)	D323	8.7–9.2	8.7–9.2
Hydrocarbon composition	D1319		
Olefins (%, maximum)		10	10
Aromatics (%, maximum)		35	35
Saturates		Remainder	Remainder

[a] From (7).
[b] ASTM = American Society for Testing and Materials, Philadelphia, Pennsylvania.
[c] For testing that is unrelated to fuel evaporative emission control, the specified range is 8.0–9.2.

B. Production-Line Testing

Testing of vehicle emissions immediately after the vehicle emerges from the production line complements prototype certification testing by giving assurance that the certified prototypes are representative of the vehicles which are being produced for sale. Procedures of this kind are in use in the state of California and are under consideration by the United States Environmental Protection Agency for application to all cars produced for sale in the United States. Assembly-line testing has also been suggested as a replacement for prototype testing.

In California, two kinds of tests are performed: (a) measurement of emissions from the vehicle while idling, performed on 100% of production, and (b) United States federal certification test on 2% of production. All vehicles are expected to pass the idle test before they can be sold. Of the vehicles undergoing the certification test procedure, 90% are required to pass in order to maintain the certification. The latter requirement applies to vehicles produced during the first two quarters following the introduction of the new model. This latter type of testing has been referred to as a quality audit.

Production-line testing can give a truer picture than prototype testing of the behavior of the vehicles actually being produced and sold, and provides detailed information on the statistical distribution of emissions from the new vehicles (*35*). There are two problems with the method. The first is the green engine effect and the second is the effect of the new car dealer's preparation on the vehicle emissions. Cars just off the production line will exhibit a greater degree of variability in their emissions over the first 50 miles than over the next several thousand miles. It has been reported that their emissions tend to drop significantly during this period (*36*). Dealer preparation can add further to the variability in the emissions. In the former case, a potential way around the green car difficulty is the use of averages and perhaps correction factors that account for any systematic bias. Some preconditioning may help in reducing increased variability or bias. This may be practical only for audit vehicles. The dealer effect, if it proves significant, probably requires a program to minimize the possibility that dealer preparation will deteriorate the car's performance from an emission point of view. Testing cars immediately after dealer preparation can be misleading, since the dealer's procedures can overload the evaporative emission control device.

C. Inspection-Lane Testing

Prototype certification and production line testing provides assurance that newly manufactured vehicles will meet United States federal emission standards. Additional testing procedures are required to assure that vehicles in use will be maintained in a condition that makes optimum use of their control systems. Such a program consists of inspection tests to identify those vehicles in the population that are high emitters followed by a required maintenance to bring the emissions back to or below stated limits. Following corrective maintenance, these vehicles presumably would be reinspected. The inspection test may consist of testing the emissions, or it may involve inspection of the engine itself to assure that the components causing high emissions are functioning properly. An alternate to either of these approaches is to avoid the inspection phase entirely and require that these vehicles periodically undergo specified maintenance procedures in authorized garages. The inspection phase of the first two alternates may be carried out in a publicly operated lane system, by licensed garages, or by stopping vehicles at road blocks set up at randomly selected locations.

While the engine parameter inspection and mandatory maintenance alternatives have been shown to lead to reductions in emissions (*37*), the emissions inspection systems provides more positive indication that the emissions of a vehicle have actually been reduced, and this appears to be the only approach under serious consideration in the United States. An ideal emission inspection test (a) could be performed quickly and accurately by relatively unskilled technicians, (b) would correlate well with the certification emission test procedure for prototype vehicles, and (c) would provide diagnostic information concerning malfunctions leading to observed high emission levels. A short test that could be performed with simple instrumentation is indicated by requirement (a). However, no short driving schedule performed on a warmed-up vehicle has been found to correlate well with a certification procedure involving a cold-start test (*38*). Thus inspection–maintenance testing has been concerned primarily with meeting the first requirement. Given a simple test, an emission standard is set at a level that will bring about a predecided failure rate (*39*). Thus, the system aims at identifying the high emitters.

To meet the third of the above-noted requirements, vehicle emissions must be tested under load, and for this, a chassis dynamometer is required. While various short cycles have been proposed, the key-mode test (*40*) and variations of it have received the greatest attention (Table V). In the key-mode, the vehicle is deliberately operated under high load conditions to put the engine under sufficient strain to show up operating malfunctions that may lead to high emission levels. The vehicle can be warmed up prior to the emission test by one repetition of the key-mode cycle. Relatively simple instrumentation is sufficient for use with this test. Its main application has been diagnosis of malfunctions leading to high hydrocarbon or carbon monoxide emissions. However, it should with

Table V Load and Speed Factors for Key-Mode Test Cycle

Vehicle weight class (pounds)	Low cruise mode		High cruise mode	
	Speed (miles/hour)	Load (horsepower)	Speed (miles/hour)	Load (horsepower)
<2800	23	5	37	14
2800–3800	30	9	45	23
>3800	33	11	49	29

little or no modification be applicable to nitrogen oxide emissions as well.

The simplest test cycle that can be performed is an idle test (*38, 39*). No dynamometer is required, and the test requires minimum time and skill. The test correlates least with certification prototype tests and provides little diagnostic information. However, its simplicity has made it widely used in inspection–maintenance programs.

Infrared analyzers employing solid-state infrared sensors and interference filters (discussed in an earlier section) are well suited to inspection–maintenance testing whether done on inspection lanes, in garages, or on the road, and a number of manufacturers market instruments designed for this application (*4*). They can be used equally well by trained inspectors or a garage mechanic. A single instrument typically measures hydrocarbon, carbon monoxide, and carbon dioxide. However, the use of an infrared instrument for measuring hydrocarbon further reduces the correlation with the United States federal prototype certification procedure. The instruments are compact, contain their own pump, an ambient temperature water trap, and coarse and fine filters. Experience has shown that it is necessary to police the calibration of instruments and their proper functioning whether the testing is done in state lanes or in private garage systems in order to gain maximum benefit from the program. Simple nitrogen oxide instruments suitable for inspection maintenance programs are not yet available. While infrared methods have the difficulties discussed in earlier sections, they are the most likely to be developed for this application. The chemiluminescent analyzer, widely used for exhaust emission measurement, is probably too complex an instrument to be successfully adapted to this use.

D. On-the-Road Emission Surveillance

The testing procedures discussed to this point do not constitute a complete system of regulatory testing procedures for automotive emissions. The prototype certification procedures and production-line testing apply to vehicles before they are in the hands of the general public. The aim of the inspection–maintenance procedures is to assure that the emissions of vehicles in use are maintained at the minimal values achieveable through good maintenance practices. What is needed in addition to these tests is further assurance that the in-use vehicles perform as was expected in granting certification based on prototype testing. In the United States, two types of testing have been used to obtain this assurance—compliance testing and surveillance testing.

In compliance testing the only in-use vehicles tested are those where there is clear indication that the vehicle has been maintained by its owner according to the manufacturer's recommendations. This follows from the fact that all prototype vehicles going through certification testing are maintained strictly according to the manufacturer's recommended procedures over the 50,000 miles of testing. Certification thus implies a properly maintained vehicle. Prior to compliance testing the vehicle is again tuned according to the manufacturers recommendations. The choice of vehicles for testing is essentially on the same basis by which the prototype vehicles were selected. The emission test procedure is identical to that followed in the original prototype certification testing. Compliance testing has tended to be confined to late-model vehicles. As such it is one step beyond production-line testing.

Surveillance testing differs from compliance testing in that the vehicles are tested in an as-received condition. The testing generally attempts to sample the overall vehicle population. The vehicles are tested in accordance with the testing that was in use at the time of their certification. Thus, surveillance testing is not a strict measure of compliance with regulations, but on the other hand, it gives a truer indication of the actual emission level of vehicles in the hands of the driving public and therefore a truer indication than compliance testing of the overall effectiveness of the certification and inspection–maintenance programs that are in effect. The primary interest in on-the-road emission surveillance is to determine compliance. If the interest were to obtain good overall indications of the behavior of the population of vehicles at large, different vehicle selection statistics would be needed as well as emission test procedures that more accurately reflect the way vehicles are driven in the region under study. More reliable vehicle emission factors for air pollution projections would result from this approach.

VI. Other Mobile Sources

The methods for measuring and characterizing the emissions from automobiles (light-duty vehicles) have been largely carried over to the measurement of emissions from the other mobile sources. The emphasis, therefore, in the following discussion will be on those features unique to the mobile source under consideration.

A. Trucks and Buses

In the United States Federal Regulations, trucks and buses are characterized as light-duty trucks and heavy-duty vehicles. Although light-duty trucks are subject to slightly different regulations, the certification test procedures (*41*) are very similar to those for light-duty vehicles. On the other hand, measurements on heavy-duty vehicles present additional problems not encountered in light-duty vehicles. Heavy-duty vehicles are not operated in the same manner. Because of their greater size, power, and need to operate at higher relative loading, the flow rates and temperatures of the exhaust gases are much greater. While an automobile engine normally lasts the life of the vehicle, a heavy-duty vehicle will customarily go through a number of engines in its lifetime. All these factors need to be taken into account. Test cycles that appropriately characterize the behavior of the large vehicles are still under development. Since a given engine can be used in a very wide variety of vehicles, certification is primarily given to the engine. The certification test procedures is therefore carried out on the engine mounted on an engine dynamometer. The present test operating schedule is a series of steady-state modes characterized by engine speed and load with different schedules for heavy-duty gasoline and diesel engines (*42, 43*). Because the test is on the engine, the mass emissions are expressed in grams emitted per unit power consumption or brake horsepower-hour.

The high flow volumes and use of an open cycle of steady-state modes make the CVS method of mass emission measurement less suited to heavy-duty engines than light-duty vehicles. As a consequence, the carbon balance method [Eq. (14)] involving fuel flow measurements at each mode has come to be used in determining the mass emissions from heavy-duty gasoline engines, while exhaust flow measurements [Eq. (1)] employing laminar flowmeters are used for heavy-duty diesel engines (*18*). For either engine type the mass emission for component X over one test cycle is then given by

$$\mathrm{BSX} = \sum_{i=1}^{n} w_i W(\mathrm{X})_i \Big/ \sum_{i=1}^{n} w_i \mathrm{BHP}_i \tag{16}$$

where BSX = brake specific emission of component X in weight per brake horsepower-hour; $W(\mathrm{X})_i$ = weight emissions per hour during the ith mode [Equation (14) for gasoline engines, Equation (1) for diesel engines]; BHP_i = brake horsepower during the *i*th mode; w_i = weighting factor for *i*th mode; n = number of modes in one cycle.

The analytical system used for heavy-duty gasoline engines is similar to that shown in Figure 8, in which the analyses are performed continuously on the undiluted exhaust. The major difference is that nondispersive infrared instruments are used to measure all the components of interest—hydrocarbons (as parts per million of hexane), carbon monoxide, carbon dioxide, and nitric oxide. A high- and low-level instrument is needed for hydrocarbons to span the concentration ranges. Because the calibration of the hydrocarbon instruments is expressed as parts per million of hexane, it is necessary to correct the concentration values obtained by a factor of $1.8 \times 6 = 10.8$ to put them on a parts per million of carbon basis, as well as to correct for the 45% low reading of nondispersive hydrocarbon instruments relative to flame ionization instruments. It is necessary to measure carbon dioxide in order to apply the carbon balance method. The continuous analysis of undiluted exhaust gives little time for conversion of nitric oxide to nitrogen dioxide, and therefore, measurement of nitrogen dioxide is not made. A humidity correction factor specific for heavy-duty gasoline engines is applied to the nitric oxide measurement (*42, 44*). The system measures all concentrations on the same basis (dry), and therefore, no further corrections need to be made to apply the carbon balance method.

The analytical system for diesel exhaust is also a continuous system. It differs from the heavy-duty gasoline engine system primarily in that a single heated flame ionization instrument is substituted for the high and low nondispersive infrared instruments. As discussed earlier, nondispersive instruments are not suitable for diesel exhaust hydrocarbon measurements. The exhaust flow measurement method requires that all the concentrations be expressed on a wet basis (no water vapor removed from the exhaust). Consequently, a correction factor is applied to the carbon monoxide and nitric oxide concentrations, since they are measured dry (*45*). The correction requires a knowledge of carbon dioxide concentration, and therefore, this component must also be measured. The humidity correction factor applied to the nitric oxide concentration is specific to diesel engines (*43, 46*).

While these methods are commonly applied at present, they do not fully reflect the advances that have been made in exhaust analysis. It is feasible to measure the emissions from either heavy-duty gasoline or diesel engines during an open operating cycle by the same method and same general analytical train. The carbon balance method can be conveniently used in both cases. Heated flame ionization analysis is desirable for both gasoline and diesel exhaust hydrocarbons measured with a continuous system to reduce loss of hydrocarbons. The chemiluminescent analyzer is a more versatile instrument than the nondispersive infrared analyzer for nitric oxide and can be used to obtain total nitrogen oxides.

A further limitation of the current testing technique is that it has been designed for testing engines and does not give a good representation of the way vehicles are used on the road. It would be very desirable to have a method that would be applicable to an engine or a vehicle operating on a chassis dynamometer. A more representative operating schedule would employ both dynamic and steady-state modes and wouldd be a closed continuous cycle. However, new measurement problems are encountered in this case. Since continuous measurement of fuel under dynamic conditions would be difficult, the carbon balance method would not be practical. Application of CVS methods customarily used in light-duty vehicles for continuous dynamic cycles is not practical for the larger engines, because the high exhaust volumes and high exhaust temperatures generated under high load conditions require extremely large systems. It would be desirable from the point of view of regulation as well as development of new control technology to have a uniform approach for all heavy-duty engine types, including gas turbine engines that are likely to come into use in the future. The vortex flowmeter discussed in an earlier section may fill this need, when used to continuously measure exhaust flow in conjunction with continuous measurement of the exhaust components. By this means mass emission measurements can be continuously determined and integrated over the test cycle.

B. Motorcycles

The characterization and measurement of emissions from motorcycles is in most respects similar to that for light-duty vehicles. However, because of the relatively small size of motorcycle engines and the use of two-stroke cycle as well as four-stroke cycle engines, there are significant differences that need to be taken into account. The emissions are expressed on a weight per unit distance traveled basis. The proposed United States Environmental Protection Agency standards are on a gram per kilometer basis (*47*). Because the interest in controlling motorcycle emissions is recent, the specific procedures are still under development. A driving schedule for use with a chassis dynamometer has been proposed. It is likely that two schedules will be adopted, one for larger engines (displacement equal to or greater than 170 cm^3) and one for smaller engines. The emission measurement will be similar to that for light-duty vehicles. The CVS method will be used to obtain mass emission measurement with the same instrument train as used with light-duty vehicles.

One potential problem that needs further resolution is contamination of the measurement system by unburned oil when two-stroke engines are tested. These engines operate on a gasoline–lubricating oil mixture. There have been observations of oil accumulation on the walls of tubing used to transport exhaust gases to the constant-volume sampling system, as well as in parts of the dilution air filter assembly. This accumulation may affect the measurement of hydrocarbon emissions.

C. Locomotives

A number of studies (*48, 49*) have been carried out to characterize the exhaust emissions from diesel-powered locomotives. The emissions of interest from diesel-powered locomotives are hydrocarbons, carbon monoxide, carbon dioxide, nitric oxide, nitrogen dioxide, oxygen, and aldehydes. The methods used are essentially those described for heavy-duty truck diesel engines.

D. Aircraft

1. Aircraft Gas Turbine Engines

Recommended practice for measuring emissions from aircraft gas turbine engines has been published by the Committee on Aircraft Emissions Measurements (E13) of the Society of Automotive Engineers (*50*). These recommendations largely have been adopted by the United States Environmental Protection Agency for their federal certification procedures (*51*). The emissions are expressed as pounds per thousand thrust pounds per hour. Mass emissions are determined using the carbon balance method. Therefore, measurement of fuel flow is required as well as the concentrations of the pollutants of interest. The engine is tested mounted on a thrust bed. Since principal concern is with contributions of aircraft to air pollution at ground levels, the test cycle is a series of steady-state modes representing taxiing, climb-out, and approach. An all-heated sample system is recommended to hold the temperature of the exhaust to 150°C entering each of the instruments. This minimizes the loss of hydrocarbons and also prevents condensation of the water. The wet measurement assures that all components are measured on the same basis for the application of the carbon balance technique, and thus no correction for loss of water is necessary as might be the case if water traps were used. The system operates on a high throughput of exhaust gas, since there is a requirement that sample transfer time be no greater than 2 seconds. The combination of rapid transport and elevated temperature appears to reduce the necessity for close temperature control. Studies (*52*) suggest that if the sample is maintained at temperatures above the dew point, there is little opportunity for loss of sample with the rapid flows. Carbon monoxide and dioxide are both determined by nondispersive infrared analysis. Hydrocarbons are determined with the heated flame ionization detector, since turbine engines employ diesel fuel. In the case of oxides of nitrogen, the Society of Automotive Engineers recommends the use of nondispersive infrared–nondispersive ultraviolet for nitric oxide and for nitrogen dioxide. The United States Environmental Protection Agency certification procedure, on the other hand, requires the use of a chemiluminescent analyzer with nitrogen dioxide converter. The Society of Automotive Engineers recommendation reflects earlier concern that the converter was unreliable.

Adequate sampling of aircraft gas turbine engine exhaust is still a significant problem, particularly for modern mixed-flow turboflow engines (*52*). It is normal practice to place a sample probe within one nozzle diameter of the turbine nozzle and take multiple samples over the cross-sectional area. Probes frequently have a number of openings so that a sample can be obtained simultaneously at several points and mixed within the probe. There are a variety of rake samplers in use in which the probe is in a form of a cross in with as many as twenty-four sampling holes (Fig. 20). The most appropriate form of sampling rake probably

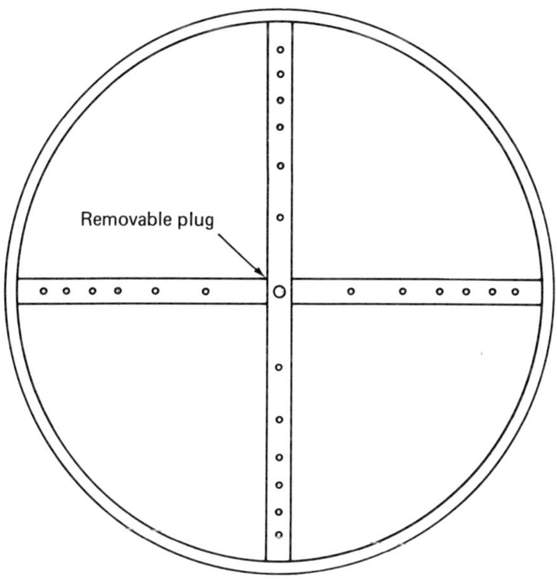

Figure 20. Sample rake for collection of aircraft turbine engine emissions.

varies from one engine to another, and as yet there has been no standardization. Experience shows that with the older jet engines, the exhaust gas is reasonably uniform over the cross section of the nozzle (*53*). However, in the newer type engines there can be a substantial variation. Because a carbon balance is used, dilution by surrounding air is not particularly important, since one is basically measuring the ratio of a given component to that of the concentration of carbon dioxide. However, with the new engines, up to fourfold variation has been found in this ratio when sampling in different quadrants (*54*). Thus, some type of a multipoint rake is required to obtain a good measure of the net behavior of the engine to avoid time-consuming scans of the nozzle cross section.

As a measure of test validity, the Society of Automotive Engineers recommends that the air–fuel ratio be calculated from the exhaust composition and compared to the air–fuel ratio determined directly from measuring air and fuel flow. If agreement is within 15%, it is suggested that the test and sampling technique are valid. Typically, the two measures are found to agree within 5–10% (*52*).

2. Piston Aircraft

As with turbine engine emission testing, the test cycle for piston aircraft consists of a series of steady-state modes designed to simulate the conditions during on-the-ground operation, take-off, and landing with each of the modes weighted for the average time spent in the mode. The engine is mounted on an aircraft engine dynamometer. The United States Environmental Protection Agency procedure (*51*) for the aircraft piston engine measures the pounds of component emitted per thousand horsepower-hours. The method of mass emission measurement in this case is based upon exhaust rather than fuel flow. The exhaust flow is determined from air and fuel flow measurements or from engine air consumption curves. The analysis is performed with a continuous system employing heated lines to a heated flame ionization detector and a chemiluminescent analyzer with converter. The sample going to the carbon monoxide and carbon dioxide instruments is passed through a cold trap to remove water. Thus, a correction is needed to put all concentrations on the same basis (*45*). Because piston engines run at near stoichiometric mixtures compared to gas turbine engines, there is substantially more water in their exhaust gases. Extra care is needed to prevent condensation of water in portions of the heated sections of the system.

Since there is an exhaust pipe, sampling is not as much of a problem as with gas turbine engines. However, when sampling from an engine with a propeller in place the probe needs to be 9–10 inches into the pipe to prevent dilution by air pulled into the pipe.

E. Boats and Ships

A variety of engines are employed in boats and ships. Diesel engines predominate and range from low- and medium-speed to high-speed turbocharged engines. Pleasure boats can involve the testing of both inboard and outboard gasoline engines. Measurement of the mass emissions from these engines introduces no new problems, if the engines can be tested in the laboratory on engine dynamometers. CVS systems can be used with gasoline engines and smaller diesels. The carbon balance method can be employed for larger engines, since marine engines typically run at steady-state conditions.

Testing marine engines in ships at sea presents some problems (*55, 56*) There is often little room and much vibration. Fuel flow is measured by installing flowmeters on the line to the engine and the return line, and taking the difference between the two readings. Power output is often difficult to measure, since few marine engines are equipped with torque meters. Generally it is necessary to resort to power curves supplied by the manufacturer. There are also occasional ships in which steam is generated for steam turbines in an oil-fired boiler. Again, the carbon balance technique can be used to obtain mass emissions.

REFERENCES

1. G. S. Springer and D. J. Patterson, "Engine Emissions—Pollutant Formation and Measurement," p. 294. Plenum, New York, New York, 1973.
2. "Continuous Hydrocarbon Analysis of Diesel Engines," SAE Recommended Practice No. J215, 1975 SAE Handbook. Soc. Automot. Eng., Detroit, Michigan, 1975.
3. B. M. Sturgis, W. F. Biller, J. W. Bosek, and S. B. Smith, "The Application of Continuous Infrared Instruments to the Analysis of Exhaust Gas." Annu. Meet. Soc. Automot. Eng., Detroit, Michigan, 1958.
4. A. J. Andreatch and J. C. Elston, "Evaluation of Idle Inspection and Maintenance Equipment Network," Paper 740134. Annu. Meet. Soc. Automot Eng., Detroit, Michigan, 1974.
5. B. Dimitriades and D. E. Seizinger, *Environ. Sci. Technol.* **5**, No. 3, 223–229 (1971).
6. J. E. Sigsby, Jr., F. M. Black, T. A. Bellar, and D. L. Klosterman, *Environ. Sci. Technol.* **17**, 51–54 (1973).
7. U.S. Code of Federal Regulations, 40CFR, Parts 85 and 86, Subpart A, July 1, 1975.

8. R. L. Beatty, L. B. Berger, and H. H. Schrenk, *U.S. Bur. Mines, Rep. Invest.* **R1-3687** (1943).
9. B. E. Saltzman, "Determination of Nitrogen Dioxide and Nitric Oxide by the Saltzman Method—Selected Methods for the Measurement of Air Pollutants," Pub. Health Serv. Publ. No. 999-AP-11. U.S. Dept. of Health, Education, and Welfare, Cincinnati, Ohio, 1965.
10. A. P. Altshuller, D. L. Miller, and S. F. Sleva, *Anal. Chem.* **33**, 621–625 (1961).
11. E. Sawicki, T. R. Hauser, T. W. Stanley, and W. Elbert, *Anal. Chem.* **33**, 93–96 (1961).
12. L. J. Papa, *Environ. Sci. Technol.* **3**, 397–398 (1969).
13. "Instrumentation and Techniques for Exhaust Gas Emission Measurement," SAE Recommended Practice No. J254, 1975 SAE Handbook. Soc. Automot. Eng., Detroit, Michigan, 1975.
14. U.S. Dept. of Health, Education, and Welfare, *Fed. Regist.* **31,** No. 61, Part II (1966).
15. U.S. Dept. of Health, Education, and Welfare, *Fed. Regist.* **33,** No. 108, Part II (1968).
16. S. H. Mick and J. B. Clark, Jr., "Weighing Automotive Exhaust Emissions," Paper 690523, Mid-Year Meet. Soc. Automot. Eng., Detroit, Michigan, 1969.
17. C. D. Paulsell, "Equivalence Testing and Evaluation of Critical Flow Venturi Constant Volume Sampler," Intern. Rep. Motor Vehicle Emissions Lab., United States Environmental Protection Agency, Ann Arbor, Michigan, 1975.
18. "Measurement of Intake Air or Exhaust Gas Flow of Diesel Engines," SAE Recommended Practice No. J244, 1975 SAE Handbook. Soc. Automot. Eng., Detroit, Michigan, 1975.
19. U.S. Code of Federal Regulations, 40CFR, Part 85, Appendix III, July 1, 1975.
20. R. Smith, A. H. Rose Jr., and R. Kruse, *Int. J. Air Water Pollut.* **8,** 427–440 (1964).
21. Olson Laboratories, "Study of Ultrasonic Effect and Tracer Gas Method for Measuring Automotive Mass Emissions," Final Rep., Proj. CAPE-22-72. Coord. Res. Counc., New York, New York, 1974.
22. D. F. White, A. F. Rodley, and C. L. McMurtrie, *in* "Flow—Its Measurement and Control in Science and Industry" (R. B. Dowdill, ed.), Vol. 1, Part III (W. E. Vannoh and H. Wayland, eds.), p. 967. Instrum. Soc. Amer., Pittsburgh, Pennsylvania, 1974.
23. U.S. Code of Federal Regulations, 40CFR, Parts 85 and 86, Subpart B, July 1, 1975.
24. D. L. Stivender, "Development of a Fuel Based Emission Measurement Procedure," Paper 710604, Annu. Meet. Soc. Automot. Eng., Detroit, Michigan, 1971.
25. U.S. Code of Federal Regulations, 40CFR, Part 85, Appendix I, July 1, 1975.
26. Systems Development Corporation, "A Survey of Average Driving Patterns in Six Urban Areas of the United States: Summary," Coord. Res. Counc., New York, New York, Air Pollution Research Advisory Committee, Proj. No. CAPE-10-68, No. PB 202 193, Nat. Tech. Inform. Serv., U.S. Department of Commerce, Springfield, Virginia, 1971.
27. M. Smith and D. M. Weston, "Construction of Chassis Dynamometer Test Cycles," Vols. 1 and 2. Motor Vehicle Mfr. Ass., Detroit, Michigan, 1961.
28. Scott Research Laboratories, "Vehicle Operations Survey," Vols. I and II. Coordinating Research Council, New York, New York; Air Pollution Research

Advisory Committee, Proj. No. CAPE-10-68, Nos. PB 210 640 and 210 844. Nat. Tech. Inform. Serv., U.S. Department of Commerce, Springfield, Virginia, 1971.

29. "Measurement of Fuel Evaporative Emissions from Gasoline Powered Passenger Cars and Light Trucks," SAE Recommended Practice No. J170a, 1975 SAE Handbook. Soc. Automot. Eng., Detroit, Michigan, 1975.
30. "Measurement of Fuel Evaporative Emissions from Gasoline Powered Passenger Cars and Light Trucks Using the Enclosure Technique," SAE Recommended Practice No. J171a, 1975 SAE Handbook. Soc. Automot. Eng., Detroit, Michigan, 1975.
31. S. Brunauer, P. H. Emmett, and E. Teller, *J. Amer. Chem. Soc.* **60**, 309 (1938).
32. "Instrumentation and Techniques for Vehicle Refueling Emission Measurement," SAE Recommended Practice No. J1045, 1975 SAE Handbook. Soc. Automot. Eng., Detroit, Michigan, 1975.
33. Scott Environmental Technology, Inc., "Investigation of Passenger Car Refueling Losses," Coord. Res. Counc., New York, New York; Proj No. CAPE 9-68, Contract CAPA 22-69-68. United States Environmental Protection Agency, Ann Arbor, Michigan, 1970.
34. Scott Environmental Technology, Inc., "Investigation of Passenger Car Refueling Losses—Second Year Program," Coord. Res. Counc., New York, New York; Proj. No. CAPE 9-68, Contract CAPA 22-69-68. United States Environmental Protection Agency, Ann Arbor, Michigan, 1972.
35. H. D. Jacoby and J. D. Steinbruner, "Clearing the Air: Federal Policy on Automotive Emission Control," Ballinger Publ. Co., Cambridge, Massachusetts, 1973.
36. F. W. Bowditch, General Motors Corporation, Warren, Michigan, 1974, (private communication).
37. "Inspection and Maintenance of Light Duty Gasoline-Powered Vehicles: A Guide for Implementation," EPA-450/2-74-005. United States Environmental Protection Agency, Ann Arbor, Michigan, 1974.
38. "Control Strategies for In-Use Vehicles," Mobile Source Pollution Control Program. Office of Air and Waste Management, United States Environmental Protection Agency, Ann Arbor, Michigan, 1972.
39. A. J. Andreatch, J. C. Elston, and R. W. Lahey, *J. Air Pollut. Contr. Ass.* **21**, No. 12, 756–763 (1971).
40. E. L. Cline and L. Tinkham, "A Realistic Vehicle Emission Inspection System," Paper No. 68-152, Ann. Meet. Air Pollut. Contr. Ass., Pittsburgh, Pennsylvania, 1968.
41. U.S. Code of Federal Regulations, 40CFR, Parts 85 and 86, Subparts C and D, July 1, 1975.
42. U.S. Code of Federal Regulations, 40CFR, Part 85, Subpart H, July 1, 1975.
43. U.S. Code of Federal Regulations, 40CFR, Part 85, Subpart I, July 1, 1975.
44. S. R. Krause, "Effect of Intake Air Humidity, Temperature and Pressure on Exhaust Emissions," Paper 710835, Ann. Meet. Soc. Automot. Eng., Detroit, Michigan, 1971.
45. "Measurement of Carbon Dioxide, Carbon Monoxide, and Oxides of Nitrogen in Diesel Exhaust," SAE Recommended Practice No. J177, SAE Handbook. Soc. Automot. Eng., Detroit, Michigan, 1975.
46. S. R. Krause, D. F. Merion, and G. L. Green, "Effect of Inlet Air Humidity and Temperature on Diesel Exhaust Emissions," Paper 730213, Ann. Meet. Soc. Automot. Eng., Detroit, Michigan, 1973.

47. United States Environmental Protection Agency, *Fed. Regist.* **40,** No. 205 (1975).
48. "Exhaust Emissions of Selected Railroad Diesel Locomotives," Southern Pacific Transportation Company, San Francisco, California, 1972.
49. C. T. Hare, K. J. Springer, and T. A. Huls, "Locomotive Exhaust Emissions and Their Impact," Publ. 74-DGP-3. Amer. Soc. Mech. Eng., New York, New York, 1974.
50. "Procedure for the Continuous Sampling and Measurement of Gaseous Emissions from Aircraft Turbine Engines," Aerospace Recommended Practice No. 1256. Soc. Automot. Eng., Detroit, Michigan, 1971.
51. U.S. Code of Federal Regulations, 40CFR, Part 87, July 1, 1974.
52. J. H. Elwood and R. H. Diek, *J. Air Pollut. Cont. Ass.* **25,** No. 8, 839–844 (1975).
53. A. F. Souza and L. R. Reckner, "Variability in Aircraft Turbine Engine Emission Measurements," EPA-460/3-74-006. United States Environmental Protection Agency, Ann Arbor, Michigan, 1974.
54. A. F. Souza, "Further Investigation into the Causes of Variability in Aircraft Turbine Engine Emission Measurement," EPA-460/3-75-011. United States Environmental Protection Agency, Ann Arbor, Michigan, 1975.
55. A. F. Souza, "A Study of Stack Emissions from Coast Guard Cutters," CG-D-13-73. United States Coast Guard, United States Department of Transportation, Washington, D.C., 1973.
56. C. L. Newton, "The DEMA Exhaust Emission Measurement Procedure for Low and Medium Speed Internal Combustion Engines," Publ. 74-DGP-9. Amer-Soc. Mech. Eng., New York, New York, 1974.

17

Emission Inventory

James R. Hammerle

I. The Air Pollutant Emissions Inventory

A. Introduction

The listing and description of air pollutant emitting sources, including an estimated pollutant emission quantification, comprises the emissions inventory. The emissions inventory provides the most elemental basis for planning air pollution reduction strategies and evaluating their results.

Initially areas with typical characteristics—economic, altitude, meterologic, population density, industrial/residential, etc.—must be defined enabling the planner to consider alternatives without regard to geographic boundaries such as rivers, mountains, or lakes. In most cases, boundaries should coincide with boundaries of local political jurisdictions, thus maintaining the ability for cooperation among local agencies.

Within a geographic area emissions can be classified as man-made or natural, controllable or uncontrollable. Although the controllable emissions are of primary interest, natural emissions must also be considered,

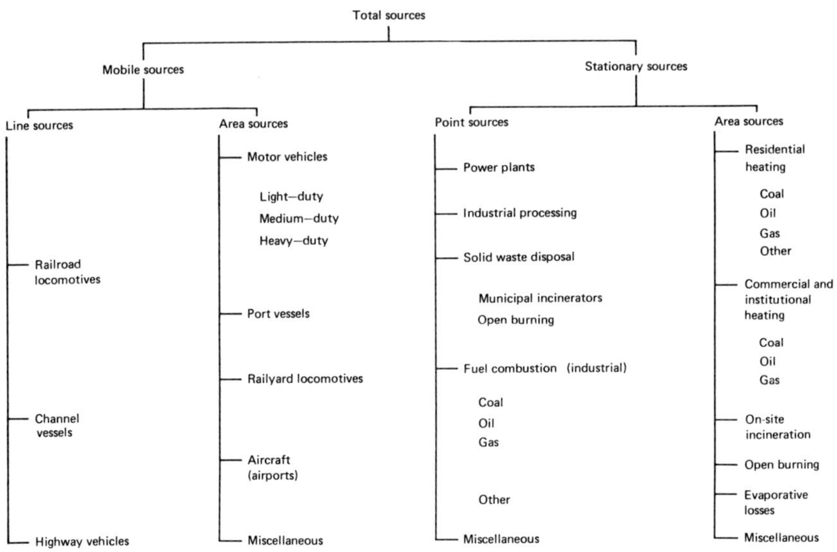

Figure 1. Emission inventory source categories.

since they may, in certain instances, constitute the major portion of the total. Emissions may be classified by source (Fig. 1), as stationary or mobile. Another method of classifying emission sources is by: (a) point sources (large stationary sources), (b) area sources (small stationary sources and mobile sources with indefinite routes), and (c) line sources (mobile sources with definite routes).

There are three classes of inventories, based upon precision and information desired: gross estimation, rapid survey, and comprehensive inventory. The gross estimation and rapid survey methods yield annual average emissions, whereas the comprehensive inventory yields annual, seasonal, daily, or hourly average emissions.

There are many ways in which data about sources can be collected and presented. However, if information from one geographic area is to be compared with that from another, it is mandatory that information about all sources conform to a single set of basic rules to assure compatibility. Common weight/volume/area units, grid coordinates, source categories, geographic identifiers, etc., must be used. Such coordination also yields ease of computerization of data, a necessity when dealing with large numbers of sources and many items of data for each source.

B. Gross Estimation Inventory

The gross estimation inventory usually considers a large geographic area for which summary statistics for fuel consumption, incineration, solvent usage, and industrial processing are available on an annual basis. Through the use of estimating procedures such as fuel consumed or vehicle distance traveled per gallon or liter of fuel, quantity of emissions per quantity processed, weight of refuse incinerated per capita, and quantity of emissions per quantity of fuel burned, the emissions can be summarized for several basic source categories. Basically the gross estimation inventory is performed in the office with use of reference documents and communication with agencies responsible for statistics on fuel use, whether it be for nationwide, state, provincial, or local jurisdictional estimation. Although the total quantities of fuels or materials consumed or processed may be relatively precise, the conversion factors for calculating emissions from them are much less precise because of the many variables in combustion and processes.

The sources of data on fuel consumption are usually from governmental statistical agencies. Conversion factors must be determined individually for the specific source categories under study. The source categories for which estimates typically can be made are shown in Table I. Since this method of determining emissions is so susceptible to error, its use should be limited, and more precise methods should be employed. With experience, nationwide estimates for source categories can be made with about four man-weeks of effort. This method has been used to estimate national emissions in the United States back to 1940 and for each year thereafter.

C. Rapid Survey Inventory

The rapid survey inventory is performed in the field for large sources and through use of reference documents for area sources.

In the rapid survey the 50 to 100 major point sources are contacted either by questionnaire or by telephone, while all other emissions are assigned to the general geographic locale where they occur; in this manner areas of high emission concentration are identified and the distribution pattern of pollutants can be estimated. The five basic source categories listed in Table I are used, however, individual point sources are grouped together as a subdivision under the appropriate category, with remaining emissions specified as being from area sources.

The rapid survey method (*1*) was designed for the limited budgetary and manpower resources of local agencies, while providing reasonable estimates of emissions for smaller study areas. The method utilizes information readily available in most communities, requiring a minimum of contact with major point sources. Technical competence in converting fuels and processes to emissions is required of personnel involved in the survey. The method is both reasonably accurate and requires a minimum

Table I Gross Estimation Inventory Sources Categories

Transportation
Highway vehicles
Gasoline
Diesel
Aircraft
Locomotives
Vessels
Off-highway vehicles
Fuel combustion in stationary sources
Coal
Fuel oil
Natural gas
Liquified petroleum gas and liquified natural gas (LPG and LNG)
Kerosene
Wood
Industrial processes
Steam/electric plants
Commercial/institutional heating
Residential heating
Industrial processes—many categories possible, dependent upon pollutant
Waste disposal
Municipal incineration
On-site incineration
Open incineration
Land fills and dumps
Miscellaneous
Forest fires
Structural fires
Coal refuse fires
Agricultural burning

of effort. Past experience has shown that approximately four to five man-weeks per moderately populated area (1,000,000 inhabitants) are required.

A community or geographic area tends to divide into areas of homogeneous fuel use, just as it tends to divide into areas of homogeneous land use. Whether a given residential area primarily burns coal, oil, or gas for residential heat is often a function of the age of the community. Geographically the older residential areas tend to be near the community's core and the newer areas tend to be in the suburbs. Various fuels have preferential use depending upon the location of the community and the nature of the user. The physical nature of residual fuel oil is such that it is usually used only in large boilers such as those for electrical generation or in large commercial establishments. Moreover, residual oil is usually a major fuel only in seaport refining areas. The chemical composition of a fuel also tends to be relatively uniform in a given locality. The coal used will usually have been mined in one major coal field and thus will contain relatively uniform percentages of ash and sulfur. Natural gas may be supplied to selected areas of a community by a gas distributing company which services only certain areas. Obtaining information about fuels is also simplified by the fact that fuels can be brought into a community and distributed by only a limited number of transportation methods and firms. By contacting these distributing firms it is usually relatively easy to obtain information on the areas of use of different fuels.

Division of a geographic area into zones should be done on the basis of common characteristics such as land use patterns, fuel use patterns, topography, or population density. Since a considerable amount of the required data may be available from census information, it is most beneficial to keep census tract size and configuration intact. When land use and zoning patterns are examined, the resulting areas are either primarily residential, commercial, or industrial. In subdividing residential areas into zones, population density as well as housing data are important in delineating those that are primarily single family residential from large apartment house areas. The differences in fuels used in these areas can therefore be easily defined. Areas of high industrial or commercial activity should not be subdivided but rather included in one zone.

Choice of the number of zones will determine their relative size and vice versa. The use of zones which have extremely large emissions totals will defeat the purpose of zone reporting, whereas the use of a multitude of small zones will complicate procedures without significantly increasing the validity or utility of results. Zones should be approximately 2–10 miles^2 (5 to 25 hectares) in area. Although one should attempt to keep zones somewhat quite similar in size and shape, the zones in outlying sections of a metropolitan area may be larger than those in the highly urbanized sectors. If the study area is a large metropolitan area, it may be necessary to increase both size and number of the zones. Irregular or elongated shapes should be avoided unless such configurations are necessary to pinpoint areas of high emissions.

The steps involved in the preparation of a reporting zone map for the rapid survey method is summarized as follows:

(a) Obtain or draw a map of the study area showing the natural features such as rivers, lakes, major interior streets, or political boundaries.

(b) Indicate the census tracts on the map. Since it is unwise to divide census tracts, the tracts should be kept intact. Subdivide the census tract into zones based on common characteristics such as land use and fuel use patterns. When necessary subdivide land use patterns on the basis of subgroup factors such as population density in residential areas.

(c) Fuel use data are then collected in the study area and allocated to the reporting zone as indicated on the prepared map. This procedure closely follows that which will be described later in the section on area source inventory data collection.

This rapid survey methodology has been utilized extensively in the United States for:

(a) Six abatement study surveys (1967–1972)*

(b) Fifty-one consultations on air quality for designation of air quality control region boundaries (1968–1970)*

(c) Eighteen transportation control, transportation strategy, and vehicle emission surveys (1968–1973)*

(d) Fifty-one real property surveys for federal facilities (1968–1971)*

(e) Thirty-three state surveys of federal facilities emissions (1968–1971)*

(f) Seventy-six emission inventory reports for assistance in preparation of state implementation plans (1966–1974) (*2*)

(g) Seventeen national emission inventories of sources and emissions of specific substances (*3*)

(h) Over 100 state-sponsored surveys for implementation plan preparation (1970–1973)

In addition six cities (St. Louis, Missouri; Ankara, Turkey; Frankfurt, Federal Republic of Germany; Oslo, Norway; Rotterdam, Netherlands; and Milan, Italy) have been surveyed as a part of the Air Pollution Pilot Study of the Committee on the Challenges of a Modern Society of the North Atlantic Treaty Organization.

Regardless of the usefulness of these surveys at the time they were performed, all of these data have been superseded by more recent information collected by the more detailed comprehensive emission inventory technique.

D. Comprehensive Inventory

Comprehensive inventory (*4*) was developed for collecting information necessary for use in mathematical atmospheric dispersion modeling. A minimum size of point source emitter selected as the lower limit for data

* These are listed in the reference list to Chapter 6, Vol. V.

collection is usually 100 tons (90 metric tons) per year of emissions. All point sources down to that level are included in the point sources inventory. All geographic areas are included in the area source inventory, and the reporting zones are constructed on a grid square basis covering the entire study area.

For each individual point source the comprehensive data usually collected include parameters such as stack height and diameter, exhaust flow, gas temperature and flow rate, coordinates for location of the stack, throughput in the process emitting the air pollutant, the sulfur or ash content of the fuel, the daily operating schedule and the existing control equipment.

The main difference between the comprehensive inventory and the rapid survey is the greater quantity of information collected for each individual point source, which permits a greatly expanded use of the point source inventory.

E. Time-Dependent Inventory

Until recently, annual average emissions were sufficient for all emission inventory uses other than dispersion modeling, and in that instance, the yearly emission rates for the sources under consideration were divided by the annual operating hours to obtain hourly emission rates. At this time, considering transportation and land use planning, air quality modeling, and other inventory usage, this method is not sufficient, and efforts must be made to collect and present emissions data on an hour-by-hour basis. This time-dependent inventory was used by the United States Environmental Protection Agency (USEPA) for the Regional Air Pollution Study in St. Louis, Missouri.

The concept of a "typical" time period (season, day, hour) simplifies data collection by reducing the number of time periods considered. For example, two typical seasons might be November to March and April to October, i.e., a heating and a nonheating season. Typical days might be weekdays and weekends. Within each category, production, etc., remains the same within a season. Typical hours might be 0700 hours to 0900 hours, and 1600 hours to 1800 hours; or 0900 hours to 1600 hours, and 1800 hours to 0700 hours, which can categorize diverse traffic and industrial operating modes.

F. Modal Inventory

The dependence of emissions on operating modes must be considered when the emissions are not a constant function of the process originating the emissions, for example, a batch operation, or a motor vehicle whose emissions are a function of the speed of the vehicle. If the time period under study is a year, modal data are not necessary. However, for hourly emissions the stage of the process or the vehicle speed at the location and during the time period considered will greatly effect the emissions. Much more information is known about the modal variation of mobile emissions than of stationary sources.

G. Pollutants

Most emission inventory effort has been devoted to particulates and sulfur oxides, which are associated primarily with stationary sources and carbon monoxide, hydrocarbons, and nitrogen oxides, which are emitted mainly from mobile sources. Originally inventories and dispersion models were primarily concerned with particulates and sulfur oxides. However, at this time, an inventory cannot be considered complete without complete data on all the above-noted pollutants. The detailed emission inventory should contain sufficient data to specify the quantities of emissions and the compliance status of at least those pollutants listed in Figure 2.

H. Estimation of Inventory Precision Required

The allowable error for a specific set of emission data can be determined to ascertain the quantity and quality of information necessary

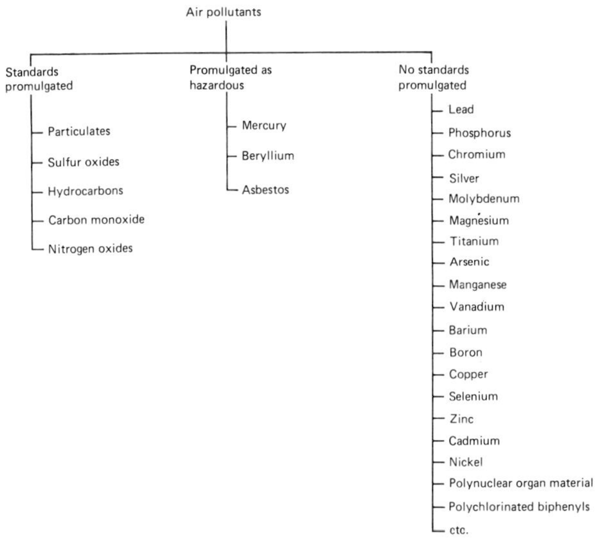

Figure 2. Example classes of air pollutants based on promulgation of ambient air quality standards by the USEPA (1975).

for calculating emissions, recognizing that each source category i contributes to the total error θ, an error proportional to its contribution to the total emissions of the particular pollutant.

When an overall error in a given set of emissions data can be specified, the allowable errors in the emission subclasses comprising the inventory can be calculated. Thus, management decisions to make improvements in the quantity and quality of emission data for these subclasses can be made in a rational manner. Determination of the allowable error for a specific subclass emission category in a given geographical area is given by

$$\sigma_i = \theta \sqrt{E_T/E_i} \tag{1}$$

where

σ_i = percent allowable error in category i
θ = percent overall error in emissions total (E_T)
E_T = total of all emissions for particular pollutant in study area
E_i = emissions of particular pollutant in the study area for category i

Overall error may be determined by the precision possible from ambient monitoring devices or the requirements of mathematical models to be used. This relationship must be applied to each pollutant separately and is quite sensitive to the number of categories considered.

Using Equation (1) and the February 1975 emissions data for the United States, the allowable error may be calculated for hydrocarbon emissions in the United States, given a total permissible error for the nation, θ = 10%, as listed in the following tabulation:

Category	E_i *(tons/year, hydrocarbons)*	σ_i ($\pm$ %)
Fuel combustion	444,000	74
Industrial processing	6,871,000	19
Incineration	971,000	50
Transportation	13,770,000	13
Other	2,104,000	34
E_T	24,160,000	

Following determination of σ for each of the five source categories, an error value for each subcategory could be determined, i.e., within the transportation category: light-duty vehicles, medium-duty vehicles, and heavy-duty vehicles, using σ_i from the transportation source category in the previous determination (13%) as θ. This computation is shown in the following tabulation:

Subcategory	E_i *(tons/year, hydrocarbons)*	σ_i (± %)
Gasoline powered light-duty vehicles	8,880,000	16
Gasoline powered heavy-duty vehicles	1,805,000	36
Gasoline powered off-highway vehicles	620,000	61
All other subcategories	2,465,000	31
E_T	13,770,000	

The permissible error (σ_i) for each of the subcategories has therefore been calculated, and can easily be used for determination of priorities of resources for data collection, selection of numbers of vehicles to be tested, etc.

Since the emissions for a sub-subcategory can be considered as a many-functioned term, it is necessary to decide how to allocate the total error among the several factors making up the emissions. The absolute error from each factor should be constant, thus the actual values measured (plants investigated, measurements performed, vehicles counted, etc.) must agree within the specified error limits as defined by Equation (1).

This procedure is being investigated in depth (*15*) with considerable emphasis being placed on the application of the results in air quality management expenditures for increasing quality and quantity of source test, ambient monitoring, and source inventory data.

I. Updating the Inventory

The emission inventory is neither a constant set of data nor a stable data base. As new facilities are added, processes modified, process rates altered, control equipment installed, new vehicles incorporated, and old vehicles removed from the population, fuels specifications changed, highways constructed, speed limits modified, etc., the emission inventory will change. Therefore, it is necessary to have some mechanism for updating the emission inventory on a regular basis, at least annually.

The updating for individual sources is most easily achieved through a control agency requirement for information submittal by the sources whenever any change which would effect the quantity of emissions takes place. However, for area and line sources it is much easier for the agency

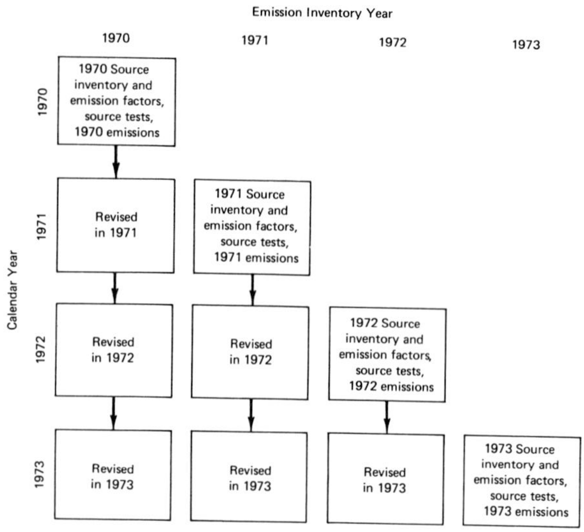

Figure 3. Record keeping for historical emission inventories.

performing the emission calculations to estimate these changes using summary statistics on fuel consumption, vehicle registration, etc.

One of the major problems in maintaining an up-to-date inventory is to induce individuals using the inventory to submit all errors and needs for modification found during such use to the persons responsible for the inventory. Many times needed changes will not be called to the attention of those responsible and the data base will have lost an excellent opportunity for upgrading.

The emission inventory also differs from most data bases in that the historical data also change. For example, in an air quality data base, new data are being added constantly but old data are not changed. However, with an annual emission summary—composed of source inventory data, emission factors data, emissions calculations, and source test data—all the items, except for source test results, may be changed for several years back into time due to changes in vehicle test procedures, new estimates of solid waste per capita, additional data to support emission factors, etc. Therefore, historical records must be kept for each year to indicate the revisions or changes in emissions factors, source inventory, and resulting emissions calculations occurring for each successive year (Fig. 3).

J. Emissions from Natural Sources

Emissions of air pollutants from natural sources such as oceans, volcanoes, swamps, marshes, and forests, are also of interest. For some pollutants on a worldwide or nationwide basis the magnitude of emissions from natural sources exceeds that of man-made emissions. Natural source emission inventory is needed to establish global balance between sources and sinks, and to estimate nonurban background concentration of pollutants. In general, in urban areas man-made sources predominate and natural sources are of concern only as they help establish background concentrations of pollutants.

II. Uses of the Inventory

A. Introduction

The two major users of an emission inventory are research and control personnel. Researchers are concerned with the relationship between ambient concentrations of pollutants and their sources, whereas regulatory personnel are charged with the achievement of air quality standards and have the responsibility and authority to collect the required data. The research uses can be further subclassified into atmospheric modeling and effects studies. The regulatory uses can be further classified into abatement and maintenance control strategies development (including land use and transportation planning) and enforcement. An additional use of the inventory is to provide data for the planning of ambient monitoring activities, an indispensible component of both research and regulatory requirements.

B. Abatement Control Strategies

The proper emission control strategy for a specific air pollution problem is dependent upon an adequate assessment of the nature and extent of the pollution in the region involved. The emission inventory indicates the major contributors (motor vehicle, industrial, processing, etc.), and this information, in turn, directs the control efforts. After strategies have been developed, they can be tested with the aid of diffusion simulation models or other systematic quantitative procedures in order to determine which strategies are capable of bringing about acceptable air quality as defined by air quality standards. If the emission inventory is updated annually, a decrease in emissions should be reflected over a period of years. This decrease would then be a measure of the effectiveness and success of the control program and could be used to indicate areas where program modification would be useful.

By constructing emission density maps of the study area, a valuable tool for metropolitan planning and zoning will be available. The maps will show the locations of high densities and therefore indicate areas in which anticipated construction of air pollutant emission sources should be carefully examined. Thus, together with other land use planning aids, the emission density map can be used to predict future growth areas and other zones in which industries of a non-air-polluting nature might be located, or in which high concentrations of population should be reduced.

Strategies and management decisions for initiating transportation plans also are derived from the inventory and the effect which the implemented plan would have on the resultant emissions. The current widespread use of environmental impact statements indicates another use of the emission inventory, especially when construction of power plants, highways, or other major sources of air pollution is under consideration.

C. Enforcement

After the emission inventory has been prepared, taking care to include the compliance status of each source with respect to each applicable control regulation, the agency charged with enforcement can make day-to-day use of it. The enforcement inventory includes *all* sources on compliance schedules, whereas most other inventory uses are primarily concerned with larger sources of air pollutants rather than sources of insignificant magnitude. It is sometimes necessary to separate the enforcement/compliance data from all others and to maintain two separate, but compatible, files, since compliance status may change rapidly and is very difficult to keep up-to-date.

An emission inventory does not maintain a continuous chain of evidence for court enforcement actions because it does not provide proof of origin of high ambient concentrations or emissions in excess of specified standards. The emission inventory rather serves as a management tool for identifying potential violators.

D. Effects Studies

Scientists and engineers concerned with effects studies are usually addressing one of the following areas.

(a) Primary effects—health of humans. The range of pollutants of interest is very broad, including not only compounds of known toxic effects, but also compounds suspect of having an effect. For this purpose detailed time resolution of emissions is not usually required since most effects are cumulative over long time periods. Historical records are of major importance in determining causal relationships of air pollution to health effects.

(b) Secondary effects—effects on plants, animals, soiling, and visibility. Plant and animal effects require long time resolution; soiling and visibility a shorter time resolution.

E. Atmospheric Modeling

The emission inventory may be used with supporting meteorological data to predict ambient air quality for a given locality. Using emission density maps, areas with high pollutant releases can be located.

The emissions model should be kept separate from the main atmospheric model to facilitate use of emissions data collected by the emission inventory.

F. Ambient Monitoring

Information concerning the location of sources and quantities of emissions in a region may be used to indicate where the highest pollutant concentrations probably exist. This knowledge will assist in locating elements of a sampling network. Samplers are normally concentrated in areas of greatest emissions or in locations where meteorological conditions result in accumulations of pollutants or atmospheric chemical reactions. In the event that a single source is believed to be primarily responsible for degraded air quality, the sampling network may be oriented in such a way that evidence of the impact of the emissions from that particular source may be obtained.

III. Emission Models

A. Concept

The concept of an emission model as an integral part of the emission inventory is relatively new, although several dispersion models include a source inventory/emission factor relationship which serves as an emission model. Any mathematical relationship combining an emission factor and a source inventory to calculate emissions is an emission model. The model must be considered a component of the emission inventory rather than of the dispersion model. Incorporation of emission models within atmospheric dispersion models prevents modification, improvement, and adjustment of the emission inventory for an area as new data become available and more sensitivity or complexity is required (e.g., hourly average rather than annual average emissions).

Spatial and temporal considerations enter into the concept of emission models. Spatial considerations relate primarily to area and line emission models since area sources or line sources with similar characteristics may need to be grouped together in order to reduce the volume of data for the area or line to be studied. Point sources for a specific source category in a study area may be grouped to permit a simplified analysis in which the mathematical relationships for calculating emissions are the same.

The most common emission model is the average annual throughput multiplied by an average emission factor and an average control efficiency to obtain average annual (controlled) emissions. More complicated models for daily or hourly average emissions are possible. The emission model for highway vehicles is more complex, since vehicle model, age, and speed must be included.

B. Spatial Considerations

1. Point Source Models

For point source models, source inventory data must be obtained for every point source of emissions. By utilization of sensitivity analysis it is possible to ascertain which sources should have continuous in-stack monitoring, which require in depth studies to provide adequate information for calculating emissions on an hourly basis, and which might be grouped together and considered to have typical operating schedules or procedures in order to calculate daily average emissions. In each case the source category becomes less specific as there is progression from continuous, to hourly, to daily, to average annual calculations and from a single point to a group of points. A point source model concept is shown (Fig. 4) in which two point sources positioned at x_1, y_1 and x_2, y_2 emit quantities E_1 and E_2, respectively.

2. Area Source Models

It is impossible to have continuous in-stack monitoring for area sources. Therefore, the most sensitive model is the grouping of similar sources into categories within a specific geographic area, collection of comprehensive data characterizing their operations, and calculation of emissions for the group as a whole using a mean emission factor for

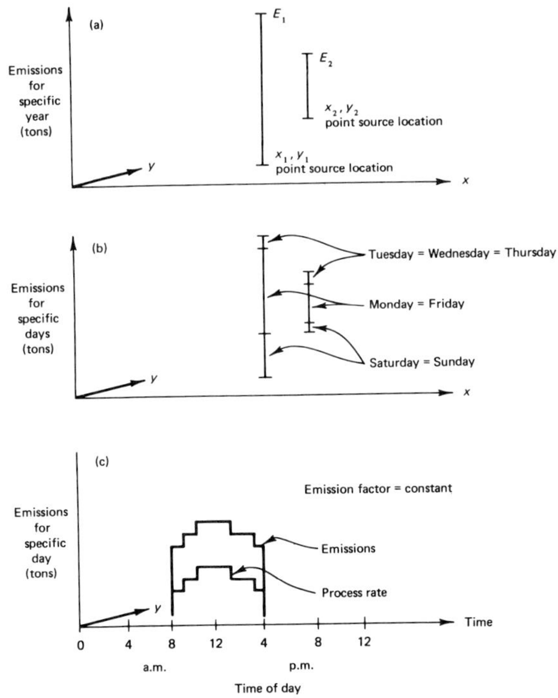

Figure 4. Point source models: (a) annual average emissions; (b) daily average emissions (with typical days); (c) hourly average emissions (for point 1).

hourly emissions. Enlarging the area, increasing the coverage of the source category (for example, considering commercial and institutional sources together, rather than separately), or accepting less comprehensive knowledge about the operations will result in less accurate estimation of emissions for the area.

An area source model demonstrating two areas A_1 and A_2 emitting E_1 and E_2 is shown (Fig. 5), in which the emissions E over the areas should be considered to be for a specific source category, for example, residential heating, because this represents a dispersion of emissions over the entire area A.

3. Line Source Models

A line source transversing a specified route and emitting a quantity in excess of a specified minimum level is composed of a number of vehicles. The route can be specified in terms of its length and location in

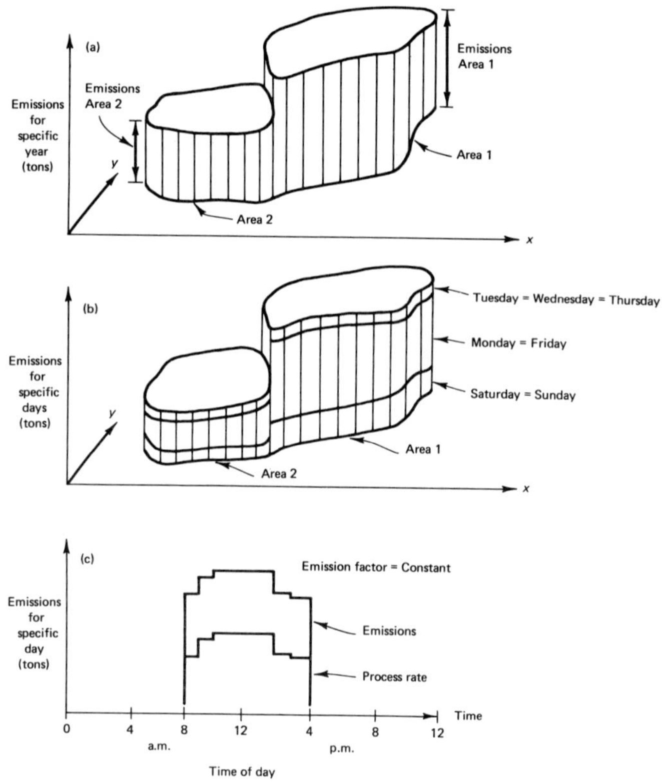

Figure 5. Area source models: (a) annual average emissions; (b) daily average emissions (with typical days); (c) hourly emissions (for area 1).

the coordinate system, with end points at x_1, y_1 and x_2, y_2, having subcomponents called "links" (Fig. 6).

In general, line source models are more complicated than point or area source models because the emissions vary over the length of the route. Although data are available to determine this variation, "constant" or average emissions over the entire length of the source are usually used in line source models. The average emissions over a line source for a specified time period may be calculated from

$$\bar{E} = \int_{x_1}^{x_2} \int_{y_1}^{y_2} \frac{E(x, y)\, dy\, dx}{\text{link length}} \tag{2}$$

where the emissions for each vehicle duty class (light, medium, and heavy) operating on each link are summed to present total emissions

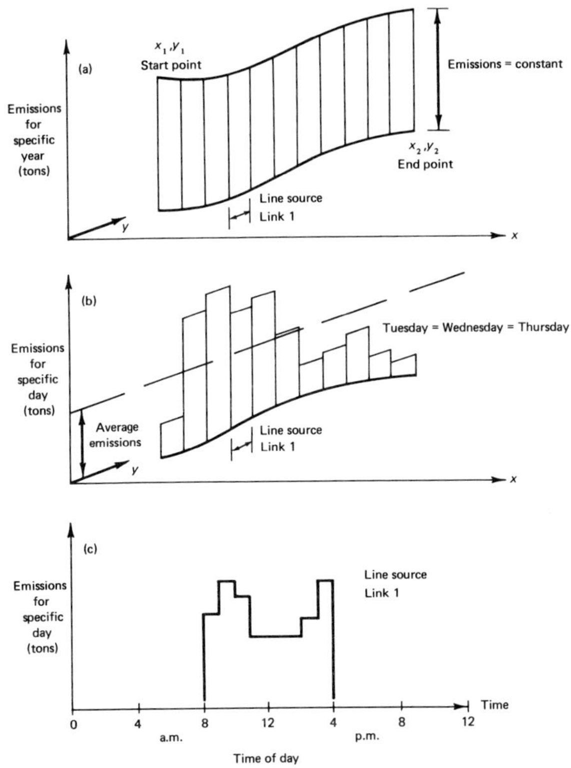

Figure 6. Line source models: (a) annual average emissions; (b) daily average emissions; (c) hourly average emissions.

for the links. Equation (2) can also be applied to railroad lines or vessel channels.

C. Temporal Considerations

1. Annual Average Models

Annual average models for point, area, and line sources are shown in the (a) portions of Figures 4–6. The emissions for the specific year must be stated, and are expressed, by pollutant, in tons/year. Gross emissions representing the total for the year are tabulated for each source considered to be a point, an area, or a line. These data are reduced to shorter time periods by averaging, apportioning, or assuming the percent of the total occurring during the time period of interest. This can lead to significant errors when comparing measured ambient concentrations with estimates of emissions obtained through the use of dispersion models. Annual average emissions are most commonly used.

2. Daily Average Models

Emissions can be specified for selected days as shown in parts (b) of Figures 4–6, which indicate that the total emissions for days may vary from day to day, but that for several days, for example, Tuesday, Wednesday, and Thursday, the daily total emission estimates may be equal. To estimate daily emissions requires a more detailed knowledge of source inventory parameters than is usually available for point and area stationary sources.

3. Hourly Average Models

Hourly average models require an intensive study of each individual point, area, and line source, with respect to source inventory data and emission factors. The use of in-stack continuous monitoring yields data especially applicable to this model. Hourly average emissions are shown in parts (c) of Figures 4–6 for the emissions from a specific source as a function of time throughout a specific day. The data are for a constant emission concentration per unit of process rate applied to a variable process rate. Theoretically, the daily average emission for a specific source and specific day is

$$\bar{E}_1 = \int_8^{16} \frac{E(t)\,dt}{16 - 8} = \frac{1}{8}\int_8^{16} E(t)\,dt \qquad (3)$$

A gross emission factor may not be suitable, so that emission factors which vary with process rate must be developed and used.

D. Modal Considerations

Modes during operations may occur consecutively or simultaneously, as in a batch operation of a stationary source or the acceleration/cruise/deceleration of a mobile source. Knowledge of the emissions as a function of the individual modes of operation permits a sensitive presentation of data for operations that vary in their rate or capacity during the period of study.

If, for example, a processing operation is composed of several modes, each of a different time length, material quantity, or throughput rate, a gross emission factor for the entire process will predict annual average emissions, but will not predict hourly emissions if the total time elapsed for the batch operations exceeds 1 hour. Other examples include the five and seven mode acceleration/cruise/deceleration automotive driving cycles which are used for determining emissions typical of characteristic vehicle operations, which take less than an hour to accomplish.

Mobile source emissions are a function of the modes which are, in turn, a function of the location of the mobile source, i.e., in traffic, on hills (low speeds), or on expressways (high speeds). In general, the larger the area under consideration the less beneficial the use of modal factors unless the cycle of operations can be clearly defined to be different from those assumed for gross emission factors.

E. Grouping within Source Categories

Sources (point, area, line) within a source category (residential heating, light duty vehicle, etc.) in a specific study area during a specified time may be grouped together and considered all to have the same operating characteristics. This simplifies calculations by identifying groups of line, area, or point sources, rather than individual sources.

The inherent difficulty in using "typical characteristics" is deciding which sources display typical characteristics, such as operating rate and operating time. Ideally, the selection of sources to group can be made by identifying as many point, area, and line sources as possible, together with emission estimates, and then performing a sensitivity analysis to indicate the minimum cutoff point for consideration of individual sources within a source category. This identifies single sources which, grouped as points, areas, or lines, within specific categories, can then be examined to ascertain further groupings such as similar operating rates, schedules, equipment configurations, etc. Following this, one value for a source parameter describes the entire class of sources within a category.

IV. Source Inventory for Source Categories

A. Classification of Categories

Source data (Table II) are required in the comprehensive inventory:

(a) To locate the source geographically and to index it with respect to political boundaries

(b) To describe the process and operating conditions of the source along with details on pollution control equipment and emission characteristics

Table II Emission Inventory Requirements

	Requirements: uses or user			
	Models			
	Atmospheric structure	*Transformation process*	*Regulatory agencies*	*Research*
Pollutants				
Hazardous material			X	X
Pollen				X
Polychlorbiphenyls				X
Benzo-*a*-pyrene				X
Trace metals			X	X
Asbestos			X	X
Sulfates				X
Fluorides				X
Heat	X			
Water vapor	X			
Particulate matter	X	X	X	X
Hydrocarbons by compounds		X		X
Hydrocarbons			X	
Peroxyacylnitrate				X
Ozone				X
Oxidants		X		X
Oxides of nitrogen		X	X	X
Carbon monoxide	X	X	X	X
Sulfur dioxide	X	X	X	X
Units				
Metric	X	X	X[a]	X
Time resolution				
Hourly average	X	X		X
Daily average			X	
Weekly average				X
Monthly average				X
Seasonal average			X	
Yearly average		X		X
Spatial resolution				
Line				
Endpoint intersections			X	
UTM		X		
Area				
Other				*b*
>1 km^2				
1 km^2	X	X		
<1 km^2			X	
Point				
0.01 km[c]	X	X	X	
0.1 km			X	X
Source				
Line		X	X	
Area	X	X	X	
Point	X	X	X	X

[a] English units are still in use in the United States and a few other countries.
[b] School districts.
[c] Error resolution.

(c) To automatically calculate emissions by means of current emission factors and compare with source estimated emissions and source tests

(d) To provide input data for mathematical modeling of ambient air concentrations

(e) To ascertain the legal status of each source with respect to control regulations and compliance schedules

B. Stationary Sources

1. Point Sources

a. Definition, General Information, and Identification. To adequately define each individual point source in a particular geographical area the following information is required:

1. Geographic code identification number—using standardized code numbers for compatibility with other data systems
2. Source number—this is a sequential identification number within each county; a source is a plant or facility
3. Point number—this is a sequential number within each source (plant or facility) used to denote individual stacks, vents, or emitting points
4. Year of record—this indicates the date of data collection or update
5. Name, address, person to contact about air pollution control, telephone number, etc.
6. Ownership of facility—to facilitate sorting of data on the basis of private, government, or public utility ownership
7. Confidentiality—an indicator to call attention to the possibility of the presence of confidential data
8. Source classification—to identify the basic process or industry activity

b. Fuel Combustion Data. The required fuel combustion data include:

1. Combustion rate—fuel consumed per hour; this amount must be tied in with the source classification code to facilitate recalculation of emissions using emission factors
2. Sulfur content, in percent
3. Ash content, in percent
4. Heat content, in millions of BTU (or kilocalories) per source classification unit

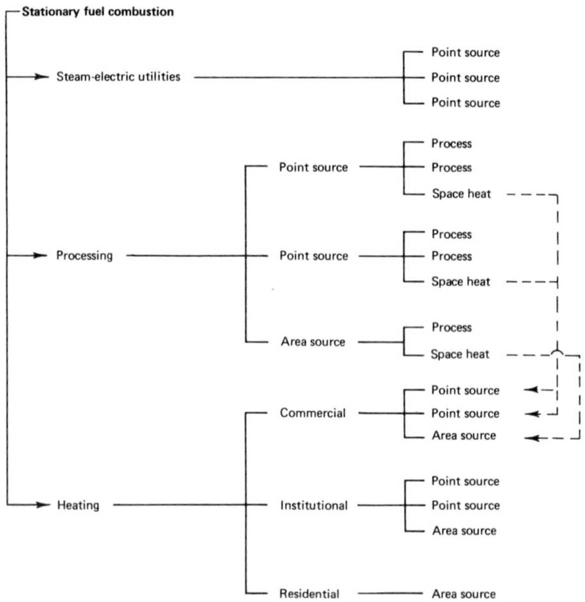

Figure 7. Fuel combustion breakdown, stationary sources.

These data are applicable primarily to boilers where external combustion of oil, coal, wood, or gas as fuels takes place (Fig. 7). Other possible combustion sources include burning coal refuse piles, structural fires, agricultural burning, and forest fires. In these latter cases, although a large area may be on fire, the source may be considered as a point.

The combustion (or process) rate may vary according to season (Fig. 8). Although the total consumption for the entire year is of most interest, it may be necessary to obtain daily consumption data, or hourly consumption data on a seasonal or monthly basis to more accurately predict emissions.

Information on the combustion of fuels must be tied together by means of an identifier, a source classification code, to clearly indicate exactly which fuel is used in what combustion process. This permits the calculation of emissions by the appropriate emission factor and the application of the proper control equipment efficiency. The use of the source classification identifier also enables summary statistics reports to be prepared providing groupings by control equipment, combustion unit, size, etc.

The stationary point source fuels commonly in use include anthracite and bituminous coal, lignite, residual, and distillate oil, natural and process gas, coke, bagasse, liquefied petroleum gas, and wood, all of which are burned in external combustion boilers.

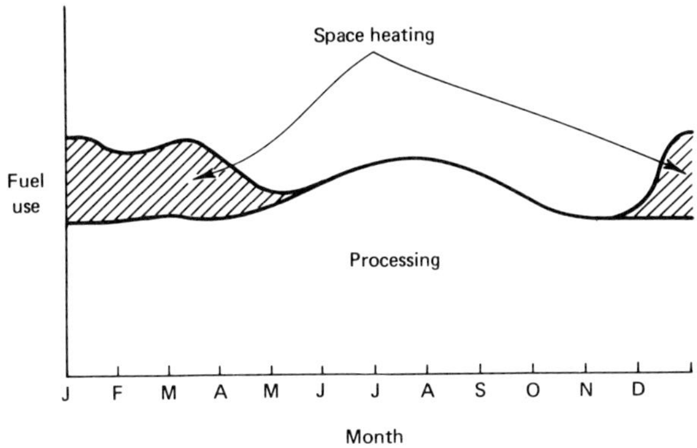

Figure 8. Monthly fuel use pattern (*1*).

For fossil fuels the sulfur and ash contents must be specified in order to determine sulfur oxides and particulates emissions. Annual boiler charging rate, maximum hourly design rate, and heat content must be available for calculation of emissions.

Further classification of the sources are required in order to summarize emissions, and the standard categories of electric/steam generation, industrial processing, and commercial-institutional are typically used for external combustion boilers.

c. Solid Waste Combustion Data. Solid waste material is incinerated usually in large municipally operated facilities. The data required for those sources are

1. Combustion rate—material quantity per hour, as required by the source classification
2. Estimated heat content, if estimation is possible

d. Process Loss Data. This particular source category covers by far the largest and most complex group of sources. The basic item of data is the throughput in units as required by the appropriate source classification.

e. Evaporative Loss Data. The major point sources of evaporative emissions are fuel storage tanks. The capacity of the storage tank, expressed as required by the source classification is necessary for emission calculation.

f. Miscellaneous Emissions Data. Information usually is grouped as follows:

1. Percent of annual throughput by season—production, consumption, throughput, etc., representing the operation of the source, proportioned among the four 3-month calendar year quarters
2. Percent of total fuel used for space heating—this amount of fuel will be assumed to produce emissions only in the winter seasons

g. Emissions Estimates. Estimates of emissions from an individual point source (in tons/year) can be made by the following methods:

1. Stack test results or other emission measurements
2. Material balance using expertise and engineering knowledge of the process
3. Emission factors
4. Rough estimates

The above methods are listed in order of preferred approach. The estimated emissions are presented as controlled emissions after passing through and taking account of the removal efficiency of the installed control equipment. The method of estimation must be specified in order to compare agency calculated emissions using given emission factors and control equipment efficiencies.

h. Modeling Information. Modeling information is required for individual point sources in order to provide input for computer analysis. The items required, in addition to those previously discussed, are

1. Coordinates of emitting point—using a standard reference system
2. Stack height—the vertical distance between the point of emission and ground level
3. Altitude of stack base—above sea level
4. Stack diameter—the inside diameter of a circular gas exit (equivalent cross-sectional area) at the point of emission
5. Stack exit gas temperature—under normal operating conditions the temperature of the exhaust stream
6. Stack exhaust flow rate—the design of maximum exhaust gas volume, or measured flow rate
7. Plume height—if all other stack data (except temperature) are zero, an estimate of the plume height must be made
8. Boiler input design capacity (before heat transfer)

i. CONTROL EQUIPMENT. The equipment controlling each individual air pollutant at the emission point and its relative efficiency in controlling each pollutant are required. In many cases there will be a primary and a secondary item of equipment, and their relative positions in the flow may effect the control efficiency.

In cases where the efficiency range of the device is unknown, make an estimate. Do not indicate equipment that normally is a part of the source since this will be considered in the emission factor. It may be anticipated that the majority of changes in the emission inventory will be in control equipment parameters.

j. COMPLIANCE DATA. The following items about each emitting point are required

1. Allowable emissions—maximum emissions, in tons per year of each pollutant that the source is legally permitted to discharge into the atmosphere under the most stringent of the following conditions:
   (i) Existing local/state/provincial/national regulations
   (ii) Imminent legislation which will legally require the source to comply with a specific emission limitation
   (iii) Official agreements with regulatory agencies whereby the emissions must be reduced to a prescribed quantity
2. Source compliance status
   (i) In compliance
   (ii) Not in compliance, on compliance schedule
   (iii) Not in compliance, not on compliance schedule
   (iv) Compliance schedule
   Date of schedule preparation
   Equipment order date
   Equipment installation date
   Equipment operation date
3. Applicable control regulations
4. Emergency control action program—certain sources are required to prepare a program for immediate reduction of emissions wherever air pollution in the area reaches prescribed conditions: this item should indicate the requirement of a program and whether the program has been submitted and approved.

k. LAND USE DATA. Several items used primarily in land use planning operations are normally included with source inventory data:

1. Normal number of operating hours per day
2. Normal operating days per week

3. Normal operating weeks per year
4. Number of employees at facility
5. Land area of facility

l. Data Confidentiality. Many sources which submit data request that all or part be kept confidential by the control agency. Legal assistance may be required in such cases to assure that the data are, in fact, legally confidential. Those items actually found to be confidential must then be protected from release, and must clearly be identified in the data system, with restrictions on their use clearly documented.

2. Area Sources

a. Definition General Information, and Identification. The significant concept for area sources is that the defined area "source" actually represents a collection of many small unidentifiable stationary points of air pollution emissions, all emitting less than the minimum level prescribed for point sources. Each small source may emit only very small quantities, but because of the great number of these sources, their collective impact is very significant, perhaps 30 to 40% of a study area's total emissions. The objective of area source calculations is to obtain an accurate estimate of this collective contribution to total emissions. Such an estimate can never be exact, because it is impossible to continuously monitor every home in a study area and determine the total residential heating emissions by summarizing individual emissions. Therefore, emissions from sources too small to be surveyed individually must be reported collectively as area sources for each reporting unit.

Each stationary area source must be uniquely identified and the source categories—fuel combustion, solid waste combustion, process loss, etc.—considered as subtotals.

b. Fuel Combustion Data. The quantity of fuel consumed by each source category must be determined. The source categories are

1. Residential heating—residential dwellings, from single family residences to multistory apartment complexes
2. Commercial-institutional heating—retail and wholesale establishments, schools, hospitals, government buildings, and other public facilities
3. Industrial—all manufacturing industries with emissions too small to qualify as point sources

Six major fuels and three less important fuels are usually considered, and the percents of sulfur and ash contents must be indicated for each in order to calculate the emissions of sulfur oxides and particulates:

1. Anthracite, or hard coal, is used in significant quantities only in areas that are within easy shipping distance from the coal fields; anthracite may be consumed by all source categories, although most is used by residential sources.

2. Bituminous coal production is more widespread than that of anthracite, therefore it is available in most areas; also included as bituminous coal are the lower grades of subbituminous coal, brown coal, and lignite. Because bituminous coal does not burn as cleanly as anthracite, a smaller proportion is used residentially. Bituminous coal is often favored for use by electric utilities, manufacturing industries, and coke producers.

3. Distillate oil includes fuel oil grades 1, 2, and 4. In addition, diesel fuel and kerosene can be considered distillate-type oils; residential and commercial-institutional sources are the largest consumers of distillate oil.

4. Residual oil includes fuel oil grades 5 and 6. In most areas, residual oil is not used by residential sources, but significant amounts may be consumed by industrial and commercial-institutional users.

5. Natural gas, consisting principally of methane but containing other hydrocarbon gases as well, is used in significant amounts by all source categories.

6. Wood may be used as heating fuel by residential sources (primarily in rural areas where wood is readily available), by industrial sources, and by commercial-institutional sources (usually in very small amounts). The chief industrial users of wood as heating fuel are those wood-processing industries such as sawmills, plywood mills, and furniture manufacturers that generate large amounts of wood waste that can be consumed in boilers.

7. Liquefied petroleum gas (LPG) consists of higher carbon-number hydrocarbons such as propane and butane. The contribution from LPG combustion to total emissions is not significant in most areas.

8. Coke is made by destructive distillation of coal to remove the volatile components. Although primarily used in the iron and steel industry, some coke may appear as an industrial area source fuel.

9. Process gas is often used at petroleum and petrochemical refining operations. "Coke oven gas," the gaseous product of destructive distillation of coal may be used industrially.

Most fuel retailers maintain sales records that can be a valuable source of information for the determination of area source fuel consumption. The information needed from fuel dealers is annual sales to each source category within the study area. The area source totals of residential and commercial-institutional fuel consumption are then simply the fuel dealers' figures minus any fuel consumed by point sources included in these source categories. It is important that all fuel dealers be contacted. The accuracy of the survey results will be significantly reduced if some fuel dealers are overlooked. It may be that not all fuel dealers will be able to furnish adequate information, or simply do not have the detailed breakdowns required. There is no assurance that fuel dealer sales accurately represent fuel consumption. Much of the fuel used may not come from retail dealers.

After determining fuels totals there may be two apportioning actions required: apportioning to smaller geographic areas and apportioning to source categories. Geographic apportioning of residential fuel use is based upon the ratio of dwelling units in the study area to the dwelling units in the larger area for which fuel statistics are available. Apportioning of commercial and institutional fuel use is based upon the ratios of population, and apportioning of industrial fuel use is based upon the ratios of manufacturing employees. Source category apportioning is more difficult.

1. Anthracite coal—the procedure for determining the use by each source category is

(i) Determine the total area source consumption of anthracite by subtracting the amount used by point sources.

(ii) Determine the residential consumption of anthracite according to the following method. The residential anthracite consumption is a product of four factors: the number of dwelling units using coal as heating fuel, the coal heating requirement factor, the average annual heating degree-days, and the correction factor for the number of rooms per dwelling unit in the study area. The heating requirement factor for coal is 0.0012 ton coal per dwelling unit per degree-F-day (1.964 kg per dwelling unit per degree-C-day). The average annual heating degree-days should be obtained from local climatological data. An average degree-day value compiled from all the reporting stations in the area for which summary fuel statistics are available for the year of interest should be used. The degree-day value given by the nearest reporting station should be used for the study area. The correction factor is the average number of rooms per dwelling unit, as given in census data for the study area divided by five—an average number of rooms per dwelling unit.

(iii) The commercial-institutional and the industrial area source consumption of anthracite can be determined by subtracting the calculated residential consumption figure from the anthracite area source total. All of the remaining anthracite should be assigned to the commercial and institutional category if no anthracite is consumed by industrial point sources. Assign 60% of the remaining amount to commercial and institutional sources and 40% to industrial sources if a significant amount of anthracite is consumed by industrial point sources.

2. Bituminous coal—the procedure is

(i) In some areas anthracite coal is not available. For these areas, all of the residential coal consumption should be considered bituminous coal, and the total consumption can be calculated according to the same procedures given for anthracite coal. For the areas in which anthracite is available, estimates of the number of dwelling units heating with anthracite and the number heating with bituminous must be made. Local coal dealers may be able to state what percent of their customers are sold anthracite, and what percent of their customers are sold bituminous. If not, the bituminous coal portion can be estimated by assuming that all bituminous coal sold by retail dealers goes to residential sources. The difference between the amount calculated by the method based on the number of dwelling units and the retail dealers figure should be considered to be the anthracite portion.

(ii) If the residential bituminous coal consumption is less than the retail dealers bituminous coal figure, assign the difference to the commercial and institutional source category. If the residential bituminous coal consumption is not less than the retail dealers figure, assume commercial and institutional bituminous coal consumption to be negligible.

3. Distillate oil—the amount consumed by residential sources estimated by using the number of dwelling units heating with fuel oil and a heating requirement factor of 0.18 gallon oil per dwelling unit per degree-F-day (1.23 liters per dwelling unit per degree-C-day), as done for anthracite.

4. Residual oil—Use the same methods as given for distillate oil. Residential sources do not usually consume residual oil. Local oil dealers should be contacted to determine what percent of their residential customers, if any, were sold fuel oil grades 5 and 6.

5. Natural gas—distribution figures are usually available from the gas suppliers.

6. Liquefied petroleum gas—assume the ratio of residential to commercial use of LPG the same as that for natural gas, and that little is used for industrial operations.

7. Wood—little published data are available for wood consumption

Table III Average Fuel Consumption for a Five-Room Residence (United States)

Fuel	*Consumption*
Natural gas	22.5 ft^3/dwelling unit degree-F-day (1.15 m^2/dwelling unit degree-C-day)
Fuel oil	0.18 gallon/dwelling unit degree-F-day (1.23 liters/dwelling unit degree-C-day)
Coal	0.0012 ton/dwelling unit degree-F-day (1.96 kg/dwelling unit degree-C-day)

by source category. Residential use can be estimated by multiplying the number of dwelling units heating with wood by the heating requirement factor of 0.0017 ton of wood per dwelling unit per degree-F-day (2.728 kg per dwelling unit per degree-C-day). Wood may also be used industrially by those wood-processing operations that are not included as point sources.

As an alternative, the data in Table III for natural gas, fuel oil, and coal may be used for residential fuel consumption.

c. Solid Waste Combustion Data. The area source solid waste category includes on-site refuse disposal activities by residential, industrial, and commercial-institutional sources. On-site incineration refers to disposal in a small incinerator. This includes backyard burners, industrial incinerators, and incinerators at food and department stores, hospitals, and schools. On-site open burning refers to unconfined burning of waste such as leaves, landscape refuse, or other rubbish. For emission inventory purposes, only solid waste that is burned is of interest. Unfortunately, very little quantitative information about on-site solid waste disposal is available.

Two methods of collecting data are available:

1. Solid waste surveys—some areas have conducted comprehensive surveys of solid waste disposal practices. If such a survey is available, it should be used to estimate solid waste quantities. Many such surveys cover only collected waste, however, and are of limited value for determination of on-site waste disposal quantities. If better information from solid waste surveys cannot be obtained, assume that 75% of the residential–commercial–institutional total was generated by residential sources. In some areas, open burning and on-site incineration are regulated and may be prohibited. If one or both of these practices is prohibited, the

corresponding combustion values should not be applied. Under such circumstances, assume that the solid waste that would normally be allotted to on-site disposal is handled by some other method that would not result in air pollutant emissions (land filling).

2. Per capita generation rate—average incineration values shown in Table IV may be used.

d. Evaporative Loss Data. This source category covers evaporative losses of hydrocarbons and other organic compounds not considered as point sources. There are three major subcategories:

1. Gasoline handling operations—losses from gasoline marketing, losses from loading storage tanks at service stations, spillage and losses from filling automobile fuel tanks

2. Dry cleaning operations—losses in cleaning plants where clothing textiles are cleaned with organic solvents

3. Printing, surface coating, metal cleaning, and degreasing operation—automobile and aircraft assembly, container, furniture, and appliance manufacture, enamel and paint application, and plastic products manufacture

Three data collection methods can be used:

1. Gasoline-handling losses can be calculated if the gasoline sales total for each study area are known. If this total has not already been determined for motor vehicle emission estimates, it should be calculated according to methods used for on-highway motor vehicles and for off-highway gasoline use.

2. Evaporative losses for solvent-use operations can best be determined by a questionnaire survey of all suspected solvent users. Information required is the amount of solvent purchased per year in tons. If figures are available only in gallons or liters, the average density of the solvents used should also be obtained and converted to tons.

Table IV Average Solid Waste Burned (United States)

Source category	*Waste burned per person-day*
Industrial on-site incineration	1.6 lb (0.72 kg)
Industrial on-site open burning	0.8 lb (0.36 kg)
Residential and commercial-institutional on-site incineration	0.65 lb (0.30 kg)
Residential and commercial-institutional on-site open burning	0.5 lb (0.23 kg)

Table V Solvent-Use Estimates (United States)

Area population	*Solvent used per capita-year*
<100,000	3 lb (1.36 kg)
100,000–500,000	8 lb (3.64 kg)
500,000–1,000,000	18 lb (8.18 kg)
>1,000,000	28 lb (12.73 kg)

3. An alternate method, which is not as accurate as a survey but much easier to employ, is to assume that solvent use is proportional to population. For dry cleaning operations, use of either of two factors, 2 pounds (0.9 kg) of solvent evaporated per person per year for colder climates, will provide reasonable estimates. Emissions from surface-coating and degreasing operations are more difficult to estimate; 30 pounds (14 kg) per person per year has been proposed, but this estimate was derived from data obtained only in large cities, where higher-than-average concentrations of industries involved in surface-coating operations would be located, and represents the total amount of solvent used by point and area sources. Thus, the graduated scale shown in Table V is recommended for estimation of other solvent-use evaporative losses.

e. Miscellaneous Sources Data. Each of the subcategories contained in this section can produce large quantities of air pollutant emissions. The techniques for collection of the required data and estimation of emissions are normally not as refined as the methods used for other area sources. The miscellaneous sources are characterized by intermittent emissions, rather than more or less continuous emissions that are characteristic of the other source categories. This means that even though the total emissions contributed by these sources may be relatively small, on particular days their contributions to the total emissions can be quite significant.

Emissions from forest fires can be very significant. Estimates of the quantity and type of growth burned should be available from forestry or agriculture departments or local fire protection agencies. The determination of tons of growth burned per acre is equally important. Local officials should be contacted for this information. If they cannot provide data, 40 tons/acre (3.64 kg/m^2) for heavily forested areas and 20 tons/acre (1.82 kg/m^2) for thinly forested or brushy areas may be used to calculate the material consumed.

Waste from logging operations is often burned under controlled conditions both to reduce the potential fire hazard in forests and to remove brush that can serve as a host for destructive insects. Because of the magnitude of this operation in certain areas, a serious air pollution problem can result if certain meteorological conditions exist when slash burning takes place. Officials of forestry departments should be contacted to provide estimates of the area burned and quantity of slash per acre. If an estimate of the quantity of slash burned per acre cannot be obtained from other sources, use 75 tons/acre. Also included under this source category are agricultural field-burning operations such as stubble burning and burning of refuse from land clearing. Estimates of acreage and tons per acre burned should be obtained. For grass burning, if a local estimate is not available, assume 2.5 tons/acre (0.23 kg/m^2). If agricultural field burning is included with slash burning, the quantity-per-acre figure should represent a weighted average for slash burning and agricultural field burning.

In some areas, orchard heaters may be used on the coldest nights of the year. Agriculture officials will usually have a breakdown of the number and types of orchard heaters that are in use. Data can also be obtained from grove operators in the area. These sources should also be able to furnish the number of nights the units were fired during the year and the approximate length of time each night the units were burning. An estimate should also be obtained of the number of units fired at any one time. The data required are the approximate number of units that are fired on each night when frost control is required, and the number of days out of the year when such operations take place.

Building fires can also produce short-term emissions of air contaminants. The best procedure for determining information for this source category is to contact local fire departments, fire protection associations, or other agencies for an estimate of the number of structural fires that occur in each jurisdiction during the year. In the absence of such information, it can be assumed that an average of four fires per 1000 population occur each year, a frequency which can be verified for the United States.

Burning coal refuse piles exist in some areas. The coal refuse piles, which were produced as a result of coal mining operations, were formed from coal refuse and rejected material containing too high a percentage of impurities to be marketed. Over the years, the coal refuse piles may have been ignited by spontaneous combustion or through careless or intentional acts of humans. If possible, local officials and owners of the coal piles should be contacted to determine if there has been any change in status of the coal fires and to estimate the extent of coal refuse fires. The total size of all burning coal refuse piles in a county should be tabulated. It is entirely possible that these piles may need to be classified as point sources in many cases.

Dust emissions can result from a variety of operations. Many of the dust emissions are caused by man, but a large portion can be the result of natural events. The principal causes of dust emissions are

1a. Vehicle travel on dirt roads—the information that should be obtained is an estimate of the number of vehicle miles or kilometers of travel on dirt roads. The highway department may have made such estimates; if not, it should at least be able to furnish the total length of dirt roads in each jurisdiction. This information can be converted to vehicle miles by estimating the traffic on dirt roads. Road departments can also be consulted to determine miles of dirt roads in the study area. The estimate of the traffic on dirt roads should be based on information that can be gained from government officials.

1b. Vehicle travel on paved roads–there is usually considerable dirt and debris on paved roads, as evidenced by the need for street cleaning and washing programs in cities. This dirt and debris comes from: dried mud from tire treads and vehicle chassis and bodies; spillage from trucks transporting dirt; rubber ground from normal wear of tires; particulate matter from the wearing of break linings; and fallout from the air of natural and man-made particulate matter from both the city and its peripheral areas. It is reentrained in the air by vehicular traffic and can be a significant part of the particulate matter at road-side air quality monitoring stations.

2. Aircraft operation on dirt airstrips—at certain small airports, flight cycles may be performed on dirt airstrips. At such locations an estimate of the number of annual landing and take-off cycles performed should be calculated, or obtained from the airport operator. This entry is made to account for dust emissions only.

3. Windblown dust and construction activities—in some areas large dust emissions are generated as a result of windblown soil. Officials of the appropriate government agency, such as the department of natural resources or department of agriculture, may be able to furnish estimates of the number of acres subject to severe wind soil erosion. Because such estimates are usually vague, reporting the total acreage to nearest 1000 acres (405 hectares) is sufficient. Also, substantial areas may be subject to construction activities that can produce dust emissions. Road construction and large building projects, such as urban redevelopment activities and large housing subdivisions, are the chief locations where significant dust emissions are produced. An estimate of the area subject to such activities can be based on information obtained from the highway de-

partment and local public works departments or similar agencies responsible for regulation of building construction activities. The total area subject to construction activities [to the nearest 1000 acres (405 hectares)] should be included in the total.

4. Rock handling and storage—usually all rock quarries and other large rock-handling operations will be included as point sources, however, area source rock-handling operations cover smaller-scale activities, such as small sand and gravel yards, stone products manufacturers, and other mineral products industries not included as point sources. The tons processed by all these small operations should be tabulated. The totals should be determined by adding the totals given on emission survey questionnaires for sources that did not qualify as point sources.

f. Emissions Estimates. The most recent gross emission estimates, in hundreds of tons per year, should be included. This provides additional information to serve as a cross check for data calculated, indicating required adjustments in the source inventory, emission factors, or the abatement and control plan and its strategies.

g. Compliance Data. At best, compliance data for area sources are limited to the applicable fuel use regulations, open burning regulations, etc., and an estimate of the percent of sources in compliance with the regulations.

h. Land Use and Other Data. Land use information includes the population of the area, the land area, and economic information helpful for predicting the anticipated growth of the area, and hence the increase in emissions. Population by census area, and some indication of the relative percents of rural, suburban, and urban land areas and populations are also very useful.

C. *Mobile Sources*

1. Area Sources

a. Definition, General Information, and Identification. Mobile area sources include all vehicles which travel in a random manner not defined as line sources. The major source categories are highway and off-highway vehicles. The emissions from these vehicles are primarily carbon monoxide, oxides of nitrogen, and hydrocarbons. Oxidants are not considered since they are a result of an atmospheric chemical reaction and do not occur as an emission.

b. HIGHWAY VEHICLE DATA. Gasoline-powered vehicles include all highway use of gasoline by light-duty vehicles [under 6000 pounds (2722 kg) gross vehicle weight] and heavy-duty vehicles [over 6000 pounds (2722 kg) gross vehicle weight]. Light-duty vehicles include automobiles and small trucks, while heavy-duty vehicles include larger trucks and buses. Two data collection methods for estimating gasoline-powered vehicle miles (or kilometers) of travel (VMT) are given in order of decreasing accuracy.

1. Measured VMT—Emissions from motor vehicles are best estimated using measured VMT. Total VMT can be obtained from traffic surveys conducted by state highway departments or local transportation agencies. Normally the total VMT will include diesel VMT as well as gasoline VMT. If actual VMT information is unavailable from these surveys, VMT can be estimated from traffic counts or traffic flow maps. This is accomplished by multiplying the traffic volume by the length of roadway. Since vehicle emission rates vary appreciably with average speed, VMT data should be obtained by roadway type. On the basis of average speed, roadways can be classified as: limited access [greater than 50 mph (80 kmph)], rural roads [45 mph (72 kmph)], suburban roads [35 mph (56 kmph)], and urban streets [25 mph (40 kmph) or less]. In addition to the total number of VMT, a breakdown of heavy-duty travel and light-duty travel should be obtained from the traffic survey. If possible the breakdown of VMT by average speed should also be determined for grid squares.

2. Gasoline sales data—some tax agencies compute gasoline sales by geographical area. The gasoline sales can be converted to VMT by multiplying by the national average distance/gallon or/liter.

To complete the source calculations by means of the first method above, gasoline VMT by roadway type are obtained for the study area; if it is not possible to determine VMT for each of the types of roads, the VMT should be broken down at least into the urban and rural components. Before attempting to calculate the gasoline-powered VMT, total diesel-powered VMT must be subtracted from total VMT, as described later. If the measured VMT method is used, the light-duty/heavy-duty fuel consumption breakdown should be calculated. This can best be accomplished by using gasoline sales data.

If the second method is used to complete the area source tabulations, total gasoline VMT for a jurisdiction, computed by assuming an average distance per gallon or per liter, is divided into heavy duty and light duty. It is assumed that approximately 10% of the total VMT in the study area is heavy-duty vehicle travel. After heavy-duty and light-duty VMT are determined, they are converted to gasoline consumption by using the mean distance per gallon or liter for light-duty vehicles and the mean distance per gallon or liter for heavy-duty vehicles. The urban–rural breakdown of VMT must be estimated.

The diesel-powered vehicle source category includes heavy-duty trucks and buses using diesel fuel. Again, two data collection methods can be used:

1. Vehicle distance—the best method is to obtain measured VMT by diesel-powered vehicles. If this information is not available, diesel VMT must be calculated from the diesel fuel sales for the jurisdiction. Diesel VMT is obtained by multiplying the diesel fuel sales by the mean distance per gallon or liter. If the first method was previously used for gasoline-powered vehicles, this method must also be used for diesel-powered vehicles to account for diesel VMT included in the total VMT.
2. Diesel fuel sales—if VMT cannot be obtained, quantity of diesel fuel sold per jurisdiction should be obtained from the appropriate state or provincial agency or oil dealers' association.

When the VMT method is used, the only quantity necessary is highway diesel fuel consumption. Diesel VMT needs to be computed only to account for the diesel portion of the total VMT.

Obtaining measured diesel VMT is by far the best method. Data may be available that indicate the percent of the total VMT by diesels. As was indicated in the methods for gasoline-powered vehicles, assume that diesel fuel purchased in the study area is consumed there. This is a grosser assumption with diesels, however, because most diesel traffic is through-traffic.

c. Off-Highway Vehicle Data. All aircraft operations are considered area sources. The required source inventory data are the number of landing-takeoff (LTO) cycles performed annually by military, civil (general aviation), and commercial aircraft. The emissions are determined from emission factors representing fuel consumption of descent from 3000 ft (914 m), taxi, and takeoff and ascent to 3000 ft (914 m) altitude. The LTO cycles can be obtained either from national compilations of statistics by the aviation regulatory agency, from military authorities, census bureaus, or from airport operators. Emissions from each aircraft category are calculated by using average emission factors based on the mix of engine and aircraft designs for each category of the aircraft fleet. As the aircraft mix within each category changes with time and as newer information becomes available on the emission rates for each aircraft engine the emission factors must be adjusted to reflect these changes.

The locomotive category includes the fuel used to heat railroad stations and workshops. The latter fuel use will be much less than that of locomotives and is difficult to separate from total railroad fuel use. Hence, the railroad category is considered to be the fuel consumption by locomotives, although it does include small amounts of fuel used for space heating. The fuels primarily consumed by railroads are distillate oil (diesel fuel) and coal, but smaller amounts of residual oil are also used. Source inventory data on fuel consumption may be obtained from national statistics or directly from the rail lines. The fuel totals can be apportioned to the study area either by track milage or by population using the techniques discussed previously. Because operations in rail yards are usually heavier than on main-line track, emissions from railroad yards may be quite significant. If yard track length cannot be obtained from rail lines, the yard track length can be roughly accounted for by doubling the measured track length in areas where it is known that there are yard operations. The population apportioning technique may be used when it is impossible to use track length. This technique assumes that most yard operations take place in large cities, an assumption that may or may not be true depending on the study area.

The vessel source category includes ocean-going ships, river vessels, and small pleasure craft used on lakes and rivers. Three data collection methods are available, each subject to considerable error since fuel is purchased in port and large amounts are consumed at sea, except for river and lake traffic. Some fuel is consumed in port for heating and dockside power.

1. Coal-powered vessels—there are still a few vessels that burn coal. No reliable method for estimation of emissions on a local basis for these vessels is currently known, because usually only nationwide fuel totals are available. Thus, information on local fuel consumption can be obtained only from estimates made by port authorities or ship operators.

2. Gasoline-powered vessels—this category includes small craft operated by inboard, outboard, or inboard–outboard motors on lakes and rivers. Gasoline consumption can be estimated by using a factor of 160 gallons (600 liters) of gasoline per boat registration per year. This factor was derived from information on average annual operating hours, average horsepower per vessel, and average fuel consumption per horsepower-hour of operation. The estimate also accounts for any nonmotorized boats that may be included in the total of registrations and for boats with small outboard motors that may not be registered. Boat registrations should be obtained from the appropriate agency, such as the Recreation, Fish, and Game Department.

3. Fuel oil (including diesel) -powered vessels—fuel consumption by these vessels includes the fuel used by large cargo and passenger ships, oil tankers, tugboats, and all other steamships and motorships that use fuel oil.

Unique apportioning methods are used for each vessel category:

1. Coal-powered vessels—because fuel consumption data on a local basis are very limited for coal-powered vessels, totals may have to be obtained by apportioning national totals. Coal use should be assigned to areas with major ports, according to tonnage handled in the ports.
2. Gasoline-powered vessels—gasoline use could be apportioned according to boat registrations. This may not be an accurate method because many boats are not used primarily in the area in which they are registered. Instead, the total gasoline consumption predicted by the total of boat registrations should be apportioned to areas on the basis of inland water surface area.
3. Fuel oil-powered vessels—if fuel consumption totals are determined from a questionnaire survey, the fuel oil consumption figures should simply be assigned to where the vessels are operated. If vessel movement data must be used, extensive apportioning measures are necessary. The apportioning method becomes somewhat involved because emissions both underway and dockside should be considered. Emissions underway include the emissions that occur when a vessel is moving under its own power through a waterway and when it is maneuvering into its dock space. The average fuel consumption rate during these periods for steamships (predominantly residual oil users) is 44 gallons per mile (104.5 liters per kilometer), and for motorships (predominantly diesel fuel consumers) the average fuel consumption rate is 19 gallons per mile (45.1 liters per kilometer). In-port, or dockside, emissions occur when a ship operates its engines or boilers when in port to provide power for the ship's utilities. Average fuel consumption rates are 1900 gallons (7200 liters) per day of residual oil for steamships and 660 gallons (2500 liters) per day for motorships. The local port authority should be able to supply a figure for the number of days, on the average, that a ship remains in port. If not, an average figure of 3 days per vessel should be used.

Farm and construction equipment uses both gasoline and diesel fuel. Off-highway equipment is usually uncontrolled from an exhaust emission standpoint.

Farm tractor gasoline fuel consumption can be calculated by applying an annual per-tractor fuel consumption rate to the number of tractors in use. The total consumption for all other off-highway gasoline-powered equipment may be estimated by assuming consumption of 13 gallons (50 liters) per person in the study area.

Farm tractor diesel fuel consumption is about 1000 gallons (3750 liters) per tractor annually. The total consumption for construction equipment is approximately 5000 gallons (19,000 liters) annually per construction employee not actually involved in building erection. All other off-highway diesel fuel consumption may be calculated by applying a factor of 7.4 gallons (28 liters) per capita annually.

For the small utility equipment source category such as outboard motors, lawn mowers, garden tractors, snowmobiles, etc. gasoline sold for automotive purposes is the usual fuel, thus the quantities are very difficult to disassociate from that used by highway vehicles.

2. Line Sources

a. Definition, General Information, and Identification. A line source is a mobile source which tranverses a specified, definite route having emissions greater than some minimum amount. Only by selecting specific highways, vessel channels, or rail lines and estimating their emissions is it possible to determine if this minimum amount has been exceeded. The minimum amount for consideration as a line source should be determined either through sensitivity analysis or by special study.

The mobile sources include light-, medium-, and heavy-duty highway vehicles (and possibly motorcycles), vessels, and locomotives. These are the same mobile sources as were considered for area sources, except that aircraft and airports are not considered to be line sources. Therefore, it must be decided in advance of inventory preparation whether line sources will be considered. The information required includes a count of vehicles and either a specific operating cycle for the route or an average value for use in application of emission factors.

b. Highways. Source inventory data in the form of daily or hourly average vehicle counts, are available from highway and transportation planning agencies. However, the vehicle counts are seldom subdivided into light-, medium-, or heavy-duty classes.

If operating cycles for the specific routes are determined, either by estimation and engineering derivation or by field measurement, the direction of travel along the routes will effect the cycle due to terrain, traffic signals, etc., therefore cycles may be direction dependent on the same route, and may vary with time of day. If average emission factors must be used, then only the length of the route is required. However, due to variation in traffic volumes along a line source, it may be necessary to subdivide the route into "links," or short lengths for which inventory data must be developed.

c. Railroads. Types other than freight yards, switching yards or station areas, constitute line sources if the magnitude of their emissions justify such consideration. The timetables of both passenger and freight trains can be obtained from station masters, and the emissions estimated in much the same manner as for highway vehicles. However, the emission factor data available do not lend themselves to modal analysis, although the power setting used by the locomotives may be determined in the course of the inventory.

d. Channels. Vessels operating in a specific channel may be considered as line sources. However, insufficient information is available for model determination due to lack of vessel test results. Source inventory information on vessel classes and counts for selected waterways is available from port and navigation authorities.

e. Modeling Information. Most atmospheric dispersion models for line sources require a minimum of information as input: (a) emissions, by pollutant, for each link; (b) coordinates of link end points; and (c) length of the link.

The models require the emissions per hour, and except when hourly values are available, annual or daily emissions must be reduced to hourly values.

D. Area Source Gridding and Apportioning of Fuels and Emissions

To perform atmospheric dispersion modeling using emissions from area sources, each small subdivided area source is considered as a point source in source-receptor analysis. Therefore study areas must be divided (or gridded) into small squares of between 1 and 100 km^2. Fuels, VMT, and other area source inventory parameters from which emissions can be calculated through the application of emission factors must then be apportioned to each square on the basis of population, vehicle activity, etc.

First it is necessary to construct the grid squares; the high emission smaller grids are in the areas of highest population density. Figure 9 shows a sample study area in which grid squares having side lengths of from 5 to 40 km are used. Also shown are the population densities of

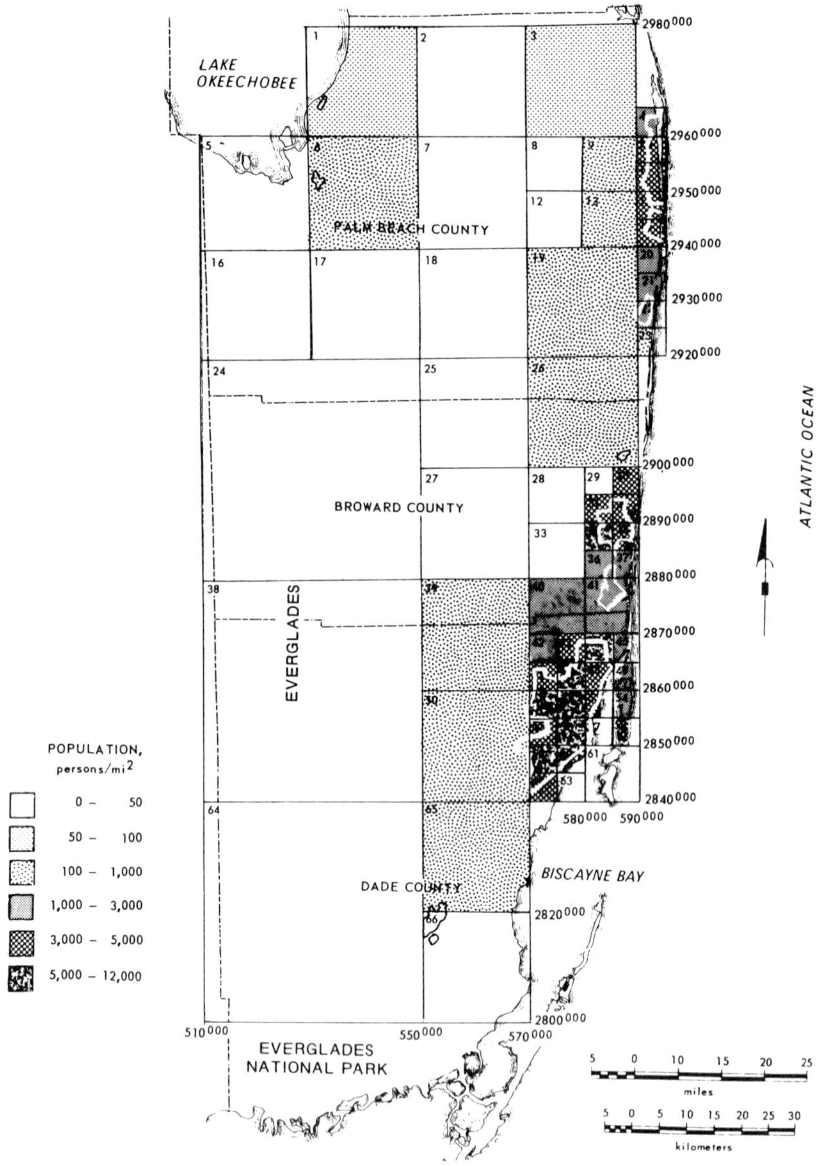

Figure 9. Grid coordinate system and population density for study area (*4*).

these grid squares, and the Universal Transverse Mercator coordinates of the squares' sides (together with their identification numbers in the northwest corner of each square).

1. Transportation Emissions

Apportioning is done on the basis of VMT in each grid square for motor vehicles, length of railroad track in each grid square for railroads, and ship traffic data for vessels. Aircraft are applied to the grid squares in which the airports are located.

If possible, comprehensive traffic-count maps from highway departments and metropolitan planning agencies should be obtained. These maps should, at least, show traffic counts (in units of vehicles per day) for all expressways, major arterial roads of major cities, and downtown streets of major cities to ensure that a substantial portion of the commuting traffic and through traffic can be accurately proportioned. The grid squares are delineated on these maps in the same manner as for census maps. The VMT on a road in a grid square is obtained by measuring the length of road between two consecutive traffic counts and multiplying this distance by the numerical average of the two counts. When measuring VMT special note should be taken of the number of VMT per grid on roads where diesel-powered vehicles would most likely travel: expressway, arterial, and downtown roads.

It is very unlikely that the traffic-count maps that are used to measure VMT will be of the same year as the intended emission inventory; the maps will, therefore, require updating. This updating should reflect the total increase in gasoline- and diesel-powered motor vehicle activity in the study area since the traffic-count map was made. Diesel VMT and gasoline VMT should be updated separately by means of the fuel consumption totals for each.

Before the traffic-count totals can be updated or the diesel-powered VMT proportioned, the number of diesel-powered VMT included in the total VMT on the traffic-count maps must be determined. The diesel-powered VMT can be estimated by multiplying the percent of the total VMT in the study area that are diesel-powered by the VMT of each grid that were measured from roads having diesel traffic. The diesel percentages should also take the year of the traffic survey into consideration. Remaining diesel-powered VMT should then be apportioned to grid zones on the basis of those diesel-powered VMT already distributed by the traffic-count maps.

The last step before calculating emissions is to distribute the gasoline-powered VMT that remain after the traffic-count maps have been used. These undistributed VMT may be apportioned on the basis of population because most remaining motor vehicle activity would occur in residential areas.

In many cases vehicle counts will be unavailable and the grid squares

Table VI Average Vehicle Counts by Street Class (United States)

Street class	*Vehicle density, vehicles/day*
Major	17,000
Industrial	14,000
Commercial	10,000
Residential	1,000
Other	1,000

must be laid over a road map of the study area. The total length of road in each grid square are calculated on the basis of the following street classes: major, industrial, commercial, residential, and other; the average vehicle counts as given in Table VI can then be applied in determining the VMT for the grid square.

The following factors must be taken into account in computing emissions based on VMT: sources of pollutants within the vehicle, degree of emissions control, and average speed for each type of road in the zone.

Pollutants emanate from the exhaust system, engine crankcase blowby, carburetor, and fuel tank in an uncontrolled gasoline-powered vehicle. The exhaust system emissions consist of all the products of fuel combustion; the crankcase emissions are hydrocarbons that escape from the engine cylinders. The emissions from the carburetor and fuel tank are evaporative hydrocarbons. Before the proper emission factors can be applied, an average speed for each road class must be assumed. These average speeds are often available from local traffic information; if not, the average speeds shown in Table VII can be used.

For each grid square an estimate should be made of total VMT by diesel vehicles. An estimate of these VMT can be made from local traffic data and subtracted from the total VMT for a given square. If such data

Table VII Average Vehicle Speeds by Route Class (United States)

Route class	*Average speed per hour*
Urban streets	25 miles (40 km)
Suburban roads	35 miles (56 km)
Rural roads	45 miles (72 km)
Limited-access expressways	50+ miles (80 km)

are not available, 3.2% of the total VMT for an area may be considered to be traveled by diesel vehicles for the vehicle mix in the United States. This percentage will be different in countries with a different vehicle mix. When estimating diesel VMT, one should be careful not to estimate diesel travel in residential areas and on roads that are restricted to automobiles, such as roads that specify "no through trucks."

2. Evaporative Emissions

Because many hours are required to locate all service stations and place them in grids, an acceptable practice is to apportion the estimated emissions from service station operations on the basis of either population or VMT. The quantity of fuel estimated for motor vehicle use should be the basis for determining evaporative losses using emission factors.

Evaporative losses due to dry cleaning are apportioned on the basis of population in those squares with commercial activity only.

3. Solid Waste Disposal Emissions

Burning in dumps and in some incinerators is considered a point source emission and is listed according to the coordinates of the source. Emissions that must be apportioned as area sources include on-site incineration and on-site open burning. The grids in the jurisdiction are rated on the basis of land use, population density, and industrial and commercial development. The rating system could be established by estimating the percentage of each grid that is devoted to residential, commercial, and industrial use. The percentages for each category are totaled for the jurisdiction, and the ratio of the percentage in the grid to the jurisdictional total is used to distribute the emissions for each appropriate source category. Source information gathered according to the previously discussed methods gives the percentage of solid waste that is industrial, commercial, or residential in origin. Once the quantities of solid waste in each category have been allocated to grids, emission factors can be applied to the solid waste disposal methods to determine emissions.

4. Residential Heating Emissions

If large apartment complexes use oil or coal as a fuel, it will be necessary to survey and pinpoint these complexes for more accurate pollutant calculations. These large sources are then easily located on the grid system as point sources; the area source classification remains. After residential fuel consumption data are obtained, it is necessary to obtain aid from local officials in apportioning these oil, coal, and gas data to the correct areas. When this type of information is not available, it is necessary to estimate the number of dwelling units in a given study area that is using each type of fuel, based on the same proportion as the larger census area. Further apportioning will be necessary to reduce these data to a grid basis.

5. Modeling Data

If the emissions for each area source category (by pollutant) are tabulated for each grid square, a total by pollutant can be presented. These data, together with the following, for each square, comprise that required for input to the dispersion model: (a) southwest coordinates of each grid square; (b) side length of grid square; (c) average stack height for stationary sources—usually estimated at 30 ft (9.14 m); and (d) mean elevation above sea level.

The preceding methodology for apportioning area fuels and emissions is based upon gross fuels totals for jurisdictions. Alternatively, typical source category operating characteristics can be specified for each individual grid square. For example, a typical operating schedule for residential heating can be developed which will yield time-dependent emissions. This will then provide diurnal area source emissions for more sophisticated modeling of dispersion.

V. Emission Factors for Source Categories*

A. Definition

Since continuous in-stack monitoring is expensive and feasible for only the very largest of sources, it is necessary to characterize source categories by generalized emission factors as a function of the process or combustion rate.

An emission factor is a statistical average of the rate at which a pollutant is released to the atmosphere as a result of some activity, such as combustion or industrial production, divided by the rate of that activity. For example, assume that in the production of 260,000 tons of ammonia per year, 26,000 tons of carbon monoxide are emitted to the atmosphere. The emission factor for carbon monoxide in the production of ammonia

* See Ref. (*6*).

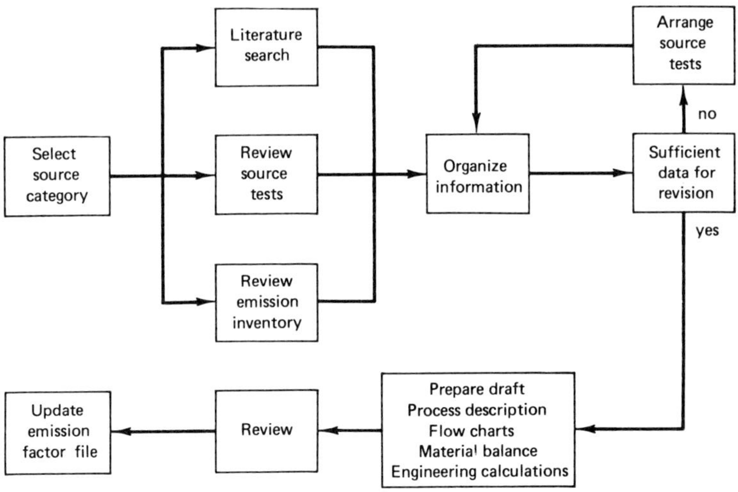

Figure 10. Emission factor methodology.

would therefore be 200 pounds of carbon monoxide released per ton (100 kg per metric ton) of ammonia produced.

Emission factors are determined by a spectrum of techniques including: detailed source testing involving many measurements, single measurements, engineering analysis of process material balances, and engineering appraisals of a given process (Fig. 10).

In general, emission factors are not precise indicators of emissions for a single process and are, therefore, more valid applied to a large number of processes.

B. Stationary Sources

Since the emission factor is, at best, a gross estimate of the emissions per unit processed, burned, etc., the emissions can only be calculated on an annual average basis from annual production figures. The factor is a composite of all the subprocesses comprising a major process, and does not take into account the matter of start-up/shutdown or batch operations (except on an average batch size basis).

The development of an emission factor from subprocesses requires an intimate knowledge of all the modes of batch and continuous operations, the emission factors associated with each and the time period during which each takes place. Modal operations (Fig. 11) must be combined to provide an average factor.

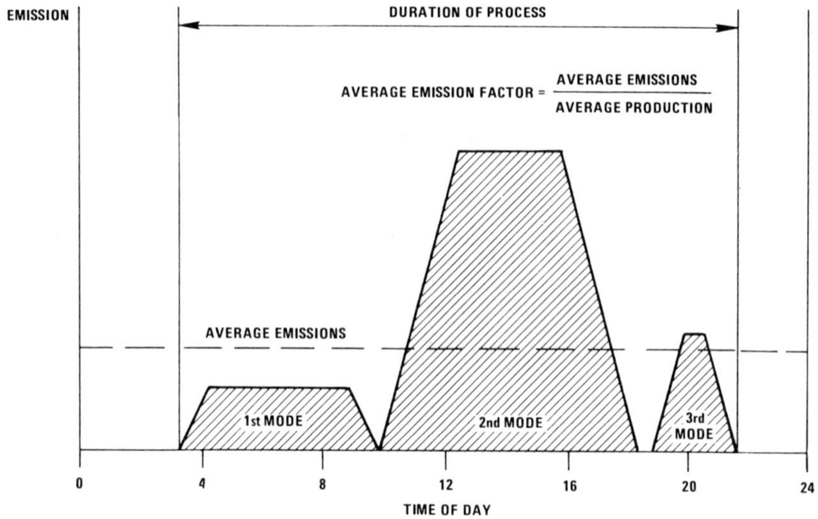

Figure 11. Average emission factor development procedure.

C. Mobile Sources

Mobile source emission factors are expressed as emissions per VMT, except for aircraft emissions which are a function of the landing/take-off cycle [up to/down from 3000 ft (914 m) altitude]. Operating or ambient temperature, as a function of the time of year, may effect combustion characteristics and therefore emission factors. The altitude of a mobile source effects its emissions owing to atmospheric pressure. Vehicle performance may be effected by fuel differences from onę location to another—additives and octane ratings vary geographically according to climatic differences.

Characteristics of the vehicle population in a specified area—such as vehicle age and model mixes—are considered under the source inventory rather than in the emission factor.

There are two modal considerations acceleration/steady-state/deceleration modes and operating characteristics modes—such as with air conditioner operation and trailer pulling.

Vehicle test driving cycles may be run on the road or on a dynamometer (Fig. 12). The emission rates during the various modes of the cycle are measured by continuous sampling techniqus and a speed-emission matrix is prepared (Fig. 13). Any driving cycle can be derived, the emission factors can be taken from the basic matrix and substituted into the driving cycle, and the emissions can be calculated, without field or laboratory testing. This procedure permits derivation of emissions for many custom-

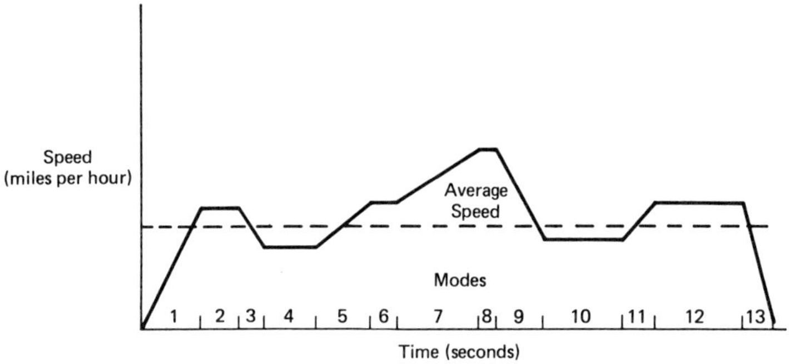

Figure 12. A 13-mode driving cycle.

developed driving cycles. Following the calculation of emissions from the driving cycle, the results may be averaged to obtain emissions per unit distance for area source calculations (Fig. 14).

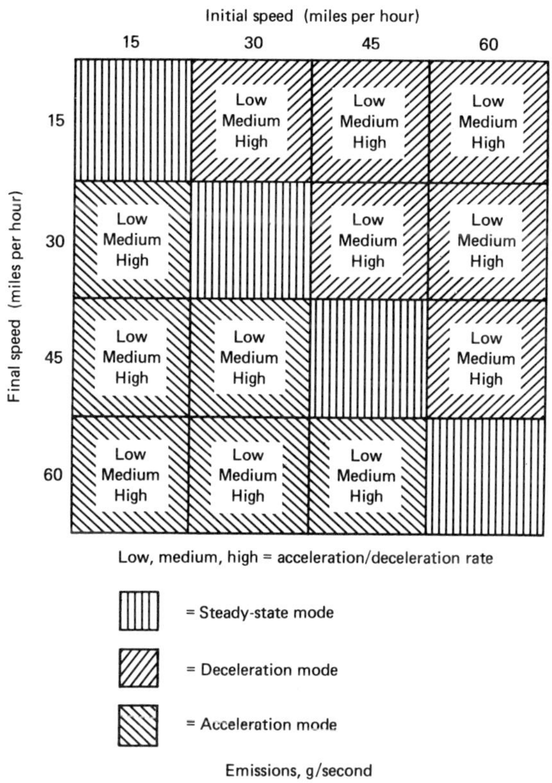

Figure 13. Speed emissions matrix.

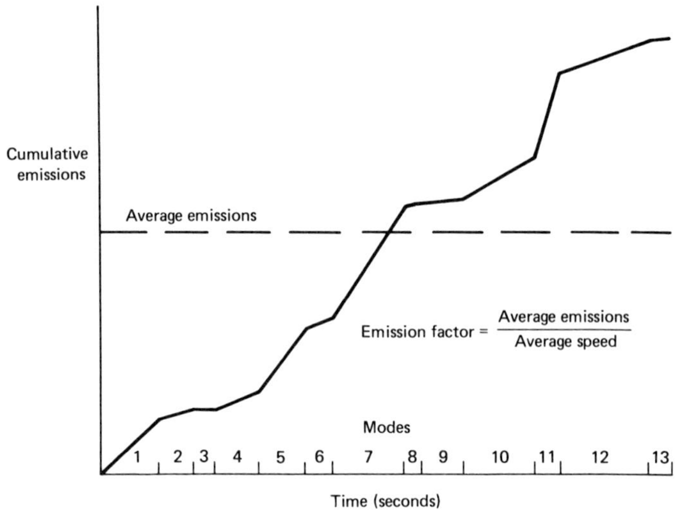

Figure 14. Emissions from a 13-mode driving cycle.

D. Control Equipment

The control equipment on an emitting source must be considered when calculating actual emissions, since a stationary source emission factor calculates uncontrolled emissions. Control equipment efficiencies are incorporated into the emissions calculation as follows:

$$E_c = (1 - \eta_c/100)\,(EF)\,(P) \tag{4}$$

where

E_c = Controlled emissions, same units as EF
η_c = Control device efficiency, percent
EF = Emission factor
P = Production rate, same units as EF

Mobile source emission factors include controls built into the vehicle.

E. Quality Ratings of Emission Factors

Both qualitative and quantitative approaches may be used to express the confidence with which emission factors can be used.

A qualitative evaluation may be expressed by grading the emission factors from A through E, where

A = excellent, factor based on field measurements of a large number of sources

B = above average, factor based on a limited number of field measurements

C = average, some reference publications available providing documentation

D = below average, little background information available

E = poor, basically the factor is an estimate based on material balance calculations

This method is satisfactory for indicating those factors which have significant source testing data support. These quality grades cannot be used in computing a statistically significant confidence level, although a quantitative evaluation can be made if sufficient source test data are available (7). Such evaluation utilizes programmed error estimating algorithms to compute the imprecision which exists in the emission estimates for each point and area source, and then sums these errors to compute the overall imprecision in each source category as well as the inventory as a whole, for any specified geographic area. Precision analysis can be used as a management tool to point out where the largest sources of error exist in any given inventory and where inventory improvement is needed.

F. Data Handling for Emission Factors

An emission inventory data handling system should include a system for handling source test data for development of emission factors and a file for storage of emission factors on a pollutant and source category basis. The source test data system can be computerized if the amount of data on hand is sufficient to justify it. A computerized source classification code file is an absolute necessity to allow an average factor to be modified as new data become available. An auxiliary control equipment efficiency file should be developed to be used in conjunction with emission factor and source inventory files to calculate controlled emissions.

VI. Automatic Data Processing

A. Introduction

The volume of data involved with a comprehensive inventory, or even with a gross estimation inventory which must be updated annually, necessitates that automatic data handling procedures be used. Manual systems and semiautomated systems have been developed and successfully used by many local and state agencies; however, with the availability of computer facilities it is no longer cost effective to implement an emission inventory data handling system without making full use of a computer.

When more than one man-week per month of professional time, in addition to clerical time, is expended in making calculations, manipulating data, and preparing data summaries and analyses, a control agency should develop a computer-based data processing system. If more than one employee spends full time processing data, machine processing is justifiable. If a control agency intends to use elaborate ambient monitoring data acquisition systems such as those that employ telemetry, or to use advanced analysis techniques such as atmospheric dispersion modeling, it is essential that it develop a workable computerized emission inventory data handling system.

The steps required to establish a system are

1. Define in detail the manual system currently in use, and define specific problems in the system and how a computer can be used to alleviate them
2. Define each separate process in the new system, taking care to demonstrate the man–machine interaction (e.g., after defining a computerized validation program, one must define how the output of the program will be used and what other manual checks the data will receive)
3. Arrange for keypunch services and computer time
4. Arrange for a programmer to write an engineering validation program that will list the data appearing on the data cards, "flagging" or marking off suspicious readings if they show:
   (a) Locational coordinates outside of boundaries of the state/province or city
   (b) Stack height less than 500 ft (150 m) or greater than 1200 ft (400 m)
   (c) Stack diameter more than 20% of stack height
   (d) Stack temperature not in 77° to 2000°F (36° to 1575°C) range
   (e) Plume height less than 200 ft (65 m)
   (f) 120% of design capacity exceeded
   (g) Control equipment matching established values.

   Several of these values may be checked for percentage change from previously collected data (b, c, f, g) whereas other data items are checked for maximum limits (f, g), minimum limits (c, e), or ranges (a, b, d)
5. Arrange for a programmer to write a program that will print out data or summaries in a format that can be photocopied and immediately incorporated in periodic reports
6. Arrange for programming to edit data for proper identifiers to facilitate computer processing

If an agency's capability is developed along these lines, professional personnel need only ensure that raw data are accessible to the keypunch operators, review computer output to validate the data, and interpret the analyzed data. No other involvement with routine data handling will be necessary, and valuable professional manpower will be free for more important work. Hazards to be avoided in introducing a computerized data handling system include:

1. Attempting to develop a more elaborate system than is required. By far the best approach is to begin with the simplest possible system, make it operational, and then add increased capability later if it becomes necessary.

2. Providing insufficient direction to computer personnel. If information is not perfectly clear to computer personnel, time and money will be wasted on development of ineffective programs.

3. Not requiring sufficient documentation. If the programmer who establishes the formats and writes the programs should no longer be available to the agency, the entire system may be rendered useless. For this reason, the formats and programs used in developing a computerized data processing system must be clearly and completely described, and completely documented so that it is understandable by any programmer familiar with the computer language in which the program is written.

B. Computerized System

Programs, formats, and identifiers comprise the software system; the computer and peripheral equipment used for storage of data, telecommunications equipment, and the terminal for access of the computer comprise the hardware system (Fig. 15). The data themselves are in computerized files referred to as the data bank. The internal processing

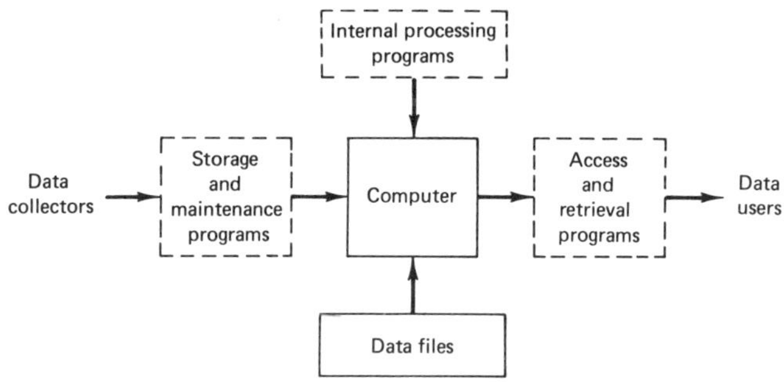

Figure 15. Simplified computerized emission inventory data system.

programs are those which perform mathematical operations each time a user accesses the data files to obtain summaries, etc., which are not stored but that are calculated each time requested. The user files, those accessed each time a retrieval is requested, are usually stored on tapes or disks and are mounted on the computer when a retrieval is requested. In an interactive system the files are "on-line" continuously.

C. Computerized Files

The computerized emission inventory system files can be divided into identification files and data files (Fig. 16).

The identification files are necessary for cross-referencing all of the data files and for providing numerical values for codification of data. The identification files used include the following:

1. Geographical—numbers assigned to each different geographical area under consideration

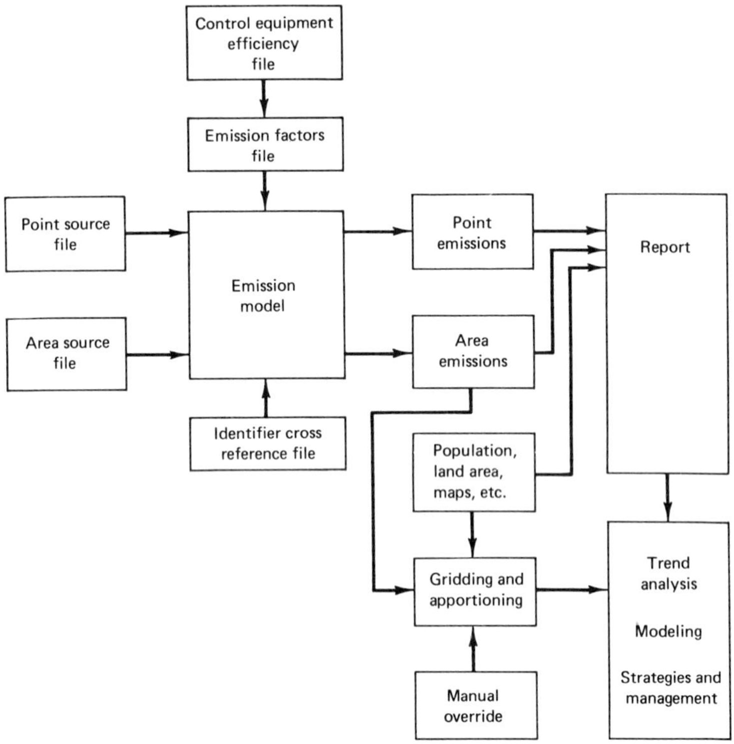

Figure 16. Computerized emissions data system.

2. Source category—numerical identification of each individual source category to be considered

3. Control equipment—value assigned to each separate control device

4. Pollutant—codes assigned to individual air pollutants

5. Compliance status—code for various stages in progress toward the compliance with control regulations.

The data files contain the information about sources and other information which may be both voluminous and subject to much updating or change. In comprehensive source inventories it is impossible to have a variable data format (one in which each data value has an associated identification) since all computer card space would be taken up by identification; therefore, it is necessary to utilize a fixed format method of data recording and storage. In fixed format, the field of data on the coding sheet or punched card is predefined to contain a specific item of data.

The data files are usually the following:

1. Point source file—all source information predetermined to be collected for each source

2. Area source file—all information to be collected on each geographical area to be considered an area source

3. Line source file—all information on each line source

4. Emission factors file—the emission factor for each pollutant for each source category

5. Control equipment efficiency file—the efficiency assumed for each pollutant control device that might be used

6. Other files—it may occur that after the system has been designed and implemented other data must be made available through the computerized system; in this case, new files must be created to handle this unanticipated data.

In the first three data files described, four subfiles should be used to store general data (names, addresses, etc.), production/consumption data for calculating emissions, modeling data, and compliance information. Segregation of these files permits ease of maintenance and expansion of the system to meet additional needs.

An additional method may be used, that of "unformatted" data files, in which the specific information is coded in a predetermined order separated by semicolons on the data card or coding sheet. The absence of items of data is indicated by no numbers being entered between semicolons.

D. Editing and Maintenance

As data are received from the collectors of information as punched cards, magnetic tape, or coding forms, validation and identification of errors must take place in a program which inspects all data applying a predetermined set of criteria. These criteria can be as comprehensive as required, and the error messages as descriptive as necessary for interpretation by the personnel assigned to correct identified errors.

All data must pass the editing procedure prior to storage in the data bank. Typical edit checks are prohibition of blanks, assurance of right or left justification, comparison of coded data with predetermined maxima or minima, cross-checking of values to assure compatibility, comparison of grid coordinates with geographical boundary lines, and comparison of emission estimates with calculated values using emission factors. Data which do not pass editing must be corrected or "validated," that is, specifically stated to be correct by the persons collecting and submitting the data.

The creation of the computerized data files and updating of the files, including the identification files, constitute files maintenance: this includes procedures for adding, deleting, and changing data.

E. Retrievals

The outputs from a computerized emission inventory take two forms: statistics, which are directly from the files or generated each time requested, and counts, which quickly provide status reports on the quantity of data in the files.

Summary statistics should be available for the following groupings: (1) smallest to largest geographic area; (2) specific source categories; (3) pollutants; (4) point, area, or line sources, in order of size; and (5) controlled/uncontrolled emissions. In addition, sorting capability should be available to rank the data in order, such as largest point source to smallest, and to bracket values, such as within specific source categories.

Counts are necessary for status reports such as (1) number of sources by source category or geographic areas; (2) missing data items; (3) identification numbers assigned; and (4) compliance status.

F. Software

Both batch processing and interactive access to the system should be considered (Fig. 17). Batch processing usually takes place at the convenience of computer operators who have a production schedule; interactive access capability permits the user to instantaneously access the data

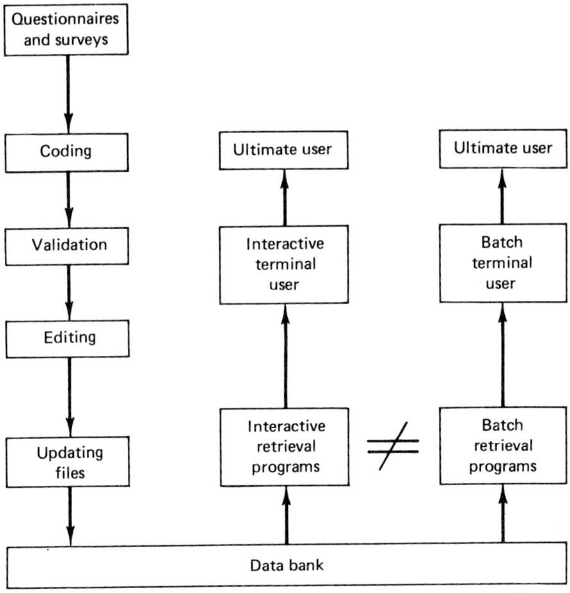

Figure 17. Data flow and retrievals for emissions data system. Some data may be accessable by both retrieval methods or by one method only.

bank, receive output, modify the request, and again receive output. Batch processing is done for large volumes of data, whereas interactive requests should be made only when relatively small amounts of data are required.

Program documentation usually consists of six items:

1. Program description—specification of the design of the program (what it does), restrictions and limitations, the procedures used, any formulas included, options available, unique features and a list of related programs

2. Flow chart showing the detailed logic of the program (should be complete enough that someone not familiar with the program can make changes)

3. Input/output description—For punched card input, the following elements are needed: control card description, data card description, end card description, and card deck setup. The control card defines input/output options—magnetic tape or direct access storage devices, file design, sequence, header, and end records. A copy of the output description and format, which can be magnetic tape, punched cards, or printed copy, should be provided

4. Operating instructions for filling out and keypunching data, control, and end cards, the input deck, and file or peripheral devices; indicat-

ing frequency and length of run in terms of volume, such as 5 minutes/per 1000 cards: describing the proper disposition of input and output when the program is finished; and providing instructions for computer operators for normal operation and on procedures for an interrupted or aborted run

5. Suggestions, warnings, and changes made
6. Program listing printout of the cards comprising the program

G. Hardware

The prime concern of the system user is the terminal at which the data is submitted or retrieved. Several terminals are available: (1) printers—data printed on sheets of paper, either by low speed typewriter or medium high speed line printer, and (2) cathode ray tube (CRT) display—data appear on television tube.

The requests for data may be entered on the keyboard of typewriter or CRT terminal or by means of punched cards fed into a card reader. Processing of requests may be performed by the computer in either batch mode or interactively, with output routed back to the desired output terminal.

H. Confidentiality

Data may be legally ruled confidential by the agency charged with supplying the data to the data processors, and by the agency processing and disseminating the data. The method for determination of confidentiality is first to have the data sources identify the data requested to be withheld from public use and then to verify that these requests meet legal criteria. It must be ascertained whether data not identified as confidential by the source have been ruled to be confidential by the courts or become confidential by a change in the statutes.

Confidential data must be purged from the files (using a confidential identifier in the file), placed in a separate file, and integrated with other data only when necessary for generating summary reports. With nonautomated emission inventory systems, the handling of confidential data is relatively simple. However, because of the ease of accessing computerized files, control of confidential data must be considered in the design of the computerized data system.

I. Users

Users might be grouped into two classes: users of the data bank and ultimate users of the output from the system. Shortening the data communication line from the data bank to the ultimate user requires planning and analysis of the system during development and operation. Comprehensive design efforts will result in output capability which does not require a trained programmer to activate. Interactive terminals are used easily after only several hours of training. Well-documented batch retrieval programs for line printers are mastered in a somewhat longer time. However, a trained computer programmer will be required if specialized outputs, not already a part of the existing library of programs, are required. The ultimate user of data should be provided with direct access to the data via terminals without requiring intermediary personnel from specialized processing or management groups.

J. Files Security

The computerized data and program files must be protected from destruction. This is most easily accomplished by maintaining a duplicate set of tapes or disks in a location remote from the master file, which is usually adjacent to the computer. This means that each time an update occurs, a duplicate is prepared and stored at another location, safe from fire, theft, or malicious damage. Another alternative is to store the old file and the updating data in another location, with the new file (old file plus update integrated) remaining with the computer. In case of destruction, the old file and update can be used to generate another master file. It is further suggested that data banks should not be updated by users, but rather by a central processing group responsible for data integrity.

K. Questionnaires

The questionnaire (*4*) is the most widely used and expeditious method of collecting point source emission inventory information. The use of the questionnaire can produce problems, especially if the "questionnaire survey" is not properly conducted. These problems may be grouped into two categories: questionnaire design, and survey administration.

Proper questionnaire design includes establishment of a suitable format, selection of the appropriate questions, proper wording, and authorship of an appropriate cover letter. The basic rule is to design the questionnaire and the letter for the person who will be asked to complete it. The respondent may have neither the benefit of a technical background in air pollution, nor the assistance of an engineer in filling out the forms.

The questionnaire should be as simple and functional as possible. Time will be saved if its format is such that a keypunch operator can punch the information directly from the questionnaire onto cards. Only information that can be used should be solicited; the questions should be well spaced to ensure that they are easy to read, yet placed on as few pages as possible because most people tend to avoid long questionnaires. When writing the questions, use terminology with which the recipient will be familiar. Each question should be self-explanatory or accompanied by clear directions. All information that will be utilized should be asked for initially, thus avoiding requests for additional data at a later time. A control agency may wish to coordinate the emission inventory with the establishment of a permit or a source registration system.

The questionnaire should be accompanied by a letter of transmittal which indicates the study area, study purpose, the legal authority, return address and contact for questions, and due date. Before the questionnaire is due to be returned, a second letter may be sent out reminding the recipient of the need for the completed questionnaire. The second letter is sent as a reminder only and should be worded so as not to be construed by the recipient as harassment on the part of the agency. When the completed questionnaire has been returned, a short letter of appreciation should be sent to the party returning the questionnaire.

The optimum procedure in distributing and processing the questionnaire will minimize the amount of time required by technical staff and maximize use of secretarial, clerical, and keypunch help.

The concept of questionnaire use in conjunction with a computerized data system must be clearly understood. The questionnaire is the communication between the source and the staff engineer. The data coding form is the communication between the staff engineer and the computerized data processor. It has been shown many times that the source cannot effectively utilize a coding form to communicate directly with a computerized data processor. This communications problem must be solved by prior planning with respect to the coding form, the instruction sheet processing guidelines, instructions for questionnaire coding, guidelines for the authorized use of the questionnaire, and procedures for forwarding the coded data to the processors.

VII. Data Presentation

A. Introduction

Emission inventory methodology is designed to facilitate grouping of point and area source data in multiple arrangements. Three groupings are basic "sorting" or indexing methods for an emission inventory: (1) emissions according to source category, (2) emissions according to pollutant, and (3) emissions within a specific geographical area.

Any of these groupings can be used to display the emission data in a multitude of ways suitable for a variety of uses. For example, emissions of hydrocarbons may be automatically listed in order of relative contributions from each point source. Another combination of groupings could tabulate emissions of all pollutants from pulp mills or chlor-alkali plants in an area. Summaries and tabulations of emissions data may include:

1. Emission tabulation by source category by geographic area
2. Stationary and mobile fuel balances within an area
3. Solid waste balances
4. Contribution of specific sources to emission totals
5. Point source versus area source contribution
6. Emission density maps for stationary and mobile source emissions arranged by pollutant
7. Installed pollution control equipment
8. Indexes of process sources emitting specific pollutants
9. Time fluctuations of emissions

The inventory system is primarily intended to accumulate data on pollutant emissions; however, the corresponding process data for each source includes extractable information on addresses, number of control devices, and other plant factors. For example, names and addresses of all organizations emitting a specific pollutant may be needed for a mailing list. In such a case the following procedure would be used: (a) select the appropriate source category; (b) instruct the computer to tabulate all sources with the given category code in the study area, and (c) instruct the system to print out names and addresses for all sources.

The inventory system should be designed to determine annual emissions of pollutants. However, as the data base becomes more complete and emission factors become more accurate, the need will arise to determine variance of emission rates with time. This determination can easily be made by basing the automatic calculations of emissions on production rate data, thus allowing estimates of daily, weekly, and seasonal variations in emission rates to be made.

It must be emphasized that the main advantages of a large computerized data base are: (1) the availability of relatively small quantities of detailed information contained in the large data base, and (2) the ability to obtain short summaries from the large data base.

B. Tables and Graphs

Pie charts are easily constructed and can simplify comparison of data on various emissions and their sources. Because the charts compare emis-

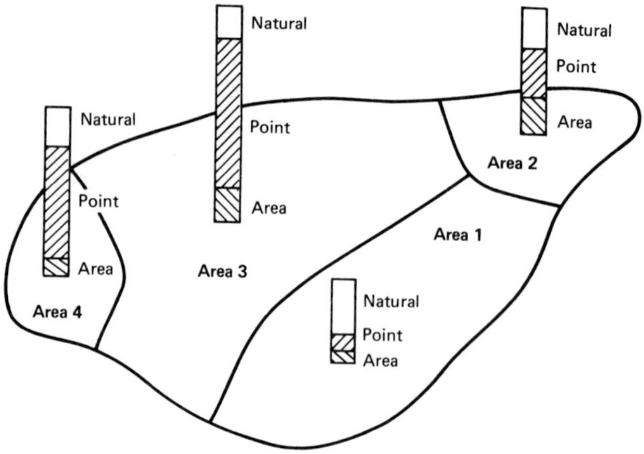

Figure 18. Comparison of natural–area–point sources emissions of hydrocarbons.

sions on a relative basis in percentages and fractions of total emissions, the total emissions should also be listed. With the totals and percentages, calculations of quantities of emissions are easily made. Furthermore, visual comparison of the relative quantities of emissions is possible if the total area of each chart corresponds to the quantity of each pollutant emitted. Another method is a graphical presentation of the distribution of natural–area–point emissions in selected geographical areas (Fig. 18).

C. Emission Density Maps

Emission density maps (*4*) offer one of the best methods of showing source contributions and concentrations of emitted pollutants by geographic area. The inventory techniques described allow emissions to be collected on a study area basis, with the emissions apportioned to each grid to construct an emission density map. A map of the study area is divided into a grid pattern in summarizing the emissions of each pollutant from all sources.

The area source emissions are apportioned to the grids and the components of line sources are allocated to the grid squares in which they occur. Point source emissions, as located by their coordinates, are added to the appropriate squares, and the resultant emission density map, similar to a population density map (Fig. 9), is obtained. If a record is kept of the quantities of area, line, and point source emissions added to each square, the total emissions per square can be calculated.

Following a tabulation of grid square area and total emissions, the emissions per unit area can be calculated. Following the procedures of basic statistics, if a frequency histogram is prepared and the appropriate density intervals are chosen, a pictorial presentation of the grid squares, with crosshatching representing the emission density distribution, can be drawn.

D. Error in the Emission Inventory

The error in the emission inventory for a specific pollutant can be calculated (7) from

$$e = [(R_1 e_1)^2 + \cdots + (R_n e_n)^2]^{1/2} \tag{5}$$

where

e = error, percent

$R_i = \dfrac{E_i}{\sum_{i=1}^{\eta} E_i}$ = emission proportion, percent, for source category i

E_i = emissions for source category i

e_i = error for source category i

The error in percent for a specific source category results from the combined errors (algebraic sum) in emission parameters (emission factors and control equipment efficiencies) and source inventory parameters (material composition and throughput):

$$e_i = [e_{\mathrm{EF}}]_i + [e_{\mathrm{CEE}}]_i + [e_{\mathrm{MC}}]_i + [e_{\mathrm{MT}}]_i \tag{6}$$

where

e_{EF} = error in the pollutant emission factor for the source category

e_{CEE} = error in the pollutant control equipment efficiency for the source category

e_{MC} = error in material composition for the pollutant source category

e_{MT} = error in material throughput rate for the pollutant source category

The error in an emission factor can be calculated with 95% confidence from source test results:

$$\pm e_{\mathrm{EF}} = \pm 2s/\bar{x}\ (100) \tag{7}$$

where s is the standard deviation, and $\bar{x}$ is the average uncontrolled emission rate.

If source test results are insufficient for statistical analysis, estimation of error is probably not possible, given material balance data, although further study in this area may yield useful results.

Error in control efficiency may arise from rounding to significant figures (three significant figures for control efficiencies greater than 90% and one significant figure for low efficiency equipment), decrease from design efficiency due to power availability, deterioration, fuel specifications change, operating conditions change, etc. In general, the rounding error is simple to calculate, but the control efficiency error due to deviation from design is much more complex.

The error in material throughput may result from the reporting precision of the emission inventory survey or coding forms (significant figures), and from errors in knowledge of throughput or weighing errors at the source (assumed to be normally distributed and therefore insignificant in the aggregate).

The error in material composition results from improper laboratory analysis procedures and also may be considered normally distributed and insignificant. However, when considering small geographic areas, the throughput and composition data should be examined carefully since the errors may not be normally distributed and may contribute significant error to the total.

E. Coordinate Systems

Emissions must be reported geographically with point sources identified and located precisely and area sources apportioned to specific grids. For inventories in different localities to be compatible, a uniform reporting system must be used.

Various grids systems utilizing coordinate systems have been developed for locating geographic points; the most common are (1) latitude and longitude and (2) universal transverse mercator (UTM).

The latitude–longitude system is well known. However, in the UTM system (*8*) instead of using the equator as the axis of the mercator, the earth's axis is rotated 90° and a meridian running north–south is used as the axis from which all measurements are made. In this manner a meridian at any point on the globe may be selected so that an approximately 2000 mile (3218 km) wide area of unskewed map covers any part of the world. The system also maximizes the distance measurable from one pair of meridians and false origins with a reasonable degree of accuracy. The globe is subdivided into 60 north–south zones, each 6 deg wide with a central transverse mercator. The false origin for each zone in the northern hemisphere is 500,000 m west of the point where the central meridian crosses the equator. Using the UTM system, grid zones may be drawn for air pollution emission inventories. The advantages of this system are (1) UTM grids are continuous and are not hindred by political subdivisions; (2) UTM grids are uniform; (3) the system is used world wide; and (4) the system is becoming the base repository for a growing body of technical information.

REFERENCES

1. G. Ozolins and R. Smith, "Rapid Survey Technique for Estimating Community Air Pollution Emissions," AP-29. United States Department of Health, Education, and Welfare, Public Health Service, Washington, D.C., 1966.
2. Emission Inventories: Air Pollution Technical Documents (APTD) United States Department of Health, Education, and Welfare, Public Health Service, Washington, D.C. *Dated 1966:* Northwest Indiana (68-4); *Dated 1968:* Dayton, Ohio, (876), Buffalo, New York (877), Cincinnati-Hamilton-Middletown, Ohio (878), Denver, Colorado (880), Indianapolis, Indiana (881), Seattle-Tacoma, Washington (886), St. Louis, Missouri (887), Pittsburgh, Pennsylvania (895); *Dated 1969:* Atlanta, Georgia (811), Houston-Galveston, Texas (818), New Orleans, Louisiana (823), Phoenix-Tucson, Arizona (827), Chittenden County, Vermont (839), Springfield, Ohio (840), Memphis, Tennessee (882), Milwaukee, Wisconsin (883), Providence and Pawtucket, Rhode Island and New Bedford and Fall River, Massachusetts (884), San Antonio, Texas (885), Toledo, Ohio (888), Louisville, Kentucky (894), United States Environmental Protection Agency, Research Triangle Park, North Carolina; *Dated 1970:* Albuquerque, New Mexico (809), Beaumont-Port Arthur, Texas (810), Charlotte, North Carolina (812), Cheyenne, Wyoming (813), Columbus, Ohio (814), El Paso, Texas (815), Fargo, North Dakota and Moorhead, Minnesota (816), Hawaii (817), Las Vegas, Nevada (819), Merrimack Valley, Massachusetts and New Hampshire (820), Miami-Ft. Lauderdale-West Palm Beach, Florida (821), Mid-Connecticut-Lower Pioneer Valley (822), Omaha, Nebraska (826), Portland, Maine (828), Reno, Nevada (829), Salt Lake City-Provo-Ogden, Utah (830), Treasure Valley, Idaho (833), Raleigh-Durham-Chapel Hill, North Carolina (834), Virgin Islands (835), Willamette Valley, Oregon (836), Baton Rouge, Louisiana (866); *Dated 1971:* Alaska (671), Nebraska (712), Missouri (732), New Jersey (742), Alabama (746), South Carolina (747), Kansas (748), Vermont (758), Minnesota (771), Oklahoma (773), Idaho (786), Montana (791), New Mexico (792), Puerto Rico (793), Louisiana (794), Arkansas (795), South Dakota (798), North Dakota (799), Wyoming (801), Billings, Montana (808), New York State Southern Tier-West (824), Oklahoma City, Oklahoma (825), San Juan, Puerto Rico (831), Souix Falls, South Dakota (832), Kansas (1131); *Dated 1972:* Dallas-Ft. Worth, Texas (695), Ohio (756), New Hampshire (1342); *Not APTD:* Kansas—Point Sources—National Emissions Data System (4503-74-007); *Dated 1973:* Iowa—National Emission Data System (4503-74-004), Utah—National Emission Data System Compilation (4503-74-005), Oregon (4503-74-006); *Dated 1974:* California (excluding Los Angeles County)—Point Sources (4503-74-017), Los Angeles County, California—Point Sources (4503-74-018).
3. National Emission Inventories of Sources and Emissions: United States Environmental Protection Agency, Research Triangle Park, North Carolina, Air Pollution Technical Document (APTD). *Dated 1971:* Arsenic (1507), Berryllium (1508),

Manganese (1509), Mercury (1510), Vanadium (1511); *Dated 1972:* Copper (1129), Selenium (1130), Zinc (1139), Barium (1140), Boron (1159); *Dated 1973 Not APTD:* Titanium (4503-74-008); Molybdenum (4503-74-009); Magnesium (4503-74-010); Silver (4503-74-011); Chromium (4503-74-012); Phosphorus (4503-74-013).

4. "A Guide for Compiling a Comprehensive Emission Inventory" (revised), APTD 1135. United States Environmental Protection Agency, Research Triangle Park, North Carolina, 1973.
5. "Weighted Sensitivity Analysis," 4503-74-022 A,B. United States Environmental Protection Agency, Research Triangle Park, North Carolina, 1975.
6. "Compilation of Air Pollutant Emission Factors," AP-42 and Supplements. United States Environmental Protection Agency, Research Triangle Park, North Carolina, 1973.
7. "Source Inventory Emission Factor Analysis," Contract 68-02-1350. United States Environmental Protection Agency, Research Triangle Park, North Carolina, 1975 (in progress).
8. C. H. Deetz and D. S. Adams, "Elements of Map Projection," Spec. Publ. No. 68. United States Department of Commerce, Washington, D.C., 1945.

Subject Index

A

F

L

M

P

T

U

ENVIRONMENTAL SCIENCES

An Interdisciplinary Monograph Series

ARTHUR C. STERN, editor, AIR POLLUTION, Second Edition, Volumes I–III, 1968; Third Edition, Volumes I, III, 1976; Volumes II, IV, V, in preparation

L. FISHBEIN, W. G. FLAMM, and H. L. FALK, CHEMICAL MUTAGENS: Environmental Effects on Biological Systems, 1970

DOUGLAS H. K. LEE and DAVID MINARD, editors, PHYSIOLOGY, ENVIRONMENT, AND MAN, 1970

KARL D. KRYTER, THE EFFECTS OF NOISE ON MAN, 1970

R. E. MUNN, BIOMETEOROLOGICAL METHODS, 1970

M. M. KEY, L. E. KERR, and M. BUNDY, PULMONARY REACTIONS TO COAL DUST: "A Review of U. S. Experience," 1971

DOUGLAS H. K. LEE, editor, METALLIC CONTAMINANTS AND HUMAN HEALTH, 1972

DOUGLAS H. K. LEE, editor, ENVIRONMENTAL FACTORS IN RESPIRATORY DISEASE, 1972

H. ELDON SUTTON and MAUREEN I. HARRIS, editors, MUTAGENIC EFFECTS OF ENVIRONMENTAL CONTAMINANTS, 1972

RAY T. OGLESBY, CLARENCE A. CARLSON, and JAMES A. MCCANN, editors, RIVER ECOLOGY AND MAN, 1972

LESTER V. CRALLEY, LEWIS T. CRALLEY, GEORGE D. CLAYTON, and JOHN A. JURGIEL, editors, INDUSTRIAL ENVIRONMENTAL HEALTH: The Worker and the Community, 1972

MOHAMMED K. YOUSEF, STEVEN M. HORVATH, and ROBERT W. BULLARD, PHYSIOLOGICAL ADAPTATIONS: Desert and Mountain, 1972

DOUGLAS H. K. LEE and PAUL KOTIN, editors, MULTIPLE FACTORS IN THE CAUSATION OF ENVIRONMENTALLY INDUCED DISEASE, 1972

Merril Eisenbud, ENVIRONMENTAL RADIOACTIVITY, Second Edition, 1973

James G. Wilson, ENVIRONMENT AND BIRTH DEFECTS, 1973

Raymond C. Loehr, AGRICULTURAL WASTE MANAGEMENT: Problems, Processes, and Approaches, 1974

Lester V. Cralley, Patrick R. Atkins, Lewis J. Cralley, and George D. Clayton, editors, INDUSTRIAL ENVIRONMENTAL HEALTH: The Worker and the Community, Second Edition, 1975

A
B 6
C 7
D 8
E 9
F 0
G 1
H 2
I 3
J 4